Ardipithecus kadabba

The Middle Awash Series

Series Editor
Tim White, University of California, Berkeley

University of California Press Editor
Charles R. Crumly

Homo erectus: *Pleistocene Evidence from the Middle Awash, Ethiopia,* edited by W. Henry Gilbert and Berhane Asfaw

Ardipithecus kadabba: *Late Miocene Evidence from the Middle Awash, Ethiopia,* edited by Yohannes Haile-Selassie and Giday WoldeGabriel

Ardipithecus kadabba

Late Miocene Evidence from the Middle Awash, Ethiopia

EDITED BY YOHANNES HAILE-SELASSIE AND GIDAY WOLDEGABRIEL

UNIVERSITY OF CALIFORNIA PRESS Berkeley Los Angeles London

University of California Press, one of the most distinguished university presses in the United States, enriches lives around the world by advancing scholarship in the humanities, social sciences, and natural sciences. Its activities are supported by the UC Press Foundation and by philanthropic contributions from individuals and institutions. For more information, visit www.ucpress.edu.

The Middle Awash Series, Volume 2

University of California Press
Berkeley and Los Angeles, California

University of California Press, Ltd.
London, England

Library of Congress Cataloging-in-Publication Data

Ardipithecus kadabba : late miocene evidence from the Middle Awash, Ethiopia / edited by Yohannes Haile-Selassie, Giday WoldeGabriel.
p. cm. — (The Middle Awash series)
Includes bibliographical references and index.
ISBN 978-0-520-25440-4 (cloth : alk. paper)
1. *Ardipithecus kadabba*—Ethiopia—Middle Awash. 2. Fossil hominids—Ethiopia—Middle Awash. 3. Human remains (Archaeology)—Ethiopia—Middle Awash. 4. Paleoanthropology—Ethiopia—Middle Awash. 5. Middle Awash (Ethiopia)—Antiquities.
I. Haile-Selassie, Yohannes, 1961– II. WoldeGabriel, Giday.
GN282.73.A73 2008
569.90963— 2008004004

Manufactured in the United States
16 15 14 13 12 11 10 09
10 9 8 7 6 5 4 3 2 1

The paper used in this publication meets the minimum requirements of ANSI/NISO Z39.48-1992 (R 1997) (*Permanence of Paper*). ∝

Cover illustration: Lingual view of ASK 3/400, *Ardipithecus kadabba* right upper canine from Asa Koma. Photograph courtesy of David Brill © copyright 2003.

Contents

List of Contributors

Yohannes Haile-Selassie
Department of Physical Anthropology
Cleveland Museum of Natural History
Cleveland, Ohio

Giday WoldeGabriel
Earth Environmental Sciences Division
Los Alamos National Laboratory
Los Alamos, New Mexico

Stanley H. Ambrose
Department of Anthropology
University of Illinois, Urbana-Champaign

Mesfin Asnake
Ministry of Mines and Energy
Addis Ababa, Ethiopia

Raymond L. Bernor
College of Medicine
Department of Anatomy
Laboratory of Evolutionary Biology
Howard University
Washington, DC

Faysal Bibi
Department of Geology and Geophysics
Yale University
New Haven, Connecticut

Michael T. Black
Phoebe A. Hearst Museum of Anthropology
University of California, Berkeley

Jean-Renaud Boisserie
Unité Paléobiodiversité et Paléoenvironnements and Département Histoire de la Terre
Muséum National d'Histoire Naturelle
Paris, France; and
Human Evolution Research Center
Museum of Vertebrate Zoology
University of California, Berkeley; and
Institut International de Paléoprimatologie, Paléontologie humaine: Evolution et Paléoenvironnements
Université de Poitiers, France; and
Department of Integrative Biology
University of California, Berkeley

David DeGusta
Department of Anthropological Sciences
Stanford University
Palo Alto, California

Stephen R. Frost
Department of Anthropology
University of Oregon, Eugene

Ioannis X. Giaourtsakis
Ludwig-Maximilians-Universität München
Department of Geo- and Environmental Sciences
Section of Paleontology
Munich, Germany

William K. Hart
Department of Geology
Miami University
Oxford, Ohio

Leslea Hlusko
Department of Integrative Biology
University of California, Berkeley

F. Clark Howell
Human Evolution Research Center
Museum of Vertebrate Zoology
University of California, Berkeley

Thomas Lehmann
Palaeontology Section
Transvaal Museum
Pretoria, South Africa

Leah E. Morgan
Department of Earth and Planetary Science
University of California, Berkeley

Cesur Pehlevan
University of Ankara
Faculty of Letters
Department of Anthropology
Ankara, Turkey

Paul R. Renne
Berkeley Geochronology Center
Department of Geology and Geophysics
University of California, Berkeley

Lorenzo Rook
Department of Earth Sciences
University of Firenze, Italy

Haruo Saegusa
Institute of Natural and Environmental Sciences
University of Hyogo
Sanda, Hyogo, Japan

Denise F. Su
Human Evolution Research Center
Museum of Vertebrate Zoology
University of California, Berkeley; and
Department of Integrative Biology
University of California, Berkeley

Gen Suwa
Department of Biological Sciences
The University of Tokyo, Japan

Elisabeth S. Vrba
Department of Geology and Geophysics
Yale University
New Haven, Connecticut

Henry B. Wesselman
Department of Anthropology
American River College
Sacramento, California; and
Department of Anthropology and
Sierra College Natural History Museum
Sierra College
Rocklin, California

Tim White
Human Evolution Research Center
Museum of Vertebrate Zoology
University of California, Berkeley; and
Department of Integrative Biology
University of California, Berkeley

Foreword

In the 1980s, the late Professor J. Desmond Clark and Professor Tim D. White from the University of California at Berkeley set up a multidisciplinary team to lead geological, paleoanthropological, and archaeological surveys, focusing on the eastern side of the Awash River in Ethiopia. As part of that project, in 1995 Yohannes Haile-Selassie and Giday WoldeGabriel initiated intensive paleontological and geological surveys along the western margin of the Middle Awash study area. In spite of very difficult fieldwork, this rich compendium shows how important and successful their research was for the knowledge of the African late Miocene (5.2 to 5.8 Ma) faunas (2,800 specimens from 23 fossiliferous localities), a crucial and dramatic period in the story of hominid origins. I am very honoured and proud to write these few words as the foreword to such a masterpiece.

First of all, I want to underline that the two editors Yohannes Haile-Selassie and Giday WoldeGabriel are Ethiopian scientists already known worldwide for their scientific accomplishments. With this volume they demonstrate their ability to bring this field work to full fruition, collecting geological and paleontological data, then gathering together around them the most qualified and comprehensive multidisciplinary team to study, to analyze, and to render all the important results presented in the chapters that follow. This masterpiece of work in paleontology honors not only them, but also Ethiopia, their country, and their mentor Professor T.D. White. It is my great pleasure to give to all of them my warmest and friendly congratulations.

With the richness and the magnitude of all its geological and paleontological results, this published work shall be an essential reference volume for all researchers working to better understand hominid origins, evolution, and paleoenvironments. This monograph is a new cornerstone in the history of our human story. All over the world in 2009, scientists will celebrate the bicentennial of Charles Darwin's birthday, the famous father of a rational scientific theory for evolution ("On the Origin of Species by Means of Natural Selection," 1859). At the beginning of the third Millennium, it is necessary to remember that in 1871, Charles Darwin in his work "The Descent of Man and Selection in Relation to Sex" suggested that close relationships existed between man and the African apes (Gorilla and Chimpanzee), and it was consequently more probable that our first ancestors would have lived in Africa than anywhere else. Darwin's prediction is now confirmed by molecular phylogeny and by paleontological evidence.

As early as 1967, Allan Wilson and Vincent Sarich of the University of California at Berkeley demonstrated our close genetic proximity with the chimpanzees (less than 2% of difference). It means *de facto* that we share a common ancestor: the chimpanzees (= Panids) are the sister group of Humans (=Hominids).

Numerous human fossil remains have been successively unearthed during the quest for our fossil ancestors, first in Europe, in the end of the 19th century, then in Asia with *Homo erectus*. Raymond Dart subsequently described the first Australopithecine (*Australopithecus africanus*), the Taung child, dated from 2 to 2.5 Ma, in South Africa in 1925. Numerous discoveries followed in eastern Africa, from 2 to 3.6 Ma: *Paranthropus boisei* (L. Leakey, in 1959); and *P. aethiopicus* (Arambourg & Coppens, in 1963). Then in the 1970s, Lucy (3.2 Ma) was found in 1974 and her relatives were named as *Australopithecus afarensis* (Johanson, White & Coppens, in 1978). Through these discoveries, researchers became clearly convinced that our story was deeply rooted in Africa.

More recently, and for the first time, three earliest hominids were unearthed from the African Late Miocene. These are the Ethiopian *Ardipithecus kadabba* (5.2–5.8 Ma and its Pliocene relative *A. ramidus*, 4.4 Ma); its contemporaneous neighbour the Kenyan *Orrorin* (*ca.* 6 Ma); and in Chad (Central Africa, 2500 km west of the Rift) *Sahelanthropus tchadensis* (nicknamed Toumaï, 7 Ma). All these Late Miocene hominids were probably bipedal and must have lived in woodlands of mosaic landscapes. This current volume describes and contextualizes the Ethiopian hominid, *Ardipithecus kadabba* Haile-Selassie, 2001. It is a work that sets a new cornerstone which confirms brilliantly the prediction of our African origin made by Charles Darwin in 1871.

Finally I want to emphasize that this monograph, with two Old World Ethiopian editors (Haile-Selassie and WoldeGabriel) and published in the New World (University of California Press) attests to an African human origin deeply rooted in time. This marks a marvellous global symbol of enlightenment at a time when, once again, we are confronted with a powerful medieval way of thinking that deliberately ignores the scientific discoveries accumulated since the 19th century. Neocreationists cleverly invented a pseudoscientific concept, their belief known as "intelligent design" (ID). In 2007, they promoted ID through a worldwide publicity campaign, even dispatching by mail to research and teaching institutions worldwide a luxurious colour photo album titled "Atlas of Creation" which purports to show that evolution never happened. In contrast, this monograph on the new Ethiopian fossils is based on science rather than belief. The chapters in this book provide the data, the irrefutable evidence that is now a large part of a clear, great, and powerful scientific answer to the question of our origins.

Michel Brunet
Institut International de Paléoprimatologie,
Paléontologie humaine: Evolution et Paléoenvironnements
Université de Poitiers, Paris,
August 2008

Series Preface

The Middle Awash valley of Ethiopia is a unique natural laboratory for the study of human origins and evolution. Sediments measuring more than a kilometer in thickness lie exposed here on the floor and margins of the Afar Rift. They provide an unparalleled composite geological, paleontological, and archaeological record of the human past. This is Earth's longest and deepest record of early hominid occupation, environment, technological development, and evolution.

The Middle Awash study area is a paleoanthropological resource very different from the nearby richly fossiliferous, stratigraphically simple, and temporally limited deposits at Hadar, where Australopithecus afarensis was found in the 1970s. The Middle Awash also differs from the more continuous depositional sequence of the Omo Shungura Formation of southern Ethiopia and from the time-compressed and spatially constrained strata of Olduvai Gorge in Tanzania. In contrast, the Middle Awash affords a series of radioisotopically calibrated "windows" opening on different time slices of the deep past, rather than a continuous accumulation of Miocene through Holocene deposits. Here, in a single valley in the Horn of Africa, it is now possible to sample dozens of biological lineages, including our own, through geological time.

The study area occupies the southwestern corner of the Afar Depression where it sits atop an active segment of the African rift system. Here, crustal extension through the last six million years created shifting centers of fluviatile and lacustrine sedimentary deposition. Today, the modern landscape of the Middle Awash is a tectonically and geomorphologically created patchwork of eroding sediments. These deposits yield the remains of ancient organisms and their environments.

The paleoanthropological importance of the Middle Awash was first revealed in the late 1960s and early 1970s by the pioneering work of geologist Maurice Taieb. Taieb and colleagues focused their efforts on the Hadar fossil field 75 kilometers to the north. Meanwhile, Jon Kalb and his Rift Valley Research Mission in Ethiopia extended Taieb's preliminary Middle Awash surveys. Their field investigations ended in 1978. In 1981 the late Berkeley professor J. Desmond Clark invited me to join him in initiating multidisciplinary work on the geological, archaeological, and paleontological resources of this unique part of the Afar. The Middle Awash project was born.

This project is an ongoing multidisciplinary effort to elucidate human origins and evolution. A carefully planned program of exploration, focused research, and resource management has maximized results. Middle Awash field investigations can be loosely divided into three stages for each of the many fossiliferous packages in the study area. Exploration identifies and constrains the package. Focused research then establishes its contents and chronostratigraphic relationships, generating most primary data. Long-term management then allows additional data recovery as erosion and excavation continue. Meanwhile, ground-truth information is synthesized with space- and aerial-based imagery to guide further fieldwork, while other field-based data are subjected to laboratory analyses.

This integrated research strategy matches ongoing problems of human evolutionary studies with the unique resources of the rich and complex Middle Awash repository. Hundreds of team members from nearly 20 different nations play key roles in this process. Paleoanthropology, by definition, is multidisciplinary. In this historical science the objects of investigation are often unique, fragile, and irreplaceable, derived from contexts that are erased during their extraction. This is a science best done deliberately, carefully, and thoroughly. At 25 years, the Middle Awash research effort is both complex and protracted compared to most laboratory-based efforts in modern science. Project success is owed to this long-term perspective and to sustained funding for simultaneous research on multiple fronts.

In its efforts to illuminate African prehistory and paleontology, including the origin and evolution of hominids and their technologies, the Middle Awash project encompasses three basic areas of research: geology, archaeology, and paleontology. A fourth dimension, crucial to the project, is capacity building. During the last century, paleoanthropological research in Ethiopia was traditionally conducted by foreign-based expeditions, with little or no scientific collaboration by Ethiopian institutions or individuals. From its nascence, the Middle Awash project has dedicated itself to developing local scientific personnel and infrastructure that today characterize internationally prominent Ethiopian field and laboratory paleoanthropology.

The Middle Awash project gives priority to constructing an accurate time-stratigraphic framework for its fossil discoveries and to nesting these discoveries into paleoenvironmental contexts. As of the time of this writing, the Middle Awash research project has recovered nearly 300 hominid specimens from 15 separate temporal horizons. Many of these fossils are pivotal to understanding the evolution of our family, genus, and species. The recovered hominid remains are but a tiny fraction of the paleobiological evidence amassed by the Middle Awash project to date. Totals as of this writing (2009) are more than 19,000 cataloged vertebrate specimens; more than 1,800 geological samples; and thousands of lithic artifacts. All recovered fossils and artifacts are permanently housed in the collections of the National Museum of Ethiopia. These data, all painstakingly extracted from radioisotopically calibrated and stratigraphically controlled sedimentary contexts, constitute an exceptional record of Africa's past.

This progress has been realized by a global scientific consortium of involved laboratories and facilities. The research has been conducted by team members working under challenging field and laboratory conditions in roles as diverse as translator, archaeologist, Ethiopian government representative, mechanic, geochronologist, cook, paleobotanist, guide, fossil preparator, and dozens more. A full listing of fieldwork participants and primary laboratory researchers resides at the web site of the Middle Awash Project (http://middleawash.berkeley.edu). Middle Awash research results are realized through the support of institutions within Ethiopia and beyond. The continuous financial support of the National Science Foundation and the Institute of Geophysics and Planetary Physics at Los Alamos National Laboratory in New Mexico, with additional assistance from many other organizations and individuals, is gratefully acknowledged (see acknowledgments to each volume).

The Middle Awash project, like most scientific endeavors, uses peer-reviewed publications as the primary means by which its data are shared. Publication of the primary data generated by any such large multidisciplinary project is a formidable and essential undertaking. In paleoanthropology, the most important discoveries are traditionally first announced in high-impact journals and then, after more detailed analysis, published in specialty journals and monographs. In envisioning how to present the most significant discoveries of the Middle Awash study area in book form, we had the advantage of more than a century of scholarly publication in this field.

The Middle Awash Series concept launched with the 2008 publication of our volume on the geological background and paleontological content of the Daka Member of the Bouri Formation, edited by Henry Gilbert and Berhane Asfaw. The series will proceed as each subsequent, edited, stand-alone volume features the original research results of collaborating teams of project scientists who work together to illuminate a particular temporal period of paleoanthropological significance. Forthcoming volumes detail the project's discoveries of early anatomically near-modern *Homo sapiens idaltu* from Herto Bouri, and the context and anatomy of *Ardipithecus ramidus*. Additional volumes will be added to document the project's active ongoing field research.

These volumes share similar production values, organization, and methods of coverage. Within each volume, richly illustrated chapters contributed by project scientists are organized by topic. The accounts of the fossils, particularly the hominid fossils, go beyond mere anatomical description and are explicitly comparative. This series places on permanent record the definitive accounts of the most major discoveries of the Middle Awash research project.

To take advantage of the opportunities opened by the rise of information technology, we have taken steps to integrate the series with online digital resources. We have established a Middle Awash web site (middleawash.berkeley.edu) that features an accessible, user-friendly portal to project activities and accomplishments, including a full bibliographic listing of all paleoanthropologically relevant work published on the area since the earliest Italian geological explorations in the 1930s.

Another feature of the electronic interface to the Middle Awash discoveries is its specimen-level presentation. Modern digital informatics has allowed a proliferation of web-based faunal lists and other compendia that are increasingly used in meta-analyses ostensibly designed to explore global relationships between data sets as diverse as proxies of global climatic change and fossil evidence for biological evolution. Vertebrate paleontology and paleoanthropology have both witnessed an explosion of uncritical uses of secondary- and tertiary-level faunal lists and other accounts to explore these relationships. These analyses, and the conclusions based upon them, are only as good as the primary data upon which they are constructed.

Specimen-level catalog detail must form the empirical foundation of any such synthetic investigations. However, the necessary comprehensive detail on individual specimens and their provenience is traditionally lacking from project-level syntheses in paleoanthropology. Therefore, we have endeavored to accompany each of the volumes in the Middle Awash Series with full and free electronic access to specimen-based catalog detail for each and every collected vertebrate fossil. Furthermore, our accompanying web site archive releases, with the publication of each respective volume in the Middle Awash Series, digital photographic coverage of all cataloged fossils and special micro-CT-generated animations for selected hominid specimens. We hope these efforts to integrate and archive scholarly printed and digital resources will move paleoanthropology forward into a new century of data sharing.

This series is dedicated to the late F. Clark Howell (see tributes at herc.berkeley.edu/fc_howell_memorial), whose global vision, detailed knowledge, and passion for paleoanthropology inspires all the participants of the Middle Awash research project.

Tim White
Human Evolution Research Center
The University of California at Berkeley
March 2009

Was the oldest *Homo sapiens* pliocene or miocene, or yet more ancient? In still older strata do the fossilized bones of an Ape more anthropoid, or a Man more pithecoid, than any yet known await the researches of some unborn paleontologist? Time will show. But, in the meanwhile, if any form of the doctrine of progressive development is correct, we must extend by long epochs the most liberal estimate that has yet been made of the antiquity of Man.

—T. H. Huxley, *Man's Place in Nature and Other Essays*
(London: J. M. Dent & Sons, 1863), p. 150

It should be recalled that the whole of the African Pliocene, a span of over ten million years, is still either almost or entirely unknown. Late Tertiary hominoids surely occupied the more central area of the continent, although fossiliferous deposits of this time range are still unknown. Until some evidences of hominoid varieties are forthcoming from the upper Neogene of sub-Saharan Africa, any hypotheses of hominid origins will lack support.

—F. C. Howell, The Villafranchian and human origins, *Science* 130 (1959), p. 842

Whether, therefore, the hominid line had become separated from the common ancestor with the Pongids as early as the mid/Tertiary must remain in doubt until more complete fossil material becomes available.

—J. Desmond Clark, *The Prehistory of Africa* (New York: Praeger, 1970), p. 52

. . . all that has been collected so far would be but an element in the vast mosaic of African mankind, and my work hitherto and what still remains for me to do merely an overture to the far vaster themes that lie ahead.

—R. A. Dart (with D. Craig), *Adventures with the Missing Link*
(New York: Harper and Brothers, 1959), p. 24

Preface

The quest to elucidate human origins, particularly the period approximating the great evolutionary division between the lineages leading to modern chimpanzees and modern humans, was long frustrated by the lack of fossil evidence. In the 1990s the discovery of a new genus of early hominid, *Ardipithecus*, finally began to truly illuminate the late Miocene African hominid record. This volume is designed to contribute to this progress in revealing the origins of the Hominidae.

Integrated, multidisciplinary investigations of the Adu-Asa Formation exposed along the western margin of the Middle Awash study area in northeastern Ethiopia have revealed late Miocene records of volcanic and tectonic processes and rift evolution, as well as diverse fauna and flora. Despite limited geographic coverage, the late Miocene (5.2 to 5.8 million years) Asa Koma and Kuseralee Members of the Adu-Asa and Sagantole Formations, respectively, yielded the early hominid *Ardipithecus kadabba* fossils among 2,760 contemporary vertebrate remains from diverse taxa collected between 1990 and 2006 and reported in detail in this volume. These fossils were all recovered from spatiotemporally controlled contexts within geological deposits exposed by tectonics and erosion. Their geochronological, sedimentary, and paleoenvironmental contexts are reported here.

The inaugural volume of this Middle Awash series, edited by W. Henry Gilbert and Berhane Asfaw, presented extensive data about *Homo erectus* in Africa during the Pleistocene. In contrast, this volume documents the geological and paleontological evidence from the much older late Miocene Adu-Asa Formation of the western margin and from the rift-bound Kuseralee Member of the Sagantole Formation in the Central Awash Complex, located about 20 kilometers to the east. Yohannes Haile-Selassie initially described the faunal assemblages from these two members in his Ph.D. thesis, submitted to the University of California at Berkeley in 2001.

The volume provides a detailed description of the earliest hominid *Ardipithecus kadabba* and its context. This is the first complete description of an early hominid older than the 4.4 Ma *Ardipithecus ramidus* (recovered from the nearby Central Awash Complex; Chapter 7). The other taxa accompanying the *Ar. kadabba* remains are detailed within the other chapters of this volume, which offer a comprehensive description and evaluation of the most complete dataset available on the vertebrate evolutionary history and geological record of the terminal Miocene in eastern Africa.

Relatively few hominid remains were recovered from the Asa Koma Member of the Adu-Asa Formation, particularly compared with the number of early Pliocene *Ar. ramidus* remains collected from the Aramis Member of the overlying Sagantole Formation. Outcrops of Middle Awash late Miocene fossil localities are small and isolated because of dense faulting and the accumulation of alluvium and colluvium along the faulted slopes and displaced blocks of these ancient sediments. However, the rich and diverse vertebrate assemblages described here shed light on evolutionary patterns and on paleoecological conditions during the terminal Miocene of eastern Africa. The integrated geological, paleontological, and paleoenvironmental records detailed here provide a foundation for exploring new fossiliferous stratigraphic sections along the rift margin that predate the Adu-Asa Formation.

Yohannes Haile-Selassie
Cleveland, Ohio
Giday WoldeGabriel
Los Alamos, New Mexico
July 2007

Acknowledgments

The research and results described in this volume would not have been possible without the support and guidance of the Middle Awash research project, which incorporates the work of many scientists conducting field and laboratory investigations. Permission to conduct the research reported here was granted by the Authority for Research and Conservation of Cultural Heritage, Ministry of Culture and Tourism, Ethiopia. We are grateful for the assistance of the many officials involved. We also thank the National Museum of Ethiopia for the facilitation of field and laboratory studies. The many Antiquities Officers who have served with the project are acknowledged for their help.

In addition to the authors of the chapters in this volume, we thank the following individuals for their contributions to understanding the late Miocene deposits of the Middle Awash: Alemu Ademassu, Awoke Amzaye, Alemayehu Asfaw, Berhane Asfaw, Tadewos Assebework, Yonas Beyene, Raymonde Bonnefille, David Brill, Michel Brunet, José Miguel Carretero, Tadewos Chernet, Sylvia Cornero, Brian Currie, Garniss Curtis, Alban Defleur, Solomon Eshete, Kebede Geleta, Ann Getty, W. Henry Gilbert, Erksin Güleç, John Harris, Grant Heiken, Ferhat Kaya, Leonard Krishtalka, Tonja Larson, Bruce Latimer, Meave Leakey, Owen Lovejoy, Ayla Sevim, Lisa Smeenk, Scott Simpson, Solomon Teshome, Robert Walter, Samson Yosef, and Liu Wu.

Research in the remote Afar depression would not have been possible without the support of a variety of individuals responsible for the logistics of organizing and maintaining field camps under difficult conditions. No science would be possible without the support of the collectors, cooks, camp managers, and drivers for the project. A full listing of the approximately 600 people engaged in the project's field activities during the last quarter-century can be found at middleawash.berkeley.edu.

The Afar Regional Government of Ethiopia and the Afar people of the Middle Awash region, particularly the communities that live along the western margin and foothills, greatly facilitated and contributed to the successes of the geological and paleontological studies and discoveries presented in this monograph. In particular, sheiks Omar Hussein, Ebrahim Helem, Abdella Hussein, and Hussein Gaas of Ena-Ito are among the people who are notably acknowledged for coordinating the labor force throughout the years. We thank Ahamed Elema and his family for their unwavering support, and the late Neina Tahiro for his friendship.

Most of the work described here was supported by grants from the National Science Foundation (NSF) of the United States (grants BNS 80-19868, BNS 82-10897, SBR-9318698, SBR-9512534, SBR-9521875, SBR-9632389, BCS-9714432, and BCS-9910344). The Revealing Hominid Origins Initiative (RHOI; NSF-HOMINID-BCS-0321893), co-directed by the late F. Clark Howell and by Tim D. White, did not support Middle Awash field or laboratory studies but helped to fund many of the investigators who worked on late Miocene collections and participated in the RHOI Analytical Working Groups, which have proven fundamentally important in the timely and accurate completion of various chapters of this volume. The Institute for Geophysics and Planetary Physics (IGPP) at the Los Alamos National Laboratory in New Mexico provided crucial support for the field geology and laboratory geochemical work associated with the project during the last 15 years. The National Geographic Society provided a vehicle for fieldwork in 1982. Additional research funding is identified in the following paragraphs on a chapter-by-chapter basis.

We are grateful to Chuck Crumly for providing guidance throughout the process of completing this volume and to Scott Norton and Francisco Reinking, who handled the submission package. We acknowledge all those at the University of California Press who contributed to the production of this volume. We would like to thank the anonymous reviewers for their comments on individual chapters and the overall presentation.

We are particularly indebted to Tim White for his selfless and untiring support and guidance for the completion of this volume. We thank members of the Human Evolution Research Center (HERC) for providing essential help—Kyle Brudvik, Jason Crosby, Brianne Daniels, Anneke Janzen, Ben Mersey, Sarah Moon, and Denise Su. We are especially grateful to Denise Su for facilitating the completion of this volume by providing substantial and invaluable editorial assistance.

The cooperation of the National Museums of Kenya (Meave Leakey and Emma Mbua) in giving access to the original, then unpublished, Lothagam material is very much appreciated. We also thank the Iziko South African Museum in Cape Town (Margaret Avery and Derek Ohland) for access to the original fossil specimens from Langebaanweg.

The following individuals and institutions made specific contributions to the individual chapters:

Chapter 2, "Stratigraphy of the Adu-Asa Formation": Access to Electron Microprobe and other support were provided by the Earth and Environmental Sciences Division of the Los Alamos National Laboratory (see above) and by the Institute of Geophysics and Planetary Physics (IGPP) of the University of California at Los Alamos National Laboratory (LANL).

Chapter 3, "Volcanic Record of the Adu-Asa Formation": The research was primarily supported by NSF Middle Awash field grants (see above) and by the Institute of Geophysics and Planetary Physics (IGPP) of the University of California at Los Alamos National Laboratory (LANL). WKH also acknowledges support from the Miami University Hampton Fund for International Faculty Development and the Miami University Committee on Faculty Research. Access to Electron Microprobe and other support from the

Earth Environmental Sciences Division at LANL facilitated the laboratory analysis. Tim and Darin Snyder are thanked for comments on an earlier version of this manuscript.

Chapter 5, "Small Mammals": Yohannes Haile-Selassie is acknowledged for his initial identification of the taxa. Valuable assistance in the systematic analysis was provided by Percy Butler, Michel Brunet, Christiane Denys, Daphne Hills, F. Clark Howell, Paula Jenkins, Wim Van Neer, Laurent Viriot, and Alisa Winkler. Research was partially funded by the National Science Foundation (ID 0521538) to HBW. The Natural History Museum at Sierra College, Rocklin, California, contributed directly to research efforts.

Chapter 6, "Cercopithecidae": Terry Harrison and Michelle Singleton are thanked for some of the data on humeri and molars, respectively. Berhane Asfaw and Tim White invited us to study the material. This research was supported by the L. S. B. Leakey and Wenner-Gren Foundations as well as the New York College of Osteopathic Medicine (to SF).

Chapter 7, "Hominidae": C. Owen Lovejoy and the late F. Clark Howell are thanked for providing constructive discussions. We thank M. Brunet for sharing unpublished and partially published information on *Sahelanthropus* and for discussions. We thank S. Semaw and S. Simpson for sharing unpublished and partially published information on Gona *Ardipithecus* materials. We thank B. Senut and M. Pickford for information and for observation of *Orrorin* casts. We thank L. de Bonis for access to *Ouranopithecus* casts and originals. The micro-ct analysis was done in collaboration with R. T. Kono. We thank the following individuals and their insitutions for access to the modern and fossil comparative materials: B. Latimer and L. Jellema, Cleveland Museum of Natural History, United States; W. Van Neer and W. Wedelen, Royal Museum of Central Africa, Tervuren, Belgium; J. de Vos, Naturalis, Leiden, The Netherlands; M. Yilma, National Museum of Ethiopia; M. Leakey and E. Mbua, National Museums of Kenya; F. Thackeray, Transvaal Museum, South Africa; S. Moya-Sola and M. Keilor, Crusafont Institute of Paleontology, Sabadell, Spain; W. Liu, I.V.P.P., Beijing, China; B. Engesser, Natural History Museum, Basel, Switzerland; and E. Heizmann, Natural History Museum, Stuttgart, Germany.

Chapter 8, "Carnivora": We thank Louis de Bonis for reviewing the final manuscript and Tim White for constructive discussions throughout the writing of this chapter.

Chapter 10, "Suidae": John Harris, Meave Leakey, and Scott Simpson are thanked for discussions on suid phylogeny, and Tim White is thanked for his constructive discussions on suid phylogeny, evolution, and for his earlier works on the family Suidae, which laid the background for this chapter.

Chapter 11, "Hippopotamidae": J. Surault, A. Foray, and E. Lavertu in Addis Ababa, Ethiopia, and G. Florent in Poitiers, France, are thanked for their kind help, as well as M. Brunet, F. C. Howell, and P. Vignaud for their advice and support. We thank Denise Su for providing helpful comments. JRB's research was funded by the Ministère Français de l'Education Nationale et de la Recherche (Université de Poitiers), the Mission Paléo-anthropologique Franco-Tchadienne, the Fondation Fyssen (postdoctoral research grant), the Ministère des Affaires Etrangères (program Lavoisier and SCAC, French Embassy in Ethiopia), and the Foundation Singer-Polignac (postdoctoral research grant).

Chapter 14, "Rhinocerotidae": For valuable assistance and discussion we would like to thank B. Asfaw, T. D. White, F. C. Howell, R. Bernor, W. H. Gilbert, J.-R. Boisserie, Z. Alemseged, F. Kaya, A. Louchart, and D. F. Su. For facilitating access to material under their care, we are indebted to P. Tassy, C. Saigne, C. Lefevre, J.-G. Michard, L. Viva (MNHN); J. Hooker, A. Currant, R. Sabin (BMNH); L. van de Hoek Ostende, J. de Vos (RMNH); A. Rol (ZMA); K. Heissig (BSPG); F. Schrenk, G. Plodowski (SMF); O. Sandrock (HLMD); R. Frey, D. Schreiber (SMNK); G. Höck, F. Spitzenberger, B. Herzig (NHMW); G. Rabeder, D. Nagel (IPUW); L. Kordos (MAFI), M. Dermitzakis, G. Theodorou (AMPG); G. Koufos (LGPUT); E. Güleç, A. Sevim (AUABL), and L. Gordon (USNM). Support for comparative studies was provided to IXG by the European Commission's Research Infrastructure Action (EU-SYNTHESYS: GB-TAF-574, FR-TAF-1226, NL-TAF-2513, FR-TAF-2545) and the European Science Foundation (ESF-EEDEN/2003/EX05).

Chapter 15, "Proboscidea": John Hooker and Andy Currant (Natural History Museum, London, UK), Margaret Avery and Graham Avery (South Africa Museum), Emma Mbua (National Museum of Kenya, Nairobi) are thanked for access to collections in their care. Georgi Markov provided papers and pictures of late Miocene–Pliocene gomphotheres from Europe and discussed the relationship between African and European gomphotheres. Martin Pickford provided pictures of unpublished material of late Miocene proboscideans from Lukeino, Kenya. Bill Sanders provided discussions and suggestions. Denise Su provided helpful comments and edits. Financial support was provided by the National Science Foundation and the Japanese Ministry of Education Culture, Sports, Science and Technology (Grant-In-Aid: #11691176, #14255009, #17540445).

Chapter 16, "Tubulidentata": Support for the research was provided by a National Research Foundation (NRF) Postdoctoral Fellowship.

Chapter 17, "Paleoenvironment": Antoine Louchart, Allison Murray, and Kathy Stewart provided unpublished data. Peter Andrews, Ray Bernor, Faysal Bibi, and Tim White reviewed this manuscript and provided helpful comments. DFS was funded by the Human Evolution Research Center at the University of California, Berkeley. Stable isotope mass spectrometry in the Environmental Isotope Paleobiogeochemistry Laboratory was supported in part by a NSF instrumentation grant and the University of Illinois Research Board.

Chapter 19, "Paleobiogeography": National Science Foundation (EAR-0125009) supported this research and funded the computer graphics facility at Howard University. LR was supported by a CNR-NATO Outreach Fellowship. L. S. B. Leakey Foundation and the National Geographic Society supported RLB's fieldwork in Hungary and Germany and LR's fieldwork in Italy, which contributed substantially to the background of this study.

1

Introduction

YOHANNES HAILE-SELASSIE AND GIDAY WOLDEGABRIEL

The Middle Awash late Miocene vertebrate fossil assemblage comprises two distinct paleontological collections, dated to 5.2 and 5.8 Ma. The combined assemblage exhibits unparalleled taxonomic diversity, large overall sample sizes, and refined spatial and temporal frameworks. The late Miocene evolutionary history of African vertebrates and the origin of the modern African fauna have been poorly understood until recently because of the lack of adequate fossil evidence from the critical 5–7 Ma time period. In the last decade, our understanding of the late Miocene evolutionary history of African vertebrates has improved as a result of the fossil discoveries at Lothagam (Kenya; Leakey and Harris 2003a), the Tugen Hills (Kenya; Hill 2002), Toros Menalla (Chad; Brunet et al. 2002), Manonga (Harrison 1997a), Langebaanweg (South Africa; Hendey 1982), Uganda (Pickford et al. 1993), and the Middle Awash (Ethiopia; Haile-Selassie 2001a). Sampling the last two million years of the Miocene, the data from these sites are crucial for understanding the origin of the modern African fauna.

Historical Background

The first European explorers crossed the Middle Awash study area (Figure 1.1) in the late 1920s, led by a British mining engineer (Nesbitt 1935). Despite widespread occurrences of fossils and stone tools in the area, the expedition made no mention of paleontological or archaeological records. Neither did the Italian geologists who worked in the Middle Awash region in 1938 (Gortani and Bianchi 1973). As part of his Awash River basin survey during the late 1960s, Taieb described fossils and artifacts, named localities, and drew stratigraphic sections for the Middle Awash area (Taieb 1974). However, by the early 1970s, Taieb and his research collaborators focused their attention on the Hadar area, about 70 km to the north of the Middle Awash (Figure 1.1). The detailed geological and paleontological investigations that they conducted at Hadar led to the discovery of the first hominid fossils in the Afar (Taieb et al. 1976; Johanson et al. 1982b). In the 1990s, more studies were conducted at Hadar (Kimbel et al. 1994) and at Gona (Semaw et al. 1997, 2005), located between the Middle Awash and the Hadar research areas. This research, particularly that ongoing in the "Dikika" area (formerly known as South Hadar), has produced additional fossils (Alemseged et al. 2006; Wynn et al. 2006).

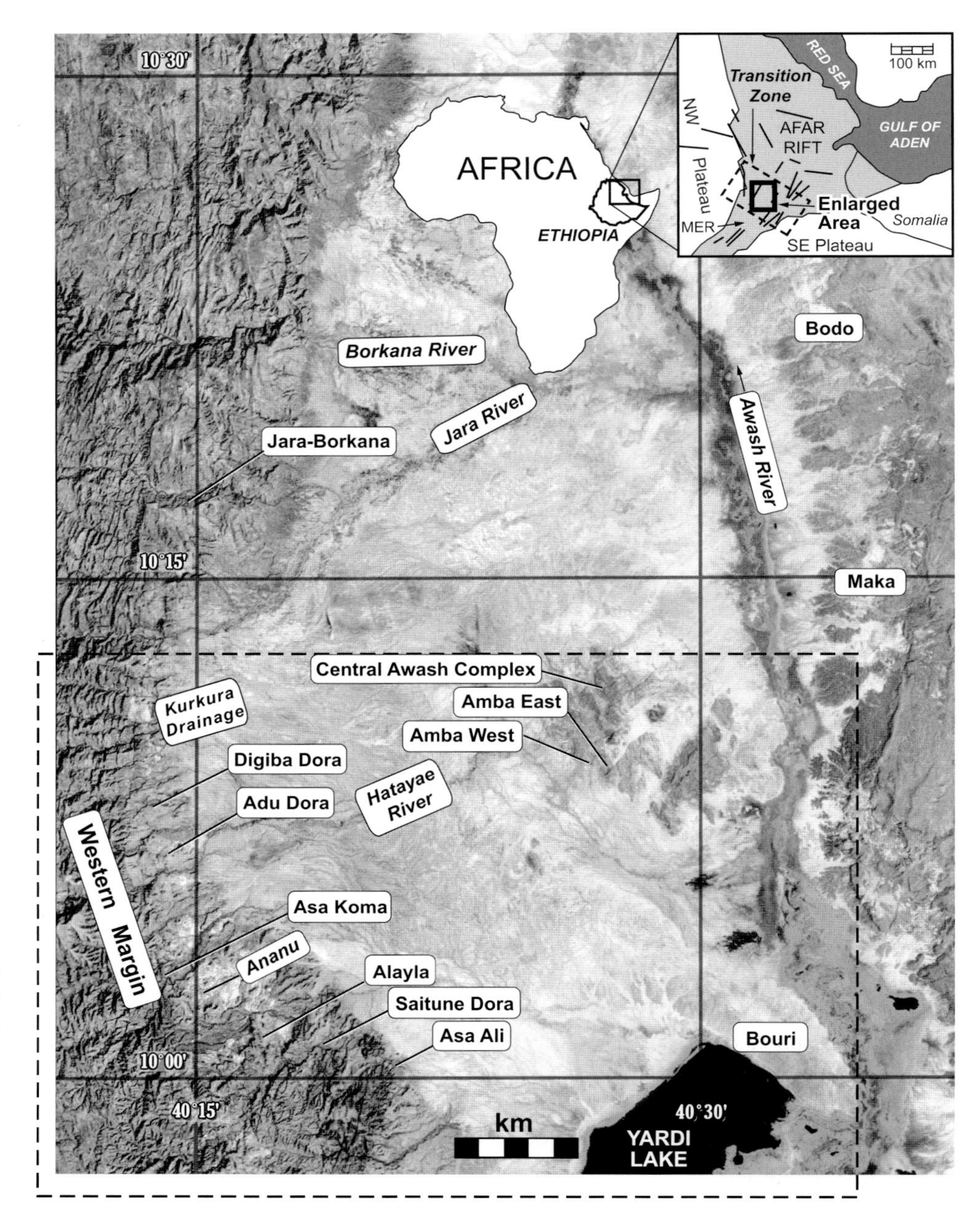

FIGURE 1.1

Satellite imagery showing the location of vertebrate collection areas of the Middle Awash study area (adapted from WoldeGabriel et al. 2001). The area within the dashed lines is shown in detail in Figure 2.5.

Jon Kalb split from the Hadar project to establish the Rift Valley Research Mission in Ethiopia (RVRME) after the discovery of the first Hadar hominid fossils in 1973. Kalb and colleagues conducted geological and paleontological reconnaissance within the rift on both sides of the modern Awash River, making the first paleontologically inspired visits to the adjacent western rift margin of the Middle Awash study area (Kalb 1993, 2000, and references therein). The RVRME project discovered the first hominid remains at Bodo in the Middle Awash study area in 1976 (Conroy et al. 1978; Kalb et al. 1980). The RVRME project was terminated in 1978, and the results of its fieldwork were subsequently published (Kalb et al. 1982a–e, and references therein).

Kalb and colleagues (1993) published a revised stratigraphic sequence and nomenclature for the entire southern Afar Rift based on aerial photograph interpretation, field lithology observations, and biochronology. In ascending stratigraphic order, Kalb and colleagues assigned the diverse volcanic and sedimentary units of the Middle Awash study area to the Miocene Adu-Asa, Pliocene Sagantole and Matabaitu, and Pleistocene Wehaitu Formations. Moreover, Kalb and colleagues assigned the Miocene and Plio-Pleistocene stratigraphic sequences of the Middle Awash study area, the Pliocene Hadar Formation (Taieb et al. 1976; Tiercelin et al. 1980), and the late Miocene Chorora Formation exposed along the southeastern rift margin and adjacent rift floor of the transition zone (Sickenberg and Schonfeld 1975) to the "Awash Group." The Chorora Formation, which crops out along the margin and foothills of the southeastern escarpment of the southern Afar and the Main Ethiopian Rifts, has no temporal or lithological continuity with the younger units of the Middle Awash study area, located more than 150 km to the north. According to Salvador (1994), the term "Group" is applied most commonly to two or more contiguous or associated formations, sharing important and diagnostic lithologic properties. There is a stratigraphic link between the Pliocene deposits of the Hadar and Middle Awash fossiliferous sedimentary and volcanic units via the Sidi Hakoma Tuff (SHT) tephra stratigraphic marker (Walter and Aronson 1982; White et al. 1993).

Unfortunately, Kalb's first attempts at time-stratigraphic segmentation were compromised by the lack of temporal resolution. No chronological or geochemical analyses were conducted on the abundant interbedded volcanic rocks. As a result, despite largely photo-interpretive mapping of the Middle Awash study area at a 1:60,000 scale (Kalb et al. 1982d), the volcanic and tectonic processes that led to rifting and to the accumulations of thick volcanic and sedimentary deposits could not be characterized, for lack of temporal and spatial controls, until our project's work began in the 1980s.

The Middle Awash research project was formed by the late J. Desmond Clark and Tim White, both of the University of California at Berkeley. They assembled a multidisciplinary team to conduct geological, paleoanthropological, and archaeological reconnaissance surveys, initially focusing primarily on the eastern side of the Middle Awash study area. The research project completed the first field season with archaeological and paleontological discoveries at Bodo, Hargufia, Maka, and Belohdelie (Asfaw 1983; Clark et al. 1984; White 1984; Clark et al. 1993). Detailed geological field investigations, coupled with geochemical and geochronological analyses, established temporally and spatially controlled stratigraphic sections for the first time by using interbedded volcanic rocks (Harris 1983; Clark et al. 1984; Hall et al. 1984). The sedimentological processes and paleoenvironmental records on the east side of the study area were also described and documented (Williams et al. 1986).

In 1981, brief reconnaissance surveys were made on the west side of the Middle Awash at Bouri, Adu Dora, and Aramis (see Figure 1.2). After this initial reconnaissance work, no Middle Awash fieldwork was allowed between 1982 and 1990. However, beginning in the autumn of 1990, detailed studies of Pliocene and Pleistocene deposits on the east side

FIGURE 1.2

Our first reconnaissance of the Middle Awash late Miocene deposits was in early November 1981, when a small team in a single vehicle crossed the Awash River by raft at Debel Plantation. The first camp was in the dust of an abandoned Afar hamlet on the floodplain of the Awash River. Photograph by Tim White, November 7, 1981.

of the Middle Awash were resumed. These generated geochemical and geochronological data that facilitated correlations among several widely separated stratigraphic sections (White et al. 1993; Clark et al. 1994). Although reconnaissance of the study area's western side was initiated in 1981 by Clark and White, work on the eastern side of the Awash River took precedence until 1992, when geological and paleontological investigations on the western side of the Middle Awash study area were resumed. These initially focused on the rift-bound, structurally domed Central Awash Complex (CAC) and the Bouri Horst (see Figures 1.3 and 1.4).

The rugged terrain of the western rift margin of the Middle Awash study area had been penetrated by 1981 foot and vehicle surveys. From a campsite along the Hatayae River in 1992 the research team extended these surveys deeper into the study area's western margin. However, the 1992 discovery of the early hominid *Ardipithecus ramidus* at Aramis in the CAC caused the project's geologists and paleontologists to focus their full efforts on associated strata there. Archaeological work on younger deposits continued through the 1990s in the Aduma and Bouri areas (Yellen et al. 2005; de Heinzelin 2000; Gilbert and Asfaw 2007). Field studies, coupled with geochemical and geochronological analytical results from several volcanic strata, established detailed chronostratigraphic sections of major outcrops within the CAC. A local chronostratigraphy of late Miocene and early Pliocene volcanic and fossiliferous fluvial and lacustrine sedimentary rocks was established (WoldeGabriel et al. 1994, 1995; Renne et al. 1999). Moreover, similar investigations within the Bouri Horst established a detailed Plio-Pleistocene chronostratigraphic sequence for that uplifted structural block (de Heinzelin et al. 1999; Clark et al. 2003; de Heinzelin 2000).

As the research project intensely focused its activities in the CAC in the early 1990s, it became apparent that sediments exposed by faulting and erosion on the western margin of the study area would be important to document in detail. It would be particularly

FIGURE 1.3

View to the east across the Asa Issie Pliocene locality (village in the distance). The Hatayae Graben is marked by the taller acacia trees in the middle distance. The center of the Central Awash Complex (CAC) is seen on the near horizon, capped by horizontal outcrops of the Gawto basalt that overlie the Kuseralee Member at sites such as Amba (center frame). Ado Fila hill (undated) is to the right of frame, and the eastern side of the study area forms the far horizon. The Pliocene site of Aramis, located above the Gawto basalt and Haradaso Member, is to the left, out of the frame. Photograph by Tim White, December 20, 2005.

important to test Kalb's hypothesis that sediments at the base of the CAC succession were coeval with deposits centered 25 km to the west.

Initial reconnaissance surveys in the western rift margin were followed by detailed studies at selected sites along the strike of the frontal fault blocks after 1993. This work characterized and ultimately defined the diverse major lithological units of the western margin of the study area. It also established accurate and detailed chronostratigraphic sections and nomenclature according to the International Stratigraphic Code (Hedberg 1976; WoldeGabriel et al. 2001). The detailed geological field investigations, coupled with the sampling and analysis of volcanic rocks, established temporal and spatial constraints for several fossiliferous sedimentary outcrops in the CAC and along the strike of the rift margin. The initial results of these various investigations, particularly those involving accompanying hominid remains, have been presented elsewhere (White et al. 1994; WoldeGabriel et al. 1994, 1995; Asfaw et al. 1999; de Heinzelin et al., 1999; Renne et al. 1999; Clark and Schick 2000; de Heinzelin 2000; Haile-Selassie 2001b; Wolde-Gabriel et al. 2001; Asfaw et al. 2002; Clark et al. 2003; Hart et al. 2003; White et al. 2003; Haile-Selassie et al. 2004b).

Post-1994 Survey of the Western Margin

In an effort to locate new fossiliferous deposits, the authors have conducted numerous foot and vehicle surveys along the western margin since 1995 (Figure 1.5). These surveys often required very long foot transects across rugged terrain. However, they resulted in the location and recovery of rich paleontological assemblages. As shown in Table 1.1, about two dozen vertebrate localities (VP) from 14 defined collection areas along the western margin of the Middle Awash study area have yielded, at the time of this writing, approximately three thousand collectible, identifiable fossil vertebrate specimens.

FIGURE 1.4
View to the east across Amba East Vertebrate Paleontology Locality 1, a locality of the CAC. Alemayehu Asfaw is the fossil collector, surveying the surface of outcrops of the Kuseralee Member, just below the whitish horizon of the Kuseralee Sands. The Gawto Member basalts, dated to 5.2 Ma, are up section to the left. Photograph by Tim White, November 14, 1995.

During five field seasons (1995–1999) Yohannes Haile-Selassie and Giday WoldeGabriel attempted to survey all exposed patches along a 50 km^2 area between an Afar village called Enayto in the south and the Kurkura drainage in the north (Figures 1.1 and 1.6). The westernmost parts of some of the exposures were accessible only after 6 to 7 hours of walking. Such long transects were made numerous times within the core area, and these survey activities were subsequently extended to the north. In the core survey area, efforts resulted in the designation of five previously unknown paleontological collection areas, two of which yielded fossil hominid remains (see Table 1.2 for details).

The first vertebrate paleontology locality designated in 1995 was Digiba Dora, north of the Adu Dora area previously worked by the RVRME (see Figure 1.1). Digiba Dora was relatively rich in vertebrate fauna given its small size. Most of the 1995 field season was spent on geological reconnaissance and rock sampling. In 1996, the survey and exploration extended further to the south and into the escarpment. During the 1996 field season, while conducting geological reconnaissance deep in the escarpment southwest of the area known as Ananu (Figures 1.7 and 1.8), two new localities were discovered near a small drainage locally known as Alayla (see Figure 1.1 for location). The two Alayla localities were patches of exposures separated by a small fault and hidden behind a massive fault block. In December 1997, one of the two localities yielded the first hominid specimen from the western margin (Figure 1.9). Asa Koma locality 3 was probably visited by the RVRME exploration, and was

FIGURE 1.5

Building a new road from Hada la Ella to Enayto, down the western escarpment of the Middle Awash study area in 1999. Successive faults associated with formation of the rift shoulder have made this terrain steep and thus rendered it largely inaccessible. This is one of many roads constructed by the research project to gain better access to the fossiliferous localities. Photograph by Tim White, November 7, 1999.

located one day after the initial discovery of the Alayla localities. This is one of the richest vertebrate localities to have yielded hominid remains in addition to numerous other vertebrate fossils (Figures 1.10 and 1.11). Between 1997 and 1999, new localities such as Gaysale, Ali Ferou Dora, and Ado Faro were added to the list (Table 1.1).

In December 1999, the survey was extended further to the north, particularly along the margin between the Jara and Borkana Rivers (Figure 1.1). During subsequent years, the survey was extended along the rift margin north of Dallifage as far as the Talalak, and some localities were designated (Figure 1.1, Table 1.1). This northern area is currently geologically less studied than its southern counterpart. Preliminary radiometric dates indicate that the Jara-Borkana localities might be slightly older than the localities in the Adu-Asa Formation (see Chapter 4). However, the taxonomic diversity is minimal, and vertebrate fossils are not as abundant as at late Miocene localities further south. Work in this region, and in the Gona area further to the north (Quade et al. 2004), is continuing and promises to expand the overall late Miocene record of this part of the Afar.

Objectives

The objectives of the investigations along the frontal fault blocks of the broad western rift margin of the Middle Awash study area were the following:

TABLE 1.1 Vertebrate Paleontology (VP) Localities Designated along the Frontal Fault Blocks of the Western Rift Margin and Adjacent Rift Floor of the Middle Awash Study Area Shown in Figure 1.1

Collection Area	*Locality*	*Year*	*Collection Area*	*Locality*	*Year*
Adu Dora	ADD-VP-1	1994	Bikir Mali Koma	BIK-VP-1	1992
	ADD-VP-2	1994		BIK-VP-2	1999
	ADD-VP-3	1995		BIK-VP-3	1999
	ADD-VP-4	1999	Bilta	BIL-VP-1	1999
Alayla	ALA-VP-1	1996	Digiba Dora	DID-VP-1	1995
	ALA-VP-2	1996			
	ALA-VP-3	1999	Gaysale	GAS-VP-1	1999
Ali Ferou Dora	AFD-VP-1	1998	Jara Borkana	JAB-VP-1	1999
	AFD-VP-2	2004		JAB-VP-2	2003
Amba	AME-VP-1	1992		JAB-VP-3	2004
	AMW-VP-1	1992	Kabanawa	KWA-VP-1	1996
Asa Koma	ASK-VP-1	1992	Kuseralee Dora	KUS-VP-1	1992
	ASK-VP-2	1992	Saitune Dora	STD-VP-1	1993
	ASK-VP-3	1996		STD-VP-2	1993

1. Investigate the volcanic strata, including lava flows, hydromagmatic basaltic eruptions, and distal silicic pyroclastic deposits.
2. Delineate the relations between faulting and rifting and the mechanism for the origin of the steeply dipping and rotated antithetic fault blocks.
3. Identify the major lithological units exposed by faulting along the strike of the rift margin and along deeply cut canyon walls that are transverse to the rift escarpment.
4. Describe and document the different sedimentological processes that led to the deposition of diverse sedimentary rocks.
5. Describe the current geomorphic features along the western rift margin and the adjacent rift floor and their implications for basin development.
6. Recover, describe, and document the faunal and floral assemblages within the fossiliferous sedimentary deposits.
7. Assess the paleoenvironmental records from the diverse lithologic assemblages, paleontological assemblages, and stable isotope analysis of paleosol, pedogenic carbonates, and carbonates.

Methods

Field Survey and Reconnaissance

Recent application of space-based imagery and aerial photographs has revolutionized paleontological field survey. During the 1980s the Thematic Mapper (TM) and Space Shuttle

FIGURE 1.6
View to the northwest across the Saragata River floodplain east of Enayto village. Faint light-colored patches in the rift margin represent late Miocene sediments. Large portions of the Middle Awash study area are covered with unfossiliferous modern alluvium, and the fossil sites are found in patches exposed via tectonics and erosion. Photograph by Tim White, December 16, 1997.

Large Format Camera (LFC) platforms provided imagery to locate areas of paleontological interest efficiently and accurately (Asfaw et al. 1990). The use of space-based imagery not only was accurate and efficient but also lowered time and money costs for the project. Using this imagery, we were able to target inaccessible localities. Foot survey guided by satellite and aerial photographic imagery led to discovery of new late Miocene localities, some of which yielded the hominid fossil specimens reported here. Our team used aerial photographs at the standard scale of 1:60,000, LFC photographs, and TM images (WoldeGabriel et al. 2004).

Kalb (1978; Kalb et al. 1982a–e) reported the use of aerial photographs of a scale 1:60,000 to conduct geologic and photogeologic mapping of the Middle Awash. However, no copies with plotted data were left available for subsequent researchers. Further paleontological survey conducted in the area in the early 1980s (Clark et al. 1984) continued to use 1:60,000 scale aerial photographs acquired by purchase from the Mapping Agency of the Government of Ethiopia. Beginning in 1990, space-based images (TM and LFC) were incorporated into the Middle Awash paleontological surveys. These new images were acquired from the Laboratory for Terrestrial Physics of the NASA Goddard Space Flight Center (see Asfaw et al. 1990 for details).

The Middle Awash project started using the Middle Awash Digital Map Archive (MADMA) in 1995. This consists of mosaics of digitized background images scanned from the 1:60,000 aerial photographs and digitally overlain with site data. MADMA facilitated easy access to existing data and also guaranteed accurate data recording and retrieval. The system was developed and administered by Henry Gilbert of the Laboratory for Human Evolutionary Studies (LHES) at the University of California at Berkeley (de Heinzelin et al. 2000b). The use of MADMA facilitated accurate documentation for both

TABLE 1.2 Collection Methods Used and Number of Specimens Collected at Each Late Miocene Vertebrate Locality

Locality	*1992*	*1993*	*1994*	*1995*	*1996*	*1997*	*1998*
Asa Koma Member							
ADD-VP-1			FS (2)			FS, C (20)	
ADD-VP-2			FS (1)				
ADD-VP-3				FS (1)			
ADD-VP-4							
AFD-VP-1							FS (4)
AFD-VP-2							
ALA-VP-1					FS (4)	FS (13)	
ALA-VP-2					FS (3)	FS, C, E/S (99)	FS, C (47)
ALA-VP-3							
ASK-VP-1	FS (3)						FS (7)
ASK-VP-2	FS (4)						FS (8)
ASK-VP-3					FS (2)		FS (76)
BIK-VP-1	FS (3)						
BIK-VP-2							
BIK-VP-3							
BIL-VP-1							
DID-VP-1				FS (4)		FS (54)	FS, C (57)
GAS-VP-1							
JAB-VP-1							
JAB-VP-2							
JAB-VP-3							
KWA-VP-1					FS (3)		
STD-VP-1		FS (13)	FS (15)	FS (2)	FS (2)		FS (17)
STD-VP-2		FS (4)		FS (4)			FS, C, E/S (124)
Total number of specimens							
Kuseralee Member							
AME-VP-1	FS (27)	FS (13)		FS (12)			
AMW-VP-1	FS (38)			FS (10)			
KUS-VP-1	FS (41)			FS (1)	FS (2)	FS (34)	FS (33)
Total number of specimens							

NOTE: Number in parentheses is the number of specimens collected by the indicated method.
FS = free survey; C = crawling; E = excavation; S = sieving.

delimiting locality boundaries and pinpointing recovered specimens. Although not a formal geographic information system (GIS), the MADMA was nevertheless able to capture accurate positional data from our 1981 and 1990–1995 work in the study area. With the availability of georeferenced digital imagery after 2000, the MADMA system was used to transfer aerial photograph-based locational information into a latitude/longitude system.

Two types of Global Positioning System (GPS) devices, Trimble and Magellan, were used between 1990 and 1997 to plot all vertebrate localities. The field accuracy of these devices was, at the time of their use under Selective Availability, ±50–100 m. In 1998 the project started using differential GPS (DGPS) to locate all new paleontological localities and col-

1999	*2000*	*2001*	*2002*	*2003*	*2004*	*2005*	*TNS*
							22
							1
FS (4)							5
FS (1)							1
					FS (2)		6
					FS (9)		9
FS (8)	FS (3)		FS (4)			FS (4)	36
FS, C, E/S (160)	FS, C (15)		FS, C (18)			FS (7)	349
FS (6)				FS (1)			7
FS (4)			FS (9)				23
							12
FS (78)		FS, C (25)	FS, C, E/S (333)			FS (28)	542
FS (9)			FS (2)				14
FS (1)							1
FS (17)			FS (8)				25
FS (1)							1
FS (18)					FS (29)		164
FS (1)							1
FS (6)				FS (11)	FS (8)		25
				FS (2)			2
					FS (13)		13
				FS (2)			5
FS (3)				FS (12)			61
FS (768)	FS (14)		FS (8)		FS (1)		923
							2248
FS, C (31)	FS (78)		FS (13)			FS (30)	204
FS (28)	FS (74)						150
						FS (9)	120
							474

lected fossils. The field accuracy of DGPS, based on blind-test returns to known points in the study area, is within 2 m of the original coordinates in over 95 percent of all tests. The Middle Awash faunal catalog contains individual latitude and longitude placements of all collected faunal remains where appropriate.

Geological Mapping and Sampling

The RVRME produced a geological base map of the Middle Awash area that was subsequently published by Kalb et al. (1982d). This map showed major drainages and rough

FIGURE 1.7
View to the southeast across the western margin of the Middle Awash study area. The Ayelu volcano is seen to the right of frame. Sediments of the Adu Dora Member of the Adu-Asa Formation are visible throughout the Ananu valley shown here. Despite the good outcrop, very few vertebrate fossil concentrations were found in Ananu sediments. Photograph by Tim White, November 18, 1981.

locations of geological contacts. Later publications by Kalb (1993) and Kalb (1995) presented a purportedly "refined" stratigraphy and geological distribution of the "Awash Group." However, the geological map presented in those later publications was effectively the same as that published in the early 1980s. Moreover, the fossils collected by the RVRME were never individually plotted on aerial photographs, so the precise spatial location of most of the collected specimens was unknowable from published sources or archives of the Ethiopian Ministry of Culture and the National Museum of Ethiopia.

Geologists of the Middle Awash research project began to collect rock samples from the Middle Awash study area in 1981. Most of the samples from the western margin were collected after 1995, for one of four reasons. First, a sample was collected if it was overlying or underlying a fossiliferous horizon and only if it was judged likely to be radiometrically datable (i.e., not weathered). Second, a sample was collected if it was critical for studying the regional geology and geochemical analysis for tephra correlation. Third, biogenic samples, such as diatomites, were collected for paleoecological interpretations. Fourth, selected samples were collected for paleomagnetic analysis. More than 80 geological samples were collected from the western margin through 2005. The most critical samples that overlie and/or underlie fossiliferous horizons have been dated by the $^{40}Ar/^{39}Ar$ dating method, and the results are presented in Chapter 4.

FIGURE 1.8
View to the northwest near the upper boundary of the Ananu catchments. Visible in this view are the basaltic tuffs of the Asa Koma Member, the laminated diatomite of the Adu Dora Member, and the underlying silty clays of the Saraitu Member. Note the successive normal faults dropping package successively to the right (east). Photograph by Tim White, December 2, 1992.

Geochemical Analysis

Tephra deposits interbedded within the sedimentary rocks of the western margin provide important temporal and spatial constraints across densely faulted stratigraphic sections. Several rhyolitic and basaltic tephra stratigraphic markers and basaltic lava flows exposed along fault scarps and drainage channels were collected for petrographic and geochemical analyses, and $^{40}Ar/^{39}Ar$ dating. More than 50 tephra and lava samples were analyzed using electron microprobe (EMP) and argon plasma methods. The details of the geochemical results and tephra correlation are documented in Chapter 3.

Radiometric Dating

Geological samples directly relevant to vertebrate paleontology were dated using the Single Crystal Laser Fusion (SCLF) dating method (see Chapter 4 for details). This method has an advantage over other methods because primary crystals can be easily segregated from detrital contaminants within the same volcanogenic sample. The $^{40}Ar/^{39}Ar$ analysis of the dated samples was conducted at the Berkeley Geochronology Center (BGC). Renne et al. (1999) have described the details of the dating method and facilities used, following Deino and Potts (1990), Deino et al. (1997), Renne (1995), and Renne et al. (1996).

Locality Designation

Locality nomenclature and designation protocols used by the Middle Awash project have been fully described in de Heinzelin et al. (2000b). The vertebrate paleontology (VP)

FIGURE 1.9
View to the southwest across ALA-VP-2 on the day of discovery of the *Ardipithecus kadabba* type specimen. This illustrates free survey, in which fossiliferous outcrops of sediments are traversed by collectors on foot, usually kneeling or bending over to scan the surface for fossils. Photograph by Tim White, December 16, 1997.

localities and the number of fossils collected from each of the designated late Miocene sites are given in Table 1.1.

A collection area is an area where there are fossiliferous deposits spatially isolated from other such deposits, and one or more localities are designated within it. An outcrop is designated as a VP locality when taxonomically diagnostic fossil specimens are found and when their constrained spatial and stratigraphic provenience can be established by surface position or adhering matrix. Such new localities were usually encountered either while conducting geological reconnaissance or by conducting satellite imagery and aerial photo-guided paleontological survey. The day-to-day survey routes were plotted on 1:60,000 aerial photograph overlays. GPS and (later) DGPS coordinates were recorded for each of the established localities.

Collections

Faunal Collection Methods

Vertebrate fossils reported here were collected in the field by one of three methods: free survey, crawling, and excavation and sieving. Table 1.2 summarizes what methods were used at each locality each year. Free survey was the most widely used and effective manner of collecting vertebrate fossils on large outcrops (see Figure 1.9). In the crawling method,

FIGURE 1.10
Overview of the Asa Koma drainage, looking to the north along the western margin of the Middle Awash study area. Excavations at the hominid locality ASK-VP-3 are under way at the left of frame, and outcrops of the Adu-Asa Formation extend in the middle distance. Vehicles in the center frame for scale. Photograph by Tim White, December 6, 2002.

developed and refined at CAC sites in the early 1990s, the surveyors line up shoulder to shoulder and collect every specimen found as they crawl on hands and knees across the terrain (see Figure 1.12). This method is usually applied in two cases. First, when an area within a vertebrate locality is littered with bones mainly dominated by fossil fragments of large mammals, it is difficult to distinguish small but important fossils. Crawling becomes the only option for collecting these smaller elements. Second, an area is crawled when any portion of a hominid fossil is found, so that every possible piece might be recovered. At the end of every line of crawling, usually measuring less than 50 m transects, specimens picked up by each individual collector are checked by a vertebrate paleontologist who separates collectibles from noncollectibles (as defined below). Major excavation and sieving of late Miocene localities were conducted at hominid localities AME-VP-1, ALA-VP-2, ASK-VP-3, and STD-VP-2 (Figure 1.10).

The determination of which specimens were deemed to merit collection rather than being left at the field locality varied according to the locality and the research questions addressed. Specimens collected during a free survey were usually craniodental specimens whose provenience could be clearly established and were identifiable by element and taxon at the generic level or below. During the initial field seasons, postcranial elements were collected only for primates and carnivores. Postcranial elements of bovids and other large mammals such as elephants and deinotheres were left in the field, usually for logistical

FIGURE 1.11
Aerial view to the northwest across the ASK-VP-3 hominid locality (foreground center). Note that the rise in elevation from the adjacent Afar floor results in heavier vegetation cover. Photograph by David Brill, January 11, 2003.

reasons. In the instance of Saitune Dora, however, specific circumstances described in Chapter 15 allowed us to recover hundreds of elephant postcranial elements. After 1999, selected body elements of bovids were collected from certain localities for ecomorphological analysis. At localities where hominid excavation was conducted, all visible fossils of any size were initially gathered prior to sorting and collection. This 100 percent pickup strategy ensured that each fragment was given individual consideration. Specimens complete enough to be identified at or below the generic or lower taxonomic level, and that could provide valid and useful measurements, were collected. Dental fragments comprising less than half of a tooth, even when identifiable at or below the genus, were collected in a bulk category with additional fragments belonging to the same family or order. This helped to document the taxonomic representation at a given locality. Broken postcranial specimens that could not be identified at the genus level were usually left in the field.

Previous Collections

In 1981, when Middle Awash research project commenced its survey of the western margin, the RVRME had made paleontological collections from at least five collection areas between 1975 and 1978 (Kalb 1978, 1993; Kalb et al. 1982a–e). The RVRME designated more than 50 vertebrate localities along the western margin. However, in some cases, it appears that no vertebrates were even collected from the localities (KL-87 and KL-114,

FIGURE 1.12

Crawling at the AME-VP-1/71 hominid foot phalanx discovery location. Photograph by Tim White, December 26, 2000.

for example). In other cases, only one or two specimens were collected from a locality. The major problem, however, came from ascertaining the spatial and stratigraphic provenience for most of the specimens. In some cases, such as at Asa Koma and Adu Dora, for example, there are multiple fossiliferous horizons within the same section, representing different temporal units. However, the designated localities and the fossil specimens collected from the localities do not appear to have been spatially or stratigraphically segregated or well positioned within such sections, and no records were maintained at the National Museum of Ethiopia, where the fossils were deposited.

The lack of precise provenience creates serious problems for integrating the fossils collected by the RVRME into this or subsequent studies. At Kuseralee Dora, for example, the RVRME designated 18 localities, all of which were described as part of the Adu-Asa Formation. However, later geological work by the Middle Awash project showed that there are two distinct, fault-bounded faunal units represented at Kuseralee Dora, temporally separated by ca. 800 kyr (5.2 Myr and 4.4 Myr). This shows the importance of first establishing spatial and stratigraphic frameworks for a collection area before fossil specimens are collected.

Middle Awash Project Collections

A basic description of the paleontological activities and protocols used by the Middle Awash research project is presented by de Heinzelin et al. (2000b). Between 1992 and 2005, the Middle Awash project collected a total of 2,800 specimens from 23 late Miocene localities. Each specimen was entered into field hard copy and then into a digital catalog with up to 25 categories of information, including geographic and stratigraphic position. In some cases, individual specimen spatial placement was recorded using GPS (see Figure 1.13).

FIGURE 1.13
Recording the position of *Nyanzachoerus* skeletal parts (AME-VP-1/137) immediately below the Kuseralee Sands unit of the Kuseralee Member at Amba East. The AME-VP-1 locality extends to the foot of the escarpment, its sediments below the Gawto basalt seen on the skyline to the left of Ado Fila hill, a prominent landmark in the CAC. Photograph by Tim White, December 27, 2000.

Documentation

Locality-based information was initially gathered on the project's vertebrate paleontology locality forms, and specimen-based information was initially gathered in the field catalog. All collected specimens were given specimen numbers with a three-letter area prefix based on the Afar geographic name, followed by "VP" (for vertebrate paleontology) and then by the locality and individual specimen numbers. A register of collected specimens was made in duplicate hard copy and filed (with field identifications) annually at the National Museum of Ethiopia. A single project vertebrate paleontology digital database file is maintained by the project, and, with the publication of this volume, available at middleawash.berkeley.edu.

Curation

All accessioned specimens were transported to the Paleoanthropology Laboratory at the National Museum of Ethiopia in Addis Ababa for cleaning, restoration, study, curation, and storage. Curatorial work ranged from simple labeling to detailed precision matrix cleaning and restoration of hundreds of fragments from single specimens. Electric and air scribes were used to clean matrix from specimens. Glyptal, cyanoacrylate, and polyvinyl acetate (Vinac^{TM}) were used to join broken pieces. All the specimens are now housed in the Paleoanthropology Laboratory of the National Museum in Addis Ababa, where all the specimens were studied.

Guide to Contents

This monograph contains 20 chapters addressing the geology, geochronology, paleontology, paleogeography, paleobiochronology, and paleoecology of the Middle Awash late Miocene. It synthesizes vertebrate evolution in eastern Africa during the end of the Miocene epoch, a time during which the modern African fauna started forming.

The Asa Koma Member of the Adu-Asa Formation and the Kuseralee Member of the Sagantole Formation (see Chapter 2 for details) are tightly controlled temporally, with radiometric dates of 5.54–5.77 Ma and 5.2 Ma, respectively (Chapters 2 and 4). Various analytical methods, including $^{40}Ar/^{39}Ar$ dating, EMP, argon plasma spectrometry, stable isotope, and paleomagnetic techniques were used to characterize the geological samples and to establish the geochronological framework for the geological processes and paleontological records (see Chapters 3 and 4).

Faunal descriptions begin with Chapter 5, a description of the small mammal assemblage compiled from the late Miocene of the Middle Awash. The micromammalian fauna from the late Miocene of the Middle Awash is diverse and documents at least 16 taxa in eight families. At least four new species are recognized in three families.

A sample of cercopithecid fossils recovered from the late Miocene sediments of the western margin and CAC of the Middle Awash is described in Chapter 6. Even though the sample is not large and most specimens are quite fragmentary, at least three, and possibly four, species are present. Specimens from the Kuseralee Member of the Sagantole Formation are tentatively assigned to both *Pliopapio alemui* and *Kuseracolobus aramisi* (these taxa were first described from the 4.4 Ma sediments from the Aramis Member of the Sagantole Formation; Frost, 2001b). In addition, there is a rare, larger species of colobine present within this member. Fragmentary fossils have also been recovered from the older Asa Koma Member of the Adu-Asa Formation. These are also tentatively allocated to *K. aramisi, P. alemui,* and the larger colobine. Also from the Adu-Asa Formation, a species that cannot be identified to subfamily but is smaller in dental size than either *P. alemui* or *K. aramisi* may also be present based on a single premolar. Although the sample is not large, it is a significant addition to primate paleobiology because late Miocene cercopithecid fossils are still very rare in sub-Saharan Africa.

The late Miocene deposits of the Middle Awash have yielded hominid fossil remains dating back to 5.8 Ma (WoldeGabriel et al. 2001). A total of 15 hominid specimens were recovered from late Miocene sediments in the Middle Awash between 1997 and 2005. These specimens are described in detail in Chapter 7. The specimens include both dentognathic and postcranial elements. The specimens found before 2000 were initially assigned to a subspecies of *Ardipithecus ramidus*. However, additional discoveries made between 2000 and 2002 warranted elevation of *Ardipithecus ramidus kadabba* (Haile-Selassie 2001b) to species level (Haile-Selassie et al. 2004b).

Chapter 8 describes the carnivores. The Middle Awash late Miocene carnivore fauna represents one of the most diverse in the latest Miocene vertebrate history of eastern Africa. At least 18 genera in six families have been recognized from the Asa Koma Member of the Adu-Asa Formation and from the Kuseralee Member of the Sagantole Formation. Large felids include *Machairodus* and *Dinofelis*. One small felid whose generic and specific affinity is currently unknown is also present. Canids are represented by at least one *Eucyon* species. Two hyaenid genera, *Hyaenictitherum* and *Hyaenictis*, are present. These genera are also known from other terminal Miocene sites of Africa. Ursids are represented by one genus, *Agriotherium*, which appears to be larger than specimens of *Agriotherium africanum* from Langebaanweg. Mustelids are diverse in the Adu-Asa Formation, with at least four species in four genera. This family includes the earliest record of *Plesiogulo*, a genus previously known only from younger deposits at Langebaanweg (Hendey 1978b). Viverrids are represented by three species (in three genera) within a single subfamily. This sample includes the earliest record of *Helogale*, a genus previously known only from Pliocene deposits of eastern Africa. The other viverrid genera are *Genetta* and *Viverra*. The Family Herpestidae is also represented by one new *Herpestes* species. The carnivore assemblage from the late Miocene of the Middle Awash has important implications for regional and global biogeography, and it also documents the first local appearances of some taxa that established the modern African carnivore fauna.

Bovids (Chapter 9) are represented in the late Miocene deposits of the Middle Awash by at least seventeen species in eleven genera and seven tribes, six of which still exist in sub-Saharan Africa today. Boselaphines were abundant during the late Miocene of Africa although they are now restricted to Asia. They probably went extinct in Africa at the end of the Miocene. Alcelaphines are absent in the Adu-Asa Formation and Kuseralee Member of the younger Sagantole Formation, whereas reduncines and tragelaphines are abundant. The abundance of tragelaphines and reduncines indicates a mixture of woodland and grassland biomes, and the early species in these two tribes would probably have been associated with plenty of permanent water in fluvial, deltaic, or lacustrine settings. Bovines, antilopines, hippotragines, and neotragines were also present in the Adu-Asa Formation even though they were less abundant compared to the reduncines and tragelaphines. The bovid assemblage from the older Asa Koma Member appears to be slightly different from the assemblage in the younger Kuseralee Member, possibly indicating a local faunal and environmental change at the end of the late Miocene.

Chapter 10 describes cranial and dentognathic suid remains from the late Miocene of the Middle Awash study area. These represent at least five tetraconodontine species between 5.2 and 5.8 Ma. *Nyanzachoerus syrticus* and *Nyanzachoerus devauxi* are limited to the western margin localities dated to older than 5.5 Ma. *Nyanzachoerus kanamensis, Ny. australis, Ny. waylandi,* and a new nyanzachoere species are identified from the terminal Miocene localities of the CAC, indicating a dramatic suid turnover in the late Miocene. In addition, the small tayassuid-like suid *Cainochoerus* is also present in the western margin and the CAC deposits of the Middle Awash. The Middle Awash fossil suid collection, added to other contemporaneous fossil collections in eastern Africa, clearly indicates the presence of several different tetraconodont lineages that might have arisen coincident with the increase in C_4 biomass toward the end of the Miocene.

Two assemblages of hippopotamus remains are identified in the late Miocene of the Middle Awash. These are described in Chapter 11. In the Adu-Asa Formation, specimens of Hippopotamidae are mostly represented by isolated teeth. These teeth indicate a primitive morphology common to other contemporary hippopotamids. These remains were identified as Hippopotaminae indet. In the Sagantole Formation, this family is represented by a recently described species of Hippopotamidae, aff. *Hippopotamus dulu*. This hexaprotodont hippopotamus exhibits a general morphology that is primitive, close in that respect to other Mio-Pliocene forms. However, its cranium and dentition display a distinctive association of measurements and features. This species increases the hippopotamid fossil record in eastern Africa and reinforces the hypothesis of hippopotamid endemism in each African basin as early as the basal Pliocene.

Chapter 12 describes the giraffid remains in the Middle Awash late Miocene deposits. Although rare in the faunal assemblage, there are possibly three species known from isolated teeth and limited postcrania. A *Palaeotragus* species, better known from other contemporary sites such as Lothagam, is known from the Middle Awash via a limited number of dental remains. *Sivatherium* is also present, but rare in the Middle Awash late Miocene deposits, as in other contemporaneous eastern African sites. The Middle Awash late Miocene remains tentatively referred here to the genus *Giraffa* may represent the earliest record of the genus in eastern Africa. *Giraffa* species are thus far documented largely from Plio-Pleistocene deposits. However, further recovery of more complete specimens will be necessary to confirm the first appearance of the genus in the Middle Awash.

Hipparionine horses of the *Eurygnathohippus* clade from the Middle Awash late Miocene are described in Chapter 13. *Eurygnathohippus* was restricted to the African continent from 7+ to less than 1 Ma. The hipparion material under consideration is mostly composed of isolated elements, but there are cases of more informative, more complete associated material. Mandibular cheek teeth confirm the attribution of material from all localities considered here as *Eurygnathohippus.* Although species distinction cannot be made by maxillary and mandibular cheek tooth size and morphology alone, mandibular and postcranial material, in particular metapodials and phalanges, confirm the presence of a small to medium-sized, gracile hipparion in the assemblage under consideration. The oldest material, from Jara-Borkana (6.0+ Ma), is closely comparable to the Upper Nawata (Lothagam Hill, Kenya) type material of *Eurygnathohippus feibeli* and is referred to that taxon. The remainder of this sample is referred to *Eurygnathohippus* aff. *feibeli* and is morphologically distinct from the Nawata Member robust-limbed form *Eurygnathohippus turkanense.* There is some evidence of evolution within the *E.* aff. *feibeli* assemblage, but there is currently insufficient morphological information to distinguish a new taxon.

Rhinocerotids are rare elements in the Middle Awash late Miocene (Chapter 14). However, the recovered specimens indicate the presence of at least two species of *Diceros.* The Kuseralee Member has the most complete cranium of *Diceros douariensis,* a species otherwise known only from the late Miocene of North Africa and southern Europe. The specimens from the older Asa Koma Member are tentatively assigned to *Diceros* sp. because of the fragmentary nature of the recovered specimens. However, the recovered specimens represent a more primitive form of the genus *Diceros* that is different from *D. douariensis.* The presence of *Diceros douariensis* in the Middle Awash late Miocene deposits indicates that there was a strong biogeographic relationship between North Africa and Africa south of the Sahara at the end of the Miocene.

The Middle Awash late Miocene fauna includes at least four elephantid, two gomphothere, and one deinothere species. This record documents an early diversification among the elephantids and includes the earliest record of the genus *Primelephas.* Chapter 15 addresses current issues in the evolution of proboscideans and suggests the need for a major revision in elephantid systematics, descriptions, and serial identification of isolated teeth. The gomphotheres, represented by the genus *Anancus,* clearly show phyletic evolution. The morphological changes are largely in the dentition, and these changes are subtle, making classification or recognition of a species besides *Anancus kenyensis* difficult. Deinotheres are rare in the Middle Awash late Miocene fossil record. Although they appeared in the fossil record as early as the early Miocene, they remained one of the less diverse taxa in the fossil record of Africa and Europe. Only two species, *Prodeinotherium hobleyi* and *Deinotherium bozasi,* are known in the deinothere fossil record; the latter is documented from the Middle Awash.

Although rare in the faunal collection from the late Miocene of the Middle Awash, tubulidentates are represented by at least two different species of the family Orycteropodidae (see Chapter 16). One of the two *Orycteropus* species is a Miocene-like, medium-sized species, whereas the other form is a large, modern-looking species. In general, the two *Orycteropus* species show the first evidence in favor of an African origin of modern aardvarks.

The paleoenvironment of the Asa Koma and Kuseralee Members is considered in Chapter 17. Different aspects of the large mammalian faunal communities from the Asa Koma and Kuseralee Members were examined. The analyses conducted include: (1) a detailed examination of the faunal list based on frequencies of dietary and locomotor variables, (2) the presence of indicator taxa and their relative abundances, and (3) ecomorphology of the bovids. Results from these analyses indicate that the faunal communities of the Asa Koma and Kuseralee Members were probably occupying riparian habitats dominated by woodland with areas of wet grassland. This is supported by other lines of evidence, including stable isotopes, geology, micromammals, and aves. Differences in the relative abundances of bovid taxa and oxygen isotopic values between the Asa Koma and Kuseralee Members may reflect subtle ecological distinctions in woodland and grassland representation and temperature.

New taxa usually arrive as a result of *in situ* speciation from an ancestral stock or as a result of immigration from other areas. The Middle Awash late Miocene fauna, particularly the faunal assemblage of the younger Kuseralee Member, shows that new taxa arrived in the Middle Awash as a result of both processes. Chapter 18 conducts a comparative biogeographic analysis to examine how much the Middle Awash fauna resembles faunas from other localities of similar age at the genus level. The results indicate that the Middle Awash had close biogeographic relationships with eastern and northern African, Arabian, and western Eurasian localities of similar age. It also indicates that part of the eastern African late Miocene fauna was established as a result of immigration from southern Europe, North Africa, and Arabia.

The faunal assemblages described in 12 chapters of this volume document diverse evolutionary trajectories for the late Miocene mammalian fauna of eastern Africa. A complete faunal list can be found in Table 1.3. The assemblages indicate a number of originations and extinctions in various groups adapted to different habitats. Chapter 19 looks at the biochronological and evolutionary significance of the Middle Awash faunal assemblages by looking at possible first appearance (FA) and last appearance (LA). It synthesizes the information to look into faunal turnovers and their probable causes. The fact that the Middle Awash late Miocene fauna samples two time successive horizons, dated to between 5.8 Ma and 5.2 Ma, makes it ideal to look into what factors trigger evolutionary changes in a lineage.

TABLE 1.3 Faunal Lists of the Asa Koma (5.54–5.77 Ma) and Kuseralee (5.2 Ma) Members

Taxon	Asa Koma Member	Kuseralee Member
MAMMALIA		
Insectivora		
Soricidae		
Crocidura aff. *aithiops*	x	
Crocidura aff. *dolichura*	x	
Primates		
Cercopithecidae		
Colobini		
Kuseracolobus aramisi	x	x
gen. et sp. indet. "large"	x	x
Papionini		
Pliopapio alemui	x	x
Subfamily indet.	x	x
Hominidae		
Hominini		
Ardipithecus kadabba	x	
Ardipithecus cf. *kadabba*		x
Lagomorpha		
Leporidae		
Serengetilagus praecapensis	x	
Rodentia		
Sciuridae		
Xerus sp.	x	
Muridae		
Tatera sp.	x	x
Lophiomys daphnae sp. nov.		
Lemniscomys aff. *striatus*	x	
Tachyoryctes makooka sp. nov.	x	x
Hystricidae		
Atherurus garbo sp. nov.	x	
Hystrix sp.	x	
Xenohystrix sp.	x	
Thryonomyidae		
Thryonomys asakomae sp. nov.	x	
Thryonomys aff. *gregorianus*		x
Carnivora		
Canidae		
Eucyon sp.		x
Ursidae		
Agriotherium sp.	x	x
Mustelidae		
Mellivora aff. *benfieldi*	x	x
Sivaonyx cf. *africanus*	x	
Plesiogulo botori	x	
Lutrinae indet.	x	x
Viverridae		
Genetta sp.	x	x
Viverra cf. *V. leakeyi*		x
Herpestidae		
Helogale sp.	x	
Herpestes alaylaii sp. nov.	x	
Hyaenidae		
Hyaenictitherium sp.	x	x
Hyaenictis wehaietu sp. nov.	x	x
Felidae		
Machairodus sp.	x	x
Dinofelis sp.	x	x
gen. et sp. indet.	x	
Tubulidentata		
Orycteropodidae		
Orycteropus sp. A	x	
Orycteropus sp. B	x	
Proboscidea		
Gomphotheriidae		
Anancus kenyensis	x	x
Anancus sp. indet.	x	x
Elephantidae		
Primelephas gomphotheroides gomphotheroides		x
Primelephas gomphotheroides saitunensis subsp. nov.	x	
cf. *Loxodonta* sp.		x
cf. *Mammuthus* sp.	x	
Deinotheriidae		
Deinotherium bozasi	x	x
Hyracoidea		
Procavidae		
Procavia sp.	x	

TABLE 1.3 *(continued)*

Taxon	Asa Koma Member	Kuseralee Member
Perissodactyla		
Equidae		
Hipparionini		
Eurygnathohippus feibeli	x	
Eurygnathohippus aff. *feibeli*	x	x
Rhinocerotidae		
Dicerotini		
Diceros douariensis	x	x
Artiodactyla		
Suidae		
Nyanzachoerus cf. *devauxi*	x	
Nyanzachoerus syrticus	x	
Nyanzachoerus australis		x
Nyanzachoerus kuseralensis sp. nov.		x
Nyanzachoerus waylandi		x
Cainochoerus aff. *africanus*	x	x
Hippopotamidae		
aff. *Hippopotamus dulu*		x
Hippopotaminae gen. et sp. indet.	x	
Giraffidae		
Palaeotragini		
Palaeotragus sp.	x	
Giraffini		
Giraffa sp.	x	x
Sivatherini		
Sivatherium sp.		x
Bovidae		
Boselaphini		
Tragoportax abyssinicus sp. nov.	x	x
Tragoportax sp. "large"	x	
gen. et sp. indet.	x	
Bovini		
Ugandax sp.	x	
Simatherium aff. *demissum*		x
gen. et sp. indet.	x	x
Tragelaphini		
Tragelaphus moroitu sp. nov.	x	x
Tragelaphus cf. *moroitu*	x	x
Reduncini		
Kobus cf. *porrecticornis*	x	
Kobus aff. *oricornis*		x
Redunca ambae sp. nov.		x
Zephyreduncinus oundagaisus	x	
gen. et sp. indet.	x	
Hippotragini		
gen. et sp. indet.		x
Neotragini		
Madoqua sp.	x	x
Raphicerus sp.	x	x
Aepycerotini		
Aepyceros cf. *premelampus*	x	x
Antilopini		
Gazella sp.		x
AVES[a]		
Podicipediformes		
Podicipedidae		
Podiceps sp.	x	
Pelecaniformes		
Phalacrocoracidae		
Phalacrocorax sp.	x	
Phalacrocorax cf. *carbo*	x	
Anhingidae		
Anhinga cf. *melanogaster*	x	
Ardeidae		
Ardea sp.	x	
gen. et sp. indet.	x	
Ciconiiformes		
cf. Ciconiiformes indet.	x	
Anseriformes		
Anatidae		
cf. *Plectropterus* sp.	x	
gen. et sp. indet. "small"	x	x
gen. et sp. indet.	x	
Falconiformes		
Pandionidae		
Pandion sp.	x	
Falconiformes indet.	x	
Galliformes		
Phasianidae		
Francolinus sp.	x	

[a] Aves are not described in this volume. See Louchart et al. (2008) for details.

The last chapter (Chapter 20) summarizes the importance of the Middle Awash late Miocene faunal assemblages in terms of their contribution toward our understanding of the diversity, evolutionary history, and evolutionary tempo and mode of eastern African late Miocene mammals. Moreover, it summarizes the results from the paleoenvironmental and paleobiogeographic analyses in relation to regional and global climatic changes and their impact on speciation and extinction of various mammalian taxa and the formation of the modern African mammalian fauna. Finally, it addresses future plans for continued paleontological work on the late Miocene of the Middle Awash paleontological study area, highlighting the significance of the site and the need for a long-term maintenance and protection program.

2

Stratigraphy of the Adu-Asa Formation

GIDAY WOLDEGABRIEL, WILLIAM K. HART, PAUL R. RENNE, YOHANNES HAILE-SELASSIE, AND TIM WHITE

The late Miocene Adu-Asa Formation comprises major sedimentary and volcanic units that crop out along the densely faulted and tilted frontal blocks of the western rift margin of the transition zone between the northern Main Ethiopian Rift (MER) and the southern Afar Rift (Figure 1.1). This transition zone is characterized by largely antithetically faulted, broad rift margins with numerous half grabens. The orientations of both rift margins have been drastically altered within the transition zone. The trend of the western margin was modified from NE-SW to N-S, whereas the southeastern escarpment changed from NE-SW to E-W. These changes are also reflected in the adjacent rift floor. The rift floor is funnel-shaped, and its width doubles within the transition zone compared to the adjacent rift floor of the northern Main Ethiopian Rift. This is also true with the Quaternary axial rift zone that widens northward into the central Afar Rift.

All of these structural modifications are attributed to the tectonic interactions among the boundary faults of the Red Sea and Gulf of Aden oceanic basins with the continental MER. The effects of the complex tectonic interactions of the boundary faults of the three rift basins are reflected in the volcanic and sedimentary records of the transition zone. During the late Miocene (5.2–6.3 Ma), the area occupied by the western rift margin was subjected to episodic tectonic and volcanic activities and fluvial and lacustrine sedimentation.

The Adu-Asa Formation is subdivided into the Saraitu, Adu Dora, Asa Koma, and Rawa Members. Each member is clearly defined by tephra stratigraphic markers, and each presents distinct volcanic and sedimentary lithologic assemblages. Late Miocene (6.16–6.33 Ma) basaltic lava flows underlie the Adu-Asa Formation. Volcanism ceased, followed by faulting, subsidence, and widespread lacustrine sedimentation along marginal basins, depositing the Saraitu Member. Lacustrine sedimentation peaked during the deposition of the overlying Adu Dora Member (≥5.77 Ma). The lacustrine environment was followed by intense and voluminous hydromagmatic eruptions and fluvial sedimentation during the deposition of the overlying Asa Koma Member units (5.54–5.77 Ma). The Rawa Member was deposited during the waning stages of the hydromagmatic eruptions. It represents diverse units that resulted from fluvial sedimentation in a more stable environment characterized by tectonic and volcanic quiescence. The lacustrine and phreatomagmatic basaltic tuff deposits suggest relatively wet paleoenvironmental conditions during the accumulations of the Adu-Asa Formation.

The initial timing of faulting and rifting along the broad western margin of the Middle Awash study area is still under investigation. However, middle Miocene syn-rift volcanism is reported from the middle part of the broad western rift margin (although no contemporaneous sedimentary rocks are known from the area). Thick sediments (≥50 m) that predate the Adu-Asa Formation crop out on the north side of the Jara River beneath late Miocene basalt (≥6.0 Ma). Older sediments (~10 Ma) are known from the southeastern rift margin of the transition zone. Given the complexity of tectonic activity of faulting, block rotation, and tilting, as well as the continuous record of volcanic and hydromagmatic volcanism along the broad western rift margin of the Middle Awash region, rifting appears to have started here earlier than along the southeastern escarpment.

This chapter highlights the geological and tectonic processes that led to the origin and evolution of the rift basin of the transition zone. It also provides the temporal and spatial contexts for the various paleontological and paleoenvironmental chapters that follow. The transition zone between the northern sector of the Main Ethiopian and the Afar Rifts (see Figure 1.1, inset map) is characterized by a number of distinct geological features. These features include a funnel-shaped rift floor and axial rift zone; voluminous hydromagmatic deposits along the frontal blocks of the rift margin and within the adjacent rift floor; an arcuate step-faulted broad western margin; a domed, rift-bound structure, the Central Awash Complex (CAC); the transverse Bouri horst; and numerous horst and graben structures mostly confined to the eastern half of the rift floor.

The transition zone includes both western and southeastern rift margins and the adjacent rift floor. The proto-oceanic Afar Rift (Barberi and Varet 1975) at the intersection of the active rift systems is a product of these complex geodynamic interactions. The complex volcanic and tectonic activities were also responsible for the changes in the orientations and widths of the western and southeastern rift margins and the widening of the rift floor and the axial rift zone within the transition zone. In the southern Afar Rift transition zone, the western and southeastern rift margins are broader. They consist of multiple densely spaced and tilted fault blocks rather than the singular boundary faults of the MER farther to the southwest.

The western margin of the Middle Awash study area is about 40 km wide and consists of multiple riftward-stepping antithetic fault blocks (Mohr 1987; WoldeGabriel et al. 2001). Although narrower in width, the southeastern margin along the transition zone from the northern MER to the southern Afar Rift is characterized by several segmented antithetic fault blocks, having variable orientations along the strike of the rift margin (Juch 1975). The orientations of the broad western and southeastern rift margins underwent drastic changes from NE-SW to N-S and from NE-SW to E-W, respectively (Juch 1975; Chernet et al. 1998; WoldeGabriel et al. 2001; Ukstins et al. 2002).

The width of the rift floor and the segmented right lateral, en-echelon displaced axial rift of the transition zone increases dramatically compared to its southern extension within the MER sectors. The rift floor increases in width from about 80 km in the northern sector of the MER to more than 200 km within the southern Afar Rift transition zone (Hayward

FIGURE 2.1
View of the western side of the Central Awash Complex (CAC), looking to the east across the Hatayae catchments. The oldest sediments exposed in this structure face the viewer, below the basaltic centers and flows that form the skyline. Ado Fila hill is on the horizon to the right of the frame, whereas the Aramis localities are out of frame and beyond the horizon to the left. The vehicle and dust plume serve as scale. Photograph by Tim White, December 5, 1996.

and Ebinger 1996). Similarly, the span across the active axial tectonic zone (which is characterized by densely faulted Quaternary basaltic and silicic lavas and pyroclastic deposits erupted from fissures, cinder cones, and major silicic central volcanoes) dramatically increases, from 5 to 10 km in the MER (Mohr 1967) to more than 40 km along the transition zone. The complex tectonic and magmatic processes along the rift margins and within the adjacent rift floor have resulted in the exposure of diverse Miocene and Plio-Pleistocene lithologic units along structural blocks and ravines within the transition zone of the Middle Awash study area (WoldeGabriel et al. 1994, 2001; Renne et al. 1999; de Heinzelin et al. 1999; Clark et al. 2003).

The Middle Awash study area is currently divided into eastern and western sides by the contemporary Awash River and its densely forested swampy flood plain (see Figure 1.1). On the east side of the study area, Plio-Pleistocene fossiliferous sedimentary rocks and intercalated volcanic units often occur beneath basaltic lava flows and pyroclastic deposits along fault scarps and drainages, mostly on the western part of the axial fault zone of the Afar rift floor (Clark et al. 1984, 1994; Williams et al. 1986; Kalb 1993; White et al. 1993). In contrast, the rift floor on the west side is mostly covered by recent alluvial and colluvial sediments except for late Miocene and Plio-Pleistocene rocks exposed along major fault scarps and within isolated uplifted structural blocks of the dome-shaped CAC and the NW-SE oriented Bouri horst (see Figure 1.1). The CAC (Figures 2.1 and 2.2) is

FIGURE 2.2
View of the eastern side of the CAC, looking to the west across the Awash River from a fly camp at Maka. Pliocene rocks exposed on this flank of the complex, including those at Aramis, overlie the Gawto Basalts and underlying Kuseralee Member faunas reported here. Photograph by Tim White, December 31, 1999.

dominated by late Miocene and early Pliocene basaltic lava flows and silicic and hydromagmatic basaltic tephra units interbedded within thick fossiliferous fluvial and lacustrine sedimentary deposits (WoldeGabriel et al. 1994; Renne et al. 1999). The NW-SE oriented Bouri horst consists of Plio-Pleistocene fluvial and lacustrine deposits with intercalated diatomaceous and glassy tephra layers (Kalb et al. 1982e; Kalb 1993; de Heinzelin et al. 1999; de Heinzelin 2000; Clark et al. 2003). In contrast to the rift floor, thick late Miocene and early Pliocene volcanic and fossiliferous sedimentary rocks are exposed along the strike of the densely step-faulted and rotated frontal fault blocks of the broad western rift margin (Figures 2.3 and 2.4) (Kalb 1993; WoldeGabriel et al. 2001).

This chapter highlights a decade-long geological and paleontological reconnaissance and focused investigation along the strike of the frontal fault blocks of the western rift margin of the Middle Awash study area. Despite the presence of widespread outcrops, geological investigations along the western margin were difficult because of rugged terrain covered by dense and thorny shrubs, densely step-faulted and rotated fault blocks, and deeply and steeply cut canyons. Thick colluvial deposits disrupt and obscure the lateral continuity of major lithological units along the strike of the rift margin, and local Afar clans occasionally restrict foot and vehicle access by the field team.

The volcanic and sedimentary deposits of the western rift margin were initially mapped, described, and assigned to the Adu Dora and Asa Koma Members of the Adu-Asa Formation according to field-based lithological and paleontological correlations and biochronology (Kalb et al. 1982e; Kalb 1993). Despite the widespread occurrences of datable local and regional volcanic markers, these studies did not utilize the tephra stratigraphic markers and lava flows interbedded within the sedimentary deposits to establish temporal

FIGURE 2.3
View to the south-southwest against the western margin of the Middle Awash study area. The Adu-Asa Formation sediments are visible as white patches against a dominantly volcanic terrain between Enayto in the south and Adu Dora in the north. The strip of acacias in the foreground marks the Hargunayu seasonal stream, which decapitated the Hatayae stream as a major tributary of the Awash River in 1997. Photograph by Tim White, November 28, 2000.

controls for the diverse lithological units, the volcanic and tectonic processes, and the paleontological assemblages. Rather, coarse biochronological considerations were used in attempts to place the various sedimentary packages relatively.

Subsequent detailed investigations by Middle Awash project scientists began in 1981. These were more comprehensive in their scope and depth, including geochronological and geochemical analyses of the major volcanic units that provided accurate temporal and spatial controls for the tectonic and volcanic processes and associated fossiliferous sedimentary deposits. Hundreds of representative geological samples were collected from the major volcanic and sedimentary units for laboratory analysis during the exploration as part of the integrated multidisciplinary investigations (Figure 2.5). Various analytical methods, including $^{40}Ar/^{39}Ar$ dating, electron microprobe (EMP), argon plasma spectrometry, stable isotope, and paleomagnetic techniques were used to characterize the geological samples and to establish the geochronological framework for the geological processes and paleontological records (see Chapters 3 and 4). New field information, coupled with laboratory results of the different volcanic units collected along the western margin, aided the reassignment of the lithologic units into an appropriate chronostratigraphic sequence first presented by WoldeGabriel et al. (2001) and elaborated upon in this chapter.

FIGURE 2.4
View to the southwest into the fault blocks of the western margin of the Middle Awash study area. Photograph by David Brill, December 29, 1998.

Geological Investigations along the Western Rift Margin of the Transition Zone

Before the end of the 1981 field season, Middle Awash project scientists crossed to the west side of the Middle Awash study area and conducted brief reconnaissance surveys within the rift floor and along the adjacent foothills of the frontal fault blocks of the western rift margin in the Asa Ali, Hindo Kali, Ananu, and Adu Dora areas (Figures 1.1 and 2.5). Samples from major tuff units interbedded within fossiliferous sediments were collected from some of the sections and subsequently dated.

Following the resumption of field research in the Middle Awash study area in 1990, the project conducted two successful seasons of geological, paleontological, and archaeological investigations on the east side of the rift floor of the Middle Awash study area (fall of 1990 and 1991). However, the focus of the exploratory studies shifted to the west side of the Awash River beginning in 1992. During the 1992 field season, most of the initial reconnaissance surveys were carried out on the Bouri peninsula and the CAC of the rift floor (see Figure 1.1). Preliminary surveys were also intermittently conducted within the Ananu and Adu Dora drainages of the western rift margin of the Middle Awash study area. Several volcanic samples were collected during these reconnaissance surveys.

In December 1993, additional surveys were carried out in the vicinities of Asa Ali, Saitune Dora, and Ananu, mostly in the southern half of the western rift margin. More geological samples and fossils were collected during these short exploratory trips. Further reconnaissance surveys were also carried out on the north and south sides of the Adu Dora area, located on both sides of the perennial Hatayae River. Several silicic and basaltic tuffs and fine-grained lacustrine and fluvial sediments were collected for geochemical, geochronological, and paleobotanical analyses, respectively. Abundant vertebrate fossils were also identified and recovered from the sedimentary deposits.

Beginning in the middle of the 1990s, with the avalanche of data generated from work in the CAC and at Bouri, additional work along the western rift margin within the study area was retasked. Two of the authors (GWG and YHS) led the reconnaissance and initiated more regular and detailed surveys there. This chapter summarizes the results of the geological investigations undertaken during this investigative phase.

This geological work is best viewed as reconnaissance in nature. Detailed geoscience studies (trace element and isotope geochemistry, paleomagnetics, geophysics, mapping, sedimentology, etc.) will continue in this region for decades or centuries to come. Our work was initiated to establish a framework for such continuing research, and this study should therefore be seen as an interim report rather than a full exposition of the study area's resources. It is our goal here to establish a framework for continuing work and to identify future research opportunities.

In December 1995, several exploratory traverses were conducted along the strike of the western rift margin, beginning in the Kurkura drainage basin and continuing southward as far as Asa Ali (see Figures 1.1 and 2.5). A number of new fossiliferous localities were identified and named, including Ali Ferou Dora, Dasga Dora, Ali Bida Dora, and Digiba Dora (Figures 1.1 and 2.5). Additional surveys were carried out at the Adu Dora, Saitune Dora, Asa Ali, and Rawa areas, where samples of numerous tuffs and lava flows were collected along measured sections.

During the 1996 field season, most of the detailed geological studies were restricted to the vicinity of the Adu Dora, Asa Koma, and Saitune Dora localities and adjacent terrain, except for limited reconnaissance surveys at the southern part of the study area, which led to the identification of several fossiliferous outcrops, including the Ado Faro site (Figure 2.5). New fossiliferous localities were designated in the vicinities of Asa Koma, Alayla, and Gaisale Daba (Figure 2.5). Major lithological units were identified, described, and sampled (see Figure 2.5). Selected sections were measured in the Ado Faro, Asa Ali, Sagala Ali, and Ounda Belayamo areas (Figure 2.5). In 1997, limited surveys were carried out along the western margin to recheck the geological sections at the Adu Dora, Alayla, and Saraitu localities and to collect additional fossils. It was in 1997 that the first hominid fossils were recovered from the late Miocene strata of this region.

Work along the western rift margin continued during 1998. Additional samples of basaltic and silicic tuffs were collected from measured sections in the Alayla, Asa Koma, Digiba Dora, and Saitune Dora areas. Samples of paleosol and pedogenic carbonate

FIGURE 2.5

The southern half of the Landsat Thematic Mapper imagery (Figure 1.1), showing the distribution and location of geological samples. The geographic names of the local areas are given.

Central Awash
Complex
Kuseralee Dora
Amba
West
Amba
East
Hatayae R.
Saragata R.
Bouri
Yardi
Lake

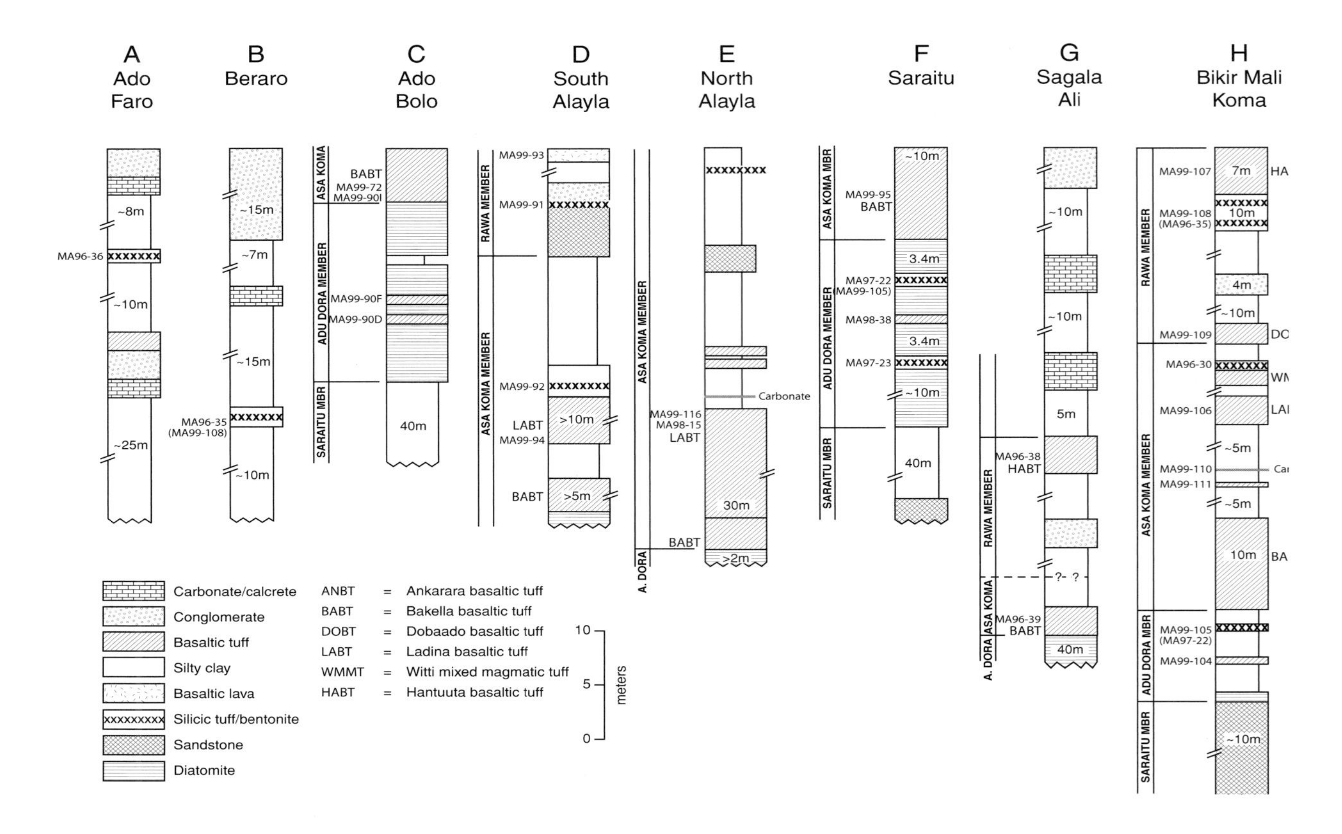

FIGURE 2.6

The lithostratigraphic sequences of the Adu-Asa Formation, showing measured sections along the western rift margin of the Middle Awash study area. Sample location, estimated thickness, stratigraphic boundaries, and major stratigraphic markers are given for each column. The locations of the measured sections are shown in Figure 1.1.

were collected from the actual hominid fossil localities for stable isotope studies to assess paleoenvironmental records.

During the 1999 field season, geological surveys continued in the Alayla and Daytoli streams and adjacent areas to determine the distribution and stratigraphic relationships of major basaltic and silicic tephra units exposed along fault scarps and stream cuts. The Ado Bolo section south of the Alayla locality was also described and measured, and several diatomite and basaltic tuff samples were collected. Other sections were also measured in the environs of Asa Koma, Adu Dora, Bikir Mali Koma, and Ananu (Figures 2.5 and 2.6).

Toward the end of the 1999 field season, the first reconnaissance surveys were conducted in the northern half of the western rift margin north of the Jara River and in the Talalak area (see Figure 1.1). The rift margin and the adjacent rift floor were explored, and important fossiliferous localities were identified. The major basaltic lava and silicic tuff units interbedded within the fossil-rich sediments were sampled to determine the age of the deposits and associated fossils. Since 2000, only a few of the most important paleontological localities at Asa Koma, Alayla, Digiba Dora, and Saitune Dora have been periodically and briefly visited, because the focus of the field research shifted back to the rift floor and to the northern and southern parts of the Middle Awash study area. However, renewed surveys during the 2004 field season between Digiba Dora and the Jara River, and in the

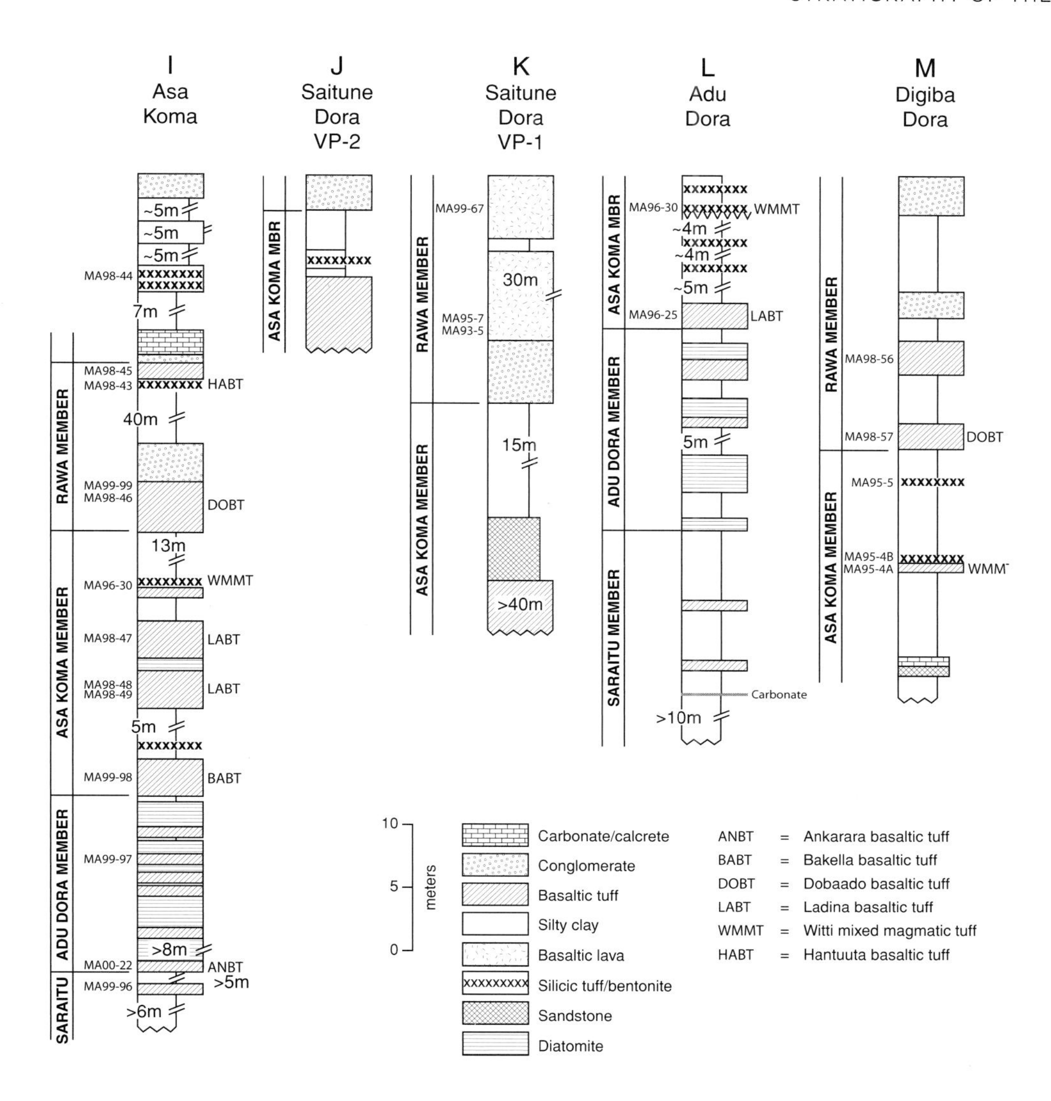

headwaters of the Bilta stream, resulted in the identification of additional sedimentary packages, some with vertebrate fossils. Numerous paleontological and archaeological sites have been designated following these exploratory and ongoing investigations.

Stratigraphy of the Adu-Asa Formation

Field studies of the major lithologic units exposed along the strike of the frontal fault blocks of the western rift margin, coupled with chemical and chronological data from distal and proximal tephra markers and lava flows, have facilitated the stratigraphic classification of the Adu-Asa Formation (WoldeGabriel et al. 2001). Kalb (1993) had grouped the fossiliferous sedimentary and volcanic rocks of the western margin of the Middle Awash study area into the Adu-Asa Formation based on mapping, lithological, and biochronological estimations. He subdivided the lithostratigraphic sequence into the Adu and Asa

FIGURE 2.7
View to the east across the Asa Koma area. The CAC is visible as dark hills (Ado Fila Mountain) below the distant horizon on the right of the frame. The Kuseralee Member sediments are exposed at the base of that uplifted structure. The Asa Koma catchments are in the foreground, with the ASK-VP-1 locality visible sandwiched between the basaltic tuff immediately above the diatomite horizon in center frame and the next, resistant, mesa-forming darker basaltic tuff above this. Dip is synthetic to the rift axis, to the southeast. Photograph taken December 22, 1996. A close-up of the triangular outcrop in the upper right of the frame is seen in Chapter 3.

Members in ascending stratigraphic order. Moreover, Kalb (1993) suggested that the Adu-Asa Formation unconformably capped late to middle Miocene (10–12 Ma) Fursa Basalts (Zanettin and Justin-Visentin 1975).

Reconnaissance geological investigations along the strike of the rift margin by the Middle Awash project revealed the complexity of the terrain induced by intense faulting, block rotation, tilting, and erosion. Detailed field investigation supplemented by $^{40}Ar/^{39}Ar$ dating, paleomagnetism, and tephra chemistry of the intercalated volcanic rocks were required to characterize the densely faulted Adu-Asa Formation accurately and to establish an appropriate chronostratigraphic framework.

Based on field studies and laboratory analyses, the lithostratigraphic nomenclature of the Adu-Asa Formation was revised and subdivided into four members using tephra markers according to the International Stratigraphic Code (Hedberg 1976). The basalts at the base of the Adu-Asa Formation, which had been described as unconformable older flows correlative to the Fursa Basalts (10–12 Ma; Zanettin and Justin-Visentin 1975), turned out to be much younger (6.0–6.8 Ma) and conformable with the overlying sedimentary deposits (WoldeGabriel et al. 2001). New $^{40}Ar/^{39}Ar$ ages and local and regional chemical correlations from tephra stratigraphic markers and paleomagnetic measurements of fine-grained sediments from measured sections facilitated the division of the diverse lithologic units of the western rift margin into four distinct members. These are the Saraitu, Adu Dora, Asa Koma, and Rawa Members, in ascending stratigraphic order (Figure 2.6). The various lithological units of each member exposed along major fault scarps and deep canyon walls of the rift margin are highlighted in the following discussions (Figures 2.7 and 2.8). Certain stratigraphic placements and regional tephra correlations are based on data presented in Hart et al. (Chapter 3). Named volcanic horizons follow the nomenclature established by WoldeGabriel et al. (2001) and Hart et al. (Chapter 3).

FIGURE 2.8
Aerial view to the north over the Adu-Asa Formation. Gaisale is to the left center of the frame. The Saraitu, Adu Dora, Asa Koma, and Rawa Members of the Formation are all visible in this view. Photograph by Tim White, January 11, 2003.

Saraitu Member

Kalb (1993) grouped into the Adu Member different sandstone and silty clay units and overlying diatomite beds. These represent the basal section of the Adu-Asa Formation. WoldeGabriel et al. (2001) subdivided the Adu Member into the Saraitu and Adu Dora Members based on their discrete lithologic and paleontological characteristics. The Saraitu Member strata are underlain by late Miocene basalts (6.2–6.3 Ma) in the Adu Dora area on the south side of the Hatayae River (Figure 2.5). This member is capped by thick (>20 m) diatomite beds exposed along the strike of the frontal fault blocks and in deeply dissected ravines of perennial and ephemeral rivers that cut across the broad rift margin (Figures 2.6F and 2.9). Because of the antithetic faulting and block rotations, deep and continuous stratigraphic sections are rarely encountered along the strike of the broad rift margin of the study area. In most cases, the displacements along the fault scarps are small and are obscured by colluvium. The base of the Saraitu Member of the Adu-Asa Formation is rarely exposed south of the Adu Dora locality. To the north, late Miocene basalts beneath the formation appear to cap >50 m of unnamed, mostly fluviatile sedimentary rocks at the JAB (Jara-Borkana) area of the western rift margin immediately to the north of the Jara River (Figure 1.1).

The main section of the Saraitu Member is located at the headwaters of the Saraitu stream on the south side of the Ananu drainage basin. It represents more than 40 m of sediments, comprising basal bedded light orange sandstone overlain by dark brown silty clays that contain mostly aquatic vertebrates (Figures 1.1, 2.5, 2.6F, and 2.9). The sequence is capped by brittle diatomite that is intruded and altered to siliceous sinter by a thick, rift-oriented basaltic dike. The base of the Saraitu Member is not exposed at the

FIGURE 2.9
View of the main section of the Saraitu Member at the base of the Adu-Asa Formation. The section is located in the Saraitu stream on the south side of the Ananu drainage system (Figure 2.5). The section is dominated by lacustrine silty clay that overlies brownish sandstone exposed at the base of the outcrop. Lacustrine deposits of diatomite and siliceous sinters of the Adu Dora Member cap the Saraitu Member in the uppermost part of the picture. Photograph taken December 9, 1999.

type section. However, strongly fractured late Miocene (6.2–6.3 Ma) basalt flows crop out beneath the basal sediments of the Adu-Asa Formation in the vicinity of the Adu Dora localities on both sides and along the Hatayae River about 10 km north of the Ananu stream (Figure 2.5). The Saraitu Member units also crop out at several locations along the strike of step-faulted blocks. For example, in the Ado Bolo area, about 5 km to the south of the type section, about 40 m of poorly exposed silty clays and sandstone crop out beneath multiple layers of diatomite with intercalated basaltic tuffs (Figure 2.6C). However, colluvium and landslide blocks obscure the base of the section. No fossils were collected from the Saraitu Member in the Ado Bolo section.

The uppermost part of the Saraitu Member is locally exposed in the Bikir Mali Koma area north of the Ananu drainage basin. About 10 m of massive, dark brown pebbly sandstone crops out directly beneath faulted diatomite beds with interbedded basaltic tuff (Figure 2.6H). Additional sedimentary units of the Saraitu Member are also exposed at Asa Koma, which exposes a complete section of the younger members of the formation. The uppermost part of the Saraitu Member at Asa Koma consists of about 12 m of multiple beds of dark brown, olive gray, and pinkish gray silty clays that contain two thin layers of basaltic tuff. The lower basaltic tuff was named the Ankarara Basaltic Tuff (ANBT) and coded MA00-22. This section occurs directly beneath diatomaceous beds on the northwest side of the Asa Koma type section (Figure 2.6I). Despite extensive faulting at the Asa Koma main section, the stratigraphic sequence was easily delineated using the diatomite deposits and interbedded basaltic tuff stratigraphic markers.

Additional Saraitu Member units crop out locally along the strike of the rift margin north of the Asa Koma type section. About 30 m of interbedded fluvial and lacustrine sediments are exposed below diatomite beds at the Adu Dora locality north of the Hatayae River. The base of the Saraitu Member section is not exposed at this locality (Figure 2.6L). However, a strongly fractured late Miocene (6.2 Ma) basaltic lava flow crops out beneath the Saraitu Member on the south side of the Hatayae River about a kilometer from the Adu Dora locality.

The Saraitu Member at the Adu Dora section contains multiple units of siltstone; sandstone; carbonate-cemented, bivalve-rich silty clay; and altered basaltic tuff, in ascending stratigraphic order. Abundant fish remains were collected from the upper half of the silty clay and sandstone beds. At the Adu Dora section, silicified diatomaceous beds cap the Saraitu Member sequence (Figure 2.6L). Although the Saraitu Member was not identified north of the Adu Dora section, the upper member of the Adu-Asa Formation is exposed at Digiba Dora (Figure 2.6M). The Adu-Asa Formation units were not recognized along the strike to the north of the Digiba Dora locality. Instead, late Miocene (6.3 Ma, MA95-1) basalt flows and altered basaltic tephra crop out along fault scarps and along canyon walls in the Kurkura River and at the Jara-Borkana locality north of the Jara River.

Adu Dora Member

The Adu Dora Member is the most distinctive lithostratigraphic unit of the Adu-Asa Formation. It is easily recognizable on aerial photographs and satellite images because of the highly reflective diatomite outcrops that form white patches along the strike of the frontal block of the western rift margin. The Adu Dora Member is composed of multiple layers of diatomaceous sediments with thin intercalated phreatomagmatic basaltic tuff and distal silicic tephra beds (Figures 2.10 and 2.11). Alternating fissile diatomite and silty clay beds occur in the lower half of the Adu Dora Member, whereas the upper part contains siliceous sinter, silty clays and sandstones, and partially altered basaltic tephra. The diatomite deposits were not characterized.

The stratigraphic units exposed at the Adu Dora section were subdivided (Hedberg 1976; WoldeGabriel et al. 2001) and comprise more than 20 m of alternating white fissile diatomite layers, silicified and cherty diatomite, silty clays, and basaltic and silicic tuffs (Figure 2.6I). The individual diatomite and massive silty clays beds range in thickness from 1 to 3 m, whereas the basaltic tuffs are generally thin (≤25 cm). At the Asa Koma section, the ANBT defines the base of the Adu Dora Member. The ANBT (MA00-22) is thin and partially altered (Figure 2.6I). At the Adu Dora type section, the Adu Dora Member is capped by the ca. 1.0 to 1.5 m thick, laminated, bedded, and partially altered Ladina Basaltic Tuff (LABT, MA96-25), containing layers of coarse- and fine-grained

FIGURE 2.10
The Adu Dora Member of the Adu-Asa Formation at the Ado Bolo section south of the Saraitu Member outcrop (Figure 2.5). The Saraitu Member sequence crops out beneath the white diatomite beds, interbedded silty clays, and hydromagmatic basaltic tuffs in the lower half of the section. The Ado Bolo section is capped by hydromagmatic basaltic tuffs that define the base of the Asa Koma Member of the Adu-Asa Formation. Photograph taken December 6, 1999.

scoriaceous fragments. However, at other sections the diatomite sequence is capped by the Bakella Basaltic Tuff (BABT), which crops out beneath the LABT stratigraphic marker. The Adu Dora Member units were not encountered at Digiba Dora to the north of the type section. However, the Adu Dora Member is exposed at several localities along the strike of the rift margin that includes the Asa Koma, Sagala Ali, Bikir Mali Koma, Saraitu, south Alayla, and Ado Bolo sections (Figures 2.5 and 2.6).

At the Asa Koma locality south of the Adu Dora Member type section, the stratigraphic sequence contains correlative lithologic units of alternating diatomite, silty clay, and basaltic tuff beds (Figure 2.6I). Unlike the type section, a number of thin (2–15 cm) basaltic tuff layers were identified within the sequence. These are intercalated with pure fissile diatomite, diatomaceous silty clays, and brown and pinkish silty clays above the ANBT. The diatomite beds are thicker (0.75–3.4 m) here compared to similar outcrops at the type section.

The top of the Adu Dora section is capped by the BABT (MA99-98), which marks the base of the overlying Asa Koma Member (Figure 2.6I). About 20 m of diatomaceous

FIGURE 2.11
Thinly laminated diatomite of the Adu Dora Member in the lower Asa Koma drainage. Photograph by David Brill, December 26, 1996.

sediments crop out at the base of the section directly beneath partially altered basaltic tuff at Sagala Ali, directly south of the Asa Koma locality (Figure 2.6G). The Bikir Mali Koma section consists of thick sedimentary units and volcanic rocks that crop out along step-faulted scarps. The upper part of the Saraitu Member and the bulk of the Adu Dora Member represent the base of the stratigraphic sequence (Figure 2.6H). Thick (~10 m), pebbly sandstone of the Saraitu Member is exposed at the base of the stratigraphic sequence directly beneath the lower silicified diatomite. Dark brown silty clay with two thin (20 cm) basaltic and silicic tuffs and a fissile diatomite overlie the silicified diatomite. The basaltic tuff (MA99-104) is dark gray and bedded, whereas the silicic tuff (MA99-105) that occurs about 3 m above it is medium gray and fine-grained. Tuff MA99-105 is chemically correlative to tuff MA93-15 from the Ananu area (Hart et al., Chapter 3). It is 30 cm thick and occurs 4 m below diatomite beds that are about 100 m upstream from the Ananu Spring. A 20 cm thick basaltic tuff (MA93-16) occurs about 4 m below the silicic tephra. The top of the Adu Dora Member at Bikir Mali Koma contains another silicified sinter, which is capped by thick (>10 m) partially altered basaltic tuff (BABT) as in the Sagala Ali section.

More diatomaceous beds capped by thick basaltic tuff crop out throughout the Ananu drainage basin to the south of the Sagala Ali and Bikir Mali Koma sections. The main Saraitu Member stratigraphic sequence at the base of the Adu-Asa Formation is located on the south side of the Ananu drainage basin, where it is overlain by about 20 m of Adu Dora Member sequence (Figure 2.6F).

At the Saraitu section at least four layers of silicified diatomite with three interbedded silicic and basaltic tuffs (MA97-22, 23, and MA98-38) constitute the Adu Dora Member (Figure 2.6F). The silicic tuffs are evenly distributed throughout the section and are moderately indurated, unlike other interbedded units. A basaltic dike is exposed in close proximity to the diatomaceous beds. At the Saraitu drainage section, baking and/or geothermal activities related to the dike intrusion were probably responsible for the silicification and indurations of the diatomite and silicic tuff units at the top of the Adu Dora Member. However, diatomite deposits replaced by siliceous sinters were also encountered in the upper part of the Adu Dora Member along the strike of the rift margin away from known dike intrusions.

Silicification and mineralization of fluvial and lacustrine deposits related to hot springs and geothermal systems, closely associated in time and space with volcanism, have commonly developed in northern Nevada (Ebert and Rye 1997; Wallace 2003) and are analogous to the hydrothermally altered Middle Awash rocks. The late Miocene Adu Dora Member lacustrine deposits were cut by major boundary faults. Meteoric water flowed along densely spaced rift-oriented faults and was heated because of the high geothermal gradient associated with the late Miocene pre-Adu-Asa Formation basaltic eruptions (6.16–6.33 Ma). Hydrothermal processes were responsible for the localized silicification of the diatomite deposits, which are replaced by siliceous sinters along bedding planes. Heat from shallow basaltic magma sources, meteoric water from the lacustrine environment and shallow groundwater, and faults and fractures related to the boundary faults facilitated the development of hydrothermal systems along the western margin of the Middle Awash study area.

The stratigraphic sequences of the Saraitu and the overlying Adu Dora Members are capped by the thick and moderately altered BABT (MA99-95). Additional outcrops of the Adu Dora Member stratigraphic units were encountered along the steep banks of the Rawa stream between the Saraitu and Alayla localities (Figure 2.5). On the north bank of the Rawa stream directly north of the Alayla locality, about 15 m of dark gray silty clay, diatomite beds with two thin (5–10 cm) vitric ash layers, diatomaceous silty clays, and sandstone capped by another vitric tuff crop out in ascending stratigraphic order. The section is cut by multiple rift-oriented fractures without apparent displacement.

At the Alayla locality and its environs, the uppermost part of the Adu Dora Member is exposed beneath thick basaltic tuffs (Figure 2.5; Figures 2.6D and E) at the bases of multiple fault blocks and ravines. Diatomite beds of variable thickness (2–5 m) occur on the east, north, and west sides of the Alayla VP localities, often in fault contact with the fossiliferous horizons. About 2 km south of the Alayla locality a thick stratigraphic sequence of the Adu Dora Member crops out at the Ado Bolo section. Here, the Adu Dora Member sequence consists of multiple layers of fissile and massive diatomite beds intercalated with diatomaceous silty clays and basaltic tuff, cropping out above limonite-stained silty clays (Figure 2.6C). Although a number of thin and altered basaltic tuff layers were noted within the section, only the thicker ones (8–20 cm) were sampled for chemical analysis. In contrast to other sections described earlier, the diatomite units are thicker (1–5 m), and

FIGURE 2.12

Panoramic view of the Asa Koma area with Asa Koma Member units exposed in the foreground. The Asa Koma vertebrate localities ASK-VP-1 and ASK-VP-3 are shown. The section in the lower half of the picture represents the main Asa Koma Member unit and consists of silty clays, bentonite, and basaltic and silicic tephra (MA96-30) overlying the whitish diatomite deposits of the Adu Dora Member in the middle part of the picture. Units of the overlying Rawa Member are outlined by the broken line. Photograph taken December 10, 1999.

no silicified diatomite beds were encountered within the Ado Bolo section. Two distinct basaltic tuffs occur within the lower half of the Adu Dora diatomaceous units at the Ado Bolo section. The lower basaltic tuff (MA99-90D) is about 8 cm thick, black, and sandy, with upward fining. In contrast, the 20 cm thick upper unit (MA99-90F) is fine-grained, laminated, and strongly indurated. The top of the Adu Dora stratigraphic sequence is capped by the thick, bedded, and coarse-grained BABT (MA99-90I), which varies in thickness (2–10 m) along the strike. To the northwest of the Ado Bolo section, thick diatomite deposits occur beneath the basaltic tuff and lava flows at Gaysale (Figure 2.5). Moreover, additional reconnaissance surveys along the strike of the rift margin to the south of the Ado Bolo and Gaisale sections showed diatomaceous outcrops to the northwest of Ado Faro (Figure 2.5). However, at the Ado Faro and Beraro sections located along the frontal fault blocks at the southern part of the study area, the stratigraphic sections contain no diatomite beds. In fact, the Beraro section contains a silicic tephra (MA96-35) that is chemically correlative to the uppermost units of the Bikir Mali Koma (MA99-108) sequence (Figures 2.6B and H; also see Hart et al., Chapter 3).

Asa Koma Member

The lithologic assemblage above the primarily lacustrine sediments of the Adu Dora Member was assigned to the Asa Koma Member (WoldeGabriel et al. 2001). The stratigraphic sequence exhibits a sharp lithologic transition from diatomite-dominated lacustrine sediments with intercalated thin basaltic tuffs of the Adu Dora Member to thick (2–4 m) multiple basaltic tuffs that are interbedded with fossiliferous fluvial sediments of the Asa Koma Member (Figures 2.12–2.14). The lithologic variation is consistent with increased

FIGURE 2.13

View to the west across the DID-VP-1 fossil locality where the Asa Koma Member of the Adu-Asa Formation is well exposed. The lower half of the picture includes the fossiliferous sediments of the DID-VP-1 locality. The resistant ledge toward the photographer is made up of carbonate and sandstone, whereas the mixed magma tephra (MA95-4 = MA96-30, WMMT or Witti Mixed Magmatic Tuff) stratigraphic marker is the white unit above the paleontologists. Units of the Rawa Member cap the section. Figures at center frame are sampling paleosol horizons for isotopic work reported in Chapter 17. Figures to the right are paleontologists crawling the surface (as described in Chapter 1) in search of collectible fossils. Photograph by David Brill, December 24, 1998.

local tectonic and volcanic activities that began at about 5.8 Ma. The Asa Koma Member main section at the Asa Koma VP localities (Figure 2.12) consists of about 40 m of varied lithologic units, including several layers of major hydromagmatic basaltic tuff, basaltic and silicic tephra from a mixed magma eruption, sandstone, sandy siltstone, orange brown sandy silty clays, medium gray silty clay, and bentonite beds. The basaltic tuff units vary in thickness from 2 to >10 m along the strike of the rift margin, probably because of postdepositional erosion from uplifted fault blocks and accumulation in fault-bounded and low-lying, elongate sediment traps. The antithetic step-faulted blocks of the broad rift margin created marginal half grabens that trapped sediments and tephra units during the evolution of the rift margin and the adjacent rift floor. In most cases, the base of the thicker (2–4 m) basaltic tuff units is totally altered to yellow-green clay-rich layers, whereas the coarser and less cemented counterparts exhibit minimal alteration.

FIGURE 2.14

View to the south across the DID-VP-1 fossil locality. The resistant horizon centered in the outcrop is the MA95-4 (WMMT) tuff. Photograph taken December 20, 1995.

At the Asa Koma type section, Saraitu Member units crop out along the footwall of a major fault scarp, whereas the overlying Adu Dora and Asa Koma Members are confined to the hanging wall on the east side of the faulted block. The BABT (MA99-98) marking the base of the Asa Koma Member is dark gray, bedded, scoriaceous, and moderately altered. The BABT is widespread along the strike of the rift margin and was identified at Adu Dora-south (MA96-27), Asa Koma (MA99-98), Saraitu (MA99-95), Ado Bolo (MA99-90i and MA99-72), and Gaysale (MA99-71) (Figures 2.5; Figures 2.6C, F, and I).

At the Asa Koma type section about 5 m of medium to dark orange-brown fossiliferous silty clay with thin bentonite beds crops out above the BABT. The orange brown silty clay is covered by the Ladina Basaltic Tuff (LABT), which ranges in thickness from 2 to ≥4 m. The LABT is exposed at both the Asa Koma (ASK) VP-1 and VP-2 fossil localities, which are about 0.5 km apart. At the ASK-VP-2 site south of ASK-VP-1, the thicker (≥4 m) LABT (MA98-48) crops out on the down-faulted block. It is a moderately altered, scoriaceous, and fine-grained bed that underlies about a meter thick fissile diatomite. However, at the ASK-VP-1 site located on the footwall of the fault block, the correlative 2 m thick LABT (MA98-49) caps an east-dipping small mesa that contains medium orange-brown fossiliferous silty clay beneath the basaltic tuff. In contrast to the VP-2 site, the LABT represents the top of the section and no other lithologic units are exposed above it. At the ASK-VP-2 locality multiple lithologic units occur. In ascending stratigraphic order (Figure 2.6I), these are white fissile diatomite (~1 m), basaltic tuff (>3 m), light gray to light brown silty clays with the intercalated Witti Mixed Magmatic Tuff (WMMT, MA96-30), bentonites, and altered basaltic tuff (Figure 2.12). The LABT (MA98-47) above the diatomite is reverse graded, with scoriaceous top and a strongly altered fine-grained base. Like the underlying BABT, the LABT marker horizon is widespread and was identified at Adu Dora (MA96-25, MA99-103), Asa Koma (MA98-47, MA98-48, and MA98-49), Bikir Mali Koma (MA99-106), and Alayla (MA96-28, MA99-94, and MA99-116) along the strike of the rift margin of the Middle Awash study area (Figures 1.1, 2.5, 2.6D, E, H, I, and L; also see Hart et al., Chapter 3).

Thick (>10 m) variegated fossiliferous silty clay units with intercalated mixed basaltic and silicic tuff and bentonite crop out above the LABT. A late Miocene age of 5.57 to 5.63 Ma was obtained on the thin (10 cm) WMMT (MA96-30) stratigraphic marker that occurs within the middle part of the fossiliferous silty clay beds (WoldeGabriel et al. 2001). The Asa Koma Member is capped by the moderately altered yellowish gray, bedded, and scoriaceous Dobaado Basaltic Tuff (DOBT, MA98-46). The DOBT marks the base of the overlying Rawa Member.

The correlations among the various basaltic tuffs of the densely faulted outcrops at the Asa Koma type section were established using chemical compositions generated by EMP analysis of discrete glass shards (Hart et al., Chapter 3). Additional lithologic units belonging to the Asa Koma Member crop out at several sections along the strike of the rift margin to the north and to the south of the type section (Figures 1.1, 2.5, and 2.6). The richest fossil horizon occurs toward the uppermost part of the Asa Koma Member section, and its minimum age is constrained by the WMMT (MA96-30) stratigraphic marker (5.57–5.63 Ma).

At the Adu Dora section north of the Hatayae River, the Asa Koma Member represents the uppermost lithologic units above the Adu Dora Member diatomite beds (Figure 2.6L). A moderately altered LABT (MA96-25), consisting of a fine-grained and laminated lower half that transitions to upper bedded coarse fragments, defines the base of the Asa Koma Member. The section (>15 m) above the basaltic tuff consists of multiple units of light gray to dark orange-brown silty clays interbedded with bentonites, and the WMMT mixed basaltic and silicic tuff marker. The late Miocene (5.57–5.63 Ma) WMMT crops out in the uppermost part of the section beneath light gray silty clays and bentonite about 3 m thick. No other lithologic units of the Adu-Asa Formation are exposed above the Asa Koma Member (Figure 2.6L) at the Adu Dora Member type section. Most of the vertebrate fossils were collected in the lower section of the Asa Koma Member in the Adu Dora area.

At the Digiba Dora locality about 5 km to the north of the Adu Dora type section, thick (>25 m) and diverse sedimentary units with interbedded silicic, hydromagmatic basaltic, and mixed magma silicic and basaltic tephra crop out along eroded fault scarps (Figures 2.5 and Figure 2.6M). The lower half of the section consists of light gray silty clay, sandstone, carbonate, and fossiliferous light brown sandy to silty clay beds in ascending stratigraphic order (Figures 2.13 and 2.14). The late Miocene (5.68 Ma) WMMT (MA95-4a and 4b) that contains the mixed silicic and basaltic tephra layer occurs in the middle section of the sedimentary units.

The upper half of the section contains thick (>5 m) silty clay, moderately altered vitric ash (MA95-5), orange-brown silty clay with two interbedded basaltic tuffs (MA98-56 and MA98-57), conglomerate, pinkish and medium gray silty clays, and conglomeratic lag deposits, in ascending stratigraphic order (Figure 2.6M). The moderately altered upper (MA98-56) and lower (MA98-57) basaltic tuffs are variable in thickness (2–2.5 m). They are bedded and laminated, with alternating layers of fine and coarse scoriaceous fragments.

FIGURE 2.15
View from the Sagala Ali horst to the north across the southeastern margin of the Asa Koma catchments. Photograph by David Brill, December 26, 1998.

The matrix of both samples is mostly altered. The lower basaltic tuff (MA98-57) is chemically similar to the DOBT, which defines the base of the Rawa Member at the Asa Koma type section. Thus, the Digiba Dora section may be mostly composed of the Asa Koma and Rawa Members of the Adu-Asa Formation. The fossiliferous area at Digiba Dora is very small, although extremely rich in its fossil content. It lies beneath the late Miocene (5.57–5.63 Ma) MA95-4 WMMT stratigraphic marker (Figure 2.6M).

The Ali Ferou Dora (AFD) area to the north of the Digiba Dora locality is relatively small (Figure 2.5). North of the Digiba Dora locality, Rawa Member units are commonly exposed along the strike of the frontal fault blocks as far north as the Kurkura River drainage system. However, the canyon walls at the Kurkura River consist of basaltic lava flows of late Miocene age (MA95-1, 6.33 Ma) that are sandwiched between older yellow-green altered basaltic tuff and younger well-sorted coarse sandstone (WoldeGabriel et al. 2001). The basalt flow is within the age range of highly fractured lava flows (MA95-22, 6.16 Ma) that underlie the Adu-Asa Formation in the general area of the Adu Dora locality on the south side of the Hatayae River (WoldeGabriel et al. 2001).

Additional Asa Koma Member lithologic units were identified along major fault scarps and drainage basins along the strike of the rift margin to the south of the Asa Koma type section at Sagala Ali, Bikir Mali Koma, Saitune Dora, and Alayla (Figures 1.1, 2.5, 2.6, and 2.15). Multiple units of volcanic and sedimentary lithologic units are exposed along the Sagala Ali horst to the south of the Asa Koma type section. The stratigraphic sequence contains diatomaceous units of the Adu Dora Member beneath thick (>10 m) yellowish brown altered BABT (MA96-39). The Asa Koma Member lithologic units above the BABT were only partially investigated because of the steepness of the section. Basaltic tuff samples were collected from the base (MA96-39) and middle (MA96-38) sections of the rift-oriented major fault scarp. The rest of the section was too steep to access. However, because of its proximity (< 2 km) to the main Asa Koma type section, it appears that both sections contain correlative lithologic units. For example, the basaltic tuffs in the middle (MA96-38) and lower (MA96-39) parts of the fault scarp are chemically correlative to the Hantuuta Basaltic Tuff (HABT, MA98-45) and the LABT (MA98-47, 48, 49) at the Asa Koma section, respectively, across from the Sagala Ali sequence.

A major fault scarp on the south side of the Sagala Ali horst exposes the thick stratigraphic sequence of the Bikir Mali Koma section. It consists of diverse sedimentary and volcanic units (Figure 2.6H). As in the Sagala Ali section, basal diatomite and diatomaceous silty clays correlative to the Adu Dora Member occur beneath thick (>10 m) altered BABT, which marks the base of the Asa Koma Member (Figure 2.6I). Diverse lithologic units consisting of light gray to light orange silty clays with intercalated bentonite, altered basaltic tuff, and a thin ostracod-bearing carbonate layer capped by altered scoriaceous LABT (MA99-106) are exposed along the strike of the hanging wall. The upper half of the stratigraphic sequence is exposed along the eastern face of the footwall, and it contains partially altered WMMT, light gray silty clay, and the DOBT (MA99-109), marking the contact with the overlying Rawa Member. Vertebrate fossils are spatially and stratigraphically restricted in this region.

At the Saraitu stream type section on the south side of the Ananu drainage system, the BABT (MA99-95), which marks the base of the Asa Koma Member, caps the Adu Dora Member diatomite sequence and interbedded silicic tephra. The bulk of the Asa Koma Member units above the BABT were removed by erosion from the section (Figure 2.6F). However, sedimentary deposits correlative to the Asa Koma Member units are commonly present at the Alayla locality to the south of the Saraitu type section (Figures 2.6D and E). The Alayla localities contain fossiliferous fluvial sediments with interbedded basaltic tuffs and bentonite units of the Asa Koma Member (Figures 1.1, 2.5, and 2.6D and E). Partially exposed diatomite beds underlie the widespread basaltic tuff that is correlative to the BABT stratigraphic marker located at the base of the Asa Koma Member (Figure 2.6I). The late Miocene (5.77 Ma) LABT (MA97-15, MA96-28, MA99-94, and MA99-116) crops out directly above the BABT (WoldeGabriel et al. 2001). The LABT is thick (>4 m), dark gray, bedded, and moderately altered, and it contains unusual ripple marks.

At the Asa Koma type section, thick (~5 m) orange-brown silty clay occurs between the LABT and BABT; however, it is absent from the Alayla north (ALA-VP-2) locality. A 3 cm thick ostracod-rich carbonate layer occurs above the LABT (Figure 2.6E). Diverse sedimentary and volcanic units similar to the outcrops of the Alayla ALA-VP-1 and ALA-VP-2 localities occur at the lower half of the stratigraphic sequence of the major fault scarp on the south side of the broad, rift-oriented Alayla graben. However, the lithologic units in the upper half of the sequence, consisting of shale, brown silty clays, sandstone, conglomerate, and bentonite, capped by late Miocene (5.5 Ma) basaltic lava flow (MA99-93), are absent from the Alayla VP localities on the north side of the basin. The late Miocene basalt lava is similar in age to the Saitune Dora basaltic flows to the southeast of the Alayla localities (Figures 2.5 and 2.6J and K). Alayla is one of the richest vertebrate fossil areas. At the Ado Bolo section to the south of the Alayla VP sites, the BABT (MA99-72 and MA99-90I) caps the Adu Dora Member diatomite sequence. However, the younger lithologic units belonging to the Asa Koma Member are absent above the BABT.

Lithostratigraphic similarities and temporal correlation of capping basalts from the south Alayla (MA99-93, 5.50 Ma) and Saitune Dora (MA93-5, MA95-7, 5.54 Ma) sections suggest that the volcanic and fossiliferous sedimentary units at the Saitune Dora (STD-VP-1 and STD-VP-2) localities belong to the Asa Koma Member stratigraphic sequence (Figures 2.5 and 2.6J and K). In contrast to most of the sections previously described, the Saitune Dora fossiliferous sediments are capped by multiple flows of late Miocene (5.54 Ma) basaltic lavas (WoldeGabriel et al. 2001). The Asa Koma Member fossiliferous units are underlain by thick, altered yellow-green basaltic tuffs. At the eastern end of the STD-VP-1 locality, massive yellowish brown to orange and reddish brown scoriaceous altered basaltic tuff beds underlie more than 5 m of massive and blocky light to dark brown gray silty clay deposits that occur beneath altered yellow-green basaltic tuff. Light greenish gray sandstone and bentonite also crop out in the graben floor west of the silty clay deposits.

The rift-oriented fault scarps on both sides of the graben between the STD-VP-1 and STD-VP-2 localities consist of thick (>35 m) pinkish and yellowish green massive basaltic tuffs that are overlain by dark brown silty clay (>15 m) and 5 m of unconsolidated and poorly sorted conglomerate with sand matrix. The section is capped by partially altered late Miocene (5.54 Ma) basaltic lava flows (WoldeGabriel et al. 2001). The section at the STD-VP-2 locality consists of thick (>6 m) layers of fossiliferous light gray to brown silty clays beneath a moderately cemented cobble conglomerate.

In contrast to the nearby sedimentary deposits that are capped by basaltic lava flows, the section at the STD-VP-2 locality is not covered by basaltic lava. The dark brown silty clay and overlying conglomerate are assigned to the Rawa Member that represents the upper units of the Adu-Asa Formation. Lithologically similar packages occur at comparable stratigraphic position above the DOBT stratigraphic marker at several sections (e.g., Asa Koma, Bikir Mali Koma, and Alayla) along the strike of the step-faulted frontal blocks of the rift margin. These are assigned to the Rawa Member (Figure 2.6). At the STD-VP-1 locality, fine-grained, light greenish gray sedimentary units are exposed beneath the reddish brown silty clays and poorly sorted conglomerate deposits. The light greenish gray sedimentary deposits appear to have been deposited in reducing environments, unlike the oxidized overlying reddish brown silty clays. The Saitune Dora area is known particularly for the incredibly rich presence of elephant remains at STD-VP-2. The Gaysale, Bilta, and Kabanawa areas are not faunally as rich as the other localities.

Rawa Member

The Rawa Member forms the uppermost lithologic sequence of the Adu-Asa Formation. It contains both volcanic and sedimentary units that were deposited during the waning stage of the hydromagmatic activities that characterized the underlying Asa Koma

Member. The Rawa Member represents the thickest (~75 m) and most diverse lithologic unit of the Adu-Asa Formation, consisting of the DOBT, conglomerate, thick (>40 m) reddish brown silty clay with pedogenic carbonates, paleosol, distal silicic tuff(s), and the HABT (MA98-45), in ascending stratigraphic order (Figure 2.6I). Basaltic lava flows (5.50 Ma) cap sedimentary successions of the Rawa Member that crop out beneath the HABT. Despite the widespread occurrence of the HABT, it was not dated to establish a minimum age for the Rawa Member because it was not associated with any important paleontological localities. Moreover, the dated basaltic lava flows that cap the south Alayla and Saitune Dora sections provide minimum ages for the underlying dark brown silty clays and conglomerate units.

A lithologically correlative coarse clastic deposit occurs at similar stratigraphic levels throughout the study area from Alayla to Digiba Dora (Figure 2.6). The DOBT at the base of the Rawa Member is about 12 m above the 5.57–5.63 Ma WMMT stratigraphic marker, which occurs in the middle part of the underlying Asa Koma Member. The south Alayla stratigraphic sequence, which contains lithologic units that are similar to the Rawa Member, is capped by a late Miocene (5.50 Ma) basaltic lava flow, providing a minimum age for the uppermost Adu-Asa Formation at this locality (Figure 2.6D). The moderately cemented cobble conglomerate, which is closely associated with the orange-brown silty clay, occurs above the DOBT to the west of the Asa Koma VP-2 locality. A similar lithologic assemblage was noted above the DOBT at several locations, including Alayla, Bikir Mali Koma, Sagala Ali, and Digiba Dora along the frontal fault blocks of the western rift margin (Figures 2.6D, G, H, and M).

The Rawa Member's main section at the Asa Koma locality (Figure 2.6I) shows a cobble conglomerate beneath thick and massive dark orange-brown silty clay and pedogenic carbonates. The uppermost part of the dark orange-brown silty clay is impregnated with calcite cement and veinlets. The Rawa Member sequence includes a thick (<10 m) fossiliferous orange to light brown silty clay that crops out above a strongly cemented cobble conglomerate in the Sagala Ali horst to the south of the Asa Koma section. It is capped by the cross-bedded and laminated dark gray unconsolidated HABT (MA96-38) exposed in the middle part of the stratigraphic sequence above the orange silty clay (Figure 2.6G). The alternating units of silty clays and carbonate layers above the HABT were not fully investigated. However, a similar stratigraphic sequence is present on the north side of the graben to the west of the Asa Koma section. There, the section contains silicic tephra above the HABT (Figure 2.6I).

The Rawa Member lithologic units at the Digiba Dora locality to the north of the Rawa Member type section contain a cobble conglomerate that is interbedded within dark brown silty clays above the DOBT (MA98-57). Multiple layers of silty clays occur above the cobble conglomerate (Figure 2.6M). At Bikir Mali Koma on the eastern side of the Sagala Ali horst, the stratigraphic sequence contains cobble conglomerate (>4 m) that is interbedded within thick (15–20 m) dark brown silty clay units above the DOBT (Figure 2.6H).

Similar correlative units were also encountered in the vicinity of the Alayla VP-1 and VP-2 locality (Figure 2.6D). For example, a major fault scarp on the eastern side of the Whydola Koma horst, about 1.0 km northwest of the Alayla VP-2 site, contains diverse lithologic units. There, the sequence consists of a basaltic tuff, fine-grained siltstone, medium gray silty clay, reddish to orange brown silty clays, cobble conglomerate, light brown silty clay, altered basaltic lava, light brown silty clays, and a basaltic lava flow (MA98-55), in ascending stratigraphic order. The base of the steep fault scarp is obscured by landslides and colluvium. Therefore, the contact between the Rawa Member and the underlying Asa Koma Member was not delineated. The major fault scarp on the south side of the broad Alayla drainage basin directly south of the Alayla VP-1 and VP-2 localities also exposes multiple basaltic tuff and sedimentary deposits correlative with the Asa Koma and Rawa Member lithologic units (Figures 2.6D, E). The rift-oriented fault block between the south side of the Alayla basin and the Saragata stream contains cobble conglomerate above vitric ash (MA99-114) that is interbedded within variegated silty clays in the middle part of a stratigraphic sequence capped by the fine- to medium-grained and spheroidally weathered basaltic lava (MA99-93; 5.50 Ma).

Reconnaissance surveys in the southern part of the Middle Awash study area identified and designated a few localities at Ado Faro, Ounda Belayamo, Beraro, and Faro (Figure 2.5). A stratigraphic sequence at the Ado Faro section contains reddish brown sediments, carbonate, conglomerate, basaltic tuff, light gray silty clay, vitric tuff (MA96-36), black silty clay, and a carbonate layer capped by conglomeratic lag deposit, in ascending stratigraphic order (Figure 2.6A). The silicic tephra sample collected from the Beraro (MA96-35) section is chemically correlative to the thick vitric tuff (MA99-108) that crops out at the top of the Rawa Member sequence at the Bikir Mali Koma section (Hart et al., Chapter 3). Thus, the bulk of the Adu-Asa Formation units are mostly exposed along the arcuate western rift margin between the Ado Faro and the Digiba Dora sections of the Middle Awash study area (Figures 1.1 and 2.5).

Post–Adu-Asa Formation sedimentary and volcanic units are widely distributed along the strike of the rift margin and were encountered at several sections to the north and to the south of the Rawa Member type section at Asa Koma. These lithologic units reflect a transition to tectonic, volcanic, and sedimentological processes that are characterized by coarse clastic deposits, pedogenesis, and pedogenic carbonate accumulations, implying different environmental conditions compared with the underlying Adu-Asa Formation lithologic units, which resulted from episodic lacustrine sedimentation and hydromagmatic eruptions with intermittent fluvial sedimentation (WoldeGabriel et al. 2001). Despite the presence of distal silicic vitric ash (MA92-04) and proximal welded ignimbrite (MA93-10) deposits above the Adu-Asa Formation units, hydromagmatic basaltic eruptions along the western rift margin of the Middle Awash study area ceased by the end of the late Miocene. However, hydromagmatic eruptions continued within the rift floor in the CAC (WoldeGabriel et al. 1994; Renne et al. 1999).

The frontal fault blocks along the rift margin at Asa Ali southeast of the Saitune Dora localities consist of thick sections of altered scoriaceous basaltic tuff capped by late Pliocene basaltic lava flows. However, a welded tuff (MA93-10) interbedded within fossiliferous sedimentary deposits exposed at the base of the section in the Hindo Kali area south of Asa Ali is early Pliocene in age (Figure 2.5) and correlates with an ignimbrite from Ounda Belayamo (MA96-37) and with ignimbrites that occur above the HABT at the Asa Koma (MA98-44) and Ananu (MA92-5) sections (Figure 2.5). Collectively these welded/ignimbritic units are designated Tuff 93-10 (Hart et al., Chapter 3). At the Ananu section, vitric tuff (MA92-6) above Tuff 93-10 is chemically correlative to the 2.5 Ma Maoleem Vitric Tuff (de Heinzelin et al. 1999) of the Hata Member, which represents the basal stratigraphic package of the Bouri Formation at Bouri (see Figure 1.1). There are Hadar-age and younger fossiliferous deposits along the rift margin further to the south, west of modern Yardi Lake. These were first probed intensively during the 2005 field season. The northwest sector of the study area, north of the Borkana river, is poor in late Miocene deposits, compared to areas south of the Jara river, because of the steep, recently faulted front scarp. Nevertheless, our preliminary explorations of the front escarpment have revealed additional fossiliferous deposits. It is clear that additional research on these younger units is warranted. In the center of the western margin of the study area, the Rawa Member, in particular, will require additional fieldwork and laboratory analyses in order to understand its internal dynamics and external correlations.

Comparison of the Stratigraphic Record of the Transition Zone to Other Rift Margins of the Afar and the Main Ethiopian Rifts

There are structural and stratigraphic similarities and differences along the strike of the western margin to the southwest and to the northeast of the Adu-Asa Formation outcrops. Semaw et al. (2005) and Simpson et al. (2007) described correlative early Pliocene to late Miocene hominid-bearing sedimentary units within the frontal fault blocks of the western rift margin in the Gona area, directly to the northeast of the Middle Awash study area. Southwest of the Adu-Asa Formation section, the frontal fault blocks of the western rift margin consist of lava flows with minor interbedded sedimentary rocks. There, Pliocene (2.54–3.76 Ma) flood basalts are exposed, unconformably capping local late Miocene basaltic lavas (6.62–10.58 Ma) along the western rift margin between the Kessem-Kebena Rivers and the Middle Awash study area (WoldeGabriel et al. 1992b; Wolfenden et al. 2004). In contrast, Precambrian and Mesozoic rocks are exposed unconformably beneath Tertiary volcanic rocks along the western and the southeastern margins of the Afar Rift, except along the margins of the transition zone. A sliver of pre-Tertiary rocks is also exposed along the western rift margin of the central sector of the MER.

In the Middle Awash study area, the base of the Tertiary sequence is not exposed along the frontal fault blocks of the western rift margin of the transition zone. Instead, late Miocene (5.3–6.33 Ma) sedimentary and volcanic rocks crop out adjacent to the rift floor, whereas early to late Oligocene volcanic rocks occur along the adjacent plateau, bordering the broad western rift margin (WoldeGabriel et al. 2001; Ukstins et al. 2002). The southeastern rift margin of the southern Afar Rift mostly consists of crystalline basement and Mesozoic sedimentary rocks that crop out unconformably beneath early Miocene basalts (Juch 1975; Kuntz et al. 1975; Juch 1980; Chernet et al. 1998). Chernet et al. (1998) reported K-Ar ages of early Miocene (23.8–24.1 Ma) on two basaltic samples from the base of the Tertiary sequence directly above the Mesozoic sedimentary rocks. These results are consistent with previous dates (22.5–23.2 Ma) from the same flows (Kuntz et al. 1975). Along the broad southeastern rift margin of the transition zone, the basaltic sequence is thicker (2,500 m) and thins toward the plateau (1,000 m). This pattern is repeated by the overlying late Miocene silicic ignimbrite flows, consistent with increased riftward volcanic eruptions, intense tectonic activities, and subsidence (Sickenberg and Schonfeld 1975; Juch 1980; Tiercelin et al. 1980).

In the central sector of the MER to the southwest of the transition zone, a sliver of pre-Tertiary rocks, consisting of crystalline basement unconformably capped by Mesozoic sedimentary rocks crops out along a major single boundary fault of the Guraghe Mountains (Arno et al. 1981; WoldeGabriel et al. 1990). The pre-Tertiary sequence is unconformably overlain by Oligocene basalt flows (31.7–32.4 Ma) that are also unconformably capped by thick Pliocene welded ignimbrite (WoldeGabriel et al. 1990). These older K-Ar ages are generally similar to new $^{40}Ar/^{39}Ar$ dates of 29.34 to 30.9 Ma obtained on volcanic rocks from the NW Plateau adjacent to the broad western rift margin of the Middle Awash study area (Ukstins et al. 2002).

As at the southeastern rift margin of the Afar Rift, the Tertiary volcanic sequence is thicker along the strike of the broad western rift margin of the Middle Awash region because of continuous volcanic and tectonic activities and subsequent subsidence along the developing rift valley. However, the volcanic and tectonic activities along the western and eastern rift margins of the transition zone were not contemporaneous.

Based on published K-Ar and $^{40}Ar/^{39}Ar$ results, late Oligocene and early Miocene volcanic rocks are commonly present adjacent to the western rift margin from the central sector of the MER as far north as the transition zone of the Middle Awash study area. For example, Oligocene basaltic and silicic rocks are exposed along both rift margins of the central sector of the MER and in the adjacent SE Plateau (Bale Mountains) (Berhe et al. 1987; WoldeGabriel et al. 1990). However, this is not the case with the earliest volcanic flows of the Afar Rift margins, because the lava flows are early Miocene in age. In the west-central Afar Rift margin, the earliest volcanic flows yielded a K-Ar age of 24 Ma and occur above the Red Series conglomerate that unconformably overlies Mesozoic and crystalline basement rocks (Bannert et al. 1970; Tiercelin et al. 1980). This age is similar to the oldest (23.8–24.1 Ma) volcanic rocks of the southeastern Afar Rift margin and shoulder (Chernet et al. 1998).

Migration of Volcanism and Patterns of Faulting within the Transition Zone

The absence of Oligocene lava flows in the western and southeastern rift margins of the transition zone suggests either that the crystalline basement and Mesozoic rocks were uplifted and were not covered by the earliest volcanic eruptions prior to the early Miocene lava flows or that the lava flows were subsequently removed by erosion. Alternatively, volcanism migrated from the northwestern to the southeastern plateau covering the intervening area during the early Miocene. Most likely, the lack of Oligocene volcanic rocks along the western and southeastern rift margins implies that volcanism probably started later in the Afar Rift region, consistent with the widespread occurrences of early Miocene basaltic eruptions that cap the pre-Tertiary rocks of the southeastern escarpment and adjacent highlands (Kuntz et al. 1975; Chernet et al. 1998). This is also generally true about the age of volcanism (≤26 Ma) in the Republic of Djibouti, which started at about 26 Ma (Chessex et al. 1980).

The early Miocene basalts unconformably capping the pre-Tertiary rocks of the southeastern escarpment probably represent the earliest record of eastward migration of volcanism from the NW highlands toward the area currently covered by the Afar Rift and the southeastern escarpment. Following the early Miocene eruptions, volcanism and tectonic activities continued to shift toward the rift margins and the adjacent rift floor over time. Such volcanic eruption patterns are uncommon within the MER. Instead, voluminous late Miocene and Plio-Pleistocene silicic and basaltic eruptions occurred within the rift floor and the adjacent rift margins and shoulders (WoldeGabriel et al. 1990; Chernet et al. 1998).

The distinct differences in volcanic eruptions and faulting patterns noted along the rift margins of the transition zone were caused by the tectonic interactions of the major boundary faults of the two oceanic rifts of the NW-SE trending Red Sea and the E-W trending Gulf of Aden with the SW-NE trending MER. The antithetically faulted broad rift margins and the adjacent wider rift floor of the transition zone are also the products of the complex dynamic interactions of major faults, controlling the three rifts. In contrast, the MER to the southwest of the transition zone is uniformly oriented (NE-SW) and has rift margins mostly defined by singular SW-NE trending major boundary faults, with the adjacent rift floor having uniform width (~80 km) along its full length.

The age decrease across the broad western rift margin toward the rift floor of the Middle Awash study area is consistent with previously reported riftward migration of volcanism (Zanettin and Justin-Visentin 1975). Similar riftward age variation (5.59–24.1 Ma) of basaltic and ignimbrite units was reported from the opposing southeastern rift margin (Kuntz et al. 1975; Chernet et al. 1998). However, in contrast to the broad western margin of the Middle Awash study area, rift-bound late Miocene to early Pliocene (4.96–6.69 Ma) silicic centers define a narrow (7–9 km) marginal graben parallel to the southeastern rift escarpment.

The transition zone is characterized by volcanic rocks of variable ages that are complexly faulted, rotated, and tilted along the broad rift margins. As a result, several rift-oriented half grabens have developed. Such structural features are rare within the MER rift margins and shoulder. Ukstins et al. (2002) reported syn-rift volcanic rocks (14.9–19.8 Ma) along these grabens of the broad western rift margin of the Middle Awash study area, but no sedimentary rocks contemporaneous with early rift basins have been identified there. Despite the presence of Oligocene volcanic rocks in the adjacent highlands and early Miocene rifting, coarse clastic deposits such as debris flows, channel conglomerates, or colluvium are generally rare in the Adu-Asa Formation sedimentary sequence.

Unlike the Adu-Asa Formation (5.2–6.3 Ma), the older Chorora Formation of the southeastern rift margin of the transition zone is bracketed between late Miocene (9.1–10 Ma) basaltic and silicic tuffs (Sickenberg and Schonfeld 1975; Tiercelin et al. 1980; WoldeGabriel et al. 2001; see Figure 1.1, inset map). The sedimentary units of the Chorora Formation were deposited along a rift-oriented, narrow (7–9 km) marginal graben located between the evolving main boundary fault of the southeastern escarpment and a rift-bound border fault that is currently marked by SW-NE trending late Miocene to early Pliocene (4.59–6.69 Ma) silicic centers (Kazmin et al. 1980; WoldeGabriel et al. 1992b; Chernet et al. 1998). Apart from the age difference, the tectonic settings, the depositional environment, and lithologic assemblages of the late Miocene sedimentary deposits of the Adu-Asa and Chorora Formations show overall similarities.

The western and southeastern rift margins of the transition zone are characterized by similar tectonic features of broad, densely faulted, riftward-dipping (antithetic) fault blocks (Morton and Black 1975; Juch 1980; Kazmin et al. 1980; Mohr 1987). The broad rift margins contain several narrow and segmented parallel half grabens along the strike of both escarpments. As stated above, the boundary faults exhibit different orientations that vary along the strike of the western and southeastern escarpments of the transition zone.

Age and Mechanism of Faulting within the Transition Zone

There is no consensus regarding how and when rifting, basin development, and the changes in the orientation and width of the broad rift margins and the adjacent rift floor in the transition zone began. Morton and Black (1975) attributed the tectonic pattern along the broad western rift margin to faulting, tilting, and rotation within the brittle upper crust, which is separated by a gradational transition zone from ductile deformation in the underlying lower crust. Mohr (1987) questioned the Morton-Black hypothesis of crustal thinning and proposed crustal dilatation driven by intrusion of dike swarms as the cause for thinning and faulting along the broad western margin of the Afar Rift.

Based on published K-Ar results (Megrue et al. 1972; Justin-Visentin and Zanettin 1974), Mohr (1987) suggested that the main structural features of the western margin of the Afar Rift were established by the middle Miocene. Published middle Miocene $^{40}Ar/^{39}Ar$ ages (14.9–19.8 Ma) of volcanic rocks that occur along the central part of the broad western rift margin of the transition zone were interpreted as syn-rift eruptions related to Red Sea Rift volcanism (Ukstins et al. 2002). However, recent field investigations along the broad western rift margin of the Middle Awash study area suggest that tectonic interactions between the boundary faults of the northern MER and the Red Sea and Gulf of Aden Rifts appear to be responsible for the distinct structural pattern that is apparent along both rift margins of the transition zone. For example, the arcuate structural pattern that is superimposed by densely spaced N-S oriented faults within the broad western margin of the Middle Awash study area is a product of the complex tectonic interaction of the boundary faults (see Figure 1.1). Tesfaye et al. (2003) described the arcuate structure as an accommodation zone, defining the transition zone between the N25°E- and N25°W-trending boundary faults of the northern sector of the MER and the Red Sea Rift, respectively. Wolfenden et al. (2004) also described the arcuate structure of the broad western rift margin as the southern boundary of the Oligocene Red Sea Rift. However, current field studies and $^{40}Ar/^{39}Ar$ results indicate that the volcanic rocks transected by the N-S oriented faults and the arcuate structure are late Miocene in age, suggesting that the tectonic patterns are younger in age (WoldeGabriel et al. 2001). Additional evidence about the age of the continuous tectonic interactions of the boundary faults within the Afar Rift is noted in the west-central part of the Afar Rift floor to the north of the transition zone. In the central Afar Rift, the SE-NW trending western boundary fault zone of the Red Sea Rift cuts across Quaternary basaltic flows and terminates the SW-NE trending axial rift zone of the MER.

Analog for the Origin of the Broad Rift Margins of the Transition Zone

The broad rift margins of the transition zone from the northern MER to the southern Afar Rift are characterized by 10 to 15 km wide and more than 100 km long half grabens (e.g., the Borkona Graben) that formed as a result of the tectonic interactions and subsequent rotations of the antithetically faulted blocks. Several of these N-S trending and right-lateral stepping grabens occur along the strike of the western margin (see Figure 1.1). These mostly closed grabens are breached on their eastern sides by major rivers that cut across the broad western rift margin and flow to the rift floor. Similar partially closed but smaller half grabens also occur along the southeastern broad rift margin (Juch 1980). The structural patterns of half grabens and horsts within the broad western rift margin probably represent a modern analog to early rifting mechanisms that slowly evolved into the greater rift basin that constitutes the modern Afar depression. The late Miocene sedimentary and volcanic rocks of the Middle Awash study area, therefore, should be seen as part of this dynamic process rather than as ancient analogs of the rift-centered deposition that is occurring along the modern low-elevation, low-topography Awash River and Yardi Lake.

Physical Paleoenvironmental Records of the Adu-Asa Formation

The paleoenvironmental conditions of the late Miocene Adu-Asa Formation were assessed from lithological and paleontological assemblages and from stable isotope analysis of paleosol and carbonates (see Su et al., Chapter 17). Based on the stratigraphic record, the western rift margin of the Middle Awash study area was subjected to complex tectonic, volcanic, and sedimentation processes during the late Miocene (5.3–6.3 Ma). Major lithologic units of altered yellow-green hydromagmatic basaltic tephra deposits along the Kurkura River canyon, which cuts the frontal fault blocks of the rift margin, and thick (≥50 m) coarse sandstone deposits at the Jara-Borkana locality north of the Jara River predate the Adu-Asa Formation. The volcanic and sedimentary records suggest that the area was subjected to several cycles of volcanism and faulting that led to subsidence and the accumulation of the thick, coarse clastic fluvial deposits prior to the accumulation of the Adu-Asa Formation sequence. By the time the Adu-Asa Formation units started to accumulate, volcanism had ceased in the area, but faulting and subsidence appear to have continued along marginal basins of the evolving rift margin of the Middle Awash region. This observation is consistent with the stratigraphic record at the base of the Adu-Asa Formation.

The Saraitu Member consists of sandstone and silty clays that transition to lacustrine deposits with aquatic fossils (Figure 2.6A). Lacustrine silty clay deposits with minor fluvial sediments continued to accumulate before thick (>20 m) diatomite beds of the Adu Dora Member started to dominate the depositional environment.

The Adu Dora Member represents a distinct lithologic sequence that is dominated by pure and fissile diatomite silica sinters. The upper part of the member contains thin (5–10 cm) beds of silicic and hydromagmatic basaltic tephra layers. The absence of major volcanic units and fluvial sedimentary deposits suggest volcanic and tectonic quiescence during this time, except for the input of distal ash layers into the basin.

However, by about 5.77 Ma a new episode of hydromagmatic eruptions changed the landscape. Several layers of thick ($\geq$3 m) basaltic tuffs interbedded with fluvial sediments dominated the Asa Koma Member of the Adu-Asa Formation. Most of these major tephra stratigraphic markers blanketed the southern half of the rift margin from Digiba Dora to Ado Bolo and beyond (see Figure 1.1). Hydromagmatic basaltic volcanism dominated the landscape for a short time interval (5.57–5.77 Ma) and started to wane with time. Despite widespread explosive hydromagmatic and extrusive volcanism, the fluvial sediments of the Asa Koma Member contain abundant vertebrate fossils, including terrestrial forms, unlike the Saraitu and Adu Dora Members, which are dominated by aquatic fossils.

A major hydromagmatic eruption occurred during the waning phase of the intense volcanic and tectonic processes that characterized the Asa Koma Member stratigraphic assemblage. Unlike the older members of the Adu-Asa Formation, the Rawa Member lithologic units formed in a different tectonic and environmental setting. The Rawa Member consists of basaltic tuffs and basaltic lava flows, conglomerate, thick ($>$40 m) reddish brown massive silty clay impregnated with calcite cement and veinlets, paleosol, and pedogenic carbonates. The cobble conglomerate ($\geq$4 m), which in most cases is interbedded within the reddish brown silty clay, represents a pulse of high-energy clastic sedimentation that was probably triggered by uplift and steep gradients in the adjacent highlands.

The conglomerate deposit in the middle part of the Rawa Member blankets a large area and forms a distinct stratigraphic marker along the strike of the rift margin from Digiba Dora to Alayla (Figure 2.6). The bulk of the reddish brown silty clay occurs above the conglomerate, and it probably represents a stable environment that was characterized by pedogenesis. The section above the reddish brown silty clay contains distal silicic ash, local basalt lava flows, basaltic tuff (HABT) that represents the last pulse of the hydromagmatic eruptions, thick calcrete and carbonates ($\geq$4 m), and welded ignimbrite interbedded within the next cycle of fluvial sedimentation. By the time the Rawa Member units were deposited, the region experienced a period characterized by volcanic and tectonic quiescence along the western rift margin, except for the depositions of distal silicic tephra.

The abundant lacustrine deposits and the overlying hydromagmatic eruptions suggest that the Adu-Asa Formation was deposited under environmental conditions that were wetter than those of today. Other paleoenvironmental evidence from stable isotope data and vertebrate fossil assemblages also indicate cool, high-altitude, and closed woodland to grassy habitats (Haile-Selassie 2001b; WoldeGabriel et al. 2001; Su et al., Chapter 17). Additional detailed studies within the rift-bound CAC about 20 km to the east of the western rift margin indicate similar cycles of uplift, faulting, basaltic lava and hydromagmatic eruptions, and fluvial and lacustrine sedimentation records and paleoenvironmental conditions between 4 and 5.2 Ma (WoldeGabriel et al. 1994; Renne et al. 1999).

FIGURE 2.16
A north-to-south microfissure in alluvium near the base of the western margin of the Middle Awash study area, north of Dallifage. Tectonism has influenced this region for the last six million years and continues to the present day (Sepulchre et al. 2006). Photograph by Tim White, December 17, 2003.

Conclusions

The Adu-Asa Formation of the Middle Awash study area represents a narrow period during the evolution of the broad western rift margin. As documented here and in the following chapters, this sediment package has yielded important information on the last six million years of tectonic and volcanic evolution in northeastern Ethiopia. There is no consensus as to when rifting started in the region. Middle Miocene (14–19 Ma) syn-rift volcanic rocks are reported from the middle part of the broad western rift margin to the northwest of the Middle Awash study area, but no contemporaneous sedimentary deposits are known from this region. The broad western rift margin appears to have been strongly modified, subsequent to its formation during the Plio-Pleistocene, by the tectonic interactions of the western boundary faults of the Red Sea and the Main Ethiopian Rifts. Within the transition zone, the western rift margin contains both SW-NE and N-S trending fault zones. In general, faults related to the Red Sea Rift postdate those of the SW-NE trending faults of the Main Ethiopian Rift. The broad western rift margin contains several parallel half grabens that also resulted from interactions among the boundary faults. Similar features are also noted along the southeastern escarpment of the southern Afar Rift, resulting from interactions of the boundary faults of the Main Ethiopian and Gulf of Aden Rifts.

In summary, the complex tectonic interactions of the boundary faults of the two oceanic rifts of the Red Sea and the Gulf of Aden with the Main Ethiopian Rift have created broad rift margins and a very wide rift floor within the transition zone. In general, faults related to the Red Sea Rift are prominent within the western margin of the transition zone and the west-central Afar Rift floor. The NW-SE trending Red Sea Rift fault zones terminate the SW-NE trending axial rift zone of the Main Ethiopian Rift.

The late Miocene Adu-Asa Formation contains the oldest hominid fossils in Ethiopia. It contains distinct lithologic units that preserve important volcanic, tectonic, sedimentological, paleontological, and paleoenvironmental records for the latest Miocene period (5.3–6.3 Ma). The Adu-Asa Formation accumulated within marginal basins of the evolving rift margin. It is dominated by lacustrine deposits in its lower half, followed by hydromagmatic eruptions in the middle part and an upper section that is generally characterized by pedogenesis during a period of relatively lessened volcanic, tectonic, and sedimentation activity. Based on the dominance of lacustrine sediments and hydromagmatic eruptions, the Adu-Asa Formation was mostly deposited during wetter paleoenvironmental conditions. Vertebrate fossils from the relatively few paleontological localities identified along the western margin provide additional paleoenvironmental indicators that are consistent with the lithological record. However, much detailed field and laboratory work remains to be done to achieve a fuller understanding of the complex geological, paleontological, and paleoenvironmental records in space and time along the western rift margin of the Middle Awash study area, an area in which active rifting continues today (Figure 2.16).

3

Volcanic Record of the Adu-Asa Formation

WILLIAM K. HART,
GIDAY WOLDEGABRIEL,
YOHANNES HAILE-SELASSIE,
AND
PAUL R. RENNE

As noted in Chapter 2, the Middle Awash region of Ethiopia resides in a unique setting within the context of the late Cenozoic Ethiopian volcanic province. This setting, the tectonic and magmatic transition zone between the northern sector of the Main Ethiopian Rift (MER) and the Afar Rift (Figure 3.1), has been extremely volcanically and tectonically active from at least the Miocene to the present. Much of this activity is focused along the western and southeastern rift margins, where intense and complex faulting and basin development and eruptions from abundant volcanic centers and complexes have resulted in the formation, preservation, and exposure of thick sequences of mixed sedimentary and volcanic deposits that partially record the tectonic and volcanic events marking the evolution of the transition zone (Kazmin and Berhe 1978; Chernet et al. 1998; Chernet and Hart 1999; Boccaletti et al. 1999; Mazzarini et al. 1999 and references therein).

An important aspect of transition zone volcanic activity (and also the vast majority of activity throughout the Afar and the central and northern sectors of the MER and the surrounding rift shoulders) is its fundamentally bimodal basalt-rhyolite/trachyte character (e.g., Walter et al. 1987; WoldeGabriel et al. 1990, 1999; Chernet and Hart 1999). The origin of this bimodality is of volcanological and petrological interest, but for the purpose of this study its significance lies in the physical and chemical characteristics of materials produced by a given volcanic eruption and the nature and style of the eruption.

In addition to mineralogical and geochemical differences between the endmembers, basalt is produced from magma that is hotter, less viscous, and typically less volatile-rich than rhyolite/trachyte. These characteristics contribute to other diagnostic properties that assist in stratigraphic and paleoenvironmental reconstructions. For example, even the most powerful explosive (pyroclastic) basaltic eruptions (phreatomagmatic or hydrovolcanic; magma and external water interaction) do not provide the extremely wide dispersal pattern common to explosive rhyolite/trachyte eruptions (e.g., Houghton et al., 2000; Pyle 2000), although eruptions of both endmembers can result in thick accumulations of medium- to coarse-grained pyroclastic flow and fall material in close proximity to the vent.

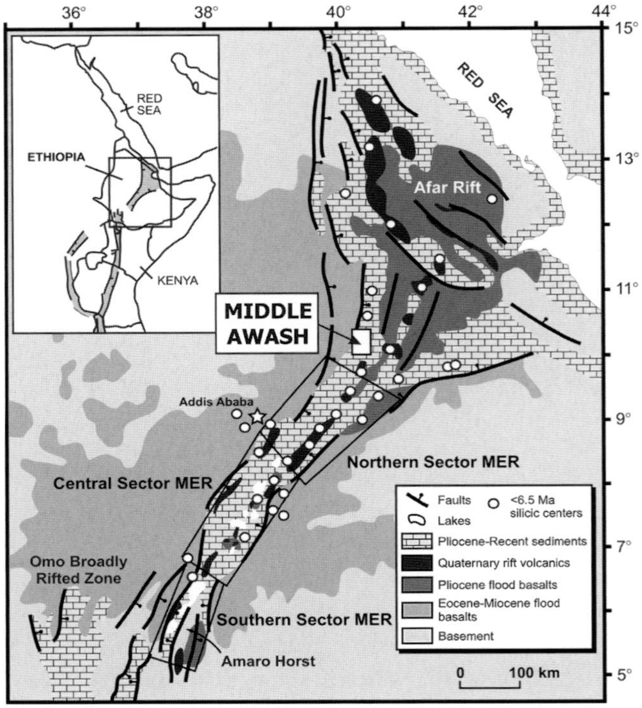

FIGURE 3.1

Generalized geological map of the Main Ethiopian and Afar Rifts and bounding plateaus. Also highlighted are major volcanic centers/complexes <6.5 Ma in age that produced voluminous silicic materials and thus are potential source regions for distal Middle Awash silicic tephra deposits. Modified after WoldeGabriel et al. (2004).

Basaltic lava flows, on the other hand, can extend for tens to hundreds of kilometers from their original vent area, whereas the high viscosity of rhyolite/trachyte magmas severely limits the distance a resulting lava flow can travel (e.g., Kilburn 2000) and typically yields a very thick accumulation of lava. Thus, in cases where very widespread accumulations of apparent silicic flow material are observed, this material typically was emplaced as a high-energy pyroclastic flow associated with an explosive origin. These general characteristics, together with more detailed examination of a volcanic deposit's physical features (thickness, grain size, bedding, alteration), can provide first-order information on relative proximity to source volcano/vent (proximal = nearby versus distal = far away), paleoenvironment (wet versus dry), depositional environment, and paleotopography/tectonic activity for a given location, particularly when used in concert with other sedimentologic, paleobiological, chronostratigraphic, and geochemical indicators (e.g., White et al. 1993; Brown 1995; de Heinzelin et al. 1999; WoldeGabriel et al. 2001; Clark et al. 2003; Quade et al. 2004; WoldeGabriel et al., 2004, 2005, Chapter 2). It should be noted at this point that, throughout the remainder of this chapter, pyroclastic deposits, regardless of composition, mode of deposition, or grain size, are referred to as tephra.

The numerous fault-bounded, highly dissected, mixed sedimentary and volcanic deposits extending along the strike of the western rift margin of the Middle Awash area were, in the recent literature, first described by WoldeGabriel et al. (2001). In keeping with previous research (Kalb 1993), the majority of these deposits were placed

FIGURE 3.2
The Adu-Asa Formation at Ananu contains multiple horizons of basaltic and vitric tephra. The rich Middle Awash volcanic record is a key to correlating and dating various paleontological localities. Photograph by Tim White, December 20, 1993.

into the Adu-Asa Formation, but new member definitions and names were presented. WoldeGabriel et al. (Chapter 2) provide a more detailed history of geological exploration and resulting stratigraphic nomenclature issues in this region. They also provide member-by-member descriptions of the main localities where detailed stratigraphic determinations and volcanic sampling have taken place since 1992. Five prominent, laterally extensive volcanic horizons were correlated along strike of the western margin, assigned names, and used to assist in the definition of the Adu-Asa Formation members based on stratigraphic observations, physical characteristics, and major element chemical characteristics (WoldeGabriel et al. 2001 and supplemental information therein). These and other volcanic horizons are discussed for their overall stratigraphic context by WoldeGabriel et al. (Chapter 2) and discussed in greater detail in the following sections.

The data presented and evaluated herein are restricted to Mio-Pliocene volcanic materials exposed along the western margin of the Middle Awash area, with a primary focus on the late Miocene deposits of the Adu-Asa Formation (Figures 3.2 and 3.3). The Mio-Pliocene deposits of the Kuseralee Member of the Central Awash Complex (CAC) Sagantole Formation (Renne et al. 1999) are separated from the western margin Adu-Asa Formation deposits by approximately 20 km, provide no lithostratigraphic or chemical links between these formations, and include vertebrate fossil localities that

FIGURE 3.3
View to the east where the Kabanawa stream cuts into the Adu-Asa Formation, exposing portions of the silty Saraitu Member at the streambed, overlain by the diatomites of the Adu Dora Member, and the basal basaltic tuff of the Asa Koma Member at the top of this local succession. Photograph by David Brill, December 26, 1996.

are younger (5.18–5.55 Ma) than those described for the western margin Adu-Asa Formation (5.50–5.77 Ma) (WoldeGabriel et al. 2004). Thus, the minor volcanic constituents of the Kuseralee Member of the Sagantole Formation are not included in this contribution.

The primary goals of this contribution are to (1) define the different types of volcanic material preserved in the Adu-Asa Formation, (2) present a working composite stratigraphic view of the Adu-Asa Formation volcanic horizons, (3) present the existing geochemical data for western margin volcanic samples, including some with uncertain chronostratigraphic context, (4) illustrate the data and observations that provided the basis for previous correlations and naming of key volcanic marker horizons, and (5) illustrate the data and observations that provide new regional correlations and hence establishment of newly named volcanic horizons. In achieving these five goals we will have provided chronostratigraphic and chemostratigraphic information that represents the base information upon which future studies must draw their comparisons.

Adu-Asa Formation and Associated Volcanic Units

Field, Laboratory, and Interpretive Methods

The general field and laboratory geological, archaeological, and paleontological methods employed by the Middle Awash research project have previously been elucidated (de Heinzelin et al. 2000a, b; WoldeGabriel et al. 2004; Haile-Selassie and WoldeGabriel, Chapter 1). Considering the large area encompassed by the western margin of the Middle Awash, the field investigation leading to identification and sample recovery of the volcanic materials described herein is best termed as detailed reconnaissance. None of these deposits have been mapped in any detail. In the context of our reconnaissance work, particular attention was paid to stratigraphic relationships, proximity and relationship to vertebrate fossil-bearing lithologies, overall unit thickness and characteristics, and choice of most appropriate material for chronologic or/and geochemical sampling. A number of the basaltic tephra units encountered illustrated complex bedding at a variety of scales; detailed description and sampling of these complexities was not undertaken.

The only analytical data reported herein are major element analyses (SiO_2, TiO_2, Al_2O_3, Fe_2O_3, MnO, MgO, CaO, Na_2O, K_2O, P_2O_5) of mafic and silicic tephra and basalt lava flow samples. The tephra data for any given sample were obtained on multiple individual glass pyroclasts by standard electron microprobe (EMP) techniques using polished thin sections or grain mounts of bulk material. All of the tephra analyses were obtained using an automated Cameca SX50 electron microprobe at Los Alamos National Laboratory following the techniques described by WoldeGabriel et al. (1994). The analytical precisions for mafic (first number) and silicic (second number) tephra glass, based on replicate analyses of standard reference materials of similar compositions, are (in weight percent): SiO_2, ±0.47, ±0.72; TiO_2, ±0.09, ±0.04; Al_2O_3, ±0.28, ±0.26; Fe_2O_3, ±0.17, ±0.11; MnO, ±0.03, ±0.02; MgO, ±0.10, ±0.01; CaO, ±0.21, ±0.03; Na_2O, ±0.09, ±0.49; K_2O, ±0.01, ±0.17; and P_2O_5, ±0.03, ±0.02. The basalt lava flow samples were reduced as whole rocks to fine powder and were analyzed for the same suite of major elements by direct current argon plasma atomic emission spectroscopy (DCP-AES) techniques at Miami University following the methods described in Katoh et al. (1999). The analytical precisions for mafic whole-rock powders, based on replicate analyses of standard reference materials of similar composition, are (in weight percent): SiO_2, ±0.37; TiO_2, ±0.02; Al_2O_3, ±0.14; Fe_2O_3, ±0.08; MnO, ±0.01; MgO, ±0.05; CaO, ±0.08; Na_2O, ±0.03; K_2O, ±0.02; and P_2O_5, ±0.04. Additional analytical work (tephra glass purification; trace and rare-earth element and Nd isotope analyses) on key Adu-Asa Formation volcanic units, particularly on purified glass separates from mafic and silicic tephra, is in progress.

In highly faulted and dissected terrains, volcanic horizons provide a critical means to correlate stratigraphic sequences over wide areas and to assemble the composite chronostratigraphic history of local and regional volcanic input into often isolated basinal depocenters (e.g., Sarna-Wojcicki 2000 and references therein). Volcanic products, particularly tephra fall deposits, are excellent stratigraphic markers and can provide both relative and absolute age constraints for associated stratigraphy (tephrostratigraphy). For a tephra layer to be useful as a stratigraphic marker it must have discernibly unique physical and chemical characteristics acquired from its eruptive event and parental magma, as well as a unique eruption age, thereby imparting characteristics that distinguish it from other tephra horizons (Sarna-Wojcicki 2000). For example, the most felsic population of quenched magma (glass pyroclasts such as shards or pumice) from individual tephra horizons provides a unique geochemical fingerprint for a given eruption (Westgate and Gorton 1981).

This logic is readily extended to more complex bimodal, multimodal, and mixed glass populations wherein the geochemical characteristics of the various components of this complexity provide a unique fingerprint (e.g., Larson et al. 1997; Hart et al. 2004). Furthermore, geochemical methods applied to basaltic tephra, focusing on glassy pyroclasts such as shards, droplets, and scoria, have also been shown to yield useful and diagnostic fingerprints (e.g., WoldeGabriel et al. 1994, 2001). In all of these cases, the geochemical signature provides one of the primary means (tephrochemistry) by which a given tephra layer and stratigraphic sequence can be correlated to other tephra horizons and their associated stratigraphic sequences. The larger the dataset of geochemical information used for comparison, the greater the certainty of assigned correlations, particularly for stratigraphically distinct tephra horizons that display very similar major element compositions. Moreover, once a tephrochemical correlation is established, any absolute age information associated with a single outcropping of this tephra unit can be applied to nearby or very distant stratigraphic sequences containing this same tephra unit (Sarna-Wojcicki 2000).

A variety of methods are employed routinely to compare the chemical characteristics of multiple tephra in order to test for potential tephrochemical correlations. These range from a qualitative assessment of geochemical characteristics as portrayed on binary, ternary, and multielement plots (e.g., Izett 1981; Brown et al., 1992; Hart et al. 1992; Preece et al. 1992; Perkins et al. 1995) to quantitative statistical treatments (e.g., Borchardt et al. 1972; Sarna-Wojcicki 1976; Stokes et al. 1992; Perkins et al. 1995). In our evaluation of the Adu-Asa Formation chemical data described in subsequent sections, we have employed the statistical approach described as an analysis of similarity (Borchardt et al. 1972), which derives a single variable (similarity coefficient, SC) denoting multivariate similarity (SC = 1 is highest degree of similarity). This approach is summarized in the following equation:

$$d_{(A,B)} = \frac{\sum_{i=1}^{n} R_i}{n}$$

where: $d_{(A,B)}$ = SC for samples A and B

i = element number
n = total number of elements
$R_i = X_{iA}/X_{iB}$ if $X_{iB} \geq X_{iA}$
$R_i = X_{iB}/X_{iA}$ if $X_{iA} > X_{iB}$
X_{iA} = concentration of element i in sample A
X_{iB} = concentration of element i in sample B

Stratigraphic and Physical Characteristics

The bimodal volcanic units investigated for this study are dominated by proximal basaltic pyroclastic fall deposits and distal silicic pyroclastic fall deposits, but they also include basalt lava flows, welded pyroclastic flow (ignimbrite) deposits, and a distinctive mixed

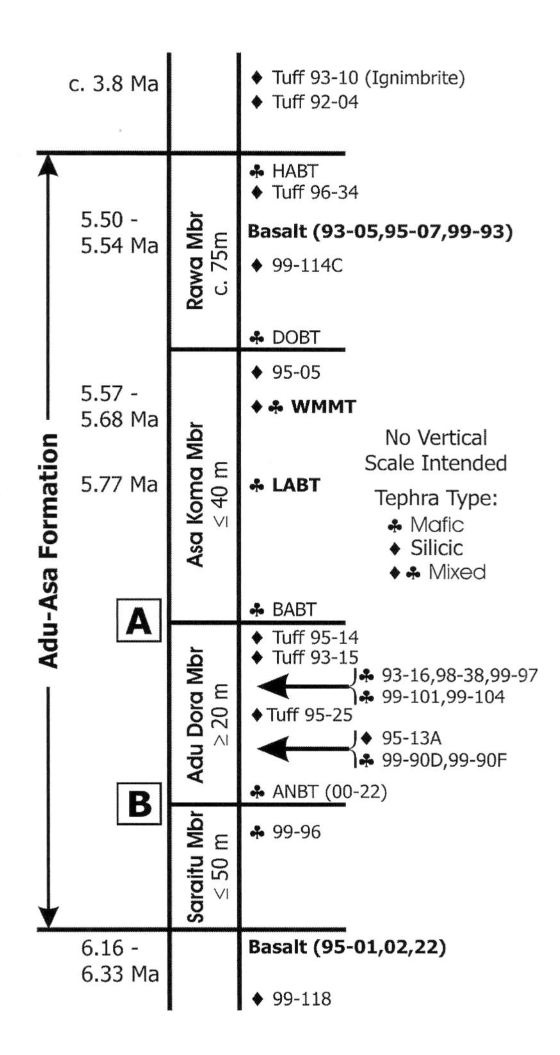

FIGURE 3.4

Schematic composite volcanic stratigraphy of the western margin area Adu-Asa Formation. This composite is derived from the stratigraphic relationships discussed and illustrated in Figure 2.6 of WoldeGabriel et al. (Chapter 2). Named volcanic units are from WoldeGabriel et al. (2001) and this chapter. All individual sample numbers carry the MA prefix. The listed ages and age ranges are for the volcanic units designated in bold type and were first reported in WoldeGabriel et al. (2001). The boxed letters A and B indicate the portions of the stratigraphic sequence represented by Figures 3.5 and 2.9, respectively.

magma pyroclastic fall deposit, the Witti Mixed Magmatic Tuff (WMMT). Individual samples that have been correlated to form a named volcanic unit (e.g., WMMT or Tuff 93-10) are designated as such and shown in descending stratigraphic order as pictured in Figure 3.4. Only those units for which stratigraphic context is available from one or more of the numerous measured sections discussed by WoldeGabriel et al. (Chapter 2) are portrayed in the composite section of Figure 3.4. Furthermore, abundant, centimeter-scale bentonite horizons are not shown in Figure 3.4, nor are they addressed in the following discussion. Thus, it is important to recognize that a more substantial record of regional volcanic activity is captured within the Adu-Asa Formation than is shown in Figure 3.4.

Considerable along-strike variability in lithology and thickness is observed for all Adu-Asa Formation members; thus, not all tephras crop out at every locality. On the other hand, certain tephra horizons are present at a sufficient number of localities, most notably many of the thick basaltic tephra units such as the Hantuuta (HABT),

FIGURE 3.5
Sharp contact between the volcanic-rich, diatomaceous Adu Dora Member (white) and the overlying basal Bakella Basaltic Tuff (BABT) of the Asa Koma Member (black) in the Ado Bolo area. Photograph by Bill Hart, December 8, 1999.

Dobaado (DOBT), Ladina (LABT), and Bakella (BABT) Basaltic Tuffs and two of the newly named silicic tephra units (Tuff 96-34 and Tuff 93-15), that composite lithostratigraphic relationships can be either precisely determined or estimated (Figure 3.4). An example of the latter is found in the ca. 20 m thick Adu Dora Member, which is readily identifiable by its fine-grained lacustrine, generally diatomaceous sedimentary deposits and its abrupt upper contact with the overlying thick (typically 3–15 m, but up to 40 m) BABT (Figure 3.5).

Although highly variable from location to location, at least four thin (15–30 cm) distal vitric silicic tephra horizons, and numerous thin (8–20 cm) distal basaltic tephra horizons are interbedded with the Adu Dora Member sediments. Figure 3.4 portrays this situation by indicating groups of tephra that occur within a particular portion of the Member without specifying specific sample-to-sample relationships.

Other relatively thin tephra horizons were sampled locally near the contact of the Adu Dora and underlying Saraitu Members (ANBT [Ankarara Basaltic Tuff], 3 cm; see Figure 2.9 for general member boundary), in the uppermost Saraitu Member (99-96, 3–5 cm), and the uppermost Asa Koma Member (95-05, 50 cm). Laterally variable but typically thicker silicic tephra horizons are observed at a number of localities; in all cases, these tephras are in the Rawa Member (99-114C, 1.5 m; Tuff 96-34, 1–10 m) or in overlying sequences (Tuff 92-04, 30 cm–10 m; Tuff 93-10, 1.5–2 m). This apparent up-section thickening of silicic tephra horizons is likely due to a combination of two factors: (1) redeposition and overthickening of distal fall units as a result of increased uplift and erosion into newly formed basinal depocenters, and (2) a shift in source eruption location to regions somewhat more proximal to the western margin. The relatively fine-grained character of all but the welded silicic tephra unit (Tuff 93-10 has pumiceous portions) strongly point to the dominance of the first factor. This same scenario likely can account for locally overthickened basaltic tephra exposures, particularly the BABT and LABT (see Figure 2.6), although certainly these deposits reflect eruptions from more proximal vents that are now missing because of erosion or are obscured by younger deposits. These major basaltic tephra marker horizons, the BABT, LABT, DOBT, and HABT (Figure 3.4) share certain common characteristics; all typically are 1–7 m thick, are bedded at varying scales, are variably altered to yellow/green to brown (palagonite), contain assemblages of juvenile (glass and crystals) and accidental (lithics and crystals) clasts, and show evidence in glassy (sideromelane) and hyalocrystalline (tachylite) pyroclasts for variable degrees of vesiculation as well as fracturing due to explosive force (hydrovolcanic fracturing). These features point to an important role for external (nonmagmatic) water-basaltic magma interactions prior to and during portions of the eruptive pulse(s) forming these tephra horizons (e.g., Heiken and Wohletz 1985). The exact nature and complexity of the eruptive phenomena and the degree to which secondary transport has affected these deposits cannot be determined without more detailed examination. However, it appears that at least the Asa Koma and Rawa Member basaltic tephra were produced predominantly by hydromagmatic (phreatomagmatic; juvenile magma + external water) processes. In this context, eruptions cycling/pulsing between wetter hydromagmatic and drier magmatic periods cannot be ruled out. That these eruptions were likely from more proximal sources can be deduced by analogy to the link between thick, hydromagmatic-dominated deposits of the Beidareem Member of the Sagantole Formation and adjacent basaltic volcanoes of the CAC (e.g., Renne et al. 1999).

Mafic Tephra: Chemistry and Correlations

Table 3.1 provides average major element data for all mafic tephra and the mafic component of the one mixed tephra (WMMT) analyzed as part of this study. Individual glass pyroclast analyses with analytical totals <96 weight percent were rare and are not utilized in the averages. Single- and multiple-glass chemical population samples are represented; discrete populations (modes) were identified by visual examination of element-element relationships such as those pictured in Figure 3.6, and through the use of relative probability calculations and plots identical to those used to evaluate Ar-Ar age data (e.g., Deino and Potts 1992). The plots illustrated in Figure 3.6 also were utilized to identify glass populations (samples) that likely represent products from the same eruption. These preliminary correlations were further tested by employing SC calculations. The results of these calculations are summarized in Table 3.2. Compared to evolved silicic magmas, mafic magmas typically show less marked elemental variability from

TABLE 3.1 Mafic Tephra Sample Average EMP Data

Sample	*Type*	*n*	SiO_2	TiO_2	Al_2O_3	Fe_2O_3*	*MnO*	*MgO*	*CaO*	Na_2O	K_2O	P_2O_5	*Total*
MA96-27	BABT	30	48.99	2.45	12.96	15.25	0.22	5.81	9.97	2.93	0.45	0.28	99.32
			0.35	0.09	0.12	0.16	0.03	0.07	0.06	0.09	0.02	0.04	0.42
MA96-39(1)	BABT	8	48.55	2.50	12.72	15.35	0.23	5.81	10.12	2.76	0.45	0.30	98.79
			0.37	0.13	0.21	0.16	0.03	0.32	0.17	0.21	0.01	0.03	0.42
MA96-39(2)	BABT	13	47.53	2.41	13.24	14.22	0.22	6.56	11.27	2.57	0.39	0.24	98.65
			0.44	0.11	0.09	0.22	0.03	0.06	0.07	0.10	0.02	0.04	0.50
MA99-102	BABT	33	49.18	2.60	12.59	15.49	0.22	5.77	10.14	2.91	0.43	nm	99.33
			0.29	0.12	0.12	0.18	0.03	0.08	0.08	0.11	0.01	nm	0.37
MA99-71	BABT	31	49.18	2.59	12.58	15.45	0.23	5.76	10.03	2.79	0.43	nm	99.04
			0.28	0.12	0.09	0.19	0.04	0.12	0.10	0.23	0.02	nm	0.37
MA99-72	BABT	32	49.34	2.60	12.53	15.53	0.24	5.75	10.01	2.87	0.44	nm	99.30
			0.30	0.11	0.11	0.19	0.03	0.09	0.11	0.13	0.01	nm	0.44
MA99-90I	BABT	27	48.93	2.63	12.79	15.60	0.23	5.94	10.64	2.91	0.44	0.32	100.42
			0.38	0.11	0.09	0.15	0.03	0.07	0.07	0.07	0.02	0.04	0.45
MA99-95	BABT	32	49.26	2.60	12.64	15.45	0.21	5.82	9.95	2.80	0.44	nm	99.19
			0.30	0.12	0.15	0.20	0.03	0.08	0.18	0.12	0.02	nm	0.73
MA99-98	BABT	34	49.29	2.63	12.74	15.61	0.22	5.87	10.22	2.94	0.44	nm	99.96
			0.27	0.09	0.08	0.15	0.03	0.06	0.09	0.09	0.02	nm	0.29
MA98-46(1)	DOBT	3	49.23	2.16	13.66	13.48	0.22	6.60	11.85	2.47	0.24	0.16	100.06
			0.24	0.03	0.29	0.36	0.02	0.12	0.52	0.16	0.04	0.03	0.32
MA98-46(2)	DOBT	3	49.05	2.69	13.76	16.23	0.24	4.44	10.43	2.63	0.41	0.25	100.13
			0.02	0.04	0.17	0.32	0.04	0.11	0.16	0.10	0.01	0.02	0.14
MA98-57	DOBT	22	48.71	2.05	13.35	13.68	0.21	6.89	11.49	2.66	0.28	0.17	99.46
			0.32	0.04	0.10	0.20	0.03	0.08	0.08	0.07	0.02	0.04	0.49
MA99-109	DOBT	26	48.23	2.05	13.41	13.76	0.21	6.88	11.77	2.66	0.28	0.20	99.45
			0.41	0.11	0.12	0.17	0.03	0.10	0.12	0.07	0.01	0.03	0.62
MA99-99	DOBT	13	49.11	2.12	13.35	13.73	0.21	6.78	11.48	2.61	0.27	nm	99.67
			0.36	0.11	0.25	0.14	0.04	0.16	0.25	0.14	0.03	nm	0.70
MA96-38	HABT	26	51.12	2.88	13.06	14.88	0.22	4.71	8.46	2.23	0.99	0.37	98.91
			0.43	0.14	0.14	0.24	0.03	0.25	0.15	0.29	0.05	0.05	0.58
MA98-45	HABT	22	51.40	3.04	12.79	15.20	0.23	4.73	8.59	2.23	0.98	0.37	99.54
			0.26	0.08	0.17	0.26	0.03	0.17	0.11	0.34	0.03	0.05	0.34
MA99-107	HABT	30	51.78	3.10	12.60	15.21	0.24	4.65	8.48	2.16	0.98	nm	99.20
			0.40	0.12	0.18	0.22	0.03	0.20	0.14	0.29	0.07	nm	0.67
MA93-12B(1)	LABT	23	45.62	4.42	12.24	14.47	0.26	5.45	10.03	2.67	1.11	2.37	98.66
			0.61	0.15	0.16	0.34	0.03	0.11	0.23	0.24	0.02	0.11	0.80
MA93-12B(2)	LABT	4	49.16	3.08	12.10	16.90	0.25	4.85	9.02	2.50	0.64	0.40	98.91
			0.30	0.10	0.03	0.20	0.04	0.17	0.13	0.24	0.02	0.02	0.33
MA96-25	LABT	26	43.33	4.42	12.07	15.24	0.29	5.56	9.87	2.94	1.07	2.41	97.20
			0.41	0.11	0.10	0.18	0.03	0.08	0.07	0.13	0.03	0.12	0.59
MA96-26	LABT	32	44.44	4.50	12.11	15.03	0.29	5.61	10.08	3.12	1.12	2.47	98.78
			0.27	0.14	0.11	0.15	0.03	0.05	0.08	0.14	0.02	0.08	0.35
MA96-28	LABT	31	42.96	4.31	12.17	15.16	0.27	5.69	9.87	3.19	1.04	2.51	97.16
			0.36	0.13	0.07	0.15	0.04	0.06	0.07	0.12	0.02	0.09	0.42
MA98-39	LABT	25	43.79	4.64	12.05	15.39	0.29	5.74	10.18	3.10	1.05	2.45	98.68
			0.33	0.07	0.08	0.26	0.03	0.08	0.06	0.21	0.03	0.11	0.45
MA98-40	LABT	31	44.12	4.65	12.09	15.13	0.29	5.83	10.23	3.25	1.05	2.55	99.18
			0.18	0.05	0.08	0.24	0.03	0.05	0.09	0.15	0.02	0.11	0.37
MA98-41	LABT	17	44.20	4.67	12.11	15.27	0.30	5.84	10.16	3.11	1.06	2.47	99.18
			0.22	0.07	0.09	0.12	0.04	0.06	0.05	0.19	0.02	0.07	0.38

TABLE 3.1 *(continued)*

Sample	*Type*	*n*	SiO_2	TiO_2	Al_2O_3	Fe_2O_3*	*MnO*	*MgO*	*CaO*	Na_2O	K_2O	P_2O_5	*Total*
MA98-47	LABT	31	43.92	4.63	12.05	15.23	0.28	5.76	10.19	2.67	1.06	2.49	98.29
			0.33	0.12	0.11	0.25	0.03	0.08	0.11	0.27	0.03	0.11	0.45
MA98-48	LABT	28	44.63	4.76	11.98	15.09	0.30	5.63	10.36	2.64	1.09	2.55	99.02
			0.14	0.09	0.12	0.17	0.02	0.06	0.06	0.25	0.02	0.11	0.36
MA98-49	LABT	31	44.67	4.74	11.95	15.10	0.29	5.66	10.18	2.57	1.08	2.46	98.69
			0.25	0.07	0.09	0.20	0.02	0.07	0.10	0.18	0.03	0.11	0.49
MA99-103	LABT	19	44.93	4.75	12.00	15.32	0.31	5.65	10.34	2.78	1.09	nm	97.17
			0.26	0.15	0.11	0.19	0.04	0.09	0.09	0.16	0.02	nm	0.34
MA99-106	LABT	35	44.46	4.67	11.99	15.33	0.30	5.74	10.29	3.10	1.06	nm	96.95
			0.27	0.17	0.08	0.12	0.03	0.06	0.09	0.17	0.02	nm	0.28
MA99-116	LABT	11	44.42	4.74	12.07	15.25	0.29	5.73	10.07	3.13	1.06	nm	96.76
			0.29	0.11	0.08	0.16	0.04	0.07	0.07	0.15	0.01	nm	0.36
MA99-69	LABT	38	44.52	4.72	12.12	15.42	0.29	5.71	10.40	2.67	1.05	nm	96.90
			0.32	0.16	0.10	0.27	0.03	0.08	0.12	0.30	0.02	nm	0.39
MA99-94	LABT	9	44.18	4.71	12.18	15.38	0.28	5.82	10.41	3.15	1.06	nm	97.16
			0.32	0.11	0.11	0.19	0.04	0.05	0.07	0.18	0.02	nm	0.39
MA93-16		36	49.66	2.24	13.30	13.40	0.18	6.25	10.52	2.94	0.40	0.27	99.16
			0.27	0.09	0.07	0.17	0.02	0.06	0.08	0.06	0.01	0.02	0.44
MA98-38		27	48.83	2.46	13.00	14.21	0.20	6.39	10.72	2.77	0.40	0.29	99.28
			0.32	0.05	0.10	0.17	0.03	0.06	0.08	0.08	0.02	0.04	0.43
MA99-101(1)		36	49.20	2.50	12.92	14.17	0.21	6.30	10.78	2.72	0.40	nm	99.20
			0.26	0.08	0.15	0.16	0.03	0.08	0.11	0.08	0.01	nm	0.41
MA99-101(2)		33	48.15	2.41	13.18	13.87	0.21	6.41	11.25	2.80	0.39	nm	98.66
			0.40	0.10	0.08	0.19	0.03	0.09	0.08	0.11	0.02	nm	0.55
MA99-104		36	49.28	2.52	12.78	14.29	0.21	6.28	10.67	2.74	0.40	nm	99.18
			0.24	0.14	0.12	0.17	0.03	0.09	0.08	0.09	0.02	nm	0.40
MA99-90D(1)		24	49.35	2.59	13.26	14.07	0.20	6.59	11.28	2.55	0.40	nm	100.28
			0.38	0.11	0.17	0.21	0.03	0.17	0.16	0.13	0.03	nm	0.48
MA99-90D(2)		4	49.60	2.56	13.23	13.33	0.19	7.06	11.97	2.30	0.31	nm	100.56
			0.17	0.04	0.19	0.39	0.03	0.22	0.21	0.16	0.06	nm	0.51
MA99-90F		31	49.32	2.53	13.19	14.41	0.21	6.41	11.17	2.75	0.41	nm	100.40
			0.24	0.11	0.08	0.16	0.03	0.08	0.08	0.07	0.01	nm	0.29
MA99-96		20	49.00	2.59	13.15	14.54	0.21	6.26	10.51	2.68	0.48	nm	99.43
			0.19	0.12	0.12	0.17	0.03	0.05	0.08	0.06	0.02	nm	0.31
MA99-97		38	49.54	2.49	13.05	14.42	0.22	6.38	10.70	2.71	0.41	nm	99.91
			0.31	0.10	0.10	0.18	0.04	0.09	0.07	0.08	0.01	nm	0.40
MA00-22	ANBT	36	49.61	2.92	12.96	15.90	0.21	5.50	9.83	2.72	0.62	0.34	100.61
			0.26	0.12	0.09	0.23	0.04	0.05	0.08	0.11	0.02	0.05	0.36
MA95-04A	WMMT	15	50.70	3.69	12.51	15.77	0.24	4.53	8.10	2.31	1.17	0.61	99.62
			0.60	0.22	0.15	0.32	0.03	0.36	0.24	0.15	0.10	0.07	0.64
MA95-04B	WMMT	7	49.89	3.89	12.69	16.21	0.24	4.68	8.35	2.06	1.06	0.55	99.63
			0.49	0.13	0.14	0.20	0.04	0.30	0.15	0.19	0.05	0.07	0.22
MA99-100	WMMT	13	49.58	3.97	12.43	16.36	0.24	4.67	8.95	1.97	1.05	0.53	99.76
			0.44	0.26	0.16	0.39	0.04	0.17	0.23	0.19	0.06	0.04	0.44

NOTE: Averages based on (*n*) number of individual glass pyroclasts and are from measured (raw) data. Total Fe expressed as Fe_2O_3 (Fe_2O_3*). P_2O_5 not measured (nm) for a number of samples. Errors are one standard deviation. Where present, individual subpopulation averages are listed as (1) and (2).

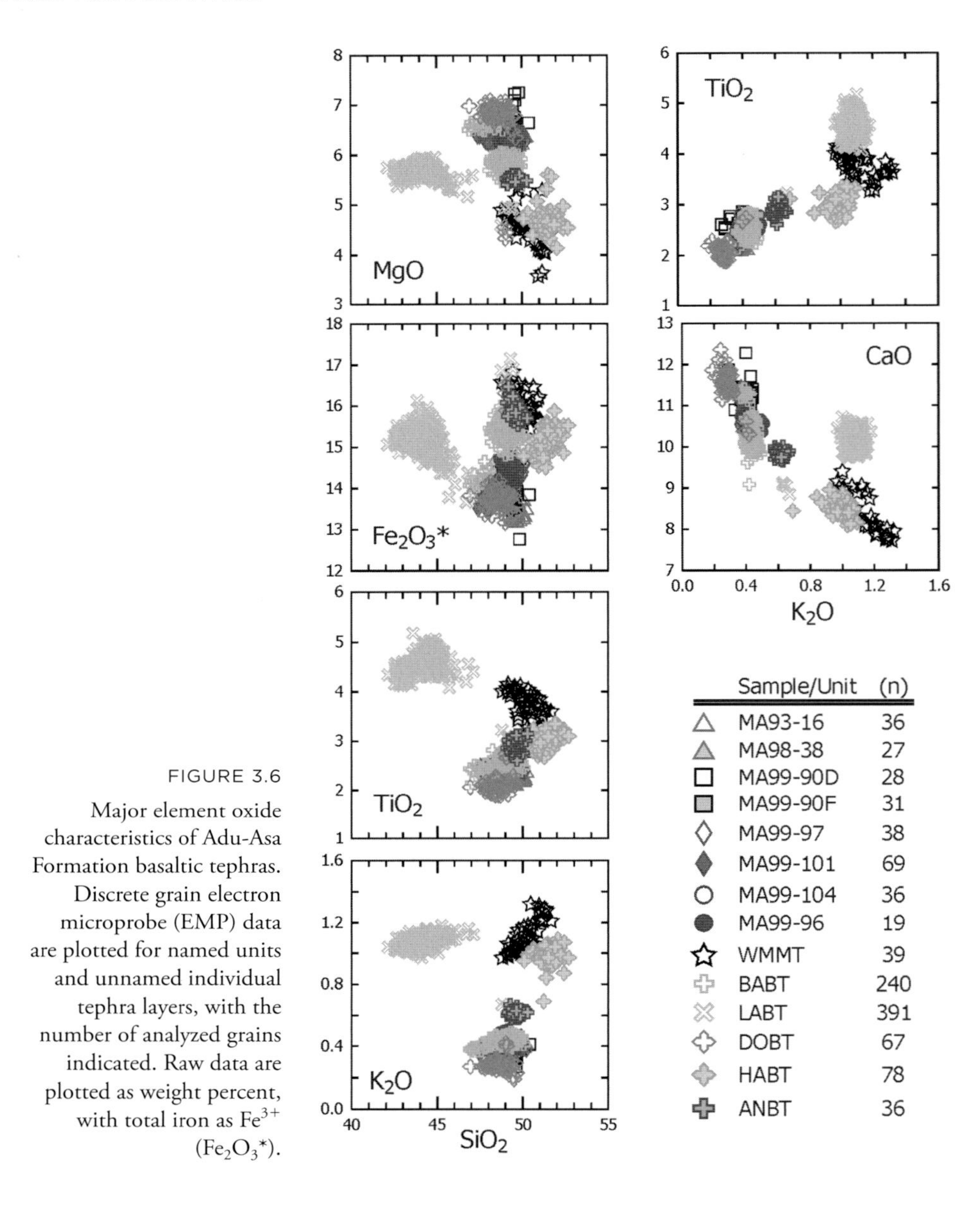

FIGURE 3.6

Major element oxide characteristics of Adu-Asa Formation basaltic tephras. Discrete grain electron microprobe (EMP) data are plotted for named units and unnamed individual tephra layers, with the number of analyzed grains indicated. Raw data are plotted as weight percent, with total iron as Fe^{3+} (Fe_2O_3*).

eruption to eruption; thus, the chemical fingerprint of a given mafic tephra horizon will often be less unique than that of a given silicic tephra horizon. Because of this, we have chosen to highlight only those mafic tephras with calculated SC values ≥0.95 as likely strong candidates for correlation. Employing the above graphical and statistical methods in concert with physical features and chronostratigraphic information leads to local and regional correlations establishing the major named mafic tephra marker horizons: the HABT, DOBT, LABT, and BABT. The overall chemical similarities between various mafic tephras found within the Adu Dora Member are noteworthy (Figures 3.6 and 3.7,

Tables 3.1 and 3.2), but field evidence is lacking to support additional formal correlations. Future analytical work on these units may provide the basis for correlations and additional tephrostratigraphic interpretations that cannot be derived from the present dataset.

The plots illustrated in Figures 3.6 and 3.7 and the summary data of Table 3.1 indicate that the mafic Adu-Asa Formation tephras are basalt to basaltic andesite in composition. The variable but uniformly elevated iron contents and the ranges of K_2O and TiO_2 are consistent with other late Miocene basaltic products and vents common to the western and southeastern margins of the Afar rift (e.g., Afar Stratoid Basalts; Hart and Walter 1983; Chernet and Hart 1999), including multiple basaltic vents within the nearby CAC (Renne et al. 1999; Middle Awash project, unpublished basalt data). Broader volcano-tectonic implications will be addressed in greater detail once trace element and isotopic data are available for the Adu-Asa Formation mafic tephra.

Silicic Tephra: Chemistry and Correlations

Table 3.3 provides average major element data for all silicic tephras and the silicic component of the one mixed tephra (WMMT) analyzed as part of this study. Individual glass pyroclast analyses with analytical totals <90 weight percent were not typically utilized in the averages, particularly when the majority of the glass in a particular sample yielded higher totals. In some cases, pervasively altered tephra samples yielded glass with analytical totals tightly constrained between 87 and 90 weight percent. These samples are reported in Table 3.3 and appear to provide robust information for relatively immobile elements such as Ti, Al, Fe, Mn, and Ca. Single- and multiple-glass chemical population samples are represented; discrete populations (modes) were identified by visual examination of element-element relationships such as those pictured in Figure 3.8, and through the use of relative probability calculations and plots identical to those used to evaluate Ar-Ar age data (e.g., Deino and Potts 1992). The plots illustrated in Figure 3.8 were also utilized to identify glass populations (samples) that likely represent products from the same eruption. These preliminary correlations were further tested by employing SC calculations. The results of these calculations are summarized in Table 3.4.

Borchardt et al. (1972) originally suggested that a SC >0.88 provided evidence for a permissive correlation. We are slightly more conservative in this regard and have highlighted all SC values>0.90 as indicative of potential correlations. As with the mafic tephra, the above graphical and statistical methods, in concert with physical features and chronostratigraphic information, provide a sound basis to propose correlations between a number of Adu-Asa Formation and post–Adu-Asa Formation silicic tephra horizons and to assign formal names to these newly established marker horizons. In ascending stratigraphic order these are Tuff 95-25 (MA95-25, 97-23), Tuff 93-15 (MA93-15, 97-16, 97-22, 97-24, 99-105), Tuff 95-14 (MA95-14, 95-21), Tuff 96-34 (MA96-34, 98-43, 99-108), Tuff 92-04 (MA92-04, 93-09, 94-18, 95-16, 95-23), and Tuff 93-10 (MA93-10, 96-37, 98-44). The identical multiple-glass populations in samples MA95-25 and MA97-23 (Figure 3.8, Tables 3.3 and 3.4) solidify their correlation and provide

TABLE 3.2 Basaltic Tephra Similarity Coefficient Summary

	96-27	*96-39(1)*	*96-39(2)*	*99-102*	*99-71*	*99-72*	*99-90I*	*99-95*	*99-98*	*98-46(1)*
96-27	■	**0.99**	0.93	0.98	0.98	**0.98**	**0.97**	**0.98**	**0.97**	0.85
96-39(1)	**0.99**	■	0.93	**0.98**	**0.98**	**0.98**	**0.97**	**0.98**	**0.98**	0.85
96-39(2)	0.93	0.93	■	0.92	0.92	0.92	0.93	0.92	0.92	0.90
99-102	**0.98**	**0.98**	0.92	■	**1.00**	**0.99**	**0.98**	**0.99**	**0.99**	0.84
99-71	**0.98**	**0.98**	0.92	**1.00**	■	**1.00**	**0.98**	**0.99**	**0.99**	0.84
99-72	**0.98**	**0.98**	0.92	**0.99**	**1.00**	■	**0.98**	**0.99**	**0.99**	0.84
99-90I	**0.97**	**0.97**	0.93	**0.98**	**0.98**	**0.98**	■	**0.98**	**0.99**	0.85
99-95	**0.98**	**0.98**	0.92	**0.99**	**0.99**	**0.99**	**0.98**	■	**0.99**	0.84
99-98	**0.97**	**0.98**	0.92	**0.99**	**0.99**	**0.99**	**0.99**	**0.99**	■	0.84
98-46(1)	0.85	0.85	0.90	0.84	0.84	0.84	0.85	0.84	0.84	■
98-46(2)	0.92	0.92	0.89	0.93	0.93	0.93	0.93	0.92	0.93	0.82
98-57	0.86	0.86	0.92	0.85	0.85	0.85	0.86	0.85	0.85	0.95
99-109	0.86	0.86	0.92	0.85	0.85	0.85	0.86	0.85	0.85	0.95
99-99	0.87	0.86	0.92	0.86	0.86	0.85	0.87	0.86	0.86	0.96
96-38	0.84	0.84	0.80	0.84	0.84	0.84	0.83	0.84	0.84	0.75
98-45	0.84	0.84	0.79	0.84	0.84	0.84	0.83	0.84	0.84	0.74
99-107	0.83	0.83	0.78	0.84	0.84	0.84	0.83	0.84	0.83	0.73
93-12B(1)	0.82	0.82	0.78	0.82	0.82	0.82	0.81	0.82	0.82	0.73
93-12B(2)	0.87	0.87	0.81	0.87	0.88	0.88	0.87	0.88	0.87	0.75
96-25	0.82	0.82	0.77	0.82	0.82	0.82	0.81	0.82	0.82	0.72
96-26	0.82	0.82	0.78	0.82	0.82	0.82	0.81	0.82	0.82	0.73
96-28	0.83	0.83	0.78	0.83	0.83	0.83	0.82	0.83	0.82	0.72
98-39	0.82	0.83	0.77	0.83	0.83	0.83	0.82	0.83	0.82	0.72
98-40	0.82	0.83	0.78	0.83	0.83	0.82	0.82	0.83	0.83	0.73
98-41	0.82	0.83	0.78	0.83	0.83	0.83	0.82	0.83	0.83	0.73
98-47	0.82	0.83	0.78	0.83	0.83	0.83	0.82	0.83	0.82	0.73
98-48	0.81	0.82	0.77	0.82	0.82	0.82	0.81	0.82	0.82	0.72
98-49	0.82	0.82	0.77	0.82	0.82	0.82	0.81	0.82	0.82	0.72
99-103	0.81	0.82	0.77	0.82	0.82	0.82	0.82	0.82	0.82	0.72
99-106	0.82	0.83	0.78	0.83	0.83	0.83	0.82	0.82	0.82	0.73
99-116	0.82	0.83	0.77	0.83	0.83	0.83	0.81	0.83	0.82	0.72
99-69	0.82	0.83	0.78	0.83	0.83	0.83	0.82	0.82	0.82	0.73
99-94	0.82	0.83	0.78	0.83	0.83	0.82	0.82	0.83	0.83	0.73
93-16	0.93	0.93	**0.95**	0.93	0.92	0.92	0.93	0.92	0.93	0.91
98-38	**0.95**	**0.95**	**0.97**	0.94	0.94	0.94	**0.95**	0.94	**0.95**	0.89
99-101(1)	**0.95**	**0.95**	**0.97**	**0.95**	**0.95**	0.94	**0.96**	0.94	**0.95**	0.89
99-101(2)	0.93	0.93	**0.99**	0.92	0.92	0.92	0.93	0.92	0.92	0.90
99-104	**0.95**	**0.95**	**0.96**	**0.96**	**0.95**	**0.95**	**0.96**	**0.95**	**0.96**	0.88
99-90D(1)	0.93	0.93	**0.98**	0.94	0.94	0.93	**0.95**	0.93	0.94	0.90
99-90D(2)	0.88	0.88	0.93	0.88	0.88	0.88	0.89	0.88	0.88	0.92
99-90F	0.94	0.94	**0.97**	0.94	0.94	0.94	**0.95**	0.94	**0.95**	0.89
99-96	**0.96**	**0.96**	0.94	**0.95**	**0.95**	**0.95**	**0.96**	**0.96**	**0.96**	0.86
99-97	**0.95**	**0.95**	**0.97**	**0.95**	**0.95**	0.94	**0.96**	**0.95**	**0.95**	0.89
00-22	0.92	0.92	0.86	0.92	0.92	0.92	0.92	0.93	0.92	0.80

98-46(2)	*98-57*	*99-109*	*99-99*	*96-38*	*98-45*	*99-107*	*93-12B(1)*	*93-12B(2)*	*96-25*	*96-26*	*96-28*
0.92	0.86	0.86	0.87	0.84	0.84	0.83	0.82	0.87	0.82	0.82	0.83
0.92	0.86	0.86	0.86	0.84	0.84	0.83	0.82	0.87	0.82	0.82	0.83
0.89	0.92	0.92	0.92	0.80	0.79	0.78	0.78	0.81	0.77	0.78	0.78
0.93	0.85	0.85	0.86	0.84	0.84	0.84	0.82	0.87	0.82	0.82	0.83
0.93	0.85	0.85	0.86	0.84	0.84	0.84	0.82	0.88	0.82	0.82	0.83
0.93	0.85	0.85	0.85	0.84	0.84	0.84	0.82	0.88	0.82	0.82	0.83
0.93	0.86	0.86	0.87	0.83	0.83	0.83	0.81	0.87	0.81	0.81	0.82
0.92	0.85	0.85	0.86	0.84	0.84	0.84	0.82	0.88	0.82	0.82	0.83
0.93	0.85	0.85	0.86	0.84	0.84	0.83	0.82	0.87	0.82	0.82	0.83
0.82	0.95	0.95	0.96	0.75	0.74	0.73	0.73	0.75	0.72	0.73	0.72
■	0.83	0.83	0.83	0.85	0.84	0.84	0.78	0.87	0.78	0.78	0.78
0.83	■	**0.99**	**0.99**	0.75	0.74	0.73	0.74	0.76	0.73	0.73	0.73
0.83	**0.99**	■	**0.98**	0.75	0.74	0.73	0.74	0.75	0.72	0.73	0.73
0.83	**0.99**	**0.98**	■	0.76	0.75	0.74	0.74	0.76	0.73	0.73	0.73
0.85	0.75	0.75	0.76	■	**0.98**	**0.98**	0.86	0.89	0.86	0.85	0.86
0.84	0.74	0.74	0.75	**0.98**	■	**0.99**	0.87	0.91	0.87	0.86	0.88
0.84	0.73	0.73	0.74	**0.98**	**0.99**	■	0.87	0.91	0.87	0.86	0.88
0.78	0.74	0.74	0.74	0.86	0.87	0.87	■	0.83	**0.97**	**0.98**	**0.96**
0.87	0.76	0.75	0.76	0.89	0.91	0.91	0.83	■	0.84	0.83	0.84
0.78	0.73	0.72	0.73	0.86	0.87	0.87	**0.97**	0.84	■	**0.98**	**0.99**
0.78	0.73	0.73	0.73	0.85	0.86	0.86	**0.98**	0.83	**0.98**	■	**0.97**
0.78	0.73	0.73	0.73	0.86	0.88	0.88	**0.96**	0.84	**0.99**	**0.97**	■
0.78	0.73	0.73	0.73	0.85	0.86	0.86	**0.96**	0.83	**0.98**	**0.98**	**0.98**
0.77	0.74	0.74	0.74	0.85	0.86	0.86	**0.96**	0.83	**0.97**	**0.98**	**0.97**
0.77	0.73	0.73	0.74	0.85	0.86	0.86	**0.96**	0.83	**0.98**	**0.98**	**0.97**
0.77	0.73	0.73	0.74	0.85	0.86	0.86	**0.96**	0.83	**0.98**	**0.98**	**0.98**
0.78	0.73	0.73	0.73	0.85	0.86	0.86	**0.97**	0.82	**0.97**	**0.98**	**0.96**
0.77	0.73	0.73	0.73	0.85	0.86	0.86	**0.97**	0.83	**0.98**	**0.98**	**0.97**
0.78	0.73	0.73	0.73	0.85	0.86	0.86	**0.97**	0.83	**0.97**	**0.98**	**0.96**
0.78	0.73	0.73	0.74	0.85	0.86	0.86	**0.96**	0.83	**0.97**	**0.98**	**0.97**
0.77	0.73	0.73	0.73	0.85	0.86	0.86	**0.96**	0.83	**0.98**	**0.98**	**0.97**
0.78	0.73	0.73	0.74	0.85	0.86	0.86	**0.96**	0.83	**0.97**	**0.97**	**0.97**
0.78	0.74	0.74	0.74	0.85	0.86	0.86	**0.96**	0.82	**0.97**	**0.97**	**0.97**
0.90	0.91	0.91	0.92	0.80	0.79	0.78	0.78	0.81	0.77	0.77	0.77
0.91	0.90	0.90	0.91	0.81	0.80	0.80	0.80	0.83	0.78	0.79	0.79
0.91	0.90	0.90	0.90	0.82	0.81	0.80	0.80	0.83	0.78	0.79	0.79
0.90	0.92	0.92	0.92	0.80	0.79	0.78	0.78	0.81	0.77	0.77	0.78
0.92	0.89	0.89	0.90	0.82	0.81	0.81	0.80	0.84	0.79	0.79	0.80
0.91	0.91	0.91	0.92	0.81	0.80	0.79	0.78	0.82	0.77	0.78	0.78
0.86	0.94	0.94	0.94	0.78	0.77	0.76	0.75	0.78	0.73	0.74	0.74
0.92	0.90	0.90	0.91	0.82	0.81	0.80	0.79	0.83	0.78	0.78	0.79
0.91	0.87	0.87	0.88	0.84	0.83	0.82	0.81	0.86	0.80	0.81	0.81
0.92	0.90	0.89	0.90	0.82	0.81	0.80	0.80	0.83	0.78	0.79	0.79
0.89	0.80	0.80	0.81	0.89	0.89	0.88	0.85	0.94	0.85	0.85	0.86

TABLE 3.2 *(continued)*

	98-39	*98-40*	*98-41*	*98-47*	*98-48*	*98-49*	*98-103*	*99-106*	*99-116*	*99-69*
96-27	0.82	0.82	0.82	0.82	0.81	0.82	0.81	0.82	0.82	0.82
96-39(1)	0.83	0.83	0.83	0.83	0.82	0.82	0.82	0.83	0.83	0.83
96-39(2)	0.77	0.78	0.78	0.78	0.77	0.77	0.77	0.78	0.77	0.78
99-102	0.83	0.83	0.83	0.83	0.82	0.82	0.82	0.83	0.83	0.83
99-71	0.83	0.83	0.83	0.83	0.82	0.82	0.82	0.83	0.83	0.83
99-72	0.83	0.82	0.83	0.83	0.82	0.82	0.82	0.83	0.83	0.83
99-90I	0.82	0.82	0.82	0.82	0.81	0.81	0.82	0.82	0.81	0.82
99-95	0.83	0.83	0.83	0.83	0.82	0.82	0.82	0.82	0.83	0.82
99-98	0.82	0.83	0.83	0.82	0.82	0.82	0.82	0.82	0.82	0.82
98-46(1)	0.72	0.73	0.73	0.73	0.72	0.72	0.72	0.73	0.72	0.73
98-46(2)	0.78	0.77	0.77	0.77	0.78	0.77	0.78	0.78	0.77	0.78
98-57	0.73	0.74	0.73	0.73	0.73	0.73	0.73	0.73	0.73	0.73
99-109	0.73	0.74	0.73	0.73	0.73	0.73	0.73	0.73	0.73	0.73
99-99	0.73	0.74	0.74	0.74	0.73	0.73	0.73	0.74	0.73	0.74
96-38	0.85	0.85	0.85	0.85	0.85	0.85	0.85	0.85	0.85	0.85
98-45	0.86	0.86	0.86	0.86	0.86	0.86	0.86	0.86	0.86	0.86
99-107	0.86	0.86	0.86	0.86	0.86	0.86	0.86	0.86	0.86	0.86
93-12B(1)	**0.96**	**0.96**	**0.96**	**0.96**	**0.97**	**0.97**	**0.97**	**0.96**	**0.96**	**0.96**
93-12B(2)	0.83	0.83	0.83	0.83	0.82	0.83	0.83	0.83	0.83	0.83
96-25	**0.98**	**0.97**	**0.98**	**0.98**	**0.97**	**0.98**	**0.97**	**0.97**	**0.98**	**0.97**
96-26	**0.98**	**0.98**	**0.98**	**0.98**	**0.98**	**0.98**	**0.98**	**0.98**	**0.98**	**0.97**
96-28	**0.98**	**0.97**	**0.97**	**0.98**	**0.96**	**0.97**	**0.96**	**0.97**	**0.97**	**0.97**
98-39	■	**0.99**	**0.99**	**1.00**	**0.98**	**0.99**	**0.98**	**0.99**	**0.99**	**0.99**
98-40	**0.99**	■	**1.00**	**0.99**	**0.98**	**0.99**	**0.98**	**0.99**	**0.99**	**0.99**
98-41	**0.99**	**1.00**	■	**0.99**	**0.98**	**0.99**	**0.98**	**0.99**	**0.99**	**0.99**
98-47	**1.00**	**0.99**	**0.99**	■	**0.98**	**0.99**	**0.98**	**0.99**	**0.99**	**0.99**
98-48	**0.98**	**0.98**	**0.98**	**0.98**	■	**0.99**	**1.00**	**0.99**	**0.99**	**0.99**
98-49	**0.99**	**0.99**	**0.99**	**0.99**	**0.99**	■	**0.99**	**0.99**	**0.99**	**0.99**
99-103	**0.98**	**0.98**	**0.98**	**0.98**	**1.00**	**0.99**	■	**0.99**	**0.99**	**0.99**
99-106	**0.99**	**0.99**	**0.99**	**0.99**	**0.99**	**0.99**	**0.99**	■	**0.99**	**0.99**
99-116	**0.99**	**0.99**	**0.99**	**0.99**	**0.99**	**0.99**	**0.99**	**0.99**	■	**0.99**
99-69	**0.99**	**0.99**	**0.99**	**0.99**	**0.99**	**0.99**	**0.99**	**0.99**	**0.99**	■
99-94	**0.99**	**0.99**	**0.99**	**0.99**	**0.98**	**0.98**	**0.98**	**0.99**	**0.99**	**0.99**
93-16	0.77	0.78	0.78	0.78	0.77	0.77	0.77	0.78	0.77	0.78
98-38	0.79	0.79	0.79	0.79	0.79	0.79	0.79	0.79	0.79	0.79
99-101(1)	0.79	0.79	0.79	0.79	0.79	0.79	0.79	0.79	0.79	0.79
99-101(2)	0.77	0.78	0.78	0.78	0.77	0.77	0.77	0.78	0.77	0.78
99-104	0.79	0.80	0.80	0.80	0.79	0.79	0.79	0.80	0.79	0.80
99-90D(1)	0.78	0.78	0.78	0.78	0.78	0.77	0.77	0.78	0.77	0.78
99-90D(2)	0.74	0.74	0.74	0.74	0.74	0.74	0.74	0.74	0.74	0.74
99-90F	0.78	0.79	0.79	0.78	0.78	0.78	0.78	0.78	0.78	0.79
99-96	0.81	0.81	0.81	0.81	0.81	0.81	0.81	0.81	0.81	0.81
99-97	0.79	0.79	0.79	0.79	0.79	0.79	0.79	0.79	0.79	0.79
00-22	0.85	0.84	0.84	0.84	0.84	0.84	0.84	0.84	0.85	0.85

NOTE: Similarity coefficients were calculated from average measured (raw) EMP data (see Table 3.1) using the following element string (all weighted equally): SiO_2, TiO_2, Al_2O_3, Fe_2O_3, MgO, CaO, and K_2O. All samples have the MA prefix (e.g., MA99-104). All SC values of 0.95 and above are designated in bold type.

99-94	*93-16*	*98-38*	*99-101(1)*	*99-101(2)*	*99-104*	*99-90D(1)*	*99-90D(2)*	*99-90F*	*99-96*	*99-97*	*00-22*
0.82	0.93	**0.95**	**0.95**	0.93	**0.95**	0.93	0.88	0.94	**0.96**	**0.95**	0.92
0.83	0.93	**0.95**	**0.95**	0.93	**0.95**	0.93	0.88	0.94	**0.96**	**0.95**	0.92
0.78	**0.95**	**0.97**	**0.97**	**0.99**	**0.96**	**0.98**	0.93	**0.97**	0.94	**0.97**	0.86
0.83	0.93	0.94	**0.95**	0.92	**0.96**	0.94	0.88	0.94	**0.95**	**0.95**	0.92
0.83	0.92	0.94	**0.95**	0.92	**0.95**	0.94	0.88	0.94	**0.95**	**0.95**	0.92
0.82	0.92	0.94	0.94	0.92	**0.95**	0.93	0.88	0.94	**0.95**	0.94	0.92
0.82	0.93	**0.95**	**0.96**	0.93	**0.96**	**0.95**	0.89	**0.95**	**0.96**	**0.96**	0.92
0.83	0.92	0.94	0.94	0.92	**0.95**	0.93	0.88	0.94	**0.96**	**0.95**	0.93
0.83	0.93	**0.95**	**0.95**	0.92	**0.96**	0.94	0.88	**0.95**	**0.96**	**0.95**	0.92
0.73	0.91	0.89	0.89	0.90	0.88	0.90	0.92	0.89	0.86	0.89	0.80
0.78	0.90	0.91	0.91	0.90	0.92	0.91	0.86	0.92	0.91	0.92	0.89
0.74	0.91	0.90	0.90	0.92	0.89	0.91	0.94	0.90	0.87	0.90	0.80
0.74	0.91	0.90	0.90	0.92	0.89	0.91	0.94	0.90	0.87	0.89	0.80
0.74	0.92	0.91	0.90	0.92	0.90	0.92	0.94	0.91	0.88	0.90	0.81
0.85	0.80	0.81	0.82	0.80	0.82	0.81	0.78	0.82	0.84	0.82	0.89
0.86	0.79	0.80	0.81	0.79	0.81	0.80	0.77	0.81	0.83	0.81	0.89
0.86	0.78	0.80	0.80	0.78	0.81	0.79	0.76	0.80	0.82	0.80	0.88
0.96	0.78	0.80	0.80	0.78	0.80	0.78	0.75	0.79	0.81	0.80	0.85
0.82	0.81	0.83	0.83	0.81	0.84	0.82	0.78	0.83	0.86	0.83	0.94
0.97	0.77	0.78	0.78	0.77	0.79	0.77	0.73	0.78	0.80	0.78	0.85
0.97	0.77	0.79	0.79	0.77	0.79	0.78	0.74	0.78	0.81	0.79	0.85
0.97	0.77	0.79	0.79	0.78	0.80	0.78	0.74	0.79	0.81	0.79	0.86
0.99	0.77	0.79	0.79	0.77	0.79	0.78	0.74	0.78	0.81	0.79	0.85
0.99	0.78	0.79	0.79	0.78	0.80	0.78	0.74	0.79	0.81	0.79	0.84
0.99	0.78	0.79	0.79	0.78	0.80	0.78	0.74	0.79	0.81	0.79	0.84
0.99	0.78	0.79	0.79	0.78	0.80	0.78	0.74	0.78	0.81	0.79	0.84
0.98	0.77	0.79	0.79	0.77	0.79	0.78	0.74	0.78	0.81	0.79	0.84
0.98	0.77	0.79	0.79	0.77	0.79	0.77	0.74	0.78	0.81	0.79	0.84
0.98	0.77	0.79	0.79	0.77	0.79	0.77	0.74	0.78	0.81	0.79	0.84
0.99	0.78	0.79	0.79	0.78	0.80	0.78	0.74	0.78	0.81	0.79	0.84
0.99	0.77	0.79	0.79	0.77	0.79	0.77	0.74	0.78	0.81	0.79	0.85
0.99	0.78	0.79	0.79	0.78	0.80	0.78	0.74	0.79	0.81	0.79	0.85
■	0.78	0.79	0.79	0.78	0.80	0.78	0.74	0.79	0.81	0.79	0.84
0.78	■	**0.97**	**0.97**	**0.96**	**0.97**	**0.96**	0.91	**0.96**	0.94	**0.97**	0.86
0.79	**0.97**	■	**0.99**	**0.98**	**0.99**	**0.98**	0.92	**0.98**	**0.96**	**0.99**	0.88
0.79	**0.97**	**0.99**	■	**0.97**	**0.99**	**0.98**	0.92	**0.98**	**0.96**	**0.99**	0.88
0.78	**0.96**	**0.98**	**0.97**	■	**0.97**	**0.97**	0.93	**0.98**	0.94	**0.97**	0.86
0.80	**0.97**	**0.99**	**0.99**	**0.97**	■	**0.97**	0.92	**0.98**	**0.96**	**0.99**	0.89
0.78	**0.96**	**0.98**	**0.98**	**0.97**	**0.97**	■	0.94	**0.99**	**0.95**	**0.98**	0.87
0.74	0.91	0.92	0.92	0.93	0.92	0.94	■	0.93	0.90	0.92	0.83
0.79	**0.96**	**0.98**	**0.98**	**0.98**	**0.98**	**0.99**	0.93	■	**0.96**	**0.99**	0.88
0.81	0.94	**0.96**	**0.96**	0.94	**0.96**	**0.95**	0.90	**0.96**	■	**0.96**	0.91
0.79	**0.97**	**0.99**	**0.99**	**0.97**	**0.99**	**0.98**	0.92	**0.99**	**0.96**	■	0.88
0.84	0.86	0.88	0.88	0.86	0.89	0.87	0.83	0.88	0.91	0.88	■

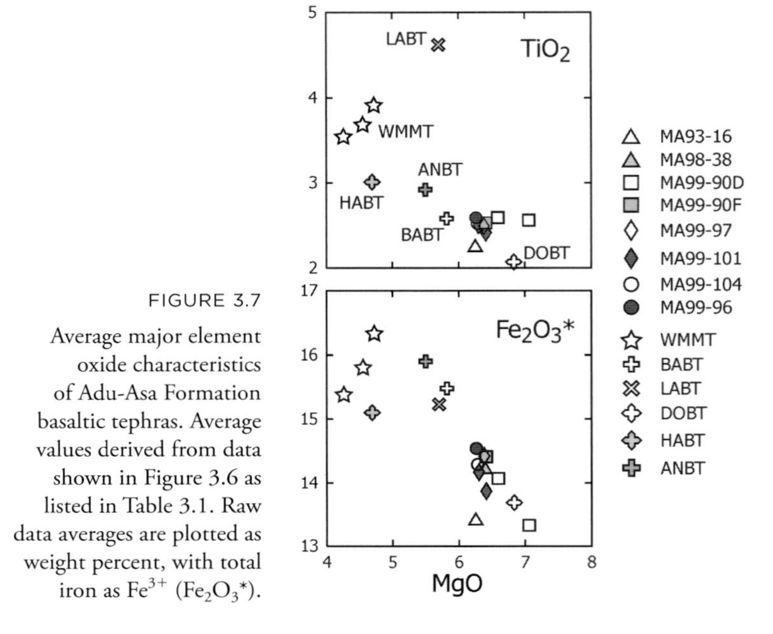

FIGURE 3.7

Average major element oxide characteristics of Adu-Asa Formation basaltic tephras. Average values derived from data shown in Figure 3.6 as listed in Table 3.1. Raw data averages are plotted as weight percent, with total iron as Fe^{3+} (Fe_2O_3*).

Tuff 95-25 with a unique chemical fingerprint. Careful examination of Figures 3.8 and 3.9 and Table 3.4 show a very strong chemical similarity between Tuff 96-34 (Rawa Member, sub-HABT) and Tuff 92-04 (post-Rawa), but the field relationships require that two stratigraphically distinct horizons exist. Furthermore, we also suggest that sample MA96-35 correlates with Tuff 96-34 (WoldeGabriel et al., Chapter 2), and that stratigraphically unconstrained samples MA95-12 and MA95-19 represent the same horizon. Future analytical work on these units may provide the basis for additional correlations, particularly between tephrostratigraphically well-constrained deposits and other isolated, local tephra occurrences and/or regional occurrences within and outside of the Middle Awash area.

The previously named WMMT (WoldeGabriel et al. 2001) is a physically and chemically recognizable unit within the upper Asa Koma Member of the Adu-Asa Formation (Figure 3.4). Like the Cindery Tuff (CT, 3.9 Ma) and Gàala Tuff Complex (GATC, 4.4 Ma) of the Sagantole Formation (Walter et al., 1987; Larson et al. 1997; Renne et al. 1999; Hart et al. 2004), the WMMT (5.6 Ma) preserves a complex mixed glass assemblage dominated by mafic and silicic endmembers but including sparse intermediate varieties. Also like the CT and GATC, the WMMT does not display the same proportions of glass chemical types everywhere, yet the glass with the most silicic endmember composition is always present and diagnostic (Figure 3.8; Tables 3.3 and 3.4). Moreover, when all major element analyses of the CT, GATC, and WMMT are plotted together, it is apparent that similar complex magmatic systems and processes are responsible for these eruptions (Hart et al. 2004). The petrogenetic and tectonic significance of these mixed magma eruptions is a focus of our ongoing Middle Awash geological and geochemical research.

TABLE 3.3 Silicic Tephra Sample Average EMP Data

Sample	*Type*	*n*	*SiO_2*	*TiO_2*	*Al_2O_3*	*Fe_2O_3**	*MnO*	*MgO*	*CaO*	*Na_2O*	*K_2O*	*P_2O_5*	*Total*
MA93-10	Tuff 93-10	30	70.29	0.36	9.97	5.61	0.20	0.00	0.27	3.41	5.41	0.02	95.54
			0.34	0.05	0.07	0.10	0.03	0.00	0.02	0.68	0.53	0.02	0.60
MA96-37	Tuff 93-10	29	70.74	0.33	9.74	5.63	0.20	0.00	0.27	4.09	4.34	0.02	95.34
			0.37	0.05	0.08	0.07	0.03	0.01	0.01	0.70	0.45	0.02	0.67
MA98-44	Tuff 93-10	29	70.22	0.36	9.55	5.71	0.19	0.00	0.29	3.68	5.02	0.02	95.05
			0.61	0.02	0.08	0.11	0.02	0.00	0.02	0.44	0.50	0.02	0.64
MA92-04	Tuff 92-04	24	70.81	0.26	9.82	4.17	0.17	0.04	0.16	2.04	4.84	0.00	92.31
			0.51	0.04	0.06	0.06	0.02	0.01	0.01	0.73	0.89	0.00	1.22
MA93-09	Tuff 92-04	29	71.03	0.27	9.92	4.08	0.18	0.05	0.17	1.55	3.22	0.02	90.49
			0.37	0.05	0.08	0.07	0.04	0.01	0.01	0.30	0.52	0.02	0.77
MA95-16	Tuff 92-04	20	71.03	0.27	9.85	4.12	0.19	0.05	0.17	1.35	3.09	0.02	90.15
			0.41	0.05	0.09	0.09	0.03	0.01	0.01	0.07	0.27	0.03	0.39
MA95-23	Tuff 92-04	27	70.82	0.28	9.88	4.12	0.19	0.04	0.18	1.69	3.85	0.02	91.07
			0.59	0.06	0.11	0.09	0.04	0.01	0.01	0.50	0.71	0.02	0.67
MA94-18	Tuff 92-04	12	70.90	0.29	9.83	4.09	0.19	0.05	0.17	1.80	2.66	0.01	90.01
			0.31	0.04	0.07	0.07	0.04	0.01	0.01	0.06	0.18	0.02	0.21
MA96-34	Tuff 96-34	33	70.30	0.28	9.50	4.13	0.18	0.04	0.17	2.73	3.78	0.01	91.12
			0.47	0.06	0.07	0.08	0.03	0.01	0.01	0.38	0.58	0.01	0.70
MA98-43	Tuff 96-34	33	70.49	0.29	9.43	4.18	0.18	0.04	0.19	1.77	5.40	0.01	91.99
			0.75	0.02	0.08	0.08	0.02	0.01	0.01	0.64	0.80	0.02	1.01
MA99-108	Tuff 96-34	21	70.29	0.30	9.39	4.18	0.17	0.03	0.18	2.36	4.24	0.01	91.14
			0.51	0.06	0.06	0.08	0.03	0.01	0.02	0.38	0.52	0.02	0.91
MA95-05		21	68.37	0.22	12.67	2.62	0.06	0.06	0.87	2.18	5.33	0.02	92.41
			0.40	0.05	0.09	0.06	0.02	0.01	0.02	0.39	0.40	0.03	0.57
MA99-114C		27	69.35	0.18	12.08	2.48	0.08	0.02	0.73	2.59	5.67	nm	93.18
			0.40	0.04	0.11	0.11	0.03	0.01	0.04	0.52	0.30	nm	0.65
MA99-112	WMMT	13	70.29	0.20	11.59	2.74	0.08	0.01	0.97	2.18	1.83	0.02	89.91
			0.31	0.06	0.06	0.09	0.02	0.01	0.02	0.09	0.11	0.02	0.34
MA93-18	WMMT	32	70.51	0.21	11.77	2.66	0.07	0.03	0.95	2.62	2.87	0.01	91.71
			0.56	0.04	0.10	0.08	0.02	0.01	0.02	0.28	0.37	0.01	0.87
MA95-04A	WMMT	6	70.43	0.17	11.58	2.57	0.06	0.01	0.87	3.05	4.75	0.01	93.49
			0.54	0.07	0.07	0.13	0.02	0.00	0.02	0.37	0.37	0.02	0.57
MA95-04B	WMMT	21	69.88	0.23	11.85	2.66	0.08	0.01	0.89	2.10	4.98	0.02	92.69
			0.49	0.06	0.09	0.10	0.04	0.01	0.04	0.47	0.47	0.02	0.94
MA96-30	WMMT	11	70.28	0.20	11.57	2.70	0.07	0.01	0.88	2.65	4.65	0.02	93.01
			0.41	0.06	0.11	0.05	0.02	0.00	0.03	0.52	0.48	0.02	0.74
MA99-100	WMMT	26	69.48	0.21	11.42	2.76	0.06	0.01	0.97	2.13	4.38	0.01	91.42
			0.39	0.07	0.07	0.09	0.02	0.01	0.03	0.43	0.59	0.02	0.87
MA95-14	Tuff 95-14	23	68.87	0.21	12.45	2.67	0.14	0.05	0.31	2.21	5.00	0.02	91.93
			0.40	0.06	0.12	0.06	0.03	0.01	0.02	0.16	0.60	0.03	0.45
MA95-21	Tuff 95-14	8	68.60	0.17	12.56	2.60	0.14	0.05	0.31	2.08	4.24	0.02	90.76
			0.31	0.06	0.16	0.08	0.02	0.01	0.02	0.13	0.71	0.03	0.49
MA93-15	Tuff 93-15	35	69.63	0.19	12.11	2.47	0.12	0.05	0.30	1.88	5.38	0.01	92.13
			0.59	0.04	0.13	0.08	0.02	0.02	0.02	0.20	0.61	0.01	1.04
MA97-16	Tuff 93-15	25	68.51	0.19	12.07	2.57	0.12	0.05	0.30	2.31	5.02	0.03	91.18
			0.61	0.05	0.11	0.08	0.02	0.01	0.02	0.21	0.15	0.02	0.69
MA97-22	Tuff 93-15	30	68.55	0.18	12.01	2.61	0.13	0.05	0.30	2.00	6.56	0.01	92.40
			0.37	0.05	0.07	0.06	0.02	0.01	0.01	0.38	0.41	0.01	0.72
MA97-24	Tuff 93-15	25	68.58	0.19	12.02	2.57	0.12	0.04	0.30	2.33	6.63	0.02	92.81
			0.46	0.06	0.10	0.08	0.03	0.01	0.02	0.40	0.39	0.03	0.64

TABLE 3.3 *(continued)*

Sample	*Type*	*n*	*SiO_2*	*TiO_2*	*Al_2O_3*	*Fe_2O_3**	*MnO*	*MgO*	*CaO*	*Na_2O*	*K_2O*	*P_2O_5*	*Total*
MA99-105	Tuff 93-15	22	69.01	0.20	11.63	2.59	0.14	0.04	0.30	1.87	5.38	nm	91.17
			0.38	0.05	0.16	0.08	0.03	0.01	0.02	0.25	0.59	nm	0.77
MA95-25(1)	Tuff 95-25	23	67.63	0.26	12.57	2.99	0.10	0.04	0.75	1.79	5.13	0.02	91.27
			0.38	0.05	0.10	0.12	0.03	0.01	0.04	0.38	0.56	0.02	0.72
MA95-25(2)	Tuff 95-25	4	66.66	0.32	12.78	3.64	0.12	0.04	0.95	1.95	5.32	0.00	91.78
			0.53	0.05	0.05	0.06	0.05	0.01	0.04	0.35	0.61	0.00	0.54
MA97-23(1)	Tuff 95-25	28	67.71	0.23	12.30	2.98	0.09	0.03	0.74	1.88	6.26	0.02	92.24
			0.42	0.04	0.07	0.09	0.03	0.01	0.03	0.38	0.41	0.02	0.70
MA97-23(2)	Tuff 95-25	4	66.58	0.27	12.45	3.50	0.11	0.04	0.92	1.34	5.42	0.01	90.63
			0.34	0.05	0.07	0.06	0.03	0.01	0.06	0.19	0.36	0.01	0.46
MA95-13A		27	68.91	0.22	12.66	1.92	0.08	0.09	0.25	2.08	5.59	0.03	91.83
			0.32	0.06	0.10	0.07	0.04	0.01	0.01	0.26	0.51	0.03	0.67
MA99-118		40	68.62	0.20	12.25	2.66	0.07	0.06	0.96	2.00	2.98	0.02	89.82
			0.44	0.05	0.09	0.07	0.03	0.01	0.02	0.27	0.40	0.02	0.82
MA95-11		17	68.66	0.23	12.57	2.50	0.10	0.05	0.69	1.86	3.43	0.02	90.12
			0.27	0.05	0.10	0.07	0.04	0.06	0.03	0.09	0.19	0.02	0.34
MA95-12		21	69.89	0.20	11.83	2.40	0.08	0.04	0.60	2.52	2.89	0.02	90.48
			0.35	0.05	0.09	0.07	0.03	0.01	0.02	0.22	0.32	0.03	0.58
MA95-19		29	69.96	0.19	11.92	2.42	0.06	0.03	0.59	2.85	5.16	0.01	93.20
			0.51	0.06	0.09	0.07	0.03	0.01	0.03	0.45	0.32	0.02	0.62
MA96-35		8	72.00	0.30	9.53	4.11	0.18	0.05	0.19	0.95	1.46	0.01	88.77
			0.39	0.06	0.04	0.11	0.03	0.03	0.01	0.06	0.30	0.02	0.33
MA96-36		25	68.28	0.17	12.09	2.72	0.10	0.05	0.67	2.89	4.18	0.02	91.17
			0.56	0.05	0.09	0.06	0.03	0.01	0.03	0.57	0.35	0.02	0.70
MA99-115		13	70.68	0.17	11.57	2.48	0.06	0.04	0.64	2.55	1.70	0.01	89.90
			0.25	0.04	0.05	0.03	0.03	0.01	0.02	0.13	0.17	0.01	0.35
MA99-122		12	71.69	0.14	11.08	1.93	0.04	0.00	0.34	2.08	2.97	0.01	90.28
			0.29	0.04	0.06	0.05	0.02	0.00	0.02	0.09	0.18	0.02	0.40

NOTE: Averages based on (*n*) number of individual glass pyroclasts and are from measured (raw) data. Total Fe expressed as Fe_2O_3 (Fe_2O_3*). P_2O_5 not measured (nm) for some samples. Errors are one standard deviation. Where present, individual subpopulation averages are listed as (1) and (2).

The plots illustrated in Figures 3.8 and 3.9 and the summary data of Table 3.3 indicate that the silicic tephras of the Adu-Asa Formation (and those just above) are rhyolites. Although difficult to assess fully because of posteruption hydration and alkali mobility, most of these rhyolites are subalkaline, with only the youngest tephras (Tuff 96-34, Tuff 92-04, Tuff 93-10) illustrating the high-Fe and low-Al signatures more characteristic of peralkaline rhyolites. A mix of subalkaline and peralkaline rhyolites with overall major element chemical signatures similar to the tephras presented here has been identified at a number of vents and vent complexes along the southeastern margin of the Afar Rift, within the transition zone between the MER and Afar rifts (Figure 3.1) (Chernet and Hart 1999). The limited chronology available on these materials suggests that they are broadly coeval with the Adu-Asa Formation (Chernet et al. 1998). In the case of the post–Adu-Asa silicic tephras, chemically similar coeval

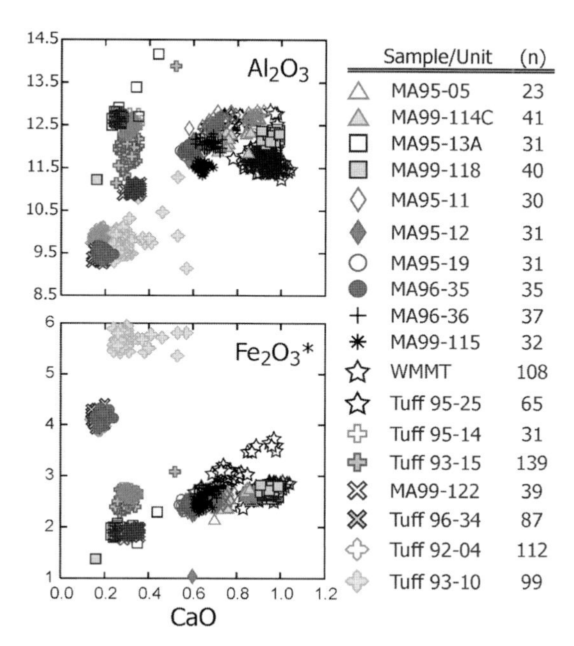

FIGURE 3.8

Major element oxide characteristics of Adu-Asa Formation and closely associated western margin silicic tephra. Discrete grain EMP data are plotted for named units and unnamed individual tephra layers, with the number of analyzed grains indicated. Raw data are plotted as weight percent, with total iron as Fe^{3+} ($Fe_2O_3^*$).

eruptions have been documented from a variety of vent regions spanning from the western Afar margin north of the Middle Awash area, south through the Addis Ababa region, extending to the central sector of the MER (Figure 3.1) (e.g., WoldeGabriel et al. 1992a, 1999; Chernet and Hart 1999; Mazzarini et al. 1999). These potential distal-proximal links will be more fully explored as we acquire complementary geochemical datasets, including Nd isotope information, for the Mio-Pliocene Middle Awash tephra and chronologically and volcanologically suitable volcanic centers (e.g., Hart et al. 1992).

Conclusions

The late Miocene Adu-Asa Formation deposits contain an excellent record of bimodal volcanism associated with episodes of local and regional rift development. Table 3.5 serves as a summary of the chemostratigraphy as currently understood. This stratigraphy has been developed through careful field investigations, selected radiometric dating, major element chemical analyses of glassy pyroclasts from mafic and silicic tephra deposits, and major element chemical analyses of whole-rock basalt lava flow material. In combination, these observations and data allow local and regional identification of, and correlation between, distinct volcanic marker horizons. Such tephrochemical and tephrostratigraphic correlations are critical because dynamic tectonic and geomorphic processes have produced a landscape characterized by abundant faulting, isolated depocenters, and patchy, yet well-preserved stratigraphic sequences.

On one end of the compositional spectrum, the numerous basaltic tephras and lava flows provide evidence of magmatic and tectonic activities occurring within or nearby

TABLE 3.4 Silicic Tephra Similarity Coefficient Summary

	93-10	*96-37*	*98-44*	*92-4*	*93-9*	*95-16*	*95-23*	*94-18*	*96-34*
93-10	■	**0.97**	**0.97**	0.78	0.80	0.82	0.82	0.82	0.79
96-37	**0.97**	■	**0.95**	0.80	0.82	0.83	0.84	0.84	0.81
98-44	**0.97**	**0.95**	■	0.77	0.79	0.81	0.81	0.81	0.79
92-04	0.78	0.80	0.77	■	**0.96**	**0.95**	**0.94**	**0.94**	**0.96**
93-09	0.80	0.82	0.79	**0.96**	■	**0.98**	**0.98**	**0.97**	**0.97**
95-16	0.82	0.83	0.81	**0.95**	**0.98**	■	**0.99**	**0.99**	**0.96**
95-23	0.82	0.84	0.81	**0.94**	**0.98**	**0.99**	■	**0.98**	**0.96**
94-18	0.82	0.84	0.81	**0.94**	**0.97**	**0.99**	**0.98**	■	**0.96**
96-34	0.79	0.81	0.79	**0.96**	**0.97**	**0.96**	**0.96**	**0.96**	■
98-43	0.82	0.84	0.82	**0.93**	**0.95**	**0.94**	**0.96**	**0.95**	**0.96**
99-108	0.80	0.83	0.80	**0.95**	**0.95**	**0.95**	**0.95**	**0.96**	**0.97**
95-05	0.49	0.50	0.49	0.55	0.55	0.54	0.54	0.54	0.54
99-114C	0.51	0.52	0.51	0.56	0.56	0.55	0.55	0.54	0.55
99-112	0.52	0.52	0.51	0.58	0.58	0.57	0.57	0.56	0.56
93-18	0.51	0.51	0.50	0.57	0.57	0.56	0.56	0.55	0.55
95-04A	0.48	0.48	0.47	0.53	0.53	0.52	0.52	0.51	0.51
95-04B	0.53	0.54	0.52	0.59	0.59	0.58	0.58	0.57	0.58
96-30	0.51	0.52	0.50	0.57	0.57	0.56	0.56	0.55	0.55
99-100	0.51	0.51	0.50	0.57	0.57	0.56	0.56	0.55	0.55
95-14	0.69	0.70	0.70	0.72	0.72	0.70	0.71	0.70	0.71
95-21	0.66	0.66	0.66	0.67	0.67	0.66	0.66	0.65	0.66
93-15	0.66	0.66	0.67	0.68	0.67	0.66	0.66	0.66	0.66
97-16	0.67	0.67	0.67	0.69	0.68	0.67	0.67	0.67	0.67
97-22	0.67	0.68	0.68	0.69	0.69	0.68	0.68	0.67	0.68
97-24	0.67	0.67	0.67	0.68	0.68	0.67	0.67	0.66	0.67
99-105	0.70	0.70	0.70	0.72	0.72	0.70	0.71	0.70	0.71
95-25(1)	0.58	0.59	0.58	0.65	0.65	0.64	0.64	0.63	0.64
95-25(2)	0.65	0.66	0.64	0.67	0.67	0.66	0.67	0.67	0.67
97-23(1)	0.56	0.57	0.56	0.63	0.62	0.61	0.61	0.61	0.61
97-23(2)	0.60	0.61	0.60	0.68	0.68	0.68	0.68	0.67	0.68
95-13A	0.62	0.63	0.60	0.64	0.65	0.64	0.64	0.63	0.63
99-118	0.50	0.50	0.49	0.56	0.56	0.55	0.55	0.54	0.55
95-11	0.56	0.57	0.56	0.62	0.61	0.60	0.60	0.60	0.60
95-12	0.54	0.54	0.53	0.58	0.58	0.57	0.57	0.56	0.56
95-19	0.52	0.52	0.51	0.56	0.55	0.55	0.55	0.54	0.54
96-35	0.82	0.84	0.82	**0.93**	**0.96**	**0.95**	**0.96**	**0.96**	**0.96**
96-36	0.54	0.54	0.54	0.59	0.59	0.58	0.58	0.57	0.58
99-115	0.50	0.50	0.50	0.54	0.54	0.53	0.53	0.53	0.53
99-122	0.53	0.53	0.53	0.52	0.53	0.52	0.52	0.52	0.51

	95-14	*95-21*	*93-15*	*97-16*	*97-22*	*97-24*	*99-105*	*95-25(1)*	*95-25(2)*
93-10	0.69	0.66	0.66	0.67	0.67	0.67	0.70	0.58	0.65
96-37	0.70	0.66	0.66	0.67	0.68	0.67	0.70	0.59	0.66
98-44	0.70	0.66	0.67	0.67	0.68	0.67	0.70	0.58	0.64
92-04	0.72	0.67	0.68	0.69	0.69	0.68	0.72	0.65	0.67
93-09	0.72	0.67	0.67	0.68	0.69	0.68	0.72	0.65	0.67
95-16	0.70	0.66	0.66	0.67	0.68	0.67	0.70	0.64	0.66
95-23	0.71	0.66	0.66	0.67	0.68	0.67	0.71	0.64	0.67

98-43	*99-108*	*95-5*	*99-114C*	*99-112*	*93-18*	*95-04A*	*95-04B*	*96-30*	*99-100*
0.82	0.80	0.49	0.51	0.52	0.51	0.48	0.53	0.51	0.51
0.84	0.83	0.50	0.52	0.52	0.51	0.48	0.54	0.52	0.51
0.82	0.80	0.49	0.51	0.51	0.50	0.47	0.52	0.50	0.50
0.93	**0.95**	0.55	0.56	0.58	0.57	0.53	0.59	0.57	0.57
0.95	**0.95**	0.55	0.56	0.58	0.57	0.53	0.59	0.57	0.57
0.94	**0.95**	0.54	0.55	0.57	0.56	0.52	0.58	0.56	0.56
0.96	**0.95**	0.54	0.55	0.57	0.56	0.52	0.58	0.56	0.56
0.95	**0.96**	0.54	0.54	0.56	0.55	0.51	0.57	0.55	0.55
0.96	**0.97**	0.54	0.55	0.56	0.55	0.51	0.58	0.55	0.55
■	**0.98**	0.53	0.54	0.56	0.55	0.51	0.57	0.55	0.55
0.98	■	0.53	0.54	0.56	0.54	0.51	0.57	0.54	0.54
0.53	0.53	■	0.86	0.89	**0.95**	**0.92**	**0.93**	**0.93**	**0.93**
0.54	0.54	0.86	■	0.89	0.86	0.88	0.89	0.89	0.85
0.56	0.56	0.89	0.89	■	**0.94**	0.88	**0.95**	**0.95**	**0.95**
0.55	0.54	**0.95**	0.86	**0.94**	■	**0.90**	**0.94**	**0.95**	**0.97**
0.51	0.51	**0.92**	0.88	0.88	**0.90**	■	0.88	**0.92**	**0.91**
0.57	0.57	**0.93**	0.89	**0.95**	**0.94**	0.88	■	**0.94**	**0.91**
0.55	0.54	**0.93**	0.89	**0.95**	**0.95**	**0.92**	**0.94**	■	**0.95**
0.55	0.54	**0.93**	0.85	**0.95**	**0.97**	**0.91**	**0.91**	**0.95**	■
0.71	0.70	0.74	0.75	0.74	0.74	0.69	0.75	0.74	0.72
0.66	0.66	0.71	0.77	0.71	0.70	0.73	0.71	0.72	0.68
0.66	0.66	0.73	0.81	0.76	0.73	0.73	0.75	0.76	0.72
0.67	0.67	0.73	0.80	0.76	0.74	0.73	0.75	0.76	0.73
0.68	0.67	0.72	0.80	0.75	0.73	0.73	0.74	0.75	0.71
0.67	0.67	0.73	0.80	0.76	0.73	0.74	0.75	0.76	0.72
0.71	0.70	0.72	0.76	0.76	0.74	0.72	0.75	0.75	0.73
0.63	0.63	0.83	0.86	0.84	0.82	0.78	0.87	0.83	0.81
0.68	0.68	0.76	0.73	0.78	0.77	0.70	0.79	0.75	0.76
0.61	0.60	0.86	**0.90**	0.87	0.85	0.80	**0.90**	0.86	0.84
0.67	0.67	0.81	0.78	0.82	0.81	0.75	0.85	0.81	0.80
0.64	0.63	0.74	0.77	0.74	0.73	0.68	0.77	0.72	0.70
0.54	0.54	**0.92**	**0.90**	**0.97**	**0.95**	0.89	**0.95**	**0.95**	**0.94**
0.60	0.60	0.85	**0.90**	0.83	0.83	0.80	0.88	0.83	0.81
0.56	0.56	0.85	**0.91**	0.89	0.87	0.84	0.89	0.89	0.85
0.54	0.54	0.87	0.89	0.84	0.87	0.88	0.84	0.88	0.87
0.99	**0.97**	0.53	0.54	0.56	0.55	0.51	0.57	0.55	0.55
0.57	0.57	0.81	**0.92**	0.85	0.82	0.84	0.84	0.85	0.81
0.52	0.52	0.86	**0.90**	0.84	0.87	**0.91**	0.83	0.88	0.87
0.52	0.51	0.67	0.69	0.65	0.67	0.74	0.64	0.68	0.68

97-23(1)	*97-23(2)*	*95-13a*	*99-118*	*95-11*	*95-12*	*95-19*	*96-35*	*96-36*	*99-115*	*99-122*
0.56	0.60	0.62	0.50	0.56	0.54	0.52	0.82	0.54	0.50	0.53
0.57	0.61	0.63	0.50	0.57	0.54	0.52	0.84	0.54	0.50	0.53
0.56	0.60	0.60	0.49	0.56	0.53	0.51	0.82	0.54	0.50	0.53
0.63	0.68	0.64	0.56	0.62	0.58	0.56	**0.93**	0.59	0.54	0.52
0.62	0.68	0.65	0.56	0.61	0.58	0.55	**0.96**	0.59	0.54	0.53
0.61	0.68	0.64	0.55	0.60	0.57	0.55	**0.95**	0.58	0.53	0.52
0.61	0.68	0.64	0.55	0.60	0.57	0.55	**0.96**	0.58	0.53	0.52

TABLE 3.4 *(continued)*

	95-14	*95-21*	*93-15*	*97-16*	*97-22*	*97-24*	*99-105*	*95-25(1)*	*95-25(2)*
94-18	0.70	0.65	0.66	0.67	0.67	0.66	0.70	0.63	0.67
96-34	0.71	0.66	0.66	0.67	0.68	0.67	0.71	0.64	0.67
98-43	0.71	0.66	0.66	0.67	0.68	0.67	0.71	0.63	0.68
99-108	0.70	0.66	0.66	0.67	0.67	0.67	0.70	0.63	0.68
95-05	0.74	0.71	0.73	0.73	0.72	0.73	0.72	0.83	0.76
99-114C	0.75	0.77	0.81	0.80	0.80	0.80	0.76	0.86	0.73
99-112	0.74	0.71	0.76	0.76	0.75	0.76	0.76	0.84	0.78
93-18	0.74	0.70	0.73	0.74	0.73	0.73	0.74	0.82	0.77
95-04A	0.69	0.73	0.73	0.73	0.73	0.74	0.72	0.78	0.70
95-04B	0.75	0.71	0.75	0.75	0.74	0.75	0.75	0.87	0.79
96-30	0.74	0.72	0.76	0.76	0.75	0.76	0.75	0.83	0.75
99-100	0.72	0.68	0.72	0.73	0.71	0.72	0.73	0.81	0.76
95-14	■	**0.94**	**0.92**	**0.94**	**0.94**	**0.93**	**0.97**	0.76	0.71
95-21	**0.94**	■	**0.91**	**0.93**	**0.96**	**0.94**	**0.94**	0.73	0.68
93-15	**0.92**	**0.91**	■	**0.98**	**0.95**	**0.97**	**0.93**	0.76	0.70
97-16	**0.94**	**0.93**	**0.98**	■	**0.97**	**0.99**	**0.96**	0.76	0.71
97-22	**0.94**	**0.96**	**0.95**	**0.97**	■	**0.98**	**0.96**	0.74	0.69
97-24	**0.93**	**0.94**	**0.97**	**0.99**	**0.98**	■	**0.95**	0.75	0.71
99-105	**0.97**	**0.94**	**0.93**	**0.96**	**0.96**	**0.95**	■	0.74	0.69
95-25(1)	0.76	0.73	0.76	0.76	0.74	0.75	0.74	■	0.84
95-25(2)	0.71	0.68	0.70	0.71	0.69	0.71	0.69	0.84	■
97-23(1)	0.77	0.73	0.77	0.77	0.75	0.76	0.75	**0.96**	0.80
97-23(2)	0.72	0.69	0.72	0.72	0.70	0.71	0.70	**0.91**	**0.93**
95-13A	0.81	0.77	0.83	0.81	0.79	0.81	0.80	0.74	0.63
99-118	0.75	0.73	0.77	0.77	0.75	0.76	0.75	0.83	0.78
95-11	0.80	0.78	0.82	0.81	0.79	0.81	0.79	**0.92**	0.79
95-12	0.77	0.75	0.81	0.80	0.78	0.79	0.79	0.82	0.69
95-19	0.75	0.75	0.80	0.79	0.77	0.79	0.76	0.78	0.66
96-35	0.70	0.66	0.66	0.67	0.68	0.67	0.71	0.63	0.68
96-36	0.79	0.82	0.82	0.82	0.82	0.83	0.79	0.88	0.75
99-115	0.71	0.76	0.76	0.75	0.75	0.76	0.74	0.78	0.66
99-122	0.70	0.75	0.74	0.73	0.73	0.74	0.72	0.59	0.51

NOTE: Similarity coefficients were calculated from average measured (raw) EMP data (see Table 3.3) using the following element string (all weighted equally); TiO_2, Al_2O_3, Fe_2O_3, MnO, and CaO. All samples have the MA prefix (e.g., MA99-122). All SC values of 0.90 and above are designated in bold type.

the Middle Awash study area. Many of the basaltic tephras preserve physical evidence for eruptions from magmatic systems that were open to external water sources prior to or during eruption. Together with sedimentological observations, this suggests that a relatively wet environment, with abundant surface water and groundwater, characterized the late Miocene western margin of the Afar Rift transition region. At the other end of the compositional spectrum, the silicic tephras primarily are distal fall deposits, presumably from sources in the Ethiopian volcanic province, but well outside the Middle Awash area. These silicic eruptions mark important episodes in the overall evolution of the MER-Afar rift systems.

97-23(1)	*97-23(2)*	*95-13a*	*99-118*	*95-11*	*95-12*	*95-19*	*96-35*	*96-36*	*99-115*	*99-122*
0.61	0.67	0.63	0.54	0.60	0.56	0.54	**0.96**	0.57	0.53	0.52
0.61	0.68	0.63	0.55	0.60	0.56	0.54	**0.96**	0.58	0.53	0.51
0.61	0.67	0.64	0.54	0.60	0.56	0.54	**0.99**	0.57	0.52	0.52
0.60	0.67	0.63	0.54	0.60	0.56	0.54	**0.97**	0.57	0.52	0.51
0.86	0.81	0.74	**0.92**	0.85	0.85	0.87	0.53	0.81	0.86	0.67
0.90	0.78	0.77	**0.90**	**0.90**	**0.91**	0.89	0.54	**0.92**	**0.90**	0.69
0.87	0.82	0.74	**0.97**	0.83	0.89	0.84	0.56	0.85	0.84	0.65
0.85	0.81	0.73	**0.95**	0.83	0.87	0.87	0.55	0.82	0.87	0.67
0.80	0.75	0.68	0.89	0.80	0.84	0.88	0.51	0.84	**0.91**	0.74
0.90	0.85	0.77	**0.95**	0.88	0.89	0.84	0.57	0.84	0.83	0.64
0.86	0.81	0.72	**0.95**	0.83	0.89	0.88	0.55	0.85	0.88	0.68
0.84	0.80	0.70	**0.94**	0.81	0.85	0.87	0.55	0.81	0.87	0.68
0.77	0.72	0.81	0.75	0.80	0.77	0.75	0.70	0.79	0.71	0.70
0.73	0.69	0.77	0.73	0.78	0.75	0.75	0.66	0.82	0.76	0.75
0.77	0.72	0.83	0.77	0.82	0.81	0.80	0.66	0.82	0.76	0.74
0.77	0.72	0.81	0.77	0.81	0.80	0.79	0.67	0.82	0.75	0.73
0.75	0.70	0.79	0.75	0.79	0.78	0.77	0.68	0.82	0.75	0.73
0.76	0.71	0.81	0.76	0.81	0.79	0.79	0.67	0.83	0.76	0.74
0.75	0.70	0.80	0.75	0.79	0.79	0.76	0.71	0.79	0.74	0.72
0.96	**0.91**	0.74	0.83	**0.92**	0.82	0.78	0.63	0.88	0.78	0.59
0.80	**0.93**	0.63	0.78	0.79	0.69	0.66	0.68	0.75	0.66	0.51
■	0.87	0.77	0.87	**0.93**	0.86	0.82	0.61	0.89	0.81	0.62
0.87	■	0.68	0.82	0.86	0.75	0.72	0.67	0.81	0.71	0.55
0.77	0.68	■	0.74	0.78	0.79	0.75	0.63	0.72	0.71	0.75
0.87	0.82	0.74	■	0.84	0.89	0.86	0.54	0.85	0.85	0.65
0.93	0.86	0.78	0.84	■	0.88	0.84	0.60	**0.91**	0.83	0.63
0.86	0.75	0.79	0.89	0.88	■	**0.95**	0.56	0.87	**0.90**	0.71
0.82	0.72	0.75	0.86	0.84	**0.95**	■	0.54	0.85	**0.94**	0.74
0.61	0.67	0.63	0.54	0.60	0.56	0.54	■	0.57	0.52	0.52
0.89	0.81	0.72	0.85	**0.91**	0.87	0.85	0.57	■	0.88	0.68
0.81	0.71	0.71	0.85	0.83	**0.90**	**0.94**	0.52	0.88	■	0.76
0.62	0.55	0.75	0.65	0.63	0.71	0.74	0.52	0.68	0.76	■

The volcanic materials summarized in Table 3.5 are associated with sediments from numerous fluvio-lacustrine depositional systems and have been utilized to provide absolute age control on abundant vertebrate taxa, including important hominid fossils. In addition, many of these volcanic horizons are widespread along the strike of the rift margin and are readily identifiable by physical or chemical attributes, thus providing ideal stratigraphic marker horizons. Identification and correlation of key volcanic horizons has been essential to the chronostratigraphic assessment of the widely spaced and highly faulted late Miocene western margin geologic exposures and to the definition of individual members within the Adu-Asa Formation. Since no other comparable chemostratigraphic

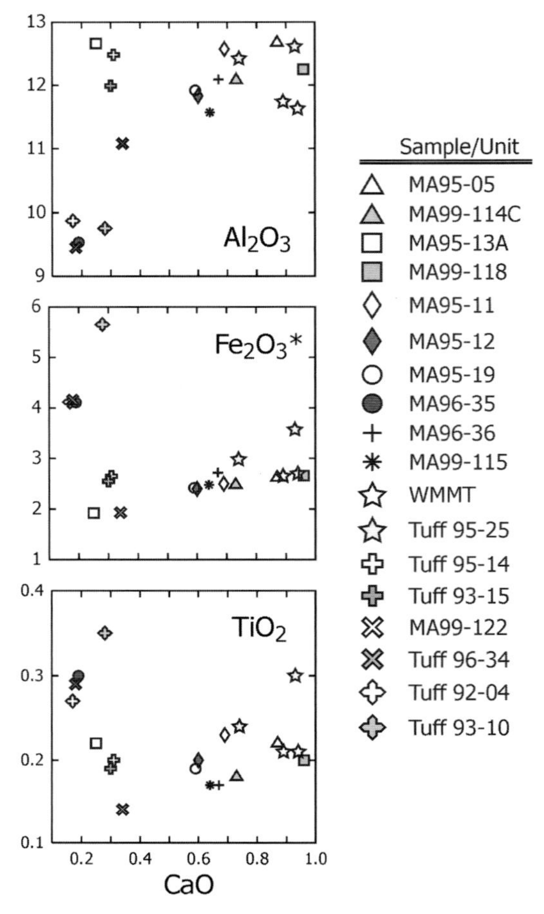

FIGURE 3.9

Average major element oxide characteristics of Adu-Asa Formation and closely associated western margin silicic tephras. Average values derived from data shown in Figure 3.8 as listed in Table 3.3. Raw data averages are plotted as weight percent, with total iron as Fe^{3+} (Fe_2O_3*).

information exists for the time period represented by the Adu-Asa Formation, the data presented herein and in WoldeGabriel et al. (Chapter 2) provide a platform for future volcanological and sedimentological investigations and the standard to which future studies will be compared.

Beginning in 1999 we have extended our studies of the western margin northward, to the Jara River and beyond. Deeper sedimentary packages have been observed in this area, vertebrate fossils have been collected from them, and geochronological and geochemical work has been initiated. We anticipate that this work, as well as more intensive sedimentological, geophysical, stratigraphic, and tephrochemical research along the western margin of the Middle Awash study area in the decades to come, will continue to elucidate the geological history of this fascinating region of Africa. This contribution is but an initial step in understanding the volcanological context and background of biological evolution in the unique setting of the Middle Awash valley.

TABLE 3.5 Summary Chemostratigraphy

Unit	*n*	*SiO_2*	*TiO_2*	*Al_2O_3*	*Fe_2O_3*	*MnO*	*MgO*	*CaO*	*Na_2O*	*K_2O*	*P_2O_5*	*Total*	*Comments*	
Post–Adu-Asa Formation														
Tuff 93-10	88	70.41	0.35	9.75	5.65	0.20	0.00	0.28	3.72	4.93	0.02	95.31	MA93-10, 96-37, 98-44	[ca. 3.8 Ma]
		0.50	0.05	0.19	0.11	0.03	0.01	0.02	0.67	0.66	0.02		Welded tuff/ignimbrite	
Tuff 92-04	112	70.92	0.27	9.87	4.12	0.18	0.05	0.17	1.68	3.64	0.01	90.91	MA92-04, 93-09, 94-18, 95-16, 95-23	
		0.47	0.05	0.09	0.08	0.03	0.01	0.01	0.50	0.94	0.02		(above HABT)	
Rawa Member, Adu-Asa Formation														
HABT	78	51.45	3.01	12.81	15.10	0.23	4.69	8.50	2.20	0.98	0.37	99.34	MA96-38, 98-45, 99-107	Top Rawa Member
		0.47	0.15	0.25	0.28	0.03	0.21	0.14	0.30	0.05	0.05			
Tuff 96-34	87	70.37	0.29	9.45	4.16	0.18	0.04	0.18	2.28	4.51	0.01	91.46	MA96-34, 98-43, 99-108	
		0.60	0.05	0.08	0.08	0.03	0.01	0.02	0.65	0.97	0.02		(below HABT)	
Basalt flows	2	46.45	1.92	15.04	12.92	0.19	7.62	9.95	2.23	0.74	0.39	97.44	MA93-05, 95-07	[5.50–5.54 Ma]
		0.02	0.03	0.15	0.01	0.00	0.17	0.11	0.02	0.09	0.08			
MA99-114C	27	69.35	0.18	12.08	2.48	0.08	0.02	0.73	2.59	5.67	nm	93.18	Below 5.5 Ma basalt (MA99-93)	
		0.40	0.04	0.11	0.11	0.03	0.01	0.04	0.52	0.30	nm			
DOBT	65	48.66	2.07	13.42	13.69	0.21	6.83	11.61	2.65	0.28	0.19	99.60	MA98-46, 98-57, 99-99, 99-109	Base Rawa Member
		0.60	0.10	0.29	0.24	0.03	0.22	0.22	0.10	0.02	0.04			
Asa Koma Member, Adu-Asa Formation														
MA95-05	21	68.37	0.22	12.67	2.62	0.06	0.06	0.87	2.18	5.33	0.02	92.41	5 m above 95-04 (WMMT)	
		0.40	0.05	0.09	0.06	0.02	0.01	0.02	0.39	0.40	0.03			
WMMT sil(f)	25	70.40	0.21	11.74	2.65	0.07	0.01	0.89	2.80	4.67	0.02	93.47	MA93-18, 95-04A, 95-04B, 96-30	[5.57 - 5.68 Ma]
		0.39	0.06	0.15	0.10	0.03	0.01	0.04	0.42	0.65	0.02			
WMMT sil(h)	83	70.00	0.21	11.63	2.70	0.07	0.01	0.94	2.26	3.54	0.01	91.39	MA93-18, 95-04A, 95-04B, 96-30,	Very heterogeneous
		0.64	0.06	0.18	0.10	0.03	0.01	0.04	0.40	1.18	0.01		99-100, 99-112	deposit typically
WMMT int	12	52.99	3.16	12.75	14.73	0.26	3.73	7.47	2.51	1.44	0.84	99.88	MA96-30, 99-100	exposed with more
		0.65	0.15	0.36	0.44	0.04	0.38	0.21	0.27	0.11	0.10			mafic and scoreaceous
WMMT maf3	9	51.05	3.54	12.57	15.38	0.24	4.26	7.92	2.43	1.25	0.68	99.32	MA95-04A, 96-30	basal portion grading
		0.42	0.16	0.38	0.27	0.04	0.53	0.18	0.28	0.05	0.07			up to finer grained
WMMT maf2	10	50.46	3.68	12.53	15.80	0.24	4.55	8.23	2.24	1.13	0.58	99.45	MA95-04A, 99-100	silicic vitric portion.
		0.42	0.23	0.16	0.23	0.04	0.27	0.24	0.15	0.08	0.05			Some outcrops
WMMT maf1	20	49.73	3.91	12.53	16.33	0.25	4.72	8.70	2.04	1.05	0.53	99.80	MA95-04A, 95-04B, 96-30, 99-100	preserve only silicic
		0.51	0.19	0.21	0.26	0.04	0.26	0.36	0.21	0.05	0.05			vitric portion.
LABT	364	44.17	4.62	12.06	15.23	0.29	5.70	10.18	2.93	1.07	2.49	98.75	MA96-25, 96-26, 96-28, 98-39,	[5.77 Ma]
		0.61	0.18	0.11	0.23	0.03	0.10	0.19	0.31	0.03	0.11		98-40, 98-41, 98-47, 98-48, 98-49,	
													99-69, 99-94, 99-103, 99-106, 99-116	

TABLE 3.5 *(continued)*

Unit	*n*	SiO_2	TiO_2	Al_2O_3	Fe_2O_3	*MnO*	*MgO*	*CaO*	Na_2O	K_2O	P_2O_5	*Total*	*Comments*	
MA93-12B	23	45.62	4.42	12.24	14.47	0.26	5.45	10.03	2.67	1.11	2.37	98.66	Stratigraphically below 92-04	
		0.61	0.15	0.16	0.34	0.03	0.11	0.23	0.24	0.02	0.11		Likely LABT variant and correlative	
MA93-12B	4	49.16	3.08	12.10	16.90	0.25	4.85	9.02	2.50	0.64	0.40	98.91		
		0.30	0.10	0.03	0.20	0.04	0.17	0.13	0.24	0.02	0.02			
BABT	227	49.15	2.58	12.69	15.48	0.23	5.82	10.13	2.87	0.44	0.30	99.69	MA96-27, 96-39, 99-71, 99-72, 99-90I, 99-95, 99-98, 99-102	Base Asa Koma Member
		0.36	0.12	0.18	0.21	0.03	0.11	0.24	0.14	0.02	0.04			
Adu Dora Member, Adu-Asa Formation														
Tuff 95-14	31	68.80	0.20	12.48	2.65	0.14	0.05	0.31	2.18	4.81	0.02	91.63	MA95-14, 95-21	
		0.40	0.06	0.13	0.07	0.03	0.01	0.02	0.16	0.71	0.03			
Tuff 93-15	137	68.90	0.19	11.99	2.55	0.13	0.05	0.30	2.07	5.80	0.02	91.99	MA93-15, 97-16, 97-22, 97-24, 99-105	
		0.67	0.05	0.20	0.09	0.03	0.01	0.02	0.36	0.81	0.02			
MA99-104	36	49.28	2.52	12.78	14.29	0.21	6.28	10.67	2.74	0.40	nm	99.18	3 m below 99-105 in upper diatomite	Multiple thin 15–30 cm basaltic tephra horizons within diatomite and between Tuff 93-15 and Tuff 95-25. These are chemically similar but cannot establish absolute correlations or absolute stratigraphy within this interval.
		0.24	0.14	0.12	0.17	0.03	0.09	0.08	0.09	0.02	nm			
MA99-101(1)	36	49.20	2.50	12.92	14.17	0.21	6.30	10.78	2.72	0.40	nm	99.20	In upper diatomite	
		0.26	0.08	0.15	0.16	0.03	0.08	0.11	0.08	0.01	nm			
MA99-101(2)	33	48.15	2.41	13.18	13.87	0.21	6.41	11.25	2.80	0.39	nm	98.66		
		0.40	0.10	0.08	0.19	0.03	0.09	0.08	0.11	0.02	nm			
MA99-97	38	49.54	2.49	13.05	14.42	0.22	6.38	10.70	2.71	0.41	nm	99.91		
		0.31	0.10	0.10	0.18	0.04	0.09	0.07	0.08	0.01	nm			
MA98-38	27	48.83	2.46	13.00	14.21	0.20	6.39	10.72	2.77	0.40	0.29	99.28	3 m below 97-22 and 3.4 m above 97-23	
		0.32	0.05	0.10	0.17	0.03	0.06	0.08	0.08	0.02	0.04			
MA93-16	36	49.66	2.24	13.30	13.40	0.18	6.25	10.52	2.94	0.40	0.27	99.16	4 m below 93-15	
		0.27	0.09	0.07	0.17	0.02	0.06	0.08	0.06	0.01	0.02			
Tuff 95-25(1)	51	67.67	0.24	12.42	2.98	0.10	0.04	0.74	1.84	5.75	0.02	91.80	MA95-25, 97-23; 6 m below 97-22	
		0.40	0.04	0.16	0.10	0.03	0.01	0.04	0.38	0.74	0.02			
Tuff 95-25(2)	8	66.62	0.30	12.61	3.57	0.11	0.04	0.93	1.65	5.37	0.01	91.21		
		0.41	0.05	0.18	0.09	0.04	0.01	0.05	0.42	0.47	0.01			
MA99-90F	31	49.32	2.53	13.19	14.41	0.21	6.41	11.17	2.75	0.41	nm	100.40	1 m above 99-90D	Stratigraphic position with respect to Tuff 95-25 and MA95-13A uncertain.
		0.24	0.11	0.08	0.16	0.03	0.08	0.08	0.07	0.01	nm			
MA99-90D(1)	24	49.35	2.59	13.26	14.07	0.20	6.59	11.28	2.55	0.40	nm	100.28	1 m below 99-90F	
		0.38	0.11	0.17	0.21	0.03	0.17	0.16	0.13	0.03	nm			
MA99-90D(2)	4	49.60	2.56	13.23	13.33	0.19	7.06	11.97	2.30	0.31	nm	100.56		
		0.17	0.04	0.19	0.39	0.03	0.22	0.21	0.16	0.06	nm			
MA95-13A	27	68.91	0.22	12.66	1.92	0.08	0.09	0.25	2.08	5.59	0.03	91.83	ca. 15 m below MA95-14	Stratigraphic position in lower Adu Dora Member uncertain.
		0.32	0.06	0.10	0.07	0.04	0.01	0.01	0.26	0.51	0.03			

ANBT	36	49.61	2.92	12.96	15.90	0.21	5.50	9.83	2.72	0.62	0.34	100.61	MA00-22	Base Adu Dora Member
		0.26	0.12	0.09	0.23	0.04	0.05	0.08	0.11	0.02	0.05			
Saraitu Member, Adu-Asa Formation														
MA99-96	20	49.00	2.59	13.15	14.54	0.21	6.26	10.51	2.68	0.48	nm	99.43	8 m below diatomite	Upper Saraitu Member
		0.19	0.12	0.12	0.17	0.03	0.05	0.08	0.06	0.02	nm			
Pre–Adu-Asa Formation														
MA95-22		50.12	3.54	12.24	15.54	0.24	4.29	8.80	2.56	1.24	0.59	99.16	Basalt flow	[6.16 Ma]
MA95-01		49.73	2.94	13.92	14.70	0.23	5.53	10.64	2.61	0.70	0.68	101.68	Basalt flow	[6.33 Ma]
MA95-02		48.26	2.87	13.85	12.47	0.27	4.92	12.93	2.54	0.61	0.55	99.27	Basalt flow	[6.33 Ma]
MA99-118	40	68.62	0.20	12.25	2.66	0.07	0.06	0.96	2.00	2.98	0.02	89.82		[>6 Ma]
		0.44	0.05	0.09	0.07	0.03	0.01	0.02	0.27	0.40	0.02			
Other Adu-Asa Formation Silicic Tephras with No or Uncertain Chronostratigraphic Context														
MA95-11	17	68.66	0.23	12.57	2.50	0.10	0.05	0.69	1.86	3.43	0.02	90.12		
		0.27	0.05	0.10	0.07	0.04	0.06	0.03	0.09	0.19	0.02			
MA95-12	21	69.89	0.20	11.83	2.40	0.08	0.04	0.60	2.52	2.89	0.02	90.48	95-12 and 95-19 are likely the same tephra	
		0.35	0.05	0.09	0.07	0.03	0.01	0.02	0.22	0.32	0.03			
MA95-19	29	69.96	0.19	11.92	2.42	0.06	0.03	0.59	2.85	5.16	0.01	93.20		
		0.51	0.06	0.09	0.07	0.03	0.01	0.03	0.45	0.32	0.02			
MA96-35	8	72.00	0.30	9.53	4.11	0.18	0.05	0.19	0.95	1.46	0.01	88.77	Likely correlation with Rawa Member Tuff 96-34	
		0.39	0.06	0.04	0.11	0.03	0.03	0.01	0.06	0.30	0.02			
MA96-36	25	68.28	0.17	12.09	2.72	0.10	0.05	0.67	2.89	4.18	0.02	91.17	Possible correlation with Rawa Member MA99-114C	
		0.56	0.05	0.09	0.06	0.03	0.01	0.03	0.57	0.35	0.02			
MA99-115	13	70.68	0.17	11.57	2.48	0.06	0.04	0.64	2.55	1.70	0.01	89.90		
		0.25	0.04	0.05	0.03	0.03	0.01	0.02	0.13	0.17	0.01			
MA99-122	12	71.69	0.14	11.08	1.93	0.04	0.00	0.34	2.08	2.97	0.01	90.28		
		0.29	0.04	0.06	0.05	0.02	0.00	0.02	0.09	0.18	0.02			

NOTE: Raw EMP data in oxide weight percent with total Fe converted from FeO to Fe_2O_3 for tephra. Raw whole-rock DCP data in oxide weight percent for basalt flows. Each tephra unit is represented by either a single sample or multiple correlated samples with data listed as averages and errors as one standard deviation based on the number of individual glass pyroclast analyses (*n*). The specific samples comprised in a given unit are listed in the Comments column. Named tephra units are those discussed in the text. Subpopulations for the WMMT are labeled as follows: sil(f), fresh silicic; sil(h), hydrated silicic; int; maf3; maf2; and maf1. Subpopulations for other units/samples are indicated by (1) and (2). All units are shown in the same composite stratigraphic context as illustrated in Fig. 3.4. Please note the specific stratigraphic information and caveats given in the Comments column.

4

Geochronology

PAUL R. RENNE,
LEAH E. MORGAN,
GIDAY
WOLDEGABRIEL,
WILLIAM K. HART,
AND
YOHANNES
HAILE-SELASSIE

$^{40}Ar/^{39}Ar$ dating has been applied extensively to volcanic units of late Miocene to Pleistocene age in the Middle Awash study area, and it has proven unequivocally to be the most important dating method applied to these deposits (Hall et al. 1984; White et al. 1993; WoldeGabriel et al. 1994; Asfaw et al. 2002; Clark et al. 2003; de Heinzelin et al. 1999; Renne et al. 1999; WoldeGabriel et al. 2001; White et al. 2006b). The character of volcanism recorded in the area evolved over this time period, such that the Miocene record up to the Gawto Basalts of the lower Sagantole Formation (Central Awash Complex) is dominated by basaltic lava flows and tuffs, whereas the Pliocene and younger eruptive units are mainly silicic tephra (see Hart et al., Chapter 3). In this chapter we summarize all available $^{40}Ar/^{39}Ar$ data for the Adu-Asa Formation, including some data not previously reported, plus some previously unpublished results from the nearby Central Awash Complex (CAC).

Methods

The methods and facilities employed for $^{40}Ar/^{39}Ar$ dating discussed herein have been described in detail by Renne et al. (1999). These methods may be broadly divided into two different approaches: (1) incremental heating, generally applied to lavas or to glass lapilli from basaltic tuffs, in neither of which xenocrysts are anticipated; and (2) single-crystal total fusion, generally applied to phenocrysts from tuffs in which the potential for xenocrystic contamination is high and analysis of multigrain samples is therefore likely to yield biased results. Previous experience in dating Middle Awash silicic tephra revealed many occurrences of such xenocrysts, particularly of 23–24 Ma age (Renne et al. 1999). Incremental heating experiments on basalt lavas were conducted on whole-rock samples from which phenocrysts (generally of olivine or pyroxene) were removed. These phenocrysts are undesirable because they dilute the potassium concentration and also may host excess ^{40}Ar.

Incremental heating experiments were performed either with an electrical resistance furnace or with a defocused Ar-ion laser. The furnace, described in detail by Sharp et al. (1996), offers the ability to measure the temperature achieved by the sample during each heating step and produces more uniform temperatures within the sample than is possible with laser heating. Laser heating, on the other hand, has the advantage of introducing relatively low background amounts of argon gas, whereas the furnace introduces large backgrounds and concomitantly large corrections at temperatures above ca. 1,000 °C.

Ages reported herein are all based on the Fish Canyon sanidine standard at 28.02 Ma (Renne et al. 1998). For comparison with results published prior to 1999, the previous ages should be increased by 0.7 percent to account for this recalibration. Uncertainties on all ages reported herein are at the level of one standard deviation and do not include contributions from uncertainties in the ^{40}K decay constants. Analytical data are summarized in Table 4.1.

Adu-Asa Formation

The following samples from the Adu-Asa Formation were dated, and the results are summarized below.

MA92-6 This unit was sampled at a fault scarp approximately 5 km N/NW of the main Ananu spring. At this locality the unit sampled is a dark-gray, calcified, poorly welded crystal vitric tuff 1.4 cm thick. Alkali feldspar phenocrysts were separated from this tuff, and five crystals were analyzed by single-crystal laser fusion. The results are shown in the age-probability diagram of Figure 4.1A. Two crystals, with ages of 5.09 ± 0.12 and 23.57 ± 0.12 Ma, are interpreted to be xenocrysts. The three remaining crystals yield consistent ages, and their weighted mean age of 2.56 ± 0.02 Ma is interpreted to represent the age of eruption and, presumably, deposition. This tephra is chemically and temporally correlative with the 2.5 Ma Maoleem Tuff of the Hatayae Member of the Bouri Formation (see WoldeGabriel et al., Chapter 2; Hart et al., Chapter 3).

MA95-11 Latitude: 10°15.252′ N, longitude: 40°18.962′ E. This 1.3 m thick crystal vitric tuff with bentonitic portions was sampled from a section in the Dasga Dora hills south of the Saragata stream. Plagioclase crystals were separated from the tuff, and 17 crystals were analyzed by single-crystal laser fusion (Figure 4.1B). Because of the relatively low K-contents and high Ca/K, the individual crystal ages are not as precise as could be obtained from alkali feldspar, but the results are internally consistent and yield a weighted mean age of 5.28 ± 0.03 Ma.

MA95-10 Latitude: 10°16.317′ N, longitude: 40°19.417′ E (Saitune Dora). At this locality the unit sampled is a >20 m thick altered basalt flow with phenocrysts of plagioclase, pyroxene, and olivine. A whole-rock basalt sample was analyzed in 12 steps by incremental laser heating. All steps define a plateau with indistinguishable ages (Figure 4.2A). The plateau age is 4.14 ± 0.07 Ma.

MA95-7 Latitude: 10°17.093′ N, longitude: 40°18.942′ E. This unit was sampled from the north bank of the Daytole stream to the north of Saitune Dora. The unit is a >5 m fractured, massive, moderately weathered, microporphyritic basalt flow. A whole-rock

TABLE 4.1 Summary of the Analytical Data

Run_ID	Power (W or °C)	^{40}Ar (mol)	$^{40}Ar/^{39}Ar$	$^{38}Ar/^{39}Ar$	$^{37}Ar/^{39}Ar$	$^{36}Ar/^{39}Ar$	$^{40}Ar^{*}/^{39}Ar$	$\%^{40}Ar^{*}$	Age (Ma)	± σ(Ma)
MA92-6	$J = 0.000387$	±0.000002								
8101-01	Fuse	4.8E–15	8.73859	0.01324	0.09864	0.00494	7.28675	83.39	5.09	0.12
8101-02	Fuse	2.2E–15	3.89090	0.01218	–0.00842	0.00033	3.79105	97.45	2.65	0.09
8101-03	Fuse	2.4E–14	34.42651	0.01197	–0.00043	0.00165	33.93737	98.58	23.57	0.12
8101-04	Fuse	4.8E–15	3.83751	0.01252	–0.00152	0.00057	3.66735	95.58	2.56	0.03
8101-05	Fuse	4.8E–15	3.71206	0.01188	–0.00140	0.00023	3.64384	98.18	2.55	0.03
MA65-11	$J = 0.001717$	±0.000006								
30641-01	Fuse	1.7E–15	1.84323	0.01176	1.50557	0.00092	1.69022	91.64	5.23	0.17
30641-02	Fuse	3.7E–15	1.88712	0.01175	1.55767	0.00110	1.68533	89.24	5.21	0.09
30641-03	Fuse	2.7E–15	1.77455	0.01198	2.14130	0.00077	1.71561	96.57	5.31	0.10
30641-04	Fuse	2.6E–15	1.90977	0.01237	1.57125	0.00104	1.72658	90.34	5.34	0.11
30641-05	Fuse	3.1E–15	1.86038	0.01205	1.82081	0.00103	1.69819	91.20	5.25	0.10
30641-06	Fuse	3.5E–15	1.84064	0.01228	1.62462	0.00093	1.69255	91.88	5.24	0.08
30641-07	Fuse	3.6E–15	1.77548	0.01240	1.84269	0.00077	1.69385	95.31	5.24	0.07
30641-08	Fuse	3.1E–15	2.23318	0.01233	1.44862	0.00209	1.73065	77.44	5.35	0.11
30641-09	Fuse	2.8E–15	1.77925	0.01250	1.91408	0.00074	1.71196	96.12	5.30	0.10
30641-10	Fuse	1.8E–15	1.78896	0.01228	1.69728	0.00058	1.75185	97.85	5.42	0.32
30641-11	Fuse	4.5E–15	2.06663	0.01204	1.43303	0.00155	1.72135	83.24	5.32	0.15
30641-12	Fuse	1.7E–15	1.82568	0.01239	1.65569	0.00070	1.74835	95.69	5.41	0.33
30641-13	Fuse	2.4E–15	1.97467	0.01220	1.94036	0.00137	1.72162	87.10	5.33	0.27
30641-14	Fuse	3.7E–15	1.81302	0.01214	1.33571	0.00057	1.74925	96.43	5.41	0.16
30641-15	Fuse	2.5E–15	1.76087	0.01262	1.83183	0.00072	1.69395	96.11	5.24	0.23
30641-16	Fuse	3.5E–15	1.87025	0.01215	1.69327	0.00084	1.75531	93.78	5.43	0.17
30641-17	Fuse	2.8E–15	2.02944	0.01223	1.88311	0.00150	1.73568	85.44	5.37	0.24
MA95-10	$J = 0.000531$	±0.000002								
31188-01A	0.3	1.5E–15	75.19289	0.04563	6.52748	0.22959	7.89033	10.45	7.54	2.81
31188-01B	0.5	1.8E–15	41.84475	0.03170	6.96096	0.12976	4.06019	9.66	3.88	1.38
31188-01C	0.7	2.3E–15	26.23309	0.02672	5.28047	0.07379	4.85568	18.45	4.64	0.70
31188-01D	0.9	2.2E–15	20.36312	0.02446	4.12244	0.05774	3.63174	17.79	3.47	0.51
31188-01E	1.2	2.4E–15	13.51678	0.01819	4.07228	0.03069	4.77651	35.24	4.57	0.32
31188-01F	1.4	2.2E–15	10.27564	0.01562	4.29869	0.01978	4.78019	46.39	4.57	0.27
31188-01G	1.6	3.1E–15	8.73903	0.01363	4.72321	0.01543	4.56223	52.05	4.36	0.17
31188-01H	1.9	3.8E–15	7.38589	0.01472	6.45261	0.01256	4.19509	56.56	4.01	0.13
31188-01I	2.3	3.7E–15	6.70348	0.01434	8.09611	0.01077	4.17495	61.95	3.99	0.13
31188-01J	2.7	1.0E–15	6.86518	0.01594	11.52509	0.01045	4.71279	68.13	4.51	0.34

TABLE 4.1 *(continued)*

Run_ID	*Power (W or °C)*	*^{40}Ar (mol)*	*$^{40}Ar/^{39}Ar$*	*$^{38}Ar/^{39}Ar$*	*$^{37}Ar/^{39}Ar$*	*$^{36}Ar/^{39}Ar$*	*$^{40}Ar^*/^{39}Ar$*	*$\%^{40}Ar^*$*	*Age (Ma)*	*$\pm\sigma$(Ma)*
31188-01K	3.5	8.7E–16	6.33631	0.01417	10.30403	0.00951	4.35825	68.32	4.17	0.33
31188-01L	5.0	1.0E–15	6.53660	0.01457	12.22373	0.01194	3.99408	60.62	3.82	0.30
MA95-7	J = 0.001783	±0.000003								
30627-01A	651	4.4E–12	490.23663	0.37253	5.00500	1.61915	12.21366	2.48	38.87	12.32
30627-01B	691	7.1E–13	66.42416	0.08606	7.44265	0.21604	3.18228	4.77	10.21	1.86
30627-01C	731	5.5E–13	28.44066	0.03123	8.91697	0.09021	2.49369	8.71	8.00	1.00
30627-01D	771	6.9E–13	22.14761	0.02516	6.78598	0.07056	1.83532	8.25	5.89	0.55
30627-01E	811	4.2E–13	8.44057	0.01600	4.13351	0.02368	1.77044	20.92	5.69	0.22
30627-01F	852	2.7E–13	6.62073	0.01494	3.00953	0.01746	1.69892	25.61	5.46	0.22
30627-01G	892	2.7E–13	10.89594	0.01794	2.98579	0.03200	1.67672	15.36	5.39	0.38
30627-01H	930	1.9E–13	7.66466	0.01618	2.45124	0.02093	1.67259	21.79	5.37	0.32
30627-01I	973	1.0E–13	4.85663	0.01478	2.70520	0.01153	1.66276	34.18	5.34	0.34
30627-01J	1010	5.8E–14	3.76271	0.01824	5.86386	0.00893	1.58652	42.00	5.10	0.45
30627-01K	1070	1.0E–13	4.45503	0.01740	16.31025	0.01417	1.55781	34.57	5.00	0.33
30627-01L	1132	3.9E–14	6.29338	0.01757	99.79829	0.04359	1.28878	19.03	4.14	1.26
30627-01M	1200	1.1E–14	10.99242	0.01868	156.05727	0.07412	1.42028	11.49	4.56	7.42
Plateau									5.54	0.17
MA95-15	J = 0.001783	±0.000003								
30628-01A	600	4.1E–16	54.19919	–0.13091	–0.37861	–0.05302	69.80596	128.83	211.63	361.51
30628-01B	652	8.3E–14	146.21340	0.11275	3.17607	0.49429	0.39785	0.27	1.28	5.86
30628-01C	691	1.6E–13	18.69008	0.02411	2.17899	0.06027	1.05213	5.62	3.38	0.42
30628-01D	732	3.2E–14	1.64353	0.01219	2.01027	0.00285	0.95859	58.27	3.08	0.16
30628-01E	762	2.8E–14	1.38006	0.01201	1.81487	0.00200	0.93187	67.47	3.00	0.15
30628-01F	793	3.3E–14	1.33679	0.01203	1.82943	0.00186	0.92971	69.49	2.99	0.13
30628-01G	821	2.6E–14	1.14182	0.01205	1.91228	0.00129	0.91043	79.67	2.93	0.14
30628-01H	862	3.5E–14	1.08253	0.01201	2.04190	0.00117	0.89526	82.63	2.88	0.10
30628-01I	898	2.2E–14	1.07177	0.01212	2.08157	0.00108	0.91489	85.29	2.94	0.15
30628-01J	941	1.9E–14	1.18266	0.01255	2.29987	0.00146	0.93212	78.73	3.00	0.19
30628-01K	984	1.8E–14	1.39578	0.01277	2.43140	0.00238	0.88234	63.14	2.84	0.24
30628-01L	1022	1.4E–14	1.34606	0.01324	2.34436	0.00217	0.88712	65.83	2.85	0.30
30628-01M	1061	4.9E–15	1.77154	0.02413	3.99951	0.00391	0.93108	52.43	2.99	1.10
30628-01N	1099	1.8E–14	1.89747	0.01611	17.95482	0.00835	0.84085	43.76	2.70	0.34
30628-01O	1150	5.5E–14	3.68682	0.01639	20.46597	0.01482	0.91791	24.54	2.95	0.22
30628-01P	1202	2.9E–14	29.93449	0.03356	17.11641	0.09923	1.97039	6.50	6.33	3.28
30628-01Q	1277	5.0E–14	174.53100	0.12973	20.18345	0.58389	3.61798	2.04	11.60	11.23
30628-01R	1353	5.7E–14	2919.40500	1.95766	21.49038	9.81231	21.88536	0.74	69.06	159.70

MA98-10	J = 0.0005322	±0.000005								
31198-01A	0.3	2.2E–15	383.14571	0.25940	3.06886	1.36396	–19.70577	–5.13	–19.02	43.73
31198-01B	0.5	2.2E–15	69.21785	0.05176	3.96291	0.22572	2.83466	4.08	2.72	2.29
31198-01C	0.8	2.5E–15	9.51076	0.01558	4.02133	0.01286	6.03967	63.34	5.79	0.21
31198-01D	1.1	1.6E–15	6.69812	0.01541	4.06014	0.00286	6.18689	92.13	5.93	0.19
31198-01E	1.4	2.2E–15	6.62950	0.01467	4.12279	0.00298	6.08526	91.55	5.83	0.16
31198-01F	1.6	3.7E–15	6.84062	0.01528	4.10358	0.00348	6.14701	89.63	5.89	0.11
31198-01G	1.9	2.3E–15	6.26891	0.01489	4.10894	0.00131	6.21744	98.92	5.96	0.13
31198-01H	2.3	4.6E–15	6.34485	0.01550	4.11574	0.00214	6.04869	95.08	5.80	0.08
31198-01I	2.7	2.8E–15	6.24996	0.01537	4.16715	0.00162	6.11156	97.53	5.86	0.11
31198-01J	3.5	2.2E–15	6.37319	0.01449	4.30640	0.00221	6.07271	95.02	5.82	0.14
31198-01K	5.0	2.8E–15	6.15873	0.01372	4.50885	0.00145	6.09826	98.73	5.85	0.11
31198-02A	0.3	2.6E–15	421.76947	0.27341	3.85536	1.40007	8.36921	1.98	8.02	39.08
31198-02B	0.5	1.9E–15	33.46927	0.03136	4.13192	0.09128	6.83774	20.37	6.55	1.05
31198-02C	0.8	5.4E–15	24.06437	0.02533	4.13958	0.06232	5.98643	24.81	5.74	0.33
31198-02D	1.1	1.7E–15	8.30823	0.01503	4.02900	0.00773	6.35468	76.29	6.09	0.24
31198-02E	1.4	3.1E–15	7.91426	0.01603	4.01877	0.00761	5.99444	75.55	5.75	0.15
31198-02F	1.6	3.9E–15	7.36006	0.01544	4.07092	0.00572	6.00421	81.37	5.76	0.11
31198-02G	1.9	3.3E–15	7.64704	0.01569	4.14601	0.00735	5.81487	75.84	5.57	0.13
31198-02H	2.3	4.1E–15	6.48388	0.01561	4.18378	0.00299	5.94122	91.39	5.70	0.09
31198-02I	2.7	3.3E–15	6.69152	0.01640	4.09652	0.00384	5.89055	87.80	5.65	0.11
31198-02J	3.5	3.9E–15	6.47342	0.01590	4.10468	0.00261	6.03950	93.05	5.79	0.09
31198-02K	5.0	7.0E–15	8.63186	0.01658	4.16766	0.00978	6.08294	70.28	5.83	0.08
MA98-11	J = 0.000531	±0.000002								
31187-20	Fuse	2.8E–15	24.59214	0.00592	–0.04344	0.00089	24.32375	98.91	23.14	0.84
31187-21	Fuse	2.4E–15	22.99566	0.01285	–0.06604	0.00207	22.37750	97.32	21.30	0.94
31187-22	Fuse	2.1E–15	16.36490	0.01183	–0.01776	0.00494	14.90382	91.08	14.22	0.68
31187-23	Fuse	2.8E–15	17.70167	0.00444	–0.00220	0.00187	17.14776	96.87	16.35	0.61
31187-24	Fuse	4.8E–15	24.73951	0.00983	–0.03853	0.00390	23.58225	95.33	22.44	0.49
31187-25	Fuse	4.1E–15	18.47458	0.01165	–0.02580	0.00294	17.60192	95.28	16.78	0.43
MA99-93	J = 0.000137	±0.000001								
33370-01A	1	4.9E–16	55.78968	0.02305	1.74805	0.16778	6.35720	11.38	1.57	3.30
33370-01B	2	1.6E–15	41.00080	0.00547	1.76294	0.04400	28.17449	68.63	6.95	0.83
33370-01C	3	5.9E–15	34.02380	0.01936	1.84034	0.02883	25.68470	75.39	6.33	0.19
33370-01D	4	1.0E–14	30.34611	0.01481	1.75701	0.02127	24.23207	79.75	5.97	0.11
33370-01E	5	1.5E–14	29.39028	0.01661	1.74731	0.01746	24.40072	82.92	6.02	0.07

TABLE 4.1 *(continued)*

Run_ID	*Power (W or °C)*	^{40}Ar *(mol)*	$^{40}Ar/^{39}Ar$	$^{38}Ar/^{39}Ar$	$^{37}Ar/^{39}Ar$	$^{36}Ar/^{39}Ar$	$^{40}Ar^{*}/^{39}Ar$	$\%^{40}Ar^{*}$	*Age (Ma)*	$\pm\sigma(Ma)$
33370-01F	6	1.5E–14	28.42251	0.01617	1.91939	0.01403	24.46465	85.96	6.03	0.07
33370-01G	7	1.7E–14	27.23550	0.01604	1.94329	0.01226	23.80236	87.28	5.87	0.06
33370-01H	8	1.8E–14	26.00269	0.01417	1.91766	0.00948	23.38645	89.82	5.77	0.05
33370-01I	10	2.6E–14	24.89196	0.01337	1.64978	0.00854	22.52714	90.40	5.56	0.04
33370-01J	12	3.3E–14	24.82880	0.01375	1.57342	0.00837	22.50768	90.55	5.55	0.03
33370-01K	14	3.2E–14	24.87466	0.01405	2.51428	0.00897	22.46689	90.16	5.54	0.03
33370-01L	16	2.2E–14	25.01556	0.01387	3.76957	0.00939	22.60506	90.12	5.57	0.04
33370-01M	18	9.1E–15	24.90323	0.01367	3.87102	0.00992	22.34320	89.48	5.51	0.09
33370-01N	21	7.8E–15	24.65895	0.01301	4.10641	0.00933	22.29779	90.16	5.50	0.10
33370-01O	25	9.2E–15	24.77691	0.01344	4.26682	0.00810	22.79552	91.73	5.62	0.09
MA99-119	$J = 0.000137$	±0.000001								
33371-01A	1	2.5E–16	845.43070	–0.17204	4.70964	0.78823	614.92860	72.49	145.82	365.64
33371-01B	2	6.5E–15	259.63900	0.15514	4.28788	0.77316	31.61024	12.14	7.79	1.52
33371-01C	3	1.5E–14	226.67290	0.13125	5.48553	0.69146	22.87508	10.05	5.64	0.80
33371-01D	4	2.2E–14	185.11070	0.11269	6.04033	0.53454	27.75832	14.93	6.84	0.52
33371-01E	5	2.4E–14	125.38380	0.07646	5.92722	0.33728	26.30482	20.89	6.49	0.34
33371-01F	6	3.0E–14	115.32690	0.06970	5.70410	0.30174	26.72936	23.08	6.59	0.24
33371-01G	7	3.6E–14	101.12950	0.06019	5.78924	0.25391	26.67420	26.27	6.58	0.19
33371-01H	8	3.1E–14	86.31792	0.04980	5.79596	0.20971	24.91672	28.75	6.14	0.19
33371-01I	10	3.1E–14	69.05452	0.03957	6.00733	0.15145	24.88992	35.89	6.14	0.14
33371-01J	12	3.2E–14	58.24176	0.03600	6.26024	0.11795	23.99693	41.02	5.92	0.12
33371-01K	14	1.8E–14	44.87331	0.02772	11.01458	0.07706	23.16920	51.23	5.71	0.12
33371-01L	16	8.2E–15	46.70477	0.02700	34.55647	0.09176	22.92931	47.89	5.65	0.22
33371-01M	18	3.3E–15	41.99219	0.02691	49.63963	0.09028	20.00893	45.97	4.94	0.43
33371-01N	21	2.2E–15	37.84037	0.02552	53.27579	0.06868	22.68608	57.68	5.59	0.59
33371-01O	25	1.3E–15	38.70922	0.02298	52.11987	0.08974	17.00998	42.32	4.20	0.99
MA92-11	$J = 0.000387$	±0.000002								
8108-02A	541	7.7E–15	1754.42000	1.18280	0.21822	5.81861	35.04506	2.00	24.33	16.39
8108-02B	600	2.5E–13	8582.29000	5.44336	1.12750	28.87491	49.88129	0.58	34.54	21.92
8108-02C	641	7.0E–13	2655.41000	1.70181	2.61868	8.87523	33.04252	1.24	22.95	6.10
8108-02D	681	1.5E–13	266.83680	0.18199	5.27564	0.87943	7.40340	2.76	5.17	0.66
8108-02E	721	1.0E–13	108.63670	0.07747	8.11624	0.34101	8.54633	7.82	5.96	0.29
8108-02F	761	4.8E–14	40.57388	0.03401	9.64782	0.11406	7.67155	18.79	5.36	0.14
8108-02G	800	2.6E–14	23.89477	0.02363	10.15412	0.05814	7.55784	31.41	5.28	0.10

8108-02H	840	2.3E–14	33.54052	0.03098	11.23868	0.09112	7.54724	22.33	5.27	0.11
8108-02I	881	1.6E–14	33.70607	0.03035	13.07846	0.09227	7.52616	22.13	5.25	0.20
8108-02J	930	2.0E–14	56.39680	0.04580	13.72243	0.16943	7.47041	13.12	5.21	0.23
8108-02K	991	1.6E–14	56.96676	0.05059	14.15243	0.17177	7.38411	12.84	5.15	0.24
8108-02L	1070	2.3E–14	76.31611	0.06293	31.34233	0.24172	7.49059	9.61	5.23	0.29
8108-02M	1161	4.3E–14	134.00440	0.13057	143.66680	0.48558	1.90700	1.29	1.33	0.47
8108-02N	1301	2.7E–14	93.14930	0.07137	140.47240	0.34634	1.94719	1.89	1.36	0.40
8108-02O	1502	2.7E–14	1059.40200	0.72254	172.07310	3.62504	1.83824	0.15	1.28	4.44
MA96-18-1	J = 0.000068	±0.0000001								
30672-01A	602	6.3E–16	–747.64550	–0.31280	5.53085	–3.22887	207.80550	–27.68	25.16	156.57
30672-01B	641	1.0E–15	4708.03400	2.13151	–4.46190	22.08258	–1811.63000	–38.61	–235.57	1432.53
30672-01C	682	1.2E–12	60220.51000	37.94025	9.57830	204.47010	–201.12120	–0.33	–24.69	63.72
30672-01D	720	9.9E–13	2701.66100	1.68880	12.66047	8.92211	66.83115	2.45	8.13	2.73
30672-01E	761	1.5E–13	252.84140	0.14684	11.61954	0.71450	43.01283	16.86	5.24	0.35
30672-01F	790	9.3E–14	226.57050	0.13294	9.41378	0.61669	45.41330	19.90	5.53	0.42
30672-01G	820	1.4E–13	526.98210	0.32007	8.61049	1.62660	47.31846	8.92	5.76	0.73
30672-01H	861	6.3E–14	217.90570	0.12380	9.51937	0.58855	45.07468	20.54	5.49	0.51
30672-01I	902	3.4E–14	119.19100	0.05682	10.33421	0.26128	43.14631	35.92	5.25	0.48
30672-01J	940	2.4E–14	102.23850	0.04428	10.95650	0.20397	43.19981	41.90	5.26	0.56
30672-01K	982	2.2E–14	127.18530	0.06127	11.28655	0.29146	42.32214	32.99	5.15	0.75
30672-01L	1020	1.7E–14	161.71180	0.09086	14.01780	0.42437	37.83229	23.15	4.61	1.27
30672-01M	1061	1.0E–14	176.51530	0.09840	22.84189	0.47128	39.76574	22.14	4.84	2.22
30672-01N	1101	1.9E–14	174.85630	0.10005	50.82643	0.45860	45.13934	24.82	5.49	1.26
30672-01O	1152	6.8E–14	215.22180	0.12805	136.73020	0.64092	40.99753	17.07	4.99	0.53
30672-01P	1202	1.7E–14	378.12310	0.21289	139.47000	1.27496	13.98205	3.31	1.70	3.38
30672-01Q	1277	1.4E–14	1459.31200	0.74256	133.09880	4.60376	121.84030	7.50	14.79	15.48
30672-01R	1352	1.5E–14	5119.45600	2.48280	122.03760	17.43653	–25.70639	–0.46	–3.14	49.32
30672-01S	1402	1.6E–14	5580.55600	3.05461	119.90330	18.58836	107.00210	1.74	13.00	50.95
30672-01T	1500	2.2E–14	455.13020	0.56275	0.97428	1.45370	25.66495	5.63	3.13	2.72
30672-01U	1502	1.4E–14	25374.22000	13.68141	–5.24535	82.10711	1106.69700	4.38	130.10	349.73

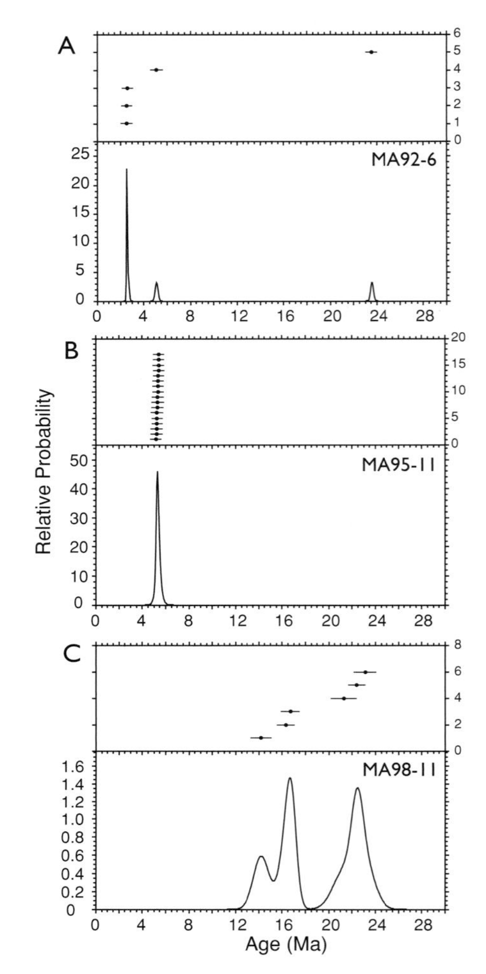

FIGURE 4.1.

Age-probability diagrams (lower panels) for units dated by single-crystal laser fusion methods. The upper panel for each indicated sample shows ranked ages (with 1σ error bars).

basalt was analyzed by incremental heating with a resistance furnace in 13 steps between 651 and 1,200 °C. The three steps below 771 °C show anomalously high ages, symptomatic of ^{39}Ar recoil artifacts (Figure 4.2B), but the remaining ten steps define a plateau with an age of 5.54 ± 0.17 Ma as reported by WoldeGabriel et al. (2001).

MA95-15 Latitude: 10°16.063′ N, longitude: 40°19.750′ E. This 5–7 m thick fractured, microporphyritic basalt flow, which is altered at the base, was sampled due west of the Asa Ali village along the east bank of the Hari Dora stream. A whole-rock basalt was analyzed by incremental heating with a resistance furnace in 18 steps between 600 and 1,350 °C. All steps define a plateau with an age of 2.95 ± 0.05 Ma (Figure 4.2C).

MA98-10 Latitude: 10°16.42′ N, longitude: 40°17.23′ E (Alayla). At this locality the unit sampled is a 7 m thick, well-bedded basaltic tuff with beds 0.2 to 4 m thick. The basaltic tuff was analyzed by 11 steps of incremental laser heating of multiple glass lapilli in two experiments. In each case, all steps defined plateaux, with ages of 5.85 ± 0.08 and

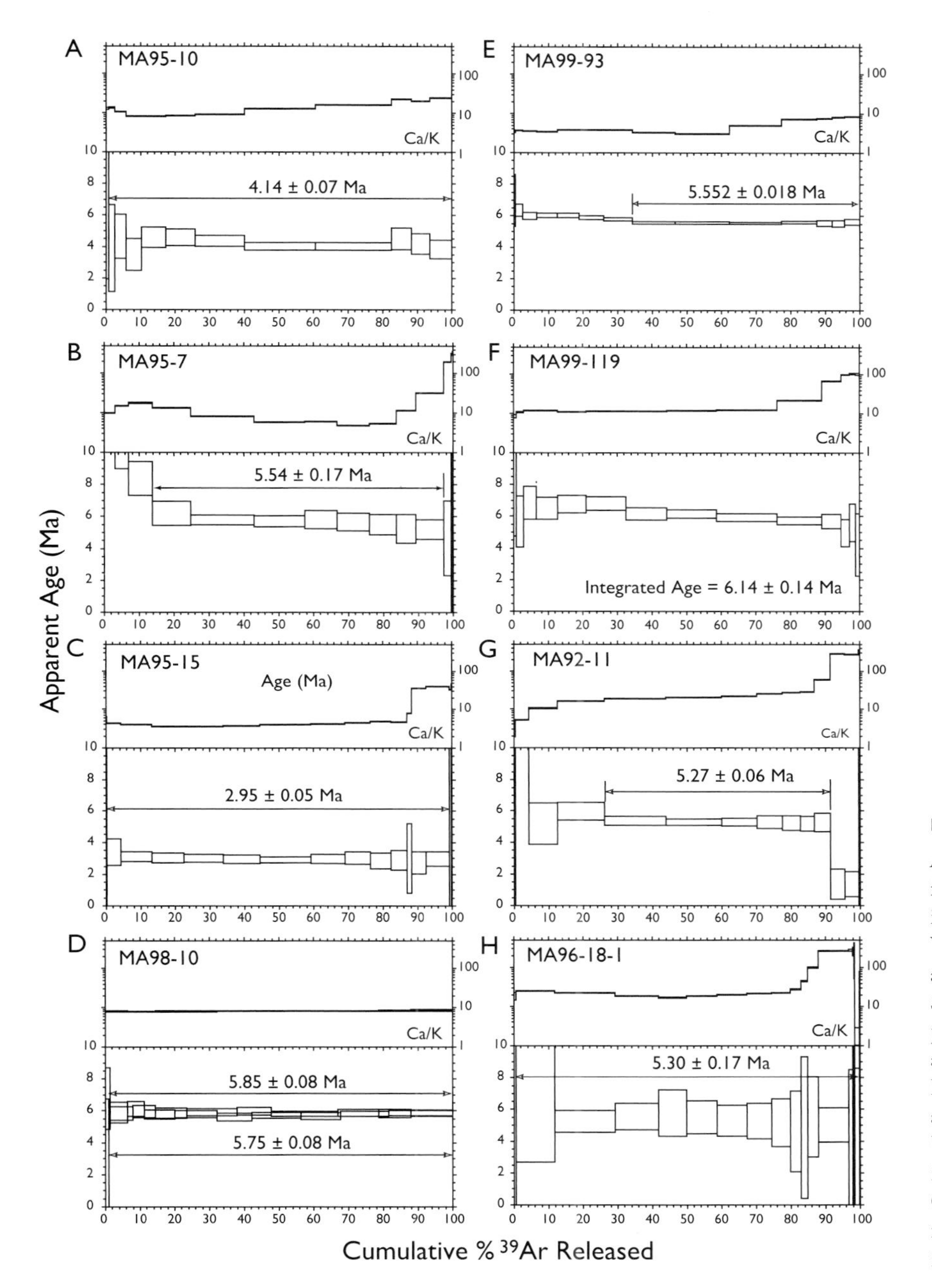

FIGURE 4.2.

Apparent-age spectra for samples analyzed by incremental heating. Vertical heights of boxes are 2σ analytical errors. Steps defining age plateaux, where relevant, are indicated by arrows. Plateau and integrated ages are given with 1σ errors. Upper panel for each indicated sample shows Ca/K ratio determined from corrected $^{37}Ar/^{39}Ar$ isotope data.

5.75 ± 0.08 Ma (Figure 4.2D). The weighted mean of the two plateau ages is 5.80 ± 0.06 Ma.

MA98-11 Latitude: 10°16.58′N, longitude: 40°17.12′E (Alayla). At this locality the unit sampled is a ~1 m thick white crystal vitric tuff, partially altered to bentonite. Alkali feldspars were separated from the tuff and analyzed by single-crystal laser fusion. Six crystals yielded ages ranging from 14.21 ± 0.68 to 23.14 ± 0.84 Ma (Figure 4.1C). The latter age, shared by two other crystals, is a common age for xenocrysts in many tuffs of the CAC (Renne et al. 1999). The younger ages are far older than permitted by ages from underlying units (i.e., the LABT = MA97-15 of WoldeGabriel et al. 2001 and MA98-10 reported herein), both dated at ca. 5.8 Ma. Thus, we consider all the crystals dated from this tuff to be xenocrysts, and no age can be determined from our data.

MA99-93 Latitude: 10°15.754′N, longitude: 40°17.284′E (South Alayla). At this locality the unit sampled is a spheroidally weathered, fine-grained basalt flow ≥4 m thick. A whole-rock sample was dated by incremental heating in 15 steps with an Ar-ion laser. The age spectrum (Figure 4.2E) is mildly discordant, with the first ~34 percent of ^{39}Ar released yielding anomalously old ages. This pattern of discordance is suggestive of ^{39}Ar recoil artifacts, but the remaining 66 percent of the gas (seven steps) defines a well-developed plateau lacking evidence of disturbance. Accordingly, the plateau age for this segment (5.55 ± 0.02 Ma) is interpreted to date the eruption of this basalt accurately. The similarity in age and stratigraphic position of this sample to basalt sample MA95-7 at Saitune Dora suggest that these two basalts may be correlative (see WoldeGabriel et al., Chapter 2; Hart et al., Chapter 3).

MA99-119 Latitude: 10°31.130′N, longitude: 40°14.944′E (Jara-Borkana). At this locality the unit sampled is a 4 m thick basalt flow capping the section. A whole-rock sample was dated by incremental heating in 15 steps with an Ar-ion laser. The resulting age spectrum (Figure 4.2F) is discordant, with initially high apparent ages decreasing monotonically after about 30 percent of the ^{39}Ar was released. This spectrum is typical of those resulting from ^{39}Ar recoil redistribution but betrays no evidence of ^{40}Ar loss due to alteration. In such cases the integrated age is likely to be meaningful, and we interpret the integrated age (6.14 ± 0.14 Ma) of this sample as being the best estimate of its eruption age.

Central Awash Complex

The following samples of basalts from the Gawto Member of the Sagantole Formation, exposed in the CAC, have been dated with the results indicated. The eruption coincides with the waning stage of the late Miocene basaltic eruption of the Adu-Asa Formation.

MA92-11 This basalt flow was sampled at Amba West, near the western limit of the CAC. At this locality the unit sampled is an aphyric, spheroidally weathered basalt that is altered at the base. A whole-rock sample was dated by incremental heating in 15 steps between 540 and 1500 °C in an electrical resistance furnace. The low-temperature steps (less than 25 percent of the total ^{39}Ar released) yield anomalously old ages, while the highest-temperature steps (25 to 91 percent of the total ^{39}Ar released) yield anomalously young ages (Figure 4.2G). This pattern of discordance is suggestive of ^{39}Ar and ^{37}Ar recoil artifacts, including possible implantation of ^{39}Ar into refractory calcic minerals such as clinopyroxene. In any case, the intermediate temperature steps yield a well-defined plateau, whose age of 5.27 ± 0.06 Ma is likely to represent the age of eruption.

MA96-18 Latitude: 10°26.264′N, longitude: 40°25.212′E (Gulubashin). This unit, a coarse-grained basalt flow with plagioclase phenocrysts, was sampled at Gulubashin Dora along the Kuseralee drainage. A whole-rock basalt was dated by incremental heating in 21 steps between 600 and 1,500 °C in an electrical resistance furnace. The age spectrum (Figure 4.2H) is completely concordant and yields a plateau over 100 percent of the ^{39}Ar released. The plateau age of 5.30 ± 0.17 Ma is interpreted as the eruption age.

Both samples MA92-11 and MA96-18 yielded ages that are indistinguishable from the age (5.186 ± 0.083 Ma) reported by Renne et al. (1999) for another sample (MA96-16) of Gawto basalt. A fourth sample (MA96-17) of Gawto basalt was also analyzed but unfortunately yielded insufficient gas to analyze because of a leaking valve in the extraction line. As other samples of Gawto basalt were dated satisfactorily, it was deemed unnecessary to perform additional analyses after repairing the problem.

Conclusions

The western margin of the Middle Awash study area comprises a tectonically complex assemblage of Miocene, Pliocene, and Pleistocene deposits, many of which contain vertebrate remains. This chapter reports results on rocks sampled within and directly above the late Miocene sediments and establishes a minimum age of >6 Ma for the oldest fossiliferous sediments in the Middle Awash. The results demonstrate that most of the basaltic volcanism recorded in the Adu-Asa Formation occurred at 5–6 Ma, thereby encompassing the ca. 5.2 Ma age of the Gawto basalts of the Sagantole Formation. It is therefore appropriate to view the two formations as temporally conformable, with the bulk of the Adu-Asa Formation, particularly its fossiliferous horizons, being both older and closer to the majority of fissural sources for the basalts and farther from the lacustrine facies represented episodically in the Sagantole Formation.

5

Small Mammals

HENRY B. WESSELMAN, MICHAEL T. BLACK, AND MESFIN ASNAKE

The last quarter-century of paleoanthropological research in Africa has greatly extended our knowledge of the earliest stages of human evolution, especially with relation to the genus *Australopithecus* and its apparent predecessors, *Ardipithecus ramidus* and *Ardipithecus kadabba,* recovered from the Middle Awash deposits (White et al. 1994, 1995; Haile-Selassie 2004b; White et al. 2006b).

Over the past decade, the search for fossilized micromammals from the hominid-bearing localities in the Middle Awash Valley has been a high priority because of their acknowledged utility in evaluating the paleoecological character of paleontological sites (e.g., Wesselman 1984; Denys and Jaeger 1986; Wesselman 1995; Denys 1999; Legendre et al. 2005). The utilization of small mammals as paleoenvironmental indicators rests on the principle of actualism, which states that the ecological requirements of a fossil taxon can be expected to be similar to those of the contemporary species with whom they share closest relationships. This principle must be qualified, however, in that the ecological tolerances of any contemporary taxon might be limited by factors that did not exist in the past. Similarly, some paleospecies may have been constrained by factors that do not exist in the present.

Many contemporary small mammals are strongly habitat-specific and live comparatively short lives within relatively small home ranges, compared to larger-bodied mammals. Coe (1972), for example, has shown that the micromammals living in the Lake Turkana Basin today are not dispersed across a wide spectrum of environments; instead, they are restricted to distinct areas of preferred habitats. Similar studies of small-bodied mammals across Africa are generating new ideas and theories about how individuals, communities, and species are integrated with their environment (Happold 2001).

The geology and geochronology of the fossiliferous localities of the Adu-Asa Formation mentioned in this chapter are covered elsewhere in this volume, as well as in Haile-Selassie et al. (2004c) and WoldeGabriel et al. (2001). Emphasis will be placed here upon the systematics and paleoecology of the small mammals recovered at these localities. All the fossil micromammal specimens from the Middle Awash discussed in this chapter are housed in the collections of the National Museum of Ethiopia, Addis Ababa; see Table 1.3 for a faunal list covering the taxa discussed. We have slightly expanded the temporal scope of this chapter in order to include important new fossils from the Haradaso Member of the

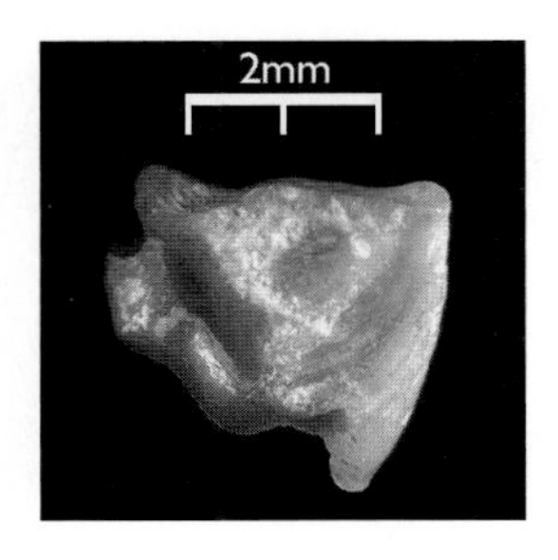

FIGURE 5.1
Procavia sp. indet. ASK-VP-3/285, left P^3 or P^4 fragment, occlusal view.

Sagantole Formation belonging to *Alilepus, Tachyoryctes,* and *Thryonomys*. These remains, dated to approximately 4.85 Ma (Renne et al. 1999), are important additions to earliest Pliocene small mammal faunas and are therefore most appropriately included here.

Hyracoidea

Procaviidae

Procavia Storr, 1780

Procavia sp.

DESCRIPTION This upper premolar fragment preserves only the distobuccal corner of the tooth crown, the metacone, which reveals the lophodont morphology typical of contemporary hyraxes (Figure 5.1). No buccal cingulum is apparent, a feature typical of *Procavia.*

DISCUSSION The late Miocene fossil hyracoids have been reviewed by Pickford (2005), including a procaviid palate with much of the dentition preserved, recovered from the Lukeino Formation and dated to about 6 Ma. As large as the western tree hyrax, *Dendrohyrax dorsalis,* the Lukeino form has been referred to as a new species: *D. samueli* Pickford, 2005.

Pickford (2005) has also described a new species of *Procavia* (*P. pliocenica*) from the early Pliocene at Langebaanweg that reveals, among other characters, the smooth ectoloph typical of *Procavia,* with weakly expressed or absent cingula on the upper molars. *Procavia pliocenica* is thought by Pickford to provide evidence of a link between the extant *P. capensis* and the genus *Heterohyrax*. The earliest known procaviids are from Nakali, Kenya (Fischer 1986), and Berg Aukas, Namibia (Rasmussen et al. 1996), and date to about 10 Ma, suggesting that this group has experienced slow evolution since that time.

In size and morphology, the Middle Awash tooth fragment ASK-VP-3/285 resembles most closely the extant species *Procavia capensis habessinica* from the Shoa province, Ethiopia, a form that is larger than both *Heterohyrax* or *Dendrohyrax*. The Asa Koma fossil is provisionally assigned to *Procavia* sp., pending recovery of more complete material.

Extant *Procavia* is typically found in drier areas of eastern Africa, extending north into Ethiopia, Sudan, Egypt, and Israel, living in rocky outcrops with trees and thickets. Its range extends up into the alpine zone of Mt. Kenya, reaching an altitude of 3,500–5,500 m (Kingdon 1971, 1997).

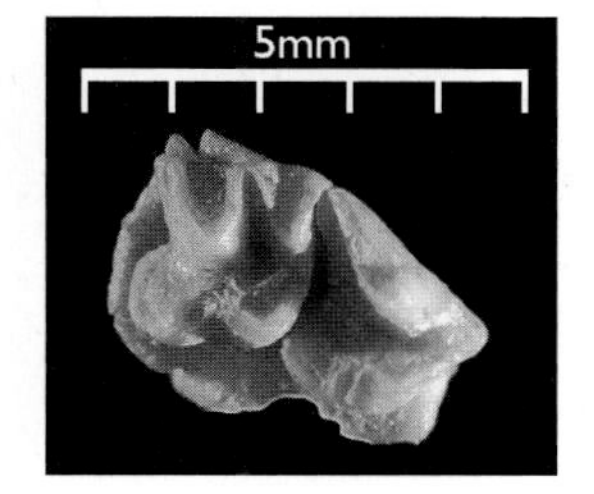

FIGURE 5.2
Crocidura aff. *aithiops*. ASK-VP-3/304, maxillary fragment with right P^4 and M^1, occlusal view.

Insectivora

Soricidae

Crocidura Wagler, 1832

Crocidura aff. *aithiops* Wesselman, 1984

DESCRIPTION The elevated P^4 (Figure 5.2) expresses the typical carnassial shearing blade, incorporating both paracone and metacone along the buccal crown margin. The

TABLE 5.1 Dental Measurements of *Crocidura*

Taxon	*Specimen*	*Tooth*	*Mandibular* *Length*	*Breadth*	*Depth at* M_2	*n*
Crocidura aff. *aithiops*	ASK-VP-3/304	P^4	2.58	2.10	nm	1
		M^1	2.05 (broken)	2.49	nm	1
Crocidura aithiops (Central Awash Complex)	Various	P^4	1.98–2.43	1.83–2.10	nm	7
		M^1	1.83–2.28	2.46–2.74	nm	10
Crocidura olivieri doriana	Various (coll. NHM)	P^4	2.86–3.23	2.46–2.94	nm	10
		M^1	1.94–2.25	2.94–3.10	nm	10
Crocidura aff. *dolichura*	ASK-VP-3/305	M_2	1.84	1.34	2.15	1
	Omo Loc. 1-617	M_2	1.45	1.05	1.60	1

NOTE: Measurements are in mm; nm = measurement was not possible.

protocone and hypocone form distinct cusps, creating a lingual half of the tooth that is expanded distally by a broad hypoconal basin and mesially by a strong anterior cingulum that runs from the protocone to a small accessory cusp anterior to the paracone.

The M^1 expresses the typical insectivorous lophodont morphology dominated by the crested arcs of the three major cusps (Figure 5.2). The paracone and metacone achieve highest elevation on the buccal side and have a deep lingual invagination between them. The hypocone sweeps posteroinferiorly from the protocone and, with the strong posterior cingulum that delineates the crown margin, forms a wide distolingual basin.

DISCUSSION The crocidurine shrews appear in Europe in the Pliocene and presumably evolved in Africa, but the Miocene record for Africa is virtually empty, causing some to wonder whether the shrew jaw recovered from Rusinga is really Miocene in age (P. M. Butler, personal communications). In South Africa, *Crocidura* does not show up until the early Pleistocene at Bolt's Farm. Therefore, the shrews from the Miocene Middle Awash deposits are of particular interest.

ASK-VP-3/304 is a large crocidurine shrew comparable in both morphology and size to the paleospecies *Crocidura aithiops* (Table 5.1), originally described from Omo Members B and F (Wesselman 1984) and also known from the Aramis, KUS-VP-2, and Asa Issie sites of the Central Awash Complex (CAC) (Wesselman and Black, unpublished data). This taxon is clearly allied with the cluster of contemporary African crocidurine species that includes the large *Crocidura olivieri doriana* from Ethiopia (formerly *C. flavescens doriana*), a relationship explored by Wesselman (1984).

Comparable in size to *Crocidura aithiops* from the CAC sites (Table 5.1), the ASK-VP-3/304 specimen is distinguishable from both *C. aithiops* and *C. olivieri* in exhibiting a considerably expanded hypocone on both the P^4 and M^1, thereby producing an enlarged basin on the distolingual sides of both teeth. In addition, while the carnassial blade of the P^4 is comparable in its breadth to *C. aithiops,* it is somewhat longer than that of *C. aithiops,* being accentuated by an enlarged accessory cusp on the mesial crown margin.

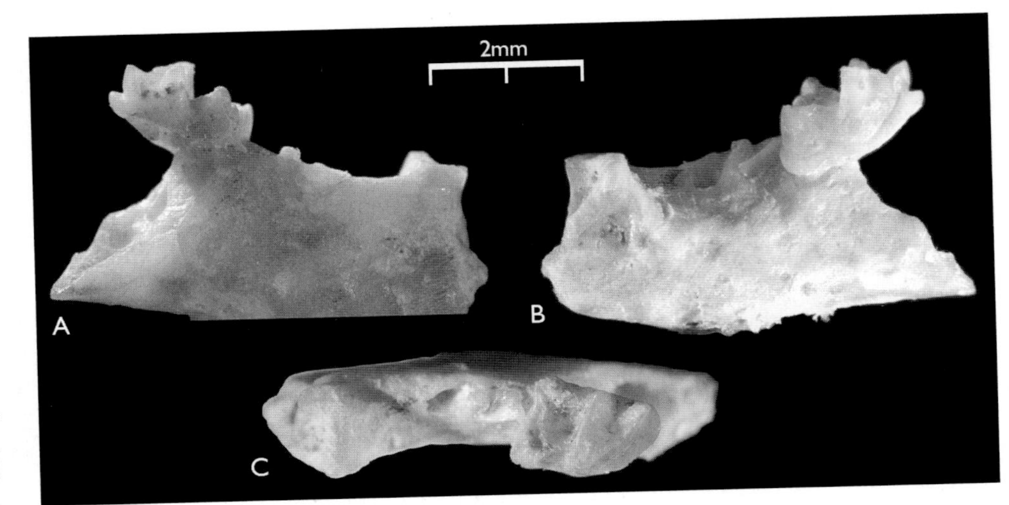

FIGURE 5.3
Crocidura aff. *dolichura*. ASK-VP-3/305, right mandible with M_2. A. Buccal view. B. Lingual view. C. Occlusal view.

ASK-VP-3/304 most likely represents a form ancestral to *Crocidura aithiops* and is referred to as *Crocidura* aff. *aithiops,* pending recovery of more complete specimens.

The contemporary large shrews in the *Crocidura olivieri/flavenscens/occidentalis* group are widely spread across a large portion of Africa—from Senegal to Egypt to South Africa—including Ethiopia, and they live in an abundance of habitats in places of sufficient moisture and cover, from mesic savannas to both highland and lowland forests (Heim de Balsac and Barloy 1966; Heim de Balsac and Meester 1977; Kingdon 1974, 1997).

Crocidura aff. *dolichura* Peters, 1876

DESCRIPTION The M_2 (Figure 5.3) displays the insectivorous dentition typical of crocidurine shrews, with the protoconid connected by shearing crests to both the paraconid and metaconid, and the resulting basin between them occupied by a lingual invagination. The distal half of the tooth is composed of the hypoconid buccally and the entoconid lingually, with both cusps joined by the lingual crest of the hypoconid. This crest runs straight across the distal margin of the tooth to merge with the entoconid and the posterolingual rib, a small knob of enamel on the posterolingual crown margin. A wide post-entoconid ledge interrupts the junction of the lingual crest and the entoconid. Mesially directed crests from both the hypoconid and entoconid form the buccal and lingual walls of a deep talonid basin. The buccal crown margin is delineated by a gracile cingulum that rises slightly between the bases of the protoconid and the hypoconid. A small lingual cingulum is confined to the base of the paraconid.

DISCUSSION This medium-sized shrew is morphologically comparable to the contemporary species *Crocidura dolichura,* which also includes as subspecies *C. d. muricauda* and *C. d. ludia* (Heim de Balsac and Meester 1977). This taxon was recovered from Omo Member B and was described as *Crocidura* aff. *dolichura*. ASK-VP-3/305 is larger than the Omo specimen (Omo Loc 1-617; Table 5.1), a possible indicator of higher altitude. Pending the recovery of more complete fossils, it is provisionally assigned to *Crocidura* aff. *dolichura*.

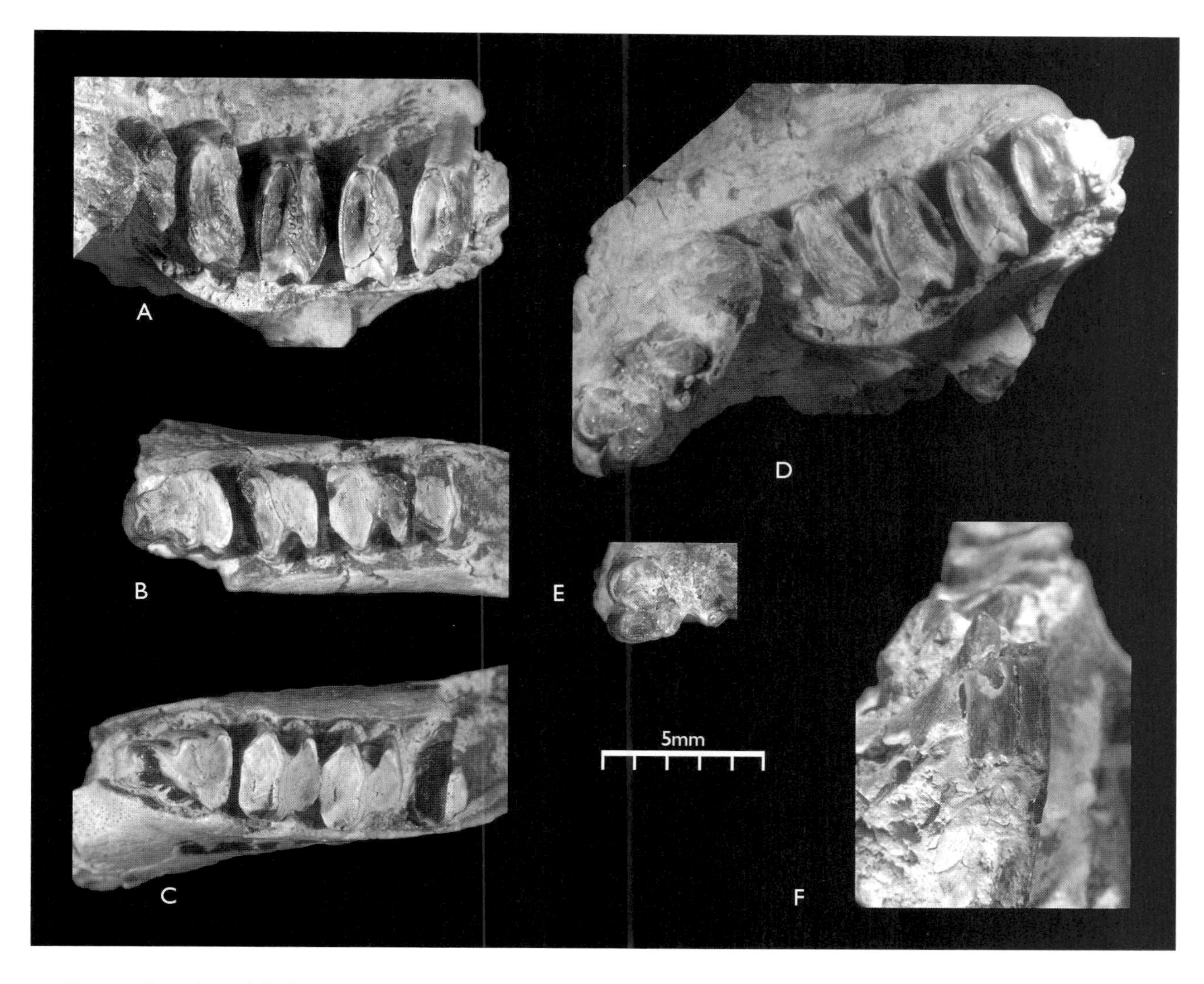

FIGURE 5.4

Alilepus sp. indet. WKH-VP-1/61. A. Occlusal view of left maxilla. B. Occlusal view of left mandible. C. Occlusal view of right mandible. D. Occlusal view of right maxilla. E. Occlusal view of upper incisors. F. Labial view of upper incisors.

Extant *Crocidura dolichura* is a species typical of the central African high forest block (Heim de Balsac and Meester 1977), and its presence in the deposits of the western margin of the Middle Awash study area argues strongly for the existence of such biotypes near the site(s) of deposition at 5.7 Ma.

Lagomorpha

Leporidae

Alilepus Dice, 1931

[*Alilepus* sp.]

DESCRIPTION The assignment of the 4.85 Ma Worku Hassan fossil (Figure 5.4; Table 5.2) to the genus *Alilepus* is permitted by the morphology of the P_3s, which, following

TABLE 5.2 Measurements of *Alilepus* sp. indet.

Taxon	*Specimen*	*Tooth*	*Length*	*Breadth*	*n*
Alilepus sp. indet.	WKH-VP-1/61	I^1	2.09	2.12	1
		P^2	1.33	3.06	1
		P^3	1.95	4.07	1
		P^4	2.12	4.00	1
		M^1	1.95	3.90	1
		M^2	1.83	3.45	1
		P_3	3.00	2.87	1
		P_4	2.63	2.97	1
		M_1	2.67	3.85	1
		M_2	nm	2.95	1
Alilepus sp. indet.[a]	Lothagam, lower Nawata (6.54 Ma)	P_3	3.33–4.00	2.92–3.42	3

NOTE: Measurements are in nm; nm = measurement was not possible.
[a]After Winkler (2003).

White (1991) and Winkler (2003), demonstrate the following characters (illustrated in Figure 5.5): the posterior internal reentrant enamel fold (PIR) extends across the occlusal surface of the crown to the midline, as does the posterior external reentrant enamel fold (PER). A narrow isthmus between them connects the trigonid to the talonid. In addition, there is a small anterior reentrant enamel fold (AR) that is absent on the Lothagam specimens, and there is a shallow anterior internal reentrant enamel fold (AIR) with smooth, thin enamel. The anterior external reentrant enamel fold (AER) is shallow compared to the PER and the PIR. The enamel is thick on the anterior edges of the PIR and PER and thin on the posterior edges.

DISCUSSION The right maxilla fragment preserves an I^1 that, in occlusal outline, resembles those of the extant *Lepus starki* Petter, 1963, an endemic form found in the central highlands of Ethiopia. The upper incisors of *L. starki* have a deep anterior groove that is not cement-filled, so the cutting edge of the tooth appears "looped"—a character that immediately separates *L. starki* from all other forms of *Lepus*.

The genus *Alilepus* is known from the late Miocene and Pliocene of North America and Eurasia and was revised by White (1991), who described its fully modern cranium and dentition. Winkler (2003) mentions an undescribed isolated leporid P_3 from the Lukeino Formation, Tugen Hills, Kenya, dated between 5.6 and 6.2 Ma (Hill 1999), which may be assignable to *Alilepus* (Table 5.2).

Winkler (2003, referring to Averianov, personal communication) discusses other fossil genera, such as *Trischizolagus* and *Serengetilagus,* that are known to have a similar P_3 morphology, which may indicate a close taxonomic relationship between the three fossil leporids. The relationship of the endemic African leporids *Bunolagus* and *Pronolagus* to *Alilepus* is currently unknown. Although the Worku Hassan fossil is assigned to *Alilepus* at this time, the relationship of this specimen to *Lepus starki* remains to be resolved.

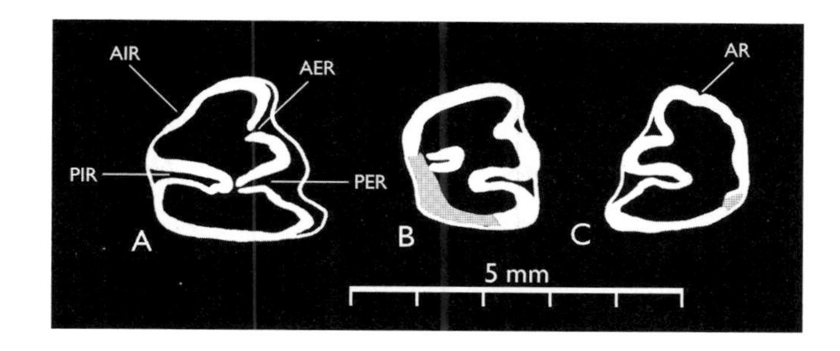

FIGURE 5.5
Leporid P_3 enamel topography terminology, after Winkler (2003: 171). AER: anterior external reentrant enamel fold; AIR: anterior internal reentrant enamel fold; AR: anterior reentrant enamel fold; PER: posterior external reentrant enamel fold; PIR: posterior internal reentrant enamel fold. A. KNM-LT 22999, *Alilepus* sp. from the Lower Member of the Nawata Formation. B. KNM-LT 23179, *Alilepus* sp. from the Lower Member of the Nawata Formation. C. KNM-LT 24963, *Serengetilagus praecapensis* from the Apak Member, Nachukui Formation.

Extant leporids in Africa favor arid grasslands and waterless plains and are also common on seasonally dry floodplain grasslands and in dry acacia scrub, especially around the fringes of clay pans (Kingdon 1974, 1997). Their presence in fossil assemblages is usually considered as a reliable indicator of aridity, and they are not typically found in woodlands or forests. *Lepus starki* is endemic to higher altitudes and is found in grasslands and moorlands of the Ethiopian plateau at 2,140–4,000 m (Yalden and Largen 1992), so this fossil taxon's presence in western margin sediments may be indicative of both altitude and a systematic relationship with *L. starki*.

Serengetilagus Dietrich, 1941

Serengetilagus praecapensis Dietrich, 1941

DESCRIPTION The attribution of the late Miocene leporid fossils from the western margin (Figure 5.6; Table 5.3) to this species is determined by the morphology of the three P_3s recovered, all of which demonstrate the morphology typical of *Serengetilagus praecapensis*. Distinguishing features include the presence of a shallow AER (see preceding discussion of *Alilepus* for these abbreviations) and a PER that crosses about one-half of the occlusal surface. The Middle Awash fossils also have a small AR on the mesial crown margin, and STD-VP-2/911 shows a small AIR. These characters are in alignment with those described for *Serengetilagus* from Lothagam, a single specimen recovered from the Apak Member of the Nachukui Formation and dated to older than 4.22 Ma (Winkler 2003, following White 1991).

DISCUSSION *Serengetilagus praecapensis* was originally described by Dietrich (1941, 1942a) with specimens collected at Laetoli by Kohl-Larsen in 1938–1939 and housed at the Museum of Natural History at Humbolt University in Berlin. Davies (1987) has also described additional remains of this taxon that were recovered from the northeastern end of Lake Eyasi and are now in the collections of the Natural History Museum (London). The upper teeth of *S. praecapensis* recovered from the Middle Awash exceed those from both Lothagam and Laetoli in their buccolingual breadths (Table 5.3). The rest of the upper and lower cheek teeth show the mesiodistally compressed, hypsodont columnar morphology typical of the leporids. The presence of this taxon in the western margin greatly extends its geographical range northward (from Tanzania through Kenya and into Ethiopia) and extends its temporal range into the late Miocene.

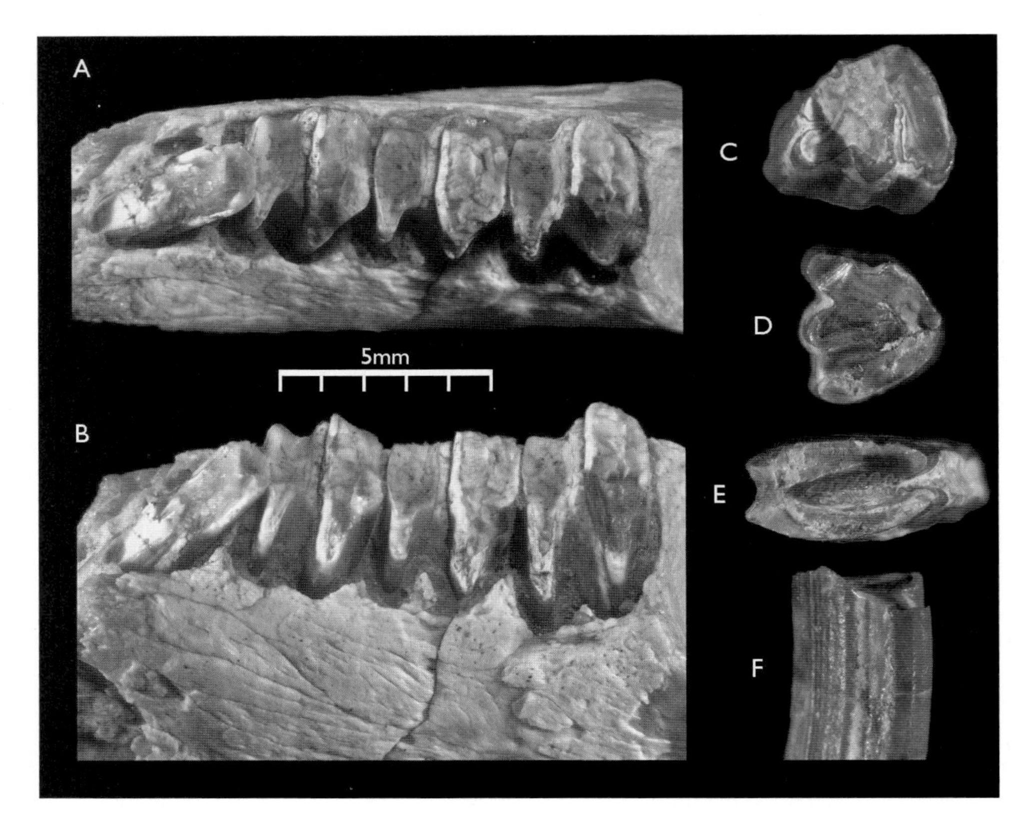

FIGURE 5.6

Serengetilagus praecapensis. A. BIK-VP-1/5, occlusal view. B. BIK-VP-1/5, buccal view. C. STD-VP-2/911, occlusal view. D. ALA-VP-2/337, occlusal view. E. ALA-VP-2/176, occlusal view. F. ALA-VP-2/337, mesial view.

Rodentia

Hystricidae

Three porcupine genera are represented in the late Miocene deposits of the Middle Awash: *Xenohystrix, Atherurus,* and *Hystrix*. The remains of the large *Xenohystrix* (previously known only from younger deposits at Langebaanweg, Makapansgat, Omo, Laetoli, and Hadar) and those of *Atherurus* are among the earliest records of these genera. *Hystrix* has also been documented at contemporaneous sites such as Lothagam (Leakey et al. 1996; Winkler 2003) and Lukeino (Hill et al. 1986; Winkler 2002), and it is ubiquitous at most African Plio-Pleistocene sites. All three of these genera have also been recovered at the late Miocene site of Lemudong'o, Kenya (Hlusko 2007b).

Atherurus Cuvier, 1829

Atherurus garbo sp. nov.

ETYMOLOGY In the Afar language, *garbo* is a word that means "forest" or "woodlands."

HOLOTYPE ALA-VP-2/172, right mandible fragment with P_4–M_1 (see Figure 5.7).

TABLE 5.3 Dental Measurements of *Serengetilagus praecapensis*

Taxon	*Specimen*	*Tooth*	*Length*	*Breadth*	*n*
Serengetilagus praecapensis[a]	ASK-VP-3/309	P^2	1.75	3.78	1
	Various	P^3	2.45–2.64	4.62–4.78	3
	Various	M^1	2.03–2.41	3.65–3.90	2
	Various	M^2	2.10–3.07	3.76–4.80	3
	Various	P_3	3.27–3.86	2.97–3.25	3
	Various	P_4	2.87–3.03	3.45–3.80	2
	Various	M_1	3.30–3.63	3.36–4.22	3
	BIK-VP-1/5	M_2	2.90	3.40	1
	STD-VP-2/858	M_3	3.60	3.78	1
Serengetilagus praecapensis	Lothagam[b]	P^3	3.14	2.57	1
	4.4 Ma	P^4	2.57	3.28	1
		M^1	2.71	3.28	1
		M^2	2.71	3.14	1
		M^3	1.71	1.71	1
	Laetoli[b]	P^3	2.08–3.83	2.71–3.50	9
	(KNM coll.)	P^4	2.57–3.00	3.00–3.58	9
		M^1	2.71–3.00	3.00–3.17	8
		M^2	2.71–2.92	2.86–2.92	3

NOTE: Measurements are in mm.

[a]In addition, two very small upper cheek teeth recovered from the Middle Awash may represent immature individuals (length 1.59, breadth 2.50).

[b]Measurements of specimens from Lothagam and Laetoli are after Winkler (2003).

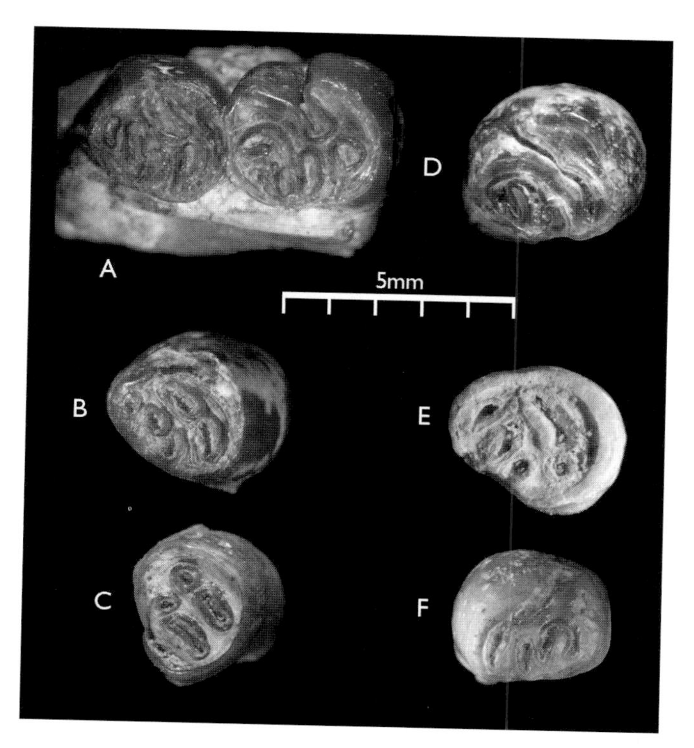

FIGURE 5.7

Atherurus garbo sp. nov. A. ALA-VP-2/172, occlusal view. B. ALA-VP-1/32, occlusal view. C. ALA-VP-2/184, occlusal view. D. ALA-VP-2/336, occlusal view. E. ALA-VP-3/259, occlusal view. F. ALA-VP-3/268, occlusal view.

TABLE 5.4 Dental Measurements of *Atherurus garbo* sp. nov.

Taxon	*Specimen*	*Tooth*	*Length*	*Breadth*	*n*
Atherurus garbo sp. nov.	Various	P^4	3.03–3.37	2.63–3.04	2
	Various	M^1	3.44–3.90	3.50	2
	Various	M^3	3.01–3.49	2.42–3.22	4
	ALA-VP-2/172	P_4	3.15	2.89	1
		M_1	3.63	2.96	1
	ALA-VP-2/175	M_2	nm	3.41	1
Atherurus africanus	Various	P^4	4.93–4.35	3.93–4.51	10
	Various	M^1	4.10–5.97	4.19–5.27	10
	Various	M^3	3.61–4.60	3.69–4.47	10
	Various	P_4	4.73–5.63	3.63–4.13	10
	Various	M_1	4.65–5.24	3.70–5.27	10
	Various	M_2	4.40–5.33	4.01–4.67	10
	Various	M_3	4.00–4.55	3.41–4.30	10

NOTE: Measurements are in mm; nm = measurement was not possible.

PARATYPES ALA-VP-1/32, left P^4; ALA-VP-2/175, left M_2 fragment; ALA-VP-2/184, right P^4; ALA-VP-2/336, left M^3; ASK-VP-3/259, left M^1; ASK-VP-3/268, right M^3; ASK-VP-3/283, right M^3; ASK-VP-3/274, left M^3; ASK-VP-3/262, left M^1 fragment (Figure 5.7).

LOCALITIES AND HORIZONS The late Miocene western margin sites of Alayla Vertebrate Localities 1 and 2 (ALA-VP-1 and ALA-VP-2) and Asa Koma Vertebrate Locality 3 (ASK-VP-3), all dated to 5.7 Ma.

DIAGNOSIS Distinguished by its size, *Atherurus garbo* is 30 percent smaller than the contemporary *Atherurus africanus* (Table 5.4). Unlike the contemporary species, the anterior half of the crown of the P_4 is narrowed and foreshortened, and the P_4 is smaller than the M_1, a tooth that is mesiodistally elongated and rectangular in occlusal outline. The anterior margin of the P_4 of *A. africanus* tends to be inflated and rounded and the M_1 is buccolingually wider, producing a tooth that is almost square in its occlusal outline.

DISCUSSION Nine isolated cheek teeth and one mandibular fragment with P_4–M_1 were recovered from the western margin deposits of the Middle Awash and are assigned to the genus *Atherurus,* the brush-tailed porcupine. The teeth of the contemporary hystricids *Atherurus africanus* and *Hystrix cristata* are similar, with a generalized brachydont/hypsodont crown morphology pierced by reentrant enamel folds (or sinuses/sinusids) and exhibiting enclosed enamel pillars or "islands." The occlusal crown patterns can vary considerably between individuals of both taxa, some being more complex, others simple. The crown morphology of both taxa also varies considerably with the state of wear. The teeth of *H. cristata* are twice the size of those of *A. africanus* and are clearly more inflated, aiding in their identification.

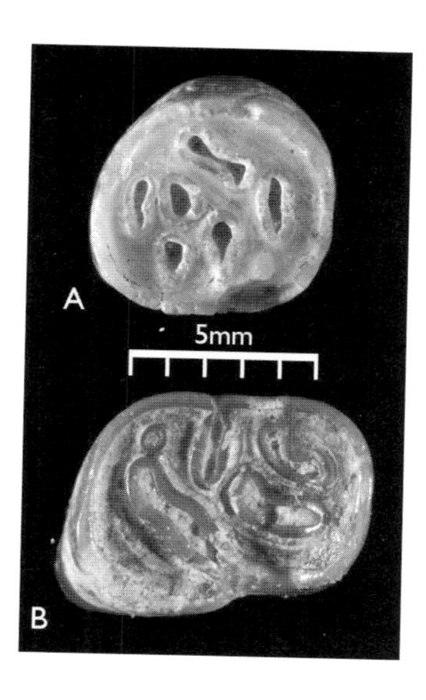

FIGURE 5.8
Hystrix sp. A. ASK-VP-1/20, occlusal view. B. DID-VP-1/102, occlusal view.

The *Atherurus* fossils from the western margin are considerably smaller than modern *Atherurus,* and they are half the size of an as-yet-undescribed *Atherurus* species from the Pliocene CAC sites at Aramis (Wesselman and Black, unpublished data). *Atherurus garbo* sp. nov. represents a species distinct from both, and its remains are among the earliest known fossil records for the genus.

Atherurus karnuliensis Lydekker, 1886, is known from the Pleistocene of India, China, and Vietnam (Van Weers 2002), indicating the extensive geographic range of this genus. The contemporary *Atherurus africanus* is a true forest-living species, found not only in forests but also in narrow woodland galleries along rivers and streams, from sea level up to 2,300 m. *Atherurus africanus* ranges from Gambia in the west through the Congo basin and into western Uganda. It is also found living in forest patches in southern Sudan and western Kenya (Kingdon 1974, 1997). The brush-tailed porcupine takes shelter in animal burrows, in termitaria, and in caves and cavities in riverbanks and is considered here to be a strong indicator of the presence of forested areas close to the sites of deposition in the late Miocene.

Hystrix Linnaeus, 1758

[*Hystrix* sp.]

DESCRIPTION The crown surface of ASK-VP-1/20 (Figure 5.8A) is heavily worn, and the reentrant lingual enamel fold (or sinus) is represented by an isolated island of enamel among a scatter of others. It has a large lingual root as well as the remains of two broken buccal roots. The mesial crown margin of DID-VP-1/102 (Figure 5.8B) is extended forward and narrows buccolingually, relative to the distal half of the crown. The crown is not worn enough to show complete islands of the buccal and lingual reentrant enamel folds, termed sinusids by Denys (1987). The mesial enamel island and the first lophid are parallel to each other, equal in size, and crescent-shaped. The posterior enamel island is

TABLE 5.5 Dental Measurements of *Hystrix*

Taxon	*Specimen*	*Tooth*	*Length*	*Breadth*	*n*
Hystrix sp. indet.	ASK-VP-1/20	L.P^4	6.32	5.85	1
	DID-VP-1/102	R.P_4	7.82	6.18	1
Hystrix cristata	Various	P^4	7.90–8.40	6.75–6.97	10
		P_4	8.99–9.87	5.83–6.67	10

NOTE: Measurements are in mm.

very small, but, as mentioned, the crown morphology of hystricids can vary considerably between individuals and with increasing wear.

DISCUSSION Both fossil specimens are small compared to most known fossil and extant *Hystrix* (Table 5.5). DID-VP-1/102 is slightly smaller than KNM-LT 24948, an isolated right P_4 from the Lower Nawata Member of the Lothagam Formation (Winkler 2003) and is closer in length to *Hystrix leakeyi* from the Laetolil Beds (Denys 1987), the smallest of the known *Hystrix* species. They are also smaller than some as-yet-undescribed *Hystrix* specimens from the CAC sites of Aramis Vertebrate Locality 1 and Kuseralee Vertebrate Locality 2, both dated to 4.4 Ma (Wesselman and Black, unpublished data).

The occlusal crown morphology of DID-VP-1/102 resembles that of LAET 75-1971, a specimen referred to *Hystrix makapanensis. Hystrix makapanensis,* however, is much larger than both the Lothagam and Middle Awash specimens. *Hystrix makapanensis* is known from Makapansgat, Laetoli, Olduvai, and Omo (Sabatier 1978; Wesselman 1984, 1995). *Hystrix cristata* is known from the Omo Beds from the Usno Formation and Members B, E, F, G, and L of the Shungura Formation. The molar crown morphology of *H. cristata* is identical to that of the contemporary species, attesting to the stability of this lineage during the Plio-Pleistocene (Wesselman 1984).

Given the diversity and intraspecific variation of known extant and extinct *Hystrix* species, it is possible to assign the Middle Awash specimen only to *Hystrix* sp. indet. until more specimens are recovered.

The relationship between the fossil and extant forms of this genus is currently unclear due to the fragmentary nature of its earlier fossil record. It is possible, however, that the Lothagam and Middle Awash *Hystrix* specimens are conspecific with specimens recovered from Lukeino (Winkler 2002) and that they represent an ancestral species which gave rise to the Plio-Pleistocene African *Hystrix*. The relationships of the large *Xenohystrix* and the small *Atherurus* to the other genera of the family have not been clarified.

Xenohystrix Greenwood, 1955

[*Xenohystrix* sp.]

DESCRIPTION ADD-VP-1/16 is an edentulous left mandible with most of the incisor preserved (Figure 5.9). The alveoli for the cheek teeth indicate that each molar had

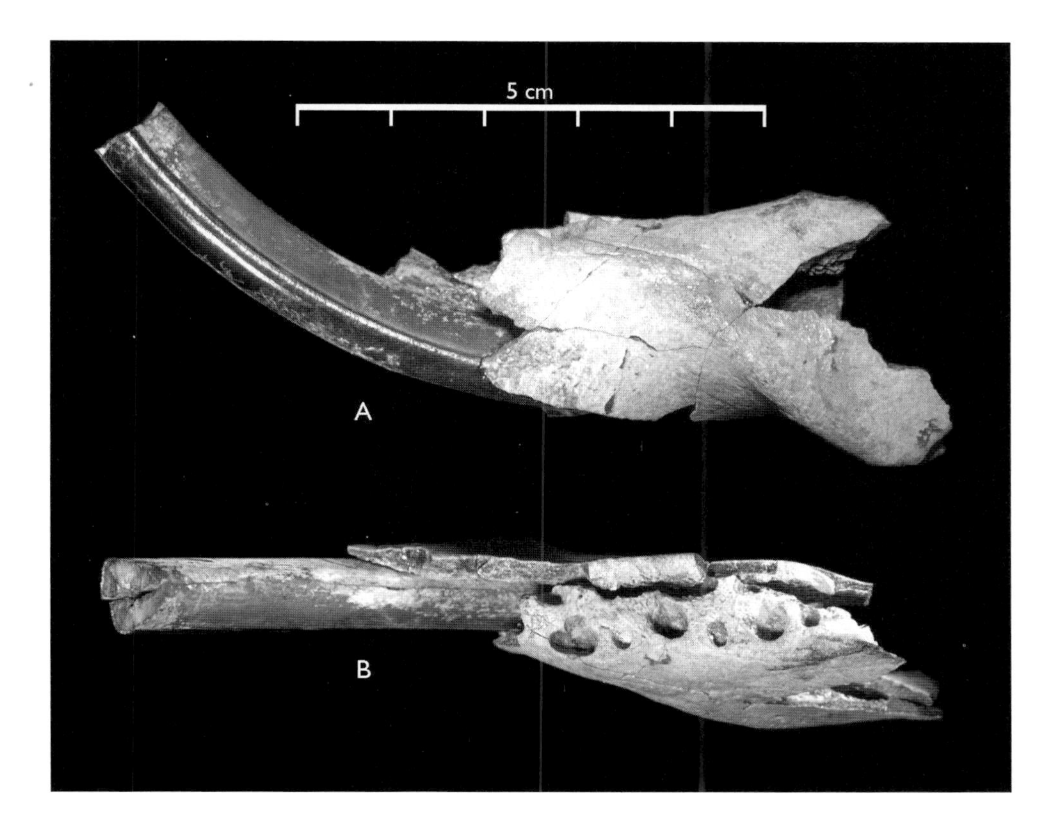

FIGURE 5.9
Xenohystrix sp. ADD-VP-1/16. A. Buccal view. B. Occlusal view.

three roots. The molars were large, and the length of the first molar can be estimated at about 10–12 mm. The M_2 is smaller than the M_1, and the M_3 is the smallest of all. The incisor is triangular in cross section, and it measures 8.75 mm mesiodistally and 9.1 mm buccolingually. On a larger *Xenohystrix* mandible from Hadar in the collections of the National Museum of Ethiopia, the lower incisor measures 9.7 mm in mesiodistal length and 10.2 mm in buccolingual breadth.

DISCUSSION Distinguished primarily by its large size, *Xenohystrix crassidens* Greenwood, 1955 has been reported from the Pliocene sites of Omo (Wesselman 1984), Hadar (Sabatier 1982), and Laetoli (Denys 1987), in addition to Langebaanweg and Makapansgat, South Africa. Its presence in the late Miocene deposits of the Middle Awash marks one of its earliest appearances in the fossil record. *Xenohystrix* is also known from a single isolated molar from Matabaietu (MAT-VP-7/6), a CAC site provisionally dated to 3.4 Ma. Pending recovery of more complete specimens, the Adu Dora fossil is assigned as *Xenohystrix* sp. indet.

Xenohystrix is an extinct taxon found in dry paleoenvironments at Laetoli, in more mesic environments at Hadar and Makapansgat (Members 3 and 4), and possibly from woodlands and/or forested conditions at Omo Member B (Wesselman 1984).

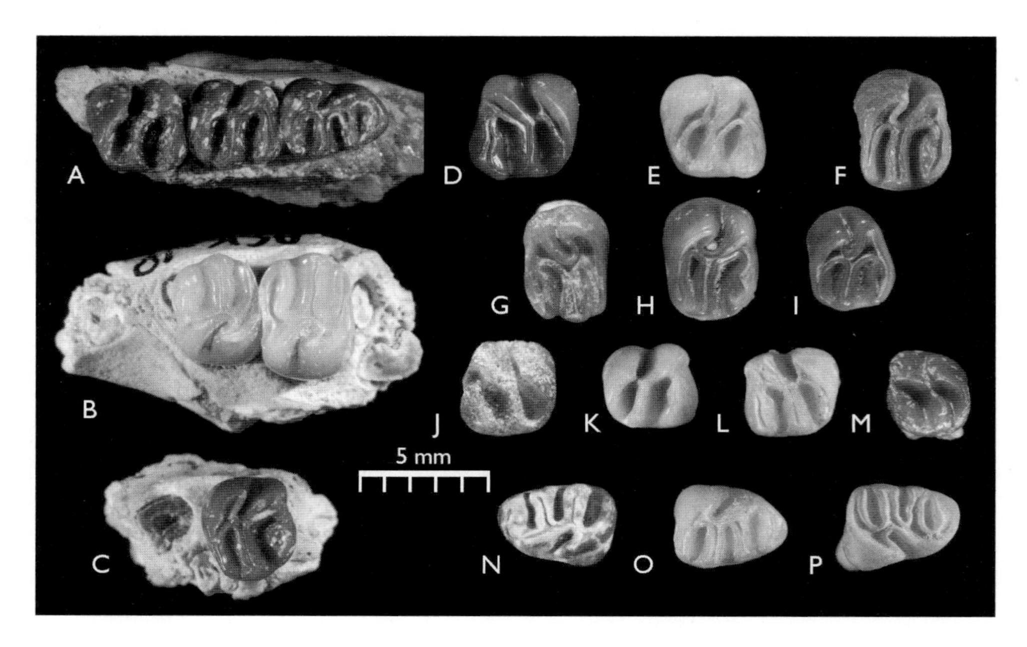

FIGURE 5.10

Thryonomys asakomae sp. nov. Cheek teeth, all in occlusal view. A. STD-VP-2/97. B. ASK-VP-3/119. C. STD-VP-2/859. D. STD-VP-1/8. E. ASK-VP-3/299. F. STD-VP-2/868. G. ASK-VP-3/282. H. ALA-VP-2/178. I. ALA-VP-2/181; J. ASK-VP-3/300. K. ASK-VP-3/281. L. ASK-VP-3/280. M. STD-VP-2/99. N. STD-VP-2/102. O. ASK-VP-3/302. P. ASK-VP-3/171.

Thryonomyidae

Thryonomys Fitzinger, 1867

Thryonomys asakomae sp. nov.

ETYMOLOGY The name *asakomae* is taken from the locality at which most of the specimens attributed to this taxon were found, Asa Koma. In the language of the Afar people, "Asa Koma" means "red hill."

HOLOTYPE STD-VP-2/97, left mandible fragment with incisor and dp_4–M_2 (see Figure 5.10A).

PARATYPES ALA-VP-2/74, right I^1 fragment; ALA-VP-2/75, right I^1 fragment; ALA-VP-2/90, cranial fragment with left I^1; ALA-VP-2/116, left I^1; ALA-VP-2/117, right I_1 fragment; ALA-VP-2/118, left I^1; ALA-VP-2/121, left I^1; ALA-VP-2/140, right mandible with dp_4 fragment and M_1; ALA-VP-2/143, right I^1 fragment; ALA-VP-2/167, left edentulous mandible; ALA-VP-2/178, right M^1; ALA-VP-2/179, left M_1; ALA-VP-2/180, right M^3; ALA-VP-2/181, right dp^4; ALA-VP-2/182, right M_3; ALA-VP-2/319, right I_1; ALA-VP-2/320, left I_1; ALA-VP-2/335, right I^1; ASK-VP-3/59, right I^1; ASK-VP-3/60, right M_1; ASK-VP-3/119, right maxilla fragment with M^2–M^3; ASK-VP-3/156, left I^1; ASK-VP-3/171, right dp_4; ASK-VP-3/174, left I_1 fragment; ASK-VP-3/255, right M_3; ASK-VP-3/267, right M^3; ASK-VP-3/275, right M_2; ASK-VP-2/276, right M_3; ASK-VP-3/280, right M_1; ASK-VP-3/281, left M_2; ASK-VP-3/282, left M^2; ASK-VP-3/289, left M^3 fragment; ASK-VP-3/299, left LM_1; ASK-VP-3/300, right M_1; ASK-VP-3/302, left dp_4; BIK-VP-3/22, left I^1; DID-VP-1/101, right I^1; STD-VP-1/8, right M_2; STD-VP-2/91,

FIGURE 5.11

Thryonomys asakomae sp. nov. incisors. All except A are pictured in occlusal view (left) and labial view (right). A. ALA-VP-2/90 (labial view only). B. ALA-VP-2/121. C. STD-VP-2/91. D. ALA-VP-2/118. E. ALA-VP-2/74. F. DID-VP-1/101. G. ALA-VP-2/116. H. ASK-VP-3/156. I. ALA-VP-2/143. J. ASK-VP-3/59.

right I^1; STD-VP-2/92, right M^3; STD-VP-2/98, right I_1; STD-VP-2/99, left M germ; STD-VP-2/100, left M^3; STD-VP-2/101, right M^3; STD-VP-2/102, left dp_4; STD-VP-2/859, right maxilla with M^3; STD-VP-2/860, right M^1 (see Figures 5.10 and 5.11).

LOCALITIES AND HORIZONS The late Miocene western margin sites of Alayla Vertebrate Locality 2 (ALA-VP-2), Asa Koma Vertebrate Locality 3 (ASK-VP-3), Bikir Mali Koma Vertebrate Locality 3 (BIK-VP-3), and Digiba Dora Vertebrate Locality 1 (DID-VP-1), all dated to 5.7 Ma, as well as the slightly younger western margin sites of Saitune Dora Vertebrate Localities 1 and 2 (STD-VP-1 and STD-VP-2), both dated to 5.6 Ma.

DIAGNOSIS The typical thryonomyid pattern is present on all the cheek teeth. The upper teeth express three transverse lophs on their occlusal surfaces. On the lower teeth, the dp_4 has four lophs, and the M_1, M_2, and M_3 all have three. The enamel is thin compared to that of the contemporary forms, and the lophs are mesiodistally compressed, a trend that creates wide, hollow synclines (uppers) and synclinids (lowers) of the reentrant enamel folds. On the dp^4, the anterior crest, the anteroloph, is crescent-shaped and lacks suggestion of a division into two cusps. The anterolophs on the M^2 and M^3 are poorly developed with thin enamel, creating wide synclines. The dp_4 is typically four-lophed, as is seen on the holotype, yet the anterior wall of the anterolophid is buccolingually narrow compared to the more expanded anterolophid of the contemporary forms. The lower molars express a minuscule anterolophid crest on the anterolingual corner of the protoconid, a feature more strongly developed on modern species.

All the I^1 fragments recovered ($n = 10$) reveal two grooves in the labial enamel surface (Figure 5.11), a species-specific character of the late Miocene thryonomyid *Paraulacodus johanesi* Jaeger et al., 1980 (see also Hinton 1933; Black 1972; López-Antoñanzas et al. 2004). The upper incisors of modern and fossil *Thryonomys* from Plio-Pleistocene sites express three grooves.

TABLE 5.6 Dental Measurements of *Thryonomys asakomae* sp. nov.

Taxon	*Specimen*	*Tooth*	*Length*	*Breadth*	*n*
Thryonomys asakomae sp. nov.	Various	I^1	3.43–4.60	3.53–4.37	10
	ALA-VP-2/181	dp^4	3.25	4.00	1
	Various	M^1	3.56–3.63	4.92–4.93	2
	Various	M^2	3.18–3.89	3.44–5.13	3
	Various	M^3	3.14–3.69	3.73–4.61	10
	Various	M_3	4.19–4.55	3.30–3.55	4
	Various	M_1	3.25–4.24	3.40–4.30	7
	Various	M_2	4.00–4.10	3.73–4.27	3
	Various	M_3	4.14–4.32	3.84–4.07	3

NOTE: Measurements are in mm. Also see measurements in Table 5.7.

DISCUSSION If the assignment of these fossils is based on the morphology of the cheek teeth, especially the four-lophed dp^4, they should be assigned to the genus *Thryonomys*. However, all the I^1 fragments from the western margin sites resemble those of *Paraulacodus johanesi,* a form originally described from the late Miocene deposits of Chorora, Ethiopia (Jaeger et al. 1980), and dated to 10.5 Ma, whose incisors exhibit two grooves. The molars of *Paraulacodus,* like those of *Thryonomys asakomae* sp. nov., have thin enamel, and wide synclines and synclinids. *Paraulacodus,* however, expresses three lophs on the dp^4.

It is also noted that one cranial fragment from Alayla (ALA-VP-2/90) bears a left I^1 fragment that displays the merest hint of a third groove taking form, suggesting a preview of the three-grooved trait that would distinguish the incisors of its descendants.

In size, the fossils from the late Miocene sites, dated to 5.7 Ma, are clearly smaller than those from the early Pliocene sites described below, dated to 4.9–5.2 Ma (Table 5.6). They are also smaller than both contemporary species, *Thryonomys gregorianus* Thomas, 1894, and *Thryonomys swinderianus* Temminck, 1827 (measurements for these taxa are presented in Table 5.7).

The fossils from the western margin sites are somewhat larger than *Thryonomys* cf. *gregorianus* from the Nachukui Formation of Lothagam (Winkler 2003) and *Thryonomys* sp. from the Wembere-Manonga Formation Inolelo Beds (Winkler 1997).

The question then arises as to whether there are two thryonomyid forms in the earlier western margin sites—*Paraulacodus* and *Thryonomys*—or whether there is a single form expressing transitional characters? It is suggested here that the latter is the case; that the late Miocene western margin sites preserve a small, early *Thryonomys* species with a four-lophed dp_4 that retains the primitive ancestral *Paraulacodus* traits of thin enamel, wide synclines and synclinids, and two-grooved incisors. As such, it represents a distinct intermediate taxon and deserves elevation as a new species.

The systematics and the phylogeny of the thryonomyids from the Eocene to the present have been revised by López-Antoñanzas et al. (2004). Thryonomyids experienced maximum diversity and widest geographic distribution during the Miocene, from southwestern Africa into Pakistan (Flynn and Winkler 1994). By the early Pliocene the family had been reduced to a single African genus, *Thryonomys,* that has been recorded from the

TABLE 5.7 Dental Measurements of *Thryonomys gregorianus* and *T. swinderianus*

Taxon	*Specimen*	*Tooth*	*Length*	*Breadth*	*n*
Thryonomys aff. *gregorianus*	WKH-VP-1/24	dp^4	3.84	4.73	1
	Various	M^1	3.53–4.11	4.96–5.27	3
	Various	dp_4	4.22–4.55	3.44–3.61	2
	Various	M_1	3.51–3.80	4.96–5.27	3
	Various	M_2	3.51–3.85	4.16–4.17	3
	Various	M_3	3.60–4.35	3.70–3.83	3
Thryonomys gregorianus (contemporary)	Various	dp^4	3.95– 4.90	4.50–4.76	10
	(coll. NHM)	M^1	3.89–4.20	4.39–5.02	10
		M^2	4.01–4.45	5.15–5.66	10
		M^3	3.96–4.47	4.27–5.22	10
		dp_4	5.18–5.44	3.95–4.34	10
		M_1	4.12–4.60	4.40–5.18	10
		M_2	4.45–4.80	5.00–5.45	10
		M_3	4.45–5.29	4.40–4.83	10
Thryonomys swinderianus (contemporary)	Various	dp^4	4.45–4.80	4.90–5.39	10
	(coll. NHM)	M^1	4.75–5.01	5.91–6.35	10
		M^2	4.70–5.58	5.76–6.70	10
		M^3	4.70–6.00	5.94–6.70	10
		dp_4	5.44–5.99	4.30–4.70	10
		M_1	5.06–5.40	5.37–5.81	10
		M_2	5.08–5.85	5.88–6.35	10
		M_3	5.95–6.50	5.01–6.43	10

NOTE: Measurements are in mm.

early Pliocene of the Chemeron Formation and Tabarin, Kenya (Winkler 1990), from the Wembere-Manonga Formation, Tanzania (Winkler 1997), from the late Pliocene Ndolanya Beds at Laetoli, Tanzania (Denys 1987), and from Members B and G of the Shungura Formation at Omo (Wesselman 1984).

The known thryonomyid genera are conservative in their dental and cranial morphology. For example, the crania of *Paraphiomys* species, known in sub-Saharan Africa after the early Miocene, are similar to the crania of the extant *Thryonomys* and *Petromus* (Lavocat 1973), and the late Miocene *Paraphiomys* from Chorora and the Lower Nawata Member of the Lothagam Formation shares dental characters with *Paraulacodus* and the extant *Thryonomys* species (López-Antoñanzas et al. 2004).

Thryonomys aff. *gregorianus* Thomas, 1894

DESCRIPTION The enamel of the fossil molars of *Thryonomys* recovered from AME-VP-1, AMW-VP-1, and WKH-VP-1 has thickened, yet the reentrant synclines (uppers) and synclinids (lowers) are still wide (Figure 5.12). These fossils are morphologically comparable to the single *Thryonomys* fossil recovered from Kuseralee Vertebrate Locality 2 of the CAC (KUS-VP-2/121, a left mandible fragment with dp_4–M_3, dated to 4.4 Ma), on which the enamel has thickened even more, approaching the condition seen on both

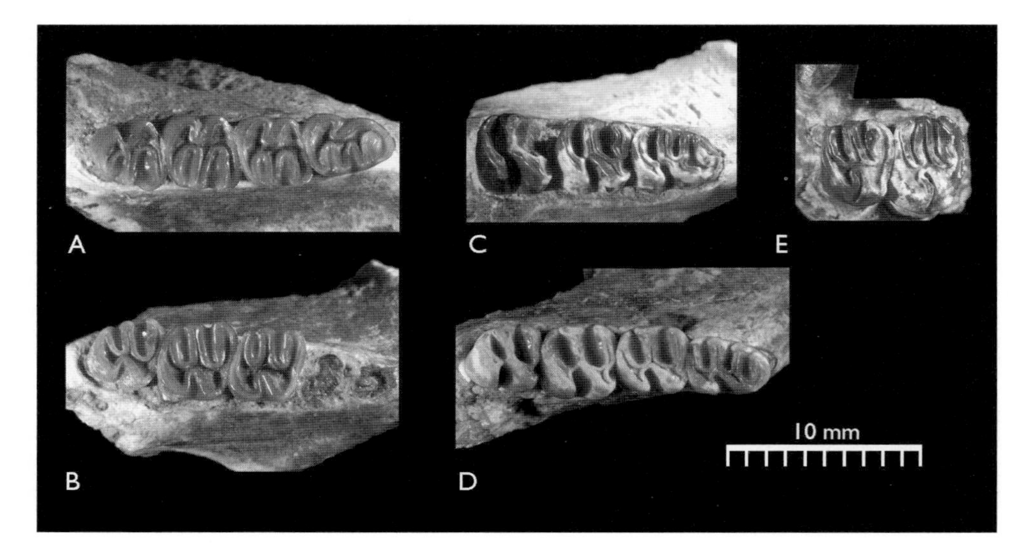

FIGURE 5.12
Thryonomys aff. *gregorianus*. All in occlusal view. A. AME-VP-1/95a. B. AME-VP-1/95b. C. WKH-VP-1/38. D. AMW-VP-1/8. E. WKH-VP-1/24.

contemporary species. On WKH-VP-1/24, a left maxilla fragment with dp^4–M^1, the minimally worn anteroloph of the dp^4 suggests the expression of two cusps. In addition, it exhibits a feature observed on many *T. gregorianus* specimens in museum collections: the reentrant enamel fold of the buccal syncline is short and straight, and is oriented directly at the middle loph (loph II; Denys 1987). By contrast, this reentrant fold tends to be curved and canted toward the anteroloph on the dp^4 of *T. swinderianus*.

In size, these molars are mesiodistally longer and buccolingually wider than those of *Thryonomys asakomae* sp. nov., approaching overlap with the smaller contemporary form *T. gregorianus*. They are provisionally attributed to that species as *Thryonomys* aff. *gregorianus* (Table 5.7).

Fossil *Thryonomys gregorianus* have also been recovered from Omo, East Turkana, and the Laetoli Ndolanya Beds, as well as from later Pleistocene deposits in the Middle Awash Valley.

The contemporary species *Thryonomys swinderianus,* the greater cane rat, is a large, semiaquatic form that is widely distributed throughout sub-Saharan Africa and is associated with mesic marshy environments and reed beds with long grass, in which it finds food and dense cover. The smaller *T. gregorianus,* the lesser cane rat, occurs from Cameroon to western Ethiopia and southward to Zimbabwe and is adapted to more open conditions, where it inhabits drier ground in moist savannas (Nowak 1991; Jenkins 2001).

Muridae

Tatera Lataste, 1882

Tatera sp.

DESCRIPTION Two specimens of *Tatera* were recovered from the late Miocene Middle Awash localities: ASK-VP-3/290 and AMW-VP-1/95 (Figure 5.13, Table 5.8). Both fossil

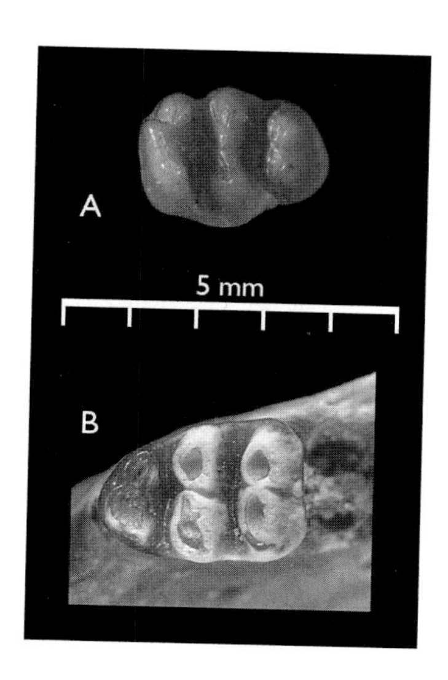

FIGURE 5.13

Tatera sp. A. ASK-VP-3/290, occlusal view. B. AMW-VP-1/95, occlusal view.

teeth preserve the typical gerbilline crown morphology, in which two cusps are incorporated into each of three transverse laminae. Both specimens also lack the longitudinal crest connections between the laminae, typical of *Gerbillus,* and are assignable to the genus *Tatera.*

DISCUSSION This gerbil is common at many eastern African sites (including the Aramis localities in the CAC, Hadar, Omo, East Turkana, Kanapoi, Olduvai, and Laetoli), and specimens from all of these sites express remarkably similar molar crown morphology. The ASK-VP-3 and AMW-VP-1 fossils are clearly larger than the Aramis *Tatera* specimens, which in turn are larger than the Omo and Hadar *Tatera.* Yet overall morphological similarity indicates affinity with contemporary *Tatera* species of the "*robusta*" group (which express molars that are narrow, lightly built, and opisthodont), and especially with *T. vicina,* recorded from Bodessa, Ethiopia, by Yalden et al. (1976), and *T. minuscula.* At this time, the paucity of the fossil material suggests that they should be assigned to *Tatera* sp.

Extant species within the genus *Tatera* are characterized by morphological and physiological adaptations to dry, open-country savanna and thicket habitats. They live on well-drained sandy soils, in which they excavate large communal burrows. They are able to survive without water for long periods and are a reliable indicator of aridity (Kingdon 1974, 1997; Petter 1971).

Lophiomys Milne-Edwards, 1867

Lophiomys daphnae sp. nov.

ETYMOLOGY This species is named in honor of Daphne Hills, curator of Mammals at The Natural History Museum (London), whose affection for *Lophiomys,* the maned rat, provided the catalyst in the identification of this paleospecies.

TABLE 5.8 Measurements of *Tatera* sp.

Taxon	*Specimen*	*Tooth*	*Length*	*Breadth*	*n*
Tatera sp.	ASK-VP-3/290	R.M^1	2.95	2.17	1
	AMW-VP-1/95	R.M_1	3.20	2.20	1

NOTE: Measurements are in mm.

HOLOTYPE ALA-VP-2/168, right mandible fragment with M_1–M_3 (see Figure 5.14).

LOCALITIES AND HORIZONS The earlier (late Miocene) western margin site of Alayla Vertebrate Locality 2 (ALA-VP-2), dated to 5.7 Ma.

DIAGNOSIS The distinctive crown morphology of the teeth of ALA-VP-2/168 suggests close taxonomic affinity with the contemporary maned rat, *Lophiomys imhausi* Milne-Edwards, 1867. *Lophiomys daphnae* sp. nov. differs from *L. imhausi,* however, in several characters. On the lower first molar, the two anterior cusps—the tF and tE—are approximated, creating a single "massif," in contrast to *L. imhausi,* in which the two cusps are divergent. The fossil M_1 presents a small posterior cingulum (cp) that is considerably more expanded on the M_1 of the contemporary species. For the second molar, both the anterior and posterior cingula are massive on *L. imhausi,* flaring into a wide tE ledge on the mesiobuccal corner of the molar crown anteriorly, and creating a large ledge on the distolingual crown corner, respectively. These cingula are small on *L. daphnae* sp. nov. The anterior cingulum of the third molar follows suit, being small on the fossil specimen, and more massive on the contemporary form. In size, the fossil teeth are 30 percent smaller than those of the contemporary species, and the enamel surface is smooth compared to that of *L. imhausi* (Table 5.9), in which the unworn areas of the enamel surface are rugose. The fossil species was also considerably smaller than its (presumed) modern descendant.

DESCRIPTION ALA-VP-2/168 is a right mandible fragment preserving all three molars at an early stage of wear. The distinctive molar crown morphology is arranged into transverse laminae, each of which is composed of two obliquely offset cusps, except for the two anterior cusps, which are approximated into a single wedge on the M_1. The laminae are connected by a central longitudinal ridge (or mure), reminiscent of the gerbilline rodents in the genus *Gerbillus* and accounting for the misidentication of this specimen as an unknown gerbilline genus and species by Haile-Selassie et al. (2004c). The M_1 presents a small posterior cingulum. The M_2 has four major cusps arranged in an oblique offset and an anterior cingulum that merges with the tE buccally, forming a small crest on the mesiobuccal crown corner. It also has a weakly developed median longitudinal ridge and a small posterior cingulum, which is oriented lingually. The M_3 is morphologically similar to the M_2; however, the M_3 tapers distally and its posterior crown margin displays a small posterior cingulum between the two distal cusps (the tB and the tA). The noticeable diastema between the M_2 and M_3 is an artifact of preservation.

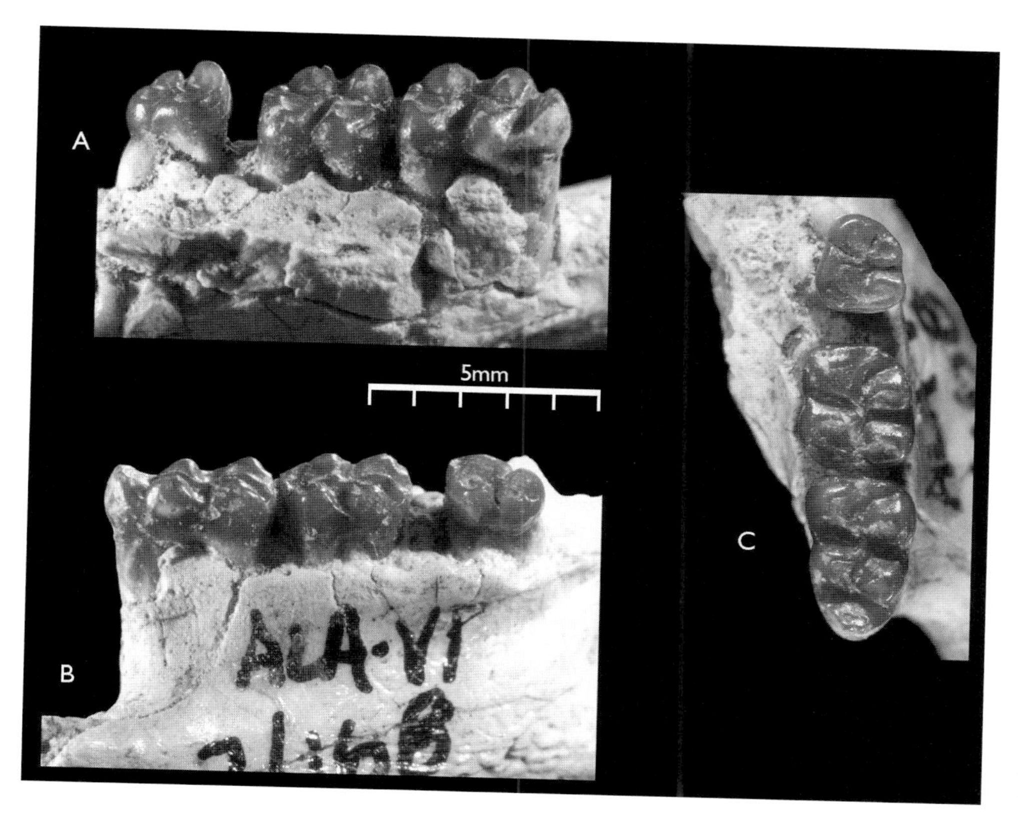

FIGURE 5.14
Lophiomys daphnae sp. nov. ALA-VP-2/168. A. Buccal view. B. Lingual view. C. Occlusal view.

DISCUSSION The molar crown morphology reveals primary affinity to the contemporary maned rat *Lophiomys imhausi,* distributed throughout eastern Africa from the Sudan, Ethiopia, and Somalia, through Kenya, Uganda, and Tanzania.

ALA-VP-2/168 is the only fossil representative of the Lophiomyinae known from eastern Africa. The closest fossil relatives of *Lophiomys* are *Microlophiomys vorontsovi* from the late Miocene of the Ukraine (Topachevskii and Skorik 1984) and *Cricetops dormitor* from the middle Oligocene of Mongolia and Kazakhstan (Wahlert 1984). Wahlert considers *Lophiomys* to be close to the Gerbillinae. We have followed Charleton and Musser (1984) in considering *Lophiomys* to be in its own subfamily, Lophiomyinae, within the family Muridae.

Lophiomys is known from Israel from subfossils and may occur in Arabia. A small microfaunal complex from the Lissafa Quarry in Morocco, dated to nearly 5.5 Ma, includes the taxon *Lophiomys maroccanus,* which Aguilar and Michaux (1990) considered to be Messinian in age and which is also known from a few Maghrebian and Spanish sites. Its presence in the corner of northwestern Africa is a testament to lowered sea levels between Spain and Morocco during the Messinian desiccation and the establishment of a land bridge across the straits of Gibraltar. *Lophiomys daphnae* sp. nov. is thus considered to be a Eurasian invader into northeastern Africa during the late Miocene.

Modern *Lophiomys* has been collected at elevations up to 3,300 m in the Ethiopian highlands (Yalden et al. 1976) and is restricted to the montane forest zone in Kenya and Uganda (Kingdon 1974, 1997). Its presence in the western margin sediments suggests

TABLE 5.9 Dental Measurements of *Lophiomys daphnae* sp. nov.

Taxon	*Specimen*	*Tooth*	*Length*	*Breadth*	*n*
Lophiomys daphnae sp. nov.	ALA-VP-2/168	M_1	3.84	2.46	1
		M_2	2.80	2.49	1
		M_3	2.16	1.97	1
Lophiomys imhausi	Various	M_1	5.15–5.80	3.05–3.42	10
		M_2	3.90–4.23	3.17–3.60	10
		M_3	3.50–3.82	2.68–2.99	10

NOTE: Measurements are in mm.

that this part of the Rift System was forested during this time and that it existed at a considerably higher altitude than it does today (Haile-Selassie 2001a; WoldeGabriel et al. 2001; Haile-Selassie et al. 2004c).

Lemniscomys Trouessart, 1881

Lemniscomys aff. *striatus* (Linnaeus, 1758)

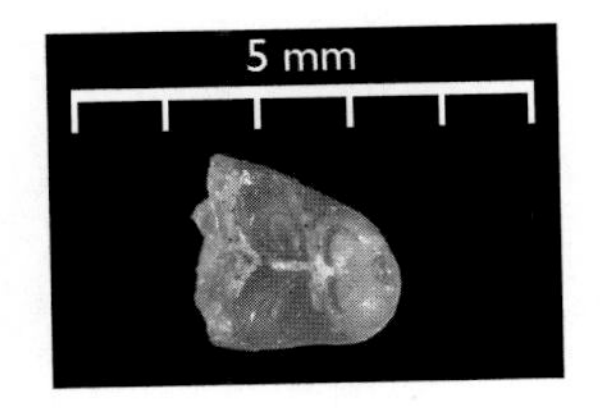

FIGURE 5.15 *Lemniscomys* aff. *striatus*. STD-VP-1/49, occlusal view.

DESCRIPTION STD-VP-1/49 (Figure 5.15) is the anterior half of an incomplete lower first molar that is tentatively assigned to *Lemniscomys* aff. *striatus* based on its crown morphology and overall size. It is at an early wear stage, and the arrangement of the anterior cusps resembles that of L1-538 from Omo Member B, which is assigned to *Lemniscomys* aff. *striatus* (Wesselman 1984). No measurements are possible.

DISCUSSION Murines are better known from the CAC sites at Aramis as well as the Nawata Formation of Lothagam (Winkler 2003). Their rarity in the Middle Awash late Miocene deposits is probably due to sampling bias. *Lemniscomys* is not present in the Aramis localities, yet it does reappear at the Asa Issie Locality 2 site (ASI-VP-2), dated to 4.0 Ma. *Lemniscomys* has been identified by one of us (HW) in the collections recovered from Kanapoi; *Lemniscomys griselda* has been tentatively identified from Makapansgat (Lavocat 1957); and *Lemniscomys barbarus* is known from the middle and late Pleistocene at Jebel Irhoud (Jaeger 1975).

Extant *Lemniscomys striatus* is a very successful murid common in grasslands, secondary forests, and woodlands, that ranges from west Africa to the Ethiopian highlands and southward to Zambia, Malawi, and northwest Angola (Misonne 1974).

Tachyoryctes Rüppell, 1835

Tachyoryctes makooka sp. nov.

ETYMOLOGY In the language of the Afar people, *makooka* indicates a sharp bend or zigzag. This name refers to the complex and highly derived pattern of enamel folding on the molars of this species.

HOLOTYPE DID-VP-1/16, a left mandible fragment with M_1–M_3 (see Figure 5.16H).

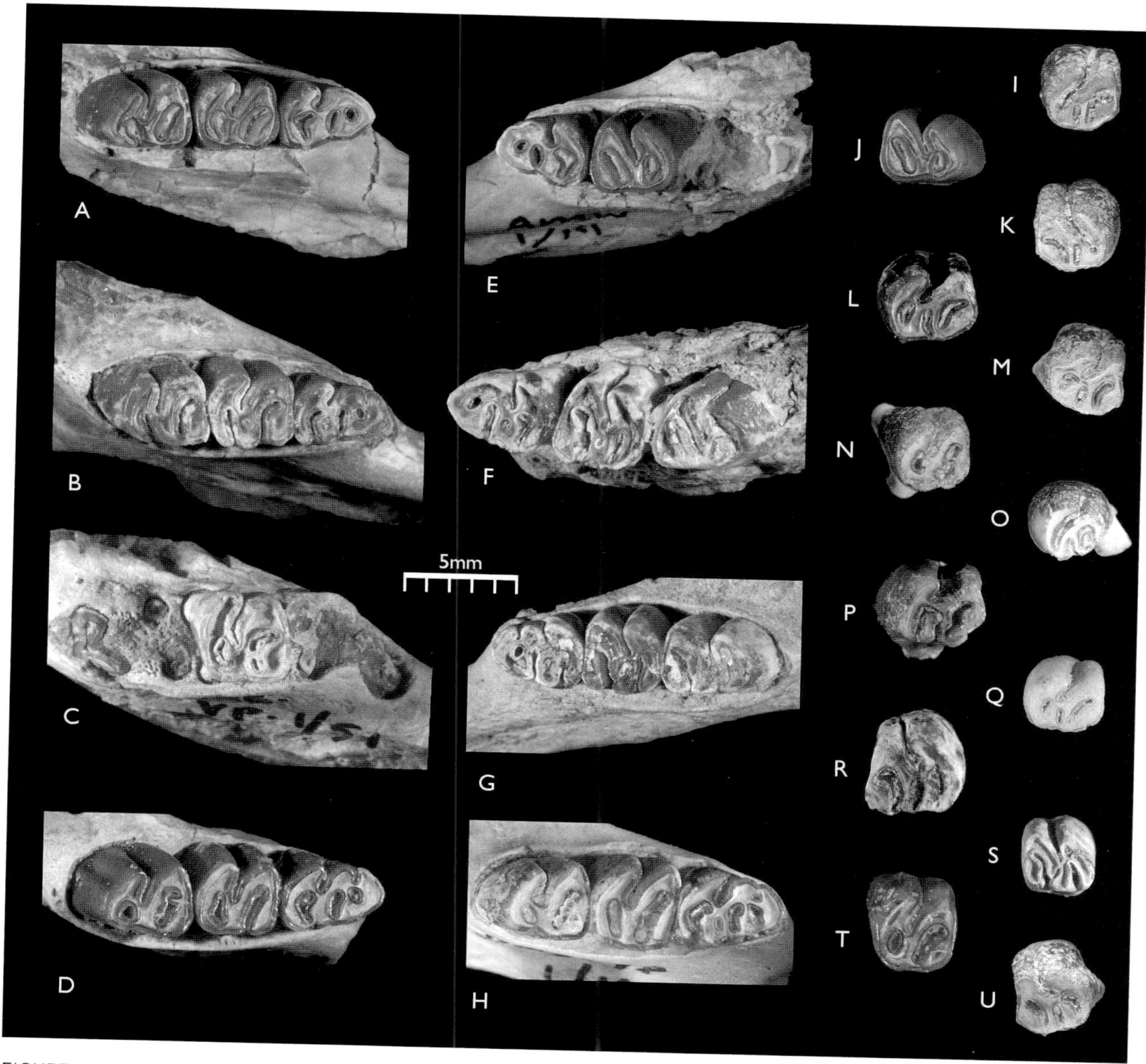

FIGURE 5.16

Tachyoryctes makooka sp. nov. All in occlusal view. A. AMW-VP-1/111a. B. AMW-VP-1/99. C. AME-VP-1/51. D. AMW-VP-1/115. E. AMW-VP-/111b. F. AMW-VP-1/59. G. AME-VP-1/97. H. DID-VP-1/16. I. ASK-VP-3/260. J. AMW-VP-1/111. K. ASK-VP-3/265. L. ALA-VP-2/174. M. ASK-VP-3/269. N. ASK-VP-3/286. O. ASK-VP-3/272. P. ASK-VP-2/173. Q. ASK-VP-3/293. R. ALA-VP-2/231. S. WKH-VP-1/53. T. DID-VP-1/103. U. ASK-VP-3/296.

PARATYPES ALA-VP-2/64, left M_1 fragment; ALA-VP-2/65, left M_2; ALA-VP-2/170, right mandible, M_1–M_3 fragments; ALA-VP-2/173, right M^3; ALA-VP-2/174, left M^2; ALA-VP-2/175, left M_2 fragment; ALA-VP-2/177, left M_1; ALA-VP-2/185, right lower M fragment: ALA-VP-2/321, right M^1; AME-VP-1/51, left mandible M_2; AME-VP-1/75, right mandible M_2–M_3; AME-VP-1/94, skull, left mandible M_1–M_3, and postcranial elements; AME-VP-1/97, right mandible M_1–M_3; AME-VP-1/136, right mandible M_1–M_3; AMW-VP-1/29, left M_3; AMW-VP-1/59, right mandible with M_1–M_3; AMW-VP-1/182, right and left mandible with M_1–M_3 and postcranial elements; AMW-VP-1/99, left mandible M_1–M_3; AMW-VP-1/111, right mandible with M_1–M_2 and left mandible with M_1–M_3; AMW-VP-1/115, left mandible M_1–M_3; AMW-VP-1/116, left mandible M_1–M_3; ASK-VP-3/172, right M_3; ASK-VP-3/173, right M_2; ASK-VP-3/254, right M^2; ASK-VP-3/256, left M_3; ASK-VP-3/257, left M_1; ASK-VP-3/258, left M_3; ASK-VP-3/260, right M^2; ASK-VP-3/261, right M^2; ASK-VP-3/263, right M_3; ASK-VP-3/264, right M_2; ASK-VP-3/265, left M^2; ASK-VP-3/266, right M_1; ASK-VP-3/269, right M^1; ASK-VP-3/270, left M_3; ASK-VP-3/271, left M_3; ASK-VP-3/272, left M^3; ASK-VP-3/276, right M_3; ASK-VP-3/277, right M_1; ASK-VP-3/278, left upper M fragment; ASK-VP-3/279, left M^1; ASK-VP-3/284, left M_2 fragment; ASK-VP-3/286, right M^2; ASK-VP-3/287, left M_1; ASK-VP-3/288, left M_3; ASK-VP-3/291, right M_2; ASK-VP-3/292, right M_2; ASK-VP-3/293, right M^2; ASK-VP-3/294, right M_1; ASK-VP-3/295, right M_3; ASK-VP-3/296, left M^1; ASK-VP-3/297, right M^1; ASK-VP-3/298, left M^2; ASK-VP-3/303, left M_1; DID-VP-1/103, right M^2; GAW-VP-1/91, right mandible M_1–M_3; GAW-VP-1/96, right maxilla fragment M^2–M^3; GAW-VP-1/119, left mandible, M_1–M_2; GAW-VP-1/120, right mandible M_1–M_3; GAW-VP-3/33, right M^1 germ; KUS-VP-1/2, left M_3; KUS-VP-1/31, right M^2; STD-VP-2/94, left M_2; STD-VP-2/862, left M_3 fragment; STD-VP-2/903, right M_3; STD-VP-2/908, left upper M fragment; WKH-VP-1/53, left M^3 (Figure 5.16).

LOCALITIES AND HORIZONS Western margin sites of Alayla Vertebrate Locality 2 (ALA-VP-2), Asa Koma Vertebrate Locality 3 (ASK-VP-3), and Digiba Dora Vertebrate Locality 1 (DID-VP-1), all dated to 5.7 Ma; the slightly younger site of Saitune Dora Vertebrate Locality 2 (STD-VP-2), dated to 5.6 Ma; CAC sites of Amba East Vertebrate Locality 1 (AME-VP-1), Amba West Vertebrate Locality 1 (AMW-VP-1), and Kuseralee Vertebrate Locality 1 (KUS-VP-1), all dated to 5.2 Ma; and the even younger CAC sites of Gawto Vertebrate Localities 1 and 3 (GAW-VP-1 and GAW-VP-3), and Worku Hassan Vertebrate Locality 1 (WKH-VP-1), all dated to 4.85 Ma.

DIAGNOSIS *Tachyoryctes makooka* sp. nov. is distinguished by its size, which exceeds that of the contemporary *Tachyoryctes splendens,* yet falls within the size range of the large *T. macrocephalus* (Table 5.10). The upper molar crowns are arranged in folded, transverse enamel lophs that are robust, hypsodont, and elongated mesiodistally, with thin enamel walls enclosing open areas of dentin that surround enamel islands, or synclines. The lingual reentrant enamel fold (or sinus) of the upper molars penetrates only to the midline of the crown surface. The posterior heel of the M^3—the metacone—is rounded and mesiodistally shortened. The lower molar crowns are similarly hypsodont and lophodont, and

TABLE 5.10 Dental Measurements of *Tachyoryctes* Species

Taxon	*Specimen*	*Tooth*	*Length*	*Breadth*	*n*
Tachyoryctes makooka (sp. nov.) (older sites)	Various	M^1	3.42–4.50	3.05–3.94	3
	Various	M^2	3.17–4.05	2.48–3.70	9
	Various	M^3	2.91–3.66	2.87–3.90	3
	Various	M_1	3.83–4.66	2.85–3.79	10
	Various	M_2	3.27–4.15	3.00–4.15	9
	Various	M_3	3.40–4.20	2.70–4.11	12
Tachyoryctes makooka (sp. nov.) (younger sites)	GAW-VP-3/33	M^1	3.53	2.75	1
	Various	M^2	3.25–3.33	2.88–3.21	2
	Various	M^3	2.90–3.25	3.06–3.24	2
	Various	M_1	3.66–4.66	2.73–3.54	12
	Various	M_2	3.15–4.07	2.98–4.31	13
	Various	M_3	3.30–4.51	2.80–3.90	12
Tachyoryctes macrocephalus (contemporary)	Various (coll. NHM)	M^1	3.37–3.64	3.70–3.95	10
		M^2	3.27–3.87	3.61–4.06	10
		M^3	4.90–5.30	3.50–4.00	10
		M_1	4.40–4.60	3.54–4.32	10
		M_2	3.60–4.30	4.20–4.75	10
		M_3	3.65–4.00	3.60–4.60	10
Tachyoryctes splendens (contemporary)	Various (coll. NHM)	M^1	2.20–3.14	2.50–3.40	20
		M^2	2.20–3.01	2.47–3.30	20
		M^3	2.80–4.16	2.25–3.10	20
		M_1	2.53–3.92	2.44–3.25	20
		M_2	2.32–3.40	2.90–3.58	20
		M_3	2.25–3.40	2.30–3.24	20

NOTE: *T. splendens* includes subspecies in the collections at the NHM labeled as *T. ruddi, T. splendens, T. s. somalicus, T. daemon, T. ankoliae, T. cheesmani, T. rex,* and *T. ibeanus.* Measurements are in mm.

the buccal reentrant enamel fold (or sinusid) penetrates only to the midline of the crown in M_2 and M_3, so that the mesolophid and hypolophid are linked by interconnecting mures; the M_1 is narrow buccolingually and mesodistally elongated, exceeding in length the M_1 of even *Tachyoryctes macrocephalus*.

DISCUSSION The fossils assignable to *Tachyoryctes makooka* sp. nov. represent a species distinct from the contemporary species *T. splendens* and *T. macrocephalus,* whose molars are more compressed mesiodistally, creating crown surfaces that are strongly flattened, and sandwiched from front to back on the upper teeth and from back to front on the lowers, with thicker reentrant enamel synclines (uppers) and synclinids (lowers) that extend across the crown to the opposite enamel wall, creating a tooth morphology that is highly derived. Unlike the fossils, the heel of the upper M^3 of the contemporary forms is elongated and complex. The fossils are also distinct from *Tachyoryctes pliocaenicus* Sabatier, 1978, described from Hadar, Ethiopia, which is considerably smaller and clearly more derived (the lower tooth row length of *T. pliocaenicus* measures 8.00 mm compared to that of the holotype of *Tachyoryctes makooka* sp. nov., which measures 12.55 mm).

The older specimens of *Tachyoryctes makooka* sp. nov. from the late Miocene sites of Alaya, Asa Koma, Digiba Dora, and Saitune Dora, dated to 5.7 Ma, are somewhat larger and more robust from those recovered from the later (early Pliocene) sites of Amba East, Amba West, and Kuseralee Vertebrate Locality 1, dated to 5.2 Ma, and those from Gawto and Worku Hassan Vertebrate Locality 1, dated to 4.85 Ma (see Table 5.10). These younger fossils are, in turn, somewhat larger than an as-yet-undescribed sample of *Tachyoryctes* fossils from the CAC sites of Aramis Vertebrate Localities 1, 6, 8, 11, and 17, all dated to 4.4 Ma (Wesselman and Black, in preparation). They all appear, however, to represent a single lineage that is diminishing in size as it travels across time, perhaps in response to decreasing altitude as a result of tectonic activity in the Ethiopian Rift.

Fossil rhizomyines are known from multiple genera from the Siwalik sequence of southwestern Asia (Jacobs 1978), and Haile-Selassie et al. (2004c) have noted the general similarity of the dental morphology of *Tachyoryctes makooka* sp. nov. to those of both the Asian genus *Kanisamys* and the earliest African rhizomyine *Nakalimys lavocati* (Flynn and Sabatier 1984), suggesting an Asiatic origin for *Tachyoryctes*. Sabatier (1978) has also noted that *Tachyoryctes pliocaeneus* from Hadar resembles the *Kanisamys* lineage of the Siwaliks. *Tachyoryctes makooka* sp. nov. may be the intermediate between the earlier *Nakalimys* and the fossil and extant African *Tachyoryctes* species, creating a *Kanisamys—Nakalimys—Tachyoryctes* lineage as suggested by Haile-Selassie et al. (2004c). If so, the Middle Awash *Tachyoryctes makooka* sp. nov. appears to establish a firm relationship between the Asian and African rhizomyines and suggests that they form a monophyletic group (see Black 1972; Jacobs et al. 1989).

With reference to the contemporary forms of *Tachyoryctes,* Hollister (1919) originally listed eight distinct species based upon consistent characters of differentiation, citing the lack of intergradation between them. Ellerman (1941) suggested the existence of 14 species, which Misonne (1974) subsumed into two: *T. macrocephalus* Rüppell, 1842, a large form endemic to the Ethiopian highlands, and *T. splendens* Rüppell, 1835, incorporating all the others. Yalden et al. (1976) claimed that *T. ibeanus* should stand as a separate species from the rest of *T. splendens*. Musser and Carleton (1993) noted that the change from 14 to two species was not based upon a careful analysis of the morphological variation characterizing the named forms, and they stated that the genus is in need of revision. The large size of *Tachyoryctes makooka* sp. nov. suggests an ancestral relationship to *T. macrocephalus*.

Modern *Tachyoryctes macrocephalus* is fossorial, endemic to the Ethiopian plateau, and an inhabitant of the Afro-Alpine moorlands and grasslands between 1,000 and 4,000 m, which it shares with other Ethiopian endemics. Yalden (1975) has provided ecological observations for *T. macrocephalus* as well as morphological distinctions separating it from *T. splendens*.

Extant *Tachyoryctes splendens* is also fossorial, ranging over the uplands of northeastern Africa, from Ethiopia and Somalia to Zaire, Rwanda, and Burundi, occurring also in the Ruwenzori range of Uganda, in Kenya, and in northern Tanzania. It is seldom found in areas with less than 500 mm annual rainfall, and it flourishes best in wet uplands, extending to 3,500 m on Mt. Kenya. This species is common on every mountainous massif in Kenya and Tanzania and is found in the Rift Valley itself only in uplifted parts between Baringo and Suswa.

Both *Tachyoryctes* species favor open grasslands with deep, fertile soils, as well as thinly treed upland savanna woodlands and higher altitude moorlands. They make short tunnel systems just below the level of the grass roots on which they feed and come to the surface more often than do other "root rats." They graze on fresh grass at the surface near their "mole-hills," which are composed of large piles of subsoil and debris.

Kingdon (1974, 1997) reveals that the population densities of *Tachyoryctes* can be astonishingly high, despite the fact that they are solitary, each living in a self-contained burrow. They are preyed upon by eagle-owls, eagles, augur buzzards, and goshawks, as well as Ethiopian wolves, serval cats, and large viverrids. At localities with elevations over 2,000 m, eagle-owls have been observed taking a great toll on these "root rats" because the birds are able to perceive the rodents working underground and drop on them, pushing long talons through the earth to grab the prey.

The presence, as well as the predominance, of *Tachyoryctes* in the western margin and CAC localities suggests that this part of the Rift Valley system existed at a considerably higher altitude during these time periods, and it presents evidence for the existence of open grasslands, upland moorlands, and montane woodlands near the sites of deposition (Haile-Selassie 2001a; WoldeGabriel et al. 2001; Haile-Selassie et al. 2004c).

Sciuridae

Xerus Ehrenberg, 1833

Xerus sp.

DESCRIPTION The P^3 is a small, stub-like tooth that has no visible occlusal crown morphology. The P^4 has a rectangular shape and is wider buccolingually than it is long (Figure 5.17). The crown outline shows slight lingual tapering, and the occlusal surface expresses the typical sciurid pattern of four semiparallel transverse ridges—the mesial cingulum, the paraloph, the metaloph, and the distal cingulum. The cingula are connected to the lingual cusps, the protocone, and the hypocone.

DISCUSSION The presence of a P^3 hints at a relationship with the smaller African bush squirrel, *Paraxerus,* and the larger African ground squirrel, *Xerus erythropus*. Both are known from the Omo Beds (Wesselman 1984), and *Paraxerus* has been reported from Laetoli (Denys 1987). A Pliocene squirrel (KB03-97-162) from Chad, *Xerus daamsi,* dated biochronologically to 4–5 Ma, also sports a minute P^3, yet this tooth is bicuspid (Denys et al. 2003), and the Alayla P^4 is clearly broader buccolingually. The large size of ALA-VP-2/186 does not support inclusion of the specimen into *Paraxerus* as reported in Haile-Selassie et al. (2004c), although taxa included in this genus can be variable.

In both the fossil specimen ALA-VP-2/186 and in the large contemporary species *X. erythropus,* the P^3 is small and peg-like. In *X. erythropus* it is tucked in close to the anterolabial corner of the P_4, whereas in the fossil specimen it is more in alignment with the midline of the P_4. Preliminary measurements reveal the fossil teeth to be considerably smaller in both length and width than those of the larger contemporary species, as

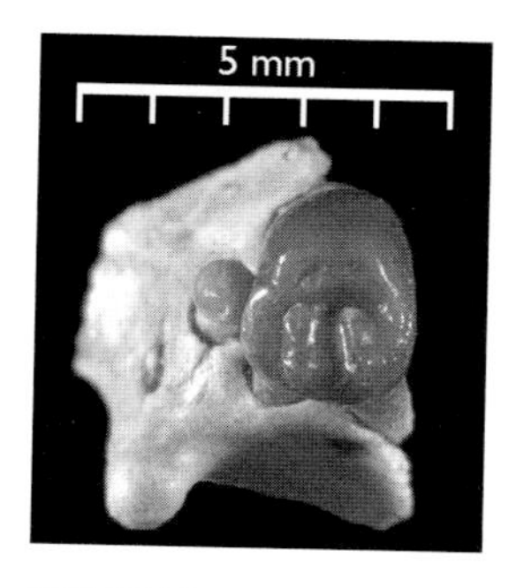

FIGURE 5.17

Xerus sp. ALA-VP-2/186, occlusal view.

TABLE 5.11 Dental Measurements of the Sciurid Genera *Xerus* and *Paraxerus*

Taxon	*Specimen*	*Tooth*	*Length*	*Breadth*	*n*
Xerus sp.	ALA-VP-2/186	P^3	0.80	0.90	1
		P^4	2.35	3.19	1
Paraxerus ochraceus	Various (Omo Member B)	P^3			4
		P^4	1.38–1.51	1.60–1.88	4
Paraxerus	Laetoli 74-304	P^3			1
		P^4	1.45	1.72	1
Xerus indet.	Omo Member F	P^3			1
		P^4	2.55	2.98	1
Xerus daamsi	KB03-97-162	P^3	Not given	Not given	1
		P^4	2.50	2.80	1
Xerus erythropus (contemporary)	Various	P^3	0.95–1.50	1.20–1.35	10
		P^4	2.70–2.96	3.55–3.67	10

NOTE: Measurements are in mm.

well as the fossilized lower teeth of *X. erythropus* from Omo Member B (Table 5.11). The teeth of *X. erythropus* are more inflated, and thus more bunodont, than those of the fossil specimen. The teeth of the contemporary *Xerus rutilis* are a closer match in both size and morphology (so too is the *Xerus* sp. indet. specimen from Omo Member F), yet the modern *X. rutilis* lacks a P^3. Cladistic analysis of the craniodental features of the Chadian specimen reveals closest affinity with *X. rutilis* as well, yet, like the Alayla specimen, *X. daamsi* expresses a P^3.

The Alayla specimen thus appears to show close affinity with both *Xerus rutilis* and *X. erythropus,* and given its age it may be ancestral to one or both. In accordance with the paucity of the sample, it is best considered as *Xerus* sp., pending recovery of more material.

The family Sciuridae has been documented to be conservative in most of its cranial aspects, and the sciurids were well-established in Eurasia and North America during the Miocene (Nowak 1991). It is also at this time that they probably entered Africa from Asia, where they exhibit considerable species diversity (Kingdon 1997). The late Miocene sciurids of Africa include *Atlantoxerus* from the Sahabi in northern Africa (Munthe 1987), and *Paraxerus* and *Xerus* from the Pliocene deposits of Omo, Ethiopia (Wesselman 1984), and Laetolil Beds of Tanzania (Denys 1987). The fossil *Paraxerus* and *Xerus* specimens from Omo are very similar to the extant genera, and it is probable that each fossil genus gave rise to the extant forms.

Modern *Xerus erythropus* is distributed across Africa, from Mauritania to southwestern Ethiopia and Kenya and southward to northern Tanzania. Extant members of *Xerus rutilis* are confined to northeastern Africa (from Somalia and Ethiopia into Sudan and then southward into Kenya, Uganda, and northeastern Tanzania), and the species lives in more arid country than does *X. erythropus,* which is primarily found in dry open woodlands and Sudanic savannas between the Sahara and the tropical forests (Amtmann 1971), although the two are sympatric where their ranges intersect. Like the viverrid *Helogale,* also found in the western margin deposits, contemporary *Xerus* are often found living in termitaria (Kingdon 1971, 1997).

Conclusions

It is suggested that a mosaic of biotypes was present during the late Miocene in the Middle Awash study area. More mesic situations were proximal to the site(s) of deposition, becoming more xeric distally.

The presence of the porcupine *Atherurus* and the crocidurine shrews suggests that there were forests and well-developed mesic woodlands present, supported by a high water table, high rainfall, or both. This forest may have taken the form of riverine and lake margin forest, spreading laterally across the floodplain and hilly rift margins, possibly grading into more deciduous woodlands distally. WoldeGabriel et al. (2001) and Haile-Selassie et al. (2004c) have confirmed that the Adu-Asa Formation is characterized by thick fluvial and lacustrine units indicating the presence of lakes and river systems, in turn confirmed by carbon isotope data suggesting the presence of woodland to grassy woodland habitat at many sites. The presence of *Thryonomys* is indicative of mesic, wet long-grass and marshland aquatic environments close to the site(s) of deposition.

Species present in the assemblages that suggest the presence of dry acacia savannas and more open savanna woodlands include *Procavia, Xerus, Tatera, Serengetilagus, Lemniscomys*, and possibly *Thronomys gregorianus*. Still drier scrub or even arid steppe habitats must have been present, as attested primarily by the presence of *Procavia, Xerus, Tatera, Serengetilagus,* and *Alilepus*. It may be that these dry, open environments were at greater distance from the sites of deposition or possibly closer to the rocky broken country of the fault zones or even the rift margins themselves. Some of these differences also reflect the long temporal span of the combined assemblage.

Several of the described taxa have descendants that typically take refuge in termitaria. These include *Xerus erythropus, Atherurus,* and the dwarf mongoose *Helogale* among the microcarnivores, suggesting the possibility that termitaria were present near the sites of deposition as well. Termites typically prefer drier, well-drained soils of open grasslands, but termitaria are also found in tropical forests and deciduous woodlands as well as more open savanna woodlands, depending on the termite species and the presence of preferred substrate types.

Although the western margin localities are found at altitudes ranging from 560 to 760 m today, several of the fossil micromammal taxa have descendants that are typically found at high altitudes—in mesic montane forests and in high-altitude uplands and moorlands—suggesting the possibility that the fossil localities of the late Miocene were more elevated in the past, above 2,000 m in altitude or even higher at 4.9–5.7 Ma (Haile-Selassie, 2001a; WoldeGabriel et al. 2001; Haile-Selassie et al. 2004c). These include, first and foremost, the prevalence of *Tachyoryctes* and the presence of *Lophiomys,* with support from *Crocidura, Procavia, Atherurus,* and possibly *Xenohystrix*. Chapter 17 integrates the small mammal faunal indications with a wider range of other paleoenvironmental indicators.

6

Cercopithecidae

STEPHEN R. FROST, YOHANNES HAILE-SELASSIE, AND LESLEA HLUSKO

A total of 107 cercopithecid fossils has been recovered from the late Miocene of the Middle Awash. Of these, the majority were collected by the Middle Awash research project between 1991 and 2003. Seven were collected by the Rift Valley Research Mission in Ethiopia (RVRME) between 1976 and 1978. In spite of the relatively small sample and its fragmentary nature, a minimum of three species are represented (Haile-Selassie 2001a): one papionin and two colobines. Only two of these can be allocated to genus. The papionin is the medium-sized *Pliopapio alemui* Frost, 2001. Colobines include *Kuseracolobus aramisi* Frost, 2001 as well as a second larger species. In addition to these three, there is a potential fourth taxon, smaller than all of the above forms, which cannot be allocated to subfamily.

Pliopapio alemui is best known from its type locality: the early Pliocene areas of Aramis, Kuseralee, and Sagantole in the Middle Awash, all dated to 4.4 Ma. There the taxon is known from a complete male skull as well as several maxillary and mandibular specimens and a large dental sample (Frost 2001b). There are a few isolated teeth that may belong to this species from the Adgantole Member of the Sagantole Formation (Frost 2001a). Additionally, there is a sizable sample of isolated papionin teeth from the Haradaso Member that are the same size (Frost, unpublished data). This undescribed material, approximately 4.9 Ma, provides a temporal bridge between the more complete material of the Aramis and Kuseralee Members.

In the late Miocene, *P. alemui* is known by a reasonably well-preserved associated maxilla and mandible from the Kuseralee Member along with other more fragmentary material tentatively assigned to this taxon. A few fragmentary specimens from the Adu-Asa Formation are tentatively assigned to *P. alemui*. This evidence provides an expanded temporal range for *P. alemui* from 5.2 to 4.4. Ma, with tentative identifications extending the possible range to as early as 5.7 and as young as 4.2 Ma.

Kuseracolobus aramisi is also best known from Aramis, where it is represented by several maxillae, mandibles, and a large dental sample (Frost 2001b). A larger sample of isolated teeth and a partial skeleton may be the same species from the Haradaso Member. There are also a few isolated teeth from the Adgantole Member, and a larger apparent descendant species *K. hafu* is also known from the 4.1 Ma Asa Issie locality

(Hlusko 2006). A well-preserved mandibular corpus and symphysis from the Kuseralee Member can be assigned to this species, along with a number of more fragmentary specimens that also have *K. aramisi* affinities. There is a small series of colobine specimens from the Adu-Asa Formation identified here as cf. *Kuseracolobus aramisi*. The temporal range of *K. aramisi* therefore extends from 5.2 Ma to 4.4 Ma, with tentative identifications potentially pushing the first appearance date of *K. aramisi* to 5.7 Ma and last appearance up to 4.2 Ma.

A larger colobine species is represented by a single molar from the Kuseralee Member and a few isolated teeth from the Adu-Asa Formation. Additionally, fragmentary colobine specimens from the Haradaso Member of the Sagantole Formation are also large. If all of these very fragmentary fossils represent a single taxon, then it would have a temporal range of 5.7 though 4.9 Ma.

Cercopithecids in the Adu-Asa Formation and Kuseralee Member of the Sagantole Formation are rare compared to other taxa (Haile-Selassie et al. 2004c), with total specimens collected from late Miocene Middle Awash deposits constituting only 3.8 percent of the total number of collected vertebrate specimens identifiable below the family level. This is a substantially smaller proportion of the total fauna than at Aramis (WoldeGabriel et al. 1994), where both *K. aramisi* and *P. alemui* are also found.

Taxonomic allocations in this chapter rely primarily on the limited number of more complete specimens. Isolated teeth from the same localities are assigned to taxa based on size and a lack of morphological evidence to the contrary.

Given the overlap in size between males of *P. alemui* and females of *K. aramisi,* as well as the presence of a second colobine and a potential smaller fourth taxon, isolated postcranial elements cannot be attributed to any species with certainty. However, we describe the available postcranial specimens and make taxonomic inferences for them. Assignments of postcranial material should be considered to be more tentative than those of the cranial specimens.

Cercopithecinae

Papionini Burnett, 1828

Pliopapio Frost, 2001

(= or including *Parapapio* Jones, 1937: WoldeGabriel et al. 1994, in part)

GENERIC DIAGNOSIS See Frost (2001b).

Pliopapio alemui Frost, 2001

DIAGNOSIS See Frost (2001b).

DESCRIPTION All of this material comes from the Kuseralee Member. This species is similar in overall cranial and dental size to larger species of *Macaca,* such as *M. nemestrina,* but smaller than most extant baboons (Delson et al. 2000; Frost 2001b). Cranial and dental dimensions of specimens described here are presented in Tables 6.1 and 6.2, respectively.

TABLE 6.1 Mandibular and Maxillary Dimensions for *Pliopapio alemui* and *Kuseracolobus aramisi*

	AME-VP-1/64	*KL308-1*
Maxilla		
Inferior premaxilla-maxilla suture to alveolar process at distal limit of M^3	42	
Palate breadth at M^2/M^3 contact	39	
Mandibles		
Width across canines	19	22
Width M^2/M^3 contact	33	
Corpus depth at P^4/M^1 contact	21	25
Corpus depth at M^2/M^3 contact	19	27
Corpus breadth at M^2/M^3 contact	10	14

NOTE: Measurements are in mm.

TABLE 6.2 Dental Dimensions of Miocene *Pliopapio alemui*

Specimen	*Sex*	*Tooth*	*WS*	*W*	*O*	*L*
AME-VP-1/64	F	I^1		4.6		5.1
AME-VP-1/64	F	I^2		4.6		3.8
AME-VP-1/64	F	C^1		5.8		5.5
AME-VP-1/64	F	P^3	8	5.3		4.5
AME-VP-1/64	F	P^4	7	6.3		5.0
AME-VP-1/64	F	M^1	14	7.3	6.8	7.4
AME-VP-1/64	F	M^2	10	8.5	7.5	8.6
AME-VP-1/64	F	M^3	3	7.5	6.0	8.9
AME-VP-1/64	F	I_1		4.8		3.4
AME-VP-1/64	F	I_2		4.5		4.2
AME-VP-1/64	F	C_1		5.9		3.5
AME-VP-1/64	F	P_3		4.0	8.0	5.6
AME-VP-1/64	F	P_4	7	4.3		5.5
AME-VP-1/64	F	M_1	16	5.1	5.6	7.4
AME-VP-1/64	F	M_2	13	6.7	6.4	8.6
AME-VP-1/64	F	M_3	4	6.5	5.6	10.2
KL-205-1	?	M^X		9.8	8.5	10.2
ASK-VP-3/354	?	M^X	5	8.5	7.8	8.5
KL-205-2	?	dp^4	3	6.6	5.8	7.2
STD-VP-2/15	?	P_4	3	5.1		7.2
AME-VP-1/107	?	M_2	5	7.1	7.1	9.4
AME-VP-1/107	?	M_3	1	7.2	6.4	11.5
AME-VP-1/124	?	M_3	1	7.1	6.6	10.8
AMW-VP-1/38	?	M_3	2	7.6	6.7	11.4
ASK-VP-3/362	?	M_X			6.3	6.0
KL-205-3	?	M_X	0	6.2	6.4	8.8

NOTE: WS = Wear stage (after Delson 1973); W = maximum buccolingual diameter (mesial width in the case of molars); O = other measures: distal width for molars, flange length for P_3s; L = maximum mesiodistal diameter. Measurements are in mm.

FIGURE 6.1

Female *Pliopapio alemui*, AME-VP-1/64. Maxilla: A. Dorsal view. B. Left lateral view. C. Ventral view. Mandible: D. Left lateral view. E. Occlusal view. F. Right lateral view (reversed).

Frost (2001b) estimated the mean male and female body mass for this population at between 10 and 14 kg and 7 and 10 kg, respectively, based on the early Pliocene material from the Aramis Member of the Sagantole Formation. The late Miocene material described here is similar in size to that population.

Maxilla AME-VP-1/64 is the only specimen to preserve any maxillary morphology, and provides the majority of the evidence for the diagnosis of this taxon (Figure 6.1). It is associated with a mandible, described below. The right and left alveolar processes are preserved with the canine through M^3 bilaterally (isolated right I^1–I^2 are also present), and the alveoli for the incisors are also present, so that the left and right halves contact each other anteriorly and nearly so at the back of the palate, which is nearly complete on the left side but largely absent on the right. The origins of the zygomatic processes are preserved bilaterally, though more is present on the right side. The premaxillae preserve about halfway up the piriform aperture on the right and a little more on the left, though the left side is more damaged. The lateral aspects of the rostrum are essentially absent.

Based on what is preserved, it is unlikely that there would have been maxillary fossae present on the lateral aspects of the rostrum. This morphology is similar to the male holotype ARA-VP-6/933, the female ARA-VP-1/1723 (both figured in Frost 2001), as well as *Parapapio, Papio (Dinopithecus),* many species of *Macaca,* and *Theropithecus oswaldi.* It is distinct from *Papio (Papio), Lophocebus, Cercocebus, Mandrillus, Gorgopithecus, Theropithecus brumpti,* and *T. gelada.* The base of the zygomatic process is positioned superior to the M^2-M^3 contact, with a rounded inferior surface. This is a similar position to that of ARA-VP-6/933, but slightly posterior to the position in the ARA-VP-1/1723.

The premaxilla is slightly projecting with a small diastema between the lateral incisor and canine. The roots of the central incisors cause a bulge on the anterior surface of the premaxilla, causing its inferior margin to form a rounded V shape, as in most papionins. In lateral view, the piriform aperture is inclined approximately 50 degrees relative to the occlusal plane. The incisive region appears more projecting than that of KNM-LT 24111, a rostrum of a male *Parapapio lothagamensis* (figured in Leakey et al. 2003).

The palate elevates slightly posteriorly, so that it is approximately 0.5 cm deep between the canines, but 1 cm deep between the third molars. In inferior view, the dental arcade is arranged in a fairly smooth arc from M^3 to M^3, with the M^2s being positioned most laterally. This shape is typical of female papionins.

In its overall rostral shape, this specimen is similar to the female *P. alemui* maxillae ARA-VP-1/1723 and ARA-VP-1/1007 from Aramis. The rostrum is slightly narrower (39 mm breadth at the M^1-M^2 contact) than that of ARA-VP-6/933 (42 mm), but considerably shorter, as would be expected given the level of sexual dimorphism seen in most papionins. The preserved area of the rostrum is similar in length to ARA-VP-1/1723. In comparison to KNM-LT 24111, the Amba East specimen is relatively narrower and more elongate. This is striking, given that the Lothagam specimen is male and that males have longer, narrower rostra than conspecific females.

Mandibles Associated with the maxilla, AME-VP-1/64 preserves nearly the entire corpus and dentition, lacking the right incisors, both rami and gonial areas (Figure 6.1). The symphysis has been refitted, and there are substantial areas missing on the right side and on the base bilaterally. There is also some damage to the inferior margin. AME-VP-1/107 is a left mandibular corpus with M_2-M_3 and the distal roots of M_1. The corpus is preserved inferior to the teeth and extends to the inferior margin. The gonial area and most of the ramus are missing, as well as much of the lateral cortical bone. It is substantially deeper than AME-VP-1/64 and therefore may represent a male individual (Table 6.1). AME-VP-1/202 is a fragment of a left mandibular corpus with M_1-M_2. It preserves only a small area of the corpus inferior to the teeth. AME-VP-1/124 is a large left M_3 with some of the mandibular corpus preserved around it, extending to below the roots and preserving part of the extramolar area.

In lateral profile, the symphysis as preserved on AME-VP-1/64 is sloping, approximating that of *Macaca fascicularis.* It is also similar to ARA-VP-1/563 and other materials from Aramis in its profile and depth (Frost 2001b). This is more sloping than in most papionins, but less so than in either *Parapapio lothagamensis* or ?*P. ado* (see Leakey et al.

2003 and Leakey and Delson 1987, respectively). The anterior surface has only very faint mental ridges. The region of the median mental foramen is damaged, but the part that is preserved indicates that one was originally present. The planum alveolare is moderately sloping and extends posteriorly to be even with the distal part of P_3s. The inferior torus is not preserved, but from what is present, it would have been similar in development to the superior torus, and not significantly longer as in the KNM-LT 23091 *P. lothagamensis* mandible.

The lateral surface of the corpus lacks fossae in AME-VP-1/64, AME-VP-1/107, and AME-VP-1/202. The corpus is relatively shallow and gracile and is nearly even in depth from anterior to posterior, though it does shallow slightly posteriorly, with its deepest point positioned inferior to the M_1. The mental foramen is not preserved on the left side. It is infilled by matrix on the right and located inferior to the anterior portion of the P_4 approximately two-thirds of the distance from the alveolar process to the inferior margin of the corpus. In superior view, the corpus of all specimens is narrow, with a narrow extramolar sulcus preserved on AME-VP-1/64 and AME-VP-1/124. AME-VP-1/107 has more of an extramolar sulcus, but it is still relatively narrow.

Dentition In addition to several more fragmentary dental specimens, nearly every element of the dentition is represented on the female specimen AME-VP-1/64. In general, the dentition is typical of most papionins other than *Theropithecus* and similar to that of *P. alemui* from Aramis. The upper incisors are relatively large and spatulate in morphology. The central incisor is not as large in relation to the other teeth as in *Lophocebus* or most Cercopithecini. The lateral incisors are relatively large and not caniniform. The lower incisors lack lingual enamel. The canines are typical of the family. The P_3 has a low protoconid and a mesiobuccal flange that is relatively long for a female, though significantly shorter than those of males from Aramis. The P_4 has a small mesiobuccal flange, relatively low, bunodont cusps, and a tall lingual notch. The molars are relatively high-crowned compared to many papionins, but much lower than those of *Theropithecus*. As in other papionins the cusps are relatively low, the loph(ids) are not as well-developed as those of colobines, and the lingual/buccal notches are relatively high on the lower/upper molars. The amount of basal flare is relatively low compared to most papionins, but similar to *P. alemui* from Aramis (Figure 6.2A). The M_3s of AME-VP-1/64 and AME-VP-1/107 lack *tubercula sexta,* whereas AME-VP-1/124 has a very small one.

cf. *Pliopapio alemui*

DESCRIPTION Four isolated papionin teeth from the Adu-Asa Formation are tentatively included here on the basis of their similarity in size and morphology to the *P. alemui* material from the Aramis and Kuseralee Members.

Colobinae

Kuseracolobus Frost, 2001

(= or including Colobinae sp. A. Eck, 1977: WoldeGabriel et al. 1994; in part cf. *Paracolobus* sp. Leakey, 1969: WoldeGabriel et al. 1994, in part)

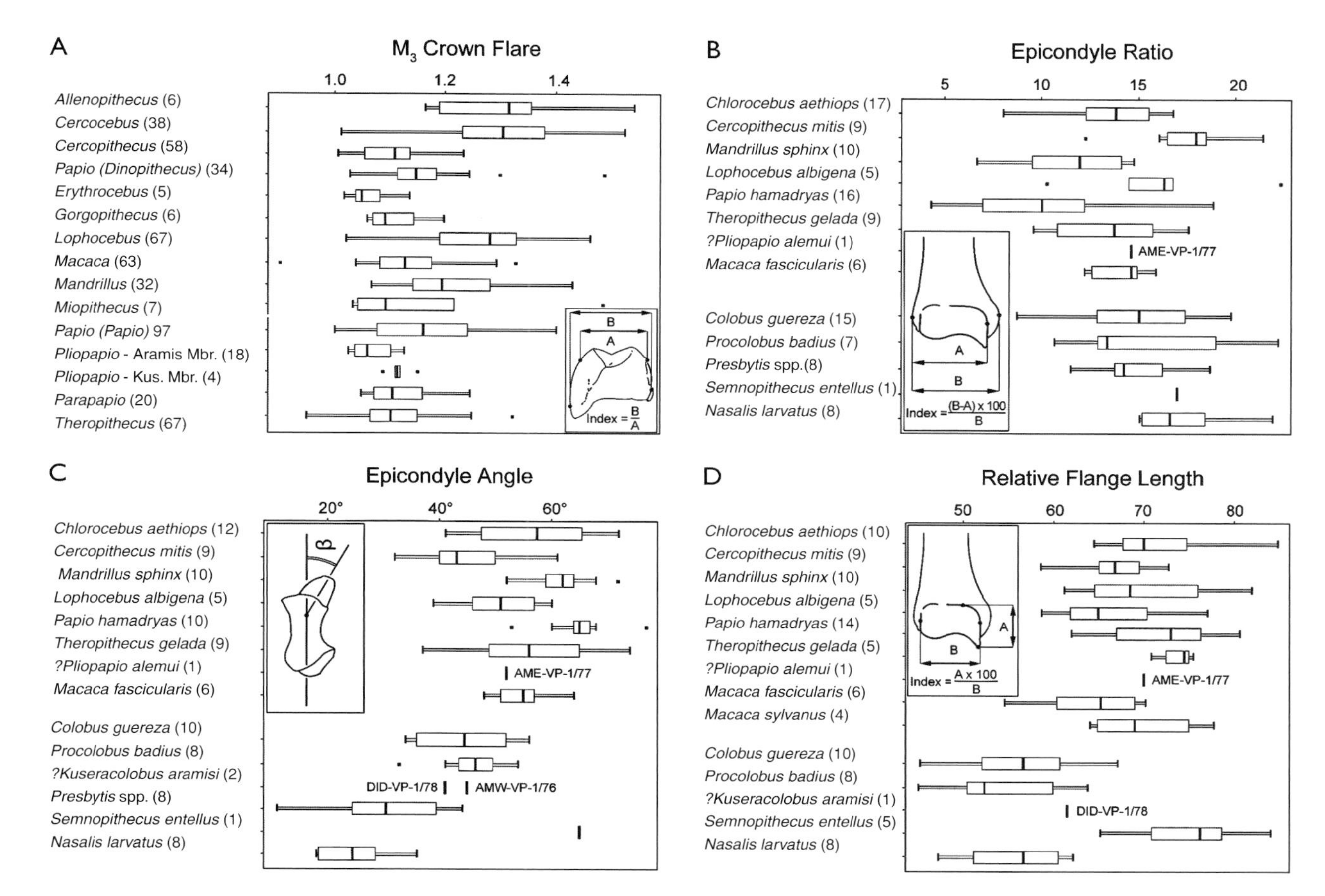

FIGURE 6.2

Box and whisker plots of comparative data. The central bar of each box represents the median, or 50th percentile. The left and right side of each box represent the 25th and 75th percentiles, respectively. The whiskers extend to the farthest observation that is less than 1.5 the length of the box. Any individuals outside of the whisker range are marked separately. Numbers in parentheses represent sample sizes. A. M_3 basal flare: the ratio of M_3 maximum mesial breadth to M_3 mesial breadth at the height of the notch. Some data from M. Singleton. B. Medial epicondyle projection: [biepicondylar breadth (*B*) – medial distal articular limit to lateral epicondyle (*A*)] × 100/ biepicondylar breadth (*B*). Some data from T. Harrison. C. Angle of medial epicondyle relative to axis of distal articular surface. Some data from T. Harrison. D. Medial trochlear flange length (*A*)/distal humeral articular breadth (*B*). Some data from T. Harrison.

GENERIC DIAGNOSIS See Frost (2001b).

Kuseracolobus aramisi Frost, 2001

(= or including Colobinae sp. A. Eck, 1977: WoldeGabriel et al. 1994; in part cf. *Paracolobus* sp. Leakey, 1969: WoldeGabriel et al. 1994, in part)

DIAGNOSIS See Frost (2001b).

DESCRIPTION All of this material comes from the Kuseralee Member. This species is similar in overall cranial and dental size to *Nasalis larvatus,* some larger subspecies of *Semnopithecus*

TABLE 6.3 Dental Dimensions of Miocene *Kuseracolobus aramisi*

Specimen	*Sex*	*Tooth*	*WS*	*W*	*O*	*L*
ASK-VP-3/363	?	I^1		(4.7)		5.0
ALA-VP-2/341	?	I^2		4.3		
ASK-VP-3/53	?	M^3	2	8.6	7.0	8.9
ALA-VP-1/26	?	M^X	10			8.9
ASK-VP-3/52	?	M^X	3	9.1	7.8	9.4
DID-VP-1/54	?	M^X	10	9.0	8.0	8.5
STD-VP-2/14	?	M^X	2	8.1	6.3	7.6
STD-VP-2/19	?	M^X	0	8.8	7.4	9.1
STD-VP-2/866	?	M^X	1	8.9	8.5	9.3
ASK-VP-3/55	?	di^1		2.3		4.0
KL-308-1	M	I_1		4.6		3.2
KL-308-1	M	I_2		4.8		3.4
KL-308-1	M	C_1		8.6		6.1
KL-308-1	M	P_3		7.9	10.8	8.3
KL-308-1	M	P_4	3	5.4		6.7
KL-308-1	M	M_1	8	6.1	6.4	8.0
KL-308-1	M	M_2	6	7.6	7.7	9.3
KL-308-1	M	M_3	4	7.6	7.0	11.7
ALA-VP-2/141	?	I_1		4.7		4.0
ASK-VP-3/170	?	I_1		5.0		4.0
ASK-VP-3/152	?	I_2		4.3		3.6
ALA-VP-2/165	?	I_2		4.2		2.8
ALA-VP-2/318	?	I_2		4.7		4.2
ASK-VP-3/353	?	M_3	7	8.2	7.8	12.4
DID-VP-1/39	?	M_3	2	8.0	7.7	12.2
KL-212-1	?	M_3	7	(7.3)	7.2	
AMW-VP-1/66	?	M_X	8	6.9		8.6

NOTE: Column headings as for Table 6.2. Numbers in parentheses are estimated. Measurements are in mm.

entellus, and *Cercopithecoides meaveae* (Delson et al. 2000; Frost and Delson 2002; Tables 6.1 and 6.3). Frost (2001b) estimated the mean male and female body mass for this species to be between 14 and 22 kg and 10 and 14 kg, respectively, based on the early Pliocene material from the Aramis Member of the Sagantole Formation. The late Miocene material described here is similar in dental and mandibular size to that population.

Mandible The only specimen to preserve the mandible is KL-308-1 (Figure 6.3). This specimen was collected by the RVRME from Amba and identified as cf. *Libypithecus* sp. by Kalb et al. (1982c). It is from a male individual and preserves the symphysis, nearly the entire left side of the corpus, and the right side posterior to the P_4. Of the dentition, the entire left side is preserved, except for the broken left canine, the right I_1, and P_4, along with the broken right canine and roots of the right P_3.

In overall morphology, KL-308-1 is similar to the mandibular material of *K. aramisi* from Aramis (Frost 2001), particularly ARA-VP-1/87 and ARA-VP-6/796, being overall deep and robust. The symphysis is short, deep, and relatively vertical in lateral view. Its

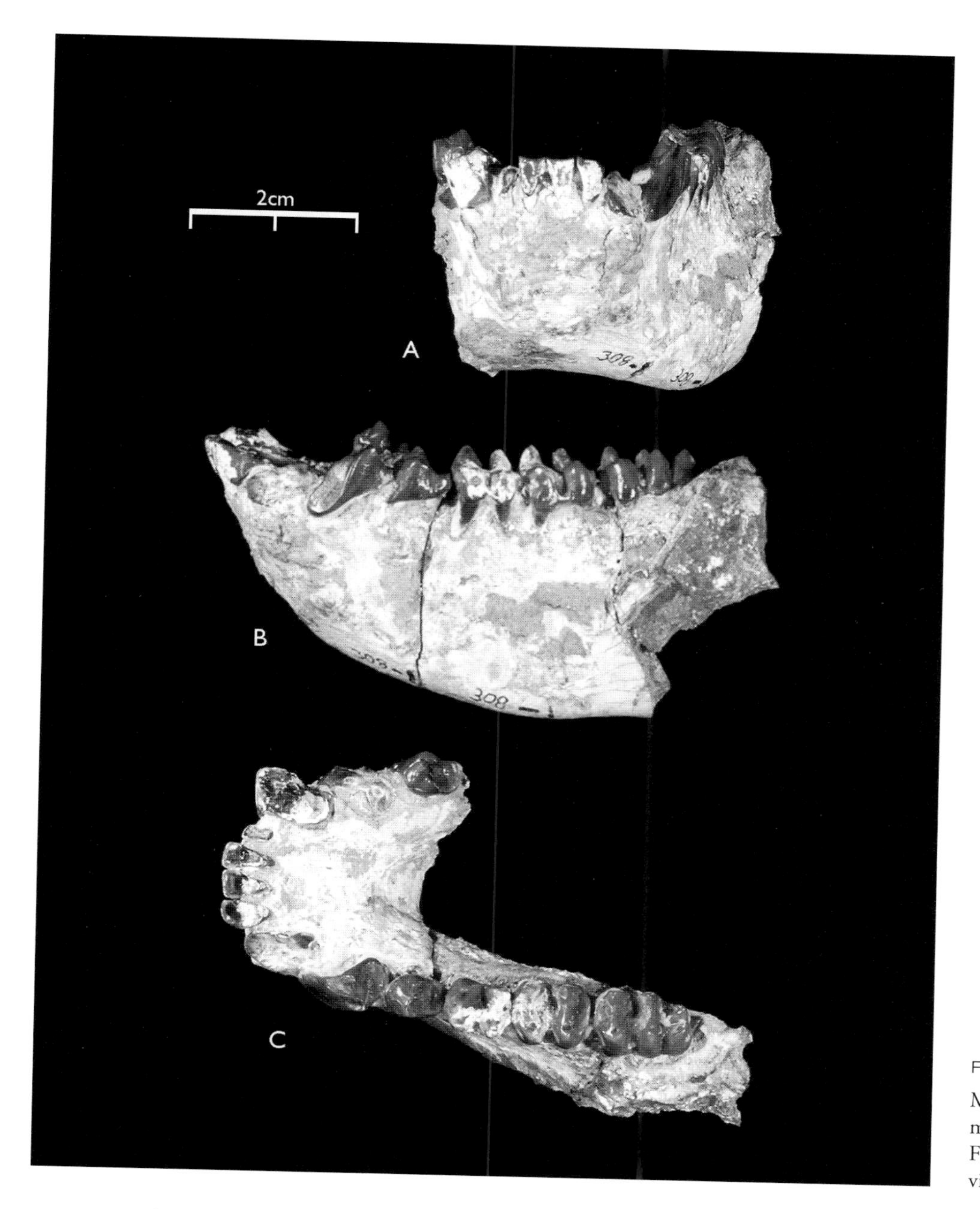

FIGURE 6.3

Male *Kuseracolobus aramisi* mandible KL-308-1. A. Frontal view. B. Left lateral view. C. Occlusal view.

anterior surface is half obscured by matrix but appears to be unmarked by mental ridges or a median mental foramen. The lateral surface of the corpus, particularly anteriorly, has slight fossae, largely because the robust and prominent lateral bulges near the inferior margin cause the widest part of the corpus to be near its inferior margin. In lateral view, the corpus deepens posteriorly from P_3 to M_3, with an inferior bulge below the M_1/M_2 contact. The corpus is also relatively deep overall, deeper than that of *Cercopithecoides* and *Procolobus*. In superior view, the corpus is broad, with a wide extramolar sulcus. Although the gonial region is absent, from what is preserved near the posterior edge of the corpus, there may have been some expansion of the gonion. This is similar to the condition in ARA-VP-1/87, *Colobus, Rhinocolobus,* and *Paracolobus mutiwa,* but generally different from other African colobines.

Dentition Within the late Miocene material for this taxon, the entire lower adult dentition is represented. In general, the morphology is typically colobine and consistent with descriptions of *K. aramisi* from Aramis (Frost 2001b). The lower incisors are small relative to the molars, and the crowns are peg-like, particularly in relation to extant *P. badius*. Lingual enamel is clearly present and of normal thickness. The lateral incisor has a distinct lateral "prong" (Szalay and Delson 1979). The P_3 has a small paraconid, in contrast to those of papionins, which generally lack it, and the talonid is relatively large and extends lingually. The P_4 has relatively tall cusps and a deep lingual notch, and a small mesiobuccal extension is present, though far shorter than that of the P_3. KL-308-1 has a small cuspule at the mesial inferior limit of the mesiobuccal flange, as occurs occasionally on colobine P_3s. There is no reduction of the P_4 metaconid. The molars are typical colobine teeth with tall, widely spaced cusps and relatively deep lingual notches. The lophids are sharp and well-developed. The distal lophid of the M_3 is narrower than the mesial lophid. The M_3 of KL-308-1 has a well developed *tuberculum sextum,* as occurs frequently in the Aramis series.

cf. *Kuseracolobus aramisi*

DESCRIPTION Isolated colobine teeth from the Adu-Asa Formation that are consistent with *K. aramisi* material from the Aramis and Kuseralee Members in size and morphology are tentatively included here. The teeth represented include upper and lower incisors and molars and a deciduous tooth. The lower central incisors from the Adu-Asa Formation are slightly larger than those from the KL-308-1 mandible or from the Aramis sample. They are not, however, out of the normal range of variation observed within most extant species. ASK-VP-3/55, a central upper incisor, shows several features that distinguish it from cercopithecine homologs. The crown narrows toward its apex and has a small lingual cingulum, and the labial surface of the root lacks a sulcus. Distinct from permanent colobine upper central incisors, the crown is labiolingually flattened and mesiodistally broad, as is the root.

Gen. et sp. indet. "Large"

DESCRIPTION There is a single specimen assigned to this taxon from the Kuseralee Member and three from the Adu-Asa Formation (Figure 6.4). There are six additional specimens from the Adu-Asa Formation that are tentatively included here as well. This sample consists entirely of isolated teeth. The Kuseralee Member specimen, AMW-VP-1/114, is a complete crown of a right upper molar, whose exact position is indeterminate as a result of poor preservation of surface detail and the general difficulty in determining tooth position in isolated colobine molars. The enamel preservation is chalky, so no evidence of interproximal wear is preserved, which would assist in refining the position of this molar. ASK-VP-3/365 is the labial aspect of the crown and the cervical 2 mm of the root of a large left upper central incisor. ASK-VP-3/366 is the distal part of a lower incisor crown, most likely of

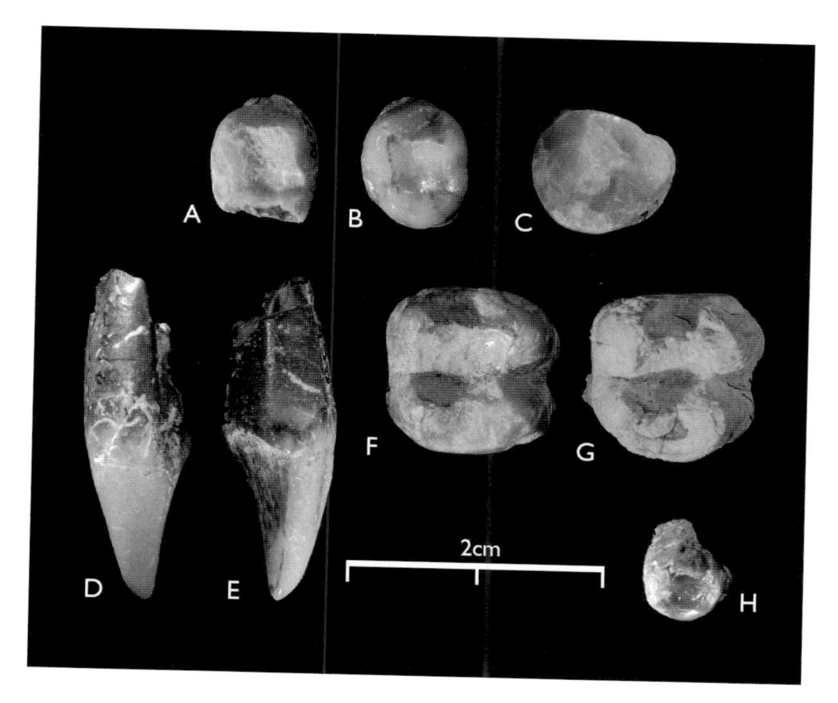

FIGURE 6.4

Teeth of Colobinae gen. et sp. indet. "Large" and Cercopithecidae subfamily indet. A. ASK-VP-3/358 (L. P_4). B. ASK-VP-3/359 (L. P_4). C. ASK-VP-3/357 (R. P^4). D. ASK-VP-3/356 (L. C^1, distal view). E. ASK-VP-3/356 (L. C^1, mesial view). F. ASK-VP-3/355 (R. M^X). G. AMW-VP-1/114 (R. M^X). H. ALA-VP-2/122 (R. P^4 of Cercopithecidae subfamily indet. A).

an I_1. ASK-VP-3/356 is a right upper canine of a female individual. ASK-VP-3/357 is a right upper fourth premolar with a low amount of wear and no obvious interproximal facets. ALA-VP-2/161 is a left upper premolar, most likely a P^3. ASK-VP-3/358 and ASK-VP-3/359 are both left lower fourth premolars, both of which had some mesiobuccal extension of the crown, which has been broken. The distal extreme of ASK-VP-3/358 is also missing as a result of damage. ASK-VP-1/19 is a left upper molar with little wear, but the tip of the protocone (and possibly that of the hypocone) is broken. ASK-VP-3/355 is a right upper molar, with a small mesial interproximal facet, but no wear facet is apparent on the distal end, which is expected given the limited occlusal wear. These specimens are larger in overall size than corresponding elements assigned to *K. aramisi,* either from the Miocene (above) or the early Pliocene at Aramis (Frost 2001b). In size they are similar to *Rhinocolobus,* larger species of *Cercopithecoides,* and the colobine from Menacer (Freedman 1957; Delson 1973; Leakey 1987). Dental dimensions are presented in Table 6.4.

ASK-VP-3/365 is most likely a large upper central incisor of a colobine, but because of damage there is some possibility of its representing *P. alemui.* While the lingual aspect is more distinct between the subfamilies, the crown does not flare apically in labial view, with the mesial and distal borders being approximately parallel. This morphology is more consistent with that of a colobine than a papionin. Although the lower incisor ASK-VP-3/366 is fragmentary, it clearly has lingual enamel and is unambiguously from a colobine. Given the lack of a "lateral prong," it is more likely to represent a lower central incisor (Delson 1973). Although its length cannot be measured, its labiolingual width is larger than any lower incisors allocated to *K. aramisi,* either from the Miocene (Table 6.3) or from the Pliocene (Frost 2001b).

TABLE 6.4 Dental Dimensions of Colobinae gen. et sp. indet. "Large"

Specimen	*Sex*	*Tooth*	*WS*	*W*	*O*	*L*
ASK-VP-3/365	?	I^1				6.5
ASK-VP-3/356	F	C^1		6.3	12.7	7.6
ALA-VP-2/161	?	P^x	5	8.2		6.3
ASK-VP-3/357	?	P^4	1	8.9		7.4
AMW-VP-1/114	?	M^X	1	10.0	9.2	9.8
ASK-VP-1/19	?	M^X	2	10.3	9.6	10.1
ASK-VP-3/355	?	M^X	4	10.2	9.8	10.0
ASK-VP-3/366	?	I_x		5.2		
ASK-VP-3/358	?	P_4	2	6.0		
ASK-VP-3/359	?	P_4	3	6.3		7.6

NOTE: Column headings as for Table 6.2. Measurements are in mm.

The upper premolar ASK-VP-3/357 is typical of colobines in morphology, with cusps that are tall relative to the talon and united by a sharp cross-loph. The talon is large and square in outline in occlusal view, with a well-developed cingulum, but no cuspules. ALA-VP-2/161, however, is more ambiguous. The cusps are not as tall, and the cross-loph is not as sharp as that of ASK-VP-3/357. Because the distinction between colobines and cercopithecines is less clear in P^3s, this difference between ASK-VP-3/357 and ALA-VP-2/161 may simply reflect the difference in tooth position. The lower P_4s have tall cusps, deep lingual notches, and large talonids. In spite of the fact that the distal end of the crown of ASK-VP-3/358 is missing, its preserved length of 7.4 mm is outside of the *K. aramisi* range from Aramis. To this preserved length, at least 1 mm would be added to achieve its original length, which would mean that it was somewhat longer than ASK-VP-3/359. Both teeth are wider than the range for *K. aramisi* from Aramis (Frost 2001b).

All of the upper molars share a similar morphology. As is typical of colobines, the cusps are widely spaced and connected by well-developed, sharp lophs. The crown shows less basal flare than in papionins, but more than in some colobines, and is similar to some *Cercopithecoides* in this regard. The smallest of the molars, AMW-VP-1/114, is 3 and 4 standard deviations above the average and more than one standard deviation above the maximum values of mesial and distal breadth, respectively, for all upper molars of *K. aramisi* from Aramis, regardless of tooth position. AMW-VP-1/114 has a well-developed protocone shelf (stage 5; Hlusko 2002). ASK-VP-1/19 is a little more ambiguous in morphology than AMW-VP-1/114 and ASK-VP-3/355. The buccal notch is a little taller, and the lophs do not appear to be as well-developed as those of the latter two teeth.

ASK-VP-3/356, a right upper canine from a female individual, is larger than those of either *K. aramisi* or *P. alemui* and falls over two standard deviations above the ranges observed in all Aramis canines for both breadth and length. It is larger than the other female canines assigned here to Cercopithecidae gen. et sp. indet. Its tentative allocation to this taxon is entirely based on size and not on any aspect of its shape.

Both ASK-VP-1/19 and ALA-VP-1/162 are less clearly colobine in their morphology than the remaining elements. Given their similarity in size to the large colobine material and the uncertain nature of their morphology, they are included here. If, however, they are not colobine, this would indicate the presence of a species of cercopithecine larger than *P. alemui*.

Gen. et sp. indet.

DESCRIPTION ASK-VP-3/504 preserves the region around the glabella, nasion, and the superior 1 cm of the maxillary portion of the interorbital pillar. The metopic suture is unfused for the preserved area. The sutures around nasion are complex, as is often the case in Colobinae. What are either the frontal processes of the maxillae or premaxillae (as sometimes occurs in the subfamily) extend superior to the nasion. The interorbital breadth is approximately 13.3 mm, which is a little larger than in the two known interorbital fragments from Aramis, where the breadth is 12.1 and 10.8 mm. Approximately 5 mm of the supraorbital rim is present bilaterally, not enough to determine whether a supraorbital notch or foramen would have been present. The supraorbital rims are approximately 6 mm thick, and a distinct ophryonic groove is present superiorly, posterior to which the frontal begins to rise.

ASK-VP-3/514 is more fragmentary than ASK-VP-3/504, with approximately 1 cm of the right supraorbital rim and 9 mm of the interorbital area (which is broken to the left of the midline, so that the full breadth would have been approximately 13 mm). The interorbital area inferior to the nasion is also absent. The supraorbital rim is 5 mm thick and lacks a supraorbital notch or foramen.

Both of these specimens are of a size compatible with larger individuals of *K. aramisi* or smaller individuals of the larger colobine. Therefore a more specific allocation is not made here. Perhaps with more complete cranial material such an allocation will be possible in the future.

STD-VP-2/921 is a small fragment of a left lower molar preserving just the area around the lingual notch, the depth of which clearly indicates that this is a colobine, but it cannot be determined whether it represents *K. aramisi* or the larger form.

Cercopithecidae

Subfamily indet. A

DESCRIPTION ALA-VP-2/122 is a lightly worn (wear stage 2) right P^3 that is smaller than the P^3s of any of the taxa previously described (buccolingual breadth 4.5 mm, mesiodistal length 5.6 mm; Figure 6.4). This specimen is smaller than the smallest female P^3 known from Aramis, ARA-VP-1/1723, a female of *P. alemui*, being approximately 1 mm, or about 20 percent, narrower (Frost 2001b). It therefore may indicate the presence of a fourth cercopithecid species in the late Miocene of the Middle Awash. It is also distinctive in its morphology; the mesial region is extended with a small cuspule (paracuspule) and a concave mesial edge that would have articulated with the canine. As a result of this extension, the mesiodistal length is comparable to that for P^3s of *P. alemui* and *K. aramisi*. There is also a mesiobuccal flange present; it occurs variably within many extant cercopithecid species. The enamel of this specimen appears to be marked by a hypoplasia as well. As is typical of cercopithecid upper premolars, three small cylindrical roots are present, though the distal-buccal root is broken near its base.

TABLE 6.5 Dental Dimensions of Cercopithecidae subfamily indet. B

Specimen	*Sex*	*Tooth*	*WS*	*W*	*O*	*L*
ALA-VP-2/162	F	C^1		5.6		7.2
ASK-VP-3/150	F	C^1		5.4		6.7
ASK-VP-3/151	F	C^1		4.4		6.2
ASK-VP-3/169	F	C^1		5.7		7.4
DID-VP-1/79	F	C^1		5.7		6.8
ASK-VP-3/54	?	P^4	0	6.1		5.5
STD-VP-2/867	?	dc^1		3.4		4.9
AMW-VP-1/85	M	C_1		11.0		5.7
ASK-VP-3/406	F	C_1		6.0		4.4
DID-VP-1/36	F	C_1		(7.2)		4.8
AME-VP-1/150	?	P_3	7	4.2	13.2	6.9
ASK-VP-3/360	?	P_4	6	5.0		6.8
ALA-VP-2/163	?	M_X				(8.2)

NOTE: Column headings as for Table 6.2. Numbers in parentheses are estimated. Measurements are in mm.

Subfamily indet. B

DESCRIPTION Nearly all of these specimens are isolated dental remains, including canines and several other teeth that are either damaged or ambiguous in their morphology and therefore do not show affinities with either subfamily. With one exception, they are all in a size range compatible with either *P. alemui* or *K. aramisi* (Table 6.5). That exception is AME-VP-1/185, a left maxillary fragment preserving most of the root of the canine and the P^3-P^4. Little of the maxillary surface is preserved, and what is present is heavily cracked, obscuring the morphology. The premolars do not show clear affinity to either subfamily, as the crowns are partially worn, broken, and infilled with matrix. For this reason, their overall sizes cannot be reliably measured but are estimated to be only slightly larger than the observed range for *K. aramisi* from the Aramis Member of the Sagantole Formation, but they are also smaller than the P^4 ASK-VP-3/357, tentatively allocated to Colobinae gen. et sp. indet. "Large." Similarly, the breadth of the upper canine is greater than that observed for either *K. aramisi* or *P. alemui* but is less than one standard deviation above the maximum observed value.

Postcranial Descriptions

The only postcranial element that might belong to *Pliopapio alemui* is AME-VP-1/77 (Figure 6.5). It preserves the complete distal articular surface of a left humerus, its epicondyles, and approximately 2 cm of the shaft proximal to this, but there is some damage to the distal extreme of the medial trochlear flange. In size, it is larger than the distal humerus of *Chlorocebus aethiops* and *Macaca fascicularis,* similar in size to *Lophocebus albigena* and *Colobus guereza,* and smaller than extant *Papio hamadryas* and *Nasalis larvatus*. It is within a range expected for perhaps a small male or larger female of *P. alemui* and is smaller than would be expected for females of *K. aramisi*.

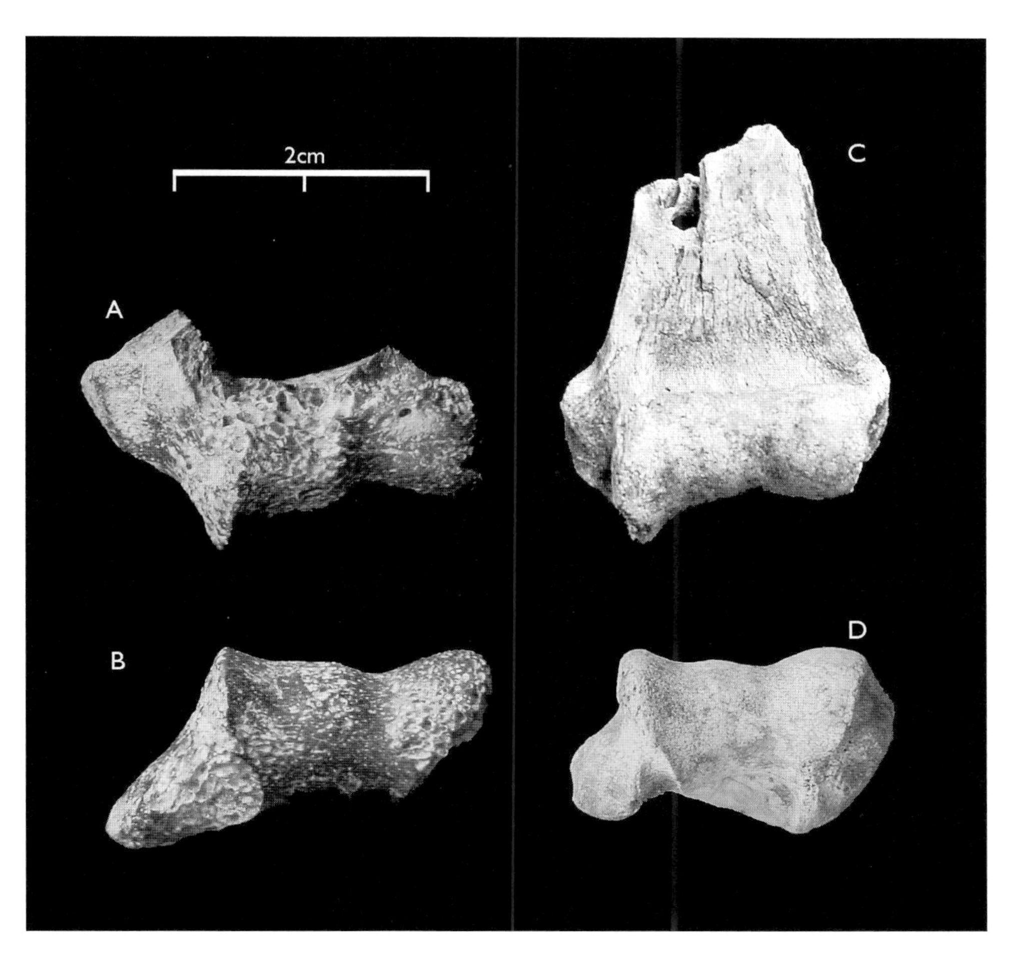

FIGURE 6.5
Miocene distal humeri from the Middle Awash. DID-VP-1/78 (?*Kuseracolobus aramisi*): A. Anterior view. B. Distal view. AME-VP-1/177 (?*Pliopapio alemui*): C. Anterior view. D. Distal view.

Morphologically, it shows several adaptations intermediate between those of terrestrial forms such as *Papio* and *Theropithecus* and more arboreal forms such as *Lophocebus* and *Colobus*. The brachio-radialis flange is moderately marked, as are those of most papionins, but differing from *Cercocebus* and *Mandrillus* (Fleagle and McGraw 2002). The supraradial fossa appears larger and deeper than the coronoid fossa. The articular surface is broad relative to the overall width of the distal end, a feature typical of most papionins and associated with a more terrestrial locomotion (Jolly 1972; Harrison 1989). The medial epicondyle is moderate in length and is retroflexed approximately 52 degrees. In comparison to humeral breadth, the medial epicondyle of AME-VP-1/77 is shorter than that of *Cercopithecus mitis* and most *L. albigena* but is also longer than in most *Papio* or *Theropithecus* (Figure 6.2B). In angle of retroflexion, AME-VP-1/77 falls in the middle of the range for most cercopithecids, being more retroflexed than most colobines (though it is in the upper part of the range for *Colobus guereza* and *Procolobus badius,* but less so than for *Papio*). It is actually in the range of overlap of most species (Harrison 1989; Frost and Delson 2002; Figure 6.2C). The capitulum is more rounded and more spherical than those of many papionins, but less so than those of most colobines. The lateral trochlear margin is marked by a slight keel. The medial trochlear keel is fairly short by cercopithecine standards but longer than those of most colobines, except for *Semnopithecus entellus* (Delson 1973; Frost and Delson 2002; Figure 6.2D).

There are two distal humeral fragments that can be tentatively allocated to *K. aramisi*. DID-VP-1/78 is a small fragment of the distal articulation and damaged medial epicondyle of a left humerus (Figure 6.5). AMW-VP-1/76 consists of a trochlea and medial epicondyle from a left humerus associated with some shaft fragments. The latter specimen would have been larger than the former, and both are larger than AME-VP-1/77. In size, they are within the range of *Nasalis larvatus* and females of *Papio hamadryas*. DID-VP-1/78 is in a size range expected for larger females or smaller males of *K. aramisi,* whereas AMW-VP-1/76 would have been in the size range for larger males.

In preserved morphology DID-VP-1/78 and AMW-VP-1/76 are similar to each other and different from AME-VP-1/77. The medial epicondyle in both specimens is long and projects medially with only moderate retroflexion, approximately 41 and 45 degrees in DID-VP-1/78 and AMW-VP-1/76, respectively (Figure 6.2C), measures equivalent to *Cercopithecus mitis, Colobus guereza,* and *Procolobus badius,* though they overlap the ranges of most taxa. The articular area of DID-VP-1/78 is relatively narrow in comparison to what is preserved of the distal end of the humerus, because of the large medial epicondyle. Even though the lateral portion is damaged, there is enough preserved to show that the capitulum was spherical and prominent. The lateral border of the trochlea is well-marked; with the zona conoidea it forms a distinct sulcus with the capitulum. The medial trochlear flange of DID-VP-1/78 is very short (Figure 6.2D), albeit damaged. The trochlea of AMW-VP-1/76 lacks its lateral edge, but the medial flange is longer than that of DID-VP-1/78 and appears more rounded in contour (perhaps as a result of damage in DID-VP-1/78).

KUS-VP-1/43 is a right astragalus with some damage to the posterior and medial limit of the trochlea. The medial and lateral margins of the trochlea are both relatively low and rounded, with the lateral margin being only a little higher than the medial margin, so that the articular surface has only a relatively shallow groove. The medial malleolar facet projects relatively far medially and is separated by a wide margin from the plantar surface, a feature of colobine astragali (Strasser 1988). The head is rotated medially out of the horizontal plane, also a feature of the Colobinae (Strasser 1988). This specimen may belong to *K. aramisi*.

There are several postcranial specimens whose affinities with any specific cercopithecid taxon cannot be determined at this time. Some common dimensions are presented in Table 6.6. These specimens are discussed in the following paragraphs.

AME-VP-1/15 is a right scapular fragment preserving the glenoid fossa, supraglenoid tubercle, base of the coracoid, and part of the acromion and spine, plus about 1.5 cm of the blade. In size, it is similar to those of extant *Theropithecus,* females of *Papio,* and males of modern *Colobus,* but smaller than those of *Papio* males. Relative to its anteroposterior length, the glenoid fossa is mediolaterally narrow (23 mm in length, 15 mm in breadth) in comparison to *Papio* and *Theropithecus,* and it is more similar in outline to *Lophocebus, Nasalis,* and *Colobus.*

ALA-VP-2/340 is a trochlear fragment of a right humerus. In size it is comparable to the other humeri just described. The fragment of the medial flange that is present appears

TABLE 6.6 Dimensions Taken on Various Postcranial Elements

Humeri	*AME-VP-1/77*	*AMW-VP-1/76*	*DID-VP-1/78*
Biepicondylar breadth	26		
Lateral epicondyle to medial edge of trochlea	22		
Distal articular breadth	18		(22+)
Medial trochlear flange length	13	(12+)	14

Femora	*ALA-VP-2/109*	*AMW-VP-1/155*
Anteroposterior head diameter	(22)	
Bicondylar breadth		33
Anteroposterior condyle length		25

Tibiae	*KUS-VP-1/24*	*KUS-VP-1/86*	*AME-VP-1/20*
Proximal anteroposterior depth	31		
Proximal mediolateral breadth	(35)		
Distal anteroposterior depth		19	17
Distal mediolateral breadth		24	20

Calcanei	*AMH-VP-1/4*	*ASK-VP-3/176*	*STD-VP-2/865*
Proximodistal tuberosity length		15	
Proximodistal proximal astragalar facet length	14	16	11
Distal length	14	13	11
Length		43	

Metapodials	*ALA-VP-2/112*	*STD-VP-2/17*	*STD-VP-1/27*	*ALA-VP-2/45*	*DID-VP-1/30*	*ALA-VP-2/113*
Proximal mediolateral breadth	7	9	7	7	12	
Proximal dorsoplantar depth	10	12	12	11		6

Phalanges	*AME-VP-1/38*	*AMW-VP-1/144*	*STD-VP-2/922*
Proximal mediolateral breadth	10	7	7
Length	20	15	

NOTE: Numbers in parentheses are estimated. Measurements are in mm.

to be short with a relatively rounded distal margin. The medial epicondyle is broken, but from what can be seen of its base, it may have been medially oriented. In both of these features, this specimen is more consistent with AMW-VP-1/76 and DID-VP-1/78 than with AME-VP-1/77. Given its fragmentary nature, no attempt at its subfamilial affinity is made here.

KUS-VP-1/25 is a proximal fragment of a right ulna missing the olecranon process and proximal half of the trochlear notch, preserving about 2 cm of the shaft. The radial facet is present, but damaged. It is approximately the right size to match AME-VP-1/77. Little functionally informative morphology is preserved. ASK-VP-2/3 is a small fragment of the olecranon process and proximal half of the trochlear notch from a right ulna. The medial and lateral margins of the trochlear notch are similar in their proximal extent, so that the superior rim is not undulated as in many terrestrially adapted species. As far as can be determined from this fragment, the olecranon process was relatively tall and angled slightly medially and does not appear to have been retroflexed.

STD-VP-2/36 is a fragment of the proximal portion of the shaft of a right radius lacking the head but preserving the tuberosity and approximately 2.5 cm of the shaft. DID-VP-1/7 is another proximal radial shaft fragment, also lacking the head but preserving the tuberosity and approximately 5 cm of the shaft. The shaft of DID-VP-1/7 has a moderate lateral curvature, flattens distally to form a sharp interosseus border with a marked flexor origin.

ALA-VP-2/109 is the head of a left femur. The surface is damaged so that little of the cortical bone is preserved. Little can be noted of its morphology, but in size it is at the larger end of the expected range for *K. aramisi*. As preserved, the anteroposterior diameter is just under 22 mm, but the true figure would be slightly greater. Among extant cercopithecids, it is larger than extant *C. guereza* but within the range for male *N. larvatus,* and it is at the lowest part of the range for male *P. hamadryas*.

AME-VP-1/155 is a distal fragment of a right femur, broken through the proximal end of the patellar sulcus. In size, it is at the top of the *C. guereza* range, within the range of female *N. larvatus* and female *P. hamadryas*. It is compatible with the estimated size for females of *K. aramisi* and males of *P. alemui*.

KUS-VP-1/24 is a proximal fragment of a left tibia. It is relatively large, being compatible with extant *Papio,* and is, therefore, approximately of a size that is expected of larger individuals of *K. aramisi,* as well as the larger colobine species. There is extensive damage to the anterior and lateral parts of the tibial plateau and lateral epicondyle. The shaft is preserved approximately 3 cm distal to the plateau. The tuberosity is relatively prominent.

KUS-VP-1/86 and AME-VP-1/20 are both distal fragments of left tibiae. The former specimen is comparable to extant *Papio,* putting it at the larger end of the expected *K. aramisi* range, or possibly the larger colobine species. AME-VP-1/20 is significantly smaller and is within the size range for extant *Colobus*. Morphologically, however, they are quite similar. For both specimens, the medial malleolus is long and the astragalar articular surface is relatively flat. On the posterior surface of AME-VP-1/20, the sulcus for the long flexor tendons is deep and well-marked, whereas that of KUS-VP-1/86 is shallow and weakly marked.

ASK-VP-3/399 is a distal fragment of a right fibula. It is relatively large in size, perhaps slightly larger than would be expected for males of *K. aramisi*.

The five calcanei preserved in the sample are all from the Adu-Asa Formation (Table 6.6). ASK-VP-3/176 is a nearly complete left calcaneus and is larger than the other calcanei recovered. It is in the size range for females of extant *P. hamadryas*. STD-VP-2/865 is a fragment of a left calcaneus, distally preserved from the proximal astragalar facet, and is also damaged on its lateral side. It is much smaller than ASK-VP-3/176 and is in the size range of both extant *C. aethiops* and *C. guereza*. STD-VP-2/863 is a fragment of a right calcaneus, preserving only the area around the sustentaculum, and probably similar in size to ASK-VP-3/176. ASK-VP-3/51 is a fragment of a left calcaneus, preserving only the area of the astragalar facets. It would have been smaller than ASK-VP-3/176 and STD-VP-2/863 and larger than STD-VP-2/865. ASK-VP-3/369 is a right calcaneal fragment that preserves the proximal astragalar facet, the proximal half of the distal facet, and the sustentaculum tali; it would also have been larger than STD-VP-2/865. Morphologically, ASK-VP-3/176 has the proximal extension of the proximal astragalar facet seen in colobines (Strasser 1988). ASK-VP-3/176 and STD-VP-2/865 have greater curvature to the proximal astragalar facet, which leads to a more vertical anterior face, and greater prominence above the tuberosity in lateral view than is the case in the others; this trait is also associated with colobine calcanei (Strasser 1988). However, given the fragmentary nature of most of this material, subfamilial affiliation is not attempted here.

ALA-VP-2/45 is a proximal fragment of a right fourth metacarpal preserving approximately 1.5 cm of the shaft. In contrast to all of the metatarsals (below), the shaft is thin and more of a flattened oval in cross section, whereas the shafts of the metatarsals are all more circular in cross section and generally more robust.

ALA-VP-2/112 is a proximal fragment of a left second metatarsal preserving approximately 1.5 cm of the shaft. STD-VP-2/17 is a proximal fragment of a left second metatarsal preserving approximately 3 cm of the shaft. STD-VP-1/27 is a proximal fragment of a right second metatarsal with approximately 2 cm of the shaft. DID-VP-1/30 is a proximal fragment of a right third metatarsal preserving approximately 1.5 cm of the shaft, but with damage to the palmar portion of the proximal facet. DID-VP-1/76 is a proximal fragment of a left third metatarsal. ALA-VP-2/113 is a proximal fragment of a right fifth metatarsal with 1.5 cm of the shaft preserved. ALA-VP-2/324 is a distal fragment of a metapodial.

AMW-VP-1/144 is a complete intermediate phalanx. It is stout and robust, making it more likely to represent a cercopithecine. It is broader than extant *C. guereza* pedal intermediate phalanges but shorter than extant *C. guereza* manual intermediate phalanges. From the size it is likely to be a pedal phalanx. In size and shape, it is within the range of extant *P. hamadryas* intermediate manual phalanges and that of *T. gelada* pedal intermediate phalanges.

STD-VP-2/922 is a proximal fragment of an intermediate phalanx.

ALA-VP-2/300 is a distal fragment of either a proximal or intermediate phalanx. Much of the shaft is preserved and is longer than would be expected for a distal phalanx. The shaft is robust and flat, making it likely to represent a cercopithecine.

STD-VP-2/883 is a small fragment of an axis vertebra preserving the dens and adjacent area of the body. In size, it is consistent with both *K. aramisi* and *P. alemui*.

STD-VP-1/42 and STD-VP-2/72 are caudal vertebrae. The former is complete, and the latter represents approximately the proximal third of a caudal vertebra. As is typical of cercopithecid caudal vertebrae, both specimens show proximal ends that are roughly pentagonal in outline. STD-VP-1/42 has an overall length of 46 mm and, judging from its size and morphology, is likely from the middle of the tail.

Finally, ASK-VP-3/67 consists of two shaft fragments, one of which appears to be a radial shaft fragment of a medium-sized cercopithecid. The other fragment is similarly sized, but it is from an uncertain element. Little diagnostic morphology is preserved on either specimen.

Postcranial Functional Morphology

We can tentatively make some functional morphological interpretations from these unassociated fossils, although we qualify these statements as provisional because they are based on some *a priori* assumptions and expectations that may be undermined if/when secure associations between craniodental and postcranial remains are made.

In the development of the medial trochlear keel, and in the length and retroflection of the medial epicondyle, the humerus that might belong to *P. alemui* shows morphology that is intermediate between that of more terrestrial forms, such as extant *Papio* and *Theropithecus,* and more arboreal forms, such as *Lophocebus* and most extant colobines other than *Semnopithecus* (e.g., Jolly 1972; Delson 1973; Harrison 1989). It is important to keep in mind that most aspects of humeral morphology related to positional behavior overlap with all of the taxa just mentioned. The humeri that might belong to *K. aramisi* show features typical of more arboreally adapted cercopithecids. In this sense, they are generally similar to those of most extant African colobines.

The astragalus that might represent *K. aramisi* shows features described by Strasser (1988) as typical of the Colobinae, and it also lacks features typical of more terrestrially adapted cercopithecoids. The trochlea has a shallow central sulcus, and the medial and lateral margins are rounded and comparatively similar in height to one another. The trochlea is also strongly wedge-shaped in superior view.

Given that only two joint complexes are possibly represented in *K. aramisi* and one in *P. alemui,* all interpretations must be treated with extreme caution until larger samples and additional anatomical regions are known. That stated, what is preserved seems to indicate some differences in locomotor behavior between these two taxa. *Kuseracolobus aramisi* appears broadly comparable to most extant African colobines in its degree of arboreality/terrestriality. *Pliopapio alemui* seems to show more adaptation to terrestrial locomotion than *K. aramisi,* but less than that seen in extant *Papio* and *Theropithecus.*

Discussion

Relative Abundances

The cercopithecid fauna reported here derives from essentially two stratigraphic intervals: the first is from the Kuseralee Member of the Sagantole Formation of the Central Awash Complex (CAC) dated to approximately 5.2 Ma (Renne et al. 1999). The other is from the Asa Koma Member of the Adu-Asa Formation from the western margin, dated to approximately 5.7 Ma (WoldeGabriel et al. 2001). Thirty-two fossil cercopithecids have been recovered from the former stratum and 75 from the latter. Of these, 13 and 30, respectively, have been diagnosed to subfamily. In the Kuseralee Member assemblage four out of the 13 assignable specimens, or approximately 31 percent, are colobines. In the Adu-Asa Formation, on the other hand, 26 out of the 30 identifiable, or approximately 87 percent, are colobines. If the postcranial assignments prove correct, then this would add one cercopithecine and two colobines to the total for the Kuseralee Member series and one colobine to the Adu-Asa Formation series. It is also important to note that the totals for members represent summations of multiple sites that may not be sampling consistent environments. While these individual collection areas vary in their relative proportions of the two subfamilies, all have so few individuals as to render proportions from individual localities highly biased by sampling error.

Comparison to Other Miocene and Early Pliocene African Cercopithecids

The earliest representatives of either of the extant subfamilies of cercopithecids are all colobines. *Microcolobus tugenensis* is known from the Ngorora Formation at Ngeringerowa, Tugen Hills, Kenya, and is dated to 9 Ma (Benefit and Pickford 1986). A single isolated colobine tooth from Wissburg, Germany, may be as old as 10 Ma (Szalay and Delson 1979). Otherwise the oldest Eurasian population is that of *Mesopithecus pentelicus* from Pikermi, Greece (Szalay and Delson 1979). The relationship of both of these taxa to either of the extant Asian or African subtribes (the Presbytina or Colobina, respectively), as well as to later African fossil forms, is unclear at this time (Szalay and Delson 1979; Delson 1994).

In the latest Miocene, fossil cercopithecids are known from several sites throughout Africa. Closest in geography and age are those from the Nawata Formation and from the lowest Nachukui Formation at Lothagam, Kenya, aged 6–7 Ma and approximately 5 Ma, respectively (Leakey et al. 2003). There are three taxa from the Nawata. The most common is the small to medium-sized papionin, *Parapapio lothagamensis*. It differs from *Pliopapio alemui* in having a shorter, broader, and more squared rostrum that lacks an anteorbital drop. Furthermore, the symphysis of *Parapapio lothagamensis* is far longer and more sloping, and the molar crowns show a greater degree of basal flare, than those of *Pliopapio alemui* (Frost 2001b; Leakey et al. 2003). There are also fragmentary papionin remains from the Apak Member of the Nachukui Formation (ca. 5 Ma) similar in size to *Pliopapio alemui* and to *Parapapio ado*. *Cercopithecoides kerioensis* is known from an uncertain stratigraphic position at Lothagam. It is smaller and different in mandibular morphology from *K. aramisi,* showing typical features of the genus (Leakey et al. 2003). There are also three species of colobine that are too fragmentary to be allocated to genus, and are designated as species A, B, and C from Lothagam. Species A and B are known from the Nawata Formation and are both smaller than *K. aramisi,* though species B is closer in size. The mandible of species A is relatively shallower than that of *K. aramisi*. Species C is from the Apak Member but is known only from very fragmentary remains. It is also smaller than *K. aramisi* in dental size. Species B and C do not preserve sufficient morphology to determine whether they are conspecific with *K. aramisi*.

Colobines of late Miocene age have also been reported from the 6 Ma locality of Lemudong'o, Kenya (Ambrose et al. 2003) and from the approximately 6–7 Ma deposits in Chad (Brunet et al. 2002). The Chadian collection has yet to be fully studied, but research on the Lemudong'o material is almost completed. This latter late Miocene cercopithecoid assemblage consists entirely of colobines. These remains are extremely fragmentary but represent at least three colobine species, two of which are relatively small in size. The best preserved taxon represents a new species of *Paracolobus* and is morphologically distinct from *Kuseracolobus aramisi* and the western margin colobines, although they are about the same size (Hlusko 2007a). The specimens of the other two colobine species do not preserve enough anatomy to determine their affinities at the generic level, although the smallest taxon is significantly smaller than *K. aramisi,* and the larger taxon is considerably bigger and in the size range of the larger colobine from the late Miocene of the Middle Awash (Hlusko 2007a). From the Tugen Hills of Kenya, late Miocene cercopithecid specimens include a colobine molar from the ca. 6.3–7 Ma Mpesida Beds and several fragmentary specimens from the ca. 6 Ma Lukeino Formation. (Gundling and Hill 2000; Pickford and Senut 2001). Two colobine M_3s comparable in size to *K. aramisi* are known from between 4 and 6 Ma in the Nkondo Formation, Uganda (Senut 1994). There are two isolated teeth of a small papionin from Langebaanweg, South Africa (Grine and Hendy 1981). They are too fragmentary to be allocated to a particular genus. Finally, there is an isolated papionin M_3 from Ongoliba, Democratic Republic of the Congo, dated to the late Miocene, which is smaller than *P. alemui* (Hooijer 1963, 1970).

North of the Sahara, the late Miocene site of Menacer (formerly known as Marceau), Algeria, thought to be 6–8 Ma on biochronological grounds, has yielded two species of fossil cercopithecids (Arambourg 1959; Szalay and Delson 1979). The more abundant of the two is an unnamed papionin similar to *Pliopapio alemui* in size, which is represented by a series of isolated teeth and one set of associated dentition. Delson (1980) tentatively allocated this species to ?*Macaca* sp., largely based on geographic grounds. The second taxon is a relatively large species of colobine close in size to the larger colobine species from the Miocene deposits of the Middle Awash. This taxon was designated ?*Colobus flandrini* (based on the holotype of *Macaca flandrini* Arambourg, 1957, which is in fact a colobine) by Delson (1973, 1980, 1994; Szalay and Delson 1979). His use of the designation ?*Colobus* was as a form genus and did not imply any special relationship to the modern genus, only that this form was likely to have been a colobine of African affinity. While quite fragmentary, the upper molars of this taxon are similar to those of Colobinae gen. et sp. indet. "Large" from the Miocene of the Middle Awash.

The holotypes of *Libypithecus markgrafi* and *Macaca libyca* are known from the terminal Miocene locality Wadi Natrun, Egypt (Delson 1973, 1980, 1994; Szalay and Delson 1979). *Libypithecus* is somewhat smaller than *K. aramisi* in size and has a longer and more prognathic rostrum. The mandible of *L. markgrafi* is unknown. The most complete specimens of *M. libyca* are two partial mandibles. They have been allocated to *Macaca* based on the dentition, an absence of mandibular corpus fossae, and geography. The symphyseal region is distinct from that of *P. lothagamensis* but similar to that of *P. alemui.*

The dentition, however, may be slightly more flaring. Both *L. markgrafi* and *M. libyca* have been tentatively identified from Sahabi, based on a few fragmentary remains (Meikle 1987). *Libypithecus* has been identified from the Middle Awash based on the mandible KL-308-1 (Kalb et al. 1982c). This specimen is here allocated to *K. aramisi* (see above).

The two cercopithecid species best represented from the late Miocene of the Middle Awash are *K. aramisi* and *P. alemui*. They are also both known in the early Pliocene from the Aramis Member and possibly from the Haradaso Member of the Sagantole Formation, also in the Middle Awash (Frost 2001b; unpublished), and have also been reported from early Pliocene deposits at Gona (Simpson et al. 2004). Neither taxon can be identified with any confidence outside of the Afar region, although there are some fragmentary specimens, mostly isolated teeth, that cannot as yet be confidently excluded (Frost 2001b). The larger colobine from the Miocene of the Middle Awash is too fragmentary to determine its affinities with other well-known forms, but colobines of similar size are known from the early Pliocene Haradaso Member of the Sagantole Formation. More specimens of similar size are found in late Miocene sediments of Menacer (Szalay and Delson 1979) as well as Pliocene localities throughout Africa and Eurasia.

Phylogenetic Implications and Relationships

Of the taxa represented in the Miocene deposits of the Middle Awash, *P. alemui* and *K. aramisi* are the better represented and therefore the most phylogenetically informative. The phylogeny of extant papionins is now relatively well-resolved with two monophyletic, and geographically disjunct, subtribes (e.g., Szalay and Delson 1979; Strasser and Delson 1987; Harris and Disotell 1998; Tosi et al., 2000, 2002, 2003). The Macacina is represented by the speciose North African and Asian genus *Macaca*, wheras the sub-Saharan Papionina consists of the genera *Papio, Theropithecus, Lophocebus, Cercocebus,* and *Mandrillus*. Unfortunately, there are no distinct morphological synapomorphies in the hard tissues that can be used to diagnose these clades.

Within the African papionins, two primary clades can be recognized: the first consists of *Papio, Theropithecus,* and *Lophocebus;* and the second consists of *Cercocebus* and *Mandrillus*. There are no known fossils that definitively represent the latter group. The position of many fossil forms relative to these extant groups is not clear. Two phylogenetic positions were proposed by Frost (2001b) for *Pliopapio*. The first places *Pliopapio* as a member of the *Papio/Theropithecus/Lophocebus* group, the second as a primitive African papionin. Evaluation of these hypotheses awaits a thorough phylogentic review. However, if our postcranial attribution to *P. alemui* is correct, it would suggest that this taxon had not adapted as strongly to a terrestrial locomotion as had either *Papio* or *Theropithecus*.

The phylogenic position of *K. aramisi* is less clear than that of *P. alemui*. The extant Colobinae are divided by most authors into two biogeographically distnct clades: the Colobina in Africa and the Presbytina in Asia (e.g., Szalay and Delson 1979; Strasser and Delson 1987; Disotell 1996, 2000). The position of most fossil forms relative to these clades is unclear, as are the relationships among extinct genera, including *K. aramisi* (e.g., Szalay and Delson 1979; Leakey 1982; Frost 2001b; Jablonski 2002). If the tentative postcranial allocations proposed in this chapter for *Kuseracolobus* are correct, then the relatively arboreal locomotor mode of *K. aramisi* would be typical of the subfamily but distinct from that of *Cercopithecoides* species with known postcrania, and possibly from *Paracolobus chemeroni* (Birchette 1981, 1982; Frost and Delson 2002).

Cercopithecid Evolution

In the Miocene, hominoids were the predominant and most diverse group of primates in the Old World, with cercopithecids being relatively rare and confined to Africa (e.g., Szalay and Delson 1979; Benefit 2000; Benefit and McCrossin 2002; Harrison 2002). By the Pliocene, the situation reversed in the Old World, and hominoids were relatively rare (e.g., Szalay and Delson 1979; Jablonski 2002).

The differences between the cercopithecoids from the late Miocene in the Middle Awash and the penecontemporaneous fossils from Lothagam indicate that, by the latest Miocene, both subfamilies of cercopithecids had attained some diversity in Africa. Therefore, the Pliocene radiation of cercopithecids appears to have actually started in the late Miocene and may not have been as rapid as previously interpreted from smaller fossil samples.

The Eurasian record at the equivalent time is extremely fragmentary, with the exception of Pikermi, Greece (Szalay and Delson 1979; Jablonski 2002), but it does indicate that colobines were present in Eurasia by the late Miocene, and cercopithecines somewhat later. As the late Miocene primate record improves, it may be possible to determine whether the radiation of cercopithecids and the extinction of non-cercopithecid catarrhines were related or separate events, as well as whether the cercopithecid radiation occurred rapidly or accumulated over a long period of time.

Conclusions

One hundred and seven cercopithecid fossils are known from the late Miocene deposits of the Middle Awash. This sample derives from two main stratigraphic levels: the Kuseralee Member of the Sagantole Formation in the CAC, dated to approximately 5.2 Ma; and from the Asa Koma Member of the Adu-Asa Formation, dated to approximately 5.6–5.7 Ma, from the western margin. Both subfamilies are present, and include a minimum of three species: *Pliopapio alemui, Kuseracolobus aramisi,* a second larger colobine, and a possible fourth species that is smaller than the others and cannot be currently allocated to a subfamily. *Pliopapio alemui, K. aramisi,* and the larger colobine are from the Kuseralee Member. These same three taxa may be represented by more fragmentary material, as well as a possible fourth smaller species known from the Adu-Asa Formation. This material extends the chronological range of both *K. aramisi* and *P. alemui* to 5.2 Ma; if the tentative allocations from the Adu-Asa Formation are correct, their ranges would extend back to 5.7 Ma.

7

Hominidae

YOHANNES HAILE-SELASSIE, GEN SUWA, AND TIM WHITE

Darwin, in the nineteenth century, suggested Africa as a continent of origin for Hominidae. His interpretation was based on anatomy and ethology because neither paleontological nor genetic evidence was available. The discovery of the first *Australopithecus* in the 1920s began a series of finds that revealed a long Pliocene record of African hominids. The discovery of *Australopithecus afarensis* in the 1970s extended this record to 3.6 Ma. Recovery of *Australopithecus anamensis* (Leakey et al. 1995, 1998; White et al. 2006b) and *Ardipithecus ramidus* (White et al. 1994, 1995) during the 1990s pushed hominid antiquity to 4.2 and 4.4 Ma, respectively.

Recent discoveries of hominid remains older than 5 Ma (*Orrorin tugenensis:* Senut et al. 2001; Pickford et al. 2002; Pickford and Senut 2005; *Ardipithecus kadabba:* Haile-Selassie 2001b; Haile-Selassie et al. 2004b; and *Sahelanthropus tchadensis:* Brunet et al. 2002, 2005) have afforded new perspectives on the origin of the hominid clade. Here we describe and assess key evidence bearing on these issues, the remains of *Ar. kadabba* from the Middle Awash study area of the Afar Rift.

The term "hominid" is used here to refer to the family Hominidae, which includes modern humans and all taxa phylogenetically closer to humans than to *Pan* (common chimpanzee and bonobo)—that is, all taxa that postdate the cladogenetic split between the lineage leading to modern humans and the lineage that led to extant chimpanzees (*sensu* Haile-Selassie 2001b; Haile-Selassie et al., 2004b; Senut et al. 2001; Pickford et al. 2002; Brunet et al. 2002, 2005; see White 2002 for further details and Wildman et al. 2003 for an alternate taxonomic view).

Numerous molecular studies on the relative and absolute timing of the human-chimpanzee split are available, and a number of possible dates for the split have been suggested (e.g., Ruvolo 1997; Deinard and Kidd 1999; Chen and Li 2001; Kumar et al. 2005; Patterson et al. 2006). Estimates vary from ca. 4–5 Ma (Sarich and Wilson 1967) to 10.5 Ma (Arnason et al. 2000), and some paleontologists suggest 12 Ma (Senut and Pickford 2004). Although it is widely agreed that such inconsistencies emanate from differences in (1) the type and quantity of the genomic infomation used, (2) analytical and statistical methodologies, and (3) fossil calibration dates, the more comprehensive molecular studies appear to suggest that the human-chimpanzee split was around 5–7 M (Chen and

Li 2001; Nei and Glazko 2002; Glazko and Nei 2003; Kumar et al. 2005; Patterson et al. 2006).

However, most of the chimpanzee-human divergence estimates depend on calibration via a poorly constrained *Pongo*-human split, with the younger *Pan* divergence estimates based on a 13 or 14 Ma date for the *Pongo* split. The latter depends on the first appearance datum of *Sivapithecus*. If the actual lineage divergence had already taken place in Africa prior to its emmigration to Asia, then a more appropriate estimate of the *Pongo* species divergence would be >17 Ma. Assuming a discrepancy between the species and average genetic (or genomic) divergence of about 2 Myr (Patterson et al. 2006), the corresponding minimal divergence date would be >19 Ma. We therefore consider a 20 Ma calibration date for the *Pongo* split to be realistic rather than an "extreme" boundary condition (contra Patterson et al. 2006), for both species and the average genetic divergences. A 20 Ma *Pongo* split would correspond to a human-chimp split of about 7.5–9.5 Ma, consistent with the hominid status attributed to *Sahelanthropus, Orrorin,* and *Ar. kadabba*.

Our descriptions and analyses are designed to introduce the major morphological features of the known *Ar. kadabba* specimens. We follow the descriptive methods used by White et al. (2000), who recommend explicit comparisons with already-described specimens. Ideally, well-known specimens of the same taxon should be referred to in such comparative descriptions. In the case of the present study, the specimens being described represent the first and only known *Ar. kadabba* fossils. The other known late Miocene to early Pliocene hominids—*O. tugenensis, S. tchadensis,* and *Ar. ramidus*—have been only provisionally described. For our comparative descriptions, we therefore primarily rely upon the more abundant and better-known early *Australopithecus* samples of *Au. afarensis* and *Au. anamensis*. Where possible, we also refer to *Ar. ramidus* homologs. Because *Ar. kadabba* is considerably more primitive than *Au. afarensis* and *Au. anamensis* in some features, such as in the C/P3 complex, we also infer possible ancestral ape conditions from the morphologies shared by extant great apes and the collective late and middle Miocene fossil ape record. An exhaustive original account of the latter is beyond the scope of the present study, but our statements are based on the combined information available in the literature and the study of original specimens or good-quality casts of *Dryopithecus, Ouranopithecus, Oreopithecus, Lufengpithecus, Gigantopithecus, Sivapithecus, Kenyapithecus, Afropithecus,* and *Proconsul*.

Following the descriptions, we compare *Ar. kadabba* to the other known late Miocene and early Pliocene hominids. Additionally, because a comparison with *Pan* (the sister taxon of the entire hominid clade) is of special interest, these comparisons and discussions are also included.

Our dental terminology primarily follows that outlined in Suwa (1990). Metric methods of the dentition and mandible follow those of White (1977) and White and Johanson (1982). The standard and other additional crown metrics are briefly defined in Table 7.1. Enamel thickness measured on selected natural fracture sections was taken with calipers under a low-powered binocular microscope. Measurements were taken only on sufficiently "radial" sections—that is, on sections that were considered to approximately exhibit local minimal distances from the enamel-dentin junction (EDJ) to the outer enamel surface (OES) (White et al. 2006b).

TABLE 7.1 Standard Crown Metrics of Teeth Attributed to *Ardipithecus kadabba*

Upper Dentition	*C^1MD*	*P^3MD*	*P^3MXMD*	*P^3BL*	*P^3MXOB*	*P^4MD*	*P^4BL*	*M^1MD*	*M^1BL*	*M^3MD*	*M^3BL*
ASK-VP-3/160		7.6	7.9	11.3	11.5						
ASK-VP-3/400	11.8										
ASK-VP-3/401								[10.3] (10.4)	11.7		
ASK-VP-3/402								[10.2] (10.3)	11.6		
ASK-VP-3/404						(7.6)					
STD-VP-2/62										10.9	12.2
STD-VP-2/63								[10.4] (10.6)	12.1		

Lower Dentition	*I_2MD*	*I_2BL*	*C_1MXOB*	*CPP*	*CHT*	*P_4MD*	*P_4BL*	*M_2MD*	*M_2BL*	*M_3MD*
ALA-VP-2/10	[6.3]	8.3	11.2	7.8	[13.4]	[8.1] (8.3)	10.0	(12.7)	11.8	(13.3)
ASK-VP-3/405[a]							(10.4)			
STD-VP-2/61			10.8	7.8	14.6					

NOTE: Measurements in mm. [ ] Dimension as worn/preserved; ( ) estimated value corrected for interproximal wear and/or damage. Abbreviations: MD, mesiodistal diameter taken along the crown's mesiodistal axis, usually across interproximal facets; BL, buccolingual diameter taken perpendicular to MD; HT, labial crown height taken from cervical line to the crown tip; MXOB, maximum oblique diameter of crown; PP, maximum diameter perpendicular to MXOB.

[a]ASK-VP-3/405 is provisionally listed as a P_4, but it may be a P^4 fragment (see the text for discussions).

For selected specimens, microcomputed tomography (micro-ct) data were obtained with the Tesco TX225-Actis system at the scanning and imaging facility at the University Museum, the University of Tokyo. For the tooth crowns, scans were taken at 130 kVp, 0.2 mA, with a copper prefilter of 0.5 mm to minimize beam hardening artifacts. Each scan was reconstructed in 512 × 512 matrix from 900 views.

Enamel thickness measurements were taken from the micro-ct data as follows:

1. Isotropic voxel volume datasets of 28 (or 42) micron size were constructed for each tooth crown from nonoverlapping serial slices.
2. The initial volume dataset was rotated and reformatted into standard orientation, with the vertical, mesiodistal, and buccolingual tooth axes determined from simultaneous examination of five orthogonal surface-rendered views.
3. Further reformatting was conducted to obtain the appropriate vertical section passing through the cusp tip and intersecting the lateral crown EDJ and OES contours at an overall "radial" situation (i.e., an approximate three-dimensional "radial" section was visually determined; see Suwa and Kono 2005).
4. Enamel thickness measures were taken on sections, each newly derived by re-rotating the original volume dataset only once, in order to avoid data corruption from successive interpolation involved in section reformatting.
5. Enamel boundaries were determined by the half-maximum height method with regard to the air-enamel interface and by visual determination of closest pixel in relation to the enamel-dentin interface. General volume rendering was done with software Analyze 6.0 (Mayo Clinic, Rochester, Minnesota, USA). Enamel thickness measurements were determined using the custom-made routines of Vol-Rugle (Medic Engineering Inc., Kyoto, Japan). Radial thickness of the lateral crown face was defined as the minimum thickness from a given point of the EDJ to the outer enamel surface, and maximum radial thickness refers to the maximum of such measures (Kono 2004; Suwa and Kono 2005).

Systematic Paleontology

Ardipithecus White, Suwa, and Asfaw, 1995

Ardipithecus kadabba Haile-Selassie, 2001

(=*Ardipithecus ramidus kadabba* Haile-Selassie, 2001)

Holotype

ALA-VP-2/10, right mandible with M3 crown, and associated left I2, lower C, P4, M1 fragment, M2, and M3 root fragment. From the Asa Koma Member of the Adu-Asa Formation, Middle Awash Ethiopia. Bracketing volcaniclastic horizons are chronometrically dated to between 5.54 6 0.17 and 5.77 6 0.08 Ma (WoldeGabriel et al. 2001).

Referred Materials

Asa Koma Member (Adu-Asa Formation) ALA-VP-2/11, distal half of intermediate hand phalanx; ALA-VP-2/101, humeral mid-shaft and proximal ulna; ALA-VP-2/349, right I^2 fragment; ASK-VP-3/160, left P^3; ASK-VP-3/78, distal humerus; ASK-VP-3/400, right upper C; ASK-VP-3/401, right M^1; ASK-VP-3/402, left M^1; ASK-VP-3/403, partial left P_3; ASK-VP-3/404, left P^4 fragment; ASK-VP-3/405, partial P4; DID-VP-1/80, proximal hand phalanx fragment; STD-VP-2/61, right lower C; STD-VP-2/62, left M^3; STD-VP-2/63, left M^1; STD-VP-2/893, left clavicle fragment.

Kuseralee Member (Sagantole Formation) *Ardipithecus* cf. *kadabba:* AME-VP-1/71, proximal foot phalanx.

The Holotype (ALA-VP-2/10; Plates 7.1–7.3)

DISCOVERY

Survey and sieving operations at Alayla Vertebrate Paleontology Locality 2 (ALA-VP-2) resulted in the collection of approximately 350 vertebrate fossils, including the holotype. The holotype of *Ardipithecus kadabba* was initially found by one of us (YHS) on December 16, 1997, in a fresh surface lag (Figures 7.1–7.4). The right mandibular corpus, preserved from the P_4 to M_3 positions and retaining the M_3 crown, was found first. Further surface survey on December 23, 1998, yielded the individual's left M_2 approximately 20 m southwest of the holotype find spot. The molar was found on the surface, transversely broken into subequal halves separated by about 1 m. In 1999 the specimen's left I_2 was recovered on the surface 3 m east of where the M_2 was found. This necessitated a recovery operation focused on the lag deposits of the locality. An area of 15 m by 30 m centered on the find spot of the second molar was stripped of surface lag, which was passed through a 1.5 mm sieve. By means of this sieve operation, the specimen's left P_4 was recovered approximately 6 m east of the M_2. The left lower canine was recovered from the sieve on December 13, 1999. A small bone fragment crossing the midline was found among fragments recovered by the sieving and reattached to the right mandibular corpus piece. Figures 7.5 and 7.6 document the various recovery operations between 1997 and 1999.

PRESERVATION

The specimen lacks abrasion on its broken edges, suggesting that it was not transported for a long distance. Anteriorly, the right corpus is preserved to just anterior to the P_4 position, where the upper portion of the preserved lingual cortical bone extends medially to the midline and conjoins the fragment of the left side. The conjoined cortical bone forms an approximately 11 mm internal segment crossing the symphyseal area corresponding to the inferiormost planum alveolare and adjacent bone above the genioglossal fossa. Posteriorly, the specimen is truncated approximately 5 mm posterior to the M_3, where its distal root apex is exposed. The preserved maximum anteroposterior length of the specimen is 61.2 mm, but both anterior and posterior breaks slant basally, so that the base of the corpus is intact only at the M_1/M_2 level for a segment of 6 to 8 mm. The posterior corpus is best preserved lingually, for a height of 27.5 mm from the M_3 alveolar margin toward the missing base. Posterobuccally, only the root of the ascending ramus is preserved opposite M_2/M_3. The corpus is less extensively preserved anteriorly but contains portions of the buccal surface below the P_4 position, including the mental foramen.

The alveolar margin is preserved between distal M_3 and M_1 position, although damage occurs at M_2 and the buccal M_1 alveolus is extensively altered by pathology. The pathology consists of a periodontal abscess centered at the distal M_1 root and extending anteriorly to the P_4/M_1 position. As a result, the corpus is swollen, especially buccally at the M_1 position, where secondary bone deposition is apparent. Trabecular bone exposed along the broken alveolar region has undergone remodeling as a result of this pathology. No crowns or roots of the right P_3, P_4, or M_1 are preserved, although portions of the P_4 and M_1 alveoli are preserved. The most likely interpretation of these alveoli is that the two anterior (mesiobuccal and distolingual) depressions are the alveolar remnants of the P_4, perhaps corresponding to a fused Tomes root system (as seen in the left P_4 root), or a variant form. Posterior to these two depressions, a small, shallower depression of the alveolar bone occurs buccally, and a deeper depression is seen distal to this. The latter is continuous lingually with what appears to be a buccolingually broad remnant of an alveolar socket, now occupied by irregular bone. It is likely that most of the M_1 was lost premortem, with perhaps small root segments retained in the pathological alveolus. The right M_2 crown is broken away, but its roots are preserved within their alveoli, with no evidence of pathology. The M_3 is almost complete, but its buccal crown face was flaked off postmortem.

The associated teeth from the left side of this individual's lower dentition consist of crowns and partial roots of I_2, lower C, P_4, and M_2. Additionally, small fragments of what appear to be the left M_1 crown and left M_3 root were also obtained. The preservation of these teeth is sketched out in the respective descriptive sections.

COMPARATIVE DESCRIPTIONS

Mandible (ALA-VP-2/10A) The mandibular body is broadly comparable in size with those of smaller examples of *Australopithecus*. In superior view, what remains of the internal mandibular contour suggests a tightly curved internal symphyseal contour comparable to the tightest curvatures seen in *Australopithecus* (e.g., A.L. 288-1). What remains of the posteriormost planum alveolare is located at mesial P_4 level, a condition common in early *Australopithecus* mandibles (MAK-VP-1/12, KNM-KP 29281).

Mandibular corpus height is approximately 32 mm lingually and 30 mm buccally (estimated) at the M_1/M_2 level, comparable with small to middle-sized examples of *Au. afarensis* (e.g. A.L. 288-1, MAK-VP-1/12) or *Au. anamensis* (KNM-KP 31713). Because most of the corpus base is missing, the inclination of the inferior corpus border relative to the alveolar margin (anterior deepening) cannot be assessed. Corpus breadths at P_4, M_1, and M_2 are 18.7, 20.2, and 18.5 mm, respectively, but breadth at M_1 (and most likely P_4) is exaggerated by pathology. An additional informative measurement is breadth at the ramus root, and this is 20.5 mm in ALA-VP-2/10. These breadth measurements of the posterior corpus of ALA-VP-2/10 are at the lowest range of *Au. afarensis* variation (e.g., close to the A.L. 198-1 condition).

The inferior corpus margin below M_1/M_2 of the Alayla mandible is narrow but rounded, similar to the configuration seen in the smaller and thinner-bodied examples

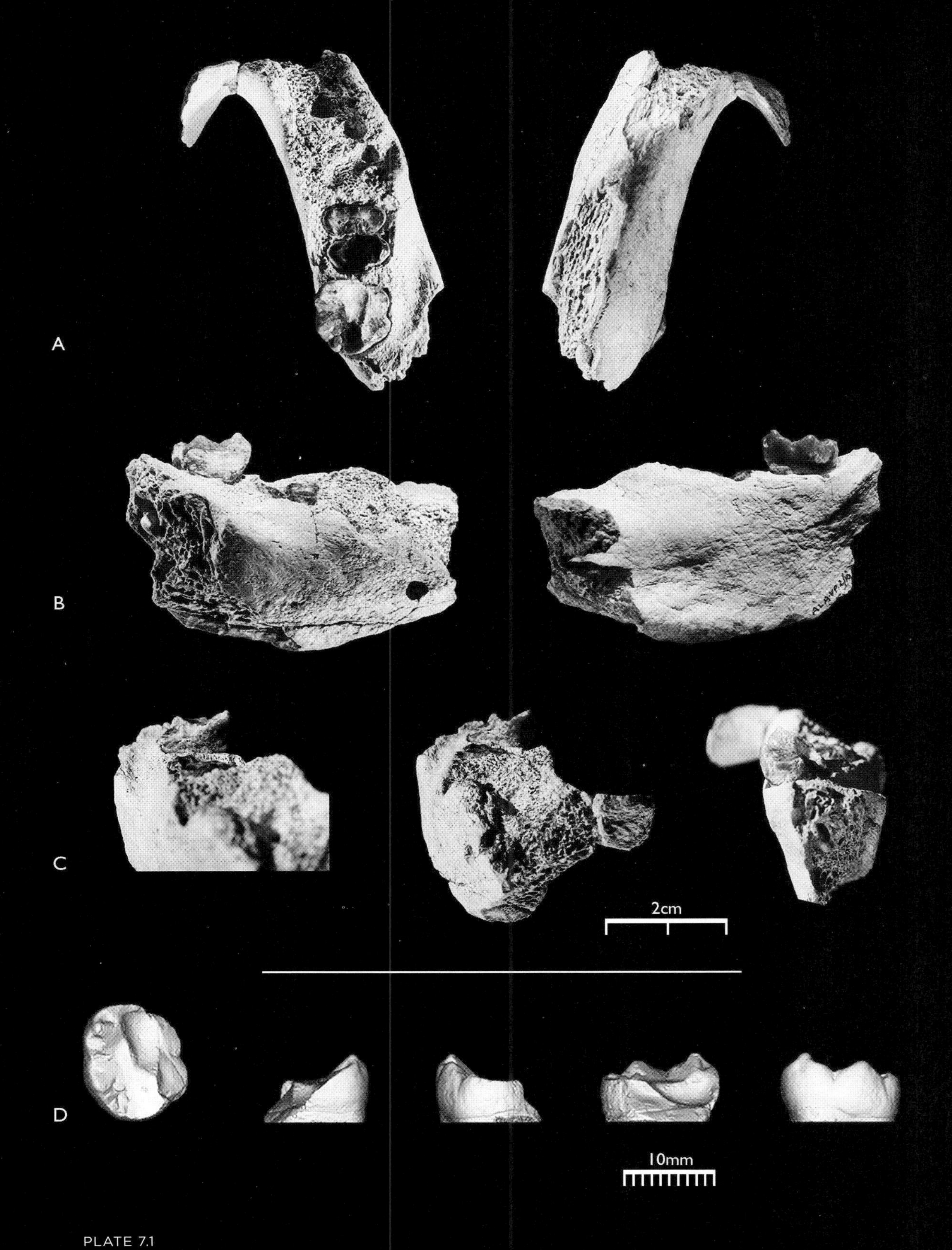

PLATE 7.1

Ardipithecus kadabba holotype specimen ALA-VP-2/10. A–C. Mandibular views. D. Micro-ct renderings of the right M_3. See text for details.

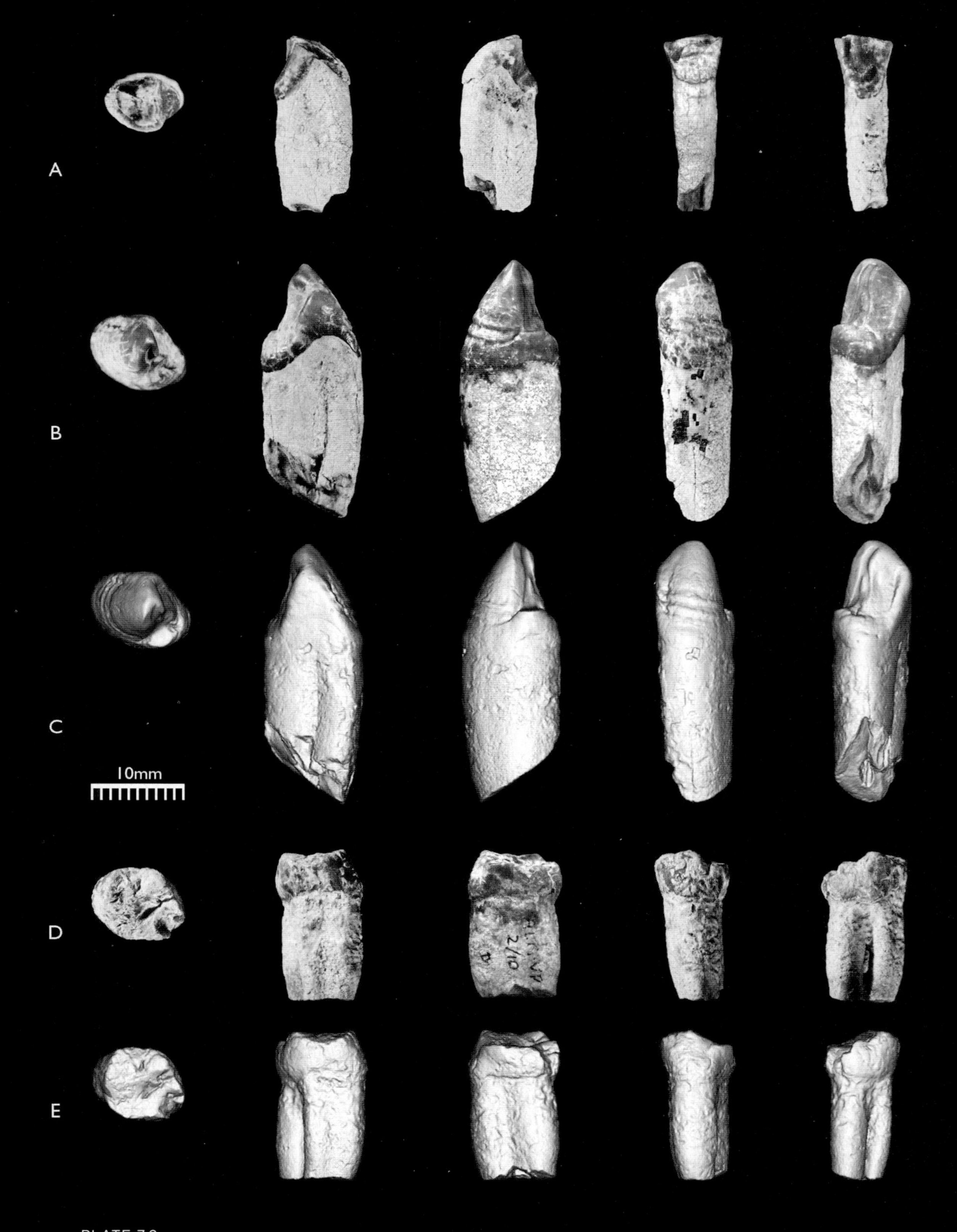

PLATE 7.2

Ardipithecus kadabba holotype specimen ALA-VP-2/10, isolated incisor, canine, and premolar. A. Left I_2. B. Left lower canine, photograph. C. Left lower canine, micro-ct renderings. D–E. Left P_4. See text for details.

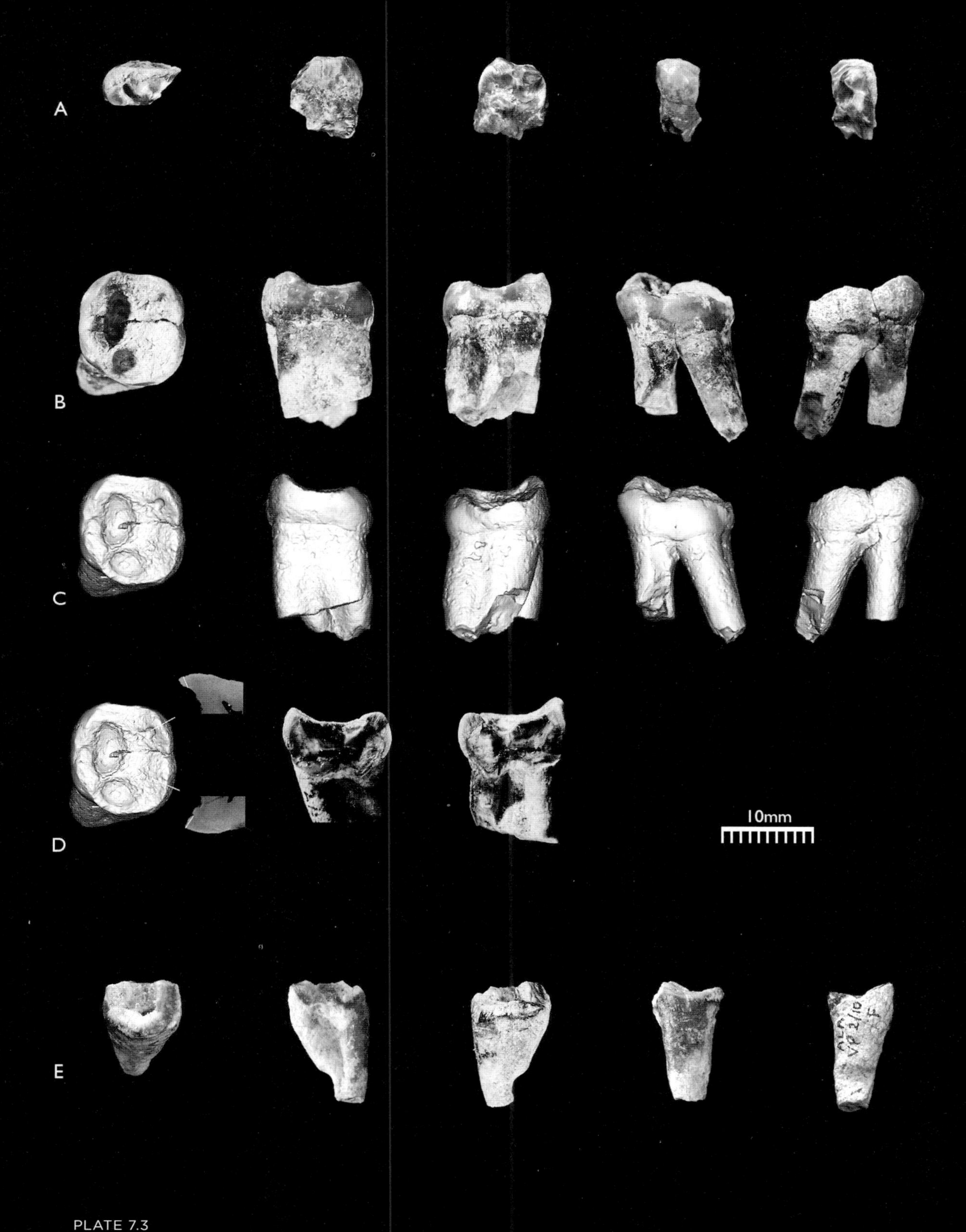

PLATE 7.3

Ardipithecus kadabba holotype specimen ALA-VP-2/10, isolated molars and molar fragments. A. Probable left M_1 crown fragment. B. Left M_2. C. Left M_2, micro-ct renderings. D. Left M_2, micro-ct rendering and photographs of unglued natural fracture sections. E. Left M_3 root fragment. See text for details.

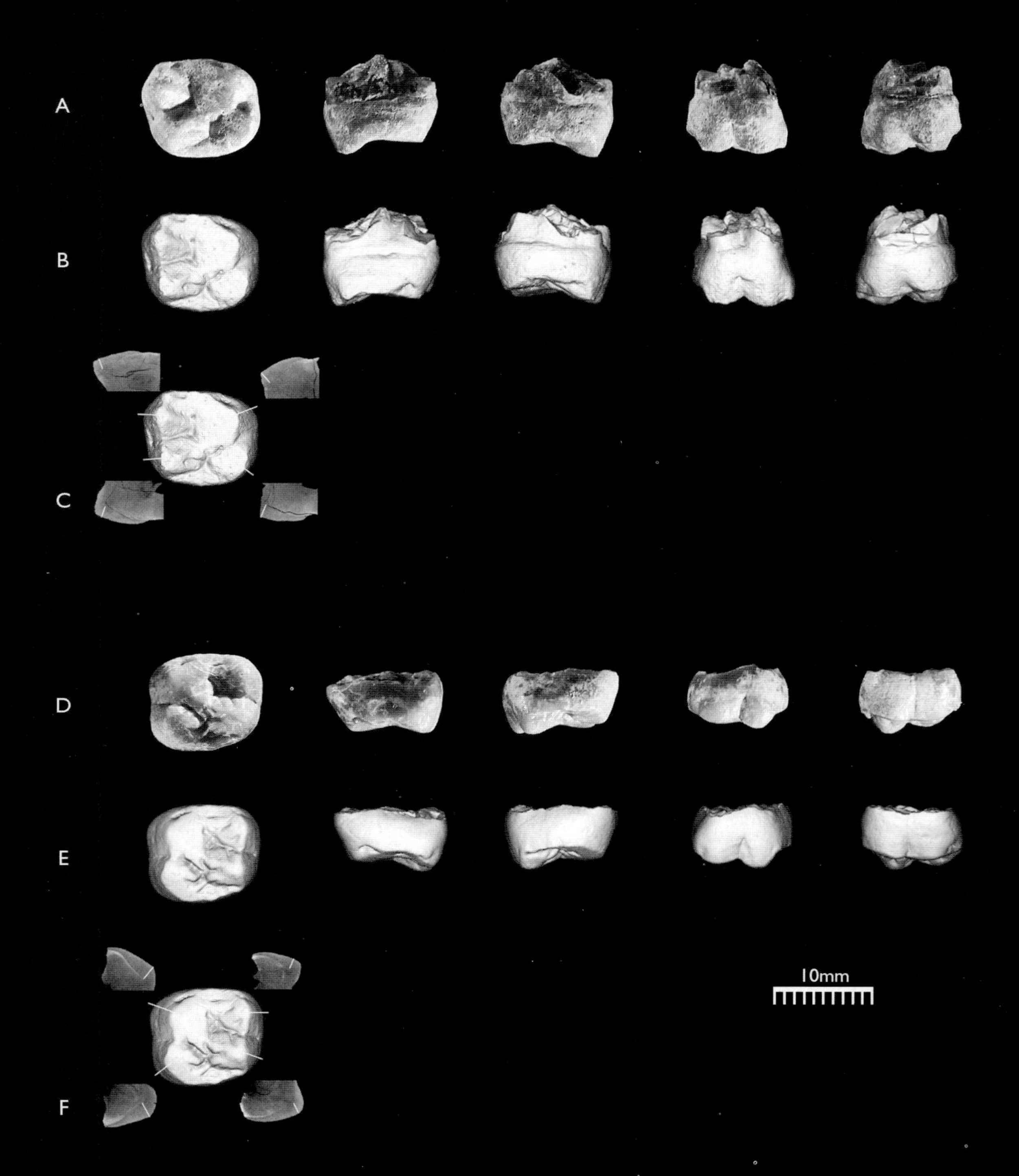

PLATE 7.4

Ardipithecus kadabba isolated molars from Asa Koma. A–C. ASK-VP-3/401 right M^1, photographs (A), micro-ct renderings (B), and micro-ct cross sections (C). D-F. ASK-VP-3/402 left M^1, photographs (D), micro-ct renderings (E), and micro-ct cross sections (F).

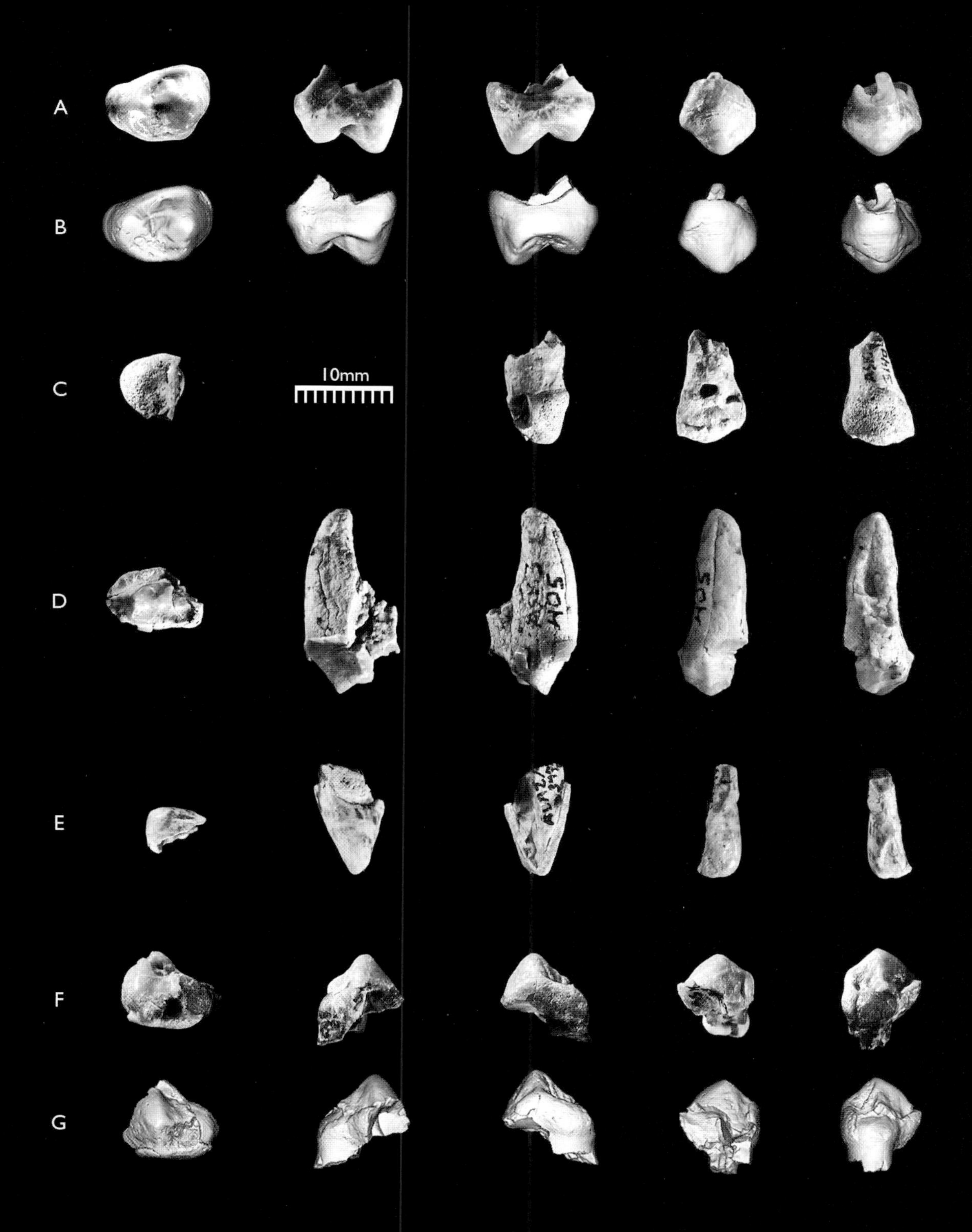

PLATE 7.5

Ardipithecus kadabba isolated premolars and incisors from Asa Koma and Alayla. A–B. ASK-VP-3/160 left P^3, photographs (A) and micro-ct renderings (B). C. ASK-VP-3/404 partial left P^4 fragment. D. ASK-VP-3/405 partial P4 fragment. E. ALA-VP-2/349 right I^2 fragment. F–G. ASK-VP-3/403 left P_3. See text for details.

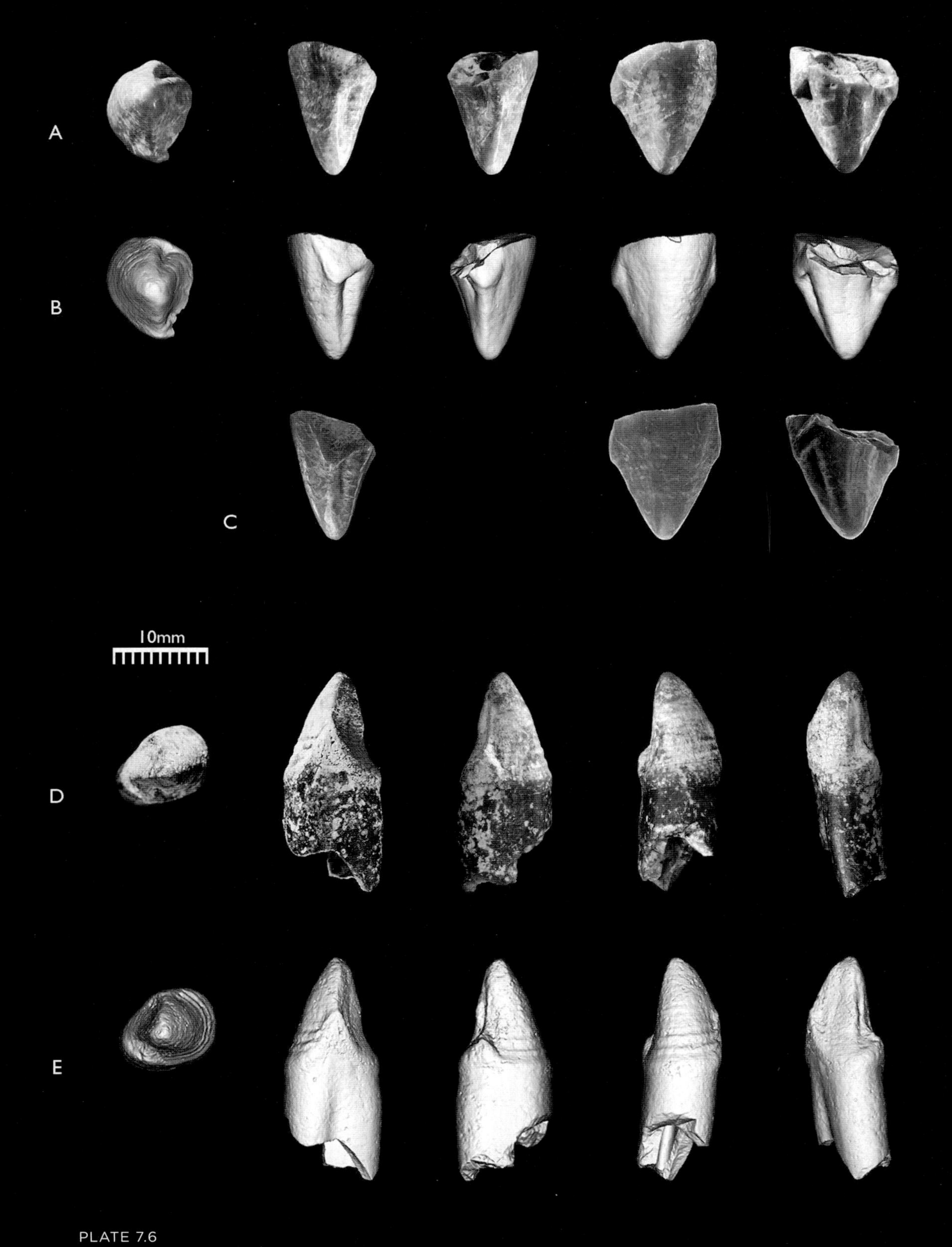

PLATE 7.6

Ardipithecus kadabba isolated canines. A–C. ASK-VP-3/400 right upper canine (micro-ct renderings shown in B). D–E. STD-VP-2/61 right lower canine (micro-ct renderings shown in E). See text for details.

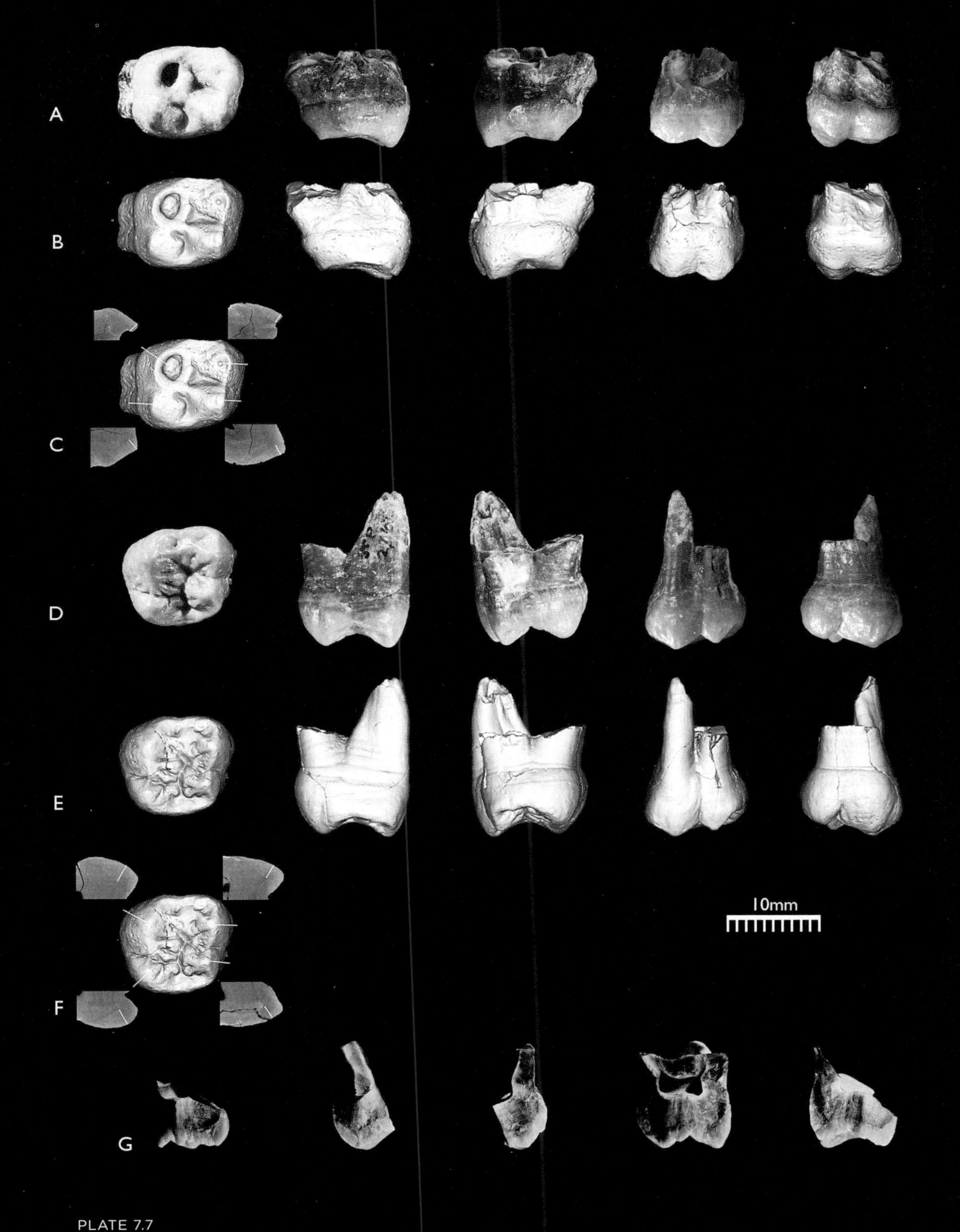

PLATE 7.7

Ardipithecus kadabba isolated molars from Saitune Dora. A–C. STD-VP-2/63 left M^1, photographs (A), micro-ct renderings (B), and micro-ct cross sections (C). D–G. STD-VP-2/62 left M^3, photographs (D), micro-ct renderings (E), micro-ct cross sections (F), and unglued natural fracture surfaces (G). See text for details.

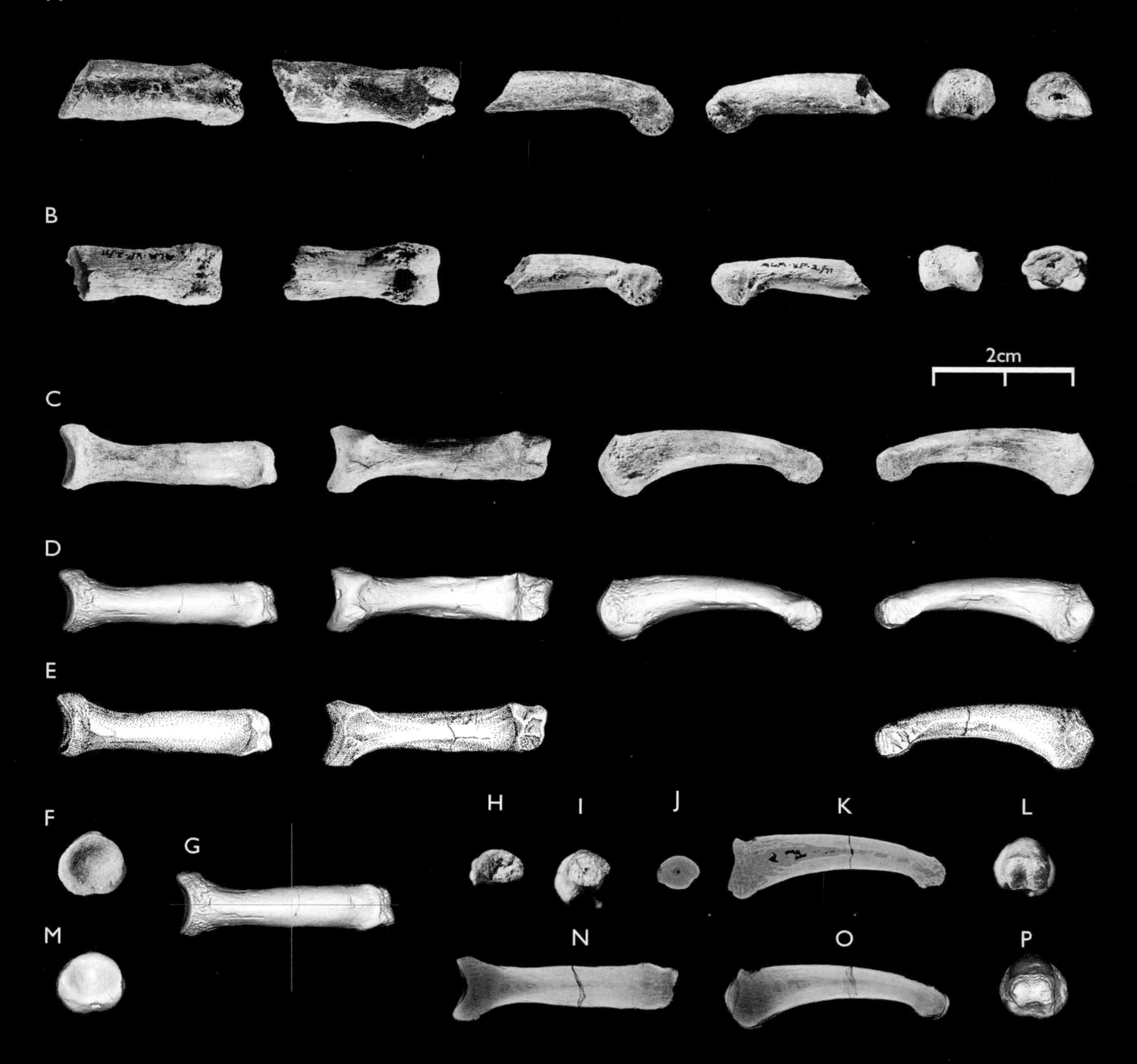

PLATE 7.8

Ardipithecus kadabba and *Ar.* cf. *kadabba* hand and foot phalanges. A. DID-VP-1/80 intermediate hand phalanx. B. ALA-VP-2/11, intermediate hand phalanx. C–P. AME-VP-1/71 proximal foot phalanx: photographs (C), micro-ct renderings (D, M, P); drawings (E); micro-ct sections (J, K); micro-ct summed-voxel projections (N, O). See text for details.

PLATE 7.9

Ardipithecus kadabba humerus and ulna fragments. A. ALA-VP-2/101 humerus shaft. B. ALA-VP-2/101 proximal ulna with shaft. See text for details.

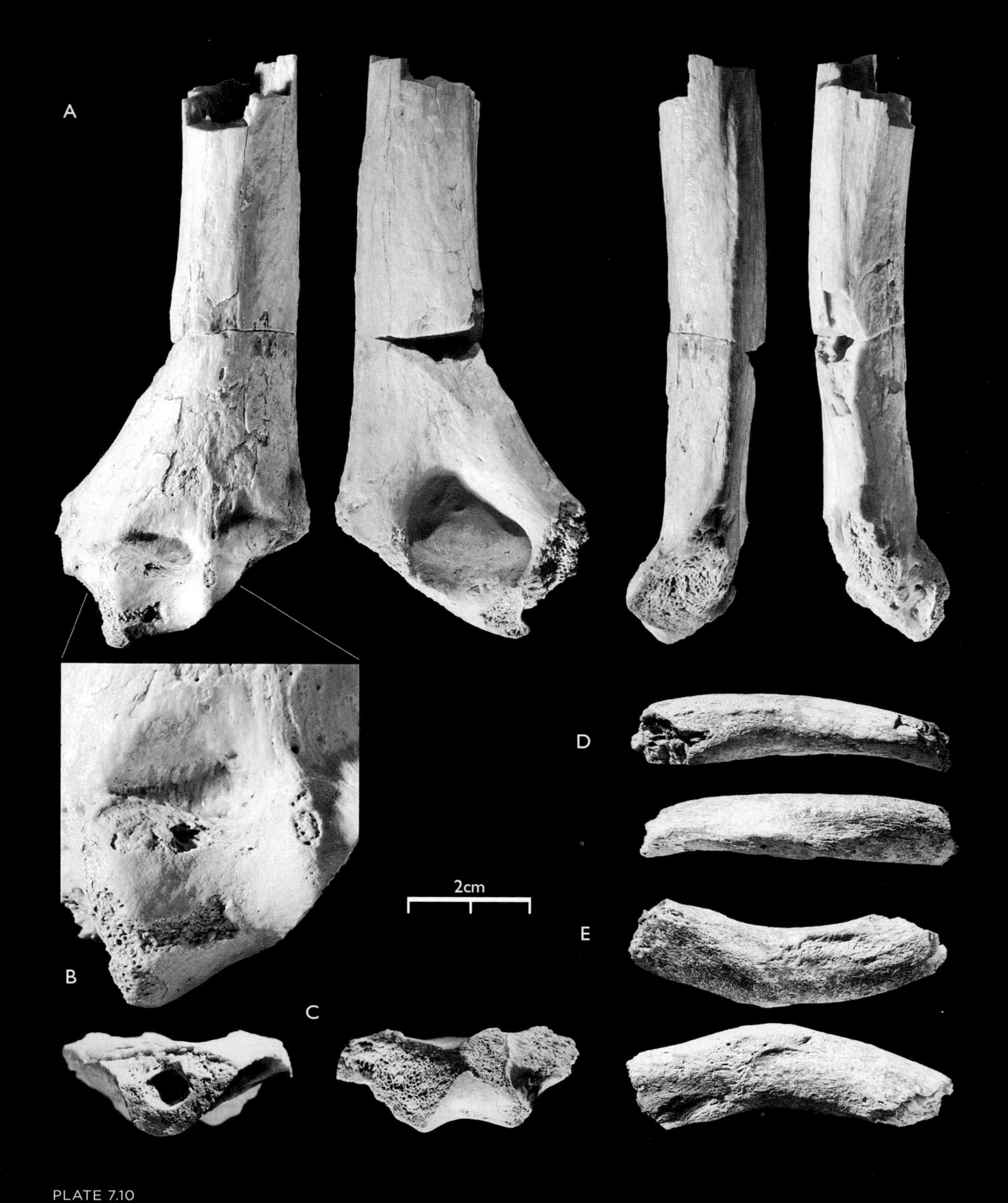

PLATE 7.10

Ardipithecus kadabba distal humerus from Asa Koma and clavicle shaft fragment from Saitune Dora. A–C. The ASK-VP-3/78 distal humerus in all six views, with the blowup (B) to show the arthritic lesions described in the text. D–E. The STD-VP-2/893 clavicular fragment in four views. See text for details.

FIGURE 7.1

View to the northwest across the ALA-VP-2 locality. Geochronologists sample the tuff horizon below vehicle while Yohannes Haile-Selassie searches for fossils. The large basalt boulder just in front of him is the discovery point for the *Ardipithecus kadabba* holotype mandible (ALA-VP-2/10) in 1997. Subsequent to that discovery, the lag of basalt cobbles and boulders was moved to expose the softer sediments to erosion, resulting in more of the individual being found across this outcrop in subsequent years—a system of locality management that continues to yield new fossils. Photograph by David Brill, November 23, 1998.

of *Au. afarensis* (e.g., A.L. 198-1). Cortex thickness can be observed in both anterior and posterior breaks, but obliquity of the break surfaces allows measurements only at limited positions. Buccal cortical thickness of the anterior break is 2.7 mm at a point 8 mm inferior to the mental foramen. Lingual cortical thickness appears to be comparable or slightly thinner. Basal cortical thickness can be estimated at 3.8 mm on the posterior break below M_2.

The root of the ascending ramus is positioned lateral to M_2/M_3, high on the corpus, about 7 mm below the M_2 alveolar margin. The extramolar sulcus is shallow and narrow (7 mm). The high placement of the ramus root corresponds to the extreme of the *Au. afarensis* range of variation (e.g., A.L. 198-1). The lateral corpus surface of the Alayla mandible is slightly hollowed below the mental foramen from P_4 to posterior M_1. However, the anterior extent of this hollow is indeterminate as a result of breakage, and the significance of its limited superior extension (thus differing from the *Au. afarensis* condition) cannot be evaluated because of the pathology affecting the lateral corpus. The mental foramen is located below the P_4 and positioned approximately midcorpus. It is oblong-shaped, with

FIGURE 7.2
View to the north across the upper catchment of the ALA-VP-2 locality. Yohannes Haile-Selassie kneels with the mandible ALA-VP-2/10, which would later be described as the holotype of *Ardipithecus kadabba,* found in the basalt cobble lag on the surface. White objects scattered on the slopes and adjacent sediment patches are other mammalian bone fragments. Photograph by Tim White, December 16, 1997.

a maximum diameter of 3 mm. It opens anterosuperiorly, as is most often the case for *Au. afarensis* and *Au. anamensis* mandibles.

The preserved medial surface of the mandibular corpus is affected by pathology only at the alveolar margin lingual to M_1. The mylohyoid line is well-defined and passes anteroinferiorly from a position ca. 7 mm behind the distal face of the M_3 to ca. 9.5 mm below the mesial M_3. The anterior portion of the pterygoid fossa is preserved below the posterior half of the mylohyoid line. It is deep and well-demarcated from the rest of the medial corpus. A similar configuration is seen in some *Au. afarensis* mandibles (e.g., MAK-VP-1/12). The submandibular fossa is shallow but distinct from the M_2 to the P_4/M_1 level. It forms a concavity close to the inferior edge of the corpus.

Lower Second Incisor (ALA-VP-2/10B; Plate 7.2) The left I_2 consists of the worn crown and about two-thirds of the root. The cervical line is preserved except at the basal labial crown, where enamel is chipped away and abraded. The labial crown surface is extensively weathered, but the lingual enamel surface is well-preserved. There are no interproximal facets (IPCFs) evident on the preserved mesial or distal faces.

The crown is mesiodistally narrow and labiolingually broad, with a mesiodistal dimension of 6.3 mm as worn and preserved. Crown dimensions are listed in Table 7.1. The unworn crown would have had a larger mesiodistal dimension, although

FIGURE 7.3

Closeup of the *Ardipithecus kadabba* mandible soon after its discovery. The second molar crown had become detached from the mandible after fossilization, as indicated by the freshly broken roots. A large sieving operation failed to recover this right crown, but other teeth from the same specimen were recovered elsewhere on the locality. Photograph by Tim White, December 16, 1997.

its dimension cannot be accurately estimated. Enough of the crown is preserved to show that lateral crown flare was probably weak, as in *Au. afarensis* and *Au. anamensis* homologs.

In occlusal view, the labial two-thirds of the crown is dominated by the broken and worn incisal surface. Preserved crown height above the cervix is 7.5 mm lingually and around 3.5 mm labially, the worn incisal surface exhibiting a strong labioapical slope. The labial and distal enamel rims surrounding this area are irregular and have polished edges, and the dentin surface is polished by wear. Therefore, the crown must have broken antemortem. The fracture accumulated wear until the individual's death.

The lingual crown face consists of a shallow lingual fossa bounded by the mesial and distal marginal ridges and the basal lingual tubercle. The mesial marginal ridge is weak and the distal marginal ridge is more distinct, similar to some examples of *Au. afarensis* (e.g., A.L. 400-1) and *Ar. ramidus*. The basal lingual tubercle is localized, but prominent, more so than in examples of *Au. afarensis* (e.g., A.L. 400-1, MAK-VP-1/12), *Au. anamensis* (KNM-KP 29287), or *Ar. ramidus*. Together with the large buccolingual dimension of this crown, the prominent basal tubercle gives the crown a strongly concave lingual profile in mesial or distal views. The lingual fossa is relatively flat and demarcated from the basal tubercle by a faint cingular furrow.

The mesial and distal cervical lines exhibit strong inverted V-profiles, the mesial cervical line being deeper than the distal, as in *Australopithecus* and *Ar. ramidus* homologs. The root exhibits a broad, shallow depression mesially and a better-defined groove distally. The buccolingual and mesiodistal dimensions of the root at the basal cervix level are 8.0 mm and 4.7 mm, respectively. Longitudinally, the root exhibits a weak mesioapical curvature. This condition is matched in known examples of *Ar. ramidus*.

FIGURE 7.4
Yohannes Haile-Selassie examines the Alayla hominid mandible by flashlight at the Dalga field camp, on the day of its discovery. This was the first late Miocene hominid to have been recovered, before either *Orrorin tugenensis* or *Sahelanthropus tchadensis*. Looking on are, clockwise from Haile-Selassie, Berhane Asfaw, Henry Gilbert, Scott Simpson, David DeGusta, Giday WoldeGabriel, and Bruce Latimer. Photograph by Tim White, December 16, 1997.

Lower Canine (ALA-VP-2/10C; Plate 7.2) The left lower canine comprises an intact crown in early to moderate wear and approximately half to two-thirds of the root. The broken labial root length is 17.5 mm. The cervical line is intact. The labial crown face is considerably weathered, whereas the other crown faces have better-preserved surface enamel. No signs of IPCFs are discernible mesially or distally. However, a small mesial IPCF cannot be ruled out because of some surface enamel weathering.

The crown outline is elongate in its mesiobuccal to distolingual axis in occlusal view, and narrow in a direction perpendicular to that. Such a strongly compressed crown contour is not characteristic of either *Australopithecus* or known *Ar. ramidus* homologs (e.g., ARA-VP-1/128). The standard crown dimensions are listed in Table 7.1.

Occlusal wear has progressed so that wear on the crown tip (or apical wear) exposes a labiolingually elongate dentin patch of about 1 mm length. Worn crown height from labial cervical line to crown tip is 13.4 mm. Crown tip wear is confined to a mesiodistally limited area of 1.5 mm, the wear surface of which is also tilted lingually. Wear continues distally and inferiorly from this confined zone of apical wear, first to a zone of oblique wear (2.5 mm in height) and then to an extensive zone of more vertical wear (4.5 mm in height). The latter wear surface covers both the distal crest and the distolingual ridge. Its dentin exposure is a continuous strip that forms a narrow inverted Y pattern. Distally, the vertical wear extends shelf-like onto the distal tubercle (cingulum), which is worn horizontally with exposed dentin. The vertical wear is due to contact with the mesial crest (or wear slope) of the upper canine. The overall wear aspect most closely resembles *Au. anamensis* and *Au. afarensis* homologs such as KNM-KP 29286, A.L. 198-1 and BMNH M.18773.

In buccal view the crown profile reveals a high mesial shoulder, steep distal wear slope, and a distinct distal tubercle. Relative to unworn crown height (a rough estimation obtained

FIGURE 7.5
Sieving at the ALA-VP-2 locality. Large boulders and cobbles are first removed from the surface and piled into cairns. The remaining silt/sand/pebble lag is then scooped into buckets, transported to the sieving location, and passed through 1.5 mm mesh. The concentrate is then picked, one pan at a time. These kinds of recovery operations are usually confined to hominid discoveries, but when micromammalian remains are present, many are recovered in this manner. Photograph by Tim White, December 7, 1999.

by projection is approximately 14.5 to 15 mm), the mesial shoulder is placed at a position of approximately two-thirds of total crown height. Such a high shoulder placement is the condition usually seen in *Au. afarensis* and *Au. anamensis* lower canines (e.g., A.L. 400-1, MAK-VP-1/12, L.H.-3, KNM-KP 29284, KNM-KP 29286). It is also seen in some *Ar. ramidus* homologs, but only rarely in female apes.

A faint vertical groove occurs adjacent to the mesial border of the buccal crown face, passing from just below shoulder height toward the cervix. The thin vertical zone of enamel defined by this groove is continuous cervically with the ill-defined basal swelling of enamel that encircles the basal buccal crown. Two strong hypoplastic lines are present on the buccal face of the crown, at a level approximately 4 mm and 5 mm above the cervical line. Another weaker line encircles the crown about 1.0 mm above this.

The distal tubercle is well-developed and projects distally by 2 mm. It extends 3.4 mm above the distal cervical line and has a buccolingual dimension of 3.2 mm as worn. It is demarcated buccally by a distinct distobuccal groove. It is separated lingually from the worn distolingual ridge by a deep, pit-like distolingual groove. Such a well-developed distal tubercle occurs in *Ar. ramidus* and *Au. anamensis* (KNM-KP 29284, KNM-KP 29286) counterparts but does not occur in ape lower canines, which exhibit weaker or virtually no distal cingular expressions. In the more incisiform *Au. afarensis* (e.g., L.H.-3) and some *Au. anamensis* (KNM-KP 31727, KNM-ER 30731, KNM-ER 30950) lower canines, the distal tubercle homolog tends to be smaller and merged into the basal portion of the better-developed distal crest.

The lingual face exhibits a dominant, fairly flat area occupying the mesial half of the lingual crown surface. A well-developed mesial vertical groove delineates a confined but

FIGURE 7.6

The 1997 discovery of the holotype of *Ardipithecus kadabba* led to revisits and maintenance of the locality. Piles of basalt cobbles are seen in this view, as well as the soft, rapidly eroding sediment below, now unarmored. These operations resulted in additional hominid remains being found, including the ALA-VP-2/101 arm bones at the large boulder between the vehicles and the sieving shade, and the lower canine from the position of the large bolder between the viewer and the shade. The original holotype mandible came from the area where the sieve is active. In this photograph, pebble lag is scooped from the hillside to the left of frame and transported in plastic buckets to the sieve location. Once sieved, the concentrate is picked by the team seated in the shade. Photograph by Tim White, December 13, 1999.

swollen marginal ridge. Such a mesiolingual marginal ridge is seen in *Ar. ramidus* and *Australopithecus* (e.g., L.H.-3, KNM-KP 29284, KNM-ER 30731) lower canines but is absent in ape canines. The flat mesiolingual face is bounded distally by the distolingual ridge, and the cleft-like distolingual groove occurs distal to it. The distolingual ridge terminates basally onto a distolingually projecting basal eminence. The latter is continuous, in a cingular-like fashion, with the distal tubercle.

The mesial crown face is triangular, with the mesial cervical line strongly convex occlusally. The mesial crown shoulder is placed approximately 5.5 mm above the deepest point of the mesial cervical line. The combined effect of a deep cervical line with a shoulder corner significantly above the cervical line produces a high placement of the mesial crown shoulder. The distal crown face is dominated by the vertical distal wear and the large distal tubercle. The distal cervical line is only very weakly convex occlusally below the distal tubercle. In mesial or distal view, relative to the vertical axis of the crown, the buccal face slopes strongly occlusally and lingually, as do female ape and *Australopithecus* lower canines, to varying degrees.

The preserved root has buccal and distal faces that form a continuously curved surface, resulting in an evenly convex cross-sectional shape. The mesial root face is overall fairly flat but has a distinct vertical groove throughout most of its preserved height, imparting a bilobate cross-sectional shape. Maximum and perpendicular root dimensions at the basal crown cervix are 11.1 mm and 7.1 mm, respectively. Radial thickness of the labial enamel is 1.2 mm measured either at the worn surface or by the micro-ct imagery.

Lower Fourth Premolar (ALA-VP-2/10D; Plate 7.2) The left P_4 consists of an intact, moderately worn crown and approximately two-thirds of the root. The broken root length is 11.2 mm buccally. The cervical line is preserved around the crown, but much of the crown and root surfaces were damaged by weathering. Both IPCFs are obscured by surface damage to the mesial and distal crown face enamel, but a planar area represents the distal IPCF on the lingual side of the distal crown face, measuring 4.5 mm in buccolingual extent. Only small segments of the original mesial IPCF are intact.

In occlusal view, the crown has a rectangular outline with rounded buccal, obtuse mesiolingual, and projecting distolingual contours. Standard crown dimensions are listed in Table 7.1. The crown is worn occlusally, with a dentin exposure of 1 mm at the protoconid cusp tip. The metaconid is faceted by wear but does not exhibit enamel perforation. A 1.5 mm to 2 mm area of the distolingual crown corner has broken away. Because the broken enamel margin is rounded and wear striae occur continuously from enamel to polished dentin, this macrofracture must have occurred premortem.

The protoconid is the largest cusp and occupies almost the entire buccal crown. The smaller metaconid is situated opposite the protoconid. The two main cusps are connected by a continuous transverse crest in moderate wear that interrupts the longitudinal groove, as seen in the *Ar. ramidus* P_4 (ARA-VP-6/1). A small, pit-like anterior fovea occurs on the worn mesial occlusal surface. Distally, the transverse groove delineates the talonid from the main cusps. The mesiodistal dimension of the talonid is greater lingually than buccally.

The talonid is delineated from the protoconid by a narrow but distinct distobuccal groove on the buccal crown face. The talonid is shelf-like and, as seen in buccal or lingual views, is situated at a position distinctly lower than the two main cusps and anterior fovea, as in the *Ar. ramidus* (ARA-VP-6/1) and *Au. anamensis* (KNM-KP 29286) P_4s, more so than in the other *Au. anamensis* (KNM-ER 20432) and comparably worn *Au. afarensis* (e.g., A.L. 400-1, A.L. 266-1) examples.

The root is of Tomes's morphology, with buccally fused mesial and distal roots and a deep vertical cleft on the mesiolingual root face. The broken root section exhibits three root canals, placed mesiolingually, buccally, and distolingually. The maximum buccolingual root dimension is 6.8 mm in the mesiobuccal root, and 10.0 mm for the fused buccal and distal root portions.

Maximum radial thickness of the lateral enamel taken at the fracture surface of the distolingual cusplet is 1.4 mm. Radial thickness of the protoconid lateral enamel is estimated at 1.6 mm at the worn dentin exposure.

Lower First Molar Fragment (ALA-VP-2/10G; Plate 7.3) The left M_1 is represented by a fragment comprising most of the protoconid and a small part of the root. Its preserved dimension is approximately 5 mm mesiodistal and 8 mm buccolingual. Dentin exposure is 1.5 mm buccolingual on the protoconid but continues and deepens distally towards the hypoconid. The mesial metaconid region exhibits a steep dentin exposure that slopes lingually, bounded by an irregular mesial enamel rim. This suggests premortem fracture and subsequent wear. The preserved mesial IPCF appears to match the distal IPCF of the left P_4, which would mean that the suggested premortem macro-fracture spanned the entire lingual region from distal P_4 to mesial M_1.

Little can be said of the size and morphology of this molar fragment, although at the distal break of the protoconid, a minute portion of buccal protoconid face begins to face distobuccally. Thus, the break is likely situated just mesial to the mesial buccal groove, suggesting that the size of the protoconid was roughly equivalent to small-sized *Ar. ramidus* homologs such as ARA-VP-1/200. The presence of a very thin, shelf-like protostylid complex at the mesial buccal groove (also seen in ARA-VP-1/200) is also suggested.

Maximum radial thickness cannot be measured on the distal broken surface of the protoconid. Radial enamel thickness estimated at the worn protoconid dentin exposure is approximately 1.5 mm.

Lower Second Molar (ALA-VP-2/10E; Plate 7.3) The left M_2 is a virtually complete and heavily worn crown with both mesial and distal roots preserved except for their apices. The cervical line is intact. The broken root lengths are approximately 11 mm for the mesial root and 12.5 mm for the distal root. The occlusal and lingual crown surfaces are weathered, but enamel of the mesial, buccal, and distal crown faces is better preserved. The mesial IPCF is flat to weakly concave, buccolingually elongate (3.7 mm wide by 1.7 mm high), and placed centrally, adjacent to the mesial occlusal margin. The distal IPCF is flat, buccolingually wide, and vertically higher than the mesial facet (3.9 mm wide by 2.2 mm high), and offset lingually from crown midline.

In occlusal view, the crown outline is a mesiodistally elongate rectangle with rounded corners and slight lingual skew of the distolingual crown contour. Both lingual and buccal crown contours are weakly bilobed. The standard crown dimensions are listed in Table 7.1.

Enamel on the buccal occlusal crown half is deeply perforated by wear. Two basin-like dentin wear surfaces are separated by a worn, narrow enamel ridge between the hypoconid and hypoconulid. The mesial of the two dentinal basins is large (6.1 mm mesiodistal, 3.5 mm buccolingual) and deep, formed by the combined protoconid and hypoconid dentin exposures. The distal dentin basin is obliquely elongate (4.1 mm maximum dimension by 2.8 mm) and corresponds to the hypoconulid and buccal portion of the distal marginal ridge. The metaconid has a scooped and moderately steep occlusal wear surface with an occlusally projecting lingual rim. A 3.4 mm by 0.9 mm, buccolingually elongate dentin wear exposure occurs along its occlusal slope. A similar but lower rim occurs at the lingual occlusal margin of the worn entoconid, but without any dentin exposure. Together, the metaconid and entoconid retain an extensive zone of phase I wear.

Although wear renders evaluations of occlusal morphology difficult, the protoconid and hypoconid appear roughly equivalent in mesiodistal dimension. The entoconid is mesiodistally larger than the metaconid. The hypoconulid is the smallest cusp. Overall, the distal three cusps appear to form a much larger portion of the crown than the two mesial cusps. Advanced wear precludes any further evaluation of occlusal morphology.

In mesial view, the buccal crown face appears to slope occlusolingually, but the degree of buccal crown flare cannot be evaluated as a result of advanced wear. A mesial buccal groove remains prominent at the worn occlusal margin, indicating the presence of an initially deep groove that incised the occlusal margin between protoconid and hypoconid. This groove continues onto the buccal surface and shallows cervically. In lingual view, the higher metaconid and lower entoconid form an asymmetric and obtuse V-shaped notch along the occlusal rim. The faint lingual groove is deeper occlusally than cervically.

The root system consists of a broader mesial root (10.5 mm) and a smaller and apically tapering distal root. The mesial root is almost vertically placed, whereas the distal root is torsioned and angled slightly distally and buccally. Root cross section shape at a position just below the cervix is similar to that of the broken root section of the right M_2, supporting the association of the isolated teeth with the ALA-VP-2/10A mandibular piece.

Maximum radial thickness of the lateral enamel was 1.8+ mm at mid-hypoconid position as worn, measured on the natural fracture surface. Enamel thickness measures of the lingual crown were obtained from the micro-ct data (the EDJ was not sufficiently imaged in the buccal side). Maximum radial thicknesses of the lateral enamel opposite the worn metaconid and entoconid were 1.38(+) mm and 1.67 mm, respectively. The "(+)" notation after the enamel thickness metric indicates that the measured value was probably close to the actual thickness of the unworn condition.

Lower Third Molar (ALA-VP-2/10A and F; Plates 7.1–7.2) The right M_3 is almost complete except for the buccal crown face, where the protoconid and hypoconid enamel and portions of the dentin are chipped off. The cervical line is well-preserved and visible, except at the damaged buccal crown face. Surface enamel is generally well-preserved, except for the occlusal enamel of the metaconid. Because of the mesiobuccal crown damage, only the lingual-most portion of the mesial IPCF is preserved.

In occlusal view, M_3 crown outline is rectangular, with rounded corners mesially and an oval, evenly rounded distal crown contour. The rounded distal contour is due to the slightly buccally placed hypoconulid and a small distolingually placed accessory cusp (C6). The measured M_3 crown dimension is given in Table 7.1. An ill-developed distal M_3 crown is shared with known *Ar. ramidus* homologs (e.g., ARA-VP-1/128). *Australopithecus afarensis* and *Au. anamensis* lower M_3s are characterized by better-developed distal accessory cusps (e.g., A.L. 266-1, A.L. 400-1, KNM-KP 29281), although minimal C6 expression (e.g., A.L. 333w-60) occurs as part of the variation.

The protoconid enamel is almost entirely perforated by wear. The worn dentin surface forms a deeply excavated concavity, the degree of which increases buccally and reaches as far inferiorly as the level of the cervical line. The maximum preserved anteroposterior length of the exposed dentin is 7.6 mm, extending onto the hypoconid. The hypoconulid is flattened by wear, but no dentin exposure is present. The metaconid and entoconid are intact. Their tips lack any blunting by wear, but wear facets occur on their occlusal slopes. On the metaconid, a steep wear facet occurs on the main occlusal and the adjacent distal accessory occlusal ridges. The facet faces lingually and slightly distally. The entoconid supports a lingually facing steep wear facet that occurs on its main and accessory occlusal ridges. These wear facets on the lingual M_3 cusps represent well-defined phase I facets in advanced wear.

The metaconid and entoconid are roughly equivalent in mesiodistal dimensions. The metaconid occupies the mesiolingual crown corner and has deep occlusal grooves situated mesial and distal to the main occlusal ridge, giving the cusp tip a nipple-like appearance. Occlusal morphology of the crown, aside from that just described, is obliterated by wear, except for the presence of a small C6 already mentioned.

In lingual view, an obtuse V-shaped notch is formed between the metaconid and entoconid cusp tips. On the lingual crown face, the lingual groove is deep occlusally and shallows cervically. A distal lingual groove between the entoconid and C6 is expressed toward the occlusal margin. Lingual crown height from the cervical line is 6.4 mm to both metaconid and entoconid cusp tips, and 4.2 mm at the lingual notch.

A strongly curved distal root is partially exposed at the posterior break of the mandibular specimen, showing a root apex bifurcated by the mandibular canal. The distal root fragment of the left M_3 (ALA-VP-2/10F) exhibits a conical structure, unlike the right M_3 root. However, the lingual face of the apical portion of this root is smoothly notched, indicating that the mandibular canal was lingually juxtaposed to the M_3 distal root.

Other Dentition

DISCOVERY

ALA-VP-2/349 is a right upper incisor fragment found by Berhane Asfaw on January 1, 2005, from the holotype locality ca. 5 m south of the holotype mandible.

STD-VP-2/61 is a right lower canine found on the surface on November 18, 1998, by Humed Mohammed, one of the local Afar workers, while a group of eight people was crawling a steep slope surface at Saitune Dora Vertebrate Paleontology Locality 2 (STD-VP-2). It was found intact among numerous bones of large mammals. STD-VP-2/61 was the first hominid to be discovered at the locality.

STD-VP-2/62 is a left M^3 recovered ca. 1 m west of STD-VP-2/61 by one of us (YHS) during group crawling on November 18, 1998. The specimen was found in four pieces, scattered within a 1 m radius. All pieces join to make up a complete crown and portions of the roots.

STD-VP-2/63 is a left M^1 recovered on December 20, 1998, during the sieving operation carried out at STD-VP-2. It was found in colluvium measuring ca. 50 cm thick, approximately 10 meters downhill from the find spots of STD-VP-2/61 and 2/62. Figures 7.7 through 7.11 record the hominid recovery operation at this locality, and Chapter 15 contains additional information about the Saitune Dora occurrence.

FIGURE 7.7
View of the STD-VP-2 hill when the first hominid teeth were discovered. They were surface finds clustered where the collection team is concentrating its efforts on the hill. Photograph taken November 18, 1998.

ASK-VP-3/160 is a left P^3, a surface discovery made on January 1, 2001, by group crawling at Asa Koma Vertebrate Paleontology Locality 3 (ASK-VP-3).

In the fall 2002 field season, a sieving operation was conducted at ASK-VP-3, where two hominid specimens (P^3 and humerus) had been found in previous years. The excavation covered an area of approximately 20 m by 30 m and involved digging into deflated lags up to 75 cm deep. Six isolated hominid teeth were recovered in the sieving operation. ASK-VP-3/400 is a right upper canine found on November 26, 2002. ASK-VP-3/401 (Plate 7.4) is a right M^1 found on November 27, 2002. ASK-VP-3/402 (Plate 7.4) is a left M^1 found on December 1, 2002. ASK-VP-3/403 is a left P_3 found on December 1, 2002. ASK-VP-3/404 is a fragment of a left P^4 discovered on December 2, 2002. ASK-VP-3/405 is a premolar fragment discovered on November 27, 2002. Figures 7.12 through 7.17 record the hominid recovery operation at this locality.

PRESERVATION

The preservation of each tooth is sketched out in the respective descriptive sections.

COMPARATIVE DESCRIPTIONS

Upper Incisor (ALA-VP-2/349; Plate 7.5) ALA-VP-2/349 is a small incisor fragment here considered part of a right I^2. As preserved, it is approximately 4 mm mesiodistal, 7 mm

FIGURE 7.8
Close-up of the area surrounding the hominid tooth fragments (indicated by individual pin flags). Note the position of the flags immediately above the proboscidean bone bed described in Chapter 15. Photograph taken November 18, 1998.

labiolingual, and 12.5 mm incisocervical. The preserved incisal edge exhibits a weakly rounded corner and an adjacent mammelon that projects further incisally. These features suggest that this fragment represents the mesial portion of an I^2. Other morphological details support this attribution. The cervical line is preserved on the mesial face, which shows an incisal convexity tending toward a broad U-shape rather than a more acute V-shape.

This tooth was either unworn or minimally polished, with lack of incisal flattening or faceting. Weathering precludes actual evaluation of incisal polish, but the adjacent enamel surface that is well preserved does exhibit some surface polish. The mesial IPCF is absent, and the combined evidence suggests either erupting or newly erupted status.

The mesial crown face is weakly convex and broad labiolingually. The preserved lingual crown exhibits a thin but distinct mesial marginal ridge defined by a groove distally, and a weakly convex lingual surface adjacent to this. A comparable morphology is seen in *Ar. ramidus* homologs, in contrast to the more spatulate and concave lingual fossa often seen in *Au. afarensis* (e.g., L.H.-3). Labial enamel thickness measured on the fracture surface is 1.0 mm.

Upper Canine (ASK-VP-3/400; Plate 7.6) ASK-VP-3/400 is a little-worn right upper canine lacking most of the root and some crown base. Only a portion of the cervical line is preserved 3 mm below the mesial crown shoulder. Otherwise, parts of the basal crown are missing, with only the buccal crown preserved close to the cervix. From the surface topography and thinness of the basal enamel, the preserved buccal face of the distal two-thirds of the crown must extend very close to the cervical line itself. In contrast, the distolingual basal crown is entirely lacking. Surface enamel is generally well-preserved,

with only scattered weathered patches. There is no mesial IPCF, while a portion (1 mm) of the distal IPCF is preserved at the basal break of the distal crown face.

Viewed occlusally, the buccal face protrudes buccally and mesially, comparable to (but slightly stronger than) *Ar. ramidus* homologs (ARA-VP-1/300, ARA-VP-6/1), suggesting perhaps a slightly more skewed mesiobuccal basal crown outline than in *Ar. ramidus*. It differs from female chimpanzee canines, which have much weaker buccal extension of the basal crown (corresponding to their mesiodistally elongate crowns). The degree of disto-lingual protrusion of the crown base cannot be assessed in ASK-VP-3/400.

The crown is little-worn, with only faint traces of polish on the crown tip, and a narrow vertical contact facet on the mesial aspect of the main mesial crest that runs from crown tip to mesial shoulder.

Maximum crown dimensions cannot be determined for this canine, but its mesiodistal dimension can be accurately measured across the preserved mesial and distal crown shoulders (see Table 7.1). Preserved canine height is 15.5 mm buccally. Crown height to the cervix was likely to have been around 16 to 16.5 mm.

In buccal view, ASK-VP-3/400 displays a high, conical crown, with steep mesial and distal crests terminating at relatively low mesial and distal crown shoulders. The mesial crest length from crown tip to mesial shoulder corner is 10.5 mm, and the distal crest length is 9.9 mm. Mesial shoulder position in buccal view was probably somewhere between

FIGURE 7.9

The pin flag (light gray) within the upper excavation unit indicates the position of the hominid clavicular fragment, from a horizon where rhinoceros skeletal elements were found *in situ*. The excavators are exposing the proboscidean bone bed found about 2 m stratigraphically below that horizon. Photograph taken December 16, 1998.

FIGURE 7.10

The northern slope of the Saitune Dora proboscidean hill partially stripped of armoring lag reveals the dense accumulation of large mammal bone just below center frame. The pin flags indicate the surface position of the hominid teeth. Large numbers of rhinocerotid fossils were found in a horizon 1 m above the proboscidean bone bed, which measures approximately 1 m thick. The slope below the proboscidean bone bed contained vast numbers of fragments of shattered bones, all of which were sieved and hand-picked in search of additional hominid pieces. Photograph by Tim White, November 23, 1998.

FIGURE 7.11

The northern slope of the STD-VP-2 locality with the proboscidean bone bed exposed just above the excavator's (YHS) head. Pin flags above this horizon indicate positions at which the hominid teeth were found on the surface. Photograph by Tim White, November 23, 1998.

FIGURE 7.12
View to the northwest across the Asa Koma drainage. ASK-VP-3 is the locality to the right of the major fault between the lighter sediments on the right and the darker sediments to the left. Most vertebrate fossils weathered out of the sediments and are trapped in the cobble lags that now armor the slope. Major sieving operations conducted at this locality recovered several specimens of *Ardipithecus kadabba*. The excavation team approaches the ASK-VP-3 hominid locality to begin maintenance recovery operations. The steepness of terrain and lack of vehicle transport close to the remote hominid recovery sites made such operations logistically difficult. Photograph by Tim White, December 10, 1999.

one-half and one-third of total crown height from the crown base, well within female ape ranges of variation, including chimpanzees. On average, however, female chimpanzees tend to have lower crown shoulders. The mesial shoulder position of *Ar. ramidus* and *Au. anamensis* upper canines range from lower than mid-crown (e.g., ARA-VP-1/300, ASI-VP-2/334, KNM-KP 35839) to mid-crown level (e.g., ARA-VP-6/1, ASI-VP-2/367). The mesial shoulder of ASK-VP-3/400 hardly protrudes mesially, in contrast to *Ar. ramidus* counterparts, which exhibit a distinct mesial projection of the shoulder angle (e.g., ARA-VP-1/300, GWM9n/P51). Faint vertical grooves occur between the shoulder corners and the main conical buccal crown face, as they do in *Ar. ramidus* canines (e.g., ARA-VP-1/300, ARA-VP-6/1).

The mesial lingual groove is well-defined and deepens toward the crown base. It terminates in a triangular pit. The groove is bounded mesially by a short marginal ridge running obliquely to the mesial crown shoulder corner, and distally by the well-developed mesiolingual ridge. The latter is a prominent but rounded and confined ridge running vertically from crown tip to basal mesiolingual crown. Whereas the mesiolingual ridge tends to flatten out basally in *Ar. ramidus* and later hominid canines (e.g., ARA-VP-6/1), this is not the case in ASK-VP-3/400. Together with its low and weakly projecting mesial shoulder corner, the mesial crest and mesiolingual ridge together define a narrow zone of potential vertical wear against an interlocking lower canine. Such a configuration is the primitive condition usually seen in apes with fully interlocking canines.

The rest of the lingual crown face is relatively flat, but with an ill-defined weak and broad accessory lingual ridge occurring at about the distal two-thirds of the lingual face. The ridge is flanked by faint vertical grooves. A marginal ridge-like structure appears to

SLOW
WHALE
CROSSING
I Brake for
Whales
www.oceanconservancy.org
Ocean
Conservancy

FIGURE 7.13

After sieving, concentrate in the screens is picked in steel pans, with each "picker" taking a small scoop of the dust-free concentrate and picking all fossil bone and tooth fragments from it. There are occasional surprises. Photograph by Tim White, December 13, 1999.

have occurred adjacent to the distal shoulder corner, but the basal crown is not sufficiently preserved to evaluate its development. The form of the distolingual basal eminence and presence or absence of cingulum-like structures cannot be assessed because of the absent crown base.

The micro-ct imagery exhibits only intermittently discernible EDJ, which shows a weak thickening of the labial enamel toward the crown tip with a maximum radial thickness of 1.15 mm.

Lower canine (STD-VP-2/61; Plate 7.6) STD-VP-2/61 is an unworn or little-worn lower canine with the complete crown and less than half of the root preserved. Maximum broken length of the root is 11.5 mm distolingually. The cervical line is intact. The enamel surface is affected by widespread light etching, and only intermittent patches of intact surface enamel remain. There are no clear signs of wear, but one small enamel patch close to the lingual crown tip exhibits polish and fine striae, indicating that this canine might have come into partial eruption. There are no signs that IPCFs were present.

In occlusal view, the crown is elongate mesiobuccal to distolingual and narrow in a perpendicular direction, similar to, but not as extreme as the condition seen in the ALA-VP-2/10 lower canine. However, because of a slightly stronger mesiobuccal basal projection and weaker distal tubercle, the occlusal crown outline of STD-VP-2/61 is an asymmetric oval with narrower distolingual and broader mesiobuccal crown. The standard crown dimensions are listed in Table 7.1.

The buccal crown face exhibits a mesial shoulder at mid-crown height, a conical buccal face narrowing toward the pointed crown tip, and a basally placed distinct distal tubercle. A 7.6 mm long, relatively sharp, and vertically steep mesial crest extends from crown tip

FIGURE 7.14
The soft sediments of the Asa Koma Member at ASK-VP-3 have eroded into an outwash colluvial fan atop a resistant basaltic tuff. The team removes the large cobbles and bone fragments by hand, then sieves the remaining gravel to remove the sand and dust, leaving a concentrate which is picked by hand. Photograph by Tim White, December 3, 2002.

to the mesial shoulder corner. The mesial shoulder placement of STD-VP-2/61 is lower than in ALA-VP-2/10 and in *Au. anamensis* and *Au. afarensis* counterparts. Comparably low mesial shoulders are commonly seen in female apes.

A faint vertical groove marks the buccal face adjacent to the mesial border. It is similar to but weaker than in ALA-VP-2/10. As with the ALA-VP-2/10 canine, the weak ridge of enamel mesial to the vertical groove is continuous basally with a zone of enamel thickening that encircles the buccal crown base to the distal tubercle. The lower half of the buccal face has three wide, horizontal hypoplastic lines centered at approximately 2.5, 3.8, and 5.3 mm from the cervical line. Other weaker hypoplastic lines occur as high as the level of the mesial crown shoulder. The more distinct hypoplastic lines on the buccal crown face are manifested on the lingual crown face as much thinner lines.

The distal tubercle is distinct, but weaker than in ALA-VP-2/10, projecting distally by 1.5 mm. It is 4.4 mm high from the highest point of the weakly convex distal cervical line. It is demarcated mesiobuccally by a short but distinct distobuccal groove. Lingually, the distal tubercle is separated from the distolingual ridge by a distolingual groove that continues to the distolingual basal eminence.

The lingual face is dominated by a fairly flat mesiolingual area, bounded distally by the distolingual ridge and mesially by the mesial crown border. In contrast to the ALA-VP-2/10 lower canine, a fold-like marginal ridge does not occur at the mesial margin of the mesiolingual face. Rather, only a faint expression of an enamel elevation is seen. It is continuous with the basal lingual crown in a cingulum-like fashion. Such a configuration is often seen in female ape canines. The distolingual ridge of the lingual face is sharp and runs almost vertically from the tip of the crown to the lingual crown base just mesial to the distal tubercle. It fans out onto the distolingual basal eminence. Another, weaker vertical crest

FIGURE 7.15

View from the east across the ASK-VP-3 locality during recovery operations. Fossils derived from sediment below the MA96-30 volcanic ash, dated to 5.63 Ma, in the right center of frame washed out atop the basaltic tuff at the base of the Asa Koma Member and were trapped in cobble lag. This lag was excavated (figures to the left) and sieved (figures to the center), and the concentrate was hand-picked to extract fossil remains, including hominid teeth and many small mammal specimens. Note the large fault to the left of the striped hill, with the horizontal dark basaltic tuff to the west uplifted relative to the fossiliferous sediments. Photograph by Tim White, November 29, 2002.

passes immediately distal to the distolingual crest and terminates at the center of the distal tubercle. This is the distal crest homolog of later, more incisiform hominid canines. The faint expression of the distal crest in STD-VP-2/61 is well within female ape ranges of variation (including chimpanzees) and attests to the primitive morphology of this canine. A distinct groove occurs between the faint distal crest and the distolingual ridge, but such structures would quickly be worn away by contact with the mesial face of the upper canine.

The triangular mesial face exhibits a cervical line that is not so occlusally convex as in ALA-VP-2/10. The mesial shoulder corner is set 4.5 mm above the deepest invagination of the cervical line. The combined effect is a mesial shoulder placement at approximately mid-crown height in buccal view, much lower than in the ALA-VP-2/10 holotype described above. The distal crown face exhibits only a very weakly occlusally convex cervical line below the distal tubercle. In mesial or distal views, relative to the vertical axis of the crown, the buccal face slopes lingually, but slightly less so than in ALA-VP-2/10.

FIGURE 7.16
Picking pan with concentrate containing a hominid upper molar from ASK-VP-3 (nearly in the center of the photograph, above the thumb). Photograph by Tim White, November 27, 2002.

The root is relatively flat distally and evenly convex buccally. The mesial root face is bilobed, with a distinct vertical groove as in ALA-VP-2/10. Maximum and perpendicular dimensions of the root at the basal crown cervix are 10.4 mm and 7.3 mm, respectively.

Maximum radial thickness of the labial enamel is estimated at 1.0 mm from the micro-ct data, with some uncertainty stemming from the faintly imaged EDJ.

Upper Third Premolar (ASK-VP-3/160; Plate 7.5) ASK-VP-3/160 is a little-worn left P^3 crown with very small root segments retained on portions of the mesial and distal tooth. The entire crown is virtually intact, mostly broken just at, or close to, the cervix. The cervical line itself is preserved mostly on the mesial and distal faces. Surface enamel is generally well-preserved but has intermittent patches of damage from weathering. Both mesial and distal IPCFs are preserved. The small mesial IPCF is oval, 2.0 mm wide by 1.0 mm high, and situated just below the occlusal margin of the buccal-most mesial marginal ridge. The distal IPCF is 2.5 mm wide by 1.5 mm high, and centrally positioned close to the occlusal margin of the distal crown face.

In occlusal view, the crown exhibits an asymmetric outline, with a mesiobuccally developed buccal contour, a distolingually slanting mesial border, and a narrow and rounded lingual contour continuing to the weakly convex distal crown. This degree of P^3 crown contour asymmetry is matched in homologs of *Ar. ramidus* and *Au. anamensis* (e.g. ARA-VP-6/1, KNM-KP 30498). Ape P^3s are generally even more asymmetric, but a comparably weak asymmetry occurs in some apes, such as in *Ouranopithecus* and as individual variation within chimpanzees. Standard crown dimensions are listed in Table 7.1.

Wear is at an early stage, with no dentin exposures on either of the cusp tips. A planar wear facet occurs on the distoocclusal slope of the paracone, but the paracone tip itself is only rounded by polish. A larger, distoocclusally facing facet occurs at and distal to the

FIGURE 7.17
Aerial view looking west across the ASK-VP-3 hominid locality. The larger ring in lower frame center contains the sieving piles, the smaller rings the fragmentary faunal remains recovered during the operation. The thin rock lines indicate the limits of the underlying basaltic tuff. Additional sieving at this site will probably produce more hominid remains. Photograph by Tim White, January 11, 2003.

protocone, extending to the distal marginal ridge. A narrow and steep wear facet occurs on the mesial aspect of the mesial protocone crest (lingual portion of the mesial marginal ridge), possibly indicating contact with a transverse P_3 crest.

The paracone is the dominant of the two main cusps in both mesiodistal and buccolingual dimensions. Mesiodistally, it is situated centrally, and the protocone tip is set slightly mesial, although the latter appears to be exaggerated by wear. The occlusal fovea consists of the triangular anterior fovea and much larger posterior fovea (or talon). A weak but distinct ridge on the mesial aspect of the buccal anterior fovea passes mesiovertically from the paracone tip. Two thin ridges extend lingually from the paracone and buccally from the protocone, meeting at the longitudinal groove. The more mesial of these ridges forms the posterior margin of the anterior fovea, which is a steeply inclined, shallow depression bounded mesially by a thin mesial marginal ridge. A deep, narrow, and buccolingually elongate groove occurs between the two sets of crests, at the middle of the occlusal fovea, separated by the longitudinal groove into larger buccal and smaller lingual portions. Posterior to this, the occlusal fovea is marked by a few accessory ridges and grooves. The longitudinal groove is distinct in the anterior two-thirds of the crown but becomes obscured in the talon towards the distal marginal ridge.

In buccal view, the mesial protocone crest is slightly longer and more steeply inclined than the distal. Such a condition is common in apes but is also matched in *Ar. ramidus* (ARA-VP-6/1) and some *Australopithecus* (e.g., L.H.-3, KNM-KP 30498) examples. A faint mesiobuccal groove (or depression) occurs at the mesial margin of the buccal face. A zone of hypoplastic lines occurs about 3 mm from the preserved crown base and extends occlusally to about mid-crown height.

In mesial view, the prominent paracone projects occlusally, much more so than the protocone, as in known examples of *Ar. ramidus* (e.g., ARA-VP-6/1). The slope of the buccal and lingual faces is also comparable to the ARA-VP-6/1 condition. The shallow and mesially facing anterior fovea is a consequence of the low (below mid-crown) and weak mesial marginal ridge and prominent transverse crest. Such a morphology is (so far) unmatched by the homologs of *Ar. ramidus* and *Australopithecus* but is common in apes, including chimpanzees. The basal mesial face is depressed, contributing to a basal crown outline at the cervix that is concave mesially.

The resolution of the EDJ with the micro-ct data is poor, and maximum radial thickness of the paracone lateral enamel is estimated at 1.3 mm.

Lower Third Premolar (ASK-VP-3/403; Plate 7.5) ASK-VP-3/403 is a little-worn partial left P_3 lacking the mesiobuccal crown base and most of the enamel of the lingual cusp. The roots are missing except for a small portion immediately below the lingual cusp. The cervical line is preserved (or its position can be inferred) for the preserved distolingual crown. The enamel surface is etched on the distal crown face but is better preserved buccally and occlusally. A shallow oval depression about 2.5 mm wide is set on the distal face and probably represents the distal IPCF, although the facet surface itself is almost completely obliterated by etching. Crown breakage precludes evaluation of presence or absence of the mesial IPCF.

In occlusal view, the premolar is narrow lingually and broad buccally. The mesiolingual crown contour exhibits a strong concavity between the protoconid and metaconid, in contrast to the more linear to weakly convex contour seen in the known examples of *Ar. ramidus* (ARA-VP-1/128, ARA-VP-6/1). Although the mesiobuccal basal crown is completely missing, this suggests a highly asymmetric crown outline. Standard measurements are not possible because of crown damage, but buccal crown length (length taken along the protoconid mesial and distal crests) can be estimated at approximately 9.3 mm.

The crown is slightly worn, with a polished occlusal surface and a wear facet on the mesiobuccal crown face adjacent to the protoconid cusp tip and mesial protoconid crest. This facet was most likely caused by contact with the distolingual surface of an interlocking upper canine.

The protoconid is the dominant cusp, whereas the metaconid is barely expressed as a distinct entity. A thin transverse crest descends from the tip of the protoconid to the metaconid. The metaconid homolog likely formed a faint projection before wear, breaking the descending contour of the transverse crest but not forming an actual projection above it. Such a weak expression and low position of the metaconid are matched in examples of *Ar. ramidus* and *Au. anamensis* (e.g., ARA-VP-6/1, KNM-KP 29286, KNM-KP 29287) and in female apes, whereas unicuspid *Au. afarensis* P_3s tend to express the metaconid homolog at a higher crown level (e.g., A.L. 288-1, A.L. 128-23). The prominent transverse crest forms a high divide between the larger posterior and the smaller, more restricted anterior fovea. The anterior fovea is positioned at the mesiobuccal angle of the occlusal surface, apparently defined only by the buccal segment of the mesial marginal ridge.

In buccal view, the occlusally projecting protoconid exhibits a slightly longer mesial than distal protoconid crest, as seen in some *Ar. ramidus* and *Australopithecus* P_3s (e.g., KNM-KP 29286, L.H.-3). A distinct distobuccal groove demarcates a vertical enamel ridge on the distal margin of the buccal face.

Micro-ct imagery of this tooth was obtained but did not reveal sufficiently reliable EDJ resolution for enamel thickness measures to be made.

Upper and Lower Fourth Premolars (ASK-VP-3/404, ASK-VP-3/405; Plate 7.5) ASK-VP-3/404 is a fragment of a left P^4. It preserves most of the crown lingual to the longitudinal groove, as well as a very minute sliver of the mesial occlusal surface situated buccal to the longitudinal groove. A small segment of the lingual root is preserved for 6.5 mm below the cervical line. The latter is intact around the preserved crown. The enamel surface is heavily etched, with only small patches of original enamel surface retained. A faint remnant of the mesial IPCF is preserved on the buccal end of the mesial break. Occlusal wear rounded the mesial and distal protocone crests, but without dentin exposure. These worn crests are less steep and slightly shorter mesially than distally, but this asymmetry is weaker than is usually the case in later hominid P^4s (e.g., ARA-VP-6/1, KNM-ER 30745, A.L. 200-1).

We had previously referred to ASK-VP-3/405 as a partial upper premolar (Haile-Selassie et al. 2004b), preserving much of what appears to be the distal half of the right P^4 crown and most of the buccal root, which measures 14.8 mm in height. The crown is damaged and worn, making observations difficult. Assuming the right P^4 attribution, the preserved distolingual portion of the crown consists of a deeply cupped dentin exposure, indicating a more advanced wear stage than in ASK-VP-3/404. What remains of the mid-buccal crown consists of a worn enamel slope, thereby suggesting strong differential wear between the buccal and lingual crown. However, according to this positional interpretation, the lingual basal crown and lingual root would both be unusually small in buccolingual dimension. This is manifested in the small buccolingual dimension of the root itself, which is 9.4 mm at the cervix, more comparable to lower P_4s (and identical to the same dimension of the ALA-VP-2/10 P_4).

The alternative possibility is that ASK-VP-3/405 is a partial left P_4 with unusual wear and root morphology. According to this interpretation, the preserved mid-crown occlusal morphology is readily interpreted as representing portions of the transverse crest and its worn distal slope. But the deeply cupped dentin exposure of the distolingual crown would be unusual, perhaps involving a history of distolingual macro-fracture as seen in the ALA-VP-2/10 P_4. The root system would also be unusual but might have been a variant of a Tome's root type, with a distal root bifurcated into buccal and lingual segments as in the L.H.-24 and *"Au. bahrelgazhali"* lower premolars.

Upper First Molar (ASK-VP-3/401, ASK-VP-3/402, STD-VP-2/63; Plate 7.4) ASK-VP-3/401 is a lightly worn complete right M^1 crown lacking most of the root. The cervical line is mostly preserved but is damaged on the lingual crown base. Etching affects surface enamel throughout the crown, with original enamel surface retained only as small intermittent patches. A buccolingually elongate mesial IPCF (about 4 mm wide by 1.7 mm high) occurs adjacent to the occlusal margin, slightly buccal to midcrown, but it is partially obscured by etching. The distal IPCF, further obscured by etching (and also about 4 mm buccolingual by 2 mm high), is more oval-shaped and positioned more centrally, slightly below the occlusal margin.

In occlusal view, crown outline is close to rhomboidal with an obtuse distobuccal corner and a slightly greater buccolingual dimension mesially than distally. The lingual crown contour is very weakly bilobed between the protocone and hypocone. Standard crown dimensions are given in Table 7.1.

Both protocone and hypocone occlusal surfaces are flattened by wear. Dentin exposure occurs on the protocone, where a small pinhole exposure of 0.3 mm diameter is present. The paracone and metacone exhibit worn occlusal surfaces, but the cusps remain salient.

The protocone is the largest cusp, followed by the metacone and then subequal paracone and hypocone. The buccal cusps are set slightly mesial to the lingual cusps. Occlusal wear precludes an assessment of detailed occlusal morphology, but its general form is similar to known examples of later hominids (e.g., *Ar. ramidus, Au. anamensis,* and *Au. afarensis*). The paracone and metacone are moderately salient, and the coarse occlusal ridges tend to "fill" the occlusal fovea, as in the upper molars of *Ar. ramidus, Au. anamensis,* and *Au. afarensis*. The anterior fovea is diminished by wear to a short groove at the mesiobuccal corner. A worn epicrista occurs buccally, bounding the anterior fovea. The central fovea is demarcated distally by a low crista obliqua, which is incised by, but continuous across, the longitudinal groove. Traces of a small distoconule are seen on the distal marginal ridge. A conspicuous accessory occlusal ridge fills the buccal portion of the posterior fovea.

On the buccal crown face, the paracone and metacone are separated by a shallow fissure terminating in a thin shelf-like structure at mid-crown height. Both paracone and metacone saliency is reduced by wear, but the subequally developed distal paracone and mesial metacone crests retain an obtuse V-shaped notch of moderate depth. Lingually, the protocone and hypocone are separated by a deep fissure that shallows cervically. At the mesiolingual corner of the crown, the protocone lateral crown face bears a small pit-like structure, representing a weakly expressed Carabelli's trait. In mesial view, the lingual protocone face tapers occlusally, more so than the buccal paracone face. The hypocone's lingual slope is steeper than that of the protocone, such that occlusally the hypocone protrudes more lingually than the protocone. In mesiodistal view, in relation to crown breadth, crown height to the level of the marginal ridges is lower (4.1 mm from the little worn distal marginal ridge to cervical line) than in *Australopithecus* M^1s (e.g., L.H.-3, L.H.-6).

The root system, broken circa 2 to 3 mm from the cervical line, shows fused lingual and distobuccal roots. Furcation between the mesiobuccal and lingual roots already occurs at this basal-most position.

Micro-ct imagery of ASK-VP-3/401 revealed sufficiently discernible EDJ surfaces, from which the maximum radial thicknesses of the lateral enamel were measured opposite the main cusps. These thickness values were 1.18 mm at the paracone, 1.23 mm at the metacone, 1.41(+) mm at the worn protocone, and 1.48(+) mm at the worn hypocone.

ASK-VP-3/402 is a lightly worn, virtually complete left M^1 crown broken at and just above the cervical region. Only small segments of the cervical line itself are preserved at the mesial and distal crown faces. The crown surface is generally well-preserved, with only intermittent patches of etched surface enamel. The mesial IPCF is a buccolingually elongate oval (3.4 mm wide by 1.7 mm high) set slightly buccal to mid-crown, just below the occlusal margin. The distal IPCF, 3.5 mm wide by 1.8 mm high, is placed lingually at mid-crown level.

In occlusal view, crown outline is close to rhomboidal with an obtuse distobuccal corner and a slightly greater buccolingual dimension mesially than distally. The lingual crown contour is more bilobed than in ASK-VP-3/401. The standard crown dimensions are given in Table 7.1.

The protocone and hypocone are largely flattened and faceted, but without enamel perforations. The paracone and metacone are less worn, and support smaller, ill-defined occlusal facets. Despite retaining some cusp saliency, the paracone exhibits a small pinhole dentin exposure of 0.2 mm diameter.

The protocone is the largest cusp, followed by the metacone and then subequal paracone and hypocone. The size difference between metacone and paracone is less than in ASK-VP-3/401. The buccal cusps are set slightly mesial to the lingual cusps. As with ASK-VP-3/401, the general occlusal morphology of ASK-VP-3/402 is similar to known examples of later hominids (e.g., *Ar. ramidus, Au. anamensis,* and *Au. afarensis*). The anterior fovea is represented by a narrow, 2 mm long groove, bounded distally by an epicrista that extends buccolingually across the longitudinal groove. The central fovea is demarcated distally by a prominent crista obliqua, its protocone and paracone segments abutting at a level distinctly above the central fovea. As in ASK-VP-3/401, a conspicuous accessory occlusal ridge fills the buccal portion of the posterior fovea. A smaller accessory ridge occupies the distolingual posterior fovea, and the distal marginal ridge exhibits a faint crenulation that hardly qualifies as a distoconule.

On the buccal crown face, the groove between the paracone and metacone is simple and does not terminate in a shelf-like structure. In ASK-VP-3/402, metacone saliency is emphasized over that of the paracone, and the tooth consequently exhibits a worn V-shaped buccal notch that is steeper distally than in ASK-VP-3/401. Lingually, the protocone and hypocone are separated by a deep fissure that shallows cervically. ASK-VP-3/402 also resembles ASK-VP-3/401 in its pit-like Carabelli's trait and low crown height (4.2 mm from the little-worn mesial marginal ridge to cervical line). The buccal and lingual crown face slopes of ASK-VP-3/402 are similar to the condition described above for ASK-VP-3/401, with slightly less disparity between slopes of the protocone and hypocone lingual faces.

Micro-ct imagery of ASK-VP-3/402 revealed sufficiently discernible EDJ surfaces, from which the maximum radial thicknesses of the lateral enamel could be measured opposite the main cusps. These thickness values were 1.16 mm at the paracone, 1.17 mm at the metacone, 1.61(+) mm at the worn protocone, and 1.66(+) mm at the worn hypocone.

STD-VP-2/63 is a moderately worn left M^1 with the entire crown and up to 6 mm of the root preserved. The cervical line is intact thoughout the crown. The crown surface is affected by enamel damage due to etching. Surface enamel is generally better preserved on the lateral crown faces than occlusally. The mesial IPCF is buccolingually elongate (4.8 mm wide by 2.8 mm high), slightly concave, and placed buccal to midcrown, adjacent to the occlusal margin. The distal IPCF is a lingually placed irregular oval (4.2 mm wide by 2.7 mm high).

In occlusal view, crown outline is close to rhomboidal, with an obtuse distobuccal corner and a slightly greater buccolingual dimension mesially than distally. The lingual crown contour is weakly bilobed between protocone and hypocone, this feature slightly stronger than in ASK-VP-3/401 and similar to ASK-VP-3/402. Standard crown dimensions are given in Table 7.1.

The protocone is worn flat and exhibits a large dentin exposure of circa 3 mm mesiodistal by 2.0 mm buccolingual diameter. The dentin exposure forms a deeply excavated cavity, to a degree not seen in *Australopithecus* molars but known in orangutans and, more rarely, in chimpanzees. The hypocone is flattened to the same level as the protocone but does not exhibit enamel perforation. The paracone retains weak buccal margin saliency and a dentin exposure of 0.5 mm diameter. The metacone is still salient and does not show dentin exposure.

The protocone is the largest cusp. In contrast to ASK-VP-3/401 and ASK-VP-3/402, paracone and metacone are subequal in size, followed by the hypocone. The buccal cusps are set slightly mesial to the lingual cusps. Although the fine details of the occlusal morphology of STD-VP-2/63 are obliterated by wear, as was the case with ASK-VP-3/401 and ASK-VP-3/402, the general occlusal structure is similar to known examples of later hominids (e.g., *Ar. ramidus, Au. Anamensis,* and *Au. afarensis*). The central fovea appears somewhat deeper than, and crista obliqua development is intermediate between, the ASK-VP-3/401 and ASK-VP-3/402 conditions. The thin distal marginal ridge suggests lack of any distoconule expression.

On the buccal crown face, the buccal groove is simple as in ASK-VP-3/402. Metacone saliency is retained despite moderate overall crown wear, and the worn, V-shaped buccal notch exhibits a steep distal slope as in ASK-VP-3/402. The lingual groove is preserved and fades basally. The form of the Carabelli's trait cannot be evaluated because of wear. The degree of occlusal taper exhibited by the lingual and buccal crown faces is stronger in STD-VP-2/63 than in ASK-VP-3/401 or ASK-VP-3/402. STD-VP-2/63 shares low crown height with the other two molars. Crown height from the minimally worn distal marginal ridge to cervical line is 4.3 mm.

An enamel bulge encircles the lingual crown base. A single horizontal hypoplastic line occurs at mid-height of the buccal crown face. This line is faintly traced onto the mesial and distal faces but is not discernible lingually.

The root system, broken circa 4 to 6 mm from the cervical line, shows fused lingual and distobuccal roots. Furcation between the mesiobuccal and lingual roots already occurs at this level.

Micro-ct imagery of STD-VP-2/63 revealed sufficiently discernible EDJ surfaces, from which maximum radial thicknesses of the lateral enamel were measured opposite the main cusps. These thickness values were 1.12(+) mm at the paracone, 1.16 mm at the metacone, 1.23+ mm at the worn protocone, and 1.18+ mm at the worn hypocone.

Upper Third Molar (STD-VP-2/62; Plate 7.7) STD-VP-2/62 is a minimally worn left M^3 with intact crown and some root. Most of the mesiobuccal root is preserved (up to 12 mm height). Only the cervical 4–5 mm of the other roots is preserved. The cervical line is preserved around the crown. Etching of the crown surface is slight, affecting only small patches of damaged surface enamel. A faint mesial IPCF (2 mm buccolingual by 1 mm) occurs on the mesial face, centered just below the occlusal margin.

In occlusal view the crown outline is a rounded irregular trapezoid, with the distal crown shorter buccolingually than the mesial crown, and with a curved distal contour. Crown dimensions are given in Table 7.1.

STD-VP-2/62 had barely started to wear, as judged from the small wear facets seen mostly on the protocone but also on the few occlusal ridges of all other cusps.

The protocone is by far the largest cusp, followed by the paracone. The subequal metacone and hypcone are even smaller. The buccal cusps are set slightly mesial to the lingual cusps. All four cusps are well-developed, including the metacone and hypocone, both of which possess prominent cusp tips. The two buccal cusps exhibit sharper cusps due to their more acute rise from the depth of the occlusal fovea. The occlusal surface is crenulated by accessory occlusal ridges, particularly marked on the protocone and paracone. However, these ridges are coarse and lack the finer and more numerous secondary crenulations seen in chimpanzee and orangutan molars. Thus, the overall occlusal appearance of this M^3 is similar to that of later hominids (e.g., *Ar. ramidus, Au. anamensis,* and *Au. afarensis*) given substantial intraspecific variation. The mesial fovea is irregularly developed, truncated just buccal to mid-crown by what appears to be a branch of the epicrista. The latter is continuously developed buccolingually, interrupting the longitudinal groove. The crista obliqua is not well-defined, and two weak accessory occlusal ridges of the protocone terminate at the longitudinal groove opposite a more prominent buccal segment of the crista obliqua (of the metacone). A well defined, buccolingually elongate posterior fovea is bounded distally by a distal marginal ridge without any distoconule expression.

On the buccal crown face, the paracone and metacone are separated by a shallow fissure that gradually fades cervically. The salient paracone and metacone cusps form a buccal notch that is as acute as in the previously described M^1s, albeit with a shorter metacone mesial crest due to the smaller size of that cusp. Crown height above the cervical line is 6.7 mm and 6.3 mm to paracone and metacone cusp tips, respectively. Lingually, the protocone and hypocone are demarcated by an occlusally deeper and cervically shallow lingual groove. The Carabelli's trait is represented only by a very faint, short vertical groove at the mesiobuccal crown face.

In mesial view, unlike the M^1 crowns, STD-VP-2/62 has a more vertical buccal and a strongly curved lingual crown face slope. Crown height in mesial view is low (3.9 mm from the little-worn mesial marginal ridge to the cervical line) and about the same as in the previously described M^1s. Crown height from the little-worn distal marginal ridge to the cervical line is 4.3 mm. The distal crown face is rounded vertically, whereas the mesial face is flat as in the M^1s. A prominent hypoplastic line encircles the crown at approximately 1.5 mm above the cervical line. Similar lines occur on the root surface. The root system consists of lingual, mesiobuccal, and distobuccal roots.

Micro-ct imagery of STD-VP-1/62 revealed sufficiently discernible EDJ surfaces, from which the maximum radial thicknesses of lateral enamel were measured opposite the main cusps. These thickness values were 1.71 mm at the paracone, 1.58 mm at the metacone, 2.01 mm at the protocone, and 1.88 mm at the hypocone.

Dentognathic Discussion

The *Ar. kadabba* material is compared in the following discussion with *Sahelanthropus tchadensis, Orrorin tugenensis, Ardipithecus ramidus, Australopithecus anamensis, Australopithecus afarensis,* the Lothagam (KNM-LT 289) and Tabarin (KNM-TH 1350) mandibles, and a comparative sample of modern and Miocene apes. A comprehensive comparison with modern and fossil apes is beyond the scope of the present analysis and discussion, and it must be noted that some of the comparative taxa have not yet been themselves fully described. The circumscribed comparisons that we present here are focused primarily on *Pan,* the living sister taxon of Hominidae, and on a broader comparative background when this appears to be useful in evaluating the *Ar. kadabba* remains.

Mandible

Comparative assessment of late Miocene and early Pliocene hominid mandibles is limited by small sample sizes and incomplete description. Although the type specimen of *Orrorin tugenensis* is a mandible with molars, morphology of the mandible was not described except for the mention of a deep corpus (35.5 mm) at the M_3 position (Senut et al. 2001). It also appears from published pictures that the specimen has undergone serious expanding matrix distortion (EMD; White 2003). Meaningful comparison of the ALA-VP-2/10 and BAR 1000'00 mandibles is therefore not possible at this time.

Four mandibular specimens have been assigned to *Sahelanthropus tchadensis:* TM 266-01-060-2, TM 266-02-154-1, TM 292-02-01, and TM 247-01-02 (Brunet et al. 2002, 2005). Brief descriptions are available for the latter three specimens. TM 266-02-154-1 is described as possessing a relatively thick corpus, 20.0 mm at M_1 (Brunet et al. 2002), with wide extramolar sulcus. TM 292-02-01 has a thinner corpus, 14.5 mm at M_1 (Brunet et al. 2002) but is apparently somewhat damaged. Corpus breadth of the distinctly abraded TM 247-01-02 is reported as greater than 16.1 mm at the M_1 position (Brunet et al. 2005). The mental foramen is reported to be at or below mid-corpus height in these specimens.

The Lothagam and Tabarin mandibles, both attributed to the 4–5 Ma time interval, have been described and discussed in some detail. Historically, the Lothagam mandible was compared with the earliest well-known hominid fossils at the time of analysis, first with *Au. africanus,* then with *Au. afarensis* (White 1986; Kramer 1986; Hill and Ward 1988; Hill et al. 1992), and most recently *Au. anamensis* (Leakey and Walker 2003). Both the Tabarin and Lothagam specimens have been considered to be within the range of variation of *Au. afarensis* in corpus size, robusticity, and most aspects of their morphology, although in our own estimates the Tabarin mandible has a smaller corpus breadth than the smallest *Au. afarensis* specimen (A.L. 198-1). More recently, aspects such as subalveolar lingual concavity, narrow corpus base, inferred anterolateral flare of the lateral corpus, expression of a low inferior transverse torus, and anteriorly opening mental foramen (one of three specimens) have been noted as features also characteristic of the known *Au. anamensis* mandibular specimens (Ward et al. 2001).

The Lothagam and Tabarin mandibles have been provisionally attributed to *Au. afarensis* by some (e.g., Kramer 1986; Ward and Hill 1987; Hill et al. 1992), whereas others considered the Lothagam mandible to be insufficiently diagnostic and hence not assignable to Hominidae (White 1986; Leakey and Walker 2003). The latter conservative view stems from the possibility that the observed morphological similarities of the Lothagam (and Tabarin) and *Au. afarensis* / *Au. anamensis* mandibles are primitive retentions. Candidate derived features suggestive of hominid attribution are the relatively thick mandibular corpus and the high (circum mid-corpus) position of the mental foramen. The only *Ardipithecus ramidus* mandible reported so far is described to have a rather low and broad corpus with an anterosuperiorly opening mental foramen located at mid-corpus (Semaw et al. 2005).

In the following discussion of mandibular morphology, we evaluate the phenetic similarities of the ALA-VP-2/10 mandible with the earliest known *Australopithecus* examples, and with *Pan,* in order to establish the candidate derived features (or such tendencies) that suggest attribution of this holotype specimen to Hominidae. The comparative metrics and observations are summarized in Tables 7.2 and 7.3.

The anterior corpus of the Alayla mandible is mostly not preserved. Lingually, the preserved posterior surface of the symphyseal area forms a sloping, shelf-like bony surface indicating the presence of a superior transverse torus that extended posteriorly to P_4 level, as in *Au. anamensis* and *Au. afarensis* examples (e.g., KNM-KP 29281, MAK-VP-1/12, A.L. 288-1). In ALA-VP-2/10 there also exists the posterior continuation of a rounded inferior transverse torus that extended to below the lingual M_1 position. This structure is well-expressed and curves anterolingually, suggesting that the symphysis was likely posteroinferiorly rounded and sloping, as typically seen in the *Au. anamensis* mandibles (strongest in KNM-KP 29281). However, the mandibular arcade was apparently narrower anteriorly than in KNM-KP 29281, as judged from the tight internal symphyseal curvature. The weaker subalveolar concavity of the posterior medial corpus of the Alayla mandible is another difference from the known *Au. anamensis* examples and the Lothagam mandible. The *Sahelanthropus* mandible, TM 292-02-01, is also described as possessing a posteriorly sloping symphysis with weakly developed superior and inferior transverse tori. In the mandibles of *Pan,* although these features may individually be seen, they usually do not occur together. On the other hand, some Miocene ape female mandibles (e.g., GSP 4622) are known to exhibit similar anterior and lingual corpus morphologies.

Corpus height of the Alayla mandible can be measured only at the M_1/M_2 level. Its value is slightly below the *Au. afarensis* mean and lower than that of the Lothagam mandible (Table 7.2, Plot 7.1). Corpus breadth of the Alayla mandible at M_1 cannot be compared with the other material, as a result of pathological modification around that area and loss of much of the basal margin. However, corpus breadth at M_2 (18.5 mm) or at ramus root (20.5 mm) was smaller than in *Au. afarensis* and *Au. anamensis* mandibles, occupying the lowest end of the *Au. afarensis* range of variation (e.g., A.L. 198-1) (Plot 7.1). Both the Lothagam and *S. tchadensis* (TM 266-02-154-1) mandibles have a slightly broader corpus

TABLE 7.2 Mandibular Corpus Dimensions

		Corpus Height					Corpus Breadth						Corpus Shape (br/ht)				
		P_4	P_4/M_1	M_1	M_1/M_2	M_2	P_4	P_4/M_1	M_1	M_1/M_2	M_2	AT RAMUS ROOT	P_4/M_1	M_1	M_1/M_2	$M_2BR/M_1/M_2HT$	M_2
Ar. kadabba	ALA-VP-2/10				30.0						18.5	20.5				61.7	
Lothagam[a]	KNM-LT 289				32.0	31.4				19.3	20.4	23.0				63.8	65.0
Tabarin[a]	KNM-TH 1350				26.0	26.5					17.5	21.0				67.3	66.0
Au. anamensis[a]	KNM-KP 29281		34.5	34.0	34.0			17.8	18.5	19.0	21.0	23.5	51.6	54.4	55.9	61.8	
	KNM-KP 29287		41.0	40.5				19.7	20.0	21.5	22.5	25.0	48.0	49.4			
	KNM-KP 31713		29.0	29.0	29.0			17.8	18.0	19.3		22.0	61.4	62.1	66.6		
	Mean		34.8	34.5				18.4	18.8	19.9		23.5	53.7	55.3			
Au. afarensis[a,b]	*n*	19	20	22	19	20	22	22	24	21	21	10	20	21	14	17	19
	Mean	36.9	35.7	34.2	32.8	31.4	19.1	18.9	19.8	20.2	21.9	24.0	52.7	57.6	63.0	67.1	70.0
	SD	4.6	4.6	4.0	3.6	3.8	2.4	2.1	2.3	2.2	3.1	2.3	5.2	5.8	7.0	7.3	7.7
P. troglodytes[c]	*n*			20	63				20	63				20	63		
	Mean			28.8	27.8				14.2	13.6				49.5	49		
	SD			2.7	2.7				1.3	1.03				5.5	5		

NOTE: Measurements are in mm.

[a]Lothagam, Tabarin, Kanapoi, and *Au. afarensis* metrics of Maka and Laetoli specimens taken on the originals.

[b]*Au. afarensis* metrics of Hadar specimens from Kimbel et al. (2004), except for corpus breadth at ramus root of the Hadar 1970s collection taken on the originals.

[c]*P. troglodytes* values are from Ward et al. (2001) and Kramer (1986) for the $M_1$and M_1/M_2 positions, respectively.

TABLE 7.3 Mandibular Ramus Root Morphology

	Extramolar Sulcus Width	*Anterior Ramus Border Position*		
		ANTERIOR BORDER HITS MOLAR ROW	ANTERIOR EXTENT OF EXTRAMOLAR CONCAVITY	ANTERIOR TERMINATION OF ROOT EMINENCE
Ar. kadabba				
ALA-VP-2/10	7.0	2	4	6
Taxon indeterminate				
KNM-LT 289	6.0	3	4	6
KNM-TH 1350	8.0	3	5	8
Au. anamensis				
KNM-KP 29281				8
KNM-KP 29287				9
KNM-KP 31713				9
Au. afarensis				
n	7	7	8	13
Mean	8.8	3.1	4.8	9.0
SD or range	1.1	1–4	2–6	6–10
P. troglodytes				
n	10	20	20	9
Mean	6.9	2.5	3.9	7.9
SD or range	0.8	1–4	2–5	7–10

NOTE: Extramolar sulcus widths are in mm. Anterior ramus border postions: 1, distal M3; 2, mid-M3; 3, mesial M3; 4, M2/M3; 5, distal M2; 6, mid-M2; 7, mesial M2; 8, M1/M2; 9, distal M1; 10, mid-M1; 11, mesial M1. *Au. afarensis* sample consists of Hadar 1970s collection and Maka. *P. troglodytes* based on Cleveland Museum of Natural History specimens.

at the M_2 position than the Alayla mandible, corresponding to the middle to lower range of *Au. afarensis/anamensis* variation. Chimpanzee mandibles overlap in range of variation with *Au. afarensis* in corpus height but do so minimally in corpus breadth.

Corpus shape (breadth/height) of the Alayla mandible is estimated to be at the middle to lower end of the *Au. afarensis* range of variation (Plot 7.1). The Lothagam and Tabarin mandibles exhibit higher corpus shape values, at or below the *Au. afarensis* mean. The Chad mandibles cannot be evaluated due to insufficient preservation. Chimpanzee mandibular bodies tend to be relatively thinner, but with some overlap in range of variation with *Au. afarensis/anamensis*. Miocene ape genera (e.g., *Ouranopithecus,* Koufos 1993; *Sivapithecus.* Pilbeam et al. 1980; and *Proconsul,* Andrews 1978) include individual mandibles that overlap in range of variation of corpus robusticity with that of *Au. afarensis/anamensis.*

The anterior root of the ALA-VP-2/10 ramus is placed relatively posteriorly and high on the corpus, with a shallow and narrow extramolar sulcus and comparatively weak lateral prominence (Table 7.3, Plot 7.2). *Australopithecus afarensis* and *Au. anamensis* mandibles tend to exhibit a more robust ramus root morphology, with a tendency for a more anterior root placement, broader extramolar sulcus, and stronger lateral prominence extending

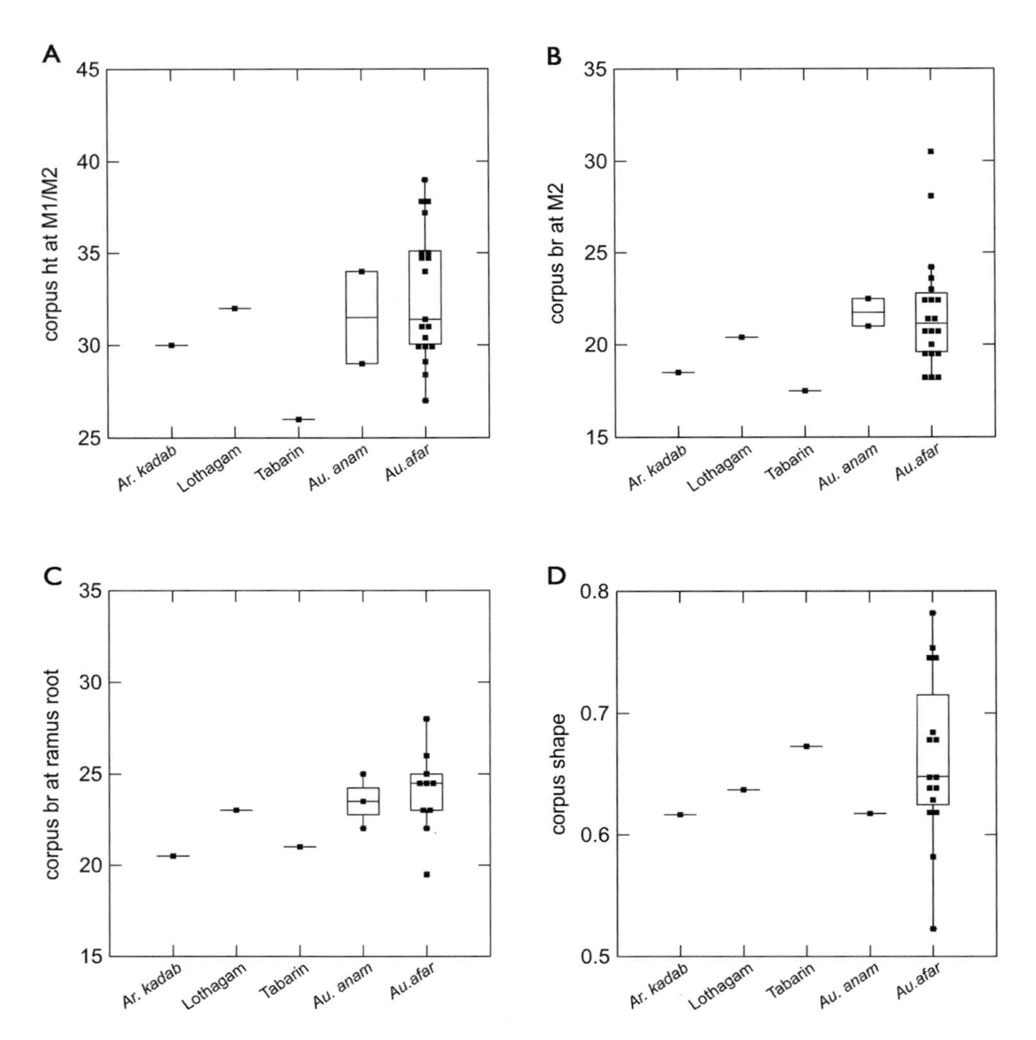

PLOT 7.1

Mandibular corpus measurements (in mm). A. Mandibular corpus height at M1/M2 position. B. Mandibular corpus breadth at M2 position. C. Mandibular corpus breadth at ramus root. D. Mandibular corpus shape (breadth at M2 divided by height at M1/M2). Abbreviations: *Ar. kadab, Ar. kadabba* (ALA-VP-2/10); *Lothagam,* KNM-LT 289; Tabarin, KNM-TH 1350; *Au.anam, Au. anamensis; Au.afar, Au. afarensis;* see Table 7.2 for further sample details. Box plots show the median (horizontal line), central 50 percent range (box margins), range (vertical line) within inner fences (1.5 times box range from box margins), and outliers. Each filled square represents an individual specimen.

more anteriorly. Only the thinnest mandibular bodies of *Au. afarensis* (e.g. A.L. 198-1) exhibit corpus proportions and ramus root robusticity comparable to the ALA-VP-2/10 condition.

The *Sahelanthropus* and Lothagam mandibles are also broadly comparable to the ALA-VP-2/10 condition, although TM 266-01-154-2 has a wider, and the Lothagam mandible a narrower, extramolar sulcus. In the Chad specimen, the ramus root is placed slightly more inferiorly on the corpus than in ALA-VP-2/10, but these differences are slight. The Tabarin mandible exhibits a broader extramolar sulcus and stronger lateral prominence. Chimpanzee mandibles tend to share the relatively high and posterior placement of the ramus root with ALA-VP-2/10, although posterior placement is often even more extreme (distal M_3) in *Pan.* The tendency for a narrow extramolar sulcus and weaker expression of the lateral prominence also represents similarities of *Pan* with the Alayla mandible.

The position of the mental foramen is comparable in ALA-VP-2/10 and the *Sahelanthropus* mandibles, in being placed about mid-corpus height. The mental foramen of the

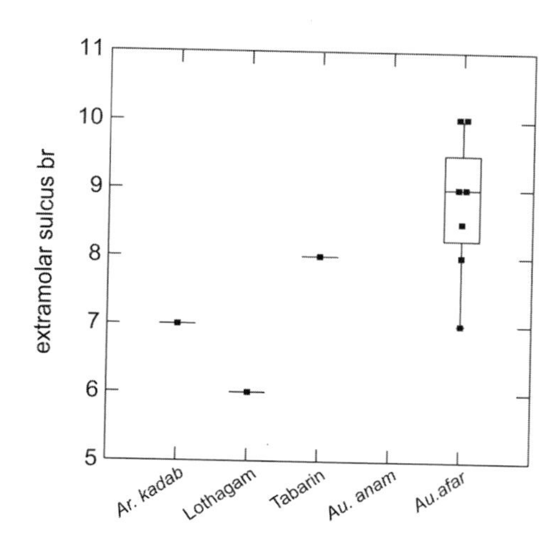

PLOT 7.2

Extramolar sulcus breadth (in mm). Abbreviations: *Ar.kadab, Ar. kadabba* (ALA-VP-2/10); Lothagam, KNM-LT 289; Tabarin, KNM-TH 1350; *Au.anam, Au. anamensis; Au.afar, Au. afarensis;* see Table 7.3 for further sample details. Box plots as in Plot 7.1. Each filled square represents an individual specimen.

Lothagam mandible is placed most likely slightly above mid-corpus height. The circum mid-corpus placement of the mental foramen is thus shared between ALA-VP-2/10, the *Sahelanthropus* mandibles, and those of *Au. anamensis,* whereas the *Au. afarensis* mandibles have been characterized by a tendency for the mental foramen to lie slightly below mid-corpus height (White and Johanson 1982; White 1986; Kimbel et al. 1994). Mental foramen position in Miocene and modern apes is known to be distinctly lower than mid-corpus height, and the *Au. afarensis* condition has been interpreted as a primitive feature perhaps related to canine root size. However, the *Sahelanthropus* mandible TM 292-02-01 exhibits a long canine root extending close to the anterior corpus base (Brunet et al. 2005), whereas its mental foramen exits relatively higher. Our observations on the position of the mandibular canal in modern apes and humans suggest an alternative interpretation. Because the mandibular canal is juxtaposed to the premolar crypts during development, the vertical offset between the canine root apex and the mental foramen seen in the *Sahelanthropus* mandible (and inferred for the Alayla and Lothagam mandibles) may be caused by growth patterns involving the relative size and placement of the developing canine and premolars prior to eruption and final root formation. In that case, the circum mid-corpus height position may be a derived feature related to dental development patterns, although some Miocene apes (and hominids) show a tendency for a deepening of the mandibular base below the mental foramen, perhaps related to masticatory robusticity. Further understanding of mandibular corpus growth patterns in relation to mandibular canal and mental foramen position is needed to better assess the validity of this hypothesis.

In summary, the preserved mandibular morphology of ALA-VP-2/10 is most similar to the few known *Au. anamensis* examples. The similarities include the general size range and shape of the corpus, the inferred posterior extension of the symphysis, and mid-corpus position of the mental foramen. However, the few known *Au. anamensis* and most *Au. afarensis* mandibles tend to have a relatively broader corpus, especially at M_2 and across the ramus root. This greater breadth is because of the anteriorly placed and laterally flaring ramus root of *Australopithecus* mandibles, resulting in a broader extramolar sulcus and corpus buttressing that extends anteriorly to the M_1 position. Conversely, relative to modern apes (including *Pan*) and most Miocene apes, the Alayla mandible exhibits a relatively broad corpus, which may be either a basal hominid characteristic or a primitive retention. The mid-corpus position of the mental foramen is also a candidate derived hominid feature. Spanning the 4 to 6 Ma time period, the known mandibular specimens of *Sahelanthropus, Ar. ramidus,* Lothagam, and Tabarin all exhibit morphologies broadly compatible with that exhibited by the ALA-VP-2/10 mandible. As a group they are similar to, but tend to be smaller, slightly thinner, and with weaker lateral flare of the ramus root than, the *Au. afarensis* and *Au. anamensis* modal conditions. The *Orrorin* mandibular specimen cannot be evaluated because of the lack of detailed information.

Dental Crown Diameter and Proportions

The general size and proportions of the *Ar. kadabba* dentition are summarized in Tables 7.4 and 7.5 and Plot 7.3.

The *Ar. kadabba* dental elements are close in size to the known *Ar. ramidus* condition, although the incisors may be an exception. The single known *Ar. kadabba* incisor, represented by the buccolingual dimension, is closer to the *P. troglodytes* mean. Relative incisor size of this individual is distinctly larger than in the available examples of *Ar. ramidus, Australopithecus* or gorillas, but smaller than the chimpanzee condition (Plot 7.4). The labiolingually broad basal I_2 crown is paralleled in the Lothagam I_1, which is also within the chimpanzee range of variation. The damaged *Sahelanthropus* and *Orrorin* I^1s have labiolingual values reported at the mid-lower chimpanzee range of variation, while their mesiodistal dimensions are difficult to estimate. The combined but very limited available evidence therefore suggests that the circum 6 Ma hominids may be characterized by relatively larger incisors than seen in either *Ar. ramidus* or *Australopithecus.*

Canine size of *Ar. kadabba,* as represented by maximum basal crown diameter, lies broadly at the mid-range of the female chimpanzee condition (Plot 7.5). The ALA-VP-2/10 dental set allows a direct evaluation of relative canine size. Relative to P_4 or M_2 size, basal canine size of the Alayla individual lies at the lower part of the chimpanzee range and is larger than the female gorilla and bonobo mean conditions (Plot 7.6).

Compared with the *P. troglodytes* dentition, both *Ar. ramidus* and *Ar. kadabba* postcanine teeth are characterized by slightly broader premolars and larger M2 and M3 sizes (Plot 7.3). The three known M^1s of *Ar. kadabba* are close to the chimpanzee mean in mesiodistal crown length but exhibit a tendency for a broader crown shape (Plot 7.7). Of the late Miocene (ca. 6 Ma) sample available for comparisons, the Lothagam (upper Nawata) and *Orrorin* M3s are similar in size with that of *Ar. kadabba.* The known *Sahelanthropus* and *Orrorin* molars include examples that are somewhat larger, but within range of what would be expected within a single taxon (e.g., Plot 7.8).

A multivariate summary of the size and proportions of the ALA-VP-2/10 lower dentition is shown in Table 7.6 and Plot 7.9. The dominant information summarized by the first two principal components are postcanine or general size (PC 1) and incisor and canine size (PC 2), but the two plots reflect other subtle overall size and proportions. The canine to M_2 analysis shows *Ar. kadabba* to be broadly comparable to the

TABLE 7.4 Summary Statistics of the Maxillary Crown Dimensions

Taxon		I^1MD	I^1BL	I^2MD	I^2BL	C^1MAX	P^3MXMD	P^3MXOB	P^4MD	P^4BL	M^1MD	M^1BL	M^2MD	M^2BL	M^3MD	M^3BL
Ar. kadabba	*n*					1	1	1	1		3	3			1	1
	Mean					11.8	7.9	11.5	7.6		10.4	11.8			10.9	12.2
Ar. ramidus[a]	*n*	1	2	1	1	4	2	2	3	3	2	1	2	2	2	2
	Mean	10.0	7.85	7.1	6.8	11.2	7.95	11.7	7.7	11.0	10.25	12.2	11.95	14.25	10.8	13.15
	SD		0.49			0.37	0.64	1.13	0.61	0.35	0.35		0.21	0.21	0.85	1.20
Au. anamensis[b]	*n*	1	6	1	2	8	3	5	4	3	10	9	10	8	11	10
	Mean	12.3	8.5	5.8	6.5	11.1	9.2	12.9	8.9	13.5	11.5	13.1	13.0	14.9	12.0	14.2
	SD		0.71		0.71	1.06	0.31	0.86	1.12	1.00	0.66	0.78	0.95	1.14	0.96	1.10
Au. afarensis[c]	*n*	6	8	6	9	14	10	12	22	16	16	13	13	14	15	14
	Mean	10.7	8.4	7.5	7.2	10.6	8.6	12.4	9.1	12.6	12.0	13.3	12.9	14.6	12.3	14.3
	SD	0.75	0.71	0.54	0.60	1.07	0.69	0.55	0.65	0.78	1.02	0.97	0.70	0.71	1.34	1.10
P. troglodytes[d] (sex-combined)[e]	*n*	84	96	80	103	96	108	109	101	103	134	133	107	109	87	87
	Mean	12.5	9.6	9.0	8.7	13.4	8.0	10.7	7.3	10.2	10.6	11.3	10.6	11.7	9.7	11.1
	SD	0.78	0.82	0.65	0.73	2.21	0.62	0.76	0.46	0.65	0.61	0.63	0.65	0.77	0.75	0.80

NOTE: All measurements are in mm. MD, mesiodistal diameter; BL, buccolingual diameter; MXMD, maximum mesidodistal diameter of P^3 taken at buccal crown; MXOB, maximum oblique diameter of crown. UCMAX is larger of UCMD and UCMXOB when both are available.

[a] Sample includes published specimens from Aramis (White et al., 1994) and Gona (Semaw et al., 2005), with minor adjustments made for methodological consistency.

[b] Sample includes Kanapoi, Allia Bay, and Assa Issie, all original measurements.

[c] Sample includes Hadar (Johanson et al., 1982, with minor adjustments; Kimbel et al., 2004), and original measurements of Laetoli, Maka, Omo, and Turkana.

[d] *P. troglodytes* from the Cleveland Museum of Natural History and the Royal Museum of Central Africa, Tervuren, original measurements.

[e] Sex-segregated statistics for *P. troglodytes* C^1MAX: Females, $n = 49$, mean = 11.6, SD = 0.80; males, $n = 47$, mean = 15.3, SD = 1.56.

TABLE 7.5 Summary Statistics of the Mandibular Crown Dimensions

Taxon		I_1MD	I_1BL	I_2MD	I_2BL	C_1MXOB	P_3MXOB	P_4MD	P_4BL	M_1MD	M_1BL	M_2MD	M_2BL	M_3MD	M_3BL
Ar. kadabba	*n*				1	2		1	1			1	1	1	
	Mean				8.3	11.0		8.3	10.0			12.7	11.8	13.3	
Ar. ramidus[a]	*n*	1		1	1	3	3	4	4	6	6	4	4	4	4
	Mean	6.0		6.6	6.6	10.4	10.3	7.9	9.3	10.9	10.3	12.8	11.6	12.3	11.2
	SD					0.70	1.12	0.71	0.43	0.22	0.26	0.35	0.25	0.71	0.62
Au. anamensis[b]	*n*	4	2	2	3	7	6	6	7	9	10	8	8	7	6
	Mean	6.7	8.0	8.5	8.3	10.4	11.9	9.0	10.8	12.4	11.8	14.6	13.2	15.0	13.3
	SD	0.56			0.53	0.91	0.82	1.11	0.90	0.82	0.93	0.92	0.99	1.27	0.60
Au. afarensis[c]	*n*	5	7		7	13	21	24	21	28	24	33	29	26	24
	Mean	6.7	7.3		8.0	10.7	11.6	9.9	11.1	13.2	12.6	14.5	13.4	15.2	13.4
	SD	0.91	0.34		0.71	1.05	1.07	1.04	0.83	0.72	0.73	1.19	0.94	1.12	0.90
P. troglodytes[d] (sex-combined)[e]	*n*	92	110	84	105	96	111	108	107	129	125	108	108	86	87
	Mean	8.2	8.8	8.5	9.1	12.4	11.2	7.9	8.7	11.1	9.7	11.8	10.5	10.9	10.0
	SD	0.61	0.67	0.62	0.70	1.81	0.79	0.57	0.66	0.57	0.57	0.69	0.68	0.71	0.65

NOTE: All measurements are in mm. MD, mesiodistal diameter; BL, buccolingual diameter; MXOB, maximum oblique diameter of crown.

[a]Sample includes published specimens from Aramis (White et al. 1994) and Gona (Semaw et al. 2005), with minor adjustments made for methodological consistency.

[b]Sample includes Kanapoi, Allia Bay, and Assa Issie, all original measurements.

[c]Sample includes Hadar (Johanson et al. 1982b, with minor adjustments; Kimbel et al. 2004), and original measurements of Laetoli, Maka, Omo, and Turkana.

[d]*P. troglodytes* from the Cleveland Museum of Natural History and the Royal Museum of Central Africa, Tervuren, original measurements.

[e]Sex-segregated statistics for *P. troglodytes* C_1MXOB: Females, $n = 48$, mean = 11.0, SD = 0.83; males: $n = 48$, mean = 13.9, SD = 1.31.

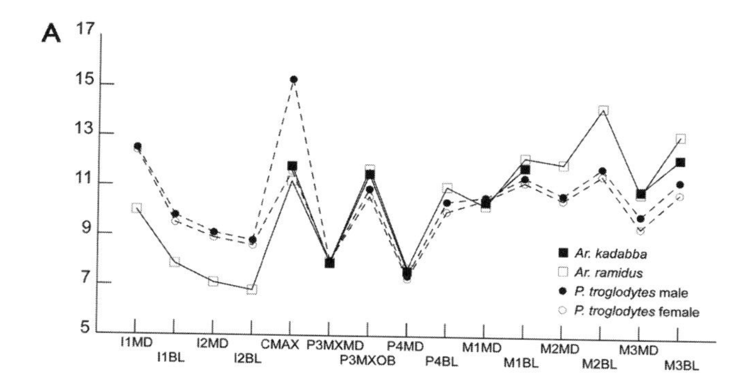

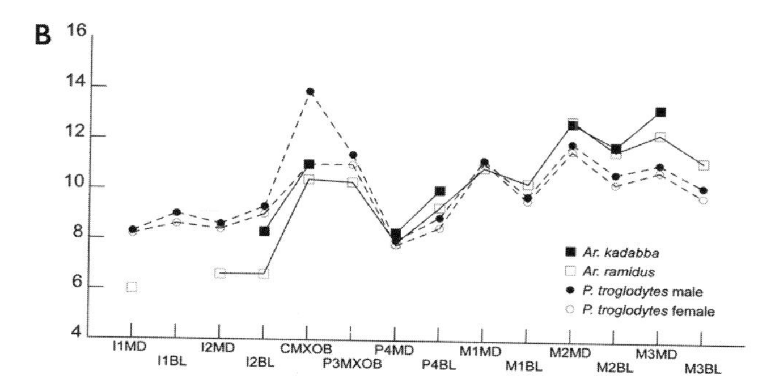

PLOT 7.3

Dental crown size of *Ar. kadabba, Ar. ramidus,* and chimpanzees (taxon means in mm). A. Maxillary dentition (see Table 7.4 for details of the metrics and samples). B. Mandibular dentition (see Table 7.5 for details of the metrics and samples).

known *Ar. ramidus* examples, occupying the margins of the female chimpanzee range of variation. When the I_2 is included in the analysis, the position of the Alayla holotype dentition becomes slightly separated in multivariate space from both chimpanzees and *Ar. ramidus*.

Incisor Morphology

Little can be said of the *Ar. kadabba* incisors, which are represented by a single I_2 with substantial wear. The possibility of *Ar. kadabba* being characterized by a larger incisor size than seen in either *Ar. ramidus* or later *Australopithecus* has been noted above. However, whether this might have involved only the labiolingual diameter or also the mesiodistal diameter cannot be concluded from the available materials. The preserved lingual morphology exhibits a narrow basal tubercle and lack of strong lingual fossa features, and therefore differs from the modal chimpanzee and bonobo conditions. The *Pan* morphology consists of a mesiodistally elongate crown, including a large mesiodistal basal crown diameter; a projecting lingual tubercle delineated by distinct cingular grooving; and a conspicuous and incisally continuous median lingual ridge. Gorilla lower incisors show weaker expressions of these features, whereas the Alayla and Lothagam lower incisors lack such signs. The single *Ar. ramidus* I_2 available for comparison is the ARA-VP-1/128 worn crown. The morphology of the lingual tubercle and adjacent regions of the ARA-VP-1/128 I_2 is similar to that of the Alayla I_2 but with a narrower mesiodistal dimension and weaker protrusion of the basal tubercle, resulting in a less concave incisocervical basal crown profile.

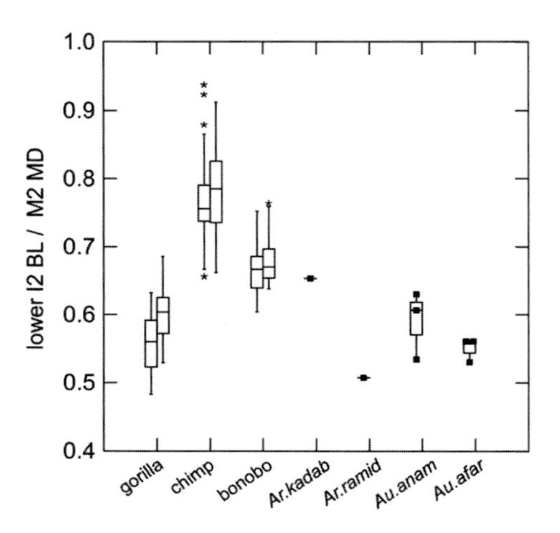

PLOT 7.4

Relative mandibular I_2 size (labiolingual breadth divided by M_2 mesiodistal length). Abbreviations: gorilla, *G. gorilla;* chimp, *P. troglodytes;* bonobo, *P. paniscus; Ar.kadab, Ar. kadabba* (ALA-VP-2/10); *Ar.ramid, Ar. ramidus; Au.anam, Au. anamensis; Au.afar, Au. afarensis.* Box plots as in Plot 7.1. Sex-segregated in the modern apes, left side females, right side males. See Table 7.5 for further details of the samples. Sample sizes of the modern apes are as follows: *G. gorilla* females (n = 19), males (n = 14); *P. troglodytes* females (n = 45), males (n = 45); and *P. paniscus* females (n = 25), males (n = 21). Each filled square represents an individual specimen.

C/P3 Complex Morphology

Elements of the C/P3 complex of *Ar. kadabba* are represented by one partial upper canine, two lower canines, and a partial P_3 crown. The combination of primitive and derived features seen in these specimens has been discussed in some detail by Haile-Selassie (2001b) and Haile-Selassie et al. (2004b).

The key primitive features that allowed recognition of *Ar. kadabba* as a species distinct from *Ar. ramidus* were the tall upper canine (ASK-VP-3/400) with a longer and more basal terminating mesial crest (mesial shoulder lower and with less flare), the P_3 (ASK-VP-3/403) with a more asymmetric occlusal view outline and weaker expression of the anterior fovea, and a lower canine with variation that includes a relatively low mesial shoulder and lack of fold-like mesial marginal ridge (STD-VP-2/61).

Morphological affinities with known *Ardipithecus* and *Australopithecus* homologs were seen, such as variation in the lower canine that includes a high mesial shoulder position, large distal tubercle, and fold-like mesial marginal ridge (ALA-VP-2/10). A fold-like buccal segment of the P_3 mesial marginal ridge (ASK-VP-3/403) is also a morphological detail shared with later hominids.

Aspects of these morphologies are further quantified and summarized in the following paragraphs. Canine crown height relative to basal dimensions does not discriminate between hominids and apes, but several aspects are relevant to the interpretation of the *Ar. kadabba* material (Plot 7.10). Using a roughly estimated upper canine crown height of 16 mm, ASK-VP-3/400 does not exhibit the particularly tall crown proportion seen in the modern African ape males (especially *Pan*). A relatively tall crown seems to be a condition characteristic of even female upper canines of chimpanzees and bonobos. The *Orrorin* upper canine is proportionally lower-crowned, as are the two examples from *Ar. ramidus.*

The relative positions (high or low) of the mesial and distal shoulders can be numerically expressed as a ratio of mesial and distal crest length to either crown height or maximum basal dimensions (Plot 7.10). It can be seen that the *Ar. kadabba* condition corresponds to the lower range of modern and some of the Miocene ape female canines. We have not had the chance to evaluate the mesial and distal crest lengths of the *Orrorin* upper canine

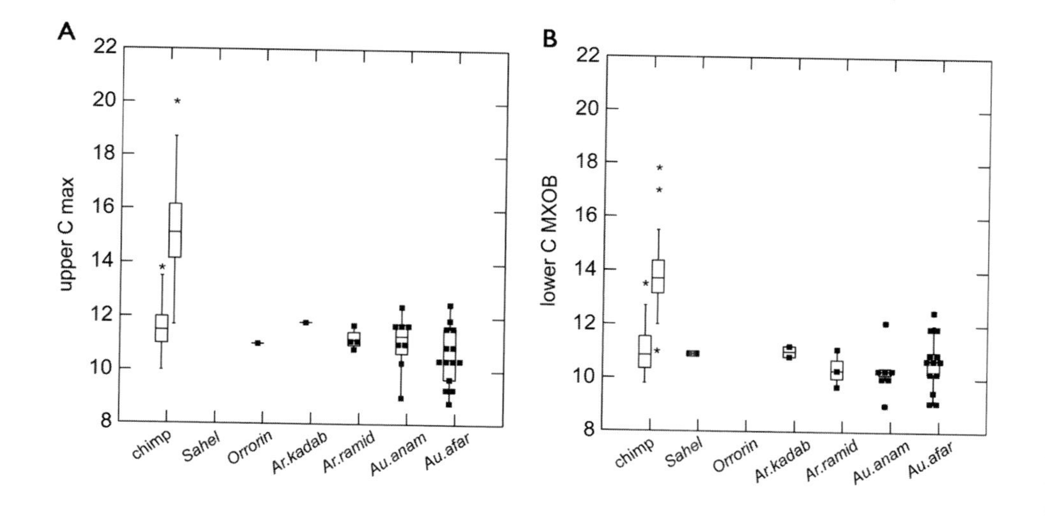

PLOT 7.5

Maxillary and mandibular canine size (in mm). A. Maxillary canine, maximum dimension. B. Mandibular canine, maximum oblique dimension. Abbreviations: chimp, *P. troglodytes;* Sahel, *Sahelanthropus; Orrorin, Orrorin; Ar.kadab, Ar. kadabba; Ar.ramid, Ar. ramidus; Au.anam, Au. anamensis; Au.afar, Au. afarensis.* See Tables 7.4 and 7.5 for further details of the samples. Box plots as in Plot 7.1. Sex-segregated in *P. troglodytes*, left side females, right side males. Sample sizes: maxillary canine, female (n = 49), male (n = 47); mandibular canine, female (n = 48), male (n = 48). Each filled square represents an individual specimen.

metrically, but the published photographs suggest a relatively longer crest than seen in ASK-VP-3/400. The *Ar. ramidus* upper canines tend have a shorter mesial crest, comparable to the more derived *Australopithecus* condition (White et al. 1994, 2006b). The crown shoulders of the *Ar. ramidus* upper canines are also more flared (protruding mesially or distally) than in ASK-VP-3/400 or the *Orrorin* upper canine. The *Sahelanthropus* partial upper canine exhibits a buccolingually narrow and near-vertical wear strip at the distal crest that extends basally to a very low position (Brunet et al. 2002), inferring a distal shoulder position as low as or lower than that seen in ASK-VP-3/400. This wear pattern differs from the condition often seen in female apes, in which the scalloped wear surface faces more lingually from honing against the mesiobuccal face of the P_3. Finally, at least two species of fossil ape, *Gigantopithecus blacki* and *Oreopithecus bambolii,* exhibit female upper canine variation that includes a high shoulder position, well into the hominid range of variation. However, sexual dimorphism in canine size and shape, as well as honing function, is retained in these taxa.

A notable morphological detail of the ASK-VP-3/400 upper canine is the strong expression of the mesiolingual ridge and adjacent deep and narrow mesiolingual groove. A similar morphology is seen in the *Orrorin* upper canine, and such a structural pattern is often more extremely expressed in Miocene apes. Exceptions occur among modern apes most often in *Pan,* in which the same features are often only faintly expressed, perhaps in association with their buccolingually narrow upper canine crown. The same mesiolingual ridge/groove complex is seen in later hominid canines in modified form. These structures become progressively weak from *Ar. ramidus* to *Au. anamensis* to *Au. afarensis* and later hominids. This "gracilization" of the mesiolingual crown structure, combined with high crown shoulders and distinct expression of the marginal ridge, results in a more spatulate, or "incisiform," rather than tusk-like upper canine. Thus, the *Ar. kadabba* upper canine exhibits the primitive structure compatible with a honing C/P_3 complex, presumably retained from the ancestral ape condition.

A similar analysis can be made with the lower canines by using roughly estimated crown height (14.5 mm) and mesial crest length (5.5 mm) of the ALA-VP-2/10 canine.

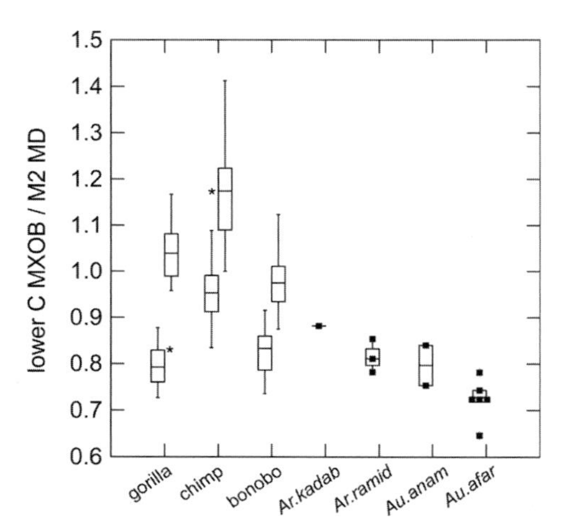

PLOT 7.6

Relative mandibular canine size (maximum oblique breadth divided by M_2 mesiodistal length). Abbreviations: gorilla, *G. gorilla;* chimp, *P. troglodytes;* bonobo, *P. paniscus; Ar.kadab, Ar. kadabba* (ALA-VP-2/10); *Ar.ramid, Ar. ramidus; Au.anam, Au. anamensis; Au.afar, Au. afarensis.* See Table 7.5 for further details of the samples. Box plots as in Plot 7.1. Sex-segregated in the modern apes, left side females, right side males. Sample sizes of the modern apes are as follows: *G. gorilla* females (n = 24), males (n = 22); *P. troglodytes* females (n = 45), males (n = 48); and *P. paniscus* females (n = 25), males (n = 22). Each filled square represents an individual specimen.

As with the upper canine, crown height relative to basal dimension does not distinguish between hominids and apes (Plot 7.11). Again, chimpanzee and bonobo female canines appear to be derived in their proportionally tall canine crowns, but it is the male gorilla, orangutan, and some Miocene ape lower canines that show the relatively tallest crowns. Mesial shoulder placement can be evaluated from the ratio of mesial crest length to either crown height or maximum basal dimensions (Plot 7.11). The two lower canines of *Ar. kadabba* represent a range comparable to female ape conditions. However, the high position of the ALA-VP-2/10 mesial shoulder corresponds to the extreme end of the female ape range of variation, a condition commonly seen in later hominids.

Further morphological details of significance occur in the distolingual lower canine crown. In the ancestral ape structural condition, the distolingual ridge is emphasized and the distal crest is de-emphasized. The distal crest can be virtually unexpressed or only weakly expressed (as in STD-VP-2/61), so that it would disappear even with minimal wear. In contrast, the well-developed distolingual ridge contributes in defining the planar zone of distal crown that wears vertically against the mesiolingual crown of the upper canine.

In the more "incisiform" lower canine there is a change of emphasis so that the distal crest is well-defined from cusp tip to distal basal crown. This occurs as part of normal variation in the canines of some of the ape species (e.g., bonobos, *Ouranopithecus, Gigantopithecus*). A distolingual groove of various depth forms between the distal crest and the distolingual ridge. When the distal lingual ridge/groove is weak, as is often the case in later species of *Australopithecus* and *Homo,* a spatulate lingual fossa extends from mesial to distal crown margins.

Another related feature is the distal tubercle of the lower canine. The evolutionary significance of the distal tubercle, however, is complex. Although the tubercle tends to be absent or weak in the large male lower canines of most ape species, a "heel"-like structure is well-developed, for example, in the lower canines of cercopithecid monkeys and hylobatids. This is associated with developed honing wear, and the "heel"-like structure supports an oblique wear zone that appears to function as a guide for the upper canine tip, as

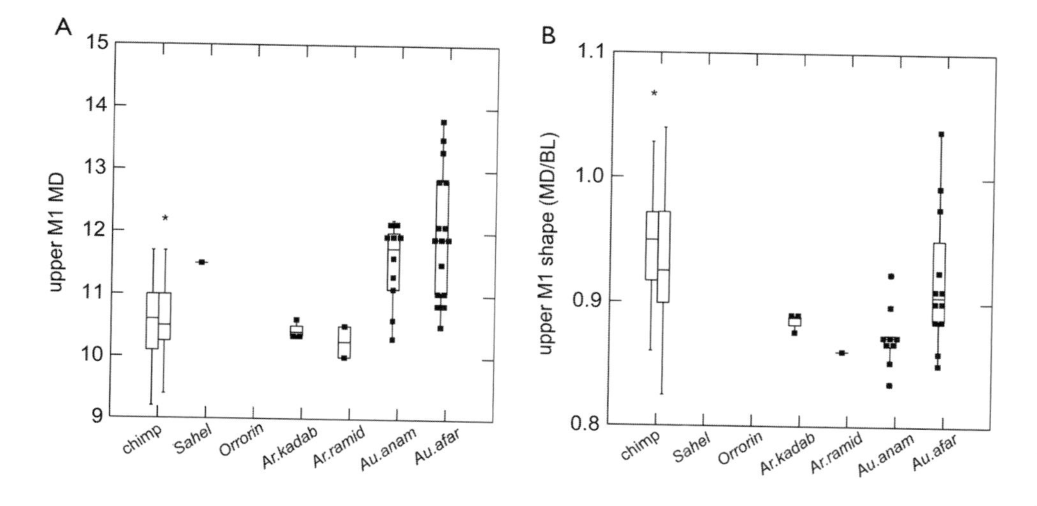

PLOT 7.7

Maxillary first molar size and shape. (A) Mesiodistal length (in mm). (B) Shape (mesiodistal length divided by buccolingual breadth). Abbreviations: chimp, *P. troglodytes; Sahel, Sahelanthropus; Ar.kadab, Ar. kadabba; Ar.ramid, Ar. ramidus; Au.anam, Au. anamensis; Au.afar, Au. afarensis*. See Table 7.4 for further details of the samples. Box plots as in Plot 7.1. Sex segregated in *P. troglodytes*, left side females, right side males. *P. troglodytes* sample sizes: MD length, females (n = 68), males (n = 61); shape index, females (n = 67), males (n = 61). Each filled square represents an individual specimen.

the upper canine starts to occlude against the sectorial P_3. In many ape canines a similar oblique wear zone is commonly formed at the distal basal crown of the lower canine (or on the distal tubercle when it is present, as noted by Haile-Selassie [2001b]), but this is not necessarily associated with the development of the tubercle itself. Rather, the distal tubercle is better expressed in female apes and in early hominids, in the form of a cingulum-like structure (STD-VP-2/61). In later hominds, it merges with the distal crest and forms the distal crown shoulder (e.g., L.H.-3, MAK-VP-1/12, and later *Australopithecus* species). Thus, in the progressively spatulate lower canine of hominids, the distal tubercle functions as a part of the distal delineation of the lingual fossa.

In modern and Miocene apes, a lower canine with a distinct distal crest and distal tubercle is seen as a part of normal variation in the females, and apparently more frequently in species that are characterized by relatively small canines, such as bonobos, *Ouranopithecus,* and *Oreopithecus.* We have already seen that the lower canine of *Ar. kadabba* retains the primitive condition of a well-developed distolingual crest and weak distal crest and combines this with a strong but cingulum-like distal tubercle. The STD-VP-2/61 canine is such an example, whereas the exposed dentin pattern of the worn ALA-VP-2/10 distal crown suggests a weak but continuous distal crest and distinct distolingual groove. A similar range of primitive morphology extends its presence into the *Au. anamensis* hypodigm, where the distal crest must have been minimal (KNM-KP 29286) or weak but continuous (KNM-KP 29284). However, in *Au. anamensis,* a more "incisiform" morphology is also a known part of the species variation (e.g., KNM-KP 31727, KNM-ER 30731), and this morphology becomes dominant in *Au. afarensis* and later hominids.

The two *Sahelanthropus* lower canines are too worn or damaged for a metric evaluation of crown and mesial shoulder height. However, the TM266-02-154-2 canine exhibits a large distal tubercle and wear pattern similar to the ALA-VP-2/10 condition (Brunet et al. 2002). The canine of the TM292-02-01 mandible is reported to combine a distinct distal tubercle and a low mesial shoulder position (Brunet et al. 2005), apparently even lower than in STD-VP-2/61. An additional similarity of the *Sahelanthropus* and *Ar. kadabba* lower canines is crown cross-sectional shape. The two known lower canines of *Ar. kadabba*

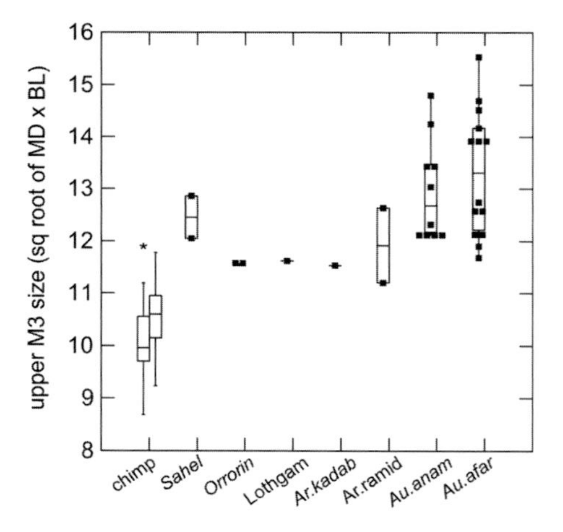

PLOT 7.8

Maxillary third molar size (mesiodistal length times buccolingual breadth). Abbreviations: chimp, *P. troglodytes; Sahel, Sahelanthropus; Orrorin, Orrorin;* Lothagam, Lothagam upper Nawata; *Ar.kadab, Ar. kadabba; Ar.ramid, Ar. ramidus; Au.anam, Au. anamensis; Au.afar, Au. afarensis.* See Table 7.4 for further details of the samples. Box plots as in Plot 7.1. Sex-segregated in *P. troglodytes*, left side females (n = 42), right side males (n = 45). Each filled square represents an individual specimen.

have a strongly compressed basal crown, also seen especially in the TM266-02-154-2 lower canine. A compressed basal lower canine is more common in Miocene apes than in modern apes (Plot 7.12) and is one of the features used by Haile-Selassie (2001b) in differentiating *Ar. kadabba* from *Ar. ramidus*.

As discussed above and in Haile-Selassie et al. (2004b), the primitive morphology and incipient wear pattern of the canines and P_3 of *Ar. kadabba* suggest that an interlocking C/P_3 complex with potential for some honing was a part of the species variation. However, it remains to be seen whether functional honing did in fact occur in adult individuals in any significant frequencies. Alternatively, apical wear seen in the Alayla lower canine and the *Sahelanthropus* upper and lower canines, at a relatively early stage of crown wear, indicates some frequency of a lack of honing wear in adults.

Other Premolar and Molar Morphology

The other postcanine dental elements of *Ar. kadabba* include a single well-preserved P^3, fragmentary and/or worn upper and lower P4s, a worn M_2 and M_3, three M^1s, and a single M^3. This is a small collection, but sufficient to see that the overall morphological affinities are with the other early hominids, such as *Ar. ramidus* and *Au. anamensis*. Some of the more salient comparative observations that justify such a conclusion are outlined in the following paragraphs.

All elements of the preserved postcanine dentition of *Ar. kadabba* lack any signs of a gorilla-like morphology with high cusps emphasizing the development of the main occlusal crests. The *Ar. kadabba* molars also differ from chimpanzee and bonobo homologs in lacking the widely spaced positions and occlusally concave slopes of the main cusps. Morphological differences and similarities with the diversity of known Miocene apes are not easily summarized, and a comprehensive evaluation is beyond the scope of the current study. However, we believe that the following combination of features would in most cases differentiate the known *Ar. kadabba* materials from those of the varied Miocene ape taxa. We restrict this summary to the upper molars, which are better represented in *Ar. kadabba*.

TABLE 7.6 Principal Component Analysis of the Mandibular Tooth Crown Metrics

	PC1	*PC2*	*PC3*
Canine to M_2 analysis			
Eigenvalues	4.535	0.322	0.089
% contribution	90.7	6.4	1.8
PC loadings			
C_1MXOB	0.86	−0.51	0.03
P_4MD	0.97	0.05	−0.20
P_4BL	0.97	0.12	0.18
M_2MD	0.98	0.10	−0.09
M_2BL	0.97	0.18	0.08
I_2 to M_3 analysis			
Eigenvalues	5.829	0.835	0.138
% contribution	83.3	11.9	2.0
PC loadings			
I_2BL	0.65	−0.73	0.19
C_1MXOB	0.87	−0.38	−0.30
P_4MD	0.97	0.09	−0.01
P_4BL	0.96	0.15	0.10
M_2MD	0.98	0.15	−0.02
M_2BL	0.96	0.24	0.06
M_3MD	0.96	0.21	0.01

NOTE: Comparative samples are outlined in plot legends.

The *Ar. kadabba* upper molars are characterized by (1) lack of a strongly expressed protoconule, (2) weak cingular expression, (3) the main cusps neither peripherally placed nor crowded internally, (4) lack of tendency for high occlusal crests, and (5) lack of expression of a hypocone-metacone crest. In addition to these characteristics of the upper molars, the well preserved P^3 of *Ar. kadabba* differs from the homologs of many Miocene ape taxa in its only weakly asymmetric crown, weak mesiobuccal projection of the basal crown, and lack of buccolingual crown expansion. The former two are also features that tend to differentiate the *Ar. kadabba* P^3 from that of *Pan*.

In contrast, the above morphological features of the *Ar. kadabba* postcanine dentition are shared by *Ar. ramidus* as well as *Au. anamensis* homologs. Differences between the known *Ar. kadabba* and *Ar. ramidus* materials have previously been described by Haile-Selassie (2001b). These include a shallow anterior fovea of the P^3 (ASK-VP-3/160) and the square occlusal outline with four well-defined cusps in the single known M^3 (STD-VP-2/62).

The *Sahelanthropus* and *Orrorin* postcanine dentitions are even more sparsely represented than those of *Ar. kadabba,* so that the differences described here merely represent individually based observations. The *Sahelanthropus* P^3 (TM 266-01-462) was described in some detail by Brunet et al. (2005). It shares with the *Ar. kadabba* and *Ar. ramidus* P^3s a lack of strong asymmetry and/or mesiobuccal crown extension. It differs from *Ar. kadabba* in the deeper anterior fovea, low crown height, and strong buccal crown flare. The

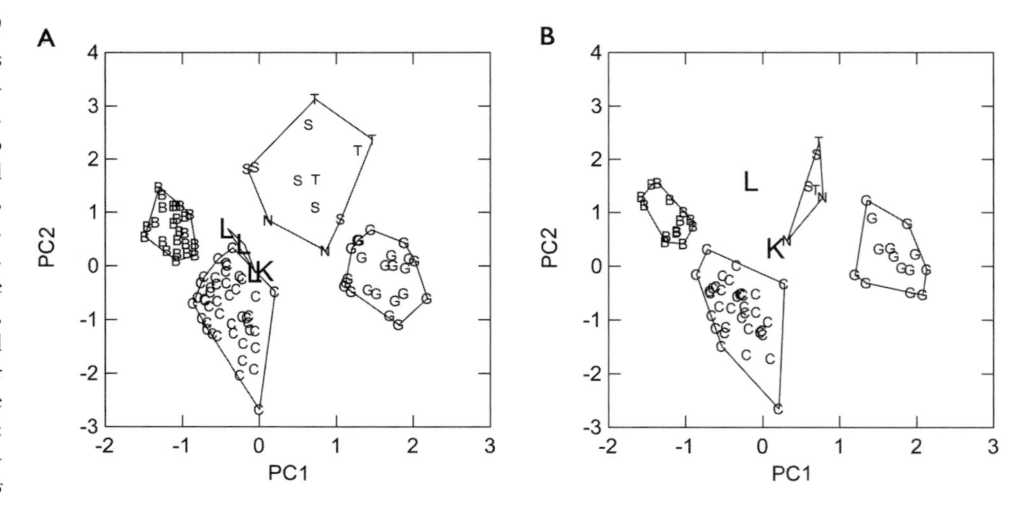

PLOT 7.9

Principal components analysis of the ALA-VP-2/10 dentition. A. Mandibular canine to M_2. Variables included were C_1MXOB, P_4MD, P_4BL, M_2MD, and M_2BL. B. Mandibular I_2 to M_3. Variables included were I_2BL, C_1MXOB, P_4MD, P_4BL, M_2MD, M_2BL, and M_3MD. See Table 7.4 for further information of the metrics. Abbreviations: K, *Ar. kadabba* (ALA-VP-2/10); L, *Ar. ramidus* (A, n = 3; B, ARA-VP-1/128); N, *Au. anamensis* (n = 2); S, *Au. afarensis* (A, n = 6; B, n = 2); T, *Au. africanus* (A, n = 4; B, n = 2); and modern ape females as follows: B, *P. paniscus* (A, n = 24; B, n = 14); C, *P. troglodytes* (A, n = 44; B, n = 35); G, *G. gorilla* (A, n = 23; B, n = 14).

mandibular postcanine dentition of *Sahelanthropus* (TM266-02-154-2) closely matches the ALA-VP-2/10 equivalent in the available P_4 to M_3 metrics. Further details are not easy to evaluate, because of damage and wear in either or both specimens, but *Sahelanthropus* and *Ar. kaddaba* appear to share a P_4 with buccal grooves and cingular features. However, although this is well-developed in *Sahelanthropus* (Brunet et al. 2002), only the distal buccal groove is distinct in the ALA-VP-2/10 P_4, as is also the case in *Ar. ramidus.* Another possible difference is the morphology of the molar cusps, described as "rounded" in *Sahelanthropus* (Brunet et al. 2002), the significance of which is difficult to evaluate with the available examples and possible effects of surface abrasion by wind-borne sand.

The *Orrorin* postcanine teeth are not described or illustrated in sufficient detail, so that a detailed comparison is not possible for the time being. The well-preserved and well-known "Lukeino molar" (KNM-LU 335) is likely to be a M_1, despite recent other opinions (e.g., Pickford and Senut 2005). No M_1 is known for *Ar. kadabba,* and the single known M_2 is too worn for adequate morphological comparisons. Pickford and Senut (2005) attribute a partial upper molar and possible M_3 to a second hominoid taxon that, they suggest, shows gorilla affinity. The major morphological reasons given for such an attribution were the large size of the upper molar and the peripheral cusp positions and wider occlusal basins of both molar specimens. However, the published intercuspal dimensions of the partial upper molar are comparable to the larger M^2 examples of *Ar. ramidus,* thus suggesting that it may also be compatible with an *Ar. kadabba*–sized post-canine dentition. Cuspal positions and crown flare are difficult to quantify and compare across small samples of variably worn molars. We are not sure how to evaluate the metrics published by Pickford and Senut (2005) (metaconid-protoconid distance, 5.4 mm; crown breadth across the cusps, 10.5 mm), since they appear to give a rather large relative intercuspal distance, in comparison to KNM-LU 335, than suggested by its mesial view photograph. We are not convinced that these molar specimens sufficiently falsify the null hypothesis of one species from the Lukeino Formation, but we await further descriptive details of the entire Lukeino dental assemblage.

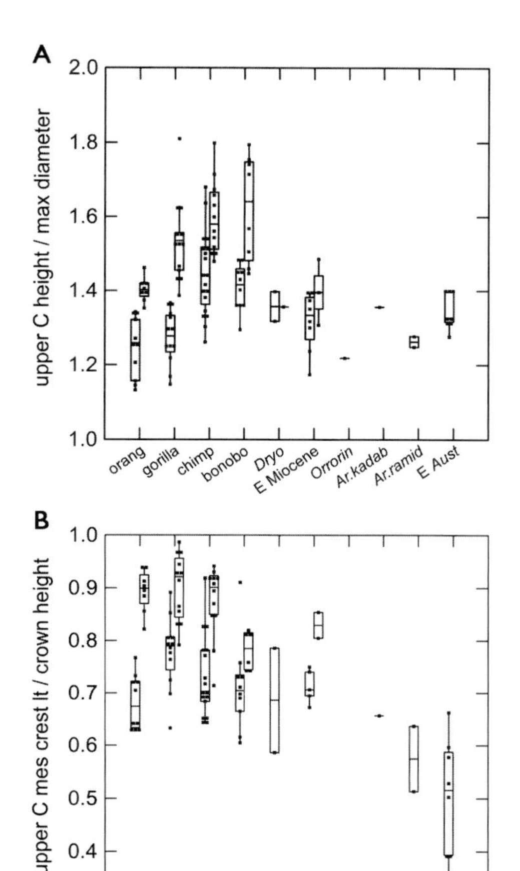

PLOT 7.10

Maxillary canine shape. A. Crown height divided by maximum basal dimension. B. Mesial shoulder position (mesial crest length divided by crown height). Abbreviations: orang, *Po. pygmaeus*; gorilla, *G. gorilla*; chimp, *P. troglodytes*; bonobo, *P. paniscus*; *Dryo, Dryopithecus laietanus*; E Miocene, *Proconsul*, *Afropithecus*, and/or *Kenyapithecus* spp. Orrorin, *Orrorin*; *Ar.kadab, Ar. kadabba*; *Ar.ramid, Ar. ramidus*; E *Aust, Au. anamensis*, *Au. afarensis*, and *Au.africanus*. Box plots as in Plot 7.1. Both modern and fossil ape taxa are sex-segregated, left side females, right side males. Each filled square represents an individual specimen.

Enamel Thickness

Enamel thickness data of the *Ar. kadabba* teeth are summarized in Table 7.7. All three *Ar. kadabba* canines examined showed a lack of, or very weak, thickening of enamel toward the canine crown tip, as was described for *Ar. ramidus* (White et al. 1994).

For the upper molars, thickness values were compiled for the comparative sample following the methodology of Suwa and Kono (2005). Using the micro-ct data set, we measured maximum radial thickness of the lateral crown faces at locations opposite the major cusps. We also measured thicknesses in naturally fractured near-radial sections of the buccal and lingual crown faces. Although this introduces some amount of noise into the data set because of the lack of strict positional control, it increases sample sizes and enables the detection of general trends across taxa, which would otherwise not be possible. We confined the natural section data to those taken at locations sufficiently close to the main cusp tips (usually closer to the cusp tip than to the intercuspal groove). Our concerns were to avoid idiosyncratic variation in thickness near grooves and to keep the vertical position of the measurement not too offset from that of the cusp tip area. We have found

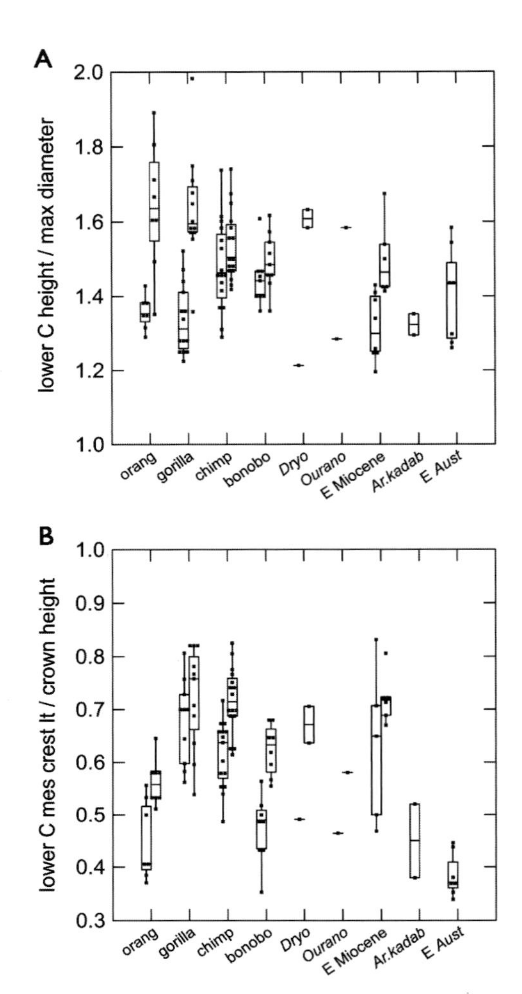

PLOT 7.11

Mandibular canine shape. A. Crown height divided by maximum basal dimension. B. Mesial shoulder position (mesial crest length divided by crown height). Abbreviations: orang, *Po. pygmaeus;* gorilla, *G. gorilla;* chimp, *P. troglodytes;* bonobo, *P. paniscus; Dryo, Dryopithecus laietanus; Ourano, Ouranopithecus macedoniensis* (based on casts housed in the University of Poitiers); E Miocene, *Proconsul, Afropithecus* and *Kenyapithecus* spp.; *Ar.kadab, Ar. kadabba;* E *Aust, Au. anamensis, Au. afarensis* and *Au. africanus.* Box plots as in Plot 7.1. Both modern and fossil ape taxa are sex-segregated, left side females, right side males. Each filled square represents an individual specimen.

that the radial sections running through the main cusps at different angles on the one hand, and near-radial sections relatively close to the cusp tips, on the other, are generally within approximately 0.15 mm of each other, whereas buccal and lingual sections closer to the intercuspal grooves can be much more variable. Thicknesses of the mesial sections are significantly smaller, and those of the distal sections are variably thick or considerably thinner than the adjacent cuspal areas. We are currently working toward a more systematic documentation of these tendencies.

In all the comparative samples we confirmed that lateral molar enamel thickness tends to be thicker lingually than buccally and thicker in the two posterior molars than in the first molars (see Suwa and Kono 2005 for a detailed presentation of human molars). The small sample of *Ar. kadabba* teeth were found to show the same tendencies and considerable variation in thickness values (Plot 7.13). Absolute values ranged from 1.1 mm to 1.7 mm at the "non-functional" side, and from 1.4 mm to 2.0 mm at the "functional" side cusps. However, this entire range is not excessive and appears to be broadly matched by a larger sample set of *Ar. ramidus* molars now under investigation. What is published of the

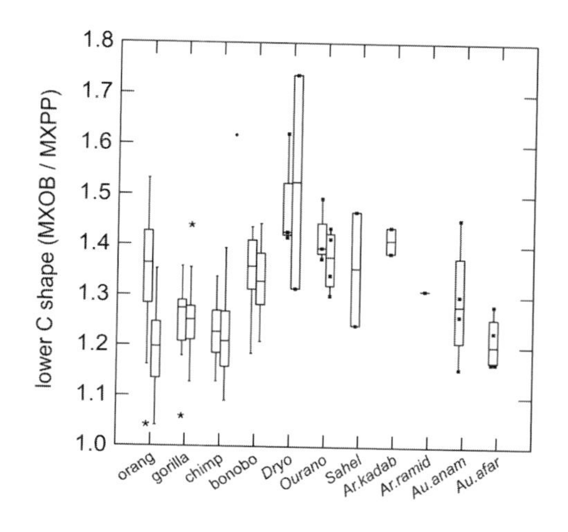

PLOT 7.12

Mandibular canine basal shape (maximum oblique breadth divided by perpendicular breadth). Abbreviations: orang, *Po. pygmaeus;* gorilla, *G. gorilla;* chimp, *P. troglodytes;* bonobo, *P. paniscus;* Dryo, *Dryopithecus laietanus; Ourano, Ouranopithecus macedoniensis* (based on casts housed in University of Poitiers); *Sahel, Sahelanthropus; Ar.kadab, Ar. kadabba; Ar.ramid, Ar. ramidus; Au.anam, Au. anamensis, Au.afar, Au. afarensis.* Box plots as in Plot 1. Both modern and fossil ape taxa are sex-segregated, left side females, right side males. Sample sizes of the modern apes are as follows: *Po. pygmaeus* females (n = 36), males (n = 20); *G. gorilla* females (n = 24), males (n = 22); *P. troglodytes* females (n = 47), males (n = 48); *P. paniscus* females (n = 28), males (n = 23). Each filled square represents an individual specimen.

Sahelanthropus (Brunet et al. 2005) and *Orrorin* (Pickford and Senut 2005) enamel thickness also appears to be broadly compatible with the *Ar. kadabba* and *Ar. ramidus* condition of enamel thickness values intermediate between extant apes and later fossil hominids, as initially described for *Ar. ramidus* but subsequently mistakenly characterized by many as having "thin enamel," as if this were a dichotomous trait (which it is obviously not).

Tooth Wear

The distinctive molar wear pattern seen in the ALA-VP-2/10 mandible was described in some detail in Haile-Selassie (2001b). It is characterized by a strong buccolingual wear differential, resulting in the retention of salient lingual cusps. There is literally no flattening of the M_3 lingual cusps, while the buccal cusps of all molars show deeply cupped and coalesced dentine exposures. The lack of a clear M_1 to M_3 wear gradient in buccal cusp wear was also described by Haile-Selassie (2001b). While a strong buccolingual wear differential is also seen in the molars of *Ar. ramidus, Au. anamensis,* and *Au. afarensis,* variable amounts of lingual cusp flattening do occur. The same is the case for the known Lothagam, Tabarin, and *Sahelanthopus* mandibular dentitions.

A close examination of the strong scooping of buccal dentin shows that there are in fact two different types of wear. One is characteristic of both *Au. afarensis* and *Au. anamensis,* and the other appears to be more typical of *Ar. ramidus* and *Ar. kadabba,* although there may be individual cases of overlap in variation. Despite the strong buccolingual wear differential seen in the two early species of *Australopithecus,* the pattern of dentin cupping is asymmetric, involving a gradual transition of enamel to dentin in the leading edge (lingual in lowers) and a stepped transition in the trailing edge (buccal in lowers) (e.g., L.H.-4, A.L. 400-1). Such a wear pattern is best interpreted as caused by abrasive wear.

The ALA-VP-2/10 mandibular molars show deep dentin wear, but the slopes are more symmetric, and the enamel margin at both leading and trailing edges tends to share the wear slope with the worn dentin surface rather than forming a "stepped" interface. The *Ar. ramidus* ARA-VP-1/128 molar row exhibits the same type of deep and buccolingually

TABLE 7.7 Summary of Enamel Thickness in *Ardipithecus kadabba*

Specimen	*Tooth Type*	*Method*	*Labial*	*Protoconid*	*Hypoconid*	*Metaconid*	*Entoconid*	*Paracone*	*Metacone*	*Protocone*	*Hypocone*
Ar. kadabba											
ALA-VP-2/10	C_1	Micro-ct/wear	(1.2)								
STD-VP-2/61	C_1	Micro-ct	(1.0)								
ALA-VP-2/10	P_4	Wear		1.6+							
ALA-VP-2/10	M_1	Wear		1.5+							
ALA-VP-2/10	M_2	Micro-ct				1.38(+)	1.67				
		Fracture			1.8+						
ALA-VP-2/349	I^2	Fracture	1.0								
ASK-VP-3/400	C^1	Micro-ct	1.15								
ASK-VP-3/160	P^3	Micro-ct						(1.3)			
ASK-VP-3/401	M^1	Micro-ct						1.18	1.23	1.41(+)	1.48(+)
ASK-VP-3/402	M^1	Micro-ct						1.16	1.17	1.61(+)	1.66(+)
STD-VP-2/63	M^1	Micro-ct						1.12(+)	1.16	1.23+	1.18+
STD-VP-2/62	M^3	Micro-ct						1.71	1.58	2.01	1.88
		Fracture							1.6		
Ar. ramidus											
ARA-VP-1/400	M^2	Micro-ct						1.34	1.24	1.54(+)	1.61
ARA-VP-1/1	M^3	Micro-ct						1.42	1.24	1.63(+)	
ARA-VP-1/182	M^3	Fracture							1.0		

NOTE: Measurements are in mm. ( ) are estimates due to poor image quality; + indicates thickness in worn condition; (+) indicates slightly worn condition considered close to actual value. Maximum radial enamel thickness opposite the major cusps from micro-ct data, or taken on near-radial natural fracture sections around the cusp tips (see text).

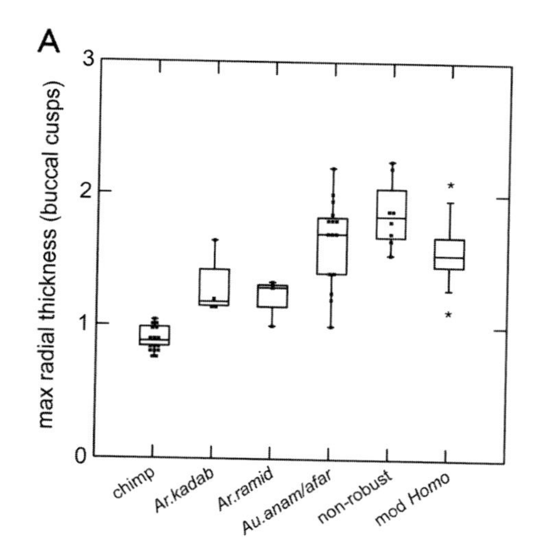

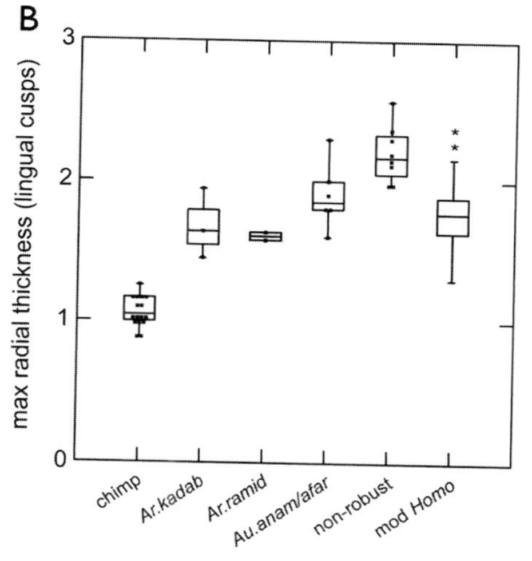

PLOT 7.13

Lateral enamel thickness of the upper molars. A. Buccal cusps. B. Lingual cusps. Abbreviations: chimp, *P. troglodytes* (n =16); *Ar.kadab, Ar. kadabba; Ar.ramid, Ar. ramidus; Au.anam/afar, Au. anamensis* (buccal, n = 10; lingual, n = 4) and *Au. afarensis* (buccal, n = 5; lingual, n = 2); non-robust, *Au. africanus* (n = 5) and eastern African non-robusts from 2.9–2.3 Ma (n = 3); mod *Homo,* modern human (n = 77). Individual data of the *Ar. kadabba* and *Ar. ramidus* specimens are given in Table 7.7. Thickness values are the maximum radial thickness of the lateral crown faces taken opposite the main cusps, as defined by Suwa and Kono (2005). When lateral thicknesses were available for both mesial and distal cusps, the average of the two was used. Enamel thickness data of the modern chimpanzee and human, *Au. africanus,* and two (out of three) eastern African non-robust molars were based on 3-dimensional micro-ct volume data. Thickness data of *Au. anamensis, Au. afarensis,* and some eastern African non-robust molars were based on natural fracture sections that were considered near-radial and sufficiently close to the cusp tip area (see the text for further details). Box plots as in Plot 7.1. Each filled square represents an individual specimen.

symmetric dentin wear and weak M_1 to M_3 wear gradient. Aside from the ALA-VP-2/10 mandibular molars, the only other *Ar. kadabba* molar of advanced wear, the STD-VP-2/63 M^1, also exhibits a scooped dentin exposure that is extremely deep. The latter suggests a significant role of "erosive" effects rather than strictly abrasive wear.

Although a comprehensive assessment of wear pattern among modern and Miocene apes is beyond the scope of this discussion, our preliminary observations suggest that among modern apes, it is *Pongo* that shows the highest frequencies of similar molar wear. Deep dentin wear with symmetric slopes is also common in the thinner-enameled Miocene apes such as *Dryopithecus,* suggesting a relationship with some aspect of frugivory, but with no simple or obvious associations with specific dentognathic morphologies. The symmetric leading and trailing edges and weak M_1 to M_3 wear gradients imply that this is not associated with an abrasive dietary regime.

Finally, we note that the detailed assessment of the ALA-VP-2/10 dental fragments has shed light on another unusual aspect of dental wear in this specimen. This is the premortem macrofracture involving loss of large chunks of tooth material at the labial side of the left I_2 and at the lingual side of the left P_4/M_1 area. The right side exhibits periodontal disease (and probable antemortem loss of M_1) and retention of salient lingual cusps in M_2 and M_3. Further fossils are needed to see whether these conditions represent an unusal wear pattern. If not, the fractures might imply some reliance on hard-object feeding, while the retention of steep phase I facets suggests the importance of softer, more pliable dietary items.

Dental Summary

In summary, the *Ar. kadabba* dentition shares many general features with that of *Ar. ramidus.* These include general size and proportions, aspects of incisor size and morphology, the C/P_3 complex morphology, postcanine size and morphology, enamel thickness, and molar wear patterns. The shared derived features of the C/P_3 complex detailed above, although subtle, securely place *Ar. kadabba* at a basal hominid position along a morphocline extending from some female ape-like form to the later hominid condition seen in *Australopithecus*. Other such candidates indicative of a shared hominid status are the relatively weak asymmetry and low height of the P^3 crown.

Some of the other shared features are likely to be primitive retentions, because the *Pan* condition is unknown or rare among Miocene and other modern apes. Such features of *Ar. kadabba* are relatively numerous and include lower incisors with simple lingual morphology and small relative size, upper canine not narrow buccolingually, upper and lower canines not particularly tall-crowned, upper canine with a strong mesiolingual ridge/groove complex, a buccolingually broad M^1 shape, the posterior molars (M2 and M3) distinctly larger than M1, some details of molar crown morphology, and an intermediate thickness of enamel. This suggests that the *Pan* dentition is derived in many ways. Still other features, or a differential expression of the same features, may be an ancestral condition shared by *Ar. kadabba* and *Pan* to the exclusion of other Miocene and modern apes. Such features appear to be less obvious, at least with the currently available samples.

Finally, the differences between the known dental elements of *Ar. kadabba* and *Ar. ramidus* were already summarized by Haile-Selassie (2001b) and Haile-Selassie et al. (2004b) and include the subtle but apparently significant differences in C/P3 morphology and some of the other detailed features that may characterize species tendencies. We believe that these differences are of the sort that does not preclude an ancestor-descendant relationship between the two taxa.

Postcranial Elements

Discovery

ALA-VP-2/11 is a hand phalanx fragment found by Solomon Eshete during a group crawling operation at ALA-VP-2 on December 29, 1997. This crawling operation was conducted after the discovery of ALA-VP-2/10 thirteen days earlier. A small sieving operation was carried out, but the remaining part of the specimen was not recovered. ALA-VP-2/101 consists of two fragments of a left proximal ulna and a humerus. It was a surface discovery by Tim White on December 6, 1999, immediately preceding the discovery of additional dental remains of ALA-VP-2/10 (Figure 7.18).

STD-VP-2/893 is a fragment of a clavicle found at STD-VP-2 in December 1998. It was found on the surface along with many other vertebrate fossils (Figure 7.9). DID-VP-1/80 is a partial hand phalanx. It was a surface recovery made on December 25, 1998, during a crawling operation at Digiba Dora Vertebrate Paleontology Locality 1 (see Chapter 2 figures). The broken proximal surface looked recent, so surface scraping and sieving were conducted in an area of ca. 20 m. The proximal half was not recovered.

ASK-VP-3/78 is a left distal humerus found at ASK-VP-3 on December 25, 1998. The specimen was found on the surface of a small secondary outwash channel along with other vertebrate fossil remains. An additional shaft fragment of the humerus was found a year later, ca. 10 m west of the initial fragment. The second fragment perfectly joined the first

FIGURE 7.18
Yohannes Haile-Selassie examines the proximal hominid ulna found at the ALA-VP-2 locality (ALA-VP-2/101). This and other discoveries led to further rubble clearing, sieving, and excavation at the locality. Photograph by Tim White, December 6, 1999.

specimen, increasing the length of the recovered specimen from 55.9 mm to 105.1 mm. This discovery eventually led to further survey of the area and the discovery of the teeth described above within ca. 20 m (see Figures 7.12 through 7.17).

AME-VP-1/71 is a foot phalanx found by Leslea Hlusko on December 4, 1999, in dark reddish brown clays ca. 2–3 m above the Kuseralee sands. The distal half was first found on the surface, and the proximal half was found nearly *in situ,* covered with concretion, within a meter or so of the initial find spot. The two pieces join perfectly. Figures 7.19 through 7.21 record the recovery circumstances. A thorough crawl and subsequent recrawls and sieving failed to produce additional remains.

Comparative Descriptions

Hand Phalanges (ALA-VP-2/11, DID-VP-1/80; Plate 7.8)

ALA-VP-2/11 is the distal two-thirds of an intermediate hand phalanx. It was a surface find positioned 20–25 m south of the holotype mandible ALA-VP-2/10, at the same horizon. The distal end is complete, and most of the shaft is preserved. The specimen does not show any chemical alteration or abrasion, and the midshaft break is recent. The preserved maximum length of the specimen is 22.9 mm. The head measures 9.1 mm mediolaterally and 6.2 mm dorsoventrally. It has a slight constriction at the neck. The base of the condyles on the palmar side shows some porosity and a depression between the condyles. The two condyles are almost symmetrical, and the areas for the attachments of the collateral ligaments are roughened. Dorsal curvature of the shaft is minimal. The palmar surface, particularly toward the proximal end of the preserved diaphysis, displays deep bilateral fossae for the flexor digitorum superficialis muscle.

FIGURE 7.19
The AME-VP-1/71 hominid foot phalanx was found on the surface of the dark brown silty clays exposed in front of the discoverer, Leslea Hlusko. The Gawto basalt, dated to 5.2 Ma, is seen on the skyline. A cercopithecid maxilla, mandible, and skull fragments (AME-VP-1/64) were previously found by H. Saegusa on the surface, and subsequently in the excavation to the right of the hominid find spot. Photograph by Tim White, December 4, 1999.

ALA-VP-2/11 is morphologically similar to most known *Au. afarensis* intermediate phalanges. The development of the bilateral fossae for the flexor digitorum superficialis muscle in ALA-VP-2/11 best matches A.L. 333w-53. However, ALA-VP-2/11 is larger than known (n = 10) Hadar intermediate hand phalanges, with an estimated length of 30–32 mm (see Figure 7.22). The maximum length of the Hadar *Au. afarensis* adult intermediate hand phalanges ranges from 21.2 mm to 26.9 mm based on a sample of eight (Bush et al. 1982). The head dimensions of ALA-VP-2/11 are also larger than the observed range for *Au. afarensis* (AP = 4.2–5.4 mm; ML = 6.8–8.2 mm, n = 8; Bush et al. 1982). However, the indeterminate serial (ray) allocation of the *Au. afarensis* and Alayla specimens reduces the value of such comparisons given the small sample sizes.

DID-VP-1/80 is the distal half of a proximal hand phalanx. The maximum preserved length of the specimen is 27.5 mm. The head dimensions are 6.5 ± 0.5 mm dorsoventrally on the preserved right condyle and 8.8 mm mediolaterally (measured at the preserved base of the head). Preserved midshaft dimensions 3.9 mm above the broken edge of the shaft are 9.1 mm mediolaterally and 5.8 mm dorsoventrally. The specimen is broken obliquely near midshaft, proximal to the ridges for the flexor retinaculum. Most of the left condyle (in dorsal view) of the distal end, and the dorsal part of the distal articular surface, are missing as a result of either abrasion or breakage. The overall

FIGURE 7.20

Closeup of the AME-VP-1/71 hominid foot phalanx. The distal end was found first, on the surface, and the remainder of the specimen, covered with calcium carbonate matrix, was found nearly *in situ*. Photograph by Tim White, December 4, 1999.

texture of the specimen suggests that it has been either fluvially abraded or chemically altered by carnivore activity, resulting in the loss of some surface detail. There is a slight constriction at the neck. The preserved part of the shaft is longitudinally slightly curved, transversely convex and smooth on the dorsal side, and flat on the palmar side. There is a small foramen (<1 mm diameter) on the palmar side along the right edge, ca. 8.3 mm below the proximal palmar base of the right condyle. The area immediately below the right condyle has a shallow longitudinal depression. The ridges for the flexor retinaculum are not well-developed. At the point of fracture, the cortex of the shaft appears to be relatively thick, with cortical thickness measurements ranging from 2.1 mm to 3.8 mm around a medullary cavity that measures 1.7 mm mediolaterally and 1.1 mm dorsoventrally at the break.

DID-VP-1/80 is fragmentary and lacks most of its measurable portions for comparative purposes. However, the preserved portion shows that morphology of the dorsal face is similar to A.L. 333-19 and to other proximal hand phalanges of *Au. afarensis* from Hadar. The palmar side in A.L. 333-19 is slightly more convex than DID-VP-1/80. The overall

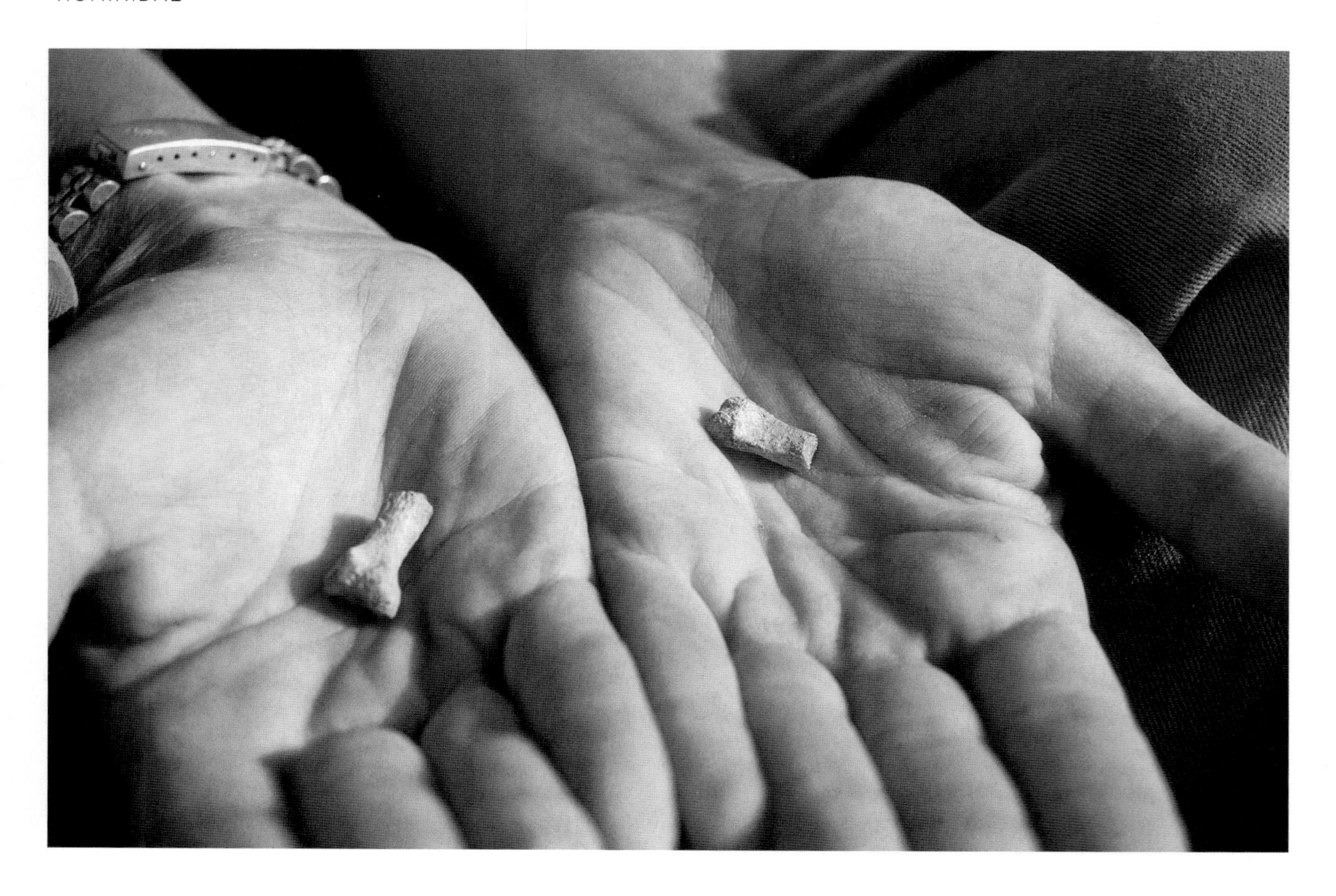

FIGURE 7.21
The two halves of the AME-VP-1/71 foot phalanx after cleaning. Photograph by Tim White, January, 2000.

degree of dorsal curvature of DID-VP-1/80, based on the preserved portion, is similar to most *Au. afarensis* known proximal hand phalanges, and some fifth-ray proximal hand phalanges of *Pan troglodytes*.

The published comparative sample from other late Miocene and early Pliocene hominids is limited to one proximal hand phalanx (BAR 349'00) of *Orrorin tugenensis* (Senut et al. 2001) and one proximal hand phalanx of *Au. anamensis* (KNM-KP 30503) (Ward et al. 2001). BAR 349'00 has not been described in detail other than publication of its dorsal view, some dimensions (length, 31.8 mm; distal breadth, 7.1 mm), and the specimen being described as "curved" (Senut et al. 2001). Morphology and dimensions of BAR 349'00 are definitely within the range seen for *Au. afarensis*. KNM-KP 30503 is also morphologically and metrically within the range of *Au. afarensis* (Ward et al. 2001). One intermediate hand phalanx (ASI-VP-2/6) from Asa Issie, Middle Awash, Ethiopia, is also described as "morphologically similar to those from Hadar, but is longer relative to its breadth" (White et al. 2006b: 887). The latter two specimens are assigned to *Au. anamensis* (Ward et al. 2001; White et al. 2006b). Like DID-VP-1/80, and all *Au. afarensis* proximal hand phalanges, none of these phalanges is assigned to a specific ray. The disadvantages with such samples have already been mentioned above.

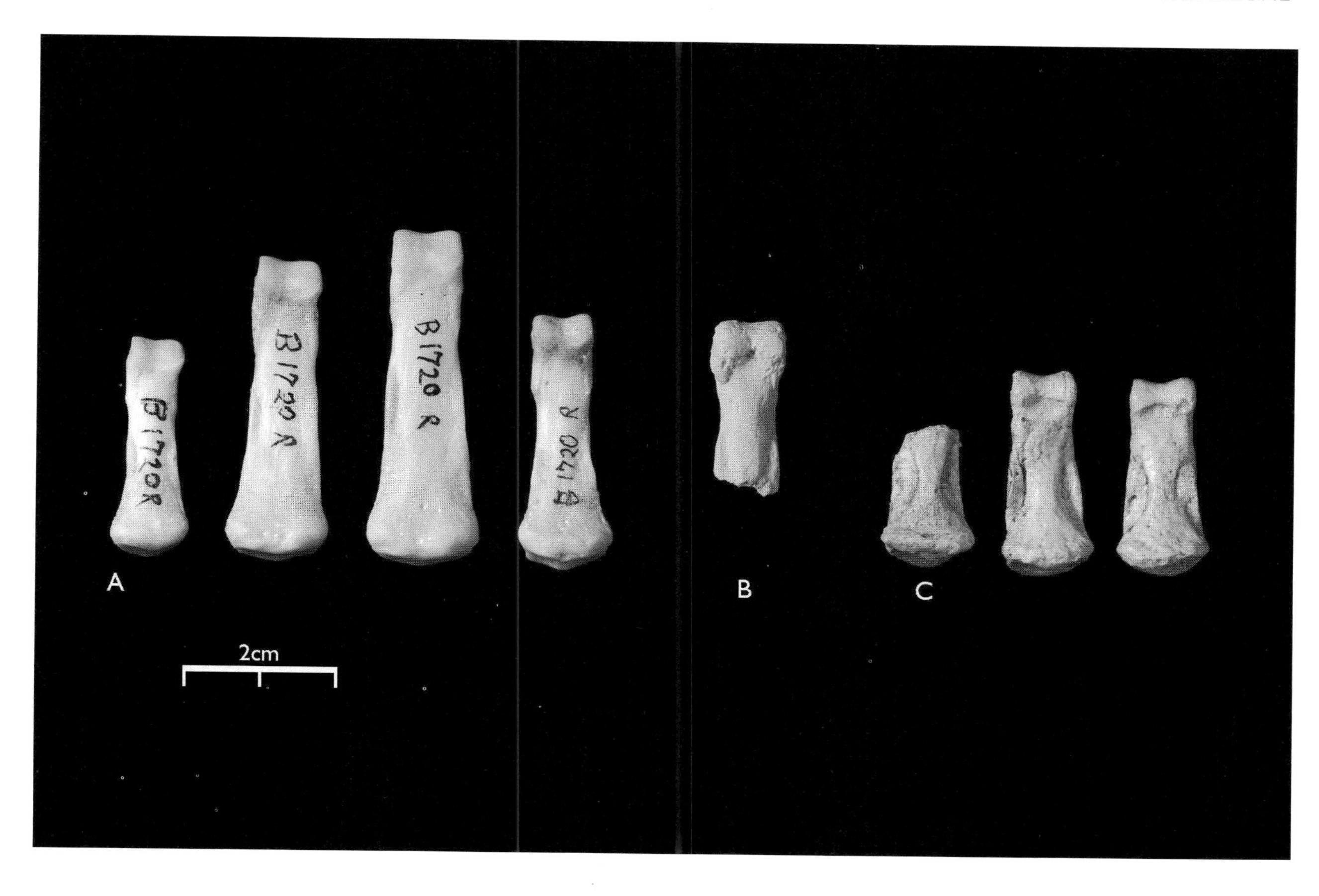

FIGURE 7.22

Palmar view of intermediate hand phalanges. A. *Pan troglodytes,* ray 2 to the left and ray 5 to the right. B. Cast of ALA-VP-2/11, aligned at the point of greatest constriction between the flexor attachments. C. *Australopithecus afarensis* adults from A.L. 333. See text for details.

Humerus and Ulna (ALA-VP-2/101, ASK-VP-3/78; Plates 7.9 and 7.10)

ALA-VP-2/101 consists of an associated left proximal ulna and a left humerus mid-shaft fragment (Figures 7.6 and 7.23). Both elements are highly weathered and abraded. The olecranon process of the ulna is missing on the proximal end. The preserved length of the ulna is 139.6 mm. The radial notch is also eroded away. The coronoid process is partially preserved. The inferior medial portion of the trochlear notch is also preserved, and it is cup-shaped. The superior third of the shaft is well-preserved, although some parts are missing and superficial bone loss is apparent. The preserved shaft is also concave anteriorly. A transverse crack goes around the proximal end at the level immediately below the coronoid process.

The distal half of the shaft lacks external cortical surface on the medial and lateral sides. The floor of the trochlear notch has a transverse ridge-like structure. The insertion area for the brachialis muscle immediately below the coronoid process is neither excavated nor medially or laterally well-marked. Rather, the medial border is a ridge-like structure that rapidly blends to the shaft proper, whereas the lateral border appears to have been the anterior border of the eroded radial notch. The superior part of the interosseous crest is a well-developed ridge-like structure, but distally it rapidly merges into the shaft proper and disappears. Posterior to the superior part of the prominent portion of the interosseous crest, there is another elevated, tubercle-like structure. A shallow longitudinal groove

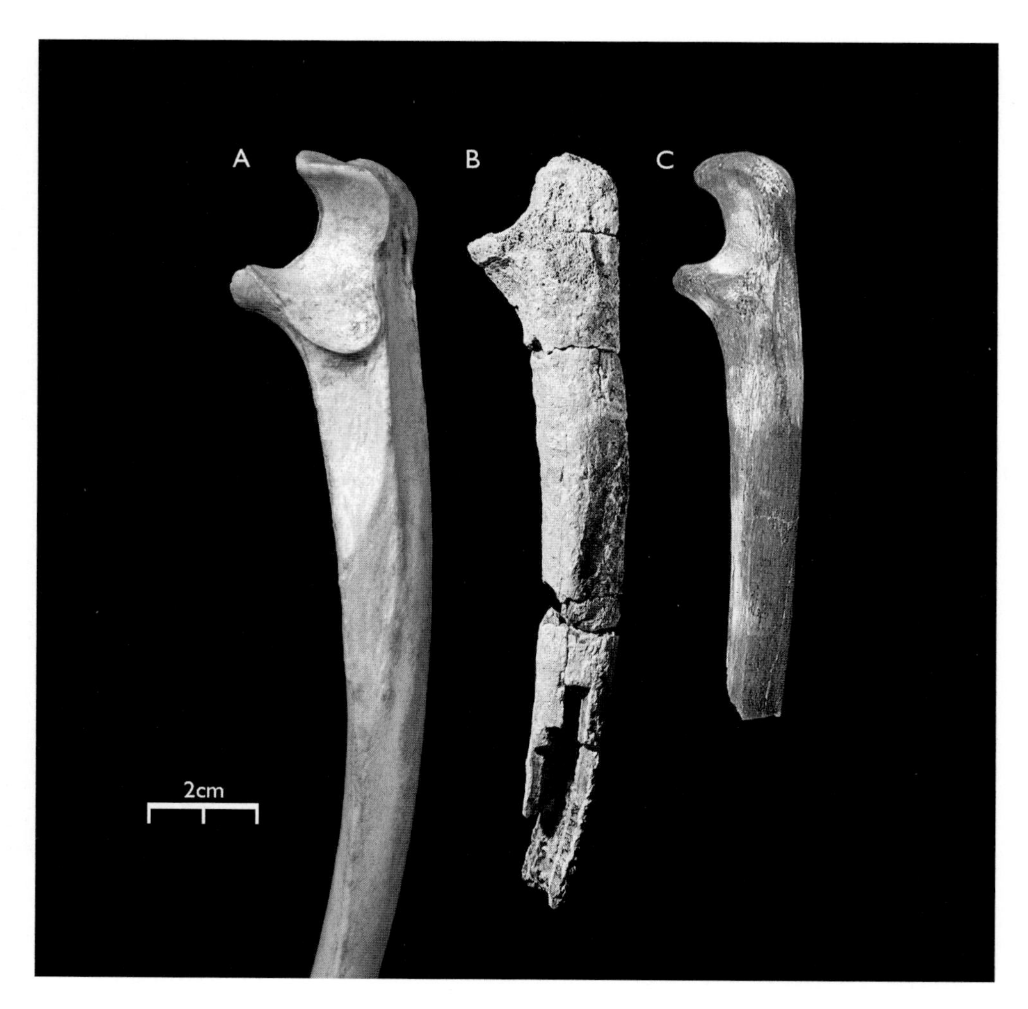

FIGURE 7.23
A. Lateral view of *Pan troglodytes* ulna (cast, reversed). B. Lateral view of the ALA-VP-2/101 left ulna. C. Lateral view of the A.L. 288 ("Lucy") ulna. See text for details.

is formed between it and the superior part of the interosseous crest. In posterior view the proximal part of the preserved shaft is rounded, whereas distally it gets flatter.

The associated (by spatial proximity and preservational characteristics) humerus is a shaft fragment from an adult left humerus. It is broken proximally below the deltoid tuberosity and lacks the distal end and some portion of the shaft immediately above it. Its preserved length is 80 mm. Cortex on both the proximal and distal broken surfaces is slightly abraded due to weathering. However, the lateral supracondylar ridge is well-developed and curves posteriorly toward its proximal end. The medial border is rounded. The broken proximal end is circular in cross section, whereas the distal broken edge has a triangular cross section and a flat posterior surface. The cortex is evenly distributed on the broken proximal surface. The broken surface at the distal end has thicker cortex on its lateral side.

ASK-VP-3/78 is a left distal adult humerus preserving most of the trochlea, part of the medial epicondyle, the entire olecranon fossa, and some of the distal diaphysis. The capitulum and inferomedial part of the trochlea are missing. There is some surface exfoliation on the anterior face of the preserved shaft immediately above the coronoid and radial fossae, but the overall bone surface is superbly preserved. The preserved length of the humerus is 105.1 mm. The maximum breadth at the preserved distal end is 42.8 mm.

Both medial and lateral supracondylar ridges are weak. The olecranon fossa is deep but not perforated. There are numerous foramina on the surface of the olecranon fossa. The lateral trochlear crest is well-developed. The medial aspect of the proximal edge of the trochlear joint surface shows arthritic subchondral erosion and bony lipping that has obscured the inferomedial edge of the coronoid fossa. The radial fossa is preserved intact and is shallower than the coronoid fossa. These two fossae are separated by a prominent ridge. The medial and lateral supracondylar ridges are equally developed. The cross section of the diaphysis at the broken edge is triangular. At the point of fracture, the medullary cavity is round in cross section.

The medial edge of the trochlea of ASK-VP-3/78 is unique in that it sweeps inferolaterally. Such a pattern is not seen in apes; for example, it sweeps either inferomedially or vertically in chimpanzees. The pathological condition (osteoarthritis) seen on this humerus is also uncommon among extinct and living hominoids in its placement and manifestation. There is an eburnated area (5.8 mm long and 10.8 mm wide) positioned on the inferior edge of the coronoid fossa (Plate 7.10). There is also some lipping on the superior edge of the eburnated area and some perforation of the articular surface.

The broken biepicondylar breadth is 52 mm, with an estimated total original breadth of 55 mm. Immediately above the olecranon fossa the humerus is 33 mm mediolaterally and 13.7 mm anteroposteriorly. The minimum breadth of the shaft at the proximal end of the specimen is approximately 22 mm. The medial and lateral pillar thicknesses measured mediolaterally immediately above the broken medial and lateral epicondyles are 10.1 and 13.9 mm, respectively. The breadth of the olecranon fossa is 20.6 mm. The medial trochlear crest height is 17.6 mm, and the minimum trochlear sagittal diameter is 13.3 mm. Thickness of the shaft cortex at the broken proximal part varies from 3.8 mm on the lateral side to 4.0 mm on the medial side. The posterior cortex thickness can only be estimated to ca. 4.0 mm. The lateral corner is twice as thick as the lateral face of the shaft. The diameter of the medullary cavity at the broken proximal edge is 7.7 mm.

The arm bones (ALA-VP-2/101) best match in overall size to smaller individuals of *Au. afarensis* such as A.L. 288-1, A.L. 322-1, and A.L. 147-38a (Figure 7.23). The ulna of ALA-VP-2/101 is absolutely smaller than ARA-VP-7/2 (associated arm bones of *Ar. ramidus;* White et al. 1994, 1995) and A.L. 438-1 (Drapeau et al. 2005). Morphologically, the ALA-VP-2/101 ulna has lost most of its surface cortical bone, and significant anatomical comparisons therefore cannot be made. However, the anterior concavity of the preserved shaft is similar to that of A.L. 288-1T and ARA-VP-7/2. Ulnar morphology of other late Miocene and early Pliocene hominids, such as *O. tugenensis, S. tchadensis,* and *Au. anamensis,* is currently unknown.

ASK-VP-3/78 is morphologically similar to similar-sized humeri of *Au. afarensis* (A.L. 322-1, A.L. 137-48a), smaller than ARA-VP-7/2 (*Ar. ramidus;* White et al. 1994, 1995) and MAK-VP-1/2 (*Au. afarensis;* White et al. 1993), but larger than A.L. 288-1m (*Au. afarensis;* Johanson et al. 1982a). Curvature of the medial epicondylar ridge of ASK-VP-3/78 and straightness of the lateral crest above the trochlea (also seen in *Au. afarensis*) match with *O. tugenensis* (BAR 1004'00; Senut et al. 2001) and *Ar. ramidus* (ARA-VP-7/2). The preserved lateral crest of KNM-KP 271, a distal humerus assigned to *Au. anamensis* (Ward et al. 2001), tends to flare laterally toward the lateral epicondyle. Olecranon fossa morphology of ASK-VP-3/78 differs from the condition seen in other hominids in its steep lateral and medial walls and slightly furrowed inferior margins. It is similar to KNM-KP 271 in this regard. However, the latter specimen differs from ASK-VP-3/78 in having extremely thick cortical bone (Ward et al. 2001). The morphology of the olecranon fossa in early hominids has been a subject of considerable functional speculation, summarized in the "Postcranial Discussion" section.

Clavicle (STD-VP-2/893; Plate 7.10)

STD-VP-2/893 is a mostly lateral fragment of a left clavicle. Its preserved mediolateral length is 53.1 mm. The acromial end is missing, and the clavicle is broken obliquely immediately medial to the attachment area of the deltoid muscle. It shows no abrasion and retains most of its surface morphological features. The conoid tubercle is well-developed. The ridge-like trapezoid line runs from its lateral end toward the acromial end. It is positioned posteriorly at the center of where the clavicle shaft bends toward the acromial end. The anteroposterior and inferosuperior dimensions at the mid-conoid tubercle level are 14.3 mm and 10.2 mm, respectively. Immediately anterior to the trapezoid line there is a deeply furrowed sulcus. The lateral end of the subclavian groove is also preserved medial to the conoid tubercle. The preserved clavicle also has a well-marked, mediolaterally elongated rugosity for the attachment of the deltoid muscle. Another ridge-like rugosity extends from the lateral end of the deltoid attachment towards the anterior edge of the acromial end. The preserved part of the shaft medial to the deltoid attachment is round in cross section, whereas the area towards the acromial end is superoinferiorly flattened.

STD-VP-2/893 shows a mosaic of features, most of which are more human-like than *Pan*-like. Its overall size and robusticity are on the lower range of human clavicles and higher end of *Pan troglodytes,* and within the range known for *Au. afarensis.* Among known early hominid fossil clavicles, the preserved morphology of STD-VP-2/893 best matches the described morphology of A.L. 438-1v (Drapeau et al. 2005), but its size, based on dimensions at the conoid tubercle, is smaller (Drapeau et al. 2005). STD-VP-2/893 is much larger and more robust than A.L. 333x-6/9 (Lovejoy et al. 1982). The latter has less pronounced muscle attachment areas. The cross section of the preserved part (including at the conoid tubercle) and positioning of the muscular attachment areas are more like those of *Au. afarensis* and modern humans than those of *Pan troglodytes.* However, STD-VP-2/893 appears to have a more developed conoid tubercle than those seen in A.L. 333x-6/9 and A.L. 438-1v. Among the early hominids, the presence of a well-developed conoid tubercle in STD-VP-2/893 could indicate a well-reinforced joint between the clavicle and the coracoid process of the scapula. However, the small and fragmentary nature of the sample does not allow valuable functional interpretation.

Pedal Phalanx (AME-VP-1/71; Plate 7.8)

AME-VP-1/71 is a complete left fourth proximal foot phalanx with minor abrasion damage around the condyles (Haile-Selassie 2001b). Ray position of this proximal phalanx was determined by comparison to the complete A.L. 333-115 foot proximal rays (Figures 7.24 and 7.25), and overall morphology of the proximal articular surface. The distal articular

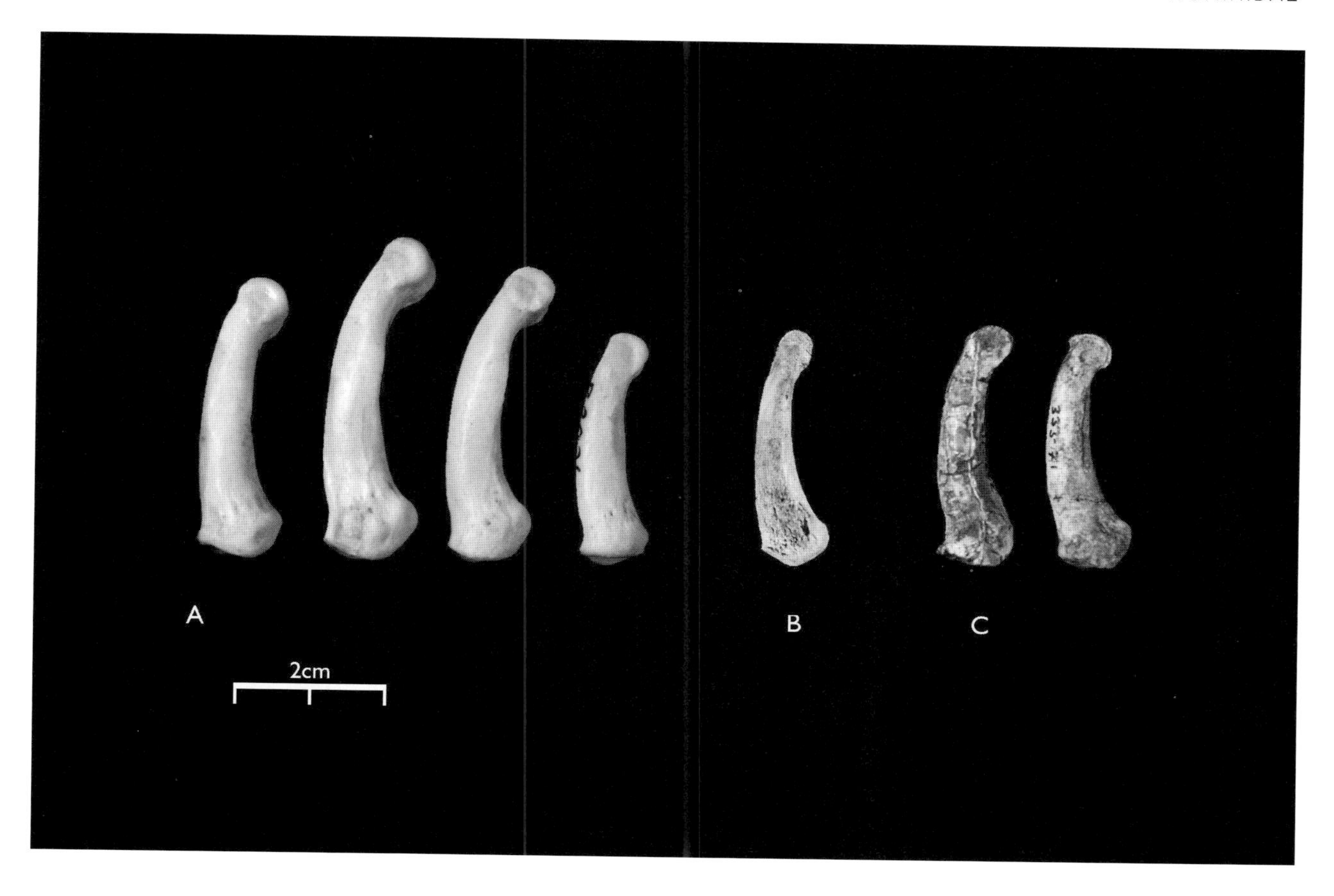

FIGURE 7.24

Side views of proximal foot phalanges. A. *Pan troglodytes,* ray 2 to the left and ray 5 to the right. B. Original AME-VP-1/71 phalanx, identified as a fourth ray. C. Casts of fourth proximal foot phalanges from the articulated Hadar *Australopithecus afarensis* foot (A.L. 333-115) and an isolated specimen (A.L. 333-71). See text for details.

surface has a trochlear form with a distinct groove between the condyles. In lateral view, the shaft is curved (plantar curvature). The distal half of the shaft is dorsoventrally compressed, whereas the proximal half is mediolaterally compressed with a prominent constriction above the base. The proximal articular surface is circular in profile and concave, with its lateral border projecting proximally more than its medial border. The proximal articular surface faces slightly proximodorsally.

The plantar tubercles for the collateral ligaments of the metatarsophalangeal articulation, particularly the medial one, are well-developed. The shaft displays on its ventral surface a strong ridge-like longitudinal structure that runs from midshaft level to the medial plantar tubercle. The crests for the fibrous flexor tendon sheath are not strong but present at midshaft level.

The comparative sample from hominid taxa older than 3 Ma is very limited. There is only one fourth proximal foot phalanx (A.L. 333-115I) positively identified and described for *Au. afarensis* (Latimer et al. 1982); there are numerous proximal foot phalanges known for *Au. afarensis* but indeterminate as to ray position (Latimer et al. 1982; Latimer and Lovejoy 1990). Hence, these specimens are less useful for comparative analysis. AME-VP-1/71 is similar to *Au. afarensis* homologs such as A.L. 333-71 and A.L. 333-115I in most of its morphological aspects, including the proximal articular surface, which faces proximodorsally, and the amount of phalangeal curvature. How-

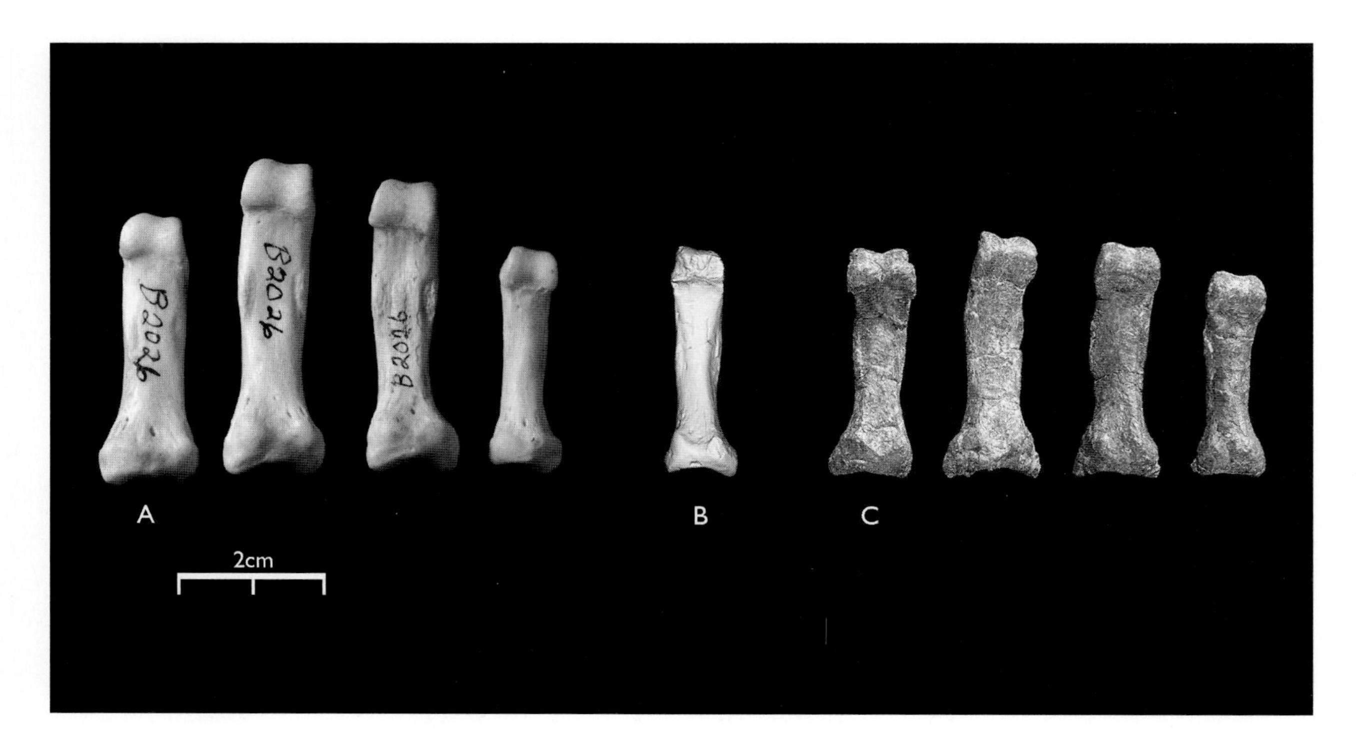

FIGURE 7.25
Plantar views of proximal foot phalanges. A. *Pan troglodytes,* ray 2 to the left and ray 5 to the right. B. Micro-ct rendering of the AME-VP-1/71 phalanx, identified as a fourth ray. C. Ray 2 through ray 5 phalanges of the articulated Hadar *Australopithecus afarensis* foot (A.L. 333-115).

ever, it is slightly smaller than either Hadar specimen in almost all dimensions. AME-VP-1/71 seems to have more dorsal canting of the proximal articular surface than the two *Au. afarensis* specimens, even though this difference is probably subsumed within the range of variation for the species. Compared to fourth proximal phalanges of *Pan troglodytes,* AME-VP-1/71 is much shorter in its length (Figures 7.24 and 7.25). AME-VP-1/71 also has a less curved diaphysis, and *P. troglodytes* lacks the dorsal canting of the proximal articular surface seen in AME-VP-1/71. The latter character is associated with toeing-off, a character uniquely associated with terrestrial bipedality (Latimer and Lovejoy 1990).

In his *Science* commentary on late Miocene hominids, Begun addressed the *Ardipithecus* phalanx from Amba East as follows (Begun 2004: 1479):

> *A. kadabba* is interpreted as a biped on the basis of a single toe bone, a foot proximal phalanx, with a dorsally oriented proximal joint surface, as in more recent hominins (6). However, the same joint configuration occurs in the definitely nonbipedal late Miocene hominid *Sivapithecus* (13), and the length and curvature of this bone closely resembles [*sic*] those of a chimpanzee or bonobo.

We are confused with this assertion. Begun's citation (13) is a paper by Rose (1986) in which three *Sivapithecus* phalanges (one with the base) are described as *hand* phalanges, not foot phalanges. We wonder whether Begun has re-identified the *Sivapithecus* phalanges as pedal (and therefore believes that he is comparing homologous elements), or whether he has misremembered the original description of these fossils. In either case, the *Ardipithe-*

cus foot phalanx appears derived in the hominid direction relative to any ape with whose phalanges we are familiar.

Postcranial Discussion

Postcranial remains attributed here to *Ar. kadabba* are limited to fragmentary elements of the forearm, two hand phalanges that are indeterminate to ray, a fragment of a clavicle, and a toe bone. However, compared to other contemporaneous, or slightly older, hominids such as *Sahelanthropus tchadensis* and *Orrorin tugenensis,* postcranial remains of *Ar. kadabba* illuminate previously unknown parts of the skeleton (e.g., intermediate hand phalanx, clavicle, and proximal foot phalanx). In this section, we highlight some of the major contributions of the remains of *Ar. kadabba* to our overall understanding of skeletal morphology of early hominids older than 4.5 Ma. We also address some previous morphological interpretations of early hominid postcrania in light of the new data derived from *Ar. kadabba.*

The curvature of the hand and foot proximal and intermediate phalanges of *Ar. kadabba* is similar to, and within the size and morphology ranges of, the known *Au. afarensis* skeletal element samples. The morphology and size of the arm bones also fit within the range seen for *Au. afarensis* and *O. tugenensis* ($n = 1$). This, in addition to the dentognathic evidence described above, further confirms the hominid status of *Ar. kadabba.*

The functional and phylogenetic importance of phalangeal curvature is a hotly debated issue in human evolutionary studies. All early hominids *(Orrorin tugenensis, Ar. ramidus, Au. anamensis, Au. afarensis),* African apes, and other primates have variably curved proximal hand and foot phalanges. The functional and evolutionary significance of curved phalanges in bipedal hominids has been interpreted by different workers who emphasize different things. Some propose functional correlation between curved phalanges and arboreality for early hominids (Stern 2000), whereas others recognize such characters as primitive retentions from a common ancestor with little or no functional value (Latimer et al. 1987).

Based on a single proximal hand phalanx and a distal humerus, Senut et al. (2001: 142) concluded that "it appears that *Orrorin* was adapted to arboreal activities." Pickford et al. (2002: 202) further elaborated Senut et al. (2001)'s conclusion by stating that "although *Orrorin* was bipedal, the morphology of its humeral shaft and the curvature of the manual phalanx, reveals that it was probably also capable of climbing trees." On the other hand, Ward et al. (2001) described the single known proximal hand phalanx of *Au. anamensis* from Kanapoi as similar to *Au. afarensis* but made no such locomotor inferences. We concur with that conservative view and suggest that the inference of arboreality in hominids based only on the curvature of isolated phalanges is unwarranted.

The arm bones of *Ar. kadabba* (ASK-VP-3/78 and ALA-VP-2/10) are, in general, similar to all other known early hominids in the midshaft curvature of the ulna and overall morphology of the distal third of the humeral shaft. The lateral wall of the olecranon fossa in ASK-VP-3/78 is relatively steep, the fossa is deep, and the lateral trochlear crest is oriented vertically or slightly inferomedially. Some of these characters are found more often in extant great apes than in early hominids or modern humans. However, some *Au. afarensis* specimens (e.g., A.L. 137-48A) show similar morphology.

The distal epiphysis and the olecranon fossa (its lateral margin and depth) of early hominid humeri have been widely discussed and various functional interpretations generated (Senut 1978, 1981; Senut and Tardieu 1985; Hill and Ward 1988; Leakey et al. 1995; Baker et al. 1998). Senut (1978, 1981) and Senut and Tardieu (1985) examined the development of the lateral trochlear crest and the position and degree of projection of the lateral epicondyle in an attempt to compile discrete and phylogenetically important morphological characters of the distal humerus. Although these studies resulted in the recognition of two "morphotypes" (human-like and ape-like), the assignment of specimens belonging to the same species to two different morphotypes reduces the validity of the method (see Hill and Ward 1988 for a critique). Moreover, although the olecranon fossa might be informative for functional interpretations, Senut and Tardieu (1985) did not discuss the morphology or phylogenetic significance of the olecranon fossa except to mention that their "robust" subgroup (one of the two ape-like submorphs) has a qualitatively shallow olecranon fossa.

There appears to be substantial variation in the depth of the olecranon fossa and expression of its lateral wall, particularly among early hominids. For example, there is substantial variation in the morphology of the olecranon fossa within the relatively well-represented *Au. afarensis* humeri (indeed, ASK-VP-3/78 also falls within this range), indicating that the significance of these features in phylogenetics or understanding of locomotor behavior in early hominids is minimal. Baker et al. (1998) suggested that the reduction in the steepness of the lateral wall of the olecranon fossa characterized all Plio-Pleistocene bipedal hominids compared to other Miocene hominoids. Early and mid- to late Miocene hominoids such as *Proconsul, Kenyapithecus,* and *Nacholapithecus* have markedly strong lateral walls of the olecranon fossa (Le Gros Clark and Leakey 1951; Leakey 1962; Pickford 1985; Conroy et al. 1992; McCrossin 1994; Nakatsukasa et al. 1998; Ishida et al. 1999, 2004; Takano et al. 2003; Nakatsukasa 2004). The locomotor behavior (quadrupedality) of these hominoids is also relatively well-understood based on their more complete skeletal remains. However, none of the known early hominid humeri has olecranon fossa morphology similar to that of these Miocene apes.

Not only do the postcranial remains of *Ar. kadabba* document previously unknown body elements for hominids older than 4.5 Ma, but their overall similarity with much younger Pliocene hominids, in addition to previous observations published on the 5.2 Ma AME-VP-1 toe bone (Haile-Selassie 2001b), corroborate previous conclusions that this species is not simply a dental hominid. However, the postcranial representation remains poor despite intensive and continuing field recovery efforts, and as a result, unequivocal evidence for habitual bipedality in *Ar. kadabba* remains to be ascertained by further discoveries in the Middle Awash or elsewhere (White 2006).

8

Carnivora

YOHANNES HAILE-SELASSIE AND F. CLARK HOWELL

Terrestrial carnivores are traditionally divided into two groups based on the morphology of the auditory region: the Arctoidea and the Canoidea. The earliest terrestrial carnivores were small, viverrid-like creatures, appearing in the fossil record during the Eocene (Martin 1989). The creodonts appeared in Africa sometime during the early Oligocene. The earliest true carnivores did not appear in Africa before the early Miocene. During the early and middle Miocene, a number of true carnivore groups appeared, including the amphicyonids, viverrids, herpestids, felids, and nimravids (Martin 1989). Sometime during the late Miocene and early Pliocene, groups such as ursids, canids, and mustelids entered Africa from Eurasia. As a result, the late Miocene fossil record of large and small carnivores in Africa is relatively speciose and represents the basis of the modern African carnivore community.

The Middle Awash carnivore fauna represents one of the most diverse in the latest Miocene of eastern Africa. At least 15 genera in seven families are recognized from the Asa Koma Member of the Adu-Asa Formation and the Kuseralee Member of the Sagantole Formation. They are described in this chapter and tabulated in Table 1.3.

Ursidae

Agriotherium Wagner, 1837

Agriotherium sp.

DESCRIPTION KUS-VP-1/17 (Figure 8.1A) is a complete right femur slightly crushed below the greater trochanter. The shaft is well-preserved, and the distal end is slightly abraded. The fovea capitis is large. The neck is short, and the muscle attachment area below the greater trochanter extends distally below the level of the lesser trochanter. The lesser trochanter is relatively small. The femur head is positioned much higher than the greater trochanter. The shaft is anteroposteriorly compressed. The medial condyle on the distal end is extended posteriorly. All these features are ursid features and show that KUS-VP-1/17 is definitely an ursid species larger than *Indarctos*. Moreover, morphology

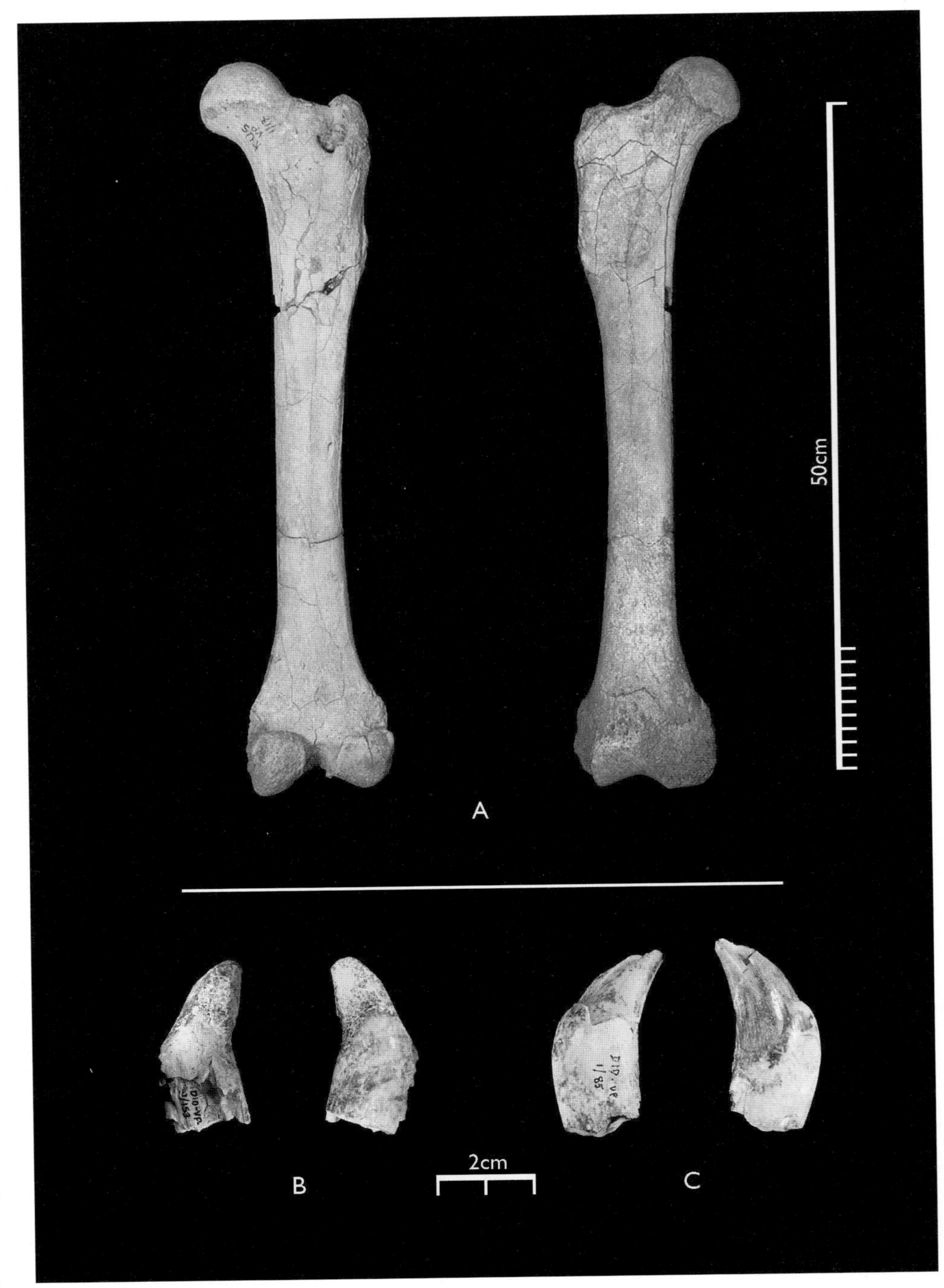

FIGURE 8.1

Middle Awash late Miocene specimens assigned to *Agriotherium* sp. A. KUS-VP-1/17, anterior and posterior views of right femur. B. DID-VP-1/153, lower canine fragment. C. DID-VP-1/85, right upper canine.

of the proximal end of the femur, added to its overall size best fits the genus *Agriotherium*. The total length of the femur is 530 mm.

DISCUSSION One of the ursid subfamilies, Agriotheriinae, is represented at the beginning of the Vallesian (early late Miocene) by the genus *Indarctos* in Europe (Thenius 1959). The subfamily includes at least four genera, *Ursavus, Ballusia, Indarctos,* and *Agriotherium* (Kurten 1966; Ginsburg and Morales 1998). Some workers recognize only the latter two genera as members of Agriotheriinae (Petter and Thomas 1986, after Thenius 1979, 1982). There are a number of species recognized in the genus *Indarctos (Indarctos salmontanus, I. vireti, I. arctoides, I. atticus, I. anthracitis, I. bakalovi,* and *I. lagrelii),* and the genus *Agriotherium* also contains numerous species identified from Eurasia, North America, and Africa. This genus seems to have undergone maximum radiation toward the end of the Miocene. The phylogenetic relationships between *Indarctos* and *Agriotherium,* and the taxonomic status of some of the species in the latter genus, have been contentious (Hendey 1977, 1980; Qiu and Schmidt-Kittler 1983; Petter and Thomas 1986). Hendey (1980) proposed an ancestor-descendant relationship between *Indarctos* and *Agriotherium* based on dental morphological similarities and larger size of the latter, supposedly a result of an evolutionary tendency toward larger size.

Petter and Thomas (1986) explained the possible presence of three phases of agriotheriine radiation during the Miocene. In the first phase, *Indarctos arctoides* was the only known species in Europe throughout the Vallesian (11.5–9 Ma) until it was replaced, in the second phase, by *Indarctos atticus,* a species that possibly arose from its predecessor. The third phase took place in the early Pliocene with the contemporaneous appearance of *Agriotherium* in Africa, Europe, Asia, and North America. According to Petter and Thomas (1986), the extinction of *Indarctos atticus* coincided with the proliferation of *Agriotherium,* and this may indicate a replacement of the former by the latter. However, it does not necessarily indicate an ancestor-descendant relationship, since *Agriotherium* was already highly diversified across continents toward the end of the Miocene. Qiu and Schmidt-Kittler (1983) argued that an ancestor-descendant relationship between *Indarctos* and *Agriotherium* (Hendey 1980) is merely based on stratigraphic occurrence and not founded on synapomorphies. Based on their study of *Agriotherium intermedium* from China, Qiu and Schmidt-Kittler (1983) concluded that *Agriotherium* may have descended from a *Hemicyon* group. As a result, the origin and affinity of *Agriotherium* remains uncertain.

There are alternative hypotheses for the phylogenetic relationships of ursids, not to mention the problems related to the position of the giant panda. However, the most parsimonious seems to be their close relationship with the amphicyonids, based on some basicranial synapomorphies (Flynn et al. 1988). What the Middle Awash *Agriotherium* tells us is that the genus was in Africa at least by about 5.2–5.8 Ma. This may support Qiu and Schmidt-Kittler's (1983) suggestion that *Indarctos* and *Agriotherium* do not have an ancestor-descendant relationship. However, the Middle Awash sample is too small to test this hypothesis.

In terms of other occurrences of the genus in Africa, Leakey et al. (1996) and Howell (1982) also reported the presence of *Agriotherium* in the Nawata Formation of Lothagam. However, Werdelin (2003b) reported that *Agriotherium* is absent from the Nawata Formation and instead refers all the specimens previously identified

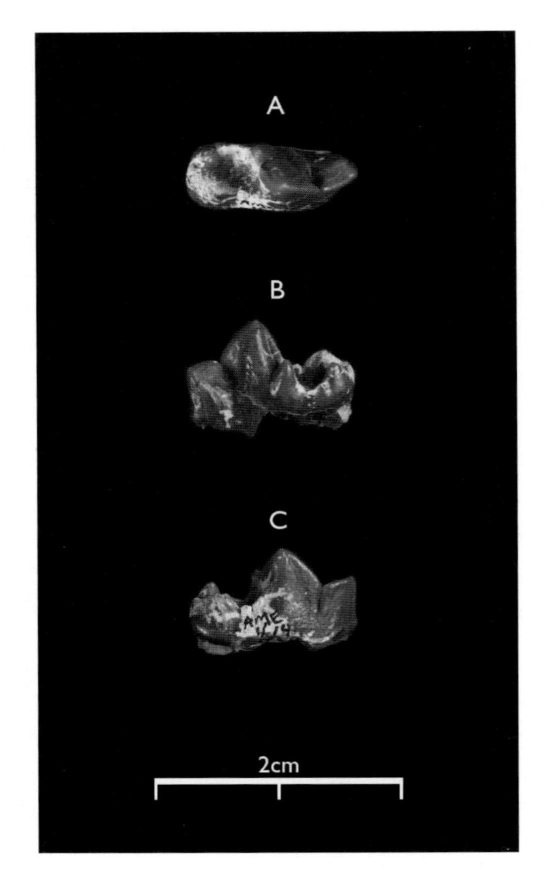

FIGURE 8.2
Specimen AME-VP-1/144 of *Eucyon* sp. A. Occlusal view. B. Lingual view. C. Buccal view.

as *Agriotherium* to Amphicyonidae. The Basal Member of the Nkondo Formation of Uganda has also yielded remains of *Agriotherium* (Petter et al. 1994). The Middle Awash fossil record shows the presence of an ursid species between 5.2 Ma and 5.8 Ma in eastern Africa. Morales et al. (2005) recently named a new *Agriotherium* species of moderate size *(Agriotherium aecuatorialis)* from deposits dated to between 4.5 Ma and 5 Ma at Nkondo, Uganda. However, the referred specimens are all dental remains and, hence, do not help in refining the taxonomic affinity of the Middle Awash specimen.

Canidae

Eucyon Tedford and Qiu, 1996

GENERIC DIAGNOSIS See Tedford and Qiu, 1996.

Eucyon sp.

The tooth AME-VP-1/144 is a right M_1 lacking the root (Figure 8.2) from Amba East Vertebrate Locality 1, Kuseralee Member, Sagantole Formation, 5.2–5.5 Ma.

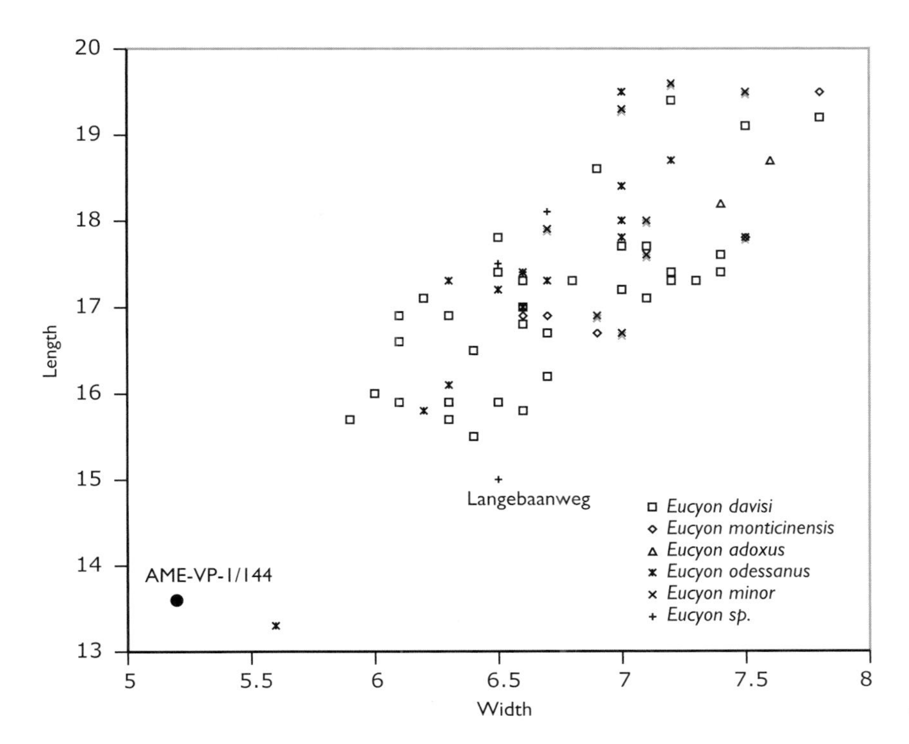

FIGURE 8.3

Plot of length/width relationship of M_1 of various *Eucyon* species.

This represents a *Eucyon* species the size of modern African *Vulpes chama* and smaller than most known species of *Eucyon.* The protostylid is positioned low, distal to the protoconid. There is a well-developed mesoconid, relatively tall and wide protoconid, and metaconid as tall as the entoconid.

The speciman appears to differ from *Eucyon davisi* (Merriam 1911), *E. monticinensis* (Rook 1992), *E. adoxus, E. odessanus* (Odintzov 1967), *E. minor, E. intrepidus* (Morales et al. 2005), and *E.* sp. (Hendey 1974) by its absolutely smaller size (Figure 8.3) and tall metaconid, although *E. intrepidus* is defined on non-comparable elements and may be conspecific.

DESCRIPTION AME-VP-1/144 is a complete right lower carnassial lacking the roots. The crown is 13.6 mm long and 5.2 mm wide. The tips of the trigon cusps do not show any sign of wear. The protoconid is the largest and tallest cusp, followed by the paraconid and the metaconid, respectively. The paraconid is about half as high as the protoconid. The metaconid is small and positioned at the distolingual corner of the protoconid. On the buccal side, a shallow groove runs down from the carnassial notch for about 3 mm. The entoconid and the hypoconid have equal height. The hypoconulid is positioned low at the distal end of a deep talonid. The valley separating the metaconid and the entoconid is deeper than the valley separating the hypoconid from the protoconid. This is a result of the presence of a protostylid between the latter two cusps.

DISCUSSION The genus *Eucyon* was erected by Tedford and Qiu (1996) to refer to the North American Hemphillian canid species previously assigned to *Canis davisi* Merriam 1911. A number of *Eucyon* species were subsequently recognized from North America (Stevens and Stevens 2003) and Eurasia (Odintzov 1967; Morales 1984; Rook 1992; Tedford and Qiu 1996; Qiu et al. 1999; Spassov and Rook 2006). The late Miocene record of canids in Africa is scant. Morales et al. (2005) reported the first occurrence of *Eucyon* in Africa from the late Miocene Kapsomin of the Lukeino Formation, Kenya. However, Hendey (1974) had previously assigned some canid specimens from Langebaanweg to *Canis* sp. (now referred to *Eucyon* by Rook 1992; Morales et al. 2005). *Canis* sp., probably a descendant of *Eucyon,* is reported from a number of African Pliocene sites including Laetoli, Tanzania (Barry 1987) and South Turkwell, Turkana, northern Kenya (Werdelin and Lewis 2000).

The Middle Awash specimen is the smallest of all known *Eucyon* lower carnassials except for one specimen (PIN 390/178) of *E. odessanus* from Odessa, which is slightly shorter but wider than AME-VP-1/144 (Figure 8.3). The M_1 of *E. intrepidus* from the contemporaneous Kapsomin of Lukeino Formation is unknown. However, the size of the M^1 suggests an M_1 much larger than AME-VP-1/144, and the crown morphology, particularly the cusp proportions, also distinguishes the Middle Awash tooth from those of all other *Eucyon* species. It is therefore possible that the new Middle Awash specimen represents a second canid lineage, but additional fossils will be required to demonstrate this.

Mustelidae

Mellivora Storr, 1780

Mellivora benfieldi Hendey, 1978b

Mellivora aff. *benfieldi*

DESCRIPTION The main cusp of the P_4 of AMW-VP-1/40 (Figure 8.4) is apically worn. The anterior and posterior accessory cusps are low but distinct. Part of the distal end of the P_4 overlaps with the anterior buccal side of the M_1. The paraconid of the M_1 is worn down to mid-protoconid level, and the protoconid is also worn on its mesial edge and apex. The talonid is small, worn flat, and looks like a single large cusp rather than a talonid. The preserved length of the P_4 and M_1 are 8.25 mm and 11.22 mm, respectively. The widths of the two teeth are 4.1 mm and 5.1 mm, respectively.

ALA-VP-2/316 is a mesiodistally elongate, unworn left M_1 lacking part of the protoconid at its base on the buccal side. The roots are not preserved. The paraconid and metaconid are not well-defined, and the platform distal to the protoconid represents the talonid. The entire crown is circled by a basal cingulum.

AMW-VP-1/56 is a left lower canine with complete crown and most of the root preserved. The structure and curvature of the root relative to the crown is similar to the Eurasian species *Simocyon primigenius* (Roth and Wagner 1954). In terms of its dimensions, however, it is longer and narrower. The preserved length and breadth of the canine are 13.74 mm and 9.14 mm, respectively. The largest lower canine of *S. primigenius* from

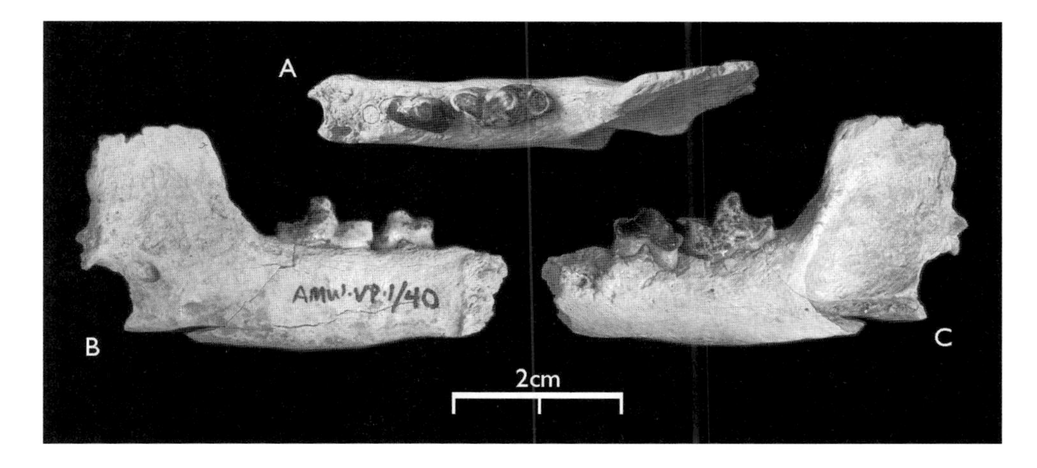

FIGURE 8.4
Middle Awash late Miocene specimens assigned to *Mellivora* aff. *benfieldi*. AMW-VP-1/40, left mandible with P_4–M_1. A. Occlusal view. B. Medial view. C. Lateral view.

Pikermi (AMPG PG 01/104) is 12.60 mm long and 9.90 mm wide. The crown height of AMW-VP-1/56 ranges from 16.70 mm on the lingual side to 18.00 mm on the buccal side. The lingual face of the tooth is flat, with sharp mesial and distal edges.

DISCUSSION Mustelids are one of the most highly diversified groups in the order Carnivora, with 65 extant species in 23 genera (Nowak 1991). However, numerous extinct mustelid genera are also known from the fossil record even though their phylogenetic relationships are poorly understood. The family Mustelidae was initially divided into five subfamilies (Simpson 1945), but recent classification recognizes at least seven subfamilies, exclusive of 18 extinct genera of the family not assigned to any of the modern subfamilies (McKenna and Bell 1997). Mustelids have a very extensive Mio-Pliocene fossil record in Africa. Werdelin (2003b) recently named two new genera and species and one mellivorine of unknown genus and species from the Nawata Formation of Lothagam. Mustelids are also diverse in the Middle Awash late Miocene fossil record and provide further evidence that mustelids were highly diversified in eastern Africa during the late Miocene. The earliest African record of *Lutra* and *Enhydriodon* is documented from Wadi El Natrun, Egypt (James and Slaughter 1974). These genera possibly made their first local appearance in eastern Africa in the late Miocene deposits of the Middle Awash (fragmentary specimens described elsewhere).

AMW-VP-1/40 is similar in size and dental morphology to L42838, the holotype of *Mellivora benfieldi* from Langebaanweg (Hendey 1978b). The species was previously known only from Langebaanweg, and its recovery from the Middle Awash indicates the species' wider geographic distribution during the late Miocene and early Pliocene. *Mellivora benfieldi* differs from *Erokomellivora lothagamensis* (Werdelin 2003b) by the lack of M_2 and overall morphology of the lower carnassial.

Hendey (1974) reported that *M. benfieldi* was possibly conspecific with the Asian Mio-Pliocene *M. punjabiensis*. However, he later concluded that these two species were not conspecific for the lack of P_1 in *M. benfieldi,* among other differences (Hendey 1978b). Hendey further stated that *M. benfieldi* is a better ancestor for the extant *M. capensis* than

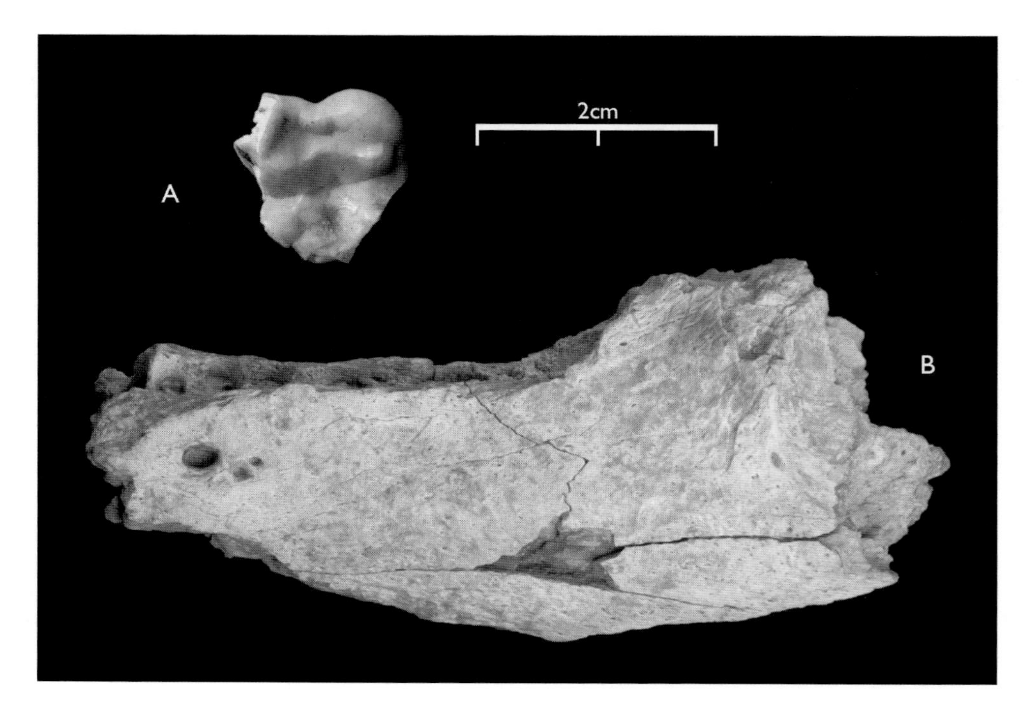

FIGURE 8.5
Middle Awash late Miocene specimens assigned to *Sivaonyx* cf. *africana*. A. ASK-VP-3/62, left P^4 fragment. B. STD-VP-2/4, left edentulous mandible.

M. punjabiensis, a species earlier thought to be an early ancestor of *M. capensis* (Schreber, 1876) (Pilgrim 1932). The Middle Awash species is more likely to be ancestral to the South African *Mellivora benfieldi.*

Sivaonyx Pilgrim, 1931

Sivaonyx africanus Stromer, 1931

Sivaonyx cf. *africanus*

DESCRIPTION ASK-VP-3/62 (Figure 8.5) is the buccal two-thirds of a left P^4. This specimen is fragmentary and most of the diagnostic occlusal morphology is not preserved. However, the metastyle and central occlusal area region are preserved. Based on what is preserved of the crown, the specimen appears to be a large premolar.

STD-VP-2/4 (Figure 8.5) is a left edentulous mandible that has all the mandibular characters referable to Lutrinae of a relatively large size such as *Sivaonyx africanus*. The second molar is single-rooted. The masseteric fossa is deep. Multiple mental foramina are situated below the level of P_3-P_4.

DISCUSSION ASK-VP-3/62 is a relatively large premolar best matched in its preserved size with BAR 1082'01 from the Kapcheberek of the Lukeino Formation (Morales et al. 2005; Morales and Pickford 2005). However, it is difficult to estimate its unbroken size or proportions of its hypocone and metacone. STD-VP-2/4 is similar to L50000 and L9138 from Langebaanweg, referred to *Enhydriodon africanus* by Hendey (1978a) and later referred to *Sivaonyx hendeyi* by Morales et al. (2005) and Morales and Pickford (2005).

Some of the similarities include the anterior extent of the masseteric fossa, the positioning of the M_2, the presence of multiple foramina, and the alveoli placement of the P_2 and P_3 relative to each other. However, STD-VP-2/4 is slightly smaller than L5000.

ASK-VP-3/62 is too small to be assigned to *Enhydriodon* species. STD-VP-2/4 is morphologically very much like *Enhydriodon africanus* from South Africa. However, the paucity of the available material does not allow detailed comparison with other known *Sivaonyx* and *Enhydriodon* species. As a result, following Morales et al.'s (2005) reclassification of the Kenyan and South African material, both Middle Awash specimens are tentatively assigned to *Sivaonyx* cf. *africanus*. This view may change with the recovery of more complete specimens from the Middle Awash, and a detailed assessment of the Kuseralee and Haradaso Member assemblages, in light of the Morales and Pickford (2005) classification.

Lutrinae gen. et sp. indet.

DESCRIPTION ALA-VP-2/106 (Figure 8.6) is a complete left M_1 lacking the mesial root and mesial base of the paraconid. The length and width of the molar are 14.9 mm and 7.8 mm, respectively. The trigonid is 8.76 mm long. The paraconid is the smallest of the trigon cusps. A parastylid is present on the lingual side of the paraconid. The protoconid and metaconid are equal in size. The mesial edges of the protoconid and metaconid form gentle slopes, whereas their distal edges are vertical. These cusps are separated by a V-shaped valley. The talonid is relatively small, and the hypoconid is positioned at the distobuccal corner. The small entoconid is positioned at the distolingual corner.

ALA-VP-2/214 (Figure 8.6) is an upper canine that possibly belongs to the same individual as ALA-VP-2/106. The tip of the canine is worn flat. The lingual face of the canine shows a cone-shaped shear facet, which extends to the root. The crown is slightly curved relative to the root, which has a shallow vertical groove on its distobuccal face.

The complete humerus STD-VP-1/2 (Figure 8.6) is large. The humeral shaft is strongly curved and has an expanded attachment area (large entepicondylar supinator wing) for the pectoralis muscle. The humeral morphological features seen in STD-VP-1/2 indicate aquatic locomotion (Willemsen 1992).

KUS-VP-1/35 is a left lower canine lacking almost one-third of the crown occlusally. The root is massive. The crown is large, measuring 15.2 mm wide at the cervico-enamel junction. The incompleteness of the specimen and absence of contemporaneous canines in the fossil record make specific assignment of this specimen difficult. It is therefore tentatively assigned only at the level of Lutrinae and may be the earliest eastern African *Enhydriodon*.

DISCUSSION Morphologically, ALA-VP-2/106 best matches the lower carnassial of NY 415'87, the holotype of *Torolutra ougandensis* (Petter et al. 1991). However, ALA-VP-2/106 is smaller, and the talonid is relatively much larger. It is larger than *Lutra lutra* and smaller than *Aonyx capensis*. ALA-VP-2/106 and ALA-VP-2/214 might belong to a small *Torolutra* (?*T. ougandensis*) or to a yet-unnamed lutrine genus and species.

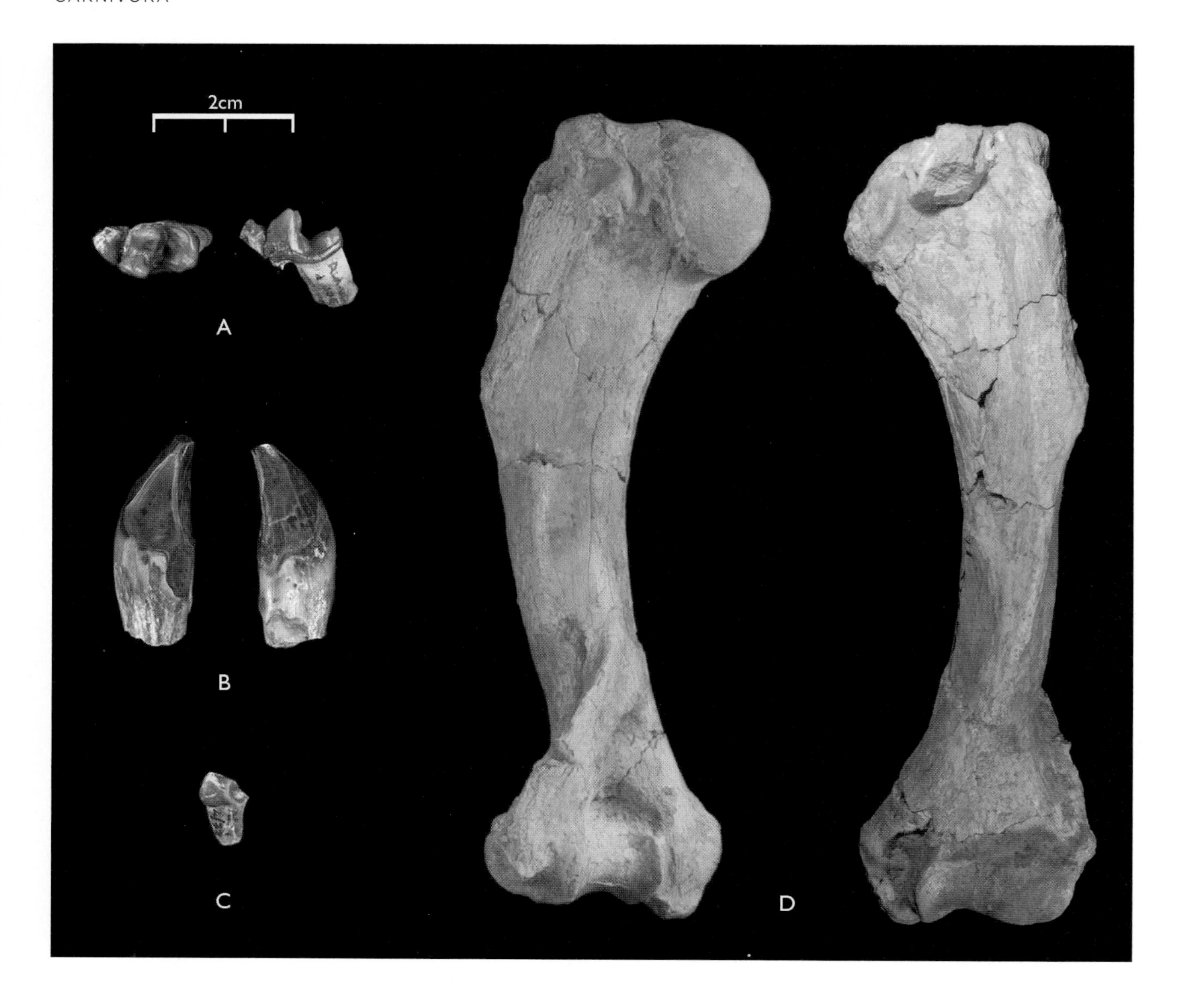

FIGURE 8.6
Middle Awash late Miocene specimens assigned to Lutrinae gen. et sp. indet. A. ALA-VP-2/106, left M_1, occlusal and buccal views. B. ALA-VP-2/214, right C^1, mesial and distal views. C. ALA-VP-2/73, left M_1 fragment. D. STD-VP-1/2, left humerus, anterior and posterior views.

Morphology of the humerus (STD-VP-1/2) is similar with European lutrines such as *Lutra simplicidens* Thenius, 1965; *Sardolutra ichnusae* Malatesta, 1977; and *Lutra trinacriae* Burgio and Fiore, 1988. However, their humeri are much smaller than STD-VP-1/2. The lack of complete fossil humeri of lutrines makes comparison with STD-VP-1/2 difficult. It is therefore tentatively assigned to Lutrinae gen. et sp. indet. However, the specimens assigned here to Lutrinae may represent more than one species.

Plesiogulo Zdansky, 1924

Plesiogulo botori Haile-Selassie et al., 2004

DESCRIPTION ADD-VP-1/10 (Figure 8.7) is a complete left M^1 lacking the buccal root. It was fully described by Haile-Selassie et al. (2004a). The crown is lingually much wider and has an anteroposterior constriction at its center. The paracone is the largest cusp. The labial side of the paracone and metacone presents a platform bound by the labial cingulum. The paraconule is not well-defined but ridge-like, running to the protocone apex. A crest descends from the paracone tip and joins the anterolabial cingulum, which runs labially and posterolabially to terminate at the base of the metacone. Another thin crest runs from the base of the paracone to the protocone, crossing a valley between the two cusps. The talon is well-developed and surrounded by a strong basal cingulum that joins the labial cingulum mesially and distally, forming a complete cingular ring. All measurements of ADD-VP-1/10 are given in Haile-Selassie et al. (2004a).

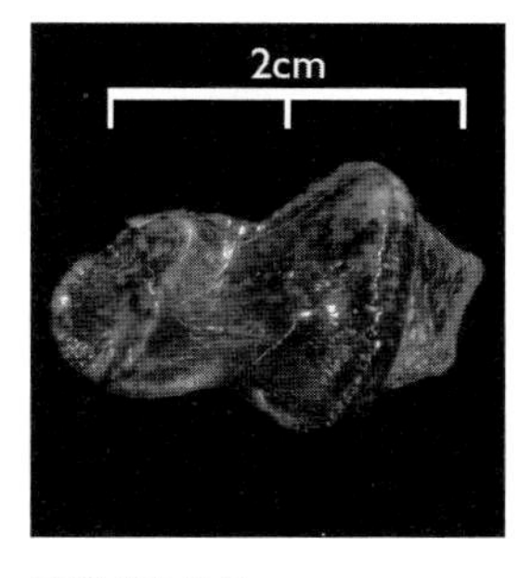

FIGURE 8.7
Middle Awash late Miocene specimen assigned to *Plesiogulo botori*. ADD-VP-1/10, occlusal view.

The occlusal crown morphology of ADD-VP-1/10 is similar to all known M^1s of *Plesiogulo*. It is indistinguishable from the upper molars of *P. crassa* Teilhard de Chardin, 1945, except that ADD-VP-1/10 is much larger. *Plesiogulo praecocidens* Kurtén, 1970 and *P. brachygnathus* (Schlosser 1903) are also smaller in size. The only close size match available for ADD-VP-1/10 is *P. monspessulanus* from Europe, even though the African specimen is still larger than the known size range for *P. monspessulanus*. Upper first molars of *P. monspessulanus* have not been reported from Langebaanweg, and hence ADD-VP-1/10 cannot be compared with the South African *Plesiogulo* (Hendey 1978b).

DISCUSSION The taxonomy and classification of the genus *Plesiogulo* have been discussed in detail by Kurtén (1970). He recognized six species of *Plesiogulo* from the late Miocene–early Pliocene fossil record of Eurasia and Africa. *Plesiogulo praecocidens* and *P. brachygnathus* were known only from the late Miocene deposits of China until Lewis (1934) reported the presence of the latter species in the Siwaliks of India. Barry (1999) has also reported *P. praecocidens* from Abu Dhabi. *Plesiogulo crassa* was previously known only from China until its recovery from the lower Turolian deposits of northern Greece (Koufos 1982). *Plesiogulo major* is documented from China and Europe, whereas *P. monspessulanus* has only been documented from Europe (Viret 1939; Morales 1984; Alcalá et al. 1994) and Africa (Hendey 1978b).

Plesiogulo botori represents one of the earliest records of the genus in Africa. The genus has also been reported from slightly older deposits at Lemudong'o, Kenya (Haile-Selassie et al. 2004a; Howell and Garcia 2007). The only African record of the genus was documented from the earliest Pliocene deposits of Langebaanweg, South Africa (Hendey 1978b). However, the size difference between the largest M^1 of *P. monspessulanus* and ADD-VP-1/10 is less than 20 percent, so a fuller assessment of the relationship between *P. monspessulanus* and *P. botori* is in order.

Viverridae

Genetta Cuvier, 1816

Genetta sp.

DESCRIPTION ADD-VP-1/17 is a right mandible fragment with intact M_1 (Figure 8.8). Dimensions of all dental remains referred here to *Genetta* sp. are given in Table 8.1. The

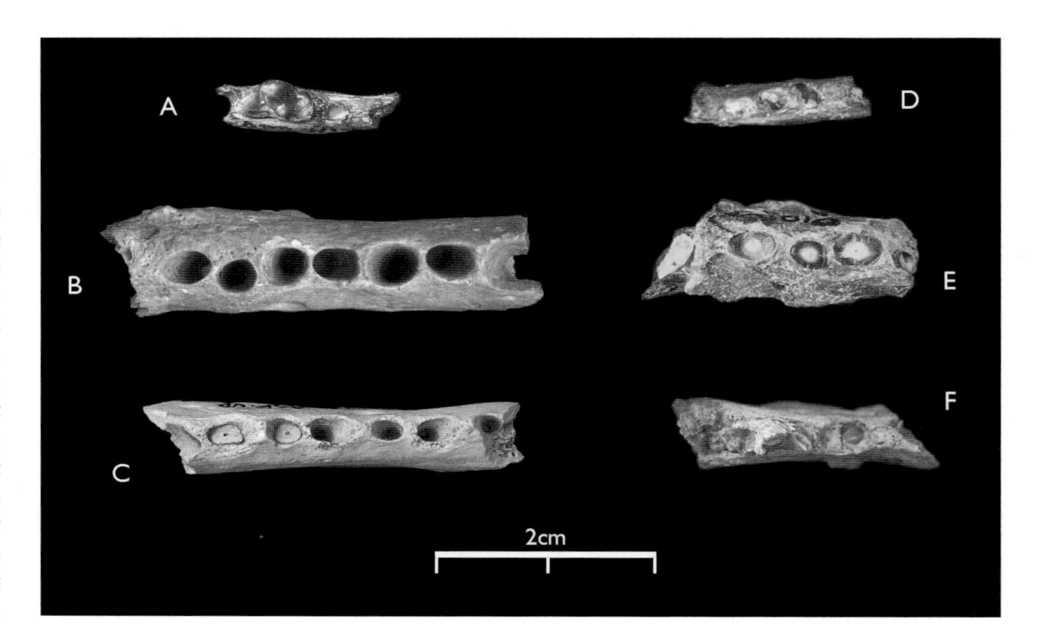

FIGURE 8.8

Occlusal views of Middle Awash late Miocene specimens assigned to *Genetta* sp. A. ADD-VP-1/17, right mandible with M_1. B. ALA-VP-2/79, left edentulous mandible. C. ASK-VP-1/14, left edentulous mandible. D. DID-VP-1/89, left edentulous mandible. E. DID-VP-1/92, left edentulous mandible. F. AME-VP-1/143, left mandible with M_1.

socket for the M_2 is single-rooted. The protoconid is the largest and tallest cusp, followed by the paraconid and the metaconid, respectively. These two cusps are slightly worn on their apices but do not show shear facets on their buccal faces. The metaconid is positioned slightly distal to the protoconid. A basal cingulum is present on the buccal side and extends from the mesial edge of the paraconid to the distal terminus of the talonid. The talonid is positioned lower and is much smaller than the trigonid. The entoconid and the hypoconid are well-developed, with the latter being substantially larger.

AME-VP-1/143 is a left mandible fragment with M_1 (Figure 8.8). The mandibular corpus is preserved from the P_3 level to posterior M_1. The protoconid is the largest and tallest cusp. The molar is similar in its morphological features to ADD-VP-1/17. However, the talonid morphology is different. The entoconid and the hypoconid are of equal size. The hypoconulid is well-defined and is positioned at the distal end of the talonid. There are two cusplets distal to the metaconid, anterior to the entoconid. The talonid is also relatively deeper than that of ADD-VP-1/17.

The four edentulous mandibles, ASK-VP-1/14, DID-VP-1/89, DID-VP-1/92, and ALA-VP-2/79 (Figure 8.8), are referred to *Genetta* sp. largely based on size. Almost all of the mandibles have multiple mental foramina and are relatively deep. None of these mandibles preserves the masseteric fossa region or the ascending ramus.

DISCUSSION Viverrids are the most primitive group in the Feloidea (Aeluroidea; Flynn et al. 1988). They retain primitive morphological features that are possibly associated with their arboreal habitus, and this makes their lower-level classification difficult. Moreover, the family includes lineages that possibly branched off independently from an earlier aeluroid stock (Flynn et al. 1988). Recent molecular systematics of Asian viverrids, excluding mongooses, showed that the subfamily Viverrinae is paraphyletic when both Asian and

TABLE 8.1 Dental and Mandibular Measurements of Specimens Referred to *Genetta* sp. from the Middle Awash Late Miocene Deposits

	M_1		Mandible Depth	
	LENGTH	WIDTH	P_4	M_1
ADD-VP-1/17	7.0	3.6	11.9	12.0
AME-VP-1/143	8.9	3.6	12.0	12.1
ASK-VP-1/14			12.2	nm

NOTE: All measurements are in mm; nm = measurement was not possible.

African viverrines are treated as a group (Veron and Heard 2000). However, the endemic Asian subfamilies Paradoxurinae and Hemigalinae form a monophyletic group to the exclusion of African civets. The African viverrids are also paraphyletic. Some civets of Madagascar, for example, share unique auditory meatus morphology with the herpestines rather than with other viverrines (Petter 1974). The Galidiinae have a specialized bullar morphology that is shared only with herpestines.

Basicranial and ear region morphology is significant in feloid (aeluroid) classification (Hunt 1987; Hunt and Tedford 1993). Hunt and Tedford (1993) argue that despite the broader morphological range, the basicranial morphology of the family Viverridae is uniform, and hence the family is a reliably monophyletic group to the exclusion of herpestines.

Viverrids are among the most poorly known mammalian families in the late Miocene fossil record of Africa. The earlier fossil record of the family shows that it was widespread in Eurasia during the Eocene and Oligocene (Hunt 1996). The African record of viverrids goes as far back as the early Miocene. However, their late Miocene record is limited to fragmentary specimens, and the evolution of modern African viverrids is far from fully understood. Recent discoveries from Lothagam and Langebaanweg and the new discoveries from the Middle Awash (although largely mandibular specimens and isolated teeth) might shed some light on the origin of extant viverrids such as *Genetta* and *Viverra*. Both of these species have been reported from the Lothagam Nawata Formation (Werdelin 2003b).

Some specimens from the Nawata Formation of Lothagam have been referred to *Genetta* sp. A and *Genetta* sp. B (Werdelin 2003b). The lower carnassial of *Genetta* sp. A (KNM-LT 25409) shows features that are different from the carnassial of ADD-VP-1/17 from the Adu-Asa Formation of the Middle Awash. In KNM-LT 25409, the paraconid is the largest cusp of the carnassial, whereas the protoconid is the largest in the case of ADD-VP-1/17. ALA-VP-2/169, a right mandible fragment with a carnassial assigned here to *Herpestes alaylaii* sp. nov. best matches the description for KNM-LT 25409. The paraconid is the largest cusp and has a deep carnassial notch. The presence of multiple foramina is apparent in both specimens. Therefore, KMN-LT 25409 might be misidentified. ALA-VP-2/169 from the Middle Awash and KMN-LT 25409 from Lothagam might be conspecific and may indicate the presence of *Herpestes* in the Lothagam Nawata Formation.

A left mandibular fragment with dentition (L11191) from the early Pliocene deposits of Langebaanweg has been referred to *Genetta* sp. (Hendey 1974). However, this specimen

FIGURE 8.9
AME-VP-1/73, left mandible with erupting M_2, assigned to *Viverra* cf. *leakeyi*. A. Occlusal view. B. Buccal view.

is much smaller than both of the Middle Awash and Lothagam specimens referred to *Genetta* sp. The Middle Awash *Genetta* sp. is also slightly larger than the extant *G. genetta* and *G. tigrina*. One large isolated left M_1 (BAR 155'01) from Kapsomin, Lukeino Formation, has also been referred to *Genetta* sp. (Morales et al. 2005). This clearly shows that the diversity of *Genetta,* or species related to *Genetta,* during the late Miocene of eastern Africa surpasses what has been inferred from the fossil record before now. However, dental plesiomorphy in small viverrids remains a major problem in the systematics of the family.

Viverra Linneaus, 1758

Viverra leakeyi Petter, 1963

Viverra cf. *V. leakeyi*

DESCRIPTION There are no known mandibular specimens of *Viverra leakeyi sensu stricto* (Petter 1963, 1987). AME-VP-1/73 (Figure 8.9) is an edentulous left mandibular piece with the M_2 still in the crypt. It is tentatively assigned to *Viverra* cf. *leakeyi* based on the size and morphological similarities of the M_2 to L12863 (one of the specimens referred to *Viverra leakeyi* from Langebaanweg; Hendey 1974). The M_2 (L = 6.62 mm, W = 4.12 mm) is slightly longer than L12863 (L = 5.9 mm, W = 4.9 mm), although slightly narrower. All four cusps are also well-defined in the AME-VP-1/73 M_2. The mesial cusps are taller than their distal homologs, which are separated by a V-shaped valley.

DISCUSSION Morales et al. (2005) referred BAR 735'02, a right M_2 from Kapsomin, Lukeino Formation, to *Megaviverra* aff. *leakeyi*. This specimen (L = 6.5 mm, W = 5.5 mm) is mesiodistally as long as the Middle Awash M_2, even though it is much wider, giving the molar a squarer occlusal outline than AME-VP-1/73. The latter is also distally narrower. The Lukeino and Middle Awash specimens are also smaller than the Langebaanweg M_2 (L = 7.5 mm, W = 6.2 mm) referred by Petter (1963) to *Megaviverra leakeyi*. However, their morphology is more or less similar, and this makes it difficult to distinguish *Viverra* from *Megaviverra*.

The presence of *Viverra leakeyi* has also been reported from younger deposits in the Omo Shungura Formation of Ethiopia (Howell and Petter 1976) and Laetoli, Tanzania (Petter 1987). Its phyletic relationships with other viverrids, however, remain uncertain.

Herpestidae

Herpestes Illiger, 1811

GENERIC DIAGNOSIS See Illiger, 1811.

Herpestes alaylaii sp. nov.

ETYMOLOGY The species name *alaylaii* is derived from the name of the locality Alayla, where the holotype was found.

HOLOTYPE ALA-VP-2/314 (right mandible with C_1–M_2; Figure 8.10).

PARATYPE ALA-VP-2/169 (right mandible fragment with M_1); ALA-VP-2/315 (right mandible with C_1–P_2 roots and P_3 crown); ALA-VP-2/225 (left M^1); ALA-VP-2/226 (right P^3); ALA-VP-2/229 (right C^1); ASK-VP-3/64 (right mandible with P_4); ALA-VP-2/221 (right proximal ulna fragment).

LOCALITIES AND HORIZONS Specimens assigned to *Herpestes alaylaii* sp. nov. were recovered from Alayla Vertebrate Localities 1 and 2 and Asa Koma Vertebrate Locality 3. They are from the Adu-Asa Formation, Asa Koma Member of the Middle Awash late Miocene. The fossiliferous deposits of the Member are radiometrically dated to between 5.54 and 5.77 Ma (WoldeGabriel et al. 2001).

DIAGNOSIS Lower premolars with proportionately tall main cusp. P_4s with well-developed anterior and posterior accessory cusps. Lower carnassial paraconid and protoconid of equal height and separated from each other by a deep V-shaped valley. Talonid one-third the length of the trigonid and positioned very low. Metaconid noticeably lower than both paraconid and protoconid.

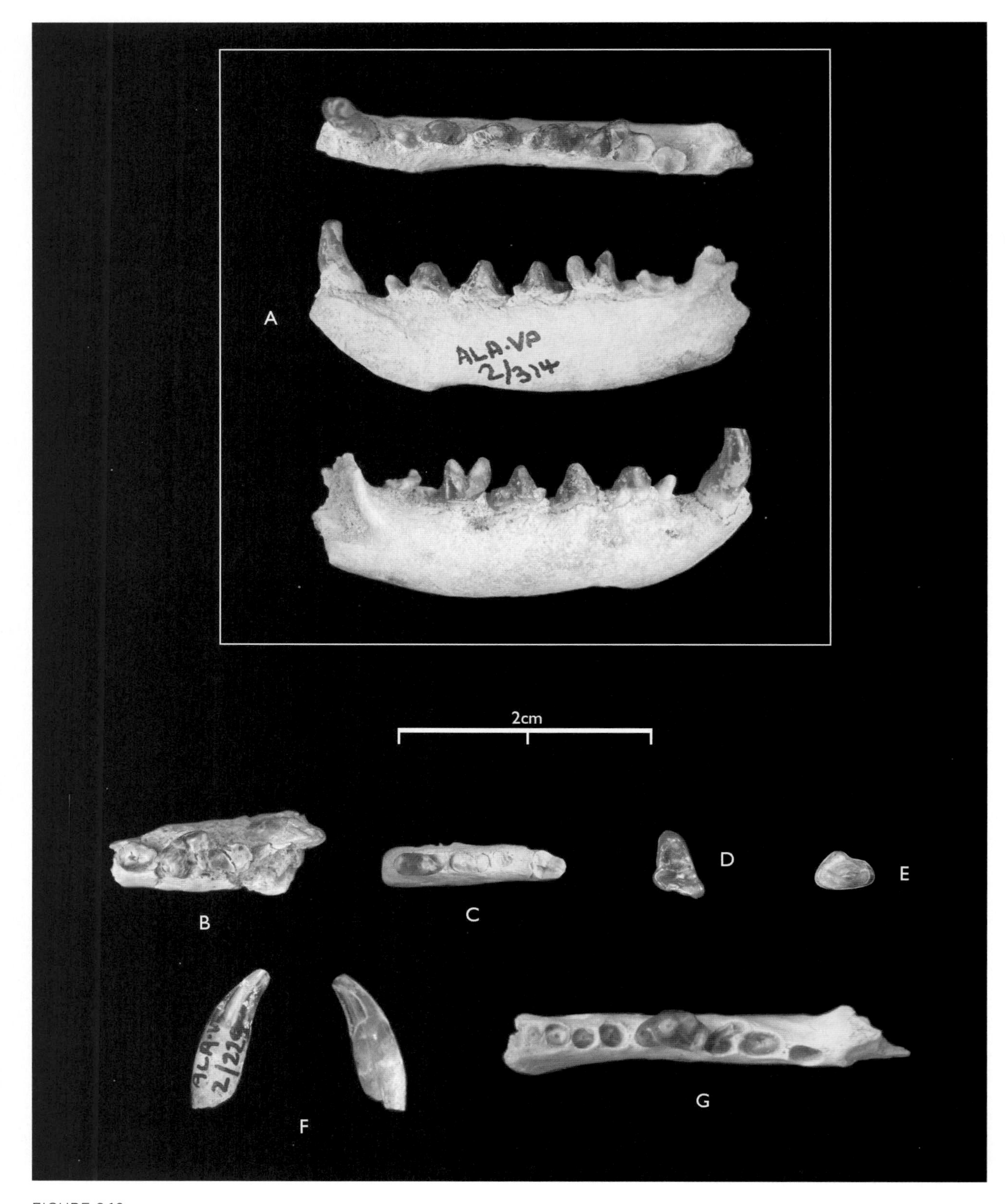

FIGURE 8.10

Middle Awash late Miocene specimens assigned to *Herpestes alaylaii* sp. nov. A. Occlusal and buccal views of the holotype ALA-VP-2/314, right mandible with C–M_2. B. ALA-VP-2/169, right mandible with M_1. C. ALA-VP-2/315, right mandible with P_3 and C–P_2 roots. D. ALA-VP-2/225, left M^1. E. ALA-VP-2/226, right P^3. F. ALA-VP-2/229, right C^1. G. ASK-VP-3/64, right mandible with P_4.

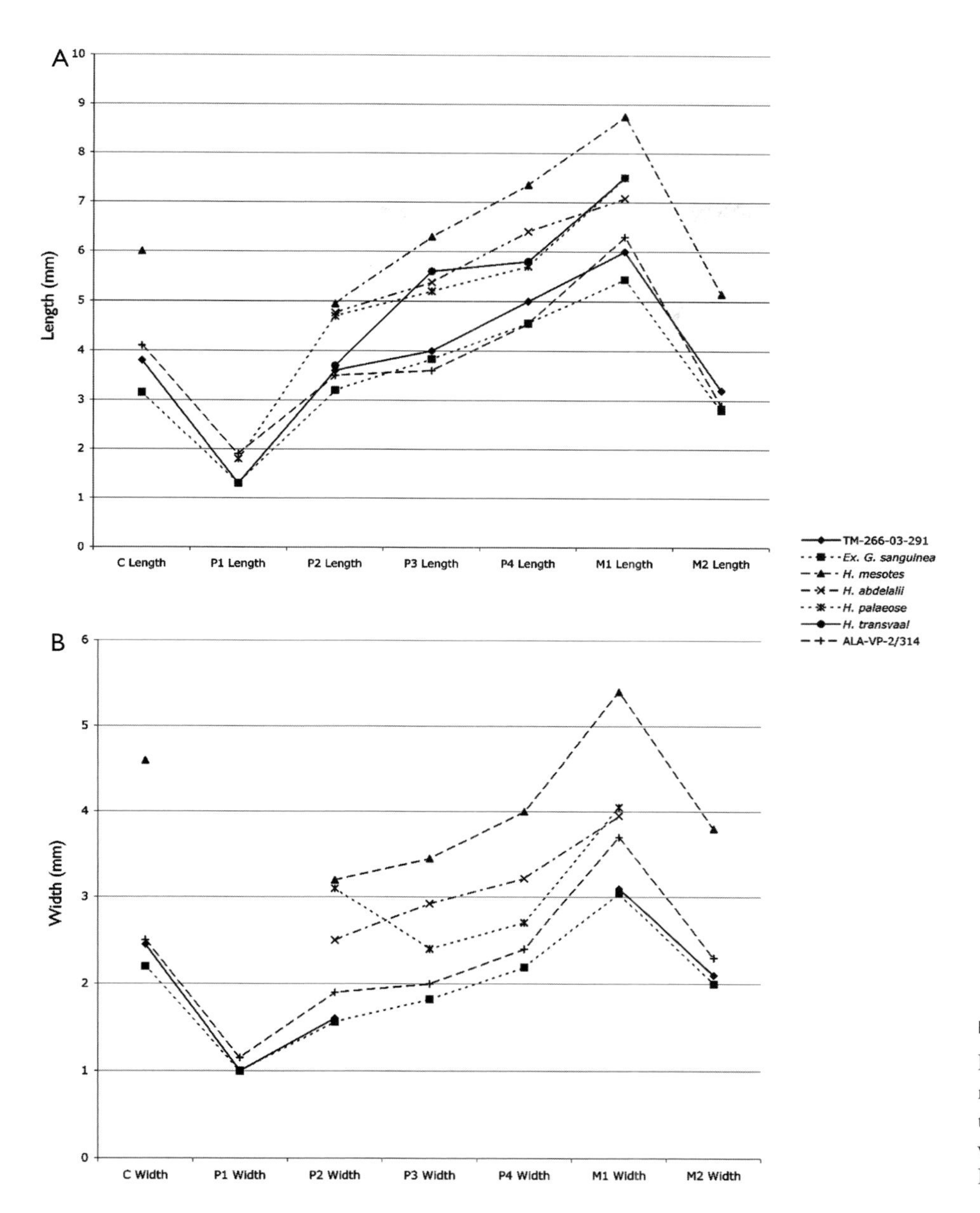

FIGURE 8.11

Plot showing the relationship between lower teeth measurements of various *Herpestes* species. A. Length. B. Width.

DIFFERENTIAL DIAGNOSIS *Herpestes alaylaii* sp. nov. is absolutely smaller than *H.* (*H.*) *ichneumon* (late-middle Pliocene) and *H.* (*G.*) *palaeoserengetensis* from Laetoli (Petter 1987). However, it is much larger than the Langebaanweg *Herpestes* (Hendey 1978b). Figure 8.11 plots the length and breadth relationships of various *Herpestes* species' lower teeth. The morphology of the P_4 also differentiates *Herpestes alaylaii* sp. nov. from all other known *Herpestes* species. *Herpestes alaylaii* sp. nov. is similar to the extant *H. pulverulentus* in its lower canine and carnassial dimensions. However, P_4 morphology and the lack of M_2 in the latter differentiate the two taxa. *Herpestes alaylaii* sp. nov. is different from *Galerella* species by the presence of P_1 and its markedly larger size. *Herpestes alaylaii* sp. nov. is different from *H. transvaalensis, H. abdelalii, H. mesotes,* and *H. palaeoserengetensis* by its markedly smaller size (Figure 8.11) and morphology of the P_4 (the relatively large posterior accessory cusp and a thick posterolingual cingulum form a talonid-like platform). Dental proportions of *H. alaylaii* sp. nov. are very similar to those of extant *G. sanguinea* and TM-266-03-291 (a fossil specimen from Toros Menalla in Chad, dated to ca. 7 Ma and referred to *G. sanguinea,* Peigné et al. 2005b). However, *H. alaylaii* sp. nov. is larger in almost all mandibular dental measurements.

DESCRIPTION Dimensions of dental remains referred to *Herpestes alaylaii* sp. nov. are given in Table 8.2. The holotype ALA-VP-2/314 (Figure 8.6) is the most complete specimen among those referred to the new species. The mandible is a complete right hemimandible lacking the ascending ramus posteriorly. Only the lower part of the masseteric fossa is preserved. The mandibular corpus is relatively shallow and slender. The corpus base behind the posterior edge of the symphysis is slightly convex inferiorly. The canine is tall, convex labially, and concave lingually. It is mesiodistally elongate. The P_1 is a small, peg-like, single-rooted tooth. It is positioned lingually relative to the rest of the tooth row. Its apex is rounded, and its distal edge forms a steep slope ending on a small distal shelf. The entire crown tilts slightly mesially. There is a diastema between the canine and the P_1.

The P_2 is long and slender. The apex of the main cusp is rounded. The anterior accessory cusp is not as well-developed as its posterior homolog. The P_3 is larger than the P_2, and its anterior accessory cusp is also well-defined. The main cusp is taller, and its distal edge forms a step at mid-crown height, descending down to the posterior accessory cusp. The P_4 is the largest of all the premolars, with an expanded posterior cingulum. The anterior accessory cusp is well-developed. The relatively large posterior accessory cusp and the thick posterolingual cingulum form a talonid-like platform.

The M_1 has paraconid and protoconid of equal height. The metaconid is slightly shorter. The three trigonid cusps are separated from each other by relatively deep valleys. The trigonid is labially buttressed due to the position of the protoconid. The talonid is low and positioned more lingually. The talonid cusps are obscured by wear. The M_2 is small and single-rooted. The paraconid is the smallest cusp, situated on the mesial edge of the trigonid. The metaconid is taller than the protoconid. These two cusps are distally connected by a thin ridge. The talonid is wider than the trigonid, and the whole crown narrows distally. The length of the M_2 is slightly less than half the size of the M_1.

ALA-VP-2/169 (Figure 8.10) is a right mandible fragment with M_1. The mandible is broken anteriorly at the P_3/P_4 level. The carnassial is at a relatively late wear stage. The talonid cusps are obliterated by wear. The trigonid is buccolingually wide and almost half the length of the crown. The talonid is buccolingually narrow. The paraconid is the largest cusp and has a deep carnassial notch.

ALA-VP-2/315 and ASK-VP-3/64 (Figure 8.10) are identical in their mandibular morphology to ALA-VP-2/314. The preserved premolars are also similar. The buccal face of ALA-VP-2/225 (left M^1; Figure 8.10) is damaged, although most of the trigon is preserved. ALA-VP-2/225 is much wider than long, and the talon is larger than the trigon. The molar has a triangular occlusal outline and is similar in this regard to extant *Herpestes* species.

TABLE 8.2 Dental Measurements of *Herpestes alaylaii* sp. nov.

	C		*P1*		*P2*		*P3*		*P4*		*M1*		*M2*	
	MD	BL	MD	BL	MD	BL	MD	BL	MD	BL	MD	BL	MD	BL
Lowers														
ALA-VP-2/314	4.1	2.5	1.9	1.15	3.5	1.9	3.6	2.0	4.55	2.4	6.3	3.7	2.9	2.3
ALA-VP-2/315	(3.0)	(2.4)	(1.5)	(1.0)	(3.2)	(1.8)	3.9	1.85						
ALA-VP-2/229	3.3	2.4												
ALA-VP-2/169											(7.4)	3.75		
ASK-VP-3/64									5.9	3.2				
Uppers														
ALA-VP-2/229	3.5	3.2												
ALA-VP-2/226							4.6	3.0						
ALA-VP-2/225											3.9	4.8		

NOTE: All measurements are in mm. Values in parentheses are estimates based on alveoli.

ALA-VP-2/229 (Figure 8.10) is a right C^1 with a slightly worn apex. The root is long relative to crown height. It has a round cross section at its base and tapers apically. It does not have much relief on the crown except for a weak distal ridge running from the tip to the base of the crown. The preserved crown height is 5.7 mm.

ALA-VP-2/226 (Figure 8.10) is a right P^3 crown lacking the lingual median root. The mesiobuccal and distobuccal roots are complete and vertically implanted. The main cusp is symmetrically positioned at the crown center. There is a low lingual median cusp that gives the crown a triangular occlusal outline. This cusp is positioned slightly more mesially. The anterior accessory cusp is small, whereas its posterior homolog is slightly larger. There is a basal cingulum around the crown.

ALA-VP-2/221 is a proximal ulna fragment tentatively assigned to *Herpestes alaylaii* sp. nov. Its size and proximal articular joint surface better match extant *Herpestes* species than any other viverrid known from the Adu-Asa Formation.

DISCUSSION The herpestids possibly diversified only in Africa, with a limited presence in the fossil record of Eurasia after the early Miocene (Schmidt-Kittler 1987). They are first documented in Eurasia from localities as old as 10 Ma. Their first appearance in the African fossil record is documented from Toros Menalla, Chad (Peigné et al. 2005b). The genus diversified in eastern Africa during the Pliocene. While absence of the genus from the Nawata Formation of Lothagam is interesting (this might be due to taphonomic reasons) some terminal Miocene herpestid specimens from the Lukeino Formation have been assigned to *Ichneumia* sp. (Morales et al. 2005).

Some workers include the mongooses as a subfamily of Viverridae (Ewer 1973; Taylor and Goldman 1993; Petter 1987; but see Taylor and Matheson 1999; Hendey 1974), whereas others recognize the mongooses as a family of their own (Herpestidae; Wozencraft 1984, 1989; Hunt and Tedford 1993; Hunt 1996; McKenna and Bell 1997; Veron et al. 2004, among others). Wozencraft (1989) recognized three subfamilies in the family Herpestidae: Galiidinae, Herpestinae, and Mungotinae, whereas Ewer (1973) recognized only Galidiinae and Herpestinae as subfamilies of Viverridae. Veron et al. (2004), on the other hand, recognized these two subfamilies in the family Herpestidae. Species that Wozencraft (1989) included in his subfamily Mungotinae were included by Ewer (1973) in the subfamily Herpestinae. The identification of herpestines is difficult in the fossil record, particularly when specimens are fragmentary and isolated teeth. Morphological distinctions often cannot be established between the herpestines and viverrines. However, since recent molecular systematics recognizes Herpestidae as a family of its own, we follow this classification in this work.

Helogale Gray, 1861

Helogale sp.

DESCRIPTION ALA-VP-2/183 (Figure 8.12A) is a complete right M_2 crown with most of the mesial and distal roots. The crown is at an early wear stage and has a rhomboidal occlusal outline. The trigonid occupies about one-third of the total occlusal crown surface. The protoconid and metaconid are subequal in size. The paraconid is positioned at the

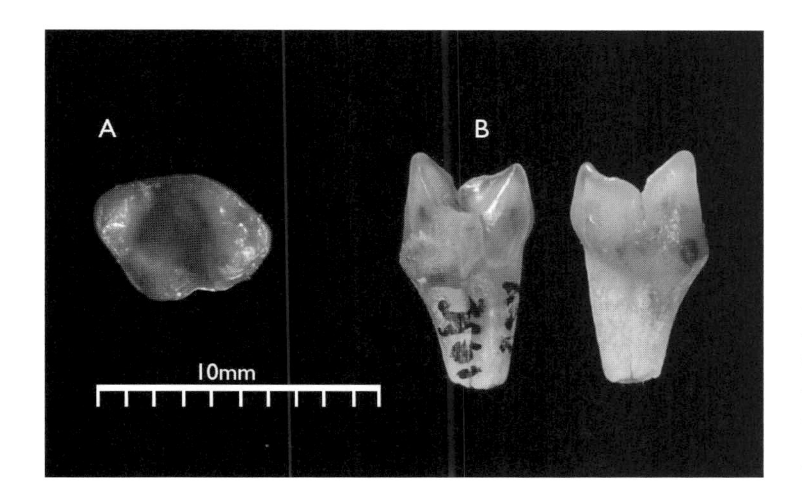

FIGURE 8.12
Middle Awash late Miocene specimens assigned to *Helogale* sp. A. ALA-VP-2/183, right M_2. B. ASK-VP-3/61, trigonid of a left M_1 with root.

center of the mesial crown edge. The talonid is positioned low and has a deep basin that opens buccally. The hypoconid and entoconid are well-developed. The distal part of the talonid is bound by a crest running between the hypoconid and the entoconid. The size of the molar (L = 3.2 mm, W = 2.4 mm) is slightly larger than *Helogale palaeogracilis* from Laetoli (Petter 1987) and within the range observed for the modern *Helogale hirtula*. It is much smaller than the M_2s of all *Mungos* species. Unfortunately no M_2s are known for *Helogale kitafe*.

ASK-VP-3/61 (Figure 8.12B) preserves only the trigonid part of a left M_1 with root. The paraconid is the tallest cusp of ASK-VP-3/61. The protoconid and the metaconid are of approximately equal size. The trigonid is positioned high, judging from the broken edge of the talonid. The trigonid basin is deep, and the metaconid and protoconid are positioned closer to each other than either of them is to the paraconid. The metaconid and protoconid are separated by a deep V-shaped valley. The protoconid and the paraconid show slight wear facetting on their buccal surfaces close to their apices. The length of the root is almost equal to the height of the paraconid. The preserved dimensions are 3.61 mm wide and 4.1 mm long.

DISCUSSION The earliest previously documented record of the genus *Helogale* was from Pliocene deposits of Omo, Ethiopia (Wesselman 1984). The specimens from the Asa Koma Member of the Middle Awash (5.54–5.77 Ma) establish the earliest record of the genus in Africa. ASK-VP-3/61 belonged to an individual larger than the Pliocene *Helogale kitafe* or the extant *H. parvula* and *H. hirtula*. ALA-VP-2/183 fits both metrically and morphologically within the range seen for the larger extant species, *H. hirtula*. The trigonid morphology of ASK-VP-3/61 also seems to be similar to all the extant and extinct species of the genus. ASK-VP-3/61 and ALA-VP-2/183 are older than the oldest known *Helogale* species (*H. kitafe*) by at least 3 million years. However, the lack of adequate sample from the Middle Awash prevents specific assignment for the two specimens below the genus level. The taxonomic status of the Middle Awash specimens may change with the recovery of more material. However, it is possible to state that the size and morphological similarities of the Middle Awash specimens to fossil and extant *Helogale* species might indicate phylogenetic relatedness. The Middle Awash specimens could thus represent the ancestral form for all the Pliocene and more recent *Helogale* species.

Felidae

Machairodus Kaup, 1833

Machairodus sp.

DESCRIPTION AME-VP-1/11 (Figure 8.13) is a complete right tibia lacking the medial side of the tibial plateau and, distally, the styloid process. The morphology of the tibia matches that of *Machairodus* spp. The shaft is bilaterally compressed proximally, and anteroposteriorly wide. The total preserved length of the tibia is 310 mm, with an estimated length no more than 320 mm. This measurement shows that it is only slightly smaller than AMPG PA 1928/91, a tibia assigned to *M. giganteus* from Pikermi, with a length of 325.3 mm (Roussiakis 2002). Immediately below the tibial tuberosity, it measures 66.5 mm anteroposteriorly and 39 mm transversely. At midshaft, it is 35.5 mm and 30 mm, respectively. At the distal end, it is 42.5 mm and 60.5 mm, respectively. The Pikermi specimen is slightly narrower anteroposteriorly at its distal end (40.6 mm; Roussiakis 2002).

AME-VP-1/21 (Figure 8.13) comprises associated postcrania including a right proximal femur, left calcaneus, left fourth metatarsal missing the midshaft portion, left second metatarsal lacking the distal one-third, and left navicular. The femur has a very deep and long intertrochanteric fossa. The neck is short but wide. The femur head and the greater trochanter are almost on the same level. The articular surface of the femur head is extended toward the intertrochanteric fossa. The head diameter is 38 mm superoinferiorly and 43 mm anteroposteriorly. The anteroposterior breadth of the greater trochanter is 46.1 mm.

The calcaneus is complete and has a typical large felid morphology. Its length is 105 mm, and its maximum breadth is 47 mm. The proximal and distal parts of the associated second metatarsal are the size of a modern African adult lion. The proximal articular surface measures 23 mm transversely and 29 mm dorsoventrally. The distal end is 20.5 mm and 21 mm, respectively. The left navicular is almost identical with 81P34A from Sahabi

---

FIGURE 8.13

Middle Awash late Miocene specimens assigned to *Machairodus* sp. A. AMW-VP-1/12, left P^4. B. STD-VP-2/21, right M_1 fragment. C. STD-VP-2/22, left M_1 fragment. D. STD-VP-2/24, left P_4 fragment. E. ALA-VP-2/213, right upper canine. F. ALA-VP-2/313, right upper canine. G. DID-VP-1/93, right astragalus. H. ASK-VP-3/29, left proximal manus phalanx. I. AME-VP-1/26, left proximal phalanx. J. ALA-VP-2/212, proximal phalanx. K. BIK-VP-1/13, proximal phalanx. L. DID-VP-1/137, proximal phalanx. M. STD-VP-2/28, right proximal femur. N. KUS-VP-1/97, left humerus. O. DID-VP-1/38, right proximal ulna. P. DID-VP-1/49, left distal humerus. Q. DID-VP-1/1, right fourth metatarsal. R. ASK-VP-3/30, right proximal third metatarsal. S. DID-VP-1/14, left fifth metacarpal. T. ASK-VP-3/32, left fifth metacarpal. U. AME-VP-1/11, right tibia. V. AME-VP-1/21, associated right proximal femur, left calcaneum, left navicular, proximal second metatarsal, proximal third metatarsal, left second metacarpal, left first metatarsal.

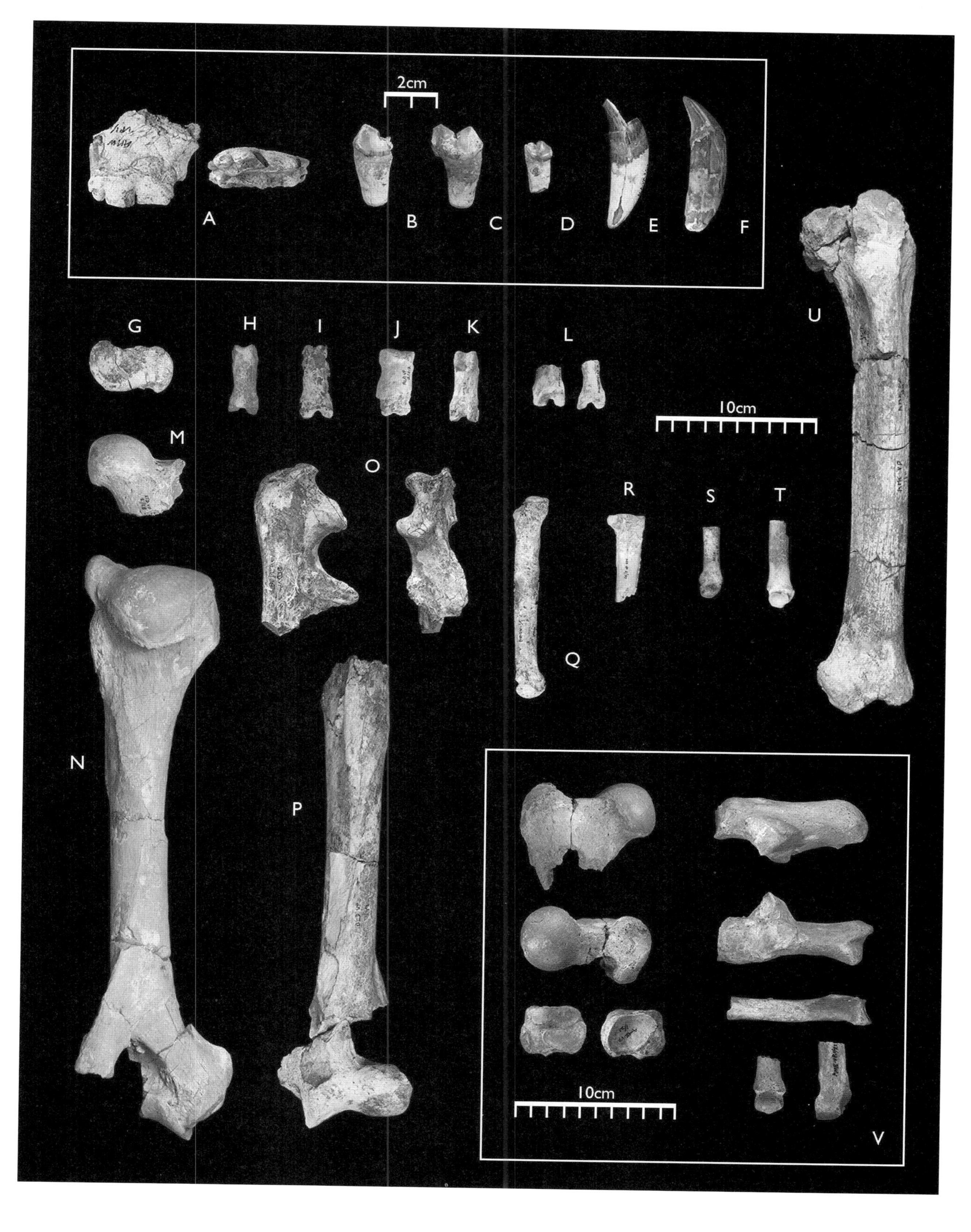
2cm
A
B
C
D
E
F
G
H
I
J
K
L
M
O
U
10cm
R
S
T
Q
N
P
10cm
V

(Howell 1987). The two specimens have similar size and morphology, and the Sahabi specimen is also referred to *Machairodus* sp.

AMW-VP-1/12 (Figure 8.13A) is a left maxilla fragment with P^4. The paracone tip is broken. It has a well-developed parastyle with a small pre-parastyle anteriorly. This carnassial shows very weak protocone reduction, in contrast to *Machairodus* spp. In overall length (34.4 mm) it is also smaller than upper carnassials of *M. giganteus* from Europe (38.2–45 mm) and KNM-LT 25405 (38.2 mm), a specimen from the Nawata Formation of Lothagam assigned to *Lokotungailurus emageritus* (W :delin 2003b). It is larger than a carnassial of *Paramachairodus maximiliani* (28.9 mm, loc. 12-M69 from Hsin An Hsien).

STD-VP-2/21 (distal half of a right M_1; Figure 8.13B), STD-VP-2/22 (distal two-thirds of a left M_1; Figure 8.13C), STD-VP-2/887 (a right P_4), and STD-VP-2/24 (distal half of a P_4; Figure 8.13D) possibly belong to the same individual based on their preservation and proximity of recovery. The right P_4 main cusp is missing most of its apical part. The anterior accessory cusp is larger than its posterior homolog. The P^3 lacks its roots. The anterior and posterior accessory cusps are well-defined, and the posterior is slightly larger, with a well-developed distolingual cingulum. The two cusps that form the blade on the M_1 are separated by a deep notch. The preserved length of the left M_1 is 19.6 mm.

ALA-VP-2/213 and ALA-VP-2/313 (Figure 8.13E, F) are lower canines. The former belongs to a subadult individual, whereas the latter is from an adult individual. ALA-VP-2/213 is serrate on both edges, which is characteristic of machairodontine lower canines other than *Dinofelis* species (Werdelin and Lewis 2001a). Its size is relatively small. It has a very long root compared to the crown size. ALA-VP-2/313 has a similar morphology but is slightly larger.

KUS-VP-1/97 (Figure 8.13N) is an almost complete left humerus lacking the capitulum and part of the trochlea. The epicondyles are preserved. The humerus is robust, with a deep bicipital groove. The humeral head points posterolaterally. It has a large greater tubercle with a deep fovea on its lateral side. The deltoid process is well-marked and its distal end extends as a crest almost to the entepicondylar foramen. The diaphysis below the head is bilaterally wide. The olecranon fossa is bilaterally elongate and very deep. The entepicondylar foramen is oval shaped. The preserved length of the humerus is 350 mm. A complete humerus from Pikermi (AMPG PG 98/26) assigned to *M. giganteus* is also reported to have a maximum length of 350 mm (Roussiakis 2002). The breadth and depth of the humerus at its proximal end are 75 mm and 91 mm, respectively. The minimum shaft diameter is 33 mm. Breadth of the distal end is more than 93 mm.

DID-VP-1/49 (Figure 8.13P) is another left humerus lacking the humeral head proximally and the area immediately above the capitulum distally. The medial epicondyle is also missing, and the midshaft is slightly chipped. It is morphologically identical to KUS-VP-1/97 even though it is smaller in size. DID-VP-1/49 may belong to a female individual. DID-VP-1/38 is a right proximal ulna fragment that possibly belongs to the same individual as DID-VP-1/49. Its proximal extremity is complete and best fits the size of DID-VP-1/49.

The hand and foot bones are assigned here to *Machairodus* based on their morphology and overall size. Measurements of most postcranial elements are given in Table 8.3.

TABLE 8.3 Postcranial Measurements of *Machairodus* sp. from the Middle Awash Late Miocene Deposits

Specimen #	Element	Length	Width	Height	*Proximal Width*		*Distal Width*	
					TRANSVERSE	ANTEROPOSTERIOR	TRANSVERSE	ANTEROPOSTERIOR
ALA-VP-1/2	R. manus proximal 3rd phalanx	43.5	nm	nm	15.6	13.3	12.6	10.5
ALA-VP-2/23	R. manus proximal 3rd phalanx	46.1	nm	nm	17.9	14.6	14.2	10.1
AME-VP-1/21	R. 2nd metatarsal	nm	nm	nm	20.5	21.1	22.9	29
AME-VP-1/21	L. calcaneum	105	46.8	44.2	29.8	36.0	nm	nm
AME-VP-1/26	L. pes proximal 5th phalanx	49.6	nm	nm	20.9	19.8	18.2	14.3
ASK-VP-3/12	L. 5th metacarpal	64.2+	nm	nm	nm	nm	nm	nm
ASK-VP-3/28	R. 4th metacarpal	37.5+	nm	nm	13.7	18.0	nm	nm
ASK-VP-3/29	L. pes proximal 2nd phalanx	47.1	nm	nm	19.4	18.5	16.7	11.5
ASK-VP-3/30	R. 3rd metacarpal	74+	nm	nm	20.2	29.9	nm	nm
ASK-VP-3/31	L. 2nd metacarpal	62+	nm	nm	19.8	27.3	nm	nm
ASK-VP-3/32	L. 5th metacarpal	56.6+	nm	nm	nm	nm	16.9	15.5
DID-VP-1/1	R. 4th metacarpal	134	nm	nm	21.2	27.4	17.9	nm
DID-VP-1/14	L. 5th metacarpal	46.8+	nm	nm	nm	—	15.7	nm
DID-VP-1/46	L. 5th metacarpal	51.8+	nm	nm	nm	—	14.4	nm
STD-VP-1/20	R. 4th metacarpal	nm	nm	nm	25.8	28.8	20.2	22.9

NOTE: All measurements are in mm; nm = measurement was not possible.

DISCUSSION Recent phylogenetic studies of the family Felidae show that felids are the sister taxon of hyaenids in the group Feloidea (Wyss and Flynn 1993). Most recent studies consider felids as the sister group of viverrids, and hyaenids as the sister group of herpestids (see Flynn et al. 2005, for example). Extinct saber-toothed cats are assigned to families of their own, Nimravidae and Barbourofelidae (Flynn et al. 1988; Wyss and Flynn 1993). All of the extant felids are members of the subfamily Felinae and form a monophyletic group (Collier and O'Brien 1985; Salles 1992). Extant and extinct felids are usually divided into two subfamilies: the Machairodontinae and the Felinae. The former subfamily has no extant relatives, and its phylogenetic relationship with felines remains unclear. The same is true when it comes to the phylogenetic relationships between extinct and extant felines. Most of the extinct large felines seem to have appeared simultaneously and remain unchanged for a long period of time (Turner 1990). The extinct *Dinofelis* and the leopard *Panthera pardus* appear to be similar in most morphological features. Moreover, *Dinofelis* seems to have appeared in the African fossil record much earlier than *Panthera,* and this combined evidence suggests a possible close phylogenetic relationship between the two taxa.

Machairodontines from the late Miocene (MN 12 and MN 13) are relatively well-known from Eurasia represented by a number of *Machairodus* species (Sardella 1993), although they appear in North Africa earlier (MN 8; Sotnikova and Noskova 2004). Their African record from the terminal Miocene is limited to fragmentary postcranial elements and isolated teeth from Sahabi (except for one partial cranium; Howell 1987), Lothagam (Werdelin 2003b), and Langebaanweg (Hendey 1974). A recent addition to this hypodigm is *M. kabir* sp. nov., a new giant machairodontine from Toros Menalla (Locality TM-266; Peigné et al. 2005a). Almost all machairodontine specimens from the eastern African late Miocene sites are also assigned to the genus *Machairodus* (but see Werdelin 2003b). However, due to the nondescript nature of most of the specimens and inconclusiveness of comparisons with European forms, assignment at the species level has always been uncertain. Machairodontines are relatively well-represented in the Plio-Pleistocene fossil record of Africa, where at least four genera are known (Hendey 1974). However, interrelationships among these genera and the exact number of different species remain unclear.

The machairodontine fossil specimens recovered from the Middle Awash terminal Miocene–basal Pliocene sites significantly augment the fossil hypodigm of the tribe. At least one species of machairodontine, possibly the size of *M. giganteus,* is known from cranial and postcranial remains. A large *Machairodus*-like felid from the Nawata formation at Lothagam is also assigned to a new genus and species, *Lokotunjairulus emageritus* (Werdelin 2003b). The Middle Awash *Machairodus* differs from the Lothagam form by being larger in its overall size and by its shorter and wider M^1. Although contemporaneous with the Middle Awash and Lothagam assemblages, no machairodontines have been reported from the Lukeino Formation, Baringo Basin, Kenya (Morales et al. 2005). The presence of multiple machairodontine species in North Africa (Howell 1987), eastern Africa (Werdelin 2003b; Haile-Selassie et al. 2004c) and central Africa (Peigné et al. 2005a) during the late Miocene is now compelling. It would indicate that the group had diversified within Africa before dispersal into Europe and rediversified while the African species gave way to smaller cats such as *Dinofelis* and *Panthera.*

Dinofelis Zdansky, 1924

Dinofelis sp.

DESCRIPTION The specimens assigned here to *Dinofelis* sp. (Figure 8.14) are mostly fragmentary and do not allow either detailed description or assignment at a specific level. However, they demonstrably indicate the presence of a *Dinofelis* species in the Middle Awash late Miocene deposits. As many as five *Dinofelis* species are considered by Werdelin and Lewis (2001a, 2005) to have existed within the Mio-Pliocene and the earlier Pleistocene of eastern Africa. The most complete specimen is AME-VP-1/1. This specimen is a left mandible with P_3, P_4, and M_1. Dental measurements are given in Table 8.4. The premolars are mesially narrow. The anterior accessory cusp of the P_3 is almost nonexistent, and the posterior cusp is small. The anterior and posterior accessory cusps of the P_4 are distinct. The main cusps of both premolars are relatively hypsodont. The M_1 protoconid is larger than the paraconid, and there is a buccal groove between the two cusps. The distal end of the protoconid has a slight bulge. The diastema between the canine and the P_3 is relatively wide. The mandible is deep at the P_4 level and shallower at the M_1 level. The posterior mental foramen is positioned low, at the level of the mesial P_3 root. The anterior mental foramen is missing, even though its posterior edge is visible on the preserved corpus. The teeth of AME-VP-1/1 are similar in absolute size to KNM-KP 30397, a complete mandible with C-M_1 from Kanapoi referred to *Dinofelis petteri* (Werdelin 2003a). However, all of the premolars and M_1 of AME-VP-1/1 are buccolingually narrower relative to their length than KNM-KP 30397. The buccolingual/mesiodistal ratios of P_3, P_4, and M_1 of AMW-VP-1/1 are 47.4, 45.1, and 44.7, respectively, whereas KNM-KP 30397 has ratios of 55.7, 47.1, and 45.7, respectively.

AME-VP-1/129 is a left edentulous mandibular fragment. It is broken anteriorly at the P_3 mesial root level and posteriorly at the mesial carnassial root. The distal P_3 root is visible in its crypt. The alveoli for the P_4 and mesial root of the carnassial are also preserved. Corpus morphology and depth at P_4 are similar to *Dinofelis* sp. indet. from the Nawata Member of Lothagam (Werdelin 2003b) and to *D. diastemata* from Langebaanweg (Hendey 1974). AMW-VP-1/124 is a right P_3 with most of the roots. The tip and distal half of the main cusp are missing. The mesial accessory cusp is well-developed. The two posterior cusplets are also well-developed as in P_4, and of equal size. The crown morphology of this tooth is slightly different from the P_3 of KNM-KP 30397 (a specimen from Kanapoi assigned to *Dinofelis petteri;* Werdelin 2003a) in having two well-defined distal accessory cusps separated from each other by a deep notch. Preserved crown dimensions are larger than the Kanapoi and Langebaanweg P_3s, and the breadth/length ratio is 53.8, similar to KNM-KP 30397 from Kanapoi (55.7). The crown of KUS-VP-1/75 is very slender and even more bilaterally compressed than the P_4s of AMW-VP-1/1 and KNM-KP 30397, with a breadth/length ratio of 40.7.

The postcranial remains referred here to *Dinofelis* sp. are very fragmentary, but their size and proportions are similar to those of *Dinofelis* species known from the African Plio-Pleistocene. *Dinofelis* is represented by limited dentognathic and postcranial remains in the Middle Awash late Miocene. The assignment of the Middle Awash postcranial elements to *Dinofelis* sp. is based on their size compared to other machairodontines. They

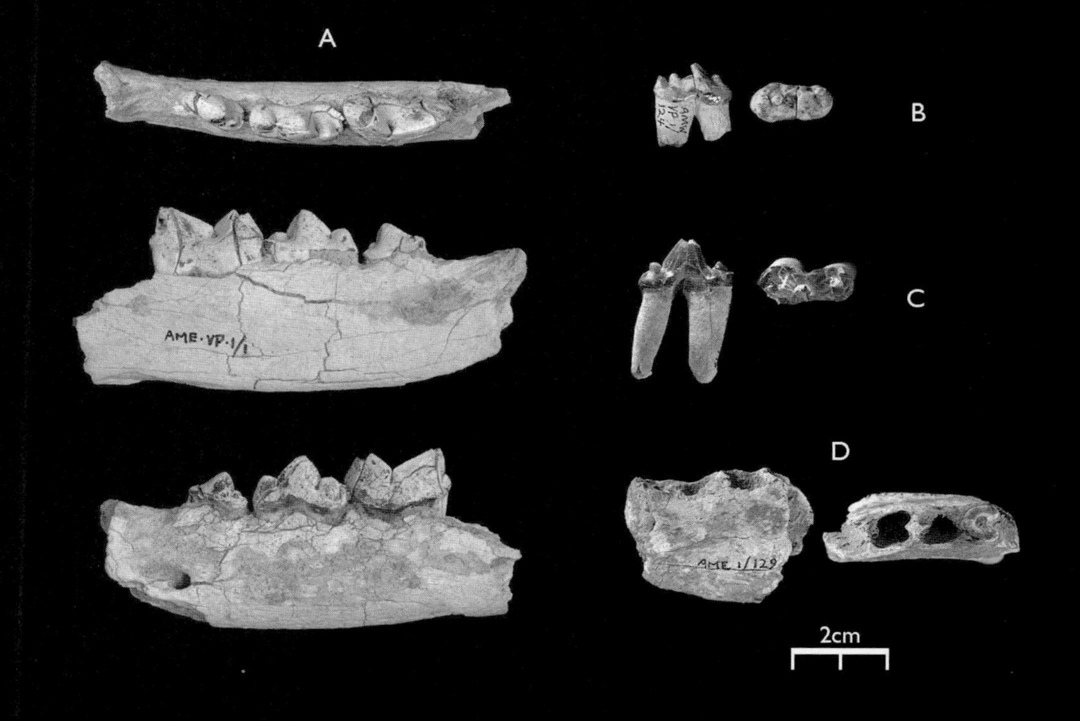
A
B
C
D
AME·VP·1/1
AME 1/129
2cm

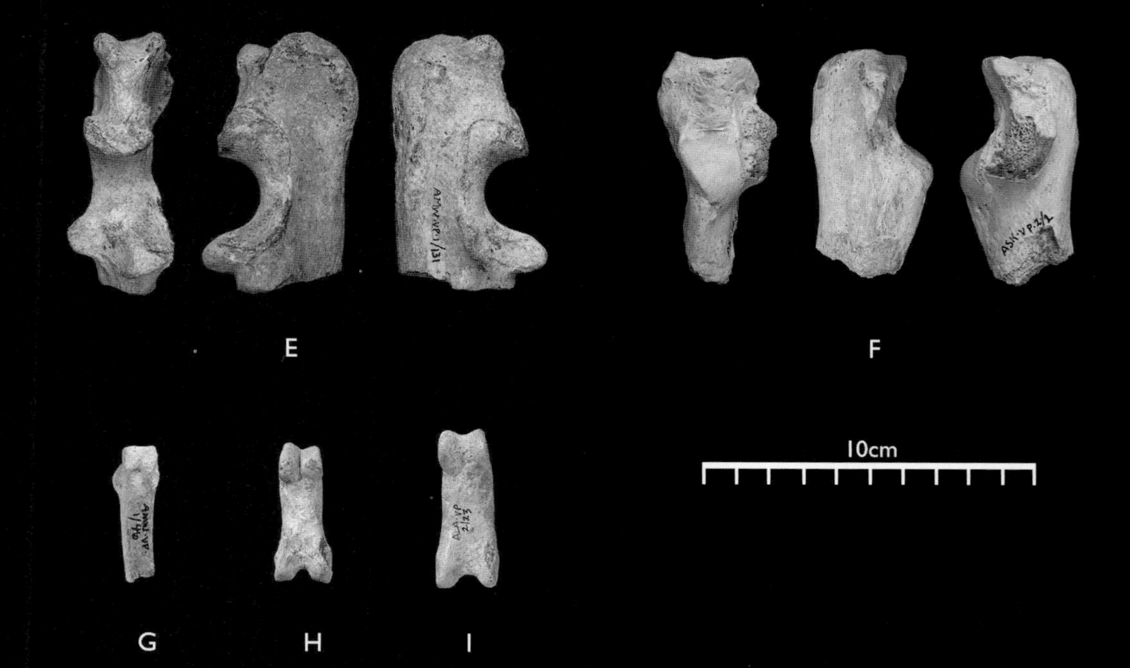
E
F
G
H
I
10cm

TABLE 8.4 Dental Measurements of *Dinofelis* sp. from the Middle Awash Late Miocene

		AME-VP-1/1	*KUS-VP-1/75*
P_3			
	MD length	13.3	
	BL width	6.3	
	Prd length	nm	
	Pad length	nm	
P_4			
	MD length	19.5	19.2
	BL width	8.8	7.8
	Prd length	nm	nm
	Pad length	nm	nm
M_1			
	MD length	22.7	
	BL width	10.1	
	Prd length	12.0	
	Pad length	10.7	
Mandible depth			
	P_4	26.5	
	M_1	24.4	
C-P_3 diastema		14.3	

NOTE: All measurements are in mm. Abbreviations: MD = mesiodistal; BL = buccolingual; Prd = protoconid; Pad = paraconid; nm = measurement was not possible.

belong to a large felid smaller than *Machairodus*. They are too small to be assigned to *Machairodus* sp. and too large to be assigned to a *Panthera*-sized felid. Their size best fits within the range of *Dinofelis petteri* from Kanapoi (Werdelin 2003a), *Dinofelis* sp. from Lothagam (Werdelin 2003b), and *Dinofelis diastemata* from Langebaanweg (Hendey 1974), although the Middle Awash form appears to be slightly smaller (Figure 8.15).

DISCUSSION Africa is now known to have been a major center of evolution for *Dinofelis,* although the genus has a broad distribution in Eurasia as well (Werdelin and

FIGURE 8.14

Middle Awash late Miocene specimens assigned to *Dinofelis* sp. A. AME-VP-1/1, left mandible with P_3–M_1. B. AMW-VP-1/124, right P_3. C. KUS-VP-1/75, right P_4. D. AME-VP-1/129, left edentulous mandible. E. AMW-VP-1/131, right proximal ulna. F. ASK-VP-2/2, left calcaneum. G. AMW-VP-1/46, left fifth metacarpal. H. ALA-VP-1/30, proximal phalanx. I. ALA-VP-2/23, proximal phalanx.

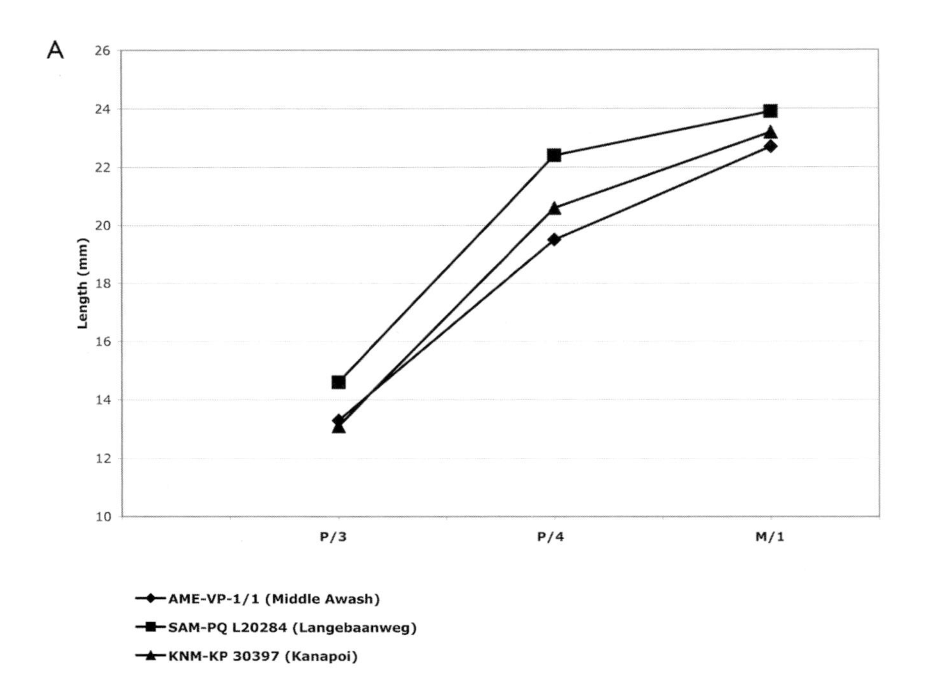

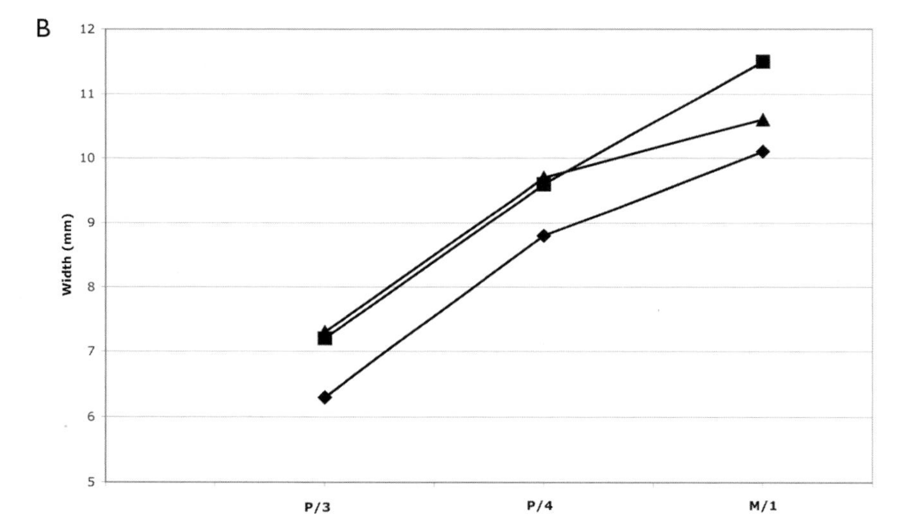

FIGURE 8.15

Plots comparing lower teeth of AME-VP-1/1 (Middle Awash), L20284 (Langebaanweg), and KNM-KP 30397 (Kanapoi). A. Length. B. Width.

Lewis 2001a). *Dinofelis* persisted in Africa until about 1 Ma (Werdelin and Lewis 2001a). The earliest *Dinofelis* in Africa was documented from Langebaanweg, South Africa (Hendey 1974) until its discovery from older deposits at Lothagam (Werdelin 2003b) and from the Middle Awash (Haile-Selassie 2001a; Haile-Selassie et al. 2004c). The Middle Awash specimens represent the earliest record of the genus in Africa together with the specimens from Lothagam (Werdelin 2003b).

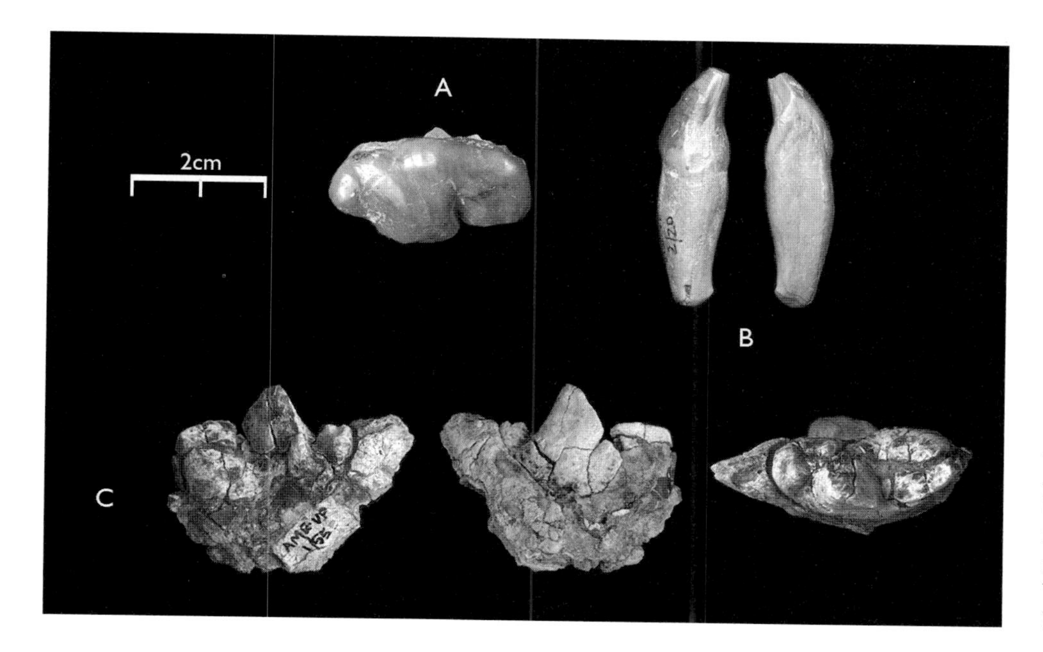

FIGURE 8.16
Middle Awash late Miocene specimens assigned to *Hyaenictitherium* sp. A. STD-VP-2/30, left P^4, buccal view. B. STD-VP-2/20, left I_3, distal and medial views. C. AME-VP-1/55, right M_1, ligual, buccal, and occlusal views.

Felidae gen. et sp. indet.

DESCRIPTION Specimens referred to Felidae gen. et sp. indet. are fragmentary and difficult to assign at a genus or species level. Recovery of more complete material is necessary to establish the specific affinity of the represented taxa. There is a small felid represented in the Middle Awash–late Miocene deposits, even though its specific affinity is currently unknown due to the small and fragmentary nature of the sample.

Hyaenidae

Hyaenictitherium Kretzoi, 1938

Hyaenictitherium sp.

DESCRIPTION STD-VP-2/30 (Figure 8.16A) is a left upper carnassial of a relatively large hyaenid. The preserved length of this tooth is 31.4 mm. Only the buccal half of the crown is preserved, and nothing can be said about the protocone or the lingual relief of the carnassial. However, in addition to its large size, there is enough morphology preserved to assign this specimen to a hyaenid other than a *Hyaenictis.* The crown base on the buccal side is inflated, and the groove between the paracone and metastyle blade is deep. The paracone is longer than the metastyle blade.

STD-VP-2/20 (Figure 8.16B) is a left I_3 (L = 9.75 mm, W = 9.12 mm) best assigned to *Hyaenictitherium.* This is mainly based on its association with STD-VP-2/30. AME-VP-1/55 (Figure 8.16C) is a right lower carnassial slightly damaged by cracks on the crown. This specimen might represent a distinctive ictithere-like but very large species, possibly larger than *H. namaquensis* from Langebaanweg (Werdelin et al. 1994). The paraconid and protoconid have comparable lengths, although the protoconid is slightly elevated. Its metaconid is very robust, and the talonid is relatively long. The lingual crest is represented by a simple salient median crest. It has a distolingually swollen entoconid and large metaconid. The talonid is bound by the entoconid and hypoconulid cusplets. The metaconid on this specimen is not so elevated as the Sahabi *Hyaenictitherium,* extending only as high as the talonid. The latter has a well-developed entoconid and hypoconid and a small hypoconulid at its distal end. AME-VP-1/55 is much longer than all the M_1 specimens referred to *Hyaenictitherium* from Sahabi. There are no lower M_1s from Lothagam referred to *Hyaenictitherium*. When more specimens are discovered, this specimen might represent a new *Hyaenictitherium* species.

A new species of *Hyaenictitherium* (*H. minimum* sp. nov.) has been identified from the Turolian-equivalent fauna of Toros Menalla, Chad (de Bonis et al. 2005). This species is absolutely smaller than the Middle Awash form. Moreover, the cusp proportions on the carnassial also differ between the two species.

The small number of specimens referred to *Hyaenictitherium* sp. from the Middle Awash does not allow a detailed phylogenetic analysis. However, the size and morphology of the M_1 indicates close relationship with *Hyaenictitherium namaquensis* from Langebaanweg, South Africa (Hendey 1978a).

DISCUSSION The diversity of hyaenids during the terminal Miocene of eastern Africa is well-documented in the Nawata Formation of Lothagam, where at least four species are identified (Werdelin 2003b). Hyaenids were also diverse in Sahabi (Howell 1987) and in slightly younger deposits at Langebaanweg (Hendey 1978a). However, the family was not as diverse in contemporary Middle Awash deposits, possibly owing to habitat differences. There are at least two species, *Hyaenictitherium* sp. and *Hyaenictis* sp. nov., in the terminal Miocene deposits of the Middle Awash.

The family Hyaenidae has three extant genera and four species distributed in Africa and parts of Asia (Nowak 1991). However, the family's fossil record shows that it was diverse during the Miocene. Recent phylogenetic analyses of the family by Werdelin and Solounias (1991, 1996), later modified by Werdelin and Turner (1996), show that there were at least 25 hyaenid genera. Werdelin and Solounias (1991) used 47 cranial and postcranial characters in their PAUP systematic analyses. Werdelin and Turner (1996) suggested that three genera (*Percrocuta, Dinocrocuta,* and *Allohyaena*) previously included in the family Hyaenidae (Howell and Petter 1985) be placed in their own family, the Percrocutidae.

The *Hyaenictitherium* sample from the Middle Awash is small and would not have any impact on the established phylogenetic relationships of the genus with other hyaenid genera. However, the most diagnostic remains are tentatively placed as a sister taxon of the exclusively African taxon, *H. namaquensis* (Hendey 1978a), based on dental similarities.

Hyaenictis Gaudry, 1861

GENERIC DIAGNOSIS See Gaudry, 1861.

Hyaenictis wehaietu sp. nov.

ETYMOLOGY *Wehaietu* is the name of the Awash River in the local Afar language. The Awash River transects the Afar Rift Valley north-south and drains to Lake Abbé.

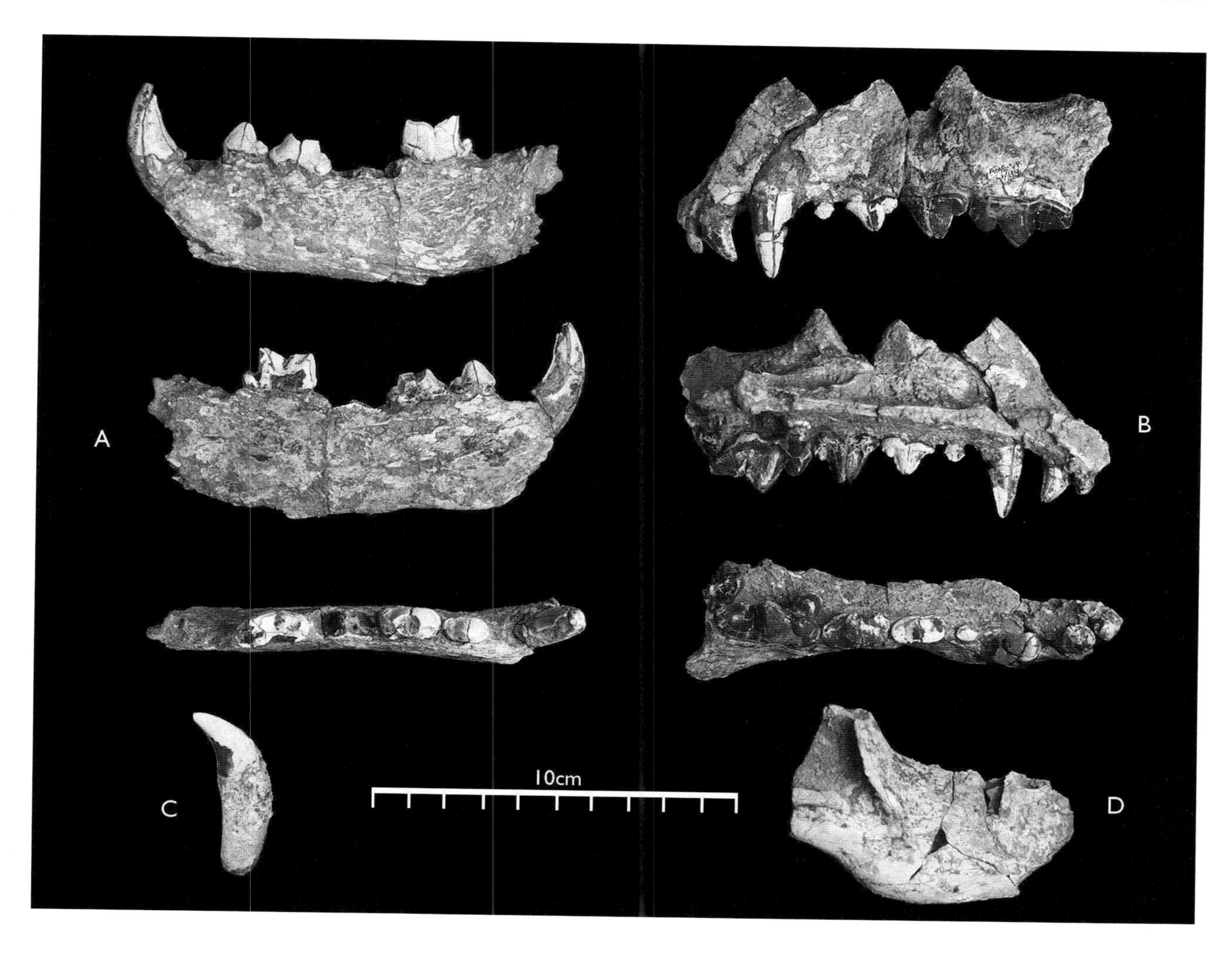

FIGURE 8.17
AME-VP-1/114, the holotype, of *Hyaenictis wehaietu* sp. nov. A. Left mandible with C, P_{2-3}, M_1, lingual and occlusal views. B. Left maxilla with I^2–M^1, lingual and occlusal views. C. Right C^1, distal view. D. Buccal view of a posterior half of the right horizontal corpus (edentulous).

HOLOTYPE AME-VP-1/114 (Figure 8.17), a left mandible with canine, P_2–P_3, M_1, a left maxilla with I^2–M^1, a right upper canine, and the posterior half of the right horizontal corpus with no teeth.

PARATYPE STD-VP-2/877 (right P_4); AME-VP-1/50 (a complete mandible with dentition and a right maxilla with P_3–M_1); AME-VP-1/16 (left mandible fragment with P_2 root and distal P_3); AME-VP-1/17 (right mandible fragment with C–P_3 roots); AMW-VP-1/64 (right P^3–P^4).

LOCALITIES AND HORIZONS The holotype was recovered from the Amba East Vertebrate Locality 1 (AME-VP-1) in the Kuseralee Member of the Sagantole Formation. Most of the paratype specimens were recovered from localities within the same member. This sequence has been radiometrically dated to 5.2 Ma. Other referred specimens include a specimen from the Asa Koma Member of the Adu-Asa Formation radiometrically dated to between 5.54 and 5.77 Ma.

DIAGNOSIS *Hyaenictis* species the size of *Ikelohyaena abronia,* metaconid on M_1 present, P_1 and M_2 absent, P_4 broader than P_3, lower canine longer than P_2.

DIFFERENTIAL DIAGNOSIS *Hyaenictis wehaietu* sp. nov. differs from *Hyaenictis gracea* by its retention of the metaconid on the M_1 and lack of P_1. It differs from *Hyaenictis almerai* in its premolar morphology. It differs from *Hyaenictis hendeyi* by its absolutely smaller size, the lack of metaconid on the *H. hendeyi* M_1, and lack of M_2 in *Hyaenictis wehaietu* sp. nov. It differs from *Ictitherium ebu* (Werdelin 2003b) by its absolutely larger size and the smaller M_1 talonid.

DESCRIPTION AME-VP-1/114 (Figure 8.17) comprises a left mandible with the canine, P_2–P_3; M_1, a left maxilla with I^2–M^1; the right upper canine; and the posterior half of the right mandibular corpus with no teeth. Dental measurements are given in Table 8.5. The left horizontal ramus is slightly abraded and broken distally at the base. The ramus is slender and deep, especially posteriorly. A single large mental foramen is positioned at mid-corpus, below P_2. The upper canine is not robust for a hyaenid. There is no P_1. The second premolar has a relatively tall main cusp with sharp anterior and posterior edges. It has no anterior accessory cusp. The posterior accessory cusp is well-defined. The crown has a generally rectangular outline and does not taper anteriorly. The P_3 lacks its apical half. The anterior accessory cusp is not developed, whereas the posterior homolog is present, although small. The crown broadens slightly distally. The main cusp has sharp anterior and posterior edges. The M_1 has a well-developed, but low, metaconid. The talonid is relatively large and composed of the entoconid and hypoconid of about the same height. The talonid is positioned lingually relative to the paraconid and protoconid and has a square outline. The protoconid is slightly more elevated than the paraconid. The mesiolingual face of the paraconid has a strong cingulum at its base.

The left hemimaxilla lacks only I^1, and most of the maxillary part above the dentition is preserved. However, as a result of abrasion, none of the sutures are visible. The maxillary surface around the canine root bulges outward, indicating a large canine root. The maxilla is hollowed posterior to the canine at the level of P^1 and P^2. The infraorbital foramen is round and positioned above the center of P^3. The inferior edge of the foramen is 21.7 mm above the base of the P^3 crown. The canine is relatively tall but slender. The P^1 is a tiny, single-rooted, incisor-like tooth with a conical crown. The P^2 is mesiodistally elongate with no anterior accessory cusp. The main cusp is tall, with convex anterior and posterior edges. The posterior accessory cusp is well-developed. A strong lingual cingulum ends at the distal end of the posterior accessory cusp. Except for the larger size, the P^3 is morphologically similar to the P^2. However, the tooth has a small anterior accessory cusp at the mesiolingual angle. The upper carnassial has a well-developed protocone positioned parallel to the parastyle. These two cusps are separated mesially by a deep groove. The protocone is situated much lower than the parastyle. The metastyle blade is much longer than the parastyle. The paracone is the largest and tallest cusp. There is a lingual cingulum running from the distal end of the protocone to the metastyle blade. The carnassial characters of *Hyaenictis wehaietu* sp. nov. show similarities with *Hyaenictis hendeyi* as described by Werdelin et al. (1994).

AME-VP-1/50 is a complete mandible lacking the right vertical ramus and the coronoid and condyloid processes on the left side. The teeth are highly worn, and none of the

TABLE 8.5 Dental Measurements of the Holotype (AME-VP-1/114) and Other Referred Material to *Hyaenictis wehaietu* sp. nov.

	AME-VP-1/114	*AME-VP-1/50*	*AME-VP-1/16*	*AME-VP-1/55*	*STD-VP-2/877*
LC_1MD	12.1	(16.2)			
BL	9.3	(11.4)			
RC_1MD	12.1				
BL	10.4				
LP_2 MD	12.4	13	(12.4)		
BL	7.5	nm	nm		
RP_2MD		13.0			
BL		8.1			
LP_3 MD	17.5		(16.5)		
BL	9.4		nm		
RP_3MD		18.0			
BL		11.1			
LP_4 MD	19.8	18.5			
BL		11.1			
RP_4MD		19.5			21.2
BL		10.9			11.3
LM_1MD	20.2	22.1			
BL	9.5	10.8			
RM_1MD		21.8		28.0	
BL		10.7		12.7	
LI^2MD	5.8				
BL	5.9				
LI^3MD	9.8				
BL	8.6				
LC^1MD	14.2				
BL	10.1				
LP^1MD	5.7				
BL	5.1				
LP^2MD	14.8				
BL	7.8				
LP^3MD	18.0				
BL	10.7				
LP^4MD	28.2				
BL	16.3				
LM^1MD	8.0				
BL	nm				

NOTE: All measurements are in mm. Values in parentheses are estimates; nm = measurement was not possible.

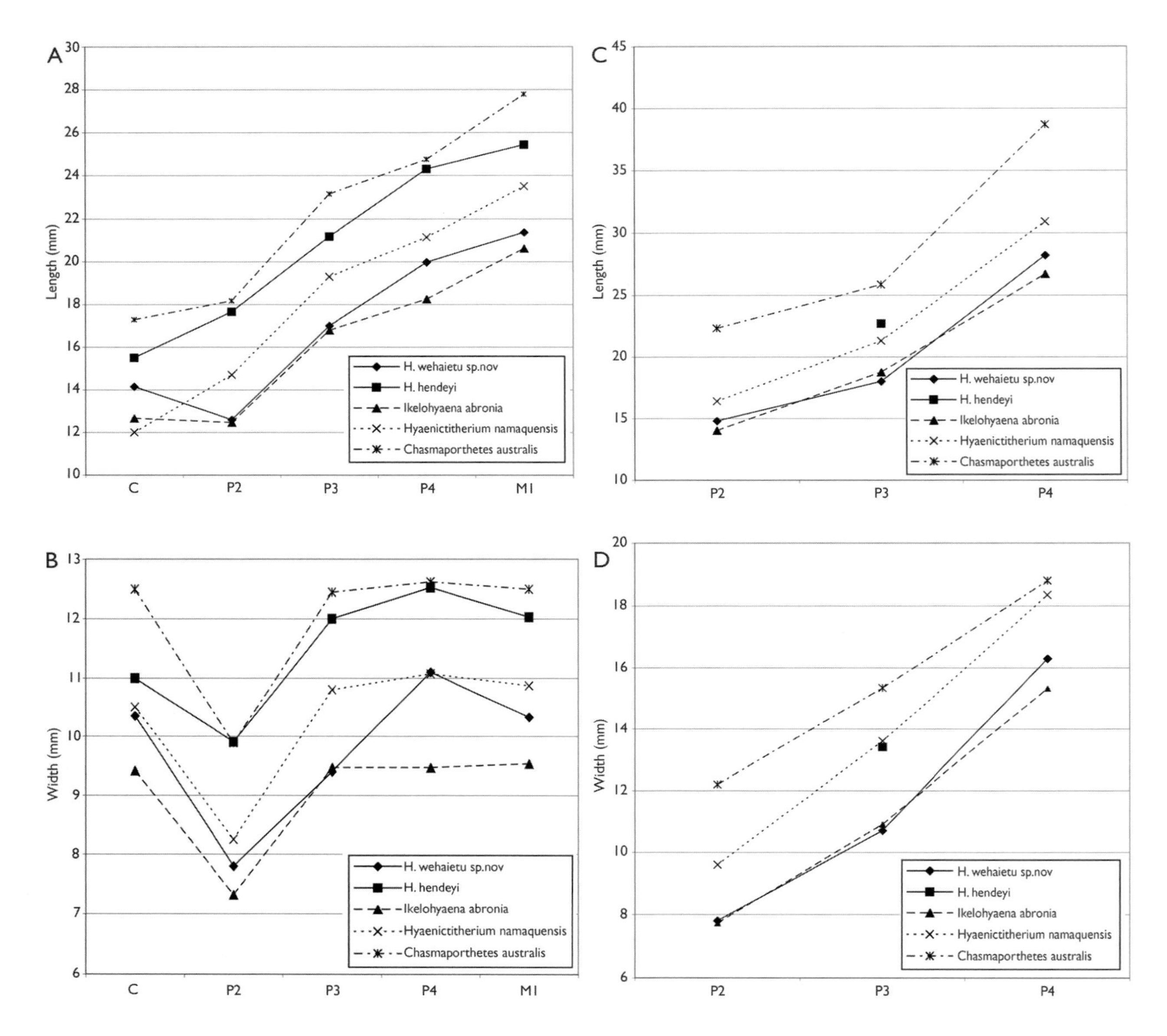

FIGURE 8.18 Plots comparing teeth measurements of various late Miocene–early Pliocene hyaenid species from eastern and southern Africa. A. Length measurements for lower teeth. B. Width measurements for lower teeth. C. Length measurements for upper teeth. D. Width measurements for upper teeth.

morphological features are visible. However, the mandible had no P_1. A single mental foramen is positioned low at the level of P_2, and the M_1 had a metaconid. This mandible is larger than AME-VP-1/114.

STD-VP-2/877 is a complete right P_4 crown lacking the roots. This specimen is comparable in its size to *H. namaquensis*. The tooth has a strong anterior crest, a very strong and salient paraconid, a well-developed posterolingual cingulum, well-developed accessory cusplets, and a strong buccal cingulum running from the anterior cusp posteriorly and connecting with the posterolingual cingulum. STD-VP-2/877 is best assigned to *Hyaenictis wehaietu* sp. nov. based on the development of the mesial and distal accessory cusps and morphology of the posterolingual cingulum, which is not as reduced as in *H. hendeyi*. This indicates that STD-VP-2/877 is more primitive than P_4s of *H. hendeyi*. Some of

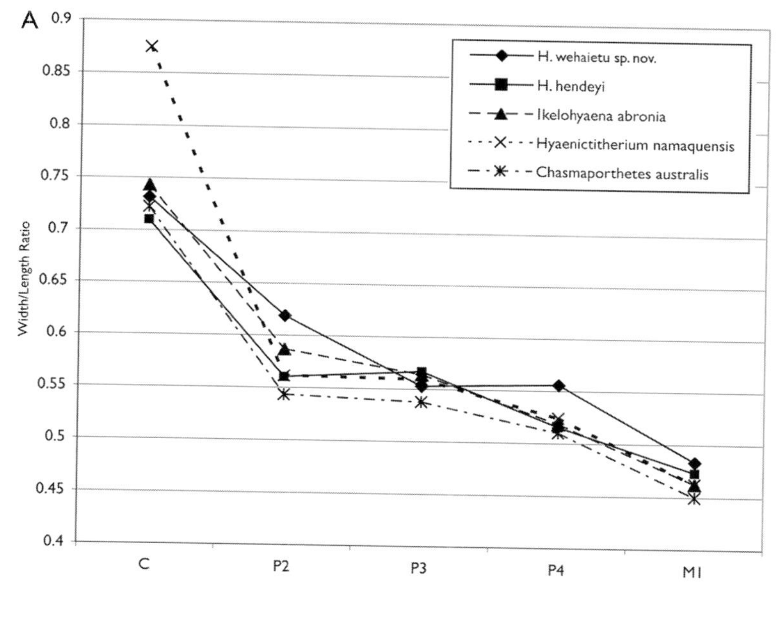

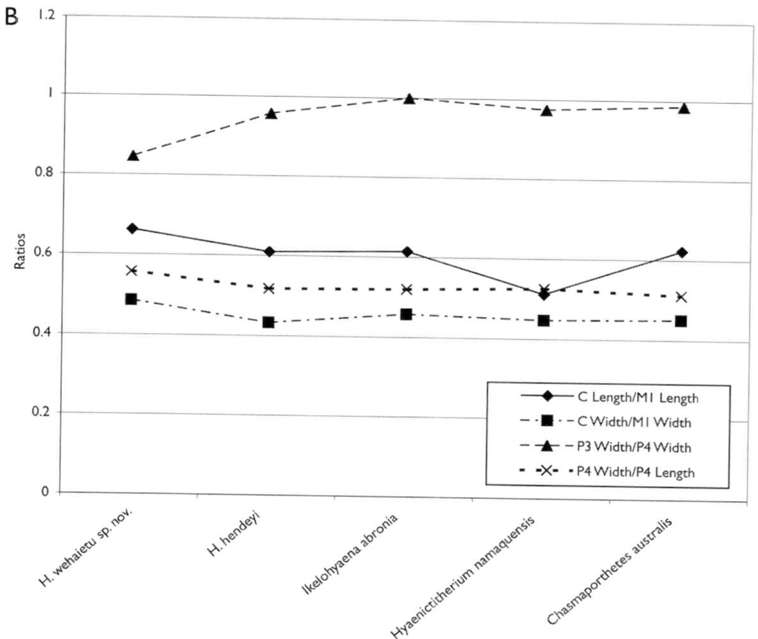

FIGURE 8.19

Plots comparing teeth measurement ratios of various late Miocene–early Pliocene hyaenid species from eastern and southern Africa. A. Width/length ratios of lower teeth. B. Ratios of lower canine length/M_1 length, lower canine width/M_1 width, and P_3 width/P_4 width.

these morphological features also resemble *Chasmaporthetes australis* from Langebaanweg (Werdelin et al. 1994), which is probably due to their close phylogenetic relationship. However, *H. wehaietu* sp. nov. is absolutely smaller in its size, and the dental size is different between the two species, *Chasmaporthetes australis* being much larger. Moreover, the posterolingual cingulum is not reduced as in *Chasmaporthetes australis*. Comparative dental measurements between *Hyaenictis wehaietu* sp. nov and other late Miocene–early Pliocene hyaenid species are presented in Figures 8.18 and 8.19.

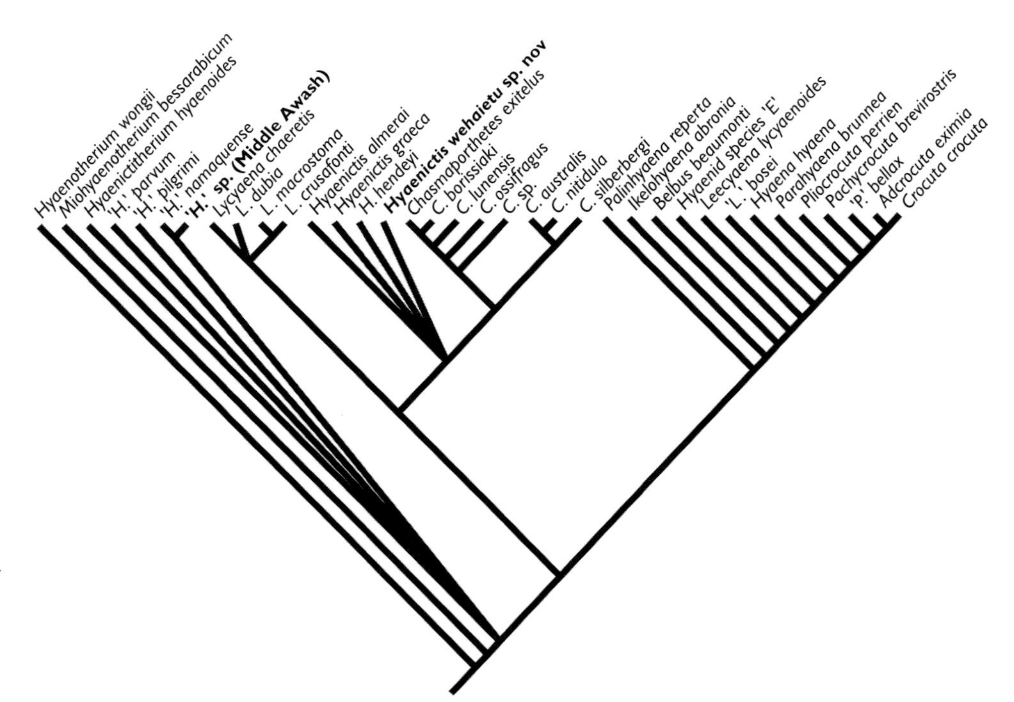

FIGURE 8.20

Cladogram showing the phylogenetic position of the Middle Awash hyaenids relative to other hyaenids. Modified after Werdelin and Turner (1996).

DISCUSSION The taxonomy and phylogenetic relationships among various African late Miocene–early Pliocene hyaenids have been addressed in detail by Werdelin et al. (1994) and Werdelin and Solounias (1990, 1991). The genus *Hyaenictis* is one of the most poorly understood hyaenid genera. Werdelin et al. (1994) suggested that it forms a monophyletic group with *Chasmaporthetes* and *Lycyaena*. These three taxa share characters such as the general narrowing of the anterior premolars, reduction of the posterolingual cingulum cusp on P_4, and the reduction of the M_1 metaconid and talonid (Werdelin et al. 1994:209; Figure 8.20).

The Middle Awash and Langebaanweg materials represent the best samples of the genus in Africa. The Lothagam dental and mandibular specimens referred to *Hyaenictis* sp. are scant. However, there is enough evidence to suggest that the specimens from Lothagam referred to *Hyaenictis* are morphologically and metrically similar to *Hyaenictis wehaietu* sp. nov. Both hypodigms lack P_1 and retain the M_1 metaconid. *Hyaenictis wehaietu* sp. nov. differs from *Hyaenictis graeca* by its retention of the metaconid on the M_1 and the lack of a P_1. *Hyaenictis wehaietu* sp. nov. and *H. hendeyi* are similar in the lack of a P_1 and the absence of a hypoconulid on the M_1 talonid. However, the two are different because the latter lacks a metaconid on its M_1. It is usually inferred that the presence of a metaconid on the hyaenid M_1 is a primitive retention for *Hyaenictis*. Its absence in *H. hendeyi* makes that species more derived than the Middle Awash and Lothagam *Hyaenictis*. Therefore, *Hyaenictis wehaietu* sp. nov. could be a possible ancestor or precursor of *H. hendeyi*. This also conforms to the younger age of Langebaanweg compared to the Middle Awash and Lothagam.

The size and morphology of the mandible and teeth of *Hyaenictis wehaietu* sp. nov. are comparable to *Hyaenictitherium parvum*. However, the retention of P_1 in the latter distinguishes *H. wehaietu* sp. nov. from *Hyaenictitherium*. The *Hyaenictitherium* recovered from the Middle Awash is also much larger based on molar size. Some dental, mandibular, and postcranial materials from the Nawata Formation of Lothagam have been referred to cf. *Hyaenictis* sp. (Werdelin 2003b). Although the postcranial elements are difficult to refer to the same taxon as the dental and mandibular specimens, the latter might belong to *Hyaenictis wehaietu* sp. nov. based on dental size and morphology, particularly in their lack of a P_1 and retention of the metaconid on the M_1.

Conclusions

The Middle Awash carnivore fauna represents one of the most diverse in the latest Miocene of Africa. Recent description of the Lothagam carnivores (Werdelin 2003b) has increased the number of known taxa in eastern Africa during the late Miocene. However, some groups such as ursids are poorly documented from this time period even though their presence in Africa around the Miocene-Pliocene boundary is well-documented at Langebaanweg (Hendey 1980). At least 15 genera in seven families are recognized from the Asa Koma Member of the Adu-Asa Formation and the Kuseralee Member of the Sagantole Formation. Three additional taxa are identified only to the subfamily level. Large felids include *Machairodus* and *Dinofelis*. One small felid of unknown generic and specific affinity is also present. The hyaenids *Hyaenictitherum* and a new *Hyaenictis* species are present. Ursids are represented by one genus, *Agriotherium,* and it appears to be larger than the *Agriotherium africanum* from Langebaanweg. Mustelids are relatively well-represented, with at least three species in three genera. This family includes the earliest record of *Plesiogulo,* a genus previously known only from Langebaanweg (Hendey 1978b) and Lemudongo (Haile-Selassie et al. 2004a). One new canid species is also recognized from the Kuseralee Member of the Sagantole Formation. This species represents the smallest known species of *Eucyon*. Viverrids are represented by two species in two genera (*Genetta* and *Viverra*); similarly, herpestids are represented by two species in two genera. This sample includes the earliest record of *Herpestes* and *Helogale* in Africa. These species were previously known only from Pliocene deposits of eastern Africa. The carnivores described from contemporaneous deposits at Lothagam show temporal overlap in some taxa. However, most of the carnivores with closed habitat preferences seem to be lacking from the Lothagam Nawata Formation while they are abundant in the Middle Awash. The carnivore assemblage from the late Miocene of the Middle Awash clearly shows greater biogeographic connections with African sites of similar age than with Eurasian ones and documents first local appearances of some taxa that established the modern African carnivore fauna.

9

Bovidae

YOHANNES HAILE-SELASSIE, ELISABETH S. VRBA, AND FAYSAL BIBI

Bovids are among the most highly diversified mammalian groups, with 137 extant species in 45 genera (Grubb 1993). Their African fossil record goes as far back as the early Miocene, although most bovid tribes existing today are believed to have originated toward the end of the Miocene (Vrba 1985a). Hence, the Middle Awash late Miocene bovid record is relevant to the study of the origin of modern African bovid communities.

Bovids are represented in the late Miocene deposits of the Middle Awash by at least 17 species in 14 genera (three as yet undetermined) and seven tribes (Boselaphini, Tragelaphini, Reduncini, Bovini, Aepycerotini, Antilopini, and Neotragini), six of which also exist in sub-Saharan Africa today. Four new species are recognized among the boselaphines, tragelaphines, and reduncines, and a new reduncine genus is also recognized. Fossil remains of Bovinae (Boselaphini, Bovini, Tragelaphini) from the Middle Awash late Miocene deposits are significant with respect to the phylogenetic relationships between the boselaphines, the tragelaphines, the extant *Bos/Bison* group, and the *Bubalus/Syncerus* group.

Boselaphines were abundant during the late Miocene of Africa, although they are now restricted to Asia. They were diverse in sub-Saharan Africa toward the end of the Miocene and probably went extinct in Africa at the onset of the Pliocene. *Tragoportax* and *Miotragocerus* appeared in Africa during the middle Miocene and diversified during the late Miocene, before going extinct at the end of the Miocene (Gentry 1999b). The presence of many boselaphine species in terminal Miocene deposits of eastern Africa therefore is not a surprise. However, this also raises the question of how these lineages suddenly went extinct. Their extinction coincides with a rise in diversity among the tragelaphines, suggesting that the boselaphines were possibly replaced ecologically by the tragelaphines. Among the Bovini, *Simatherium* appears in the Kuseralee Member for the first time. The earliest *Ugandax* appears in the earlier Asa Koma Member. These two taxa appear to be the earliest representatives of the *Simatherium-Pelorovis* and *Ugandax-Syncerus* lineages, respectively.

Reduncines are believed to have originated ca. 7–8 Ma. However, they underwent a major radiation toward the end of the Miocene, as seen from their fossil record in the Middle Awash and other contemporaneous sites in eastern Africa. At least five species were present in the Middle Awash between 5.2 Ma and 5.8 Ma, one of them representing a new genus. The abundance of isolated reduncine teeth in the Middle Awash fossil assemblage prompted a detailed study of early reduncine dentition, and two dental morphs were identified in the collection. Both morphs are present in the earlier set of assemblages dated 5.7 Ma (containing horn cores and frontlets of *Zephyreduncinus oundagaisus, Kobus* cf. *porrecticornis,* and Reduncini gen. et sp. indet. based on a single specimen) as well as in the later 5.2 Ma-old assemblages (in which *Redunca ambae* sp. nov. and *Kobus* aff. *oricornis* first appear alongside the persistent *K.* cf. *porrecticornis* lineage).

Alcelaphines are absent in the Asa Koma and Kuseralee members of the Middle Awash. Their absence, especially in the younger Kuseralee Member, is puzzling. Alcelaphines were present at the end of the Miocene not only at Langebaanweg (Kuseralee Member contemporary and where they are abundant) but also in North Africa (in the Sahabi and Wadi Natrun assemblages) and in even earlier deposits of eastern Africa (at Lothagam Nawata Member). Given their presence in contemporaneous deposits, their absence in the Middle Awash late Miocene deposits, if it signifies more than a sampling error, suggests either that crucial aspects of the habitat required by early alcelaphines were not represented in the area or that at the time (or times) of early alcelaphine geographic expansion there were topographic and/or environmental barriers that prevented alcelaphine penetration into the Middle Awash area.

Fossil gazelles are rare in the Miocene-Pliocene fossil record. The fragmentary specimens recovered from this time period are difficult to assign at the species level. The gazelle fossil remains recovered from the Sahabi, for example, were identified only at the genus level (Lehman and Thomas 1987). In addition to the nature of the fossil record of this genus, another difficulty with the classification of gazelles is due to the evolutionary history of the group itself. The genus *Gazella* is a long-lasting bovid. Its relatively conservative horn core morphology renders designation at the species level difficult except for more complete specimens (Gentry 1980). Lehman and Thomas (1987) state that the Sahabi gazelles resemble the *Gazella* species from southwestern Europe and other sites in North Africa. On the other hand, the fossil gazelles from the "E" Quarry at Langebaanweg are closely related to *Gazella vanhoepeni* from Makapansgat Limeworks (Gentry 1980). *Gazella,* although uncommon, is also part of the bovid community at Lothagam (Leakey et al. 1996), with three specimens from the Nawata Formation assigned to *Gazella* sp. indet. (Harris 2003c). The available evidence, therefore, indicates that there were at least two or three species of the genus during the terminal Miocene and basal Pliocene of Africa.

The genus *Raphicerus* belongs to the tribe Neotragini and subfamily Antilopinae. There are three extant species of this genus—*R. melanotis, R. sharpei,* and *R. campestris*—mainly confined to the southern and south central parts of Africa (Gentry 1980). The earliest record of the genus, although fragmentary, comes from the Mpesida Beds, Kenya. It has also been reported from the Lower Nawata Formation of Lothagam (Leakey et al. 1996) by one specimen assigned to *Raphicerus* sp. (Harris 2003c). An extinct form of the species, *R. paralius* Gentry 1980, was first reported from the Quartzose Sand Member (QSM) of Langebaanweg, South Africa. Subsequently, a mandibular fragment with dentition, a horn core, two metapodials, and three astragali from Sahabi have been assigned to *Raphicerus.* However, their affinity at the species level is not clear (Lehman and Thomas 1987). Morphology of the horn core, particularly the basal thickening, indicates similarity with *R. paralius* from Langebaanweg. However, the dental remains are more like the extant species *R. campestris.* As a result, Lehman and Thomas (1987) assign the remains from Sahabi to *Raphicerus* sp.

Madoqua sp. is rare in the late Miocene fossil record of Africa. It is reported from Lothagam as one specimen from the Lower Nawata and three specimens from the Upper Nawata (Harris 2003). Its presence in higher horizons is also limited to one specimen from the Apak Member (Harris 2003).

The bovid assemblages from the older Asa Koma Member and the younger Kuseralee Member of the Middle Awash have many taxa in common with other contemporaneous eastern and South African localities. However, there are also noticeable differences in the bovid assemblages from these sites. Even the two time-successive members within the Middle Awash have different composition of bovid taxa. These differences may reflect real biological effects (environmental change, evolution, or both) or sampling errors. The Middle Awash bovid record, along with evidence from other contemporaneous eastern African sites such as Lothagam and Lukeino, document a significant bovid diversity in eastern Africa toward the end of the Miocene.

Boselaphini

Systematic Paleontology

Tragoportax Pilgrim, 1937

(*Tragocerus* Gaudry, 1861; *Pontoportax* Kretzoi, 1941; ?*Gazelloportax* Kretzoi, 1941; ?*Mirabilocerus* Hadjiev, 1961; *Tragoceridus* Kretzoi, 1968; *Mesembriportax* Gentry, 1974; ?*Mesotragocerus* Korotkevich, 1982)

TYPE SPECIES *Tragoportax salmontanus* Pilgrim, 1937

OTHER SPECIES *Tragoportax amalthea* (Roth and Wagner 1854); *Tragoportax rugosifrons* (Schlosser, 1904); ?*Tragoportax curvicornis* Andree, 1926; *Tragoportax maius* Meladze, 1967; ?*Tragoportax cyrenaicus* Thomas, 1979; *Tragoportax acrae* (Gentry, 1974); *Tragoportax macedoniensis* Bouvrain, 1988.

GENERIC DIAGNOSIS Dorsal frontoparietal surface comprises a flat or slightly concave depressed area, bordered laterally by raised temporal ridges and posteriorly by a slightly raised plateau. Basioccipital often with a median crest present. Occipital only slightly broader ventrally than dorsally, with a trapezoidal rather than triangular outline. Adult male horn cores slender, with triangular to quadrangular basal cross section; sharp posterolateral keel; flattened lateral surfaces; strong mediolateral compression, with transverse diameter (DT) about 50 percent to 80 percent of anteroposterior diameter (DAP); anterior demarcations often present and variably expressed; weak to moderately strong anticlockwise torsion (assessed on the right side from the base up). Intercornual region raised and almost rectangular in outline. Premolar row length reduced to about 60 percent to 70 percent that of molar row (modified from Spassov and Geraads 2004).

Tragoportax Pilgrim, 1937

Tragoportax abyssinicus sp. nov.

ETYMOLOGY The species is named for Ethiopia, using the older Latin name of this country.

HOLOTYPE AMH-VP-1/1 (calvarium with both horn cores, Figure 9.1; left mandible with P_2–M_3, right M_2, fragmentary right M_3, and left upper molar, Figure 9.2) from the Kuseralee Member.

PARATYPE AMW-VP-1/1 (calvarium with both horn cores, left mandible with P_{2-3} and part of M_2, fragmentary left P^2 and P^3, left P^4, left M^{2-3}, fragmentary right M^3) from the Kuseralee Member.

OTHER REFERRED MATERIAL None.

LOCALITIES AND HORIZONS AMH-VP-1 and AMW-VP-1, Kuseralee Member, Middle Awash, Ethiopia, radiometrically dated to between 5.2 and 5.6 Ma (Renne et al. 1999).

DIAGNOSIS Medium-sized *Tragoportax.* The following combination of horn core characters serves to distinguish these specimens from all known species of *Tragoportax:* intermediate size; horns short and straight, with only slight lateral curvature and slight anterior curvature of the distal portions; inserted above the orbits, little inclined (ca. 55°) and little divergent (ca. 40°), with weak anticlockwise torsion (ca. 25° total) and strong mediolateral compression (DT is ca. 60 percent DAP); very prominent demarcation present along the anterior keel halfway to two-thirds distally; distal to demarcation, horn cross section is circular, while proximal to it the cross section is elongate, with flat lateral and anterior faces, convex medial face, sharp posterolateral keel, and strong anterior keel (Figure 9.3); anterior keel rugosity extends 2–3 cm proximally beyond the remainder of the horn core; lower premolar row length between 67 percent and 70 percent of molar row.

DIFFERENTIAL DIAGNOSIS *Tragoportax* (*sensu* Spassov and Geraads 2004) can generally be differentiated from species of *Miotragocerus* (*sensu* Spassov and Geraads 2004) by horn cores with less extreme mediolateral compression, and dentitions with absolutely larger molar rows and relatively shorter premolar rows, although some overlap exists. The new species can be distinguished from most species of *Miotragocerus,* including *M. valenciennesi* (senior synonym of *M. gaudryi* in Kostopoulos 2005) in these regards (Figures 9.4 and 9.5).

Miotragocerus (=Sivaceros) gradiens from the Chinji deposits (Pilgrim 1937, 1939) is similar to the Middle Awash specimens in the presence of frontal and pedicel sinuses and straight, upright horns bearing a large anterior demarcation with some anterolateral flattening of the anterior keel. Furthermore, the holotype of *M. gradiens* exhibits mediolateral horn compression that is within the range of *Tragoportax* and comparable to

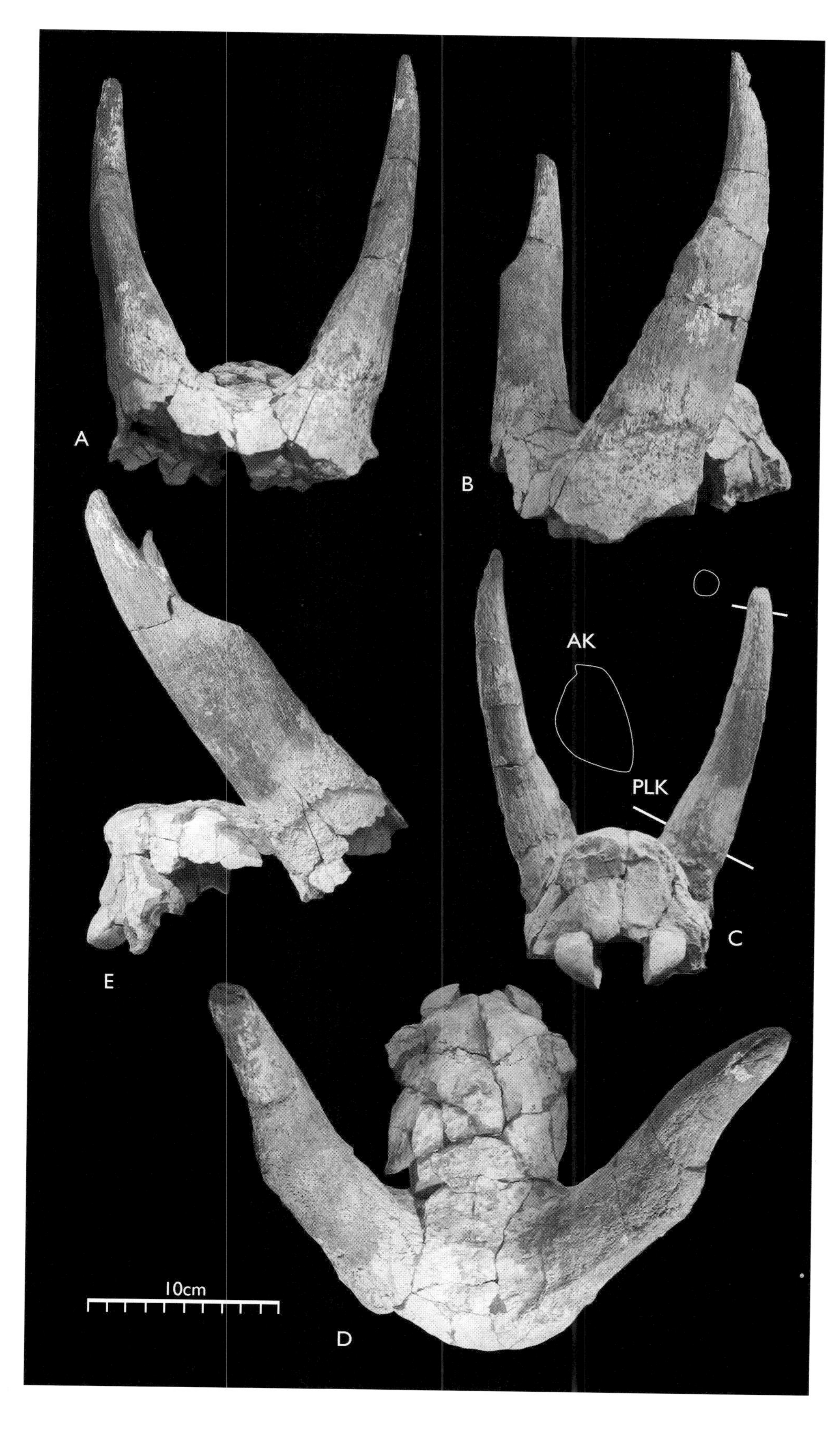

FIGURE 9.1

Cranium of AMH-VP-1/1, holotype of *Tragoportax abyssinicus*. A. Anterior view. B. Left anterolateral view. C. Posterior view. D. Dorsal view. E. Right lateral view. Horn core cross-sectional outlines are shown at scale. Abbreviations: AK = anterior keel; PLK = posterolateral keel.

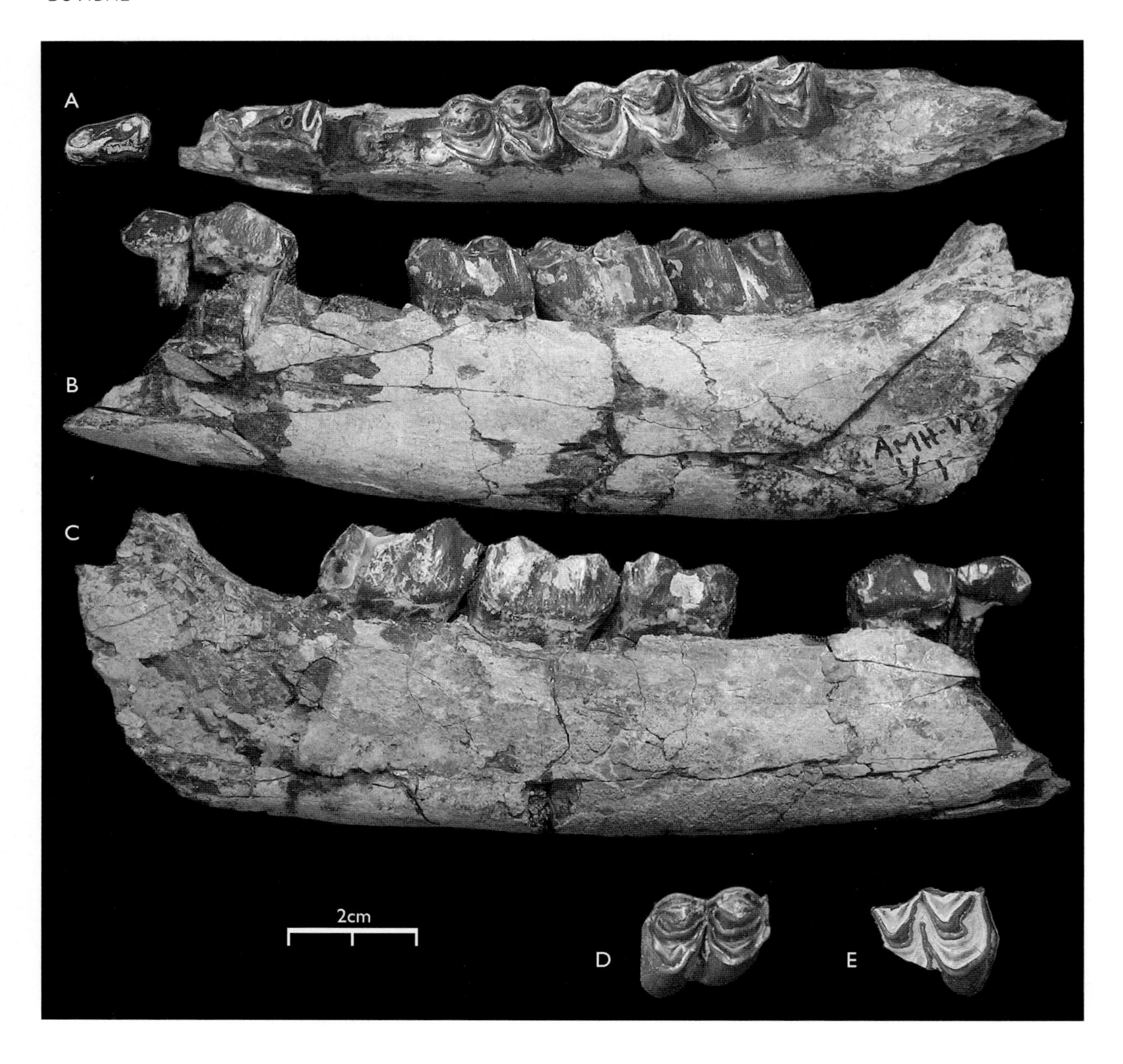

FIGURE 9.2
Left mandible and isolated teeth of AMH-VP-1/1, holotype of *Tragoportax abyssinicus*. A. Mandible, occlusal view. B. Mandible, lateral view. C. Mandible, medial view. D. Right M_2. E. Fragmentary left upper molar.

that of the Middle Awash specimens (Figure 9.5). The Middle Awash fossils differ from *M. gradiens* in having horns that are absolutely larger and longer, but relatively shorter; with greater divergence; even more prominent anterior demarcation, and marked lateral and anterior face flattening; slight lateral curvature; slight anterior curvature of the distal tips; more torsion; dorsal cranial depression present; and less convergent temporal ridges. A very likely descendant of *M. gradiens* is *Miotragocerus (=Sivaceros) vedicus* (Pilgrim, 1939) from the late Miocene Dhok Pathan deposits. This species is little known, but its short and straight horns with flat anterior face and evidence of a large demarcation

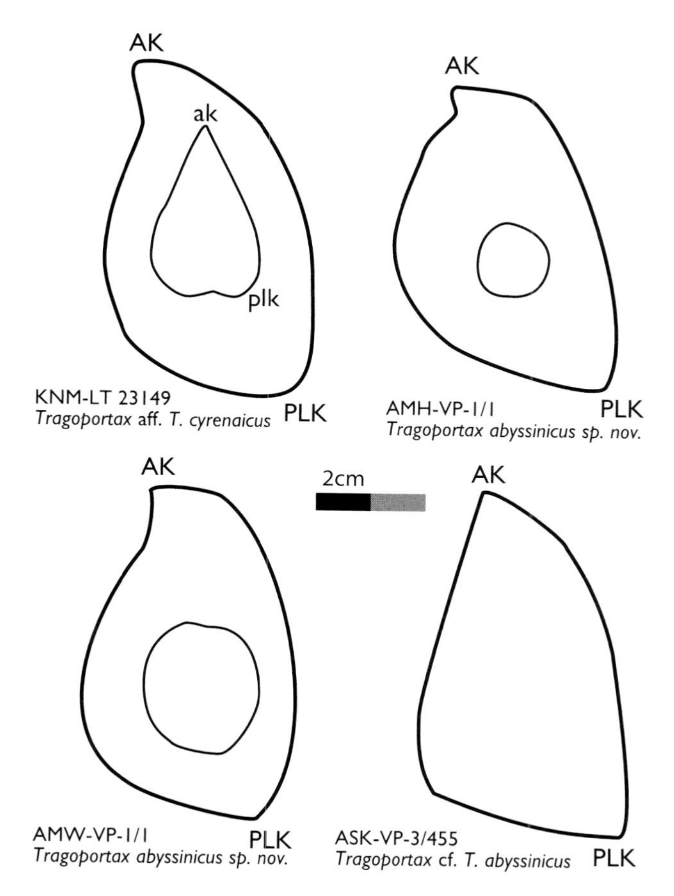

FIGURE 9.3

Basal and distal horn core cross-sectional outlines of *Tragoportax* from Lothagam (top left) and the Middle Awash. All are shown as right horn cores with the anterior keel or its corresponding position towards the top of the page. Distal cross-section outlines are inset in the basal cross section (not present for ASK-VP-3/455, which only preserves the proximal horn core). Distal cross sections were taken at approximately 50 mm (KNM-LT 23149), 15 mm (AMH-VP-1/1), and 30 mm (AMW-VP-1/1) from the horn core distal tip. Abbreviations: same as Figure 9.1.

(in some specimens) make it the fossil boselaphine most closely approaching the Middle Awash material. *Miotragocerus vedicus* is immediately distinguished from the Middle Awash fossils by its smaller size, although larger horn specimens described by Pilgrim (1939) as *Sivaceros* sp. may approach the Middle Awash material in this regard (Figure 9.5). Based on the description and illustrations provided by Pilgrim (1939), the Middle Awash specimens differ from *M. vedicus* in horn cores that are slightly less inclined, more mediolaterally compressed, and with a more flattened lateral surface; dorsal cranial depression present; temporal ridges present; and intercornual region raised, probably reflecting a greater degree of sinus development.

Solounias (1981: figures 35B and D) illustrates female specimens of *Miotragocerus monacensis* that resemble the Middle Awash species in having straight, uprightly inserted horns with a large anterior demarcation toward the distal third of the horn. Specimens attributed to the male of this species, however, have longer, curved horns with multiple small demarcations along the course of the anterior keel. Solounias (1981) describes *M. monacensis* as having pedicels with no significant sinuses, and frontals with only small sinuses. This distinguishes *M. monacensis* from the Middle Awash form, which bears large frontal and pedicel sinuses. Other characters that distinguish the Middle Awash form from

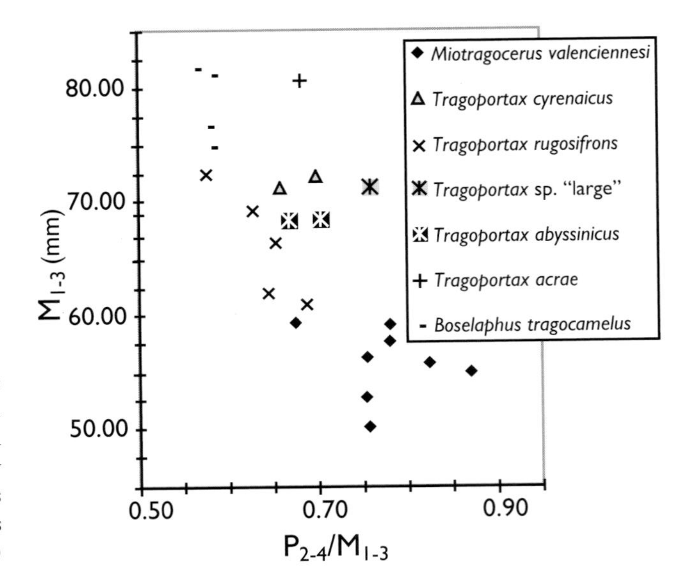

FIGURE 9.4

Proportions of the mandibular dentition in boselaphines. (Data included from Gentry 1974, 1999b; Thomas 1979; Spassov and Geraads 2004; Kostopoulos 2005)

M. monacensis include, in *T. abyssinicus,* a more rounded horn core cross section bearing a flat surface anteriorly; horns inserted farther apart; and intercornual region probably more raised. It should be noted that Solounias describes *M. monacensis* as possessing a dorsal postcornual depression, outlined by the temporal ridges, a feature used by Spassov and Geraads (2004) to diagnose *Tragoportax.* The latter authors include certain specimens identified as *M. monacensis* by Solounias (1981) in *M. gaudryi* (synonym of *M. valenciennesi* according to Kostolopoulos 2005).

Tragoportax abyssinicus differs from *T. cyrenaicus* from Abu Dhabi (Gentry 1999b) and Sahabi (Thomas 1979), as well as *T. acrae* from Langebaanweg (Gentry 1974), in having slightly smaller horn core basal size and much shorter and straighter horns. It also differs from these species in having less spiraling, less divergent horns inserted farther apart, and a less prominent intercornual torus. It resembles *T. acrae* in its lack of backward curvature of its horns, while further differing from *T. cyrenaicus* in this regard. *Tragoportax abyssinicus* further differs from *T. acrae* in the more distal placement of the horn anterior demarcation.

Tragoportax abyssinicus may be distinguished from *T.* aff. *T. cyrenaicus* from Lothagam (Harris 2003) in the very large anterior demarcation, which, if even present in the Lothagam species, is located more proximally in the Middle Awash specimens; more flattened anterior keel (proximal to demarcation); less torsion; horns without backward curvature; and much shorter horns.

Tragoportax abyssinicus differs from the small Lothagam *Tragoportax* sp. A and sp. B (Harris 2003) in having a larger size; much more prominent anterior demarcation; horns absolutely longer than sp. A; and much more polygonal horn core cross section, with flattened lateral and anterior faces.

Tragoportax rugosifrons is taken to include *T. curvicornis, T. browni,* and *T. recticornis* (Bouvrain 1988; Gentry 1999b; Moyà-Solà 1983). The new material differs from

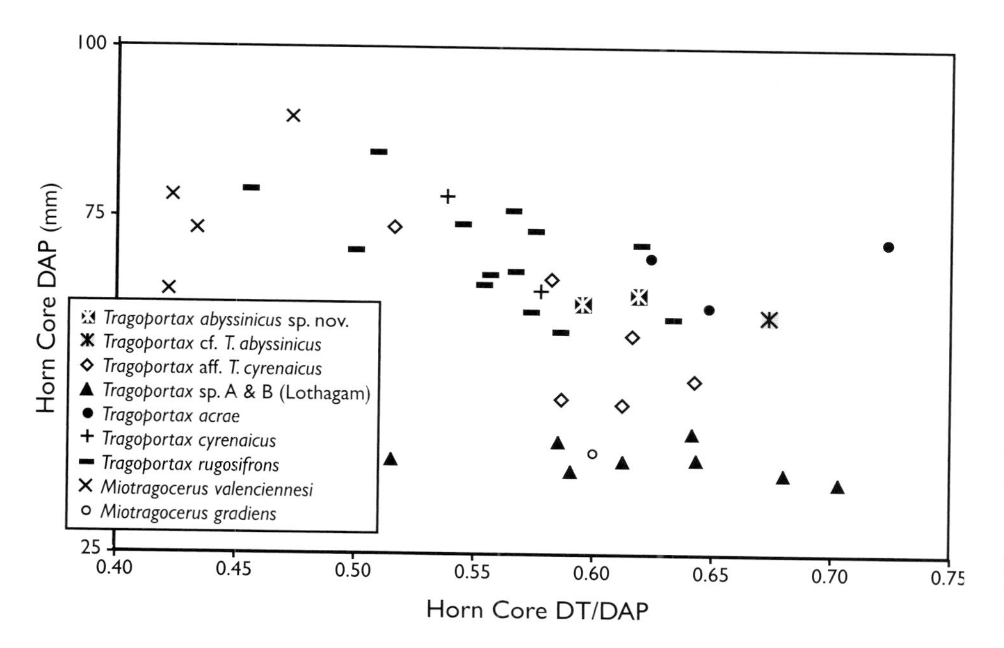

FIGURE 9.5

Basal horn core size and compression in boselaphines. Abbreviations: DAP = horn core basal anteroposterior diameter; DT = horn core basal transverse diameter. (Data included from Pilgrim 1937, 1939; Gentry 1974, 1999b; Thomas 1979; Bouvrain 1994; Harris 2003; Spassov and Geraads 2004; Kostopoulos 2005)

T. rugosifrons from Hadjidimovo (Spassov and Geraads 2004), and Prochoma and Vathylakkos (Bouvrain 1994) in horns that are straighter and shorter, with less divergence; no posterior curvature; weaker intercornual torus; and large anterior demarcation and flattened anterior face.

Tragoportax abyssinicus differs from *T. salmontanus* from the Dhok Pathan (Pilgrim 1937) in the large anterior demarcation and flattened anterior face; horns more upright and without backward curvature or significant torsion; larger basal horn core size; greater mediolateral compression; longer horns; and horns inserted more widely apart.

Tragoportax abyssinicus differs from *T. amalthea* in horns that are shorter, straighter, with smaller basal size and a large anterior demarcation. Metrics are given in Tables 9.1 and 9.2.

REMARKS The diagnosis is based on the cranial morphology of two adult specimens (AMH-VP-1/1 and AMW-VP-1/1). Although it is not possible to be sure, both specimens are probably male, given that female horns are often small and morphologically simple (e.g., Bouvrain 1988; Kostopoulos 2005; Spassov and Geraads 2004).

DESCRIPTION

Type Specimen AMH-VP-1/1 preserves much of the calvarium, including complete left and right horn cores, as well as most of the left mandible, and some isolated teeth. The cranial vault was reconstructed from over a dozen fragments, but each of these is relatively large and well-preserved, such that the reconstructed morphology is faithful to the original condition. The intercornual region is slightly raised, with a transverse ridge bordering this area posteriorly; frontal sinuses are large and strutted, extending into the pedicel and possibly proximal few centimeters of the horn core proper. The frontoparietal

TABLE 9.1 Measurements of AMH-VP-1/1, the Holotype of *Tragoportax abyssinicus* sp. nov.

Horn core divergence	45°	
Horn core inclination	55°	
Frontal biorbital breadth	As preserved = 132.44	Complete = (136.5)
Bimastoid breadth	97.69	
Occipital bicondylar width	External = 64.26	Internal = 20.65
Frontal-parietal angle	ca. 135°	
Parietal-nuchal angle	105°	
Distance posterior border of intercornual torus–exoccipital process	122.47	
Occipital height	ca. 77.5	
Horn core length along anterior keel	223	
Horn core length along posterior surface	210	
Intercornual distance	External = 143.12	Internal = ca. 62.25
Lower premolar row length	ca. 47.5	
Lower molar row length	ca. 68.40	
Horn core basal DAP	62.22	
Horn core basal DT	37.05	
Frontal thickness at intercornual midline (including sinuses)	49.46	
Torsion	25°	

NOTE: All measurements in mm except for angles as indicated. Estimated values are in parentheses.

dorsal surface comprises a depressed area that is highly textured and bounded laterally by raised temporal ridges and posteriorly by a raised and rugose plateau. A weak sagittal ridge is present, extending posteriorly from the moderately raised intercornual "torus." The occipital surface of AMH-VP-1/1 is quite broad inferiorly and narrow superiorly, approaching the condition supposedly diagnostic of *Miotragocerus* (occipital "triangular," Spassov and Geraads 2004), although not different from that seen in *Tragoportax rugosifrons (=Tragocerus browni)* from the Siwaliks (Pilgrim 1937). In dorsal view, the braincase appears narrow and long, another character suggestive of affinities to species of *Miotragocerus.* The basioccipital bears very large posterior tuberosities with prominent posterior rims, a wide and smooth area separating these from the condyles. The anterior tuberosities are small. A very prominent longitudinal crest is present, beginning anterior to these tuberosities and increasing in prominence as it crosses over the fused basioccipital-sphenoid suture. Spassov and Geraads (2004) propose that a "keeled" basioccipital is characteristic of *Miotragocerus,* but the presence of this character in AMH-VP-1/1 as well as in the Siwalik *T. rugosifrons* suggests that it is variable within these genera.

The mandible retains P_3, M_1, M_2, and most of the M_3. All teeth possess strongly wrinkled enamel. The P_2 is complete but isolated from the mandible. It is in early wear, whereas the other teeth are all moderately worn. The P_3 paraconid is small and not much separated from the parastylid. The metaconid sweeps back strongly, and a central enamel island is present between it and the long and broad entoconid. The metaconid is short but well-separated from the entoconid. The hypoconid projects lingually, but weakly. The lower molars exhibit transverse anterior flanges (although not a true "goat fold") that decrease in size from M_1 to M_3; curved (not flat) medial faces; small basal pillars probably

TABLE 9.2 Dental Metrics of Boselaphini and cf. Boselaphini

Specimen	*Taxon*	*Tooth*	*AP*	*T*	*Unworn Crown Height*
ASK-VP-1/12	cf. Boselaphini	R. M^3	19.62 (22.41 w/style)	19.03	ca. 30
ASK-VP-3/35	cf. Boselaphini	L. $P_{2 \text{ or } 3}$	13.88	7.42	9.92
ASK-VP-3/41	cf. Boselaphini	R. M^3	17.32+	18.39	ca. 22.5
ASK-VP-3/43	cf. Boselaphini	L. M^3	20.12	17.54	ca. 20.5
ASK-VP-3/333	cf. Boselaphini	R. $P_{2 \text{ or } 3}$	15.61	9.06	11.57
ASK-VP-3/528	cf. Boselaphini	L. M^2	(28.2+)	19.50	
ALA-VP-1/17	*Tragoportax* sp. "large"	R. M_3 fragment		13.81+	
ALA-VP-2/9	*Tragoportax* sp. "large"	R. M_2	22.66	14.92	ca. 26.0
ALA-VP-2/69	*Tragoportax* sp. "large"	R. M^2 fragment	21.62+		
ALA-VP-2/115	*Tragoportax* sp. "large"	L. M^2	22.24	23.63+	
ALA-VP-2/265	*Tragoportax* sp. "large"	R. M_1 or $_2$	21.26	14.38	
ALA-VP-2/266	*Tragoportax* sp. "large"	L. M_3 fragment		14.25	
ALA-VP-2/267	*Tragoportax* sp. "large"	R. M_2	21.84	14.39	
ASK-VP-3/11	*Tragoportax* sp. "large"	R. M_3	30.07	15.08	24
ASK-VP-3/37	*Tragoportax* sp. "large"	L. M_2	20.68+	15.32	
ASK-VP-3/527	*Tragoportax* sp. "large"	L. M_2	20.54+		
STD-VP-2/5	*Tragoportax* sp. "large"	R. M_1	20.62	14.23	
STD-VP-2/5	*Tragoportax* sp. "large"	R. M_2	21.83	15.11	
STD-VP-2/5	*Tragoportax* sp. "large"	R. M_3	29.08	14.37	
STD-VP-2/5	*Tragoportax* sp. "large"	R. P_2 ALV	12.79	6.95	NA
STD-VP-2/5	*Tragoportax* sp. "large"	R. P_3 ALV	18.25	8.20	NA
STD-VP-2/5	*Tragoportax* sp. "large"	R. P_4	19.92	10.74	
ALA-VP-1/16	*Tragoportax* cf. *abyssinicus* sp. nov.	L. M^3	20.10 (23.96 w/style)	20.84	ca. 21.5
ALA-VP-1/20	*Tragoportax* cf. *abyssinicus* sp. nov.	R. P_3	16.96	9.86	
ALA-VP-2/148	*Tragoportax* cf. *abyssinicus* sp. nov.	R. $M_{1 \text{ or } 2}$	19.77	12.00	
DID-VP-1/11	*Tragoportax* cf. *abyssinicus* sp. nov.	R. M_1	18.58	11.73	
DID-VP-1/11	*Tragoportax* cf. *abyssinicus* sp. nov.	R. M_2	20.36	13.67	
DID-VP-1/11	*Tragoportax* cf. *abyssinicus* sp. nov.	R. M_3	25.11	12.67	ca. 19.5
ALA-VP-2/345	*Tragoportax* cf. *abyssinicus* or sp. "large"	L. M_3	27	12.90	
ASK-VP-3/38	*Tragoportax* cf. *abyssinicus* or sp. "large"	R. M_3	26.56	13.64	
KWA-VP-1/3	*Tragoportax* cf. *abyssinicus* or sp. "large"	R. P_4	18.61	10.08	(13.5)
STD-VP-2/9	*Tragoportax* cf. *abyssinicus* or sp. "large"	L. M^2	22.49	21.72	ca. 21.5
AMH-VP-1/1	*Tragoportax abyssinicus* sp. nov.	L. M_1	18.08+	12.22	
AMH-VP-1/1	*Tragoportax abyssinicus* sp. nov.	L. M_2	19.59	13.12	
AMH-VP-1/1	*Tragoportax abyssinicus* sp. nov.	L. M_3	ca. 28.65	12.79	
AMH-VP-1/1	*Tragoportax abyssinicus* sp. nov.	L. P_2	12.69	7.25	
AMH-VP-1/1	*Tragoportax abyssinicus* sp. nov.	L. P_3	16.08	9.38	
AMH-VP-1/1	*Tragoportax abyssinicus* sp. nov.	L. P_4 ALV	16.61	7.45	NA
AMH-VP-1/1	*Tragoportax abyssinicus* sp. nov.	L. M^x fragment	>15	>18	
AMH-VP-1/1	*Tragoportax abyssinicus* sp. nov.	R. M_2	19.9	13.55	
AMW-VP-1/1	*Tragoportax abyssinicus* sp. nov.	L. M_1 ALV	20.29	11.89	NA
AMW-VP-1/1	*Tragoportax abyssinicus* sp. nov.	L. M_2 ALV		13.44	
AMW-VP-1/1	*Tragoportax abyssinicus* sp. nov.	L. M_3 part ALV		ca. 13.0	NA
AMW-VP-1/1	*Tragoportax abyssinicus* sp. nov.	L. P_2	13.51	7.83	
AMW-VP-1/1	*Tragoportax abyssinicus* sp. nov.	L. P_3	16.66	10.12	
AMW-VP-1/1	*Tragoportax abyssinicus* sp. nov.	L. P_4 ALV	18.08	8.00	NA

TABLE 9.2 *(continued)*

Specimen	*Taxon*	*Tooth*	*AP*	*T*	*Unworn Crown Height*
AMW-VP-1/1	*Tragoportax abyssinicus* sp. nov.	L. M^2	20.77	22.92	nm
AMW-VP-1/1	*Tragoportax abyssinicus* sp. nov.	L. M^3	19.81; 21.79 w/style	20.97	nm
AMW-VP-1/1	*Tragoportax abyssinicus* sp. nov.	L. P^2 or ^{3}	17.77	13.97	nm
AMW-VP-1/1	*Tragoportax abyssinicus* sp. nov.	L. P^2 or ^{3} fragment	16.81+		nm
AMW-VP-1/1	*Tragoportax abyssinicus* sp. nov.	L. P^4	13.36	16.67	nm
AMW-VP-1/1	*Tragoportax abyssinicus* sp. nov.	R. M^3	(19.7)		nm

NOTE: All measurements are in mm. Abbreviations: ALV = alveolus only; ( ) = estimated values; + = worn or broken, value is higher; nm = too worn to measure.

never taller than half of the unworn crown height; and deep valleys between the anterior and posterior lobes such that the two lateral cusps do not fuse until late in wear.

Other Key Specimens The paratype specimen AMW-VP-1/1 is also a relatively complete calvarium, although with less complete horns and dorsal cranial surface than in the holotype. The mandibular dentition in AMW-VP-1/1 is poorly preserved, yet this specimen preserves some of the upper dentition. The teeth of this species are generalized and typical of the fossil Boselaphini, or even of the primitive bovid condition. The M^2 and M^3 are complete and in mid-wear. These teeth bear very deep valleys between the anterior and posterior lobes, and the connection between the two lingual cusps is very weak. Basal tubercles (hardly "pillars") are present, associated with narrow cingulae that wrap around the lingual cusps at the crown base. Parastyles and mesostyles are strong, projecting anterolaterally and laterally, respectively, and the metastyle on M^3 is prominent and posteriorly projecting. ASK-VP-3/455 is also referred to this group based on its overall horn core size and shape similarity to AMW-VP-1/1 and to the type specimen. It also shows some similarities with the Lothagam form.

cf. *Tragoportax abyssinicus*

DESCRIPTION AMW-VP-1/1 preserves only the proximal horn core portion with pedicel. It matches *T. abyssinicus* in overall size and in the general cross-sectional shape. Its basal DAP and DT are 63.55 mm and 39.33 mm, respectively. Its complete length is ca. 190 mm. The pedicel and proximal 3 cm of the horn core proper are hollowed and strutted. This horn core specimen differs from those of *T. abyssinicus* in less basal mediolateral compression (Figure 9.5) and a basal cross section with a much wider anterior-lateral face and slightly flatter medial and posterior faces. The proximal portions of horn cores of *T.* aff. *T. cyrenaicus* from Lothagam and *T. abyssinicus* may be difficult to distinguish. ASK-VP-3/455 is probably just an older representative of *T. abyssinicus* (basal DAP = 60.3 mm; basal DT = 40.62 mm), but without knowledge of the distal horn morphology one cannot distinguish it with any certainty from the Lothagam form. The isolated dental specimens match those of *T. abyssinicus* in size and general morphology (Table 9.2).

Tragoportax sp. "large"

DESCRIPTION A second, larger, boselaphine species is present in the Middle Awash, best represented by the mandible STD-VP-2/5. The lower molars exhibit relatively large goat folds and small, columnar basal pillars reaching halfway up the crown. The P_4 metaconid is very elongated anteriorly, a character suggested to be a synapomorphy of *Tragoportax* (Spassov and Geraads 2004). It is differentiated from *T. abyssinicus* by larger overall size and a relatively longer premolar row (especially P_3 and P_4). Length of M_1–M_3 compares with that of *T. cyrenaicus*, while the degree of premolar row reduction is very low and on the order of that seen in *Miotragocerus* (Figure 9.4). Isolated dentitions are also referred to the large *Tragoportax* (Table 9.2).

cf. *Tragoportax abyssinicus* or sp. "large"

DESCRIPTION Isolated molars that, on the basis of size, cannot be distinguished from either *T. abyssinicus* or the larger species represented by STD-VP-2/5 are referred to this group. They are referred to *Tragoportax* based on general congruence of morphology with the specimens of *T. abyssinicus* and *T.* sp. "large" described earlier. Measurements are given in Table 9.2.

cf. Boselaphini Knottnerus-Meyer, 1907

DESCRIPTION These are isolated teeth that, on the basis of size and morphology, probably belong to a boselaphine. Characters identifying these teeth as boselaphine include brachydonty, pointed labial lower and lingual upper cusps that do not fuse until late wear stage, goat folds (variably expressed), weak ribs and pillars, and overall size. Measurements are given in Table 9.2.

DISCUSSION Early late Miocene boselaphines such as *Protragocerus labidotus* and *Sivoreas eremita* occur at Ngorora. Only the latter has been reported from the Ngeringerowa Formation, which is contemporaneous with Ngorora (Benefit and Pickford 1986). *Miotragocerus, Tragocerus* and *Tragoportax* are known only after 7 Ma. *Miotragocerus* is mostly known outside Africa, particularly from the Siwalik Formation of Pakistan and India, along with numerous other species of the genus *Tragoportax. Miotragocerus cyrenaicus* is the only known species of the genus documented at Sahabi (Thomas 1979; Lehman and Thomas 1987). *Mesembriportax acrae* is another boselaphine documented from Langebaanweg (Gentry 1980). Neither *M. cyrenaicus* nor *M. acrae* has been reported outside Sahabi and Langebaanweg, respectively. Gentry (1999b), however, lumps both *Mesembriportax* of Langebaanweg and *Miotragocerus* of Sahabi into the genus *Tragoportax.* The Boselaphines from Lothagam have been initially assigned to the genus *Tragocerus* (Leakey et al. 1996). However, Harris (2003) recognized four species from the Nawata Formation: *Tragoportax* aff. *T. cyrenaicus, Tragoportax* sp. A, *Tragoportax* sp. B, and Boselaphini gen. et sp. indet.

The Middle Awash *Tragoportax abyssinicus* represents a new species of the *Miotragocerus-Tragoportax* complex, a diverse group of boselaphine bovids that inhabited much of Africa and Eurasia during the late Miocene and earliest Pliocene. The morphological variation between the different species within this complex is pronounced. This, coupled with a confused nomenclatural history, has made for a lack of consensus on the systematics of these fossil boselaphines. The most recent attempt to clarify the differences between *Tragoportax* and *Miotragocerus* is that of Spassov and Geraads (2004), whose classification differs from previous ones (e.g., Solounias 1981; Moyà-Solà 1983; Bouvrain 1988; Gentry, 1999b). Regardless of which classification scheme is used, it is apparent that this successful group was subject to high degrees of endemism. This is exemplified by a comparison between the localities of Lothagam (Harris 2003) and the Middle Awash, which, although within close proximity and roughly contemporaneous, do not seem to share any of the three or four *Tragoportax* species present. At the same time, the three Lothagam *Tragoportax* species (Harris 2003) share with *T. abyssinicus* certain straightness and lack of torsion of the horn, although the Middle Awash species is distinctly differentiated from these forms by its pronounced anterior demarcation.

The horns of *T. abyssinicus* resemble those of *Miotragocerus (=Sivaceros) gradiens* and *Miotragocerus (=Sivaceros) vedicus* from the Indo-Pakistani Siwaliks (Pilgrim 1937, 1939). These similarities, if not convergent, may reflect the degree of continuity between the faunas of eastern Africa and southern Asia in Mio-Pliocene times. The similarities noted here between the morphology of *T. abyssinicus* and that of species included in *Miotragocerus* indicate that the assignment of the Middle Awash material to the genus *Tragoportax* may require later amendment, or that Pilgrim's *Sivaceros* would be better accommodated in *Tragoportax* rather than *Miotragocerus.* Gentry (1999b) posits that shorter horns, presence of an anterior demarcation, and weaker torsion are characters primitive within *Tragoportax.* Further analyses of the systematics within the *Miotragocerus-Tragoportax* complex, including *T. abyssinicus,* are forthcoming (Bibi, in prep.).

Bovini

Systematic Paleontology

Ugandax Cooke and Coryndon, 1970

Ugandax sp.

DIAGNOSIS The diagnosis is that of the only named species, *Ugandax gautieri,* and is compiled from Cooke and Coryndon (1970), Gentry (1978a), and Vrba (1987). Horn cores have rounded cross sections, inserted relatively upright, with almost no compression and weak keels. The braincase is not very shortened and is less low and wide than in *Syncerus.* Dental occlusal morphology is more derived, and the premolar row is less reduced than in contemporaneous *Simatherium-Pelorovis.*

DESCRIPTION *Ugandax* sp. from the Middle Awash late Miocene is a bovine of large size. The horn cores (Figure 9.6) are inserted above the orbits. Horns are moderately divergent (80°) and moderately inclined (35° against braincase); with strong posterolateral keel, rounded anterior keel, and rounded posterointernal corner; mediolaterally compressed with a very flat medial surface, flat posterior surface, and a slightly convex lateral face that

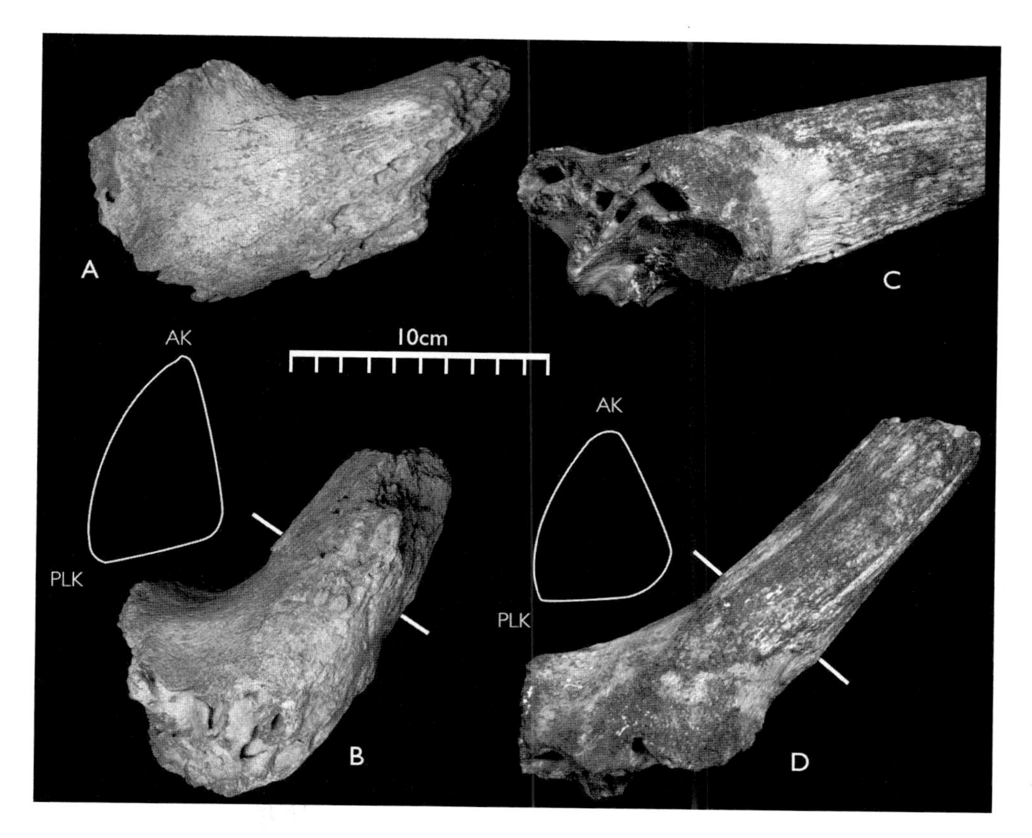

FIGURE 9.6

Horns cores assigned to *Ugandax* sp. A. ASK-VP-3/3, left horn core, dorsal view. B. Same, anterior view. C. KWA-VP-1/1, left horn core, anterolateral view. D. Same, anterior view. Cross sections are shown at scale. Abbreviations: same as Figure 9.1.

forms a triangular cross section with a rounded 90° angle at the posterointernal corner. They lack torsion. The anterior keel rugosity continues proximally along the pedicel a few more centimeters than the remainder of the horn core. The horn cores bear rugose longitudinal grooves. The frontal contains large, extensive, and strutted sinuses that extend into the pedicel and probably into the first few centimeters of horn core proper. ASK-VP-3/3, a left horn core, has DAP and DT of 72.33 mm and 58.24 mm, respectively. KWA-VP-1/1 is also a left horn core with a DAP measurement of 65.74 mm and DT of 52.59 mm.

These specimens differ most significantly from the type specimen of *Ugandax gautieri* in the horn core cross section, this being triangular in the Middle Awash specimens and rounded in *U. gautieri* (Cooke and Coryndon 1970). A substantial amount of morphological variation, particularly in horn core cross section, is common in living bovines such as *Syncerus* and *Bison,* where geographic variation is strong and female horn morphology differs from that of males. However, the range of horn core shape variation in *Ugandax* is not completely determined, hence the Middle Awash specimens are only tentatively referred to this genus. The two referred horn cores from the Middle Awash late Miocene could morphologically match certain horn cores identified as *Ugandax* from much younger deposits in eastern Africa. For example, they are similar to *Syncerus acoelotus* from Olduvai in possessing a triangular cross section and sinuses not extending deeply into the horn core, but differ from this more advanced species in the much more upright insertion of the horns.

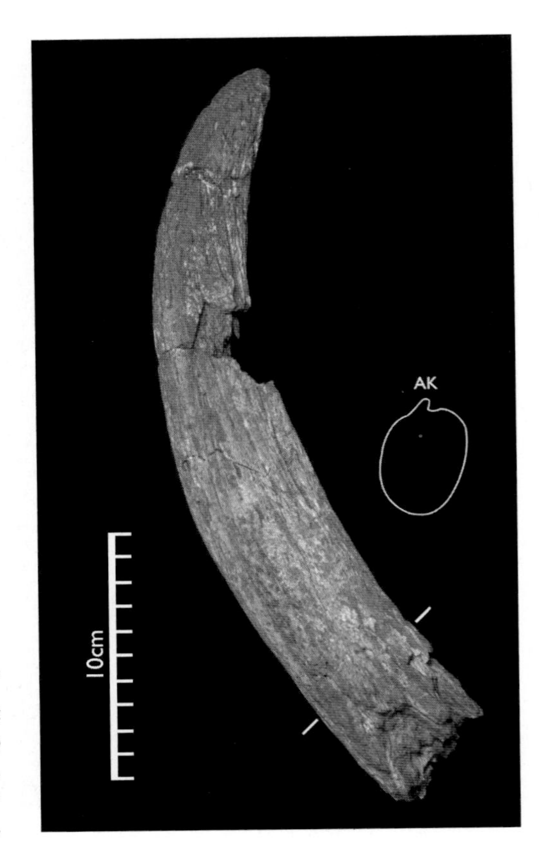

FIGURE 9.7
AME-VP-1/49, right horn core assigned to *Simatherium demissum*. Abbreviation: AK = anterior keel.

Simatherium Dietrich, 1941

Simatherium demissum Gentry, 1980

DIAGNOSIS Horn cores with strong anterior keel, rounded cross section, strong basal divergence, and strong curvature. Frontal sinuses continuing into pedicel and slightly into the horn core proper. Cheek teeth are only moderately hypsodont, with simple central cavities, moderately strong styles, large ribs (localized on lower molars but not on uppers), and weak goat folds on lower molars. P_4 shows a weakly projecting hypoconid, backward-slanting metaconid, and paraconid distinct from parastylid (abbreviated from Gentry 1980).

DESCRIPTION AME-VP-1/49 (a right horn core and right edentulous mandible preserving alveoli P_3–M_3; Figure 9.7) is the only specimen attributed to this species. The horn core is broken very close to its natural base, with a small part of the pedicel probably present. The cross section is rounded, with slight mediolateral compression. Horns are curved but lack torsion. A strong anterior keel is present, and a wide but shallow groove borders the keel ridge laterally (ventrally). Longitudinal grooving is present on the horn surfaces. Measurements of the horn core DAP and DT are 71.38+ mm and 52.79+ mm, respectively. It is 365 mm long. Dental measurements of Bovini and cf. Bovini are given in Table 9.3. AME-VP-1/49

TABLE 9.3 Dental Metrics of Bovini and cf. Bovini

Specimen	*Taxon*	*Tooth*	*AP*	*T*	*Unworn Crown Height*
ALA-VP-2/6	Bovini	R. M_3	41.41	20.11	>34
ALA-VP-2/98	Bovini	L. P_3	22.8	13.8	
ALA-VP-2/262	Bovini	L. $M^{1 \text{ or } 2}$	27.55	29.74	
ALA-VP-2/263	Bovini	L. M^1	28.19	27.31	
ALA-VP-2/264	Bovini	L. P^4	17.69	23.80	ca. 25
ALA-VP-2/344	Bovini	R. M^x	28.90+	28.98+	
AME-VP-1/6	Bovini	L. M_1	21.48	ca. 18.5	
AME-VP-1/6	Bovini	L. M_2	26.69	19.74	
AME-VP-1/6	Bovini	L. M_3	41.17	19.24	
AME-VP-1/6	Bovini	L. P_3 fragment		10.95	
AME-VP-1/6	Bovini	L. P_4	20.44	13.81	
AME-VP-1/6	Bovini	R. M_1	21.75+	18.72	
AME-VP-1/6	Bovini	R. M_2	27.5	19.79	
AME-VP-1/6	Bovini	R. M_3	43.35	18.51	
AME-VP-1/6	Bovini	R. P_2	15.65	8.79	11.28
AME-VP-1/6	Bovini	R. P_3	19.18	11.49	
AME-VP-1/6	Bovini	R. P_4	21.10	13.62	
AME-VP-1/39	Bovini	L. M_3		16.95+	ca. 35.5
AME-VP-1/49	*Simatherium demissum*	R. M_1 ALV	22.6	15.53	NA
AME-VP-1/49	*Simatherium demissum*	R. M_2 ALV	25.72	16.23	NA
AME-VP-1/49	*Simatherium demissum*	R. M_3 ALV	35.9	18.22	NA
AME-VP-1/49	*Simatherium demissum*	R. P_3 ALV	ca. 23.60	9.68	NA
AME-VP-1/49	*Simatherium demissum*	R. P_4 ALV	20.4	12.45	NA
AME-VP-1/89	Bovini	R. M_3	37.38	17.84	ca. 38
AME-VP-1/118	Bovini	L. M_2	25.46	18.14	
AME-VP-1/118	Bovini	L. M_3	36.49+	18.34	
AME-VP-1/138	Bovini	R. M_3	35.41+	14.46	ca. 33.0
AME-VP-1/157	Bovini	L. M^1	27.98	29.26	ca. 26
AME-VP-1/157	Bovini	L. M^2	28.31	27.4	ca. 34.5
AME-VP-1/157	Bovini	L. M^3		27.09+	ca. 33.0
AMW-VP-1/24	Bovini	L. M_3	38.77+	20.52	ca. 40.25
AMW-VP-1/25	Bovini	R. M^1	26.01	27.30+	
AMW-VP-1/48	Bovini	R. M_x fragment		18.28	
AMW-VP-1/88	Bovini	M^x fragment			
AMW-VP-1/102	Bovini	R. mandible fragment w/$M_{1 \text{ or } 2}$	25.25	14.65	35.15
ASK-VP-3/113	Bovini	L. $M^{1 \text{ or } 2}$	28.04+	19.72+	
ASK-VP-3/528	Bovini	R. M_3 fragment	(38.0)	19.36	ca. 36.2
ASK-VP-3/535	Bovini	L. M_3	36.30	17.97	
BIK-VP-3/25	Bovini	L. M_3 fragment	(36.5)	17.53+	
KUS-VP-1/6	Bovini	L. M_2	27.45	20.03	ca. 33
KUS-VP-1/28	Bovini	R. M_1	24.97		
KUS-VP-1/61	Bovini	R. M_3	39.68	18.63	ca. 41
KUS-VP-1/61	Bovini	R. P_2	16.8	9.09	13.62
KUS-VP-1/66	Bovini	R. $M^{2 \text{ or } 3}$	29.47	24.07	
KUS-VP-1/70	Bovini	L. $M^{2 \text{ or } 3}$	29.24	28.41	ca. 33
KUS-VP-1/87	Bovini	R. M_3	37.58	16.21	

TABLE 9.3 *(continued)*

Specimen	*Taxon*	*Tooth*	*AP*	*T*	*Unworn Crown Height*
KUS-VP-1/90	Bovini	R. mandible fragment w/M_3	38.43	16.51	
ADD-VP-1/9	cf. Bovini	R. $M^{1 \text{ or } 2}$	23.93	23.38+	
ASK-VP-3/39	cf. Bovini	R. M^x	23.53	27.46	30.0
ASK-VP-3/40	cf. Bovini	L. M^x	25.46+		
DID-VP-1/17	cf. Bovini	R. $dP^{4?}$	22.00 (25.82 w/style)	24.17	

NOTE: All measurements are in mm. Abbreviations: ALV = alveolus only; ( ) = estimated values.

is a very good match, both in size and morphology, for the type material of *Simatherium demissum* described by Gentry (1980) from Langebaanweg, South Africa.

Bovini gen. et sp. indet.

DESCRIPTION Specimens assigned to this group are largely isolated dental remains from the Adu-Asa Formation and Kuseralee Member (see Figure 9.8 and Table 9.3 for measurements). The teeth are assigned to Bovini based on characters such as large size, large and tall basal pillars, enlarged ribs, anteroposteriorly compressed upper labial and lower lingual cusps, high crowns, and the presence of cementum. Although the specimens exhibit some variation in size, there are no appreciable differences between bovine teeth from the two members. Compared to *Simatherium demissum* from Langebaanweg (Gentry 1980), these teeth accord with the smaller end of the range for that species. This places them closer to the center of the size range of *Ugandax* from Kaiso (Cooke and Coryndon 1970), Hadar (in Gentry 1980), and Lukeino (Thomas 1980).

DISCUSSION While certain late Miocene Siwaliks forms may represent the earliest bovines (Bibi, in prep.), the earliest African record of the tribe Bovini is documented from Toros-Menalla, Chad (Vignaud et al. 2002), biochronologically estimated to be between 6 and 7 Ma. Slightly younger bovine material from Lukeino was assigned to the genus *Ugandax* (Thomas 1980, 1984). These specimens were collected from the Kapcheberek Member (BPRP site 76; A. Hill, personal communication), dated to between 5.70 and 5.88 Ma (Deino et al. 2002). Thomas's recognition of *Ugandax* at Lukeino was based on a right maxillary fragment with all three molars, and another isolated upper molar. Thomas noted that the molars are not different from those of *Simatherium kohllarseni* from Laetoli (3.6 Ma), except that the Lukeino molars lacked cement. *Ugandax,* only known from Africa, is also known from the early Pliocene. *Ugandax* is considered ancestral to the *Syncerus* lineage, although *Syncerus* appears first in Member C of the Omo Shungura Formation.

Diagnosing horns of *Ugandax* is problematic because they are characterized by their primitiveness. Limited sinuses, keels, a triangular cross section (in the Middle Awash specimens), and upright insertion are characters primitive for Bovini horn cores. Similarities between

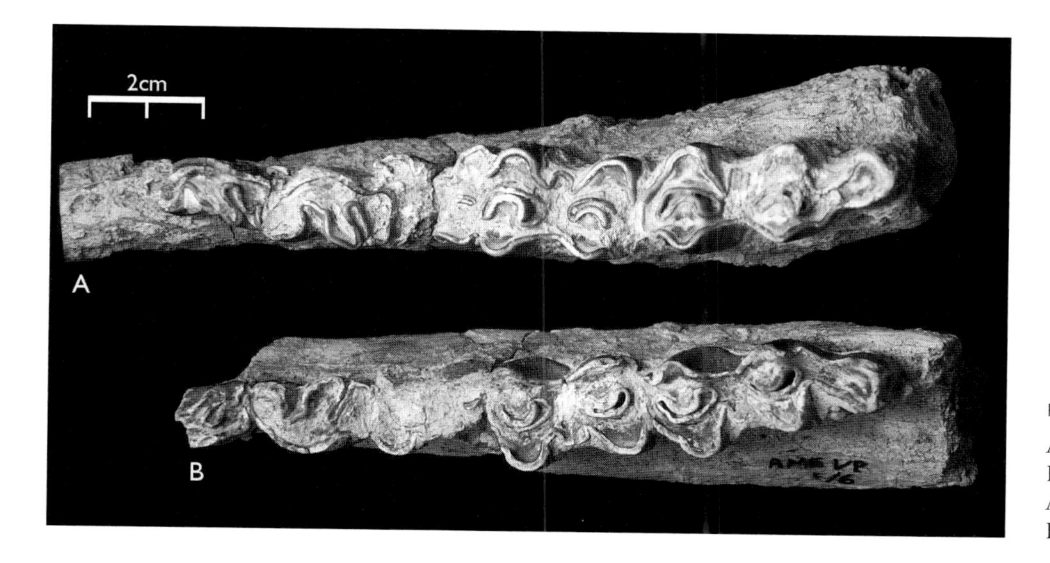

FIGURE 9.8

AME-VP-1/6, assigned to Bovini gen. et sp. indet. A. Partial right mandible. B. Partial left mandible.

Ugandax and Asian bovines such as *Proamphibos* and *Bubalus* (Cooke and Coryndon 1970; Gentry 1978a) are probably a result of the retention of primitive characters and cannot be indicative of phylogenetic affinity. Gentry and Gentry (1978) and Vrba (1987) have proposed that *Ugandax* gave rise to *Syncerus* and that *Simatherium* and *Pelorovis* constitute a second, separate, African bovine lineage. However, it is worth reiterating Gentry and Gentry's (1978) observation that *Simatherium* itself may be a descendant from a form like *Ugandax*. This is particularly possible when one considers the wholly primitive state (lacking apomorphies) of the horns in *Ugandax*. Cladistic analyses of Bovini by Geraads (1992, 1995) place *Ugandax* at the base of the bovine clade, sister to a clade uniting *Simatherium, Pelorovis,* and *Syncerus,* thus also hinting at the possible ancestral status of *Ugandax*. The age of KWA-VP-1/1 and ASK-VP-3/3 (ca. 5.7 Ma) suggests that they could be close to the last common ancestor of both the *Simatherium-Pelorovis* and *Ugandax-Syncerus* lineages.

Simatherium demissum is documented only from Langebaanweg (early Pliocene; Gentry 1980), and it is the most likely ancestor for the eastern African *Simatherium kohllarseni* from Laetoli. In describing *Simatherium demissum,* Gentry (1980) expressed uncertainty in choosing between *Simatherium* and *Ugandax* for the generic identity of this species. It remains difficult to identify the phylogenetic attributes of bovines of this age confidently. Harris (2003) attributes specimens from the Apak and Kaiyumung Members at Lothagam to *Simatherium* aff. *S. kohllarseni*. However, he notes some similarities between the Lothagam material and that of *S. demissum*. Geraads and Thomas (1994) report bovines from Mio-Pliocene deposits of Uganda, among which is a new species of *Simatherium,* bearing an affinity to, although perhaps more primitive than, *S. demissum*. These authors also leave open the possibility that this species may be related to *Ugandax gautieri*. Geraads's (1995) cladistic analysis presents *Simatherium* as paraphyletic, and Geraads suggests that *S. shungurense* may be close to the ancestry of *Syncerus*. In possessing a rounded horn core, *S. demissum* may be considered derived relative to the keeled and triangular cross section of *Ugandax* sp. from the older Adu-Asa Formation (this chapter). The identification of *S. demissum* from the Kuseralee Member of the Middle Awash, at an age almost contemporaneous with Langebaanweg, establishes the geographic range of this species extending at least from the Cape in the south to the Horn of Africa.

The bovines from the Adu-Asa Formation have primitive horn core features that are otherwise documented for early Pliocene European and Indian bovines such as *Alephis* and *Proamphibos* as well as late Pliocene taxa such as *Hemibos*. These taxa are characterized by horn cores more or less triangular in cross section (Pilgrim 1947), but this is probably a primitive character for the Bovini and may not be a reliable taxonomic indicator. The bovines from the younger Kuseralee Member of the Middle Awash, however, have horn cores with more or less rounded cross sections that retain anterior keels. Eurasian forms such as the Pliocene *Leptobos* also have horn cores that are more rounded in cross section, with a pronounced anterior keel. The primitive forms also had a posterolateral keel.

In summary, it appears that two distinct bovine lineages were contemporaneously present in the Middle Awash at around 5.3 Ma. One group is recovered entirely from the Kuseralee Member localities of Kuseralee Dora and Amba, dated to 5.2 Ma. The other group is recovered entirely from earlier localities in the Adu-Asa Formation. Vrba (1987) discussed the parallel existence of two major related bovine lineages, the *Simatherium-Pelorovis* and *Ugandax-Syncerus* lineages, in Sub-Saharan Africa during the late Neogene. The former lineage may have been associated with more open vegetation in seasonally cooler habitats.

Tragelaphini

Systematic Paleontology

Tragelaphus de Blainville, 1816

TYPE SPECIES *Tragelaphus scriptus* (Pallas 1766)

OTHER SPECIES *T. strepsiceros,* (Pallas 1766); *T. imberbis* (Blyth 1869); *T. angasi* (Gray 1849); *T. buxtoni* (Lydekker 1910); *T. spekei* (Speke 1863); *T. nakuae* (Arambourg 1941); *T. kyaloae* (Harris 1991)

GENERIC DIAGNOSIS Medium to large tragelaphines with spiraled horn cores inserted close together and having an anterior keel and sometimes a stronger posterolateral one; small- to medium-sized supraorbital pits, which are frequently long and narrow; and occipital surface tending to have a flat top edge and straight sides (Gentry 1985).

Tragelaphus moroitu sp. nov.

ETYMOLOGY The species is named *moroitu,* which in the Afar language means "spiral," in reference to the spiraling horns characteristic of all Tragelaphini.

HOLOTYPE ALA-VP-2/2 (frontlet, with almost complete left horn core, proximal half of right horn core, and occipital fragment; Figure 9.9) from the Adu-Asa Formation. Measurements are given in Table 9.4.

PARATYPES None assigned.

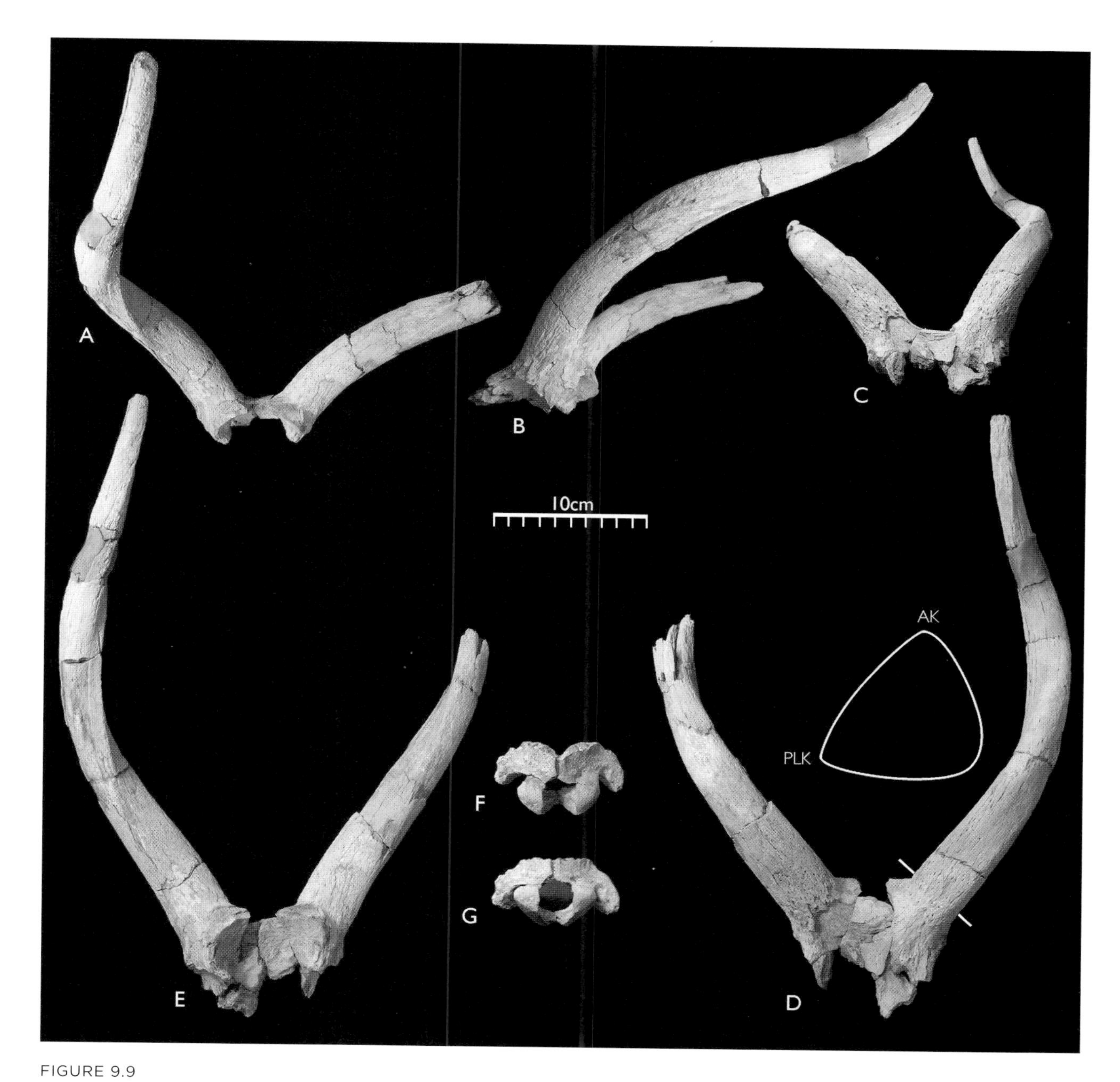

FIGURE 9.9

ALA-VP-2/2, holotype of *Tragelaphus moroitu* sp. nov., frontlet with right and left horn cores and portion of occipital. A. Horn cores, posterior view. B. Horn cores, left lateral view. C. Horn cores, anterior view. D. Horn cores, anterodorsal view. E. Horn cores, posteroventral view. F. Occipital fragment, dorsal view. G. Occipital fragment, posterior view. Left horn core cross-sectional outline is shown (not to scale). Abbreviations: same as Figure 9.1.

TABLE 9.4 Measurements of ALA-VP-2/2, the Holotype of *Tragelaphus moroitu* sp. nov., and KUS-VP-1/33

	ALA-VP-2/2	*KUS-VP-1/33*
Horn core divergence	ca. 70°	(90°)
Frontal biorbital breadth	(110)	ca. 75
Occipital bicondylar width	50.38	(115.0)
Horn core length along anterior keel	(385)	(96.0)
Horn core length along posterior face	(350)	(430); 347 preserved
Intercornual distance, internal	42.69	ca. 52.5
Intercornual distance, external	109.9	ca. 115.0
Horn core basal DAP	34.49	32.01
Horn core basal DT	36.25	35.99
Frontal thickness at intercornual midline	12.55	ca. 11.5
Torsion	ca. 270°	(270°); 255° preserved

NOTE: All measurements are in mm. Abbreviations: ( ) = estimated values.

LOCALITIES AND HORIZONS Asa Koma Member (Adu-Asa Formation) and Kuseralee Member (Sagantole Formation), Middle Awash, Ethiopia. These members are radiometrically dated to between 5.54 and 5.77 Ma and 5.2 Ma, respectively (WoldeGabriel et al. 2001; Renne et al., Chapter 4).

DIAGNOSIS A medium-sized tragelaphine characterized by horns with triangular cross section that is relatively constant throughout the extent of the horn; slight to moderate anteroposterior compression; 270° of torsion; and relatively low divergence and subdued lateral curvature in anterior view, curving outward and backward proximally, and converging in their distal halves.

DIFFERENTIAL DIAGNOSIS Horns resemble those of *T. angasi* and *T. spekei* in general form, particularly in the triangular cross section, gentle spiraling, and great degree of torsion, although being of smaller size than these living species. In overall shape, the horns of *T. moroitu* are very similar to those of *T.* cf. *spekei* from Lukeino (Thomas 1980) and *T.* sp. from Langebaanweg (Gentry 1980). *Tragelaphus moroitu* differs slightly from the Lukeino tragelaphine (horn cores KNM-LU 852 and KNM-LU 906) in having longer horns and in the presence of anteroposterior compression. *Tragelaphus moroitu* may perhaps be distinguished from the Langebaanweg tragelaphine in lesser horn divergence. Horn lengths of the Middle Awash species may also be greater. *Tragelaphus moroitu* differs from *T. kyaloae* in the lesser lateral curvature of the horns (in anterior view), and in the much more developed anterior keel, which produces a triangular cross section that is consistent throughout the length of the horn (Figure 9.10). The *T. kyaloae* holotype (KNM-WT 18673) possesses only a very faint anterior keel, and its cross section changes along the course of the horn core, being more triangular and less compressed distally, and quite oval and exhibiting marked compression proximally. The new material is much smaller than *T. nakuae,* horns of which also exhibit extreme anteroposterior compression,

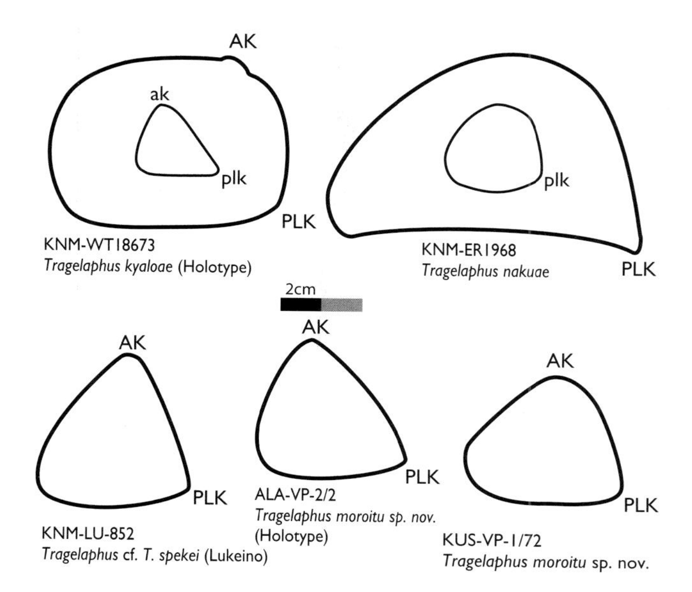

FIGURE 9.10

Basal and distal horn core cross-sectional outlines in fossil species of *Tragelaphus*. All are shown as right horn cores with the anterior keel or its corresponding position toward the top of the page. Distal cross-sectional outlines were taken at approximately 45 mm and 30 mm from the distal tips in KNM-WT 18673 and KNM-ER 1968, respectively. Abbreviations: same as Figure 9.1.

a more teardrop-shaped cross section, less spiraling, and more parallel distal portions (Figure 9.10). *Tragelaphus gaudryi, T. strepsiceros,* and *T. imberbis* differ from *T. moroitu* in having open helical spirals, ovate horn cross sections, and the tendency toward mediolateral compression. *Tragelaphus moroitu* differs from *T. scriptus* in having longer horn cores that exhibit much greater torsion. *Tragelaphus moroitu* is smaller than the two species of *Tragoportax* with which it co-occurs in the Middle Awash.

REMARKS While dental material of boselaphines and tragelaphines is not easily differentiated, tragelaphines are identified here by their less developed basal pillars and goat folds and their generally sharper labial lower and lingual upper cusps. Isolated teeth are assigned to this species on the basis of appropriate size, and the fact that it is the only tragelaphine for which horn core material has been retrieved from Middle Awash Mio-Pliocene deposits.

REFERRED MATERIAL

Cranial

Asa Koma Member, Adu-Asa Formation ALA-VP-3/5 (proximal right horn core); BIK-VP-3/10 (proximal right horn core); ALA-VP-1/9 (proximal left horn core); STD-VP-2/836 (proximal left horn core); STD-VP-1/29 (left horn core midsection fragment).

Kuseralee Member, Sagantole Formation KUS-VP-1/33 (frontlet with left horn core, parts of right horn core, part of right mastoid, part of basioccipital); KUS-VP-1/12 (left horn core with frontlet and parts of right horn core); AMW-VP-1/54 (right horn core); KUS-VP-1/18 (proximal right horn core); KUS-VP-1/72 (proximal left horn core); AMW-VP-1/4 (left and right horn cores); KUS-VP-1/99 (right horn core fragment); AME-VP-1/61 (right horn core and fragment of left horn core); AME-VP-1/112 (proximal right horn core); KUS-VP-1/63 (proximal left horn core); KUS-VP-1/36 (subadult right horn core); KUS-VP-1/65 (right proximal horn core and fragment of left horn core); KUS-VP-1/16 (proximal right horn core).

Dental

Asa Koma Member, Adu-Asa Formation ALA-VP-2/327 (left M^1or M^2); ALA-VP-1/19 (left M^2); ASK-VP-3/337 (right M_3 fragment); BIK-VP-3/14 (right M_1).

Kuseralee Member, Sagantole Formation AME-VP-1/60 (left M^1or M^2); AME-VP-1/68 (right mandible fragment w/P_3–P_4); AMW-VP-1/6 (left mandible fragment with M_1–M_3); AMW-VP-1/26 (right mandible fragment with partial M_3); AMW-VP-1/32 (left P^2–P^4); AMW-VP-1/36 (left mandible fragment with parts of M_1–M_2); AMW-VP-1/93 (right mandible with P_4–M_3); AMW-VP-1/113 (left M_3); AMW-VP-1/121 (left M^2 or M^3); AMW-VP-1/145 (right M^1 or M^2); KUS-VP-1/94 (left M^2 or M^3); KUS-VP-1/120 (left M^2).

DESCRIPTION

Type Specimen ALA-VP-2/2 is a frontlet retaining almost all of the left horn core and the proximal half of the right one, along with a portion of the occipital with the condyles, part of the nuchal plane, and the exoccipital processes. Measurements are given in Table 9.4. The left horn core is missing only a short section in its distal half, although a sliver of the missing articulation is present to guide the reconstruction. Horn morphology follows that given in the diagnosis for the species. Horn core anteroposterior compression on this specimen is minimal (DT/DAP = 1.05). The supraorbital foramina are not preserved. Bone surfaces are very weathered.

Other Key Specimens KUS-VP-1/33 comprises a frontlet and portion of parietal, retaining an almost complete right horn core and parts of the right horn core, as well as part of the right mastoid region and the basilar process at the anterior tuberosities. Measurements are given in Table 9.4. Among all the specimens from the Kuseralee Member, the horns of KUS-VP-1/33 are the least anteroposteriorly compressed (DT/DAP = 1.12). KUS-VP-1/36 is a frontlet and right horn core that is taken to be that of an immature individual based on the following: small basal size and shortness of the horn; low torsion (ca. 200°); unfused metopic and frontoparietal sutures; and thin frontal (ca. 8 mm just medial to the posterointernal keel). It is by far the most anteroposteriorly compressed of all the horn cores of *Tragelaphus moroitu* (DT/DAP = 1.37), although this should not be considered informative of the adult basal horn core morphology. Horn core metrics of remaining specimens of *T. moroitu* are provided in Table 9.5, and dental measurements are given in Table 9.6.

cf. *Tragelaphus* cf. *moroitu*

DESCRIPTION These are dental specimens only doubtfully assigned to Tragelaphini on the basis of morphology and to the new tragelaphine species on the basis of size. It is difficult to distinguish between teeth of tragelaphines and boselaphines when both are known within a fauna. Cheek teeth of tragelaphines have a tendency to possess smaller pillars

TABLE 9.5 Measurements of Horn Cores of *Tragelaphus moroitu* sp. nov.

Specimen	*Side*	*Taxon*	*Basal DAP*	*Basal DT*	*Preserved Length*	*Estimated Complete Length*
ALA-VP-1/9	L	*Tragelaphus moroitu* sp. nov.	39.68	41.06		
ALA-VP-3/5	R	*Tragelaphus moroitu* sp. nov.	34.33	41.78		
AME-VP-1/61	R	*Tragelaphus moroitu* sp. nov.	30.41	37.62	178	
AME-VP-1/112	R	*Tragelaphus moroitu* sp. nov.	31.64	36.9	83	
AMW-VP-1/4	L	*Tragelaphus moroitu* sp. nov.	37.55	42.57	310	(470)
AMW-VP-1/54	R	*Tragelaphus moroitu* sp. nov.	nm	nm	300	(430)
BIK-VP-3/10	R	*Tragelaphus moroitu* sp. nov.	32.78	41.23		
KUS-VP-1/12	L	*Tragelaphus moroitu* sp. nov.	37.58	47.7	325	(430)
KUS-VP-1/16	R	*Tragelaphus moroitu* sp. nov.	27.73	35.28		
KUS-VP-1/18	R	*Tragelaphus moroitu* sp. nov.	29.69	36.75		
KUS-VP-1/36	R	*Tragelaphus moroitu* sp. nov. (subadult)	25.81	35.36	245	(275)
KUS-VP-1/65	R	*Tragelaphus moroitu* sp. nov.	31.72	40.1		
KUS-VP-1/72	L	*Tragelaphus moroitu* sp. nov.	31.59	38.11		
KUS-VP-1/99	R	*Tragelaphus moroitu* sp. nov.	nm	nm	240	
STD-VP-2/836	L	*Tragelaphus moroitu* sp. nov.	35.42	35.69		

NOTE: Length is taken along the posterior surface. All measurements are in mm. Abbreviations: ( ) = estimated values; nm = measurement was not possible.

(if any at all), absent or weak anterior transverse flanges, more V-shaped labial lower and lingual upper cusps, and more readily forming central enamel islands on upper teeth as a result of deep separation and late fusion of the lingual cusps, although all these characters are also present in boselaphines. Measurements are given in Table 9.6.

DISCUSSION Tragelaphines first appear in the African fossil record during the late Miocene. Gentry (1978a) assigned a lower molar and a mandibular fragment with second molar from the Mpesida beds to *Tragelaphus* sp. indet. Additional dental remains recovered later were also included into this group. These dental remains mark the first appearance of *Tragelaphus* in Africa (Thomas 1980). A number of dental and cranial remains of Tragelaphini recovered from Lukeino are assigned to *T.* cf. *spekei* (Thomas 1980). A similar form has also been documented from the "E" Quarry at Langebaanweg (Gentry 1970). Tragelaphines are rare in the Nawata Formation of Lothagam. Smart (1976) reported two possibly tragelaphine species at Lothagam, one of which was identified by Thomas (1980) as *T.* cf. *spekei*. However, Leakey et al. (1996) assign all the tragelaphine remains from the Upper Nawata into one small variant of *T. kyaloae*. Some specimens from the Lower Nawata are also assigned to *Tragelaphus* sp. indet. (Harris 2003). Geraads and Thomas (1994) recently named a new species of a small tragelaphine, *T. nkondoensis,* from the Nkondo Formation of Uganda. Three fragmentary dental remains from the Ibole and Tinde Members of the Wembere Manonga Formation, Tanzania, have also been assigned to *Tragelaphus* sp. (Gentry 1997).

Tragelaphus moroitu sp. nov. includes some of the earliest known representatives of Tragelaphini. Morphological variations are noted between specimens from the Adu-Asa and Kuseralee assemblages. Horn cores from the Adu-Asa Formation (ca. 5.6 Ma) exhibit

TABLE 9.6 Dental Metrics of Tragelaphini

Specimen	*Taxon*	*Tooth*	*AP*	*T*	*Unworn Crown Height*
ALA-VP-1/15	cf. Tragelaphini cf. *Tragelaphus moroitu*	L. P	10.86	10.05	nm
ALA-VP-2/123	cf. Tragelaphini cf. *Tragelaphus moroitu*	R. M^2	16.39	17.12	ca. 19.5
ALA-VP-2/261	cf. Tragelaphini cf. *Tragelaphus moroitu*	R. M^1 fragment	nm	15.82+	nm
AME-VP-1/28	cf. Tragelaphini cf. *Tragelaphus moroitu*	L. M_3	19.9	9.22	nm
AME-VP-1/81	cf. Tragelaphini cf. *Tragelaphus moroitu*	L. M_3	21.88	9.97	20.48
AME-VP-1/99	cf. Tragelaphini cf. *Tragelaphus moroitu*	L. M^3	16.67	13.43	20.55
AME-VP-1/115	cf. Tragelaphini cf. *Tragelaphus moroitu*	R. $M_{1 \text{ or } 2}$	15.52	9.48	ca. 19.5
AME-VP-1/126	cf. Tragelaphini cf. *Tragelaphus moroitu*	R. $M^{2 \text{ or } 3}$	16.21	15.87	nm
AME-VP-1/188	cf. Tragelaphini cf. *Tragelaphus moroitu*	L. M^3	17.4	16.7	nm
AME-VP-1/204	cf. Tragelaphini cf. *Tragelaphus moroitu*	R. M_2	(15.5)	10.4	nm
AMW-VP-1/35	cf. Tragelaphini cf. *Tragelaphus moroitu*	L. $M_{1 \text{ or } 2}$	14.75	8.41	ca. 13.0
AMW-VP-1/132	cf. Tragelaphini cf. *Tragelaphus moroitu*	R. M_1	13.58	8.31	nm
AMW-VP-1/142	cf. Tragelaphini cf. *Tragelaphus moroitu*	R. M_3	15.28+	9.18	ca. 18.5
ASK-VP-3/165	cf. Tragelaphini cf. *Tragelaphus moroitu*	R. M^1	12.83	15.04	nm
DID-VP-1/35	cf. Tragelaphini cf. *Tragelaphus moroitu*	R. M^2	15.32	15.69	nm
DID-VP-1/65	cf. Tragelaphini cf. *Tragelaphus moroitu*	R. $M^{2 \text{ or } 3}$	17.94+	14.33+	nm
KUS-VP-1/91	cf. Tragelaphini cf. *Tragelaphus moroitu*	R. $M^{1 \text{ or } 2}$	14.73	15.1	ca. 18.5
KUS-VP-1/93	cf. Tragelaphini cf. *Tragelaphus moroitu*	L. M^3	16.79	15.46	ca. 18
STD-VP-1/43	cf. Tragelaphini cf. *Tragelaphus moroitu*	L. P_4	12.57	7.07	nm
ALA-VP-1/19	*Tragelaphus moroitu* sp. nov.	L. M^2	15.74	14.44	ca. 18.0
ALA-VP-2/327	*Tragelaphus moroitu* sp. nov.	L. $M^{1 \text{ or } 2}$	14.16	12.24	nm
AME-VP-1/60	*Tragelaphus moroitu* sp. nov.	L. $M^{1 \text{ or } 2}$	15.23	15.38	ca. 18.75
AME-VP-1/68	*Tragelaphus moroitu* sp. nov.	R. P_3	12.62	6.09	9.53
AME-VP-1/68	*Tragelaphus moroitu* sp. nov.	R. P_4	13.14	7.13	nm
AMW-VP-1/6	*Tragelaphus moroitu* sp. nov.	L. M_1	13.66	9.77	nm
AMW-VP-1/6	*Tragelaphus moroitu* sp. nov.	L. M_2	17.03	9.37	nm
AMW-VP-1/6	*Tragelaphus moroitu* sp. nov.	L. M_3	22.77	8.56	nm
AMW-VP-1/26	*Tragelaphus moroitu* sp. nov.	R. M_3	(24.17)	10.16	nm
AMW-VP-1/32	*Tragelaphus moroitu* sp. nov.	L. P^2	11.51	8.37	ca. 12
AMW-VP-1/32	*Tragelaphus moroitu* sp. nov.	L. P^3	9.46	8.89	ca. 12.5
AMW-VP-1/32	*Tragelaphus moroitu* sp. nov.	L. P^4	9.07	10.17	ca. 13.5
AMW-VP-1/36	*Tragelaphus moroitu* sp. nov.	L. M_1	13	9.38+	nm
AMW-VP-1/93	*Tragelaphus moroitu* sp. nov.	R. P_4	13.47	7.8	ca. 12
AMW-VP-1/93	*Tragelaphus moroitu* sp. nov.	L. M_3	24.9	10.72	ca. 19.5
AMW-VP-1/93	*Tragelaphus moroitu* sp. nov.	R. M_1	13.9	9.33	nm
AMW-VP-1/93	*Tragelaphus moroitu* sp. nov.	R. M_2	16.83	10.95	nm
AMW-VP-1/113	*Tragelaphus moroitu* sp. nov.	L. M_3	19.82+	9.35	nm
AMW-VP-1/121	*Tragelaphus moroitu* sp. nov.	L. $M^{2 \text{ or } 3}$	16.82+	16.18	nm
AMW-VP-1/145	*Tragelaphus moroitu* sp. nov.	R. $M^{1 \text{ or } 2}$	14.85+	14.05+	nm
ASK-VP-3/337	*Tragelaphus moroitu* sp. nov.	R. M_3 fragment	nm	9.18+	nm
BIK-VP-3/14	*Tragelaphus moroitu* sp. nov.	R. M_1	14.68	9.18	nm
KUS-VP-1/94	*Tragelaphus moroitu* sp. nov.	L. $M^{2 \text{ or } 3}$	16.83	14.51	18.0
KUS-VP-1/120	*Tragelaphus moroitu* sp. nov.	L. M^2	15.3	14.6	nm

NOTE: All measurements in mm. Abbreviations: ( ) = estimated values; nm = measurement was not possible.

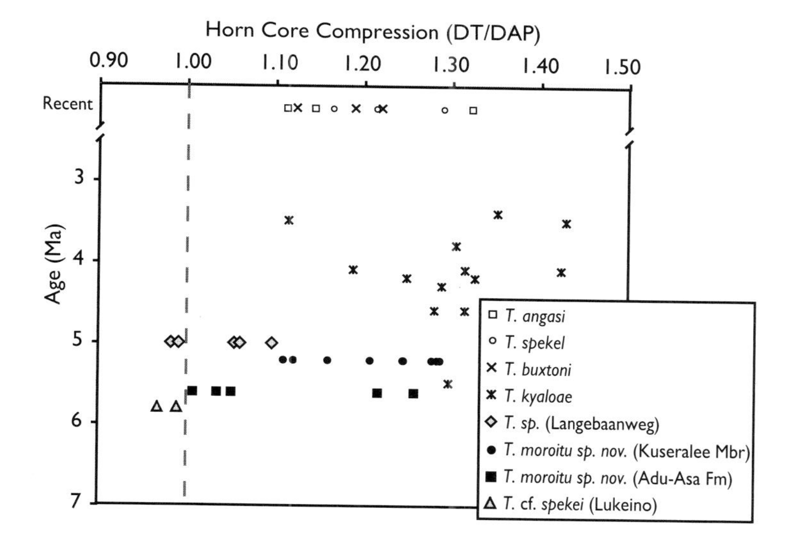

FIGURE 9.11

Horn core compression through time in *Tragelaphus*. Values above 1 indicate anteroposterior compression; those less than 1 indicate some degree of mediolateral compression. Horn cores with compression values around 1.00 (indicated by a dashed line) are practically uncompressed. Abbreviations: DAP = horn core basal anteroposterior diameter; DT = horn core basal transverse diameter. (Data included from Gentry 1980; Harris 1991c; Harris 2003)

slight anteroposterior compression and are probably shorter than those from the Kuseralee Member (ca. 5.2 Ma). The Kuseralee horn cores are marked by increased anteroposterior compression, but the degree of compression is within the lower range of specimens of *T. kyaloae* (Figure 9.11). The nature of the morphological changes between the Adu-Asa and Kuseralee assemblages, carried further, is capable of producing the morphology characteristic of *T. kyaloae,* namely increased compression with a weakened anterior keel. The distal portions of horns of *T. kyaloae* lack these advanced characters, retaining the anterior keel, triangular cross section, and lesser compression. This is evidence for a change in the ontogenetic program of the horn core that took place sometime between *T. moroitu* and *T. kyaloae*. The oldest record of *T. kyaloae* is a single horn core fragment from the Upper Nawata (6.5–5.4 Ma), whereas the taxon is more conclusively recorded from the Apak Member (~5.4–4.2 Ma) (Harris 2003; McDougall and Feibel 2003). A reexamination of the single Upper Nawata specimen, if this is its true provenance, may show that the age of *T. kyaloae* does not necessarily preclude this species from being a direct descendant of *T. moroitu*. In turn, the Lukeino specimens, with their slightly older age (between 5.88–5.70 Ma, Deino et al. 2002) and shorter and uncompressed horns, probably represent the ancestral condition from which *T. moroitu* is itself derived. Tragelaphines have not been reported from Toros Menalla (Chad; Vignaud et al. 2002). The Langebaanweg *Tragelaphus* sp. (Gentry 1980), although of about the same age as the Kuseralee Member specimens, exhibits very slight anteroposterior compression comparable to that seen at Lukeino and in specimens of *T. moroitu* from the older Adu-Asa Formation (Figure 9.11). This may be evidence for the longer retention of more primitive cross-sectional proportions in the South African form. Tragelaphines from slightly younger deposits, such as from the Haradaso Member of the Middle Awash, are crucial in testing this hypothesis. Study of the Haradaso Member faunal Assemblage is under way.

Horns of the extant nyala (*T. angasi*), sitatunga (*T. spekei*), and mountain nyala (*T. buxtoni*), although larger, are very similar in shape and torsion to those of *T. moroitu*. Horns of all three of these species exhibit the weak anteroposterior compression seen in *T. moroitu* (Figure 9.11). It is possible that *T. moroitu* is related to these living tragelaphines to the exclusion of the remainder of the tribe. However, published molecular phylogenies (e.g., Willows-Munro et al. 2005) do not support a sister relationship between these three species, suggesting that this shared horn core morphology is primitive within Tragelaphini. As such, we interpret *T. moroitu* to be a primitive tragelaphine possibly older than many of the evolutionary events within this group. Accordingly, the slightly older Lukeino tragelaphine assigned to *T.* cf. *spekei* (Thomas 1980) may be close to the most common recent ancestor of all Tragelaphini.

Aepycerotini

Aepyceros Sundevall, 1847

Aepyceros cf. *premelampus* Harris, 2003

DIAGNOSIS Anatomically similar to that of the living *A. melampus,* but the cranial vault is proportionately longer, wider, and less steeply angled. The auditory bullae are more inflated, less mediolaterally compressed, and more transversely directed. The horn cores of the males are similar to or larger than those of the living species. Horn cores arise from shorter pedicels and are located immediately above the rear of the orbit, rather than behind it, as in the living species. As in *A. melampus,* the horn cores are mediolaterally compressed at their base, with a slight keel at the posterolateral edge. The lyrate shape and torsion of the proximal portion of the horn cores are similar to those of the living species, but the distal tips converge instead of extending parallel to each other. The supraorbital fossae are larger, set in a depressed area of the frontals, and separated by an elevated intrafrontal suture (Harris 2003).

DESCRIPTION AMW-VP-1/75 (Figure 9.12) and AMW-VP-1/49, a left and right horn core base and frontlet, respectively, are the most complete horn cores referred to this taxon from the Middle Awash. AMW-VP-1/75 has a basal DAP and DT of 35.4 mm and 28.8 mm, respectively. The width across the lateral margins of the supraorbital foramina is ca. 55 mm. The same variable measures ca. 52 mm for AME-VP-1/49, which also has a distance of ca. 20.6 mm from its superior margin of the supraorbital foramen to its horn core base. Both horn cores are little compressed at the base and have transverse ridges and slight but definite anticlockwise torsion, typical of this genus. The horn cores have short pedicels and were fairly divergent and long when complete, with gradual distal taper. The postcornual fossae are shallow and round (AMW-VP-1/75) to oval (AMW-VP-1/49). The supraorbital foramina were well-separated and fairly close to the horn core base. Their preserved superior margins are nearly flush with the frontal surface above that.

ALA-VP-2/31 is a right horn core fragment. It is mediolaterally compressed with very strong anterior keel and no posterolateral keel. The cross section is oval at the base, with a sharp angle along the anterior keel. ALA-VP-2/31 has a strong anterior keel. ALA-VP-1/5 is a left horn core that has a weak anterior keel and a much stronger posterolateral keel with a longitudinal groove alongside.

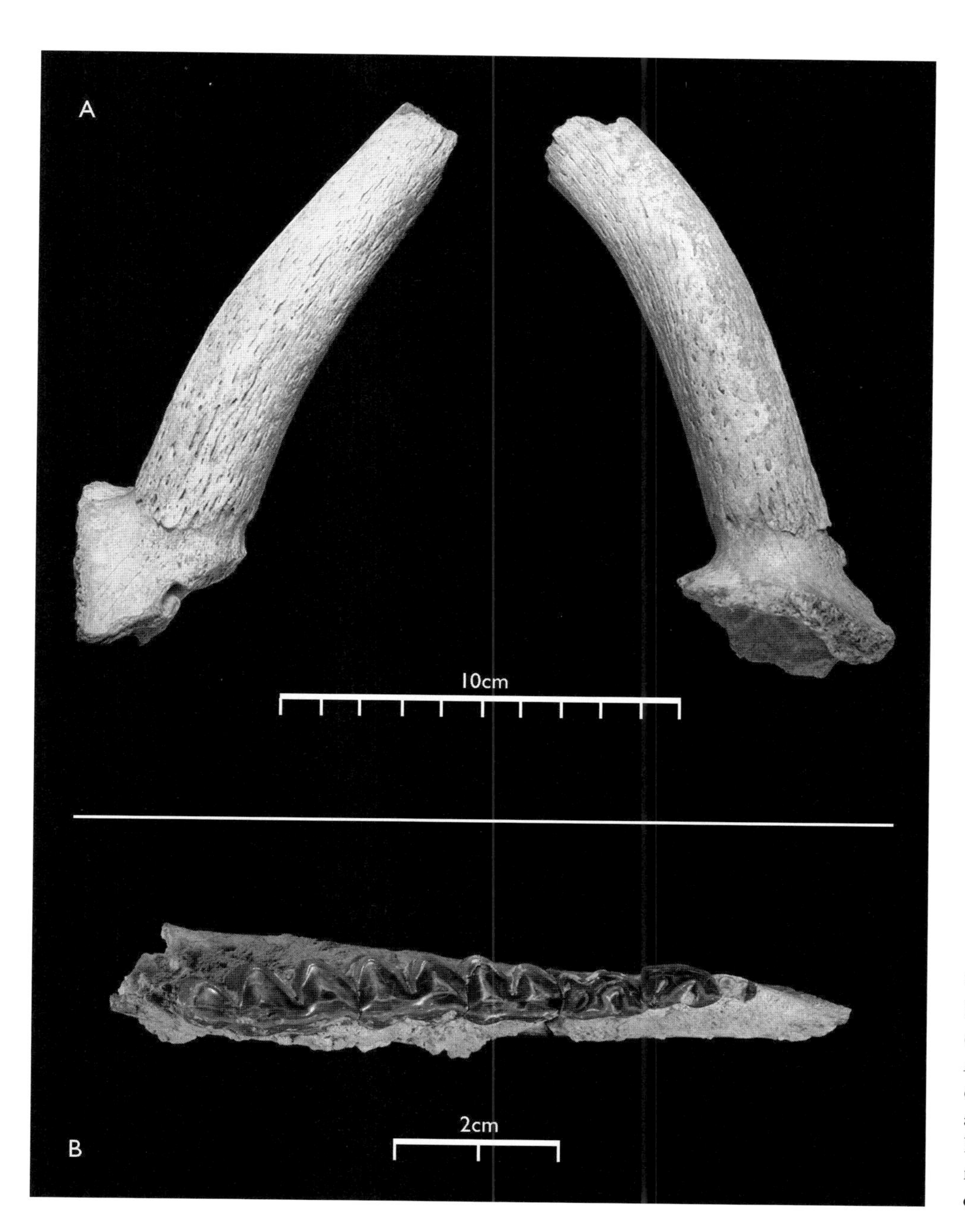

FIGURE 9.12

Specimens of *Aepyceros* cf. *premelampus*. A. AMW-VP-1/75, left horn core with frontal bone, anterior and medial views. B. AME-VP-1/187, left mandible with P_3–M_3, occlusal view.

Teeth are generally larger than those of *Tragelaphus,* even though they are morphologically similar (Figure 9.12). The buccal central rib of upper molars is very strong, and the lingual groove between the two lobes on lower molars is relatively shallow. The upper molars have enamel islands between the two lobes (mostly circular but variable). AME-VP-1/187 is the more complete mandible assigned to this species. This left mandible preserved all teeth but P_2. The base of the corpus is broken. The lingual walls of the molars are flat and the buccal ribs are V-shaped. Small parastyles are present on all molars, and the median islands tend to disappear at an early wear stage. AME-VP-1/178 is a left partial mandible preserving the M_2 and the first rib of the M_3. The M_2 is morphologically similar to the M_2 of AME-VP-1/187, and the size of the mandible is also comparable to that specimen.

The size and compression of these horn cores and other visible features agrees well with *Aepyceros premelampus* known throughout the Lothagam sequence (Harris 2003). Additional and more complete material may confirm that the Middle Awash impala belongs to this species.

Antilopini

Gazella sp. de Blainville, 1816

GENERIC DIAGNOSIS See Gentry and Gentry (1978).

DESCRIPTION KUS-VP-1/78 (Figure 9.13), a right M_2, has a well-developed parastylid. It does not show any ribbing or basal pillar. The anterior buccal lobe is asymmetrical, with the anterior face elongated. KUS-VP-1/76 (Figure 9.13), a left M_3, also has no basal pillar. The third lobe is well-developed and has a central cavity. The parastylid is not as developed as on the second molar. KUS-VP-1/57 (Figure 9.13) is another M_3 morphologically similar to KUS-VP-1/76 but less worn and with weak styles between lobes. The parastylid, however, is strong and no basal pillar is visible. All dental measurements are given in Table 9.7.

DISCUSSION The material from the Middle Awash late Miocene attributable to *Gazella* is fragmentary, comprising only isolated teeth. This is also the case at Lothagam and other contemporaneous eastern African sites. As such, detailed comparative analysis is not possible. However, it is clear that species of *Gazella* were very successful in the late Miocene, as evidenced by their wide distribution in Eurasia and Africa during this time.

Neotragini

Madoqua sp. Ogilby, 1837

DIAGNOSIS See Gentry (1987).

DESCRIPTION This species is known from both the Asa Koma and Kuseralee Members by a partial cranium (AMW-VP-1/149), a right M_3 preserving the two anterior lobes (ASK-VP-3/163), and a right M^2 (ASK-VP-3/524: MD = 7.6 mm, BL = 6.9 mm;

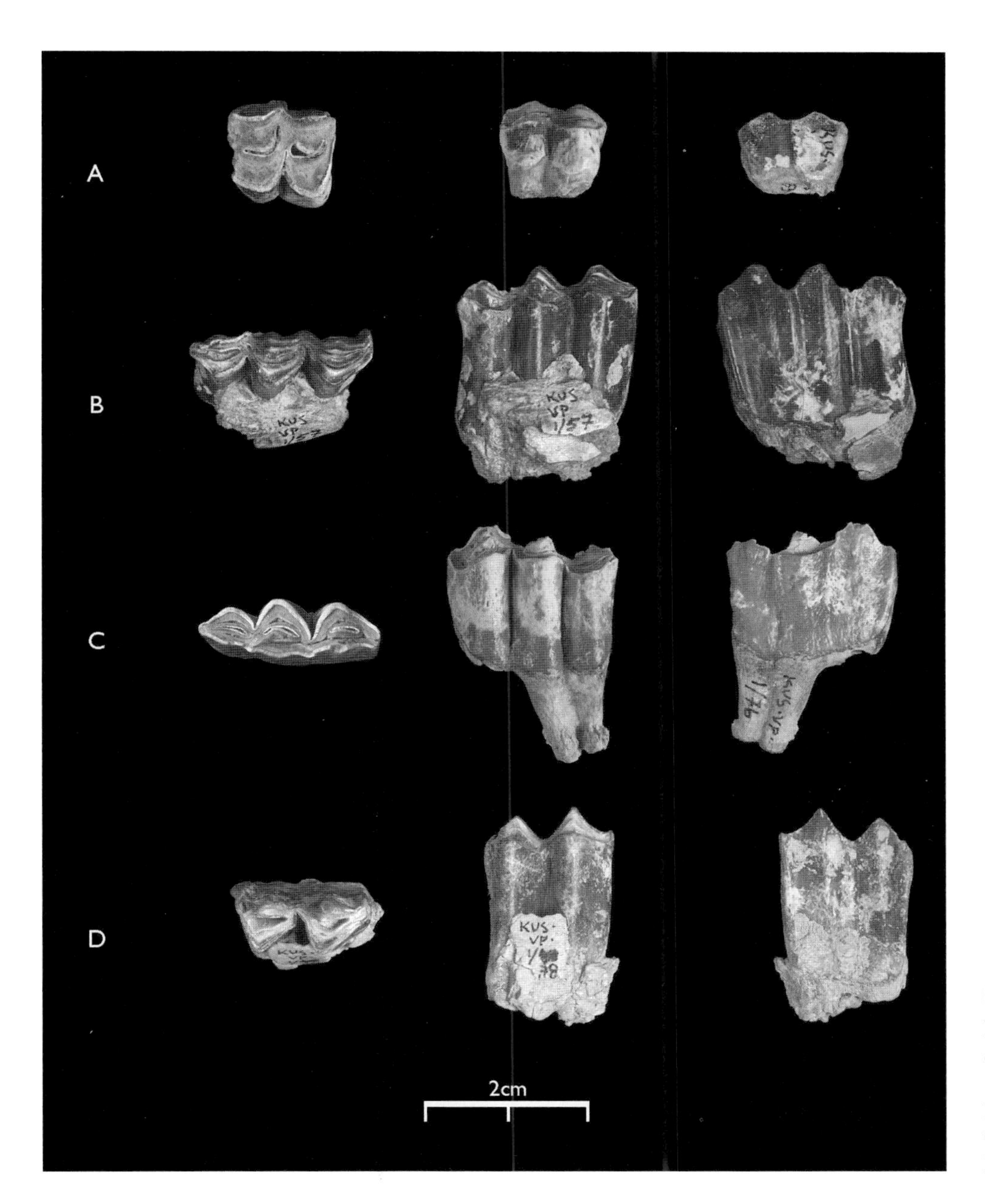

FIGURE 9.13

Isolated dental specimens of *Gazella* sp. A. KUS-VP-1/53, right M^2. B. KUS-VP-1/57, R. M_3. C. KUS-VP-1/76, L. M_3. D. KUS-VP-1/78, R. M_2.

Figure 9.14). The cranial element is a posterior half of the cranial vault around the occipital. Its size best fits the extant *Madoqua,* although it is too fragmentary to assign to any extant species. The right M_3 fragment lacks the distal lobe and the roots. Its size best matches the M_3 of KNM-LT 177 from the Upper Nawata of Lothagam, also referred to *Madoqua*. The M^2 size is also within the known range of the Lothagam sample of the genus. The sample of this small antilopine from the Adu-Asa Formation is scanty and can only be tentatively referred to *Madoqua* sp.

TABLE 9.7 Dental Metrics of *Gazella* sp.

Specimen	*Element*	*MD*	*BL*
AMW-VP-1/130	$M^{1 \text{ or } 2}$	12.2	9.3
KUS-VP-1/53	$M^{1 \text{ or } 2}$	13.4	12.8
KUS-VP-1/57	M_3	22.4	20.4
KUS-VP-1/76	L. M_3	20.9	15.0
KUS-VP-1/78	R. M_2	16.7	10.9

NOTE: All measurements are in mm.

Raphicerus sp. Smith, 1827

DIAGNOSIS See Gentry (1980).

DESCRIPTION The two horn core specimens (STD-VP-1/22 and AME-VP-1/111; Figure 9.15) referred to *Raphicerus* sp. are fragmentary. They are uncompressed, almost circular at the base, and the insertions are very upright. STD-VP-1/22 belongs to a subadult individual. Specimen AME-VP-1/111 is similar to KNM-LT 503 and KNM-LT 14125, from the Lower Nawata and Kaiyumung Members of Lothagam, respectively, also referred to *Raphicerus* sp.

Raphicerus specimens from the late Miocene come from Lothagam (only dentitions, Harris 2003) and *R. paralius* was described from Langebaanweg (Gentry 1980). The basal horn core dimensions of the Middle Awash specimens have low compression similar to those from Langebaanweg but are larger in size, falling just beyond the upper range of 13 Langebaanweg specimens. The horn core lengths are comparable. A strong difference is the more upright insertion of Middle Awash horn cores, in contrast to the very low angle of insertion in *R. paralius* (Gentry 1980). The mandibular size and morphology of the preserved teeth show clear affinity with *Raphicerus*. Harris (2003) reported that the Lothagam specimens are larger than those of *Madoqua* species and *Raphicerus campestris*, but have similar morphology. Both the Lothagam and Middle Awash dental specimens are smaller than *Raphicerus paralius* from Langebaanweg (Gentry 1980).

Reduncini

Systematic Paleontology

Kobus A. Smith, 1840

Kobus cf. *porrecticornis* Lydekker, 1878

DIAGNOSIS A small- to medium-sized reduncine about the size of the living larger *Redunca* species. Horn cores are moderately compressed and fairly strongly divergent, inserted close to each other and quite upright, with the basal cross section having the widest mediolateral diameter situated anteriorly and a flattened lateral surface. The horn cores are moderately long with a low rate of tapering from the base upward and little lyration.

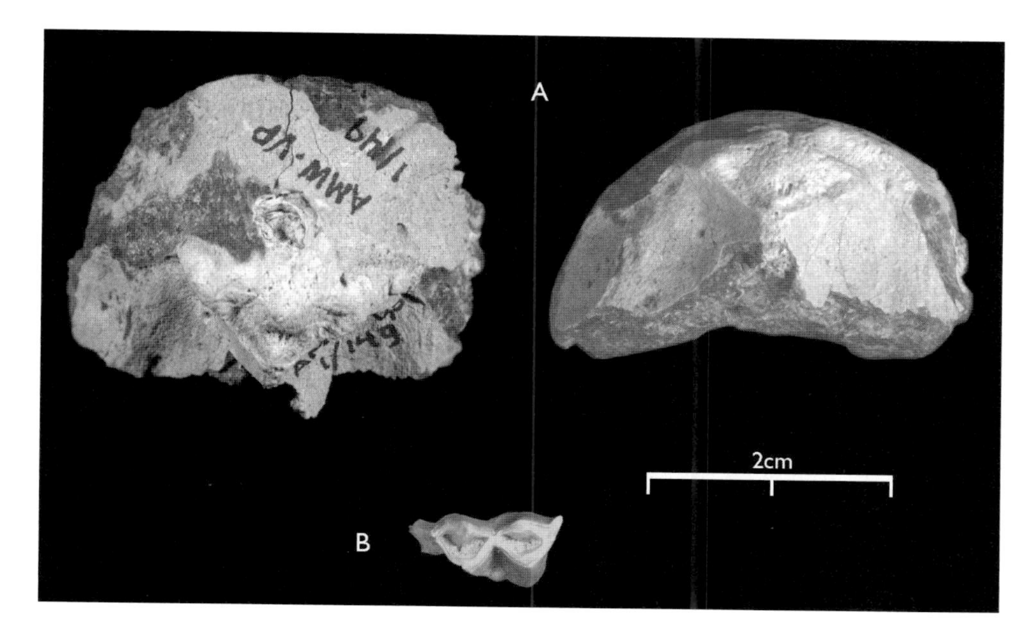

FIGURE 9.14

Specimens of *Madoqua* sp. A. AMW-VP-1/149, cranial fragment (cranial and caudal views). B. ASK-VP-3/163, R. M_3.

They lack the concave-forward course at or shortly above the horn core base that is present in many reduncines; instead, they show slight and gradual backward curvature from the upright base upward. The horn cores have strong longitudinal grooving. The transverse ridges are absent to very slight. The supraorbital foramina are situated close to each other and to the anterior horn core bases, in a moderately deep to deep pit lacking rims, toward the midline. There is a fairly wide and deep postcornual fossa (see Figure 9.16). The "craniofacial angle," between the forehead and the dorsal braincase is low, especially for such an early reduncine. The angle between the maximum horn core diameter and the midfrontal suture is from low to moderately low. An unusual (although not unique) character in Reduncini is the thickening of the frontal roof of the braincase at the level of the midfrontal suture.

REMARKS All of the more securely identified specimens come from the earlier Asa Koma Member, whereas other reduncine species predominate in younger members (discussed later).

DESCRIPTION AFD-VP-1/1 is a cranial fragment preserving both horn cores and the frontal. The orbits are damaged on both sides. The horn core tips are missing on both sides. Horn cores are inserted upright slightly apart from each other. The pedicels are relatively long above the orbits. The lateral walls of the horn cores are flattened. The right supraorbital is preserved, and its medial wall is ca. 15 mm from midline and ca. 40 mm from its superior edge to the pedicel. The postcornual fossae are positioned on the distal-lateral corners of the horn cores, long superoinferiorly, and shallow. AFD-VP-1/1 is similar to KNM-LT 189 (type specimen of *Kobus presigmoidalis* sp. nov. from the Upper Nawata of Lothagam, Harris 2003) in having long pedicels above the orbits. However,

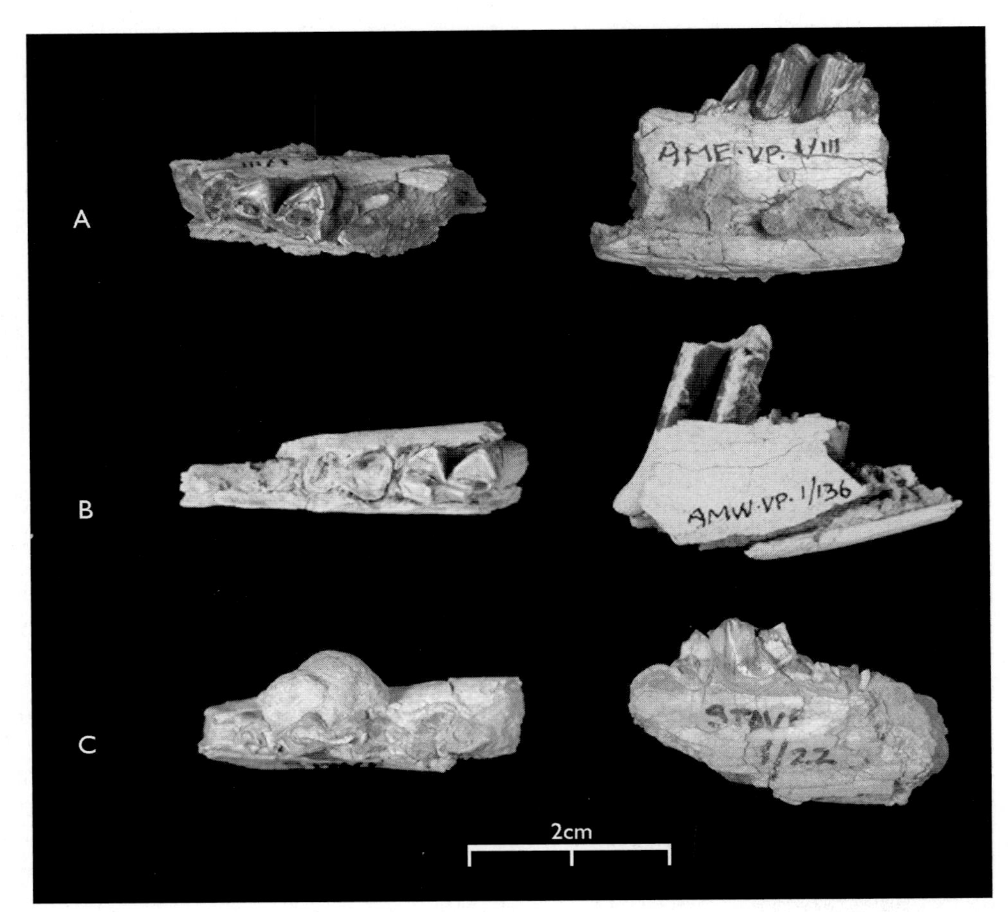

FIGURE 9.15
Dental specimens of *Raphicerus* sp. A. AME-VP-1/111, right mandible fragment with M_2. B. AMW-VP-1/136, right mandible fragment with M_1. C. STD-VP-1/22, right mandible with dp_4.

they differ in the insertion of their horn cores and bilateral compression. KNM-LT 189 is more compressed bilaterally. AFD-VP-1/1 is also larger than KNM-LT 189 in its overall size. Other Middle Awash specimens assigned to *K.* cf. *porrecticornis* are fragmentary horn cores similar in their preserved morphology to AFD-VP-1/1. Horn core measurements are given in Table 9.8.

DISCUSSION The set of characters described here for the Middle Awash specimens is consistent with those of specimens that have been assigned to the species *K. porrecticornis.* This species is known from late Miocene to early Pliocene strata in the Siwaliks of India and Pakistan (with a range of ca. 8.1–3.5 Ma; Barry 1995; Barry et al. 2002), Mpesida in Kenya (Thomas 1979; dated to ca. 6.5 Ma), Lukeino in Kenya (Gentry 1978b, 1980; Thomas 1979; dated to ca. 5.8 Ma), Baard's Quarry in the western Cape of South Africa (Gentry 1980; of uncertain but more likely early Pliocene age), and possibly from the Hadar Formation, Ethiopia (at 3.4 Ma; Gentry 1981).

Some of the material from Lothagam that Harris (2003) described as a new species, *Kobus presigmoidalis,* appears to belong to this species (in the view of one of us, ESV, who has studied all the original specimens). This includes the partial skull that Harris (2003)

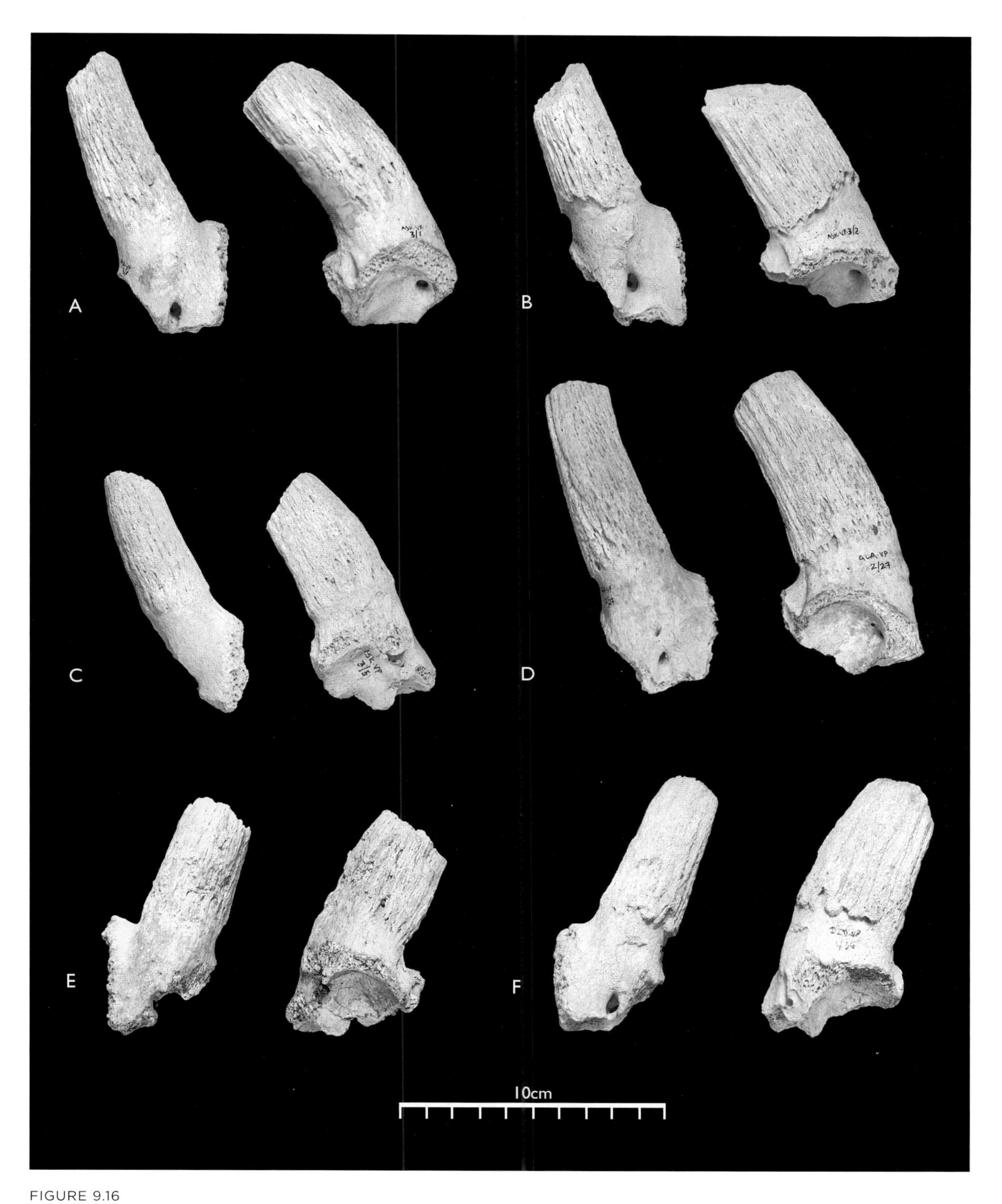

FIGURE 9.16

Horn cores assigned to *Kobus* cf. *porrecticornis*. A. ASK-VP-3/1, right horn core. B. ASK-VP-3/2, right horn core. C. ASK-VP-3/5, right horn core. D. ALA-VP-2/27, right horn core. E. KUS-VP-1/3, left horn core. F. DID-VP-1/36, left horn core.

TABLE 9.8 Horn Core Measurements of *Kobus* cf. *porrecticornis* and *Kobus* aff. *oricornis*

Specimen	*Taxon*	*MAX*	*MIN*	*LEN*	*SOF*	*S-HC*	*ADIV*	*AHMET*	*AHCBR*
ADD-VP-1/12 (R)	*Kobus* cf. *porrecticornis*	43	33		48	31	(60°)	(25°)	
ADD-VP-1/15 (R)	*Kobus* cf. *porrecticornis*	41	33						(90°)
AFD-VP-1/1 (R, L)	*Kobus* cf. *porrecticornis*	(44)	(30.5)						
ALA-VP-2/15 (L)	*Kobus* cf. *porrecticornis*	40	30			(33)	(50°)	(30°)	(75°)
ALA-VP-2/27 (R)	*Kobus* cf. *porrecticornis*	43.9	32.1		(44)	(38)	(55°)	(20°)	(105°)
ASK-VP-2/5 (R)	*Kobus* cf. *porrecticornis*	(42)	31		(40)	(33)	(55°)	(35°)	
ASK-VP-2/6 (R)	*Kobus* cf. *porrecticornis*	43.4	30.5			29.7	(50°)	(25°)	
ASK-VP-2/10 (L)	*Kobus* cf. *porrecticornis*	41.5	31						
ASK-VP-3/5 (R)	*Kobus* cf. *porrecticornis*	38.1	27.5		(36)	27.7	(50°)	(30°)	
DID-VP-1/26 (R)	*Kobus* cf. *porrecticornis*	(45.5)	(31.5)	(190)	(44)	(31.5)	(50°)	(35°)	
DID-VP-1/36 (L)	*Kobus* cf. *porrecticornis*	43	31.8		(40)	30	(47.5°)	(17.5°)	(100°)
KUS-VP-1/3 (R)	*Kobus* cf. *porrecticornis*	38	29.5		(45)	30.2	(35°)	(35°)	(70°)
AME-VP-1/4 (L)	*Kobus* aff. *oricornis*	30	25		(50)		(40°)	(90°)	
AME-VP-1/22 (R)	*Kobus* aff. *oricornis*	29	25						
AME-VP-1/32 (L, R)	*Kobus* aff. *oricornis*	(41)	(31)	(240)					

NOTE: R, L = right or left horn core; length measurements in millimeters; ( ) = estimated values; MAX = horn core maximum diameter; MIN = horn core minimum diameter; LEN = estimated length of complete horn core; SOF = width across lateral margins of supraorbital foramina; S-HC = distance from superior margin of supraorbital foramen to horn core base; ADIV = angle of basal horn core divergence; AHMET = angle of MAX to metopic suture; AHCBR = angle of horn core to braincase in lateral view.

designated as the holotype of *K. presigmoidalis,* KNM-LT 189, from the Upper Nawata Member of Lothagam, dated ca. 6.5–5.2 Ma, as well as specimens from later Lothagam strata. Harris (2003) did not compare these fossils with *K.* (*D.*) *porrecticornis* from the Siwaliks. He did note that although the Lothagam fossils "compare quite closely to *K.* aff. *porrecticornis* . . . from Lukeino . . . they differ in that the pedicel is shorter [in the Lukeino form], the horn core curves backward in a single plane whereas that of the Lothagam specimens is more lyrate (curving outward in its proximal portion and recurving medially towards its tip), and the backward curvature is slightly more pronounced in the Lothagam material" (Harris 2003:541). Harris also noted that the Mpesida *K.* aff. *porrecticornis* horn cores differ from the Lothagam form by being "smaller, less elongate, and proportionally wider" (Harris 2003:541).

In the context of degree of conspecific variation typical of living reduncine species (e.g., Vrba et al. 1994), we do not consider the differences cited by Harris (2003) as sufficient to warrant a species separate from *K. porrecticornis*. Minor differences in characters such as pedicel length and horn core course occur between subadults and adults, as well as among adults, of a single living species. Given the strong overall similarities in all comparable characters, including the unusual thickening of the frontal roof of the braincase at the level of the midfrontal suture, we suggest that the Lothagam form belongs to the same (or a very closely related) species as that from the Siwaliks and from elsewhere in Kenya, Ethiopia, and possibly also southern Africa. The differences and similarities between the *K.* cf. *porrecticornis* specimens described here and the other late Miocene–early Pliocene reduncines from the Middle Awash will be cited as we discuss those species below.

Kobus aff. *oricornis* Gentry, 1985

DIAGNOSIS See Gentry (1985).

DESCRIPTION AME-VP-1/4 (Figure 9.17) is a small left horn core base that probably belongs to the same individual as right horn core base AME-VP-1/22. The same species may be represented by the distal right or left horn core piece KUS-VP-1/58 lacking the base, and the larger AME-VP-1/32 comprising left and distal right horn core pieces. All horn core measurements are given in Table 9.8. The smaller specimens reflect a spongy bone density consistent with juvenile status whereas AME-VP-1/32 may represent the adult. The horn cores have low basal compression, a very high posterior angle of the maximum basal diameter to the midfrontal suture, absence of basal anterior swelling, presence of transverse ridges, appreciable length in relation to their small girth, and (for reduncines of such an early age) some posteromedial basal flattening, and some increased distal divergence. The presumed juvenile horn core AME-VP-1/4 curves slightly outward above the base in anterior view and forward soon above the base in lateral view. Its forehead is quite flat up to the preserved level of the supraorbital foramina, without any sign of a deep supraorbital pit. There is some horn core torsion that would have been clockwise from the base up in the right horn core. There is no sign of pedicel or basal horn core hollowing. The postcornual fossa is small, round, and quite shallow and is situated close to the horn core base. The inferred adult specimen AME-VP-1/32 has a short pedicel, no torsion or lateral flattening, some posteromedial flattening and an incipient posterolateral

FIGURE 9.17

Horn cores assigned to *Kobus* aff. *oricornis*. A. AME-VP-1/5, right horn core. B. AME-VP-1/4, left horn core.

keel, a very small and round postcornual fossa, distally increasing divergence, a single longitudinal groove preserved that extends about halfway up the horn core, and very little sigmoid curvature in lateral view.

Many of these characters are reminiscent of *Kobus oricornis* and allied forms, known from later strata in the Middle Awash and Hadar, Ethiopia, and the Turkana Basin in northern Kenya and southwestern Ethiopia, although the best known occurrences of *K. oricornis* from post–3.5 Ma strata are considerably larger. Until more complete material becomes available, an appropriate assignation to the Middle Awash late Miocene material is *K.* aff. *oricornis*. One of us (ESV) has recently assigned (via tabulation; 2006) two of these specimens (from Amba East) to *K.* cf. *basilcookei*.

Redunca H. Smith, 1827

TYPE SPECIES *Redunca redunca* Pallas, 1767.

OTHER SPECIES *Redunca arundinum* Boddaert, 1785; *Redunca fulvorufula* Afzelius, 1815; *Redunca darti* Wells and Cooke, 1956; *Redunca* cf. *R. darti* Lehman and Thomas, 1987.

According to one of us (ESV), *Kobus subdolus* (Gentry 1980) probably belongs to *Redunca*. The similarities of *Kobus ancystrocera* (Arambourg 1947) to *Redunca* species might indicate that *K. ancystrocera* also belongs to *Redunca*. An additional, yet unnamed, *Redunca* species is also present at Maka, Middle Awash, and Hadar, in the Afar region of Ethiopia.

GENERIC DIAGNOSIS *Redunca* is generally smaller than *Kobus,* with horn cores that are short, with little or no upright orientation of horn cores just above the pedicel, resulting in a low inclination angle between the posterior horn core base and the dorsal braincase. Distally, the horn cores curve upward and their tips recurve forward more strongly than is usual in other reduncines, so that they appear forwardly concave in lateral view. The horn cores have little mediolateral compression, moderate to strong longitudinal grooves, and a posteromedial flattened surface with the greatest basal diameter situated anteriorly. The supraorital pits are bounded laterally by ridges that are moderately to strongly developed. The orbital rim tends to project prominently. The braincase tends to be low and wide. The auditory bulla is typically inflated, and the anterior tuberosities of the basioccipital are large and often outwardly splayed (after Gentry and Gentry 1978; Vrba et al. 1994; and additional personal observations by ESV).

Redunca H. Smith, 1827

Redunca ambae sp. nov.

ETYMOLOGY The specific name refers to the type locality, Amba East.

HOLOTYPE The holotype is a cranium, AME-VP-1/42, with both horn cores (Figure 9.18).

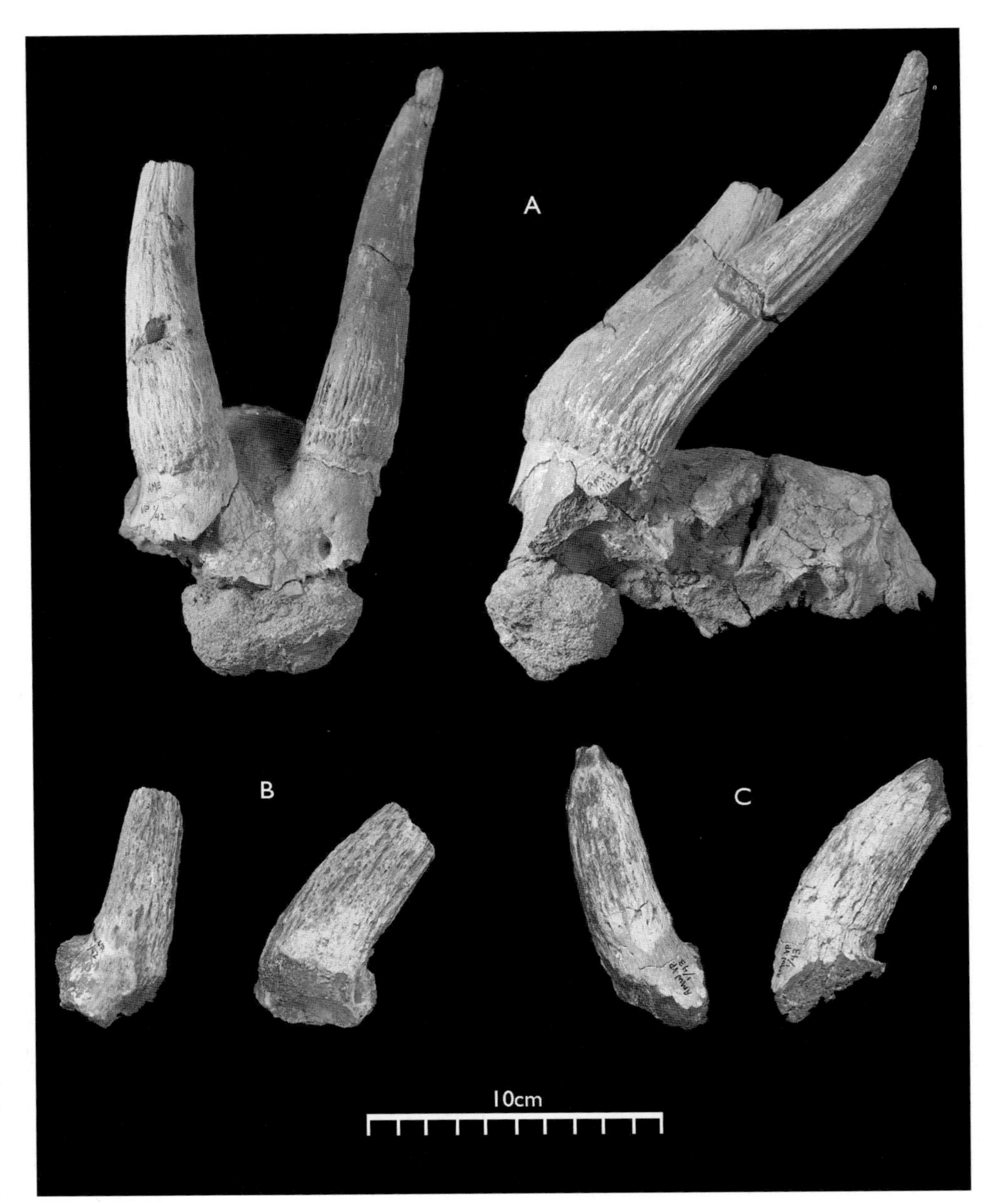

FIGURE 9.18

Specimens of *Redunca ambae* sp. nov. A. AME-VP-1/42 (holotype), cranium with right and left horn cores. B. AMW-VP-1/42, left horn core. C. AMW-VP-1/43, right horn core.

PARATYPES None.

LOCALITIES AND HORIZONS AME-VP-1, AMW-VP-1, KUS-VP-1, Kuseralee Member, Middle Awash, Ethiopia, and radiometrically dated to 5.2 Ma (Renne et al. 1999).

DIAGNOSIS A small reduncine about the size of a southern reedbuck or a small *Kobus kob*. The short horn cores are little separated at their bases and weakly divergent. They have a tendency for anterior swelling at the base, with rapid tapering towards the tip. Horn core curvature in lateral view is sigmoid, with posterior bending above the base and marked anterior recurvature toward the tip. The horn cores are moderately compressed, with definite anterior transverse ridges and many strong longitudinal grooves, a widest basal diameter situated anteriorly, some posteromedial flattening, and a maximum anteroposterior diameter that is little angled relative to the midfrontal suture. There is no hollowing in the pedicel or basal horn core. The postcornual fossae are fairly large and deep, and the orbital rims flare out prominently. The forehead area is expanded as reflected by the wide spacing of the supraorbital foramina and the relatively great distance between them and the basal horn cores. The craniofacial angle, between the forehead and the dorsal braincase, is low. The braincase is long relative to skull size, with an occipital surface that is wide relative to its height and partly laterally facing. The basioccipital is elongated and has prominent anterior tuberosities.

DIFFERENTIAL DIAGNOSIS The appropriate comparisons are with the earliest Reduncini, late Miocene to Pliocene, notably *Kobus porrecticornis* from the Siwaliks of India and Pakistan, Mpesida and Lukeino in Kenya, and elsewhere in Africa; *Zephyreduncinus oundagaisus* from slightly earlier strata in the Middle Awash (Vrba and Haile-Selassie 2006); ?*Redunca subdolus* from Langebaanweg in South Africa, which Gentry (1980) described as *Kobus subdolus* but may belong to *Redunca* (Vrba 1995b); *Redunca darti* (Wells and Cooke 1956) from Makapansgat in South Africa; and *Redunca* aff. *darti* from Sahabi (Lehmann and Thomas 1987) and allied northern and eastern African material (see Vrba 1995b).

There are strong differences in horn core shape and orientation between *R. ambae* sp. nov. and *K. porrecticornis*. *Redunca ambae* sp. nov. differs from *K. porrecticornis* in having much stronger sigmoid curvature of horn cores in lateral view with a far greater diminution rate from the base upward, and horn cores that are less upright basally in lateral view and less divergent in anterior view. It also has more widely separated supraorbital foramina than found in any *K. porrecticornis* and tends to have larger postcornual fossae. *Redunca ambae* sp. nov. variously resembles specimens assigned to *K. porrecticornis* in overall size, horn core compression and length, and in a braincase that is long and strongly angled with respect to the dorsal forehead.

Redunca ambae sp. nov. differs from *Z. oundagaisus* decisively in its much lower basal horn core compression, as well as in its less divergent and less upright horn cores and greater separation of the supraorbital foramina. However, it resembles *Z. oundagaisus* in its overall horn core size and length. *Redunca ambae* sp. nov. differs from *R. darti* in horn cores that are on average larger, more basally compressed, less divergent, and with less marked temporal ridges and less widely separated supraorbital foramina. Resemblances to *R. darti* include horn core shortness and shape in lateral view with basal anterior swelling and rapid diminution toward the tip, a long braincase that is strongly angled with respect to the dorsal forehead, prominently jutting dorsal orbits, and an anteriorly indented coronal suture at bregma. *Redunca ambae* sp. nov. differs from *Redunca* aff. *darti* from Sahabi in having longer horn cores in relation to smaller basal horn core size, and decidedly lower divergence and posterior angle of the maximum basal diameter to the midfrontal suture.

Some horn cores of ?*Redunca subdolus* have anterior swelling of the basal horn core in lateral view with rapid diminution in girth above that, as seen in *R. ambae* sp. nov. But the Langebaanweg species differs in having horn cores that are larger in basal girth and on average less compressed, more uprightly inserted, more divergent, shorter in relation to basal size, with a much weaker tendency to sigmoid curvature in lateral view, a lower incidence of transverse ridges (only 8 of 31 horn cores; Gentry 1980), and absolutely lower separation of the supraorbital foramina from the basal horn core (and also from each other) relative to basal horn core size. Nevertheless, ?*R. subdolus* may be more closely related to *R. ambae* sp. nov., with a more recent common ancestry, than the other species compared above.

Referred Material

Kuseralee Member, Sagantole Formation KUS-VP-1/11 (frontal with base of both horn cores and a partial left horn core); AMW-VP-1/42 (left horn core); AMW-VP-1/43 (right horn core).

DESCRIPTION

Type Specimen AME-VP-1/42 is an almost complete cranium of a small reduncine (see Table 9.9 for measurements). It includes a complete left horn core, a right horn core missing the distal part, parts of the frontals and both orbits, the skull roof, occipital surface, both mastoids, left occipital condyle, basioccipital, and parts of the temporal bones on both sides. The occipital region is slightly crushed and moderately deformed.

The horn cores are fairly short and little-separated at their bases. The base of each horn core has a marked swelling anteriorly, with rapid tapering towards the tip. The nearly complete left horn core in lateral view has remarkably strong sigmoid curvature for a horn core of limited length, with posterior bending above the base and anterior recurvature towards the tip. The angle of the posterior few centimeters of the horn core to the cranial roof in lateral view is low, although the very base is somewhat less inclined, and horn core divergence is low. The horn cores are moderately compressed, with definite anterior transverse ridges and many longitudinal grooves that are especially extensive and deep on the posterior and adjacent posterolateral and posteromedial surfaces, with some grooves extending from the base upward over about three-quarters of the horn core length.

The basal horn core cross section has the widest mediolateral diameter situated anteriorly, and a flattened lateral surface with a tendency to a posterolateral keel. There is also some posteromedial flattening, and a maximum anteroposterior diameter that is little angled relative to the midfrontal suture. There is no hollowing in the pedicel or basal horn core. In spite of damage, the postcornual fossa was deep and localized, and the orbits flared out prominently. The forehead area is expanded as reflected by the wide spacing of the supraorbital foramina and the relatively high distance between them and the basal horn cores. The round supraorbital foramina are situated in deep funnel-shaped pits with the wide part of the funnel anterior, and a depressed flattish frontal area between the foramina. The pits are bounded laterally by blunt ridges. The craniofacial angle, between the forehead and the dorsal braincase, is quite low, especially for such an early reduncine. The

TABLE 9.9 Measurements of AME-VP-1/42, the Holotype Cranium of *Redunca ambae* sp. nov.

		Measurement
MAX	Horn core maximum diameter	40.3
MIN	Horn core minimum diameter	32.9
LEN	Estimated length of complete horn core	(190)
SEP	Basal horn core separation	24.5
SOF	Width across lateral margins of supraorbital foramina	(46)
S-HC	Distance from superior margin of supraorbital foramen to horn core base	(32)
ADIV	Angle of basal horn core divergence	(20°)
AHMET	Angle of MAX to metopic suture	(30°)
AHCBR	Angle of horn core to braincase in lateral view	(60°)
ABRFAC	Craniofacial angle, i.e., the angle between the forehead and the dorsal braincase	(125°)
ABROCC	Parietal-occipital angle, i.e., the angle between the straight line from bregma to occiput, and the straight line from occiput to the top of the foramen magnum	(130°)
BRLEN	Braincase length from coronal suture to occiput	(73)
BASANT	Basioccipital width across the anterior tuberosities	23.5
BASPOS	Basioccipital width across posterior tuberosities;	(21)
BASLEN	Basioccipital length: anterior to posterior tuberosities	41.3
OCCHT	Occipital height from the top of the foramen magnum to occiput	37.3
MASW	Distance across mastoid exposures	(82)

NOTE: Length measurements are in mm except where indicated; ( ) = estimated values.

braincase is long relative to skull size and may have widened posteriorly in the complete state. The coronal suture is situated far behind the posterior pedicels and slightly, but definitely, indented anteriorly toward the midline at bregma, which is somewhat raised. The braincase roof is only slightly convex longitudinally. The temporal ridges are moderately strong anteriorly on the right side. The occipital surface is wide relative to its height, with a tendency to a triangular shape, and partly laterally facing. Its central spine forms an angle of approximately 130° with the dorsal braincase. The mastoid has a prominent dorsal rim and is oriented laterally. The anterior tuberosities of the basioccipital project ventrolaterally, yet they are not as widely splayed as in some reduncines. The basioccipital is markedly elongate. Its shallow midpart is laterally constricted and lacks prominent ridges flanking a marked central groove.

Other Key Specimens Features visible on the additional specimens assigned to this species agree well with those on the holotype, especially in terms of horn core compression, size, curvature, and posterior bending. The horn core and frontlet specimen KUS-VP-1/11 has a more pronounced raising at bregma. The Sahabi specimen M41945 has a markedly raised area on the braincase roof at bregma, which resembles that found in KUS-VP-1/11. Horn core dimensions of specimens referred to *Redunca ambae* sp. nov are given in Table 9.10.

TABLE 9.10 Horn Core Measurements of the Paratypes of *Redunca ambae* sp. nov.

Specimen	*MAX*	*MIN*	*LEN*	*SOF*	*S-HC*	*ADIV*	*AHMET*	*AHCBR*
KUS-VP-1/11 (L)	49.9	33.8		(45)			(32.5°)	
AME-VP-1/5 (R)	39.3	32.4	(230)	(55)	25		(40°)	
AMW-VP-1/42 (L)	36.5	27.4				(30°)		
AMW-VP-1/43 (R)	36.3	28				(40°)		(75°)

NOTE: R, L = right or left horn core; length measurements in mm; ( ) = estimated values; see Table 9.8 for other abbreviations.

Zephyreduncinus Vrba and Haile-Selassie, 2006

Zephyreduncinus oundagaisus Vrba and Haile-Selassie, 2006

DIAGNOSIS See Vrba and Haile-Selassie (2006).

DESCRIPTION Specimens referred to *Zephyreduncinus oundagaisus* have been described in Vrba and Haile-Selassie (2006), except for additional parts of AFD-VP-1/3, and will not be repeated here (Figure 9.19). However, it is important to reiterate some of the major species-level comparative assessments. *Zephyreduncinus oundagaisus* represents a small antelope with short and mediolaterally more compressed horn cores than any known species of Reduncini (Table 9.11). Compression index (CI) is less than 70 percent (see Vrba and Haile-Selassie 2006, for a complete description). Strong and deep longitudinal grooves are present on the lateral and posterolateral distal half of the horn core. The horn cores have very little anticlockwise torsion and are almost parallel superiorly; however, they show a tendency to curve anteriorly at the tip. The horn cores are obliquely inserted above the orbits and are divergent at the pedicel. The orbital walls are smooth. The supraorbital foramina are large. The postcornual fossa is relatively deep and teardrop-shaped. Pedicels are well-developed. The cross section at pedicel is an anteroposteriorly elongated oval shape. The lateral side of the horn core is almost flat, and the medial side a little convex.

DISCUSSION *Zephyreduncinus oundagaisus* has horn cores that are, on average, as short as, but absolutely more compressed than, *Kobus subdolus*. *Zephyreduncinus oundagaisus* also lacks transverse ridges that are frequently present on horn cores of *K. subdolus*. Three horn cores from Sahabi assigned to *Redunca* aff. *darti* are more rounded in cross section and longer than *Zephyreduncinus oundagaisus*. *Kobus porrecticornis,* an early reduncine from the Siwaliks, is similar to *Zephyreduncinus oundagaisus* by its horn core size and absence of transverse ridges on the horn cores. However, *Zephyreduncinus oundagaisus* differs from *K. porrecticornis* by its less divergent horn cores and higher degree of compression. *Zephyreduncinus* shares more of its characters with *Redunca* than with *Kobus,* including short horn cores with longitudinal grooving and a widest mediolateral diameter

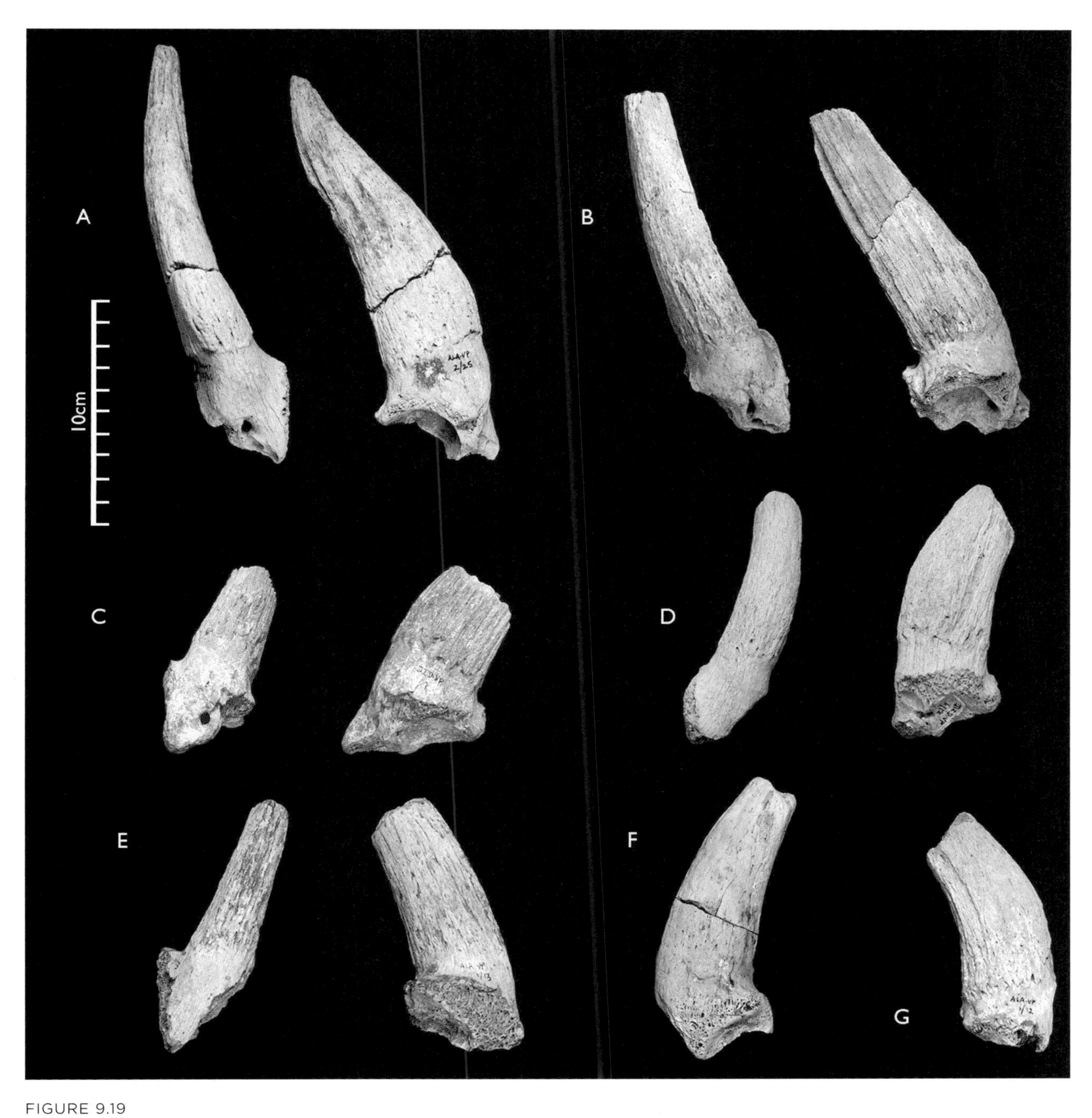

FIGURE 9.19

Horn core specimens of *Zephyreduncinus oundagaisus*. A. ALA-VP-2/25 (holotype), right horn core. B. AFD-VP-1/3, right horn core. C. DID-VP-1/9, left horn core. D. ALA-VP-1/13, left horn core. E. DID-VP-1/12, left horn core. F. ALA-VP-1/1, left horn core. G. ALA-VP-1/12, right horn core.

TABLE 9.11 Horn Core Measurements of *Zephyreduncinus oundagaisus*

Specimen	*MAX*	*MIN*	*CI*	*LEN*
ALA-VP-2/25 (R)	45.8	29.4	0.64	160
ALA-VP-1/13 (L)	43.2	28.0	0.65	—
ALA-VP-1/12 (R)	(41)	(28)	(0.68)	(155)*
ALA-VP-1/1 (R)	(43.5)	(29.5)	(0.68)	(150)*
ASK-VP-2/9 (R)	45.3	29.8	0.66	—
DID-VP-1/9 (L)	44.0	28.0	0.64	—
DID-VP-1/12 (L)	39.5	26.6	0.67	(145)*

NOTE: All measurements in mm. Abbreviations: R, L = right or left horn core; MAX = maximum basal horn core diameters; MIN = minimum basal horn core diameters; CI = horn core compression index (min/max); LEN = length of horn core; ( ) = estimated max or min to the nearest 0.5 mm; ( )* = estimated horn core length along the anterior margin based on a tracing of preserved outline in lateral view.

situated anteriorly. This suggests that *Z. oundagaisus* may have a more recent common ancestry with *Redunca*.

Reduncini gen. et sp. indet.

DESCRIPTION ALA-VP-2/7 (Figure 9.20) is larger than all other reduncine specimens from the Middle Awash 5.7 Ma-old assemblages. This horn core has a basal DAP and DT of 52.3 mm and 38.5 mm, respectively, with an estimated length of 200 mm. The width across the lateral margins of its supraorbital foramina, and the distance from the superior margin of the supraorbital foramen to the horn core base, are ca. 43 mm and 33.5 mm, respectively. The angle of basal horn core divergence, angle of DAP to metopic suture, and angle of horn core to brain case in lateral view are all estimated to be 50°, 35°, and 65°, respectively. It is consistent with Reduncini in having an anterior basal horn core that, in lateral view, curves backward strongly in a smooth arc above the pedicel, anterior recurvature toward the tip, a supraorbital foramen situated at the dorsal end of a depression—or pit—that extends forward and to the midfrontal suture, a deep postcornual fossa, and deep longitudinal grooves on its posterior and posterolateral surfaces. The horn core shows a slight tendency to anticlockwise torsion. There is some sinus development medial to the orbit. The pedicel is well-developed. The basal horn core is quite compressed, has a laterally flattened surface, and a greatest mediolateral diameter situated anteriorly. Horn core divergence is low.

ALA-VP-2/7 differs from contemporaneous specimens of *K.* cf. *porrecticornis* in its larger size, in having supraorbital foramina more closely spaced in relation to size but further separated from the horn core, in having a less upright horn core insertion relative to the dorsal braincase, and probably in having horn cores that are somewhat shorter relative to basal size. *Zephyreduncinus oundagaisus* differs from ALA-VP-2/7 in its much smaller size, in horn cores that are more compressed basally, more uprightly inserted, and on average less divergent. The new species *Redunca ambae* differs from ALA-VP-2/7 in being of a smaller size; having horn cores that are less divergent and on average less compressed and longer relative to basal size; and in having supraorbital foramina that are more widely separated relative to basal horn

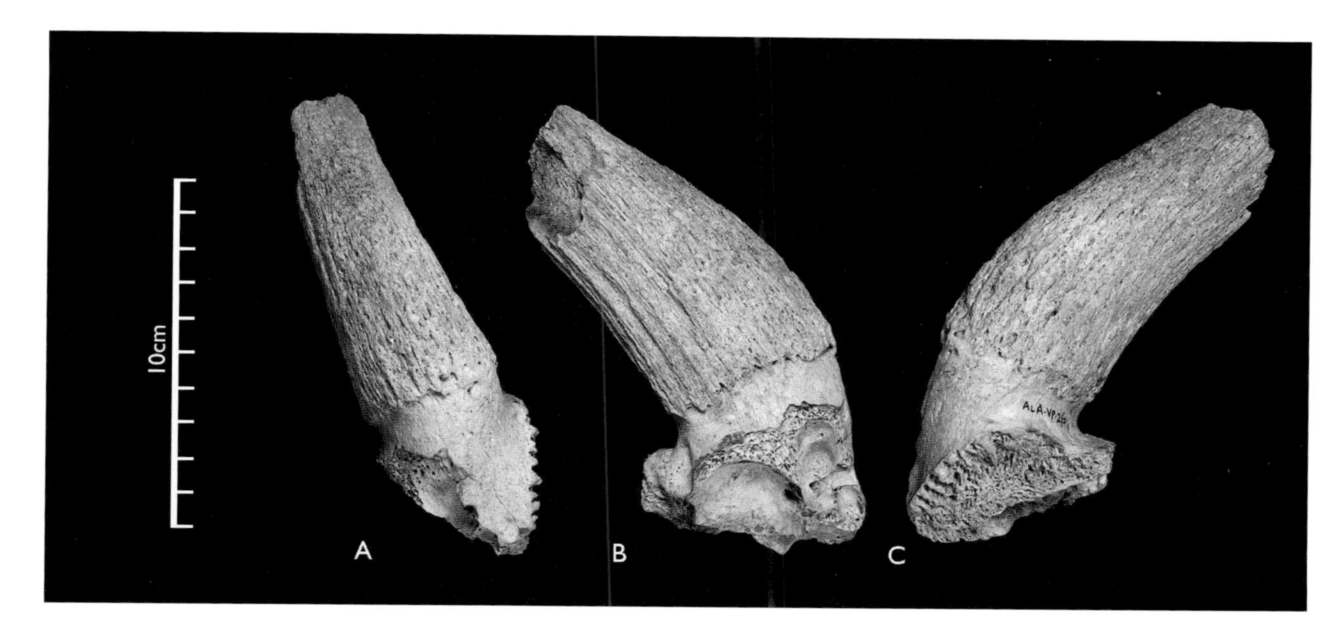

FIGURE 9.20
Horn core assigned to Reduncini gen. et sp. indet. ALA-VP-2/7, right horn core. A. Cranial view. B. Lateral view. C. Medial view.

core size and much closer to the basal horn core. The closest comparison is with ?*R. subdolus* from Langebaanweg, South Africa, which resembles ALA-VP-2/7 in overall size, horn core shortness and low divergence, and a similar posterior angle of the maximum basal diameter to the midfrontal suture. ?*Redunca subdolus* has a low incidence of transverse horn core ridges, compatible with their absence on the preserved part of the horn core of ALA-VP-2/7. ?*Redunca subdolus* differs from ALA-VP-2/7 in displaying horn cores that are inserted more uprightly, are on average less compressed, have a less well-marked horn pedicel, and have supraorbital foramina that are closer to the horn core base and absolutely closer together, especially in relation to basal horn core size. In sum, because ALA-VP-2/7 shows a combination of size and shape characters that set it apart from other reduncines, it is best left as Reduncini gen. et sp. indet. until additional fossils come to light.

REMARKS ON REDUNCINE DENTITION Numerous isolated teeth and jaw fragments with teeth of reduncines have been recovered from the basal Pliocene–terminal Miocene sites of the Middle Awash. None of these specimens were found associated with other skeletal parts identifiable at the species or genus level, although they all belong to Reduncini. Measurements for the more complete lower and upper reduncine teeth are given in Tables 9.12 and 9.13, respectively.

Two kinds of reduncine dentitions can be distinguished. Type A dentitions are overall less advanced and on average larger than those of Type B. At least some of the specimens suggest this distinction (among lower dentitions Type A includes, for example, KUS-VP-1/14 and KUS-VP-1/40; Type B includes AME-VP-1/67, AMW-VP-1/14, and KUS-VP-1/45). However, it is impossible to assign all specimens to one of these two categories. The primitive characters of Type A include the following: they are very low crowned; the medial (lingual) enamel surfaces of the anterior and posterior lobes of upper molars fuse

TABLE 9.12 Measurements of Reduncine Mandibular Dentitions

Specimen	P_{2-4}	P_2	P_3	P_4	M_{1-3}	M_1	M_2	M_3	*Ramus*
ADD-VP-1/13 (L)	(31.5)		10.5	13		12.5	16.9		(30)
			6.5	8.4		(10)	10.3		
ALA-VP-1/4 (L)							16.3		
							8.5		
ALA-VP-2/18 (L)							16.6		
							9.9		
ALA-VP-2/311 (L)						11.5	(20)		
						(8.5)	8.6		
ASK-VP-3/9 (L)								(22)	
								8.8	
ASK-VP-3/13			11.9						
			6.5						
ASK-VP-3/36 (L)							17.6		
							9.8		
ASK-VP-3/161 (R)								22.4	
								9.7	
ASK-VP-3/533 (L)								21.6	
								12.1	
ASK-VP-3/534 (R)								21.2	
								10.8	
DID-VP-1/53 (R)							17.1		
							8.3		
AME-VP-1/37 (L)								(20)	
								8.8	
AME-VP-1/67 (R)								20.4	
								7.4	
AME-VP-1/104 (L)(j)				16.8d		14.8			
				(8.5)d		7.4			
AMW-VP-1/14 (R)	(26)	(6.5)	9.6	10.3		12.9			
			6.8	8		9			
AMW-VP-1/33 (R)(j)	38.0d	7.3d	12.0d	19.1d		15.7	18.4		
		4.0d	5.7d	7.8d		7.3	7.5		
KUS-VP-1/7 (L)								23.6	32
								10.8	
KUS-VP-1/14 (L)					53.4	14.3	16	22.8	(30)
						9.3	(9.5)	(9.5)	
KUS-VP-1/40 (R)				12.5			16.2	23	30.5
				7.3			9.2	9	
KUS-VP-1/45 (R)							(15.5)		
							(10)		
KUS-VP-1/59 (L)	27.6			11.5e		11.7			
				6.6		8.4			
KUS-VP-1/104 (R)				16.8d		14.8			
				(8.5)		7.4			

NOTE: All measurements in mm; Abbreviations: d = deciduous, ( ) = estimated values; j = juvenile (at least one deciduous tooth present); L = left; R = right; Ramus = depth of mandibular ramus at a right angle to the occlusal surface at the M2/M3 junction. In the case of each tooth, the mesiodistal length is given above the buccolingual breadth, both measured at the occlusal surface.

TABLE 9.13 Measurements of Reduncine Maxillary Dentitions

Specimen	M^{1-3}	M^1	M^2	M^3
ALA-VP-2/93 (R)		11.3 10.6		
ALA-VP-2/261 (R)	(50)	(12) (12)		17.8 13
AME-VP-1/45 (L)			(16.2) 12	
AME-VP-1/53 (L)				19.7 12.6
DID-VP-1/64 (L)			(15)	17
KUS-VP-1/52 (L)				(17) (12.5)
KUS-VP-1/68 (L)			16.1 12	
STD-VP-1/6 (R)				17 11.3
STD-VP-1/16 (R)				17 11.5

NOTE: All measurements in mm. Abbreviations: d = deciduous, ( ) = estimated values; L = left; R = right. In the case of each tooth, the mesiodistal length is given above the buccolingual breadth, both measured at the occlusal surface.

only late in wear to give the typical occlusal morphology of a medially continuous enamel border with discrete central enamel cavities; basal pillars are absent on upper molars and are small and poorly developed on lower molars; the premolar row to molar row ratio is greater in Type B dentitions; goat folds on the lower molars are small, thin, and poorly developed; and lateral walls of the lower molars and medial walls of the upper ones are pointed, with little or no pinching (narrowing). The smaller Type B dentitions are more advanced in all these respects. The paraconid and metaconid on P_4 are not fused in any of the available Type A and B specimens. Both kinds of dentitions are present in the earlier set of assemblages dated at 5.7 Ma (containing horn cores and frontlets of *Zephyreduncinus oundagaisus, Kobus* cf. *porrecticornis,* and Reduncini gen. et sp. indet. based on the single specimen ALA-VP-2/7), and in the later 5.2 Ma-old assemblages (in which *Redunca ambae* and *Kobus* aff. *oricornis* first appear alongside the persistent *K.* cf. *porrecticornis* lineage).

The Middle Awash early reduncine dental material in the present 5.2 Ma assemblages can be compared closely with that from the similarly-aged "E" Quarry of the Langebaanweg site in South Africa. Gentry (1980) reported two reduncine species based on horn core and cranial material: the more abundantly represented form he described as *Kobus subdolus* (which we refer to as ?*Redunca subdolus*), and the rarer, smaller species he called *Kobus* sp. 2. Most of the dentitions that Gentry (1980) assigned to Reduncini (and are expected to include mostly the dentitions of ?*Redunca subdolus* because they are the most common) have primitive characteristics that are closely comparable in all respects cited above for the Middle Awash Type A dentitions. We noted the resemblances and possible close relationship between

R. ambae and ?*R. subdolus* above. In the 5.2 Ma strata, the less advanced and larger Type A dentitions may belong to *R. ambae,* while the more advanced, smaller Type B dentitions may include those of *Kobus* aff. *oricornis* and *K.* cf. *porrecticornis*. The association of Type A dentitions with *R. ambae* is further supported by the reduncine fossils from the slightly later ca. 4.9 Ma Gawto strata in the Middle Awash (Haradaso Member). The only reduncine species known to date from Gawto is *R. ambae,* and those dentitions are of the primitive, larger type, namely Type A. If the association in the 5.2 Ma assemblages of larger Type A teeth with *R. ambae* and smaller Type B with *K.* aff. *oricornis* and *K.* cf. *porrecticornis* is correct, then in the older 5.7 Ma strata the Type B teeth may belong to *K.* cf. *porrecticornis* and the larger, more primitive ones of Type A to *Z. oundagaisus*. We noted above that *Zephyreduncinus* may have a more recent common ancestry with *Redunca* than with the *Kobus-Menelikia* clade. The pattern we find suggests that early members of a *Zephyreduncinus-Redunca* clade had less advanced teeth than contemporaneous *Kobus* species.

DISCUSSION Kobus is one of the two genera in the tribe Reduncini. The African late Miocene record of *Kobus* is limited to fragmentary specimens recovered from the Mpesida Beds, Lukeino, Sahabi, and Langebaanweg. This genus is better known from the Nawata Formation of Lothagam. Four isolated horn cores from the Mpesida Beds and two lower molars from the Lukeino Formation have been assigned to *Kobus* aff. *porrecticornis* (Gentry 1978a; Thomas 1980). This species first appears in the type locality of the Dhok Pathan Formation of the Siwaliks sequence, which is late Miocene in age (Thomas 1980). Gentry (1980) named a new species, *Kobus subdolus,* to accommodate the reduncine remains from the "E" Quarry at Langebaanweg.

Fragments of horn cores from Lothagam (Smart 1976) and Langebaanweg (Gentry 1980) were assigned to *Kobus* sp. Recent discoveries at Lothagam, however, indicate that there are at least four species of the genus *Kobus* with different horn core sizes and morphology. Two of the species are new (*Kobus presigmoidalis* sp. nov. and *Kobus laticornis* sp. nov.), whereas one is referred to *Menelikia* and the other to Reduncini gen. et sp. indet. (Harris 2003). Unlike the Miocene and basal Pliocene, the Plio-Pleistocene record of *Kobus* is particularly rich in East Africa. Leakey et al. (1996) have also reported the genus *Menelikia* from Lothagam, although they do not indicate whether it derives from the Nawata or Nachukui Formation. It is more likely to be from the Apak Member because the genus *Menelikia* is unknown before the Pliocene.

The genus *Redunca* is mainly known in Africa after the early Pliocene, and it has three extant species in Africa. Gentry (1980) described two frontlets and one horn core discovered at Sahabi as reduncine remains belonging to the genus *Redunca* or *Kobus,* more likely the latter. Lehman and Thomas (1987), however, argue that the Sahabi reduncine remains are morphologically more like *Redunca darti* than to any other reduncine described from Langebaanweg (Gentry 1980), Lukeino, or Mpesida (Thomas 1980). As a result, Lehman and Thomas (1987) assign the Sahabi reduncine remains to *Redunca* aff. *darti*. However, they also admit Gentry's (1978a, b) statement that generic-level definitions of reduncines are difficult. If the Sahabi record of *Redunca* aff. *darti* is correct, it would represent the earliest record of the genus. *Redunca darti* was the name given by Wells and Cooke (1956) to reduncine remains from the middle Pliocene cave deposits at Makapansgat, South Africa.

Hippotragini

Systematic Paleontology

Hippotragini Retzius and Lovén, 1845

Hippotragini gen. et sp. indet.

DIAGNOSIS Large bovids whose long horn cores have a large basal area relative to skull size and moderately to well-developed basal sinuses in their horn core pedicels; horn cores unkeeled, with little divergence, and no transverse ridges; postcornual fossae shallow or absent; bulbous, lingually extending metaconids on the P_4s (after Vrba and Gatesy 1994).

DESCRIPTION The Middle Awash specimens referred to this group derive from the Kuseralee Member and represent a medium-sized bovid that was smaller than living Hippotragini, and comparable to the smallest known hippotragine material from Lothagam (Smart 1976; Harris 2003; Figure 9.21). Horn core measurements are given in Table 9.14. We consider this assemblage hippotragine, rather than reduncine or alcelaphine, because its specimens combine limited, and mainly anterior, sinus development in the pedicel; small supraorbital foramina not situated in a pit but flush on an evenly sloping frontal surface; very uprightly inserted horn cores with low divergence; basal separation, and substantial girth relative to visible adjacent skull morphology; absent to very poor development of transverse ridges on horn cores; and a straight parietofrontal suture lacking central anterior indentation. The frontal between the horn cores is hardly elevated above the dorsal orbital rim, which is thin and not prominent (as can be seen, for instance, in AMW-VP-1/3 in spite of some marginal damage). AMW-VP-1/3 is most similar to KNM-LT 23598 and KNM-LT 23131 from the Apak and Upper Nawata Members of Lothagam, respectively, assigned to *Hippotragus* sp. (Harris 2003). The top of the orbit is closer to the horn core base in KUS-VP-1/8 in spite of damage. The craniofacial angle was also relatively high in this specimen. The horn cores are very compressed at the base, and variously show some lateral flattening (e.g., in AMW-VP-1/3, AMW-VP-1/69, Figure 9.21). The horn cores appear to have been short in comparison with later hippotragines (see AMW-VP-1/3, AMW-VP-1/87, Figure 9.21). The supraorbital foramina are fairly far apart, especially in AMW-VP-1/87, and have a shallow groove anteriorly (e.g., AMW-VP-1/3). The postcornual fossae are more elongated, shallower, and less extensive than in the reduncines from the same strata, and deeper than is usual in alcelaphines.

DISCUSSION The Middle Awash hippotragine specimens are among the earliest records of the tribe. The material currently available from the Middle Awash is limited to the younger Kuseralee Member and does not indicate the presence of more than one species. The closest resemblance of the Middle Awash material is to the material from Lothagam previously referred to as Hippotragini gen. indet. sp. nov. (by Vrba and Gatesy 1994, whose cladistic results place this form as one of the three most basal lineages in Hippotragini) and as Hippotragini gen. nov. (Vrba 1995b), which Harris (2003 and elsewhere) variously included in *Hippotragus* sp., and *Damalacra* sp. A. This shows that phylogeny of early hip-

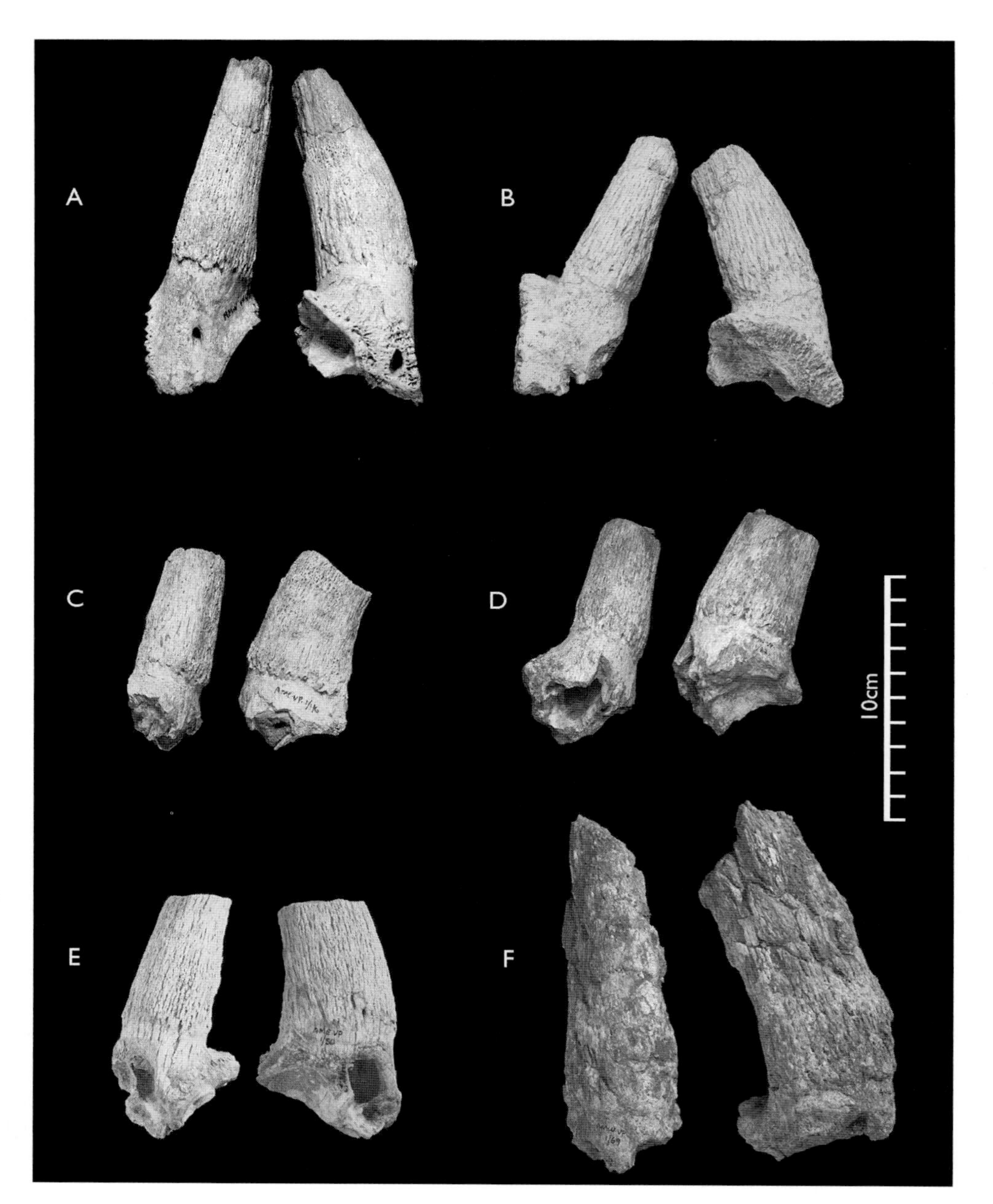

FIGURE 9.21

Horn core specimens of Hippotragini gen. et sp. indet. A. AMW-VP-1/3, left horn core. B. KUS-VP-1/8, left horn core. C. AME-VP-1/46, left horn core. D. AMW-VP-1/87, left horn core. E. AME-VP-1/56, right horn core. F. AMW-VP-1/69, horn core fragment.

potragines is far from understood. Moreover, due to the overall absence of adequate late Miocene fossil samples from other contemporaneous sites, detailed phylogenetic studies of the tribe are not possible. However, the hippotragines documented thus far from both the Middle Awash and Lothagam late Miocene are different from the early hippotragine genus *Praedamalis* in the less extensive sinuses in the pedicel and basal horn core, lower separation between supraorbital foramina, and probably also horn cores that are shorter and basally more compressed than in the earlier form (Vrba and Gatesy 1994).

TABLE 9.14 Measurements of Horn Core and Frontlet Specimens of Hippotragini gen et sp. indet.

Specimen	*MAX*	*MIN*	*LEN*	*SOF*	*S-HC*	*ADIV*	*AHMET*	*AHCBR*
AME-VP-1/46 (L)	(42.5)	(30)						
AME-VP-1/56 (L)	39.6	28.7		(46)	34			(70°)
AMW-VP-1/3 (L)	42.2	29.7	(180)	(46)	26.4	(20°)	(15°)	(70°)
AMW-VP-1/69 (R)	(52.5)	39.4						(95°)
AMW-VP-1/87 (L)	41.5	30.4		(56)	(26)			
KUS-VP-1/8 (L)	39.6	29.3		(46)	35.3	(25°)	(35°)	(60°)

NOTE: R, L = right and left horn cores; ( ) = estimated values. All measurements in mm except where indicated. See Table 9.8 for other abbreviations.

Conclusions

The Middle Awash late Miocene bovid assemblages belong to one of the few mammalian fossil sequences in Africa that sample the 5–6 Ma interval, which includes the Miocene-Pliocene transition. These assemblages document a much higher diversity in the late Miocene bovid community of eastern Africa and further hint at active species origination during the 5–6 Ma interval. Several instances of species first appearances are present, including the new genus *Zephyreduncinus* as well as *Simatherium* (the latter appearing also in the Langebaanweg assemblage). The Middle Awash bovids also share a number of taxa with other contemporaneous African sites such as Lothagam, Lukeino, Sahabi, and Toros Menalla.

The two time-successive Middle Awash horizons document first appearances of some taxa and the extinction of others, particularly at the Plio-Miocene boundary (5.33 Ma), creating a clear difference between the Asa Koma Member assemblage (5.5–5.8 Ma) and the Kuseralee Member assemblage (5.2 Ma). Exclusive to the earlier Asa Koma Member are the bovine *Ugandax* sp., a different and larger species of the boselaphine genus *Tragoportax,* and the reduncines *Zephyreduncinus oundagaisus* and Reduncini gen. et sp. indet. In contrast, the bovine *Simatherium demissum, Gazella* sp., and *Kobus* aff. *oricornis* and Hippotragini gen et sp. indet. are all confined to the Kuseralee Member. In addition, the species *Kobus* cf. *porrecticornis* is more abundant in the earlier period and, in fact, may not be represented at all in the later fauna. Thus, the change in the bovid assemblages from the Asa Koma Member (5.5–5.8 Ma) to the Kuseralee Member (5.2 Ma) is substantial.

Based on the overall assemblage compositions of the available sample, relative abundances of indicator taxa, bovid ecomorphology, and associated contextual evidence of the Asa Koma and Kuseralee Members, Su et al. (Chapter 17) documented similar paleoenvironmental conditions for both members. These results indicate that bovids associated with densely vegetated, floodplain grasslands and swampy habitats were abundant in both members, whereas those found in open habitats were not as common, but present in both members.

There are some genera that persist throughout the two members, particularly the larger-sized ones, such as *Tragoportax* and *Tragelaphus*. Although one might argue the possibility that the two species of these genera from the Asa Koma Member gave rise to the species in the Kuseralee Member, the available fossil evidence is not robust enough to infer phyletic evolution, largely because of the small sample size and lack of comparable elements.

If these preliminary findings hold as other assemblages are recovered from other sites, then it will be likely that the changes evident in the present late Miocene bovid assemblages of the Middle Awash not only indicate local faunal and environmental change between ca. 5.7 and 5.2 Ma, but that they may also reflect a general response of the larger African mammals to global climatic change. Vrba (2000) found that over the past 20 Ma the largest net temperature changes, especially cooling trends, were associated with levels of significantly elevated origination and/or extinction episodes in African larger mammals. In a more recent refinement of this study, using data for all the African larger mammals over the past 10 Ma, and a statistical model that compensates for uneven fossil preservation in time, one of the significant origination pulses was found to overlap the Messinian interval (Vrba, unpublished data). Thus, the first appearances of some bovid taxa in the Middle Awash may have been a part of a larger episode of elevated origination and extinction in African mammals between 5 and 6 Ma.

10

Suidae

YOHANNES HAILE-SELASSIE

At least seven genera are recognized in the subfamily Tetraconodontinae (Pickford 1995), yet the origin of African tetraconodontines is poorly understood. Some researchers argue that they emigrated from Asia and are unrelated to the forms already established in Africa by the early to middle Miocene. Others argue that they emerged from a hyotheriine stock (particularly from the European genus *Aureliachoerus*) with similar molar and premolar morphology (Pickford 1995). Cooke and Wilkinson (1978) suggested that the African genus *Nyanzachoerus* probably descended from a form like *Kubanochoerus massai,* a species from the middle Miocene of Africa. However, considering the enlarged premolars and other dental characters of *Nyanzachoerus,* its closer affinity seems to lie with the tetraconodontines.

The genus *Conohyus* represents a possible African tetraconodontine candidate ancestor that might have emigrated from Europe to Africa and given rise to *Nyanzachoerus.* Its fossil record in Africa is restricted to north of the Sahara. *Conohyus* is known from Europe, Africa, Asia, and the Indian subcontinent and is currently the best candidate ancestor for *Nyanzachoerus.* No *Conohyus* has been documented in Africa from deposits younger than 11 Ma. *Nyanzachoerus* might have emerged from and then replaced *Conohyus* during its African diversification late in the Miocene when at least four nyanzachoere taxa (*Ny. devauxi, Ny. syrticus, Ny. waylandi,* and *Ny. kuseralensis* sp. nov.) occupied eastern Africa.

Historically, almost all the late Miocene and early Pliocene African tetraconodont suid taxa have been referred to *Nyanzachoerus* and *Notochoerus* (Cooke and Wilkinson 1978). By the early to middle Pliocene, at least three new suine lineages (*Potamochoerus, Kolpochoerus,* and *Metridiochoerus*) appeared. Perhaps immigrating to Africa from Eurasia (Harris and Leakey 2003b), *Kolpochoerus deheinzelini* appears in Pliocene deposits of Chad and Ethiopia (Brunet and White 2001), and the origin of *Metridiochoerus* remains elusive (White et al. 2006a).

Historical Background

The tetraconodont fossil record has expanded dramatically during the last two decades of field research at late Miocene and early Pliocene sites such as Lothagam, Kanapoi, and those of the Afar. As a result, a thorough revision of the Mio-Pliocene tetraconodont suids of Africa is overdue. Before proceeding with such a revision, it is important to review the history of discovery and interpretation of African tetraconodontines.

The phylogeny of African tetraconodonts has been a subject of interest since the early 1970s. A number of phylogenetic schemes have been offered and revised for the known African tetraconodonts (e.g., White and Harris 1977; Harris and White 1979; Cooke 1978, 1987; Pickford 1988, 1989a; van der Made 1999). Most of these works agreed on basic phylogenetic relationships, but residual problems of species definition and recognition persist.

Discoveries from Sahabi, Libya, during the first half of the twentieth century led to the naming of two *Sivachoerus* species (cf. *S. giganteus* and *S. syrticus*), that were later reassigned to *Nyanzachoerus syrticus* (Leonardi 1952). Leakey (1958) named the type species *(Ny. kanamensis)* to refer to a suid mandible from Kanam, Kenya. Arambourg (1968) named another species *(Propotamochoerus devauxi = Ny. devauxi)* to refer to specimens from the late Miocene of Bou Hanifia in Algeria. Coppens (1971) named the species *Ny. jaegeri* for a mandible with teeth from Hamada Damous, Tunisia. Similar specimens from Kanapoi had already been recognized as a new species (*Ny. plicatus,* Cooke and Ewer 1972). However, the latter was required to be synonymized with *Ny. jaegeri* as a result of *Ny. jaegeri's* priority. Most of these *Nyanzachoerus* species were based on single specimens. A major exception to this rule was Cooke and Ewer's 1972 study of a large suid collection from the late Miocene of Lothagam. They named *Nyanzachoerus tulotos* based on this series and recognized an additional new nyanzachoere species from the Kanapoi deposits in northwestern Kenya, which they named *Ny. pattersoni.*

White and Harris (1977) presented the first synthesis of African suid phylogeny after the Kanapoi and Lothagam suids had been described. They presented *Ny. tulotos* as the earliest known nyanzachoere in Africa. They recognized *Ny. syrticus* as a species similar to *Ny. tulotos,* although they did not include it on their phylogenetic tree or synonymize it with *Ny. tulotos*. They hypothesized that *Ny. kanamensis* probably evolved from a form similar to *Ny. tulotos,* and that *Ny. jaegeri* evolved from *Ny. kanamensis* (Figure 10.1B).

Cooke (1978) proposed an alternative phylogeny and taxonomy as a reply to White and Harris's 1977 proposals, recognizing *Ny. syrticus* as ancestral to *Ny. tulotos*. He also differed from White and Harris by recognizing *Ny. pattersoni* as a valid species and possible ancestor of *Ny. kanamensis*. Cooke (1978) argued that the Kanam material assigned to *Ny. kanamensis* was outside the range of variation seen in the material assigned to *Ny. pattersoni*. He further argued that *Ny. pattersoni* was not a direct descendant of *Ny. tulotos* and probably evolved from an earlier stock. He also disagreed with White and Harris (1977) on the ancestor-descendant relationship between *Ny. kanamensis* and *Ny. jaegeri,* suggesting that *Ny. jaegeri* possibly diverged at an earlier stage rather than being a derivative of *Ny. kanamensis* (Figure 10.1A).

Subsequent work by Cooke and Wilkinson (1978) reported that *Ny. pattersoni* was widespread in eastern Africa, present at Hadar and the Omo Basin in Ethiopia, East Rudolf in Kenya, and in the Kaiso Formation of Uganda. They also suggested that the suid remains from the Quartzose Sand Member (QSM) and Pelletal Phosphorite Member (PPM) at Langebaanweg, South Africa, belonged to a species similar to *Ny. pattersoni* and

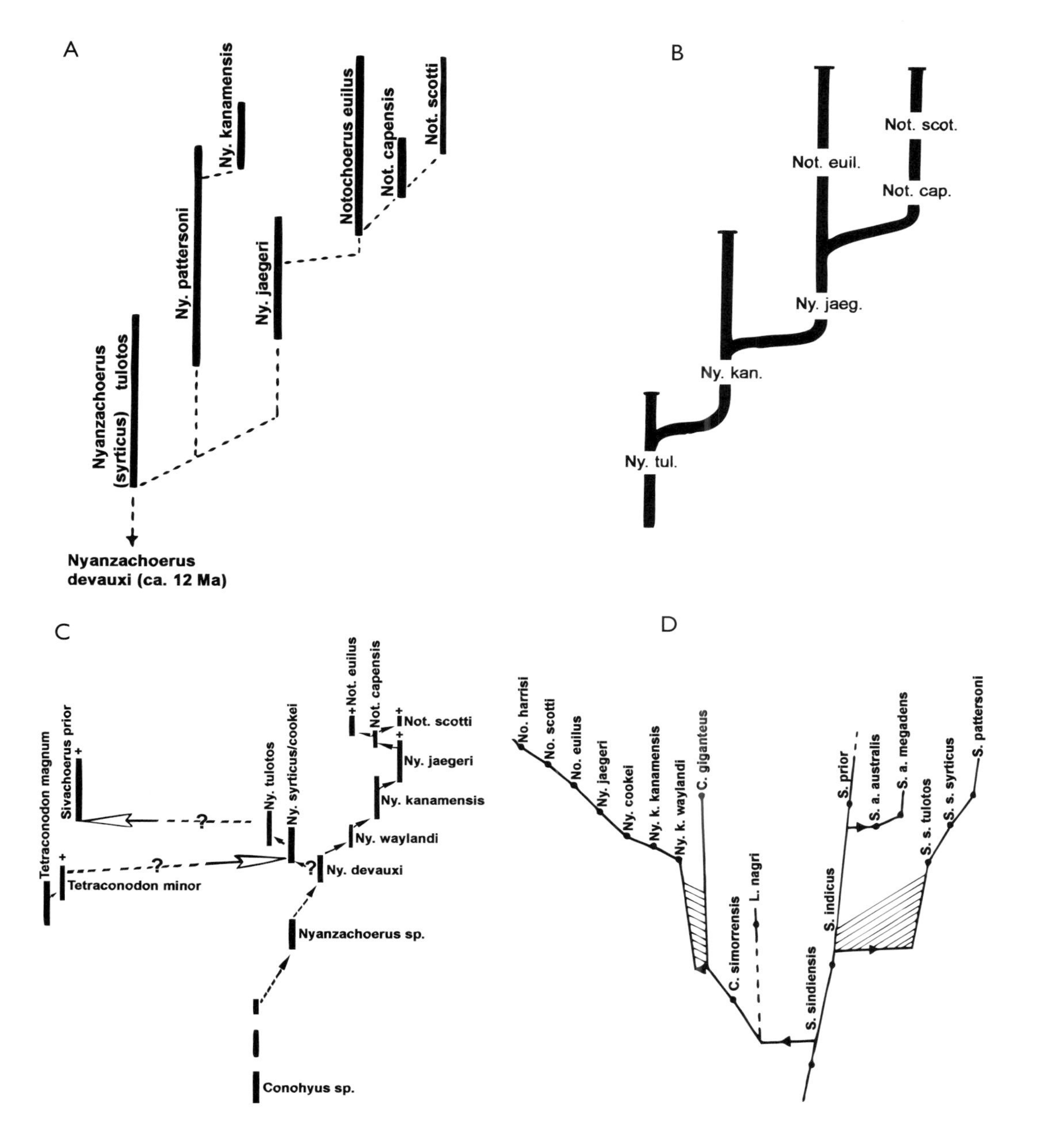

FIGURE 10.1

Proposed phylogenetic relationships among African tetracondonts.
A. Cooke 1978; Cooke and Wilkinson 1978. B. Harris and White 1979; White and Harris 1977.
C. Pickford 1985.
D. Van der Made 1999.

probably represented a slightly earlier southern race (Cooke and Wilkinson 1978:456). Cooke and Wilkinson's (1978) phylogenetic assessment of various late Miocene–early Pliocene African suid taxa showed *Ny. devauxi* as the stem species for the African nyanzachoeres, giving rise to *Ny. syrticus,* evolving to *Ny. tulotos* and, possibly simultaneously or a little later in time, giving rise to the *Ny. pattersoni* and *Ny. kanamensis* clade (Figure 10.1A). Cooke and Wilkinson also suggested that *Ny. jaegeri* was part of this early radiation, which eventually gave rise to the *Notochoerus* lineage sometime during the lower Pliocene. Harris et al. (2003) recently reclassified *Ny. jaegeri* as *Notochoerus jaegeri,* largely based on morphological similarities claimed to be shared between the mandibular symphyseal regions of *Ny. jaegeri* and *Not. euilus.*

Harris and White (1979) conducted a detailed comparative description of the African Plio-Pleistocene suids, elaborating on their 1977 synthesis. They presented a compelling argument that the third molar of suids is highly informative in discerning the evolution of mammalian groups, using this tooth heavily to infer phylogenetic relationships. They synonymized *Ny. pattersoni* with *Ny. kanamensis,* depicting the latter as the direct descendant of *Ny. tulotos.* Although Harris and White (1979) agreed with Cooke and Ewer (1972) in recognizing *Ny. tulotos* as a valid species, they rejected *Ny. pattersoni* as a valid species and instead synonymized it with *Ny. kanamensis* and included the Langebaanweg material in *Ny. kanamensis.* They also recognized *Ny. jaegeri* as the most derived nyanzachoere and inferred that it was the ancestor of all later notochoeres. After studying the material from Sahabi, Cooke (1987) revised his earlier suid phylogeny and accepted that *Ny. pattersoni* was a junior synonym of *Ny. kanamensis.* He further accepted *Ny. tulotos* as a junior synonym of *Ny. syrticus.*

Phylogenetic relationships among African suids were further revised by Pickford (1989a) after he recognized a new nyanzachoere species, *Ny. waylandi,* from the Nyaburogo Formation of Uganda, estimated to be 5–6 Ma. The *Ny. waylandi* discovery further complicated the postulated direct ancestor-descendant relationship between *Ny. syrticus* and *Ny. kanamensis* (Figure 10.1C). This species was interpreted as derived relative to *Ny. syrticus* and possibly ancestral to *Ny. kanamensis* (White 1995).

Cooke and Hendey (1992) recognized a new subspecies of *Nyanzachoerus kanamensis (Ny. kanamensis australis)* to refer to the suid remains from the QSM at Langebaanweg, South Africa. This was mainly based on the overall larger size of the South African form relative to *Ny. kanamensis* from Hadar and Kanapoi. Three years later, White (1995) suggested *Ny. kanamensis* as the possible descendant of *Ny. waylandi* (Pickford 1989a). Leakey et al. (1996) followed the synonymy of *Ny. syrticus* and *Ny. tulotos,* although Harris and Leakey (2003b) preferred to recognize the northern African and eastern African forms as two distinct subspecies: *Ny. syrticus syrticus* and *Ny. syrticus tulotos.* In doing so, however, Harris and Leakey (2003b) failed to present sufficient evidence to recognize two different subspecies for the hypodigm, and they appear to have been largely influenced by van der Made's (1999) metric-based classification.

The most recent suid systematic analysis, which basically revised previous classifications (White and Harris 1977; Harris and White 1979; Cooke 1978, 1987; Pickford 1989a, for example), was performed by van der Made (1999). He reclassified the African nyanzachoeres into two tribes: Tetraconodontini and Nyanzachoerini. He included three species in the Nyanzachoerini: *Ny. cookei, Ny. jaegeri,* and the type species, *Ny. kanamensis. Nyanzachoerus syrticus (=Ny. tulotos)* and *Ny. australis* were moved to the Asian genus *Sivachoerus* Pilgrim, 1926, and further classified into many subspecies. By doing so, van der Made not only reclassified the African nyanzachoeres into three separately evolving lineages (two related to the Asian genus *Sivachoerus* and a third descending from the European *Conohyus*) but also inflated the number of African late Miocene–early Pliocene suid taxa by naming multiple subspecies in some groups based on minor metric differences. His classification further complicated the phylogenetic

FIGURE 10.2

Tim White applies Vinac™ preservative (hardener) to the left mandible of a *Nyanzachoerus* specimen from Amba East. The mandible is *in situ* in the Kuseralee Sands and can be lifted without falling apart only after the hardener has set. Photograph by Gen Suwa, December 12, 1992.

relationships of African late Miocene–early Pliocene suids both in terms of species diagnosis and taxonomy vis-à-vis the Asian forms (Figure 10.1D). For example, he assigned the South African *Ny. australis* material to two subspecies of *Sivachoerus australis: Sivachoerus australis australis* for the material from the QSM, and *Sivachoerus australis megadens* for the material from the PPM (van der Made 1999:212–213). These systematic revisions have not received support or acceptance because of their excessive splitting and their lack of clearly diagnostic features used to distinguish one taxon from another.

Two major questions are significant to address in current African tetraconodont systematics: the distinction between *Ny. kanamensis* and *Ny. australis,* and the relationship between *Ny. jaegeri* and *Not. euilus*. As described above, there are multiple views proposed by various researchers. It is important to review them here before proceeding to address how the newly recovered Middle Awash material relates to them.

The phylogenetic relationship between *Ny. kanamensis* from eastern Africa and *Ny. australis* from southern Africa has been proposed, with slightly differing taxonomic labeling, as ancestor-descendant (Harris and White 1979; White 1995; Haile-Selassie 2001a; Harris and Leakey

FIGURE 10.3A
Nyanzachoerus cranium STD-VP-1/1 *in situ,* the first vertebrate specimen collected by the Middle Awash paleoanthropological project from this locality. Ahamed Elema, Ann Getty (discoverer), Mohammed Hamadou, and local guide.

2003b). In contrast, Cooke and Wilkinson (1978; but see Cooke 1987), interpreted *Ny. australis* as an early stock of *Ny. pattersoni* unrelated to *Ny. kanamensis*. Van der Made (1999) claimed that *Ny. kanamensis* and *Ny. australis* do not belong to the same genus and represent two separate lineages. A different view was proposed by Fesseha (1999), suggesting that *Ny. kanamensis* and *Ny. jaegeri* belong to the genus *Notochoerus*. Based on a study of Hadar suids using dental character states and quantitative analysis, Fesseha argued that *Ny. kanamensis* and *Ny. jaegeri* share more derived characters with *Not. euilus* than with *Ny. syrticus*. Fesseha (1999) also agrees with Harris and White (1979) in *Not. euilus* being a direct descendant of *Ny. jaegeri*.

Despite Cooke's (1978), and Cooke and Wilkinson's (1978), earlier argument that *Ny. jaegeri* may not have descended from *Ny. kanamensis* but rather diverged at an earlier stage, the ancestor-descendant relationship between *Ny. kanamensis* and *Ny. jaegeri* appears to be accepted by most researchers (White and Harris 1977; Harris and White 1979; Cooke and Wilkinson 1978; Harris 1983b; Cooke 1987; Pickford 1989a; White 1995; Fesseha 1999, among others). Fesseha's 1999 reassignment of *Ny. kanamensis* to the genus *Notochoerus* has received no support, although Harris et al. (2003) recognized *Notochoerus jaegeri* from Kanapoi, Kenya.

The available fossil suid sample has increased enormously since the 1980s with the discovery of specimens from sites such as Lothagam (Harris and Leakey 2003b), Kanapoi (Harris et al. 2003), and the Middle Awash (Haile-Selassie 2001a) (see Figures 10.2, 10.3). These new collections provide deeper insights into the systematics of African tetraconodont suids during the late Miocene and early Pliocene. The late Miocene Adu-Asa Formation and the Mio-Pliocene Sagantole Formation of the Middle Awash have yielded hundreds of suid specimens directly relevant to these issues, because of their good preser-

FIGURE 10.3B

A close-up of STD-VP-1/1. Photographs by Tim White, December, 19, 1993.

vation, abundance, and tight chronostratigraphic placements. These Middle Awash assemblages from time-successive members illuminate tetraconodont diversity toward the end of the Miocene and necessitates reconsideration of some of the phylogenetic relationships proposed earlier. These discoveries are presented and interpreted here.

Tetraconodontinae

Systematic Paleontology

Nyanzachoerus Leakey, 1958

GENERIC DIAGNOSIS See Leakey, 1958.

Nyanzachoerus devauxi Arambourg, 1968

Nyanzachoerus cf. *devauxi*

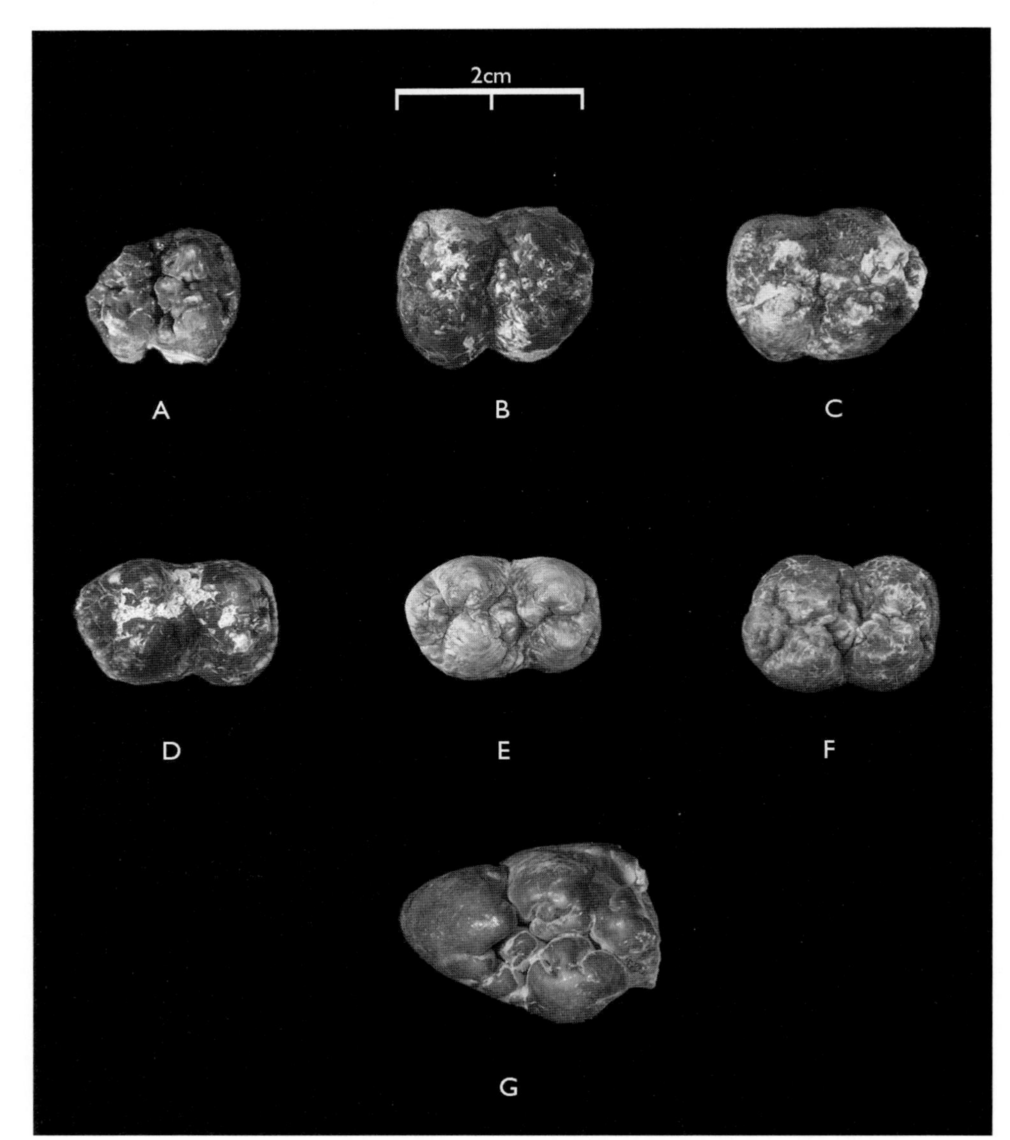

FIGURE 10.4
Occlusal views of isolated upper and lower molars of *Nyanzachoerus devauxi*. A. ALA-VP-1/35, left M^1. B. ALA-VP-2/12, left M^1. C. ALA-VP-2/86, left M^1. D. ALA-VP-1/23, right M_2. E. ALA-VP-2/80, right M_2. F. DID-VP-1/99, left M_2. G. ASK-VP-3/115, left M_3 fragment.

DESCRIPTION Arambourg (1968) recognized *Nyanzachoerus devauxi* as one of the earliest African tetraconodonts based on specimens from Bou Hanifia, Algeria. He initially described the material as *Propotamochoerus devauxi*. However, later workers referred the same material to *Nyanzachoerus devauxi* largely based on similarities with later nyanzachoeres from eastern Africa. Van der Made (1999) assigned *Ny. devauxi* as a junior synonym of *Conohyus giganteus* Falconer and Cautley, 1847. Harris and Leakey (2003b), however, retained the name *Ny. devauxi* to refer to a small nyanzachoere from the Lower Member of the Nawata Formation of Lothagam. A M_3 recovered from the Namurungule Formation (Nakaya et al. 1984), and therefore older than the oldest deposits in the Lothagam Formation, is also referable to *Ny. devauxi* (Nakaya 1994) and establishes its earliest record in eastern Africa at ca. 8.5 Ma. Given the evolutionary trend in nyanzachoeres, *Ny. devauxi* best represents the earliest tetraconodont referred to the genus *Nyanzachoerus* in eastern Africa, and a sister taxon to all other nyanzachoere species.

Nyanzachoerus devauxi is best known from Sahabi (Cooke 1982, 1987) and the Lower Nawata Member of Lothagam (Harris and Leakey 2003b). This species is represented in the Middle Awash by a number of isolated teeth (Figure 10.4). Isolated molars of *Ny. devauxi* and *Ny. syrticus* are difficult to distinguish except that the latter species generally has larger teeth. The dimensions of the teeth referred here to *Ny.* cf. *devauxi* are slightly below the range of *Ny. devauxi* from the Nawata Formation of Lothagam (Figure 10.5). However, the M^1s, most of which are unerupted crowns, are metrically and morphologically similar to specimens from Sahabi, such as 6P16B (Cooke 1987). The M_2s are also referable to *Ny.* cf. *devauxi* largely based on their size. They are too small to be included in the *Ny. syrticus* hypodigm. ASK-VP-3/115 is assigned to this species based largely on its size and morphology of the talonid. It has one small and two large cusplets on the mesiobuccal side of the terminal cusp. This is a character commonly seen in *Ny. devauxi* specimens from the Lower Nawata of Lothagam. The dental measurements are given in Table 10.1.

DISCUSSION *Nyanzachoerus devauxi* and *Ny. syrticus* are very similar in their dental morphology even though *Ny. devauxi* third molars are much smaller. Harris and Leakey (2003b:497) stated that morphological differences between the two species lie in the talon(id) of the third molars. The M^3 talon has a median cusp posterior to the metacone and hypocone, a posteromedially positioned terminal cusp, and two or three cusplets on the distobuccal corner. However, this is largely the case in both species. There does appear to be a difference in the way that the two or three cusplets on the distobuccal corner are proportioned. In *Ny. devauxi,* there are usually two major cusplets and the third, if present, is more of a basal pillar than a cusplet. In *Ny. syrticus,* there are at least three cusplets that are usually of equal size in breadth and height. An additional difference that Harris and Leakey (2003b) noted is the simplicity of the talonid on the M_3s of *Ny. devauxi.* However, the M_3s of *Ny. syrticus* from the Adu-Asa Formation show that this character is highly variable. For example, KNM-LT 26075, a M_3 from the Lower Nawata Member of Lothagam (Harris and Leakey 2003b: figure 10.45), is identical in occlusal morphology with KL164-1 (Figure 10.7C) from the Middle Awash and assigned to *Ny. syrticus.* Therefore, talonid simplicity on M_3s by itself does not appear to be a reliable character to distinguish *Ny. devauxi* from *Ny. syrticus.* However, it appears that on average, the median pillars and posterior cusplets on *Ny. syrticus* M_3s increase in size and contribute to the increase in the overall length of the molar.

Nyanzachoerus devauxi upper and lower third molars are distinctly smaller than those of *Ny. syrticus,* even though there is overlap in length between the smallest *Ny. syrticus* and largest *Ny. devauxi* third molars from Lothagam (Figure 10.5). This kind of overlap is also seen in other ancestor-descendant tetraconodont groups, such as *Ny. australis* and *Ny. kanamensis.* Indeed, the teeth from the Upper Nawata assigned to *Ny. syrticus* are much larger than those from the Lower Nawata. In light of the fact that Harris and Leakey (2003b) made the distinction between the two groups from the Lower Nawata based on the size of the third molars, it is not impossible that they assigned the males and

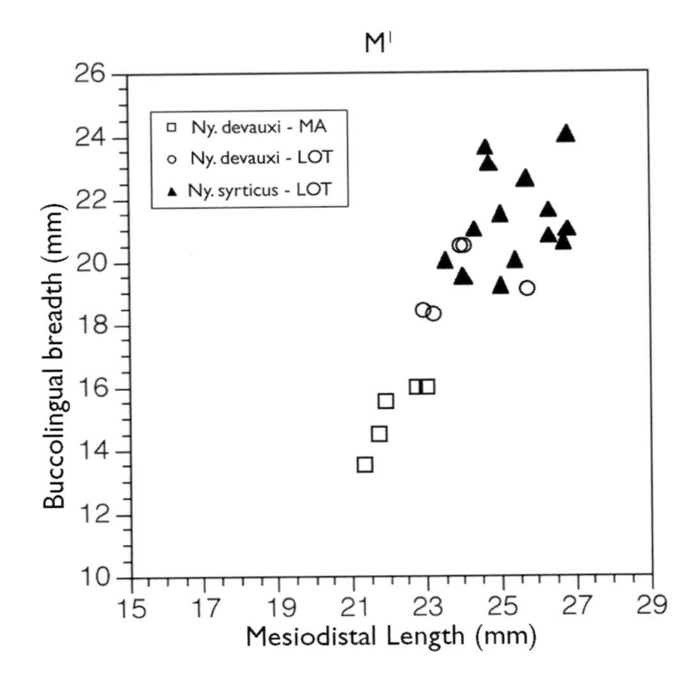

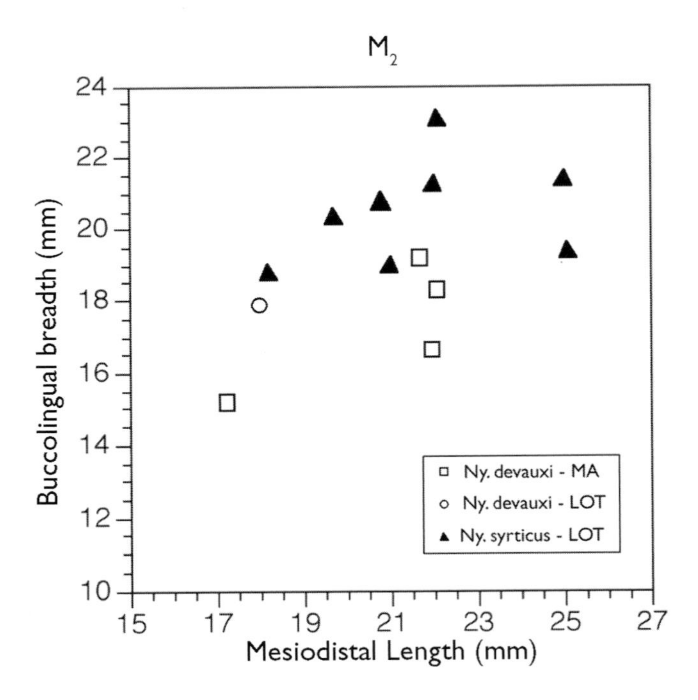

FIGURE 10.5

Bivariate plots showing comparisons of dental dimensions between *Ny. devauxi* (Lower Nawata of Lothagam), *Ny. devauxi* (Middle Awash), and *Ny. syrticus* (Lower Nawata of Lothagam). Abbreviations: MA = Middle Awash; LOT = Lothagam.

TABLE 10.1 Dental Measurements of *Nyanzachoerus* cf. *devauxi*

	P3		*P4*		*M1*		*M2*		*M3*	
Specimen	MD	BL	MD	BL	MD	BL	MD	BL	MD	BL
Uppers										
ALA-VP-2/12					22.1	18.3				
ALA-VP-2/83					21.7	19.2				
ALA-VP-2/86					21.9	17.6				
STD-VP-2/58							(24)	nm		
Lowers										
ALA-VP-2/80							21.3	13.4		
ALA-VP-2/81					(21)	(16)				
ALA-VP-2/120							22.7	16		
DID-VP-1/99							21.9	15.5		

NOTE: Values in parentheses are estimates. All measurements are in mm; nm = measurement was not possible.

females of *Ny. devauxi* to two different species, in which case *Ny. syrticus* first appears in the Upper Nawata Member, while *Ny. devauxi* persisted into the same member. Whether *Ny. syrticus* is present in the Lower Nawata or not, however, there is now good evidence from Lothagam and the Middle Awash to suggest that *Ny. syrticus* speciated by cladogenesis rather than by anagenesis. The absence of *Ny. syrticus* in the Lower Nawata would have significant bearing on the First Appearance (FA) of *Ny. syrticus* in Africa and calls for a reassessment of the taxonomic affinity of the suid specimens from the Namurungule Formation.

Nyanzachoerus syrticus Leonardi, 1952

(=*Nyanzachoerus tulotos* Cooke and Ewer, 1972; *Sivachoerus syrticus syrticus* van der Made, 1999; *Sivachoerus syrticus tulotos* van der Made, 1999; *Nyanzachoerus syrticus tulotus* Harris and Leakey, 2003)

DESCRIPTION The most complete cranium of *Nyanzachoerus syrticus* recovered from the Middle Awash is STD-VP-1/1 (Figure 10.6). This specimen consists of most of the cranium. The specimen is partially damaged by cracks and surface exfoliations that render description of detailed morphological features difficult. Most of the muzzle and skull roof are intact. The left zygoma, most of the basioccipital and foramen magnum regions, and the right superior rim of the occiput are missing. The right zygoma is preserved but medially crushed at the center. The tip of the muzzle is missing on the left side, and the left wall of the snout is crushed and distorted, with the maxillary flange missing. The maxillary flange on the right side is perfectly preserved. Both canines are broken at their roots. The right dental row is perfectly preserved from P^1 to M^3, although enamel loss is apparent on M^1 and M^2. On the left side, P^3–P^4 and the distal half of the M^3 are preserved.

The maxillary flange on the right side is small and positioned above the canine alveoli, although most of it lies posterior to the canine. The dorsal wings of the flange do not

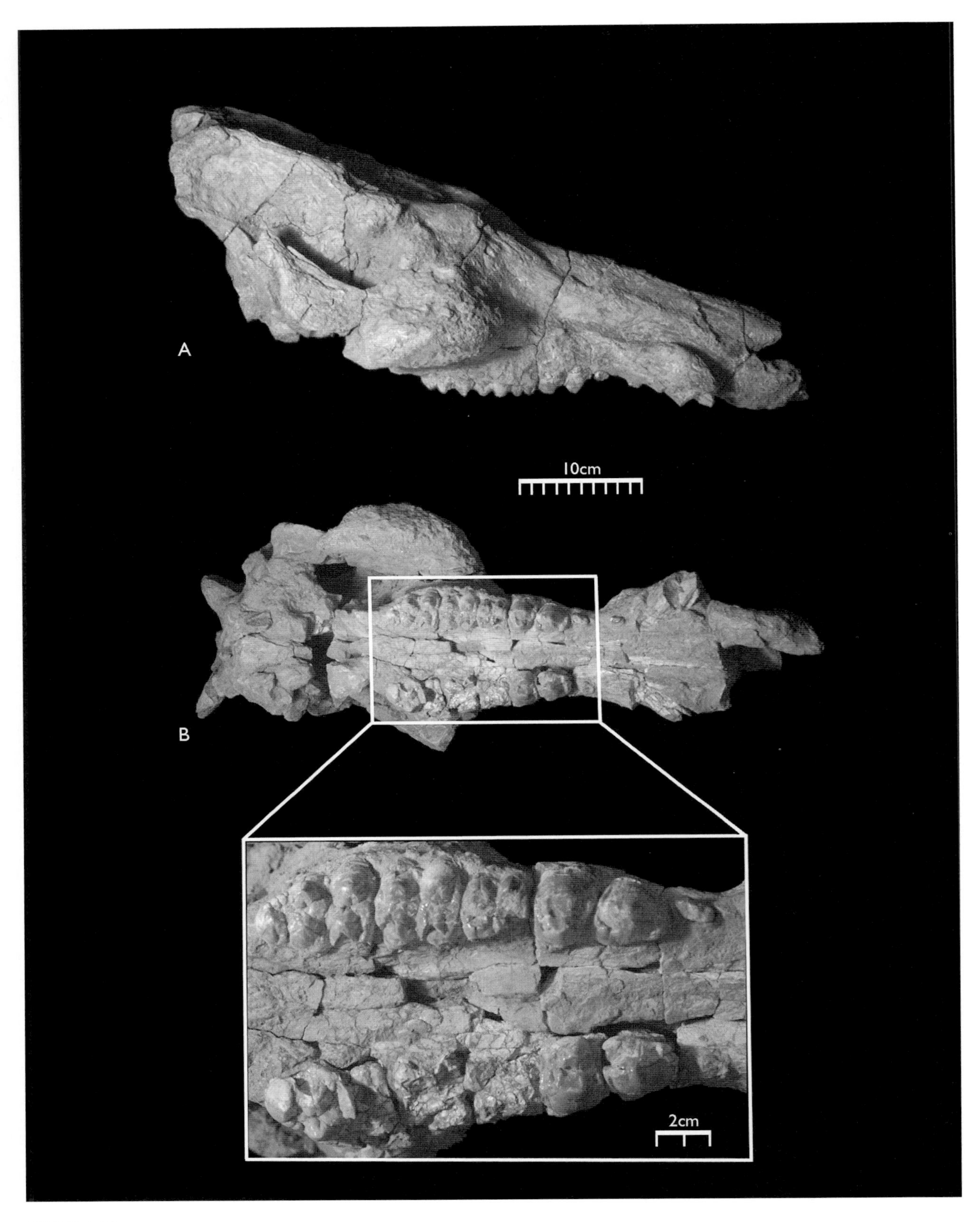

FIGURE 10.6

STD-VP-1/1, a complete cranium of *Nyanzachoerus syrticus* from the Middle Awash. A. Lateral view. B. Occlusal view.

go as high as the dorsal wall of the muzzle. The lateral wall of the right muzzle shows a strong depression, and the dorsal wall of the muzzle is rounded. The dorsolateral posterior boundary of the muzzle forms a strong ridge-like structure that extends all the way to the superior border of the occiput, forming the lateral boundaries of the cranial floor. The occiput steeply declines toward the foramen magnum. The right zygomatic process originates at the level of M^1 and does not flare outward as much as KNM-LT 316 (Cooke and Ewer's 1972, presumably male, type specimen of *Ny. tulotos*). The canine is relatively small, has a semicircular cross section, and curves downward.

Most of the dental measurements and proportions are similar to KNM-LT 316 from Lothagam. However, the vertex length of STD-VP-1/1 is much higher than that of KNM-LT 316. The vertex length of the latter might have been underestimated, since it lacks the posterior portion. The canine morphology of the two specimens also seems to be different. STD-VP-1/1 has a canine with a semicircular cross section, whereas KNM-LT 316 is described as having a canine that is dorsoventrally flattened. It is currently unclear whether these differences are related to sexual dimorphism, distortion, or variation at the specific level or above. Substantial variation within *N. syrticus* is obtained when the Lothagam specimens are combined with *N. syrticus* specimens from Sahabi (Cooke 1987). STD-VP-1/1, likely a male individual and larger than KNM-LT 316 (also presumably an adult male), is further evidence of significant size variation within the species *Ny. syrticus.*

The isolated Middle Awash teeth are assigned here to *Ny. syrticus* largely based on occlusal morphology (Figures 10.7 and 10.8) and overall size. All dental measurements of *Ny. syrticus* are given in Table 10.2. The M^3s range from 40.2 mm to 43.5 mm, which is similar to the range seen for the M^3s of *Ny. syrticus* from the Upper Nawata Member of the Lothagam Formation (40.8–45.9 mm; Harris and Leakey 2003b). The M^3s from the Lower Nawata tend to have a mesiodistally shorter dimension, with a range between 36.6 mm and 42.1 mm, and only two out of the nine specimens from the Lower Nawata overlap with the specimens from the Upper Nawata Member. The mesiodistal dimension of the M_3s ranges from 43.2 mm to 45.7 mm. This is also within the range seen for the Lothagam specimens (39.1–44.1 mm for 14 specimens from the Lower Nawata, and >43–49.7 mm for 4 specimens from the Upper Nawata; Harris and Leakey, 2003b).

DISCUSSION Cooke and Ewer (1972) conducted a systematic description of the Lothagam suids and recognized three species from the Lothagam Formation. They recognized a new species, *Nyanzachoerus "tulotos,"* to refer to one of the earliest tetraconodontines from the formation. A similar tetraconodont from North Africa had been previously assigned to the species *Sivachoerus syrticus.* Following his initial suggestion in 1982, and taking into account the morphological similarities between the specimens from Lothagam assigned to *Ny. tulotos* and specimens from Sahabi assigned to *Sivachoerus syrticus* (Leonardi 1952), Cooke (1987) merged the Lothagam *Ny. tulotos* material with the *Sivachoerus syrticus* material from Sahabi into a single species *Nyanzachoerus syrticus.* This is currently widely accepted in suid systematics.

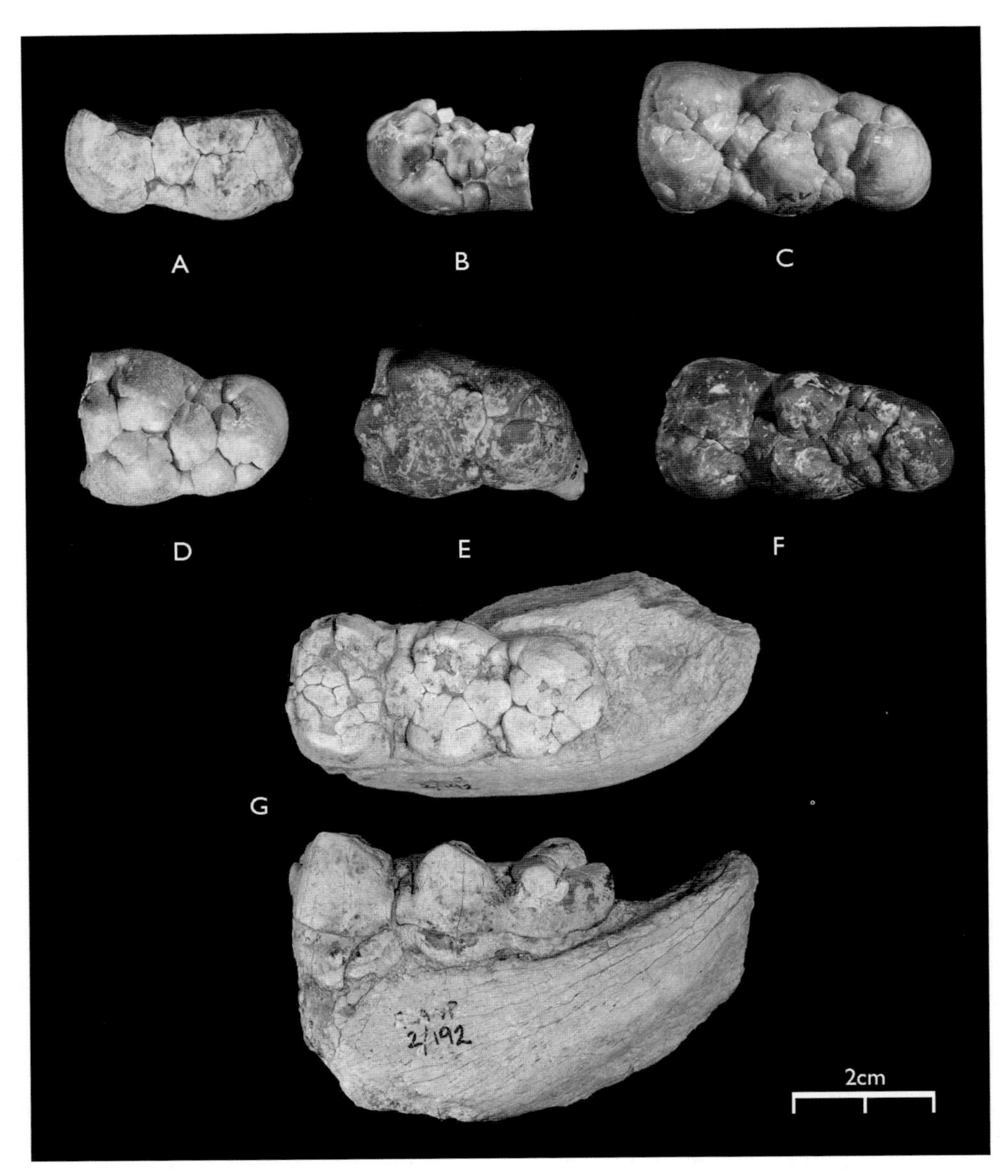

FIGURE 10.7

Occlusal views of M_3s of *Nyanzachoerus syrticus*. A. ALA-VP-2/28, left M_3. B. DID-VP-1/143, left M_3. C. KL164-1, left M_3. D. ASK-VP-3/47, right M_3. E. STD-VP-1/46, left M_3. F. KL174-1, left M_3. G. ALA-VP-2/192, right mandible with M_3, occlusal and lingual views.

The Middle Awash *Ny. syrticus* dental remains overlap in size and morphology with those from the Upper Nawata of Lothagam. The absence of *Ny. syrticus* from Middle Awash deposits younger than 5.2 Ma is worth noting, as this is also about the time that *Ny. australis* first appears in the fossil record of the Middle Awash. It is also worth noting here that some specimens collected from Lothagam in the 1960s and 1970s and reported by Harris and Leakey (2003b) as from the Upper Nawata might have problems of provenience similiar to that plaguing the Lothagam hominid mandible (KNM-LT 329). Therefore, some of the specimens assigned to *Ny. syrticus* from the Upper Nawata Member should be considered with caution. For example, the mesio-

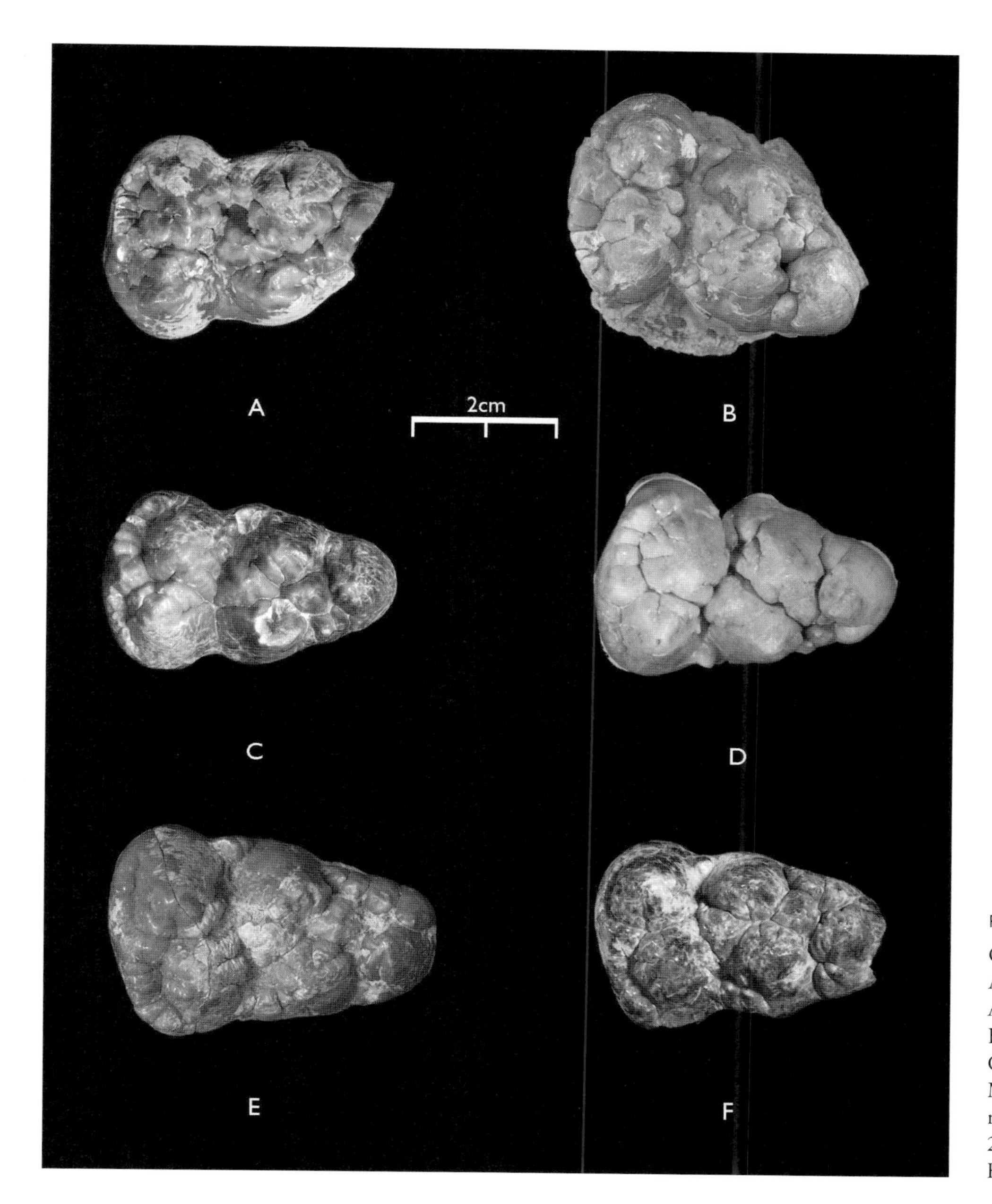

FIGURE 10.8

Occlusal views of M^3s of *Nyanzachoerus syrticus*. A. ALA-VP-1/25, right M^3. B. ASK-VP-1/11, left M^3. C. ASK-VP-3/48, right M^3. D. ASK-VP-3/538, right M^3. E. STD-VP-2/849, left M^3. F. KL173-1, left M^3.

distal dimension of the KNM-LT 388 M_3 is two standard deviations outside the range seen for those of Middle Awash *Ny. syrticus*. It falls within the range seen for *Ny. australis*. Harris and Leakey (2003b) referred two specimens (KNM-LT 309, a mandible with M_3; KNM-LT 26110, M^3) from the Apak Member to *Ny. syrticus*. Although all measurements of the M^3s from the Apak Member are estimates, it is obvious that they fall at or above the upper range of *Ny. syrticus* from the Upper Nawata. Measurements of KNM-LT 309 also show that it is mesiodistally the longest M_3 in the entire *Ny.*

TABLE 10.2 Dental Measurements of *Nyanzachoerus syrticus*

	P3		*P4*		*M1*		*M2*		*M3*	
Specimen	MD	BL	MD	BL	MD	BL	MD	BL	MD	BL
Uppers										
STD-VP-1/1 (L)	24.1	22.5	21	27						
STD-VP-1/1 (R)	26.3	22.5	19.8	25.7	21.8	20.9	26.8	28.2	43.5	30
ALA-VP-2/63	21.2	17.8	17.9	22	19.9	20.6	28.6	27.2		
ASK-VP-1/11									43.5	30.8
ASK-VP-3/10							26.5	26.1		
ASK-VP-3/48									41.6	26.3
ASK-VP-3/538									42.1	29.1
ASK-VP-3/541			18	22						
STD-VP-2/8							27.3	26.7		
STD-VP-2/849									47.2	30.4
KL173-1									40.2	27.5
Lowers										
ALA-VP-2/1							25.3	18.9		
ASK-VP-1/8			22.6	18.3						
ASK-VP-3/8							27	21.9		
STD-VP-1/28	24.6	16.6	22.8	20	20.5	16.1			45.7	27.3
STD-VP-1/31			23.1	20.6						
KL164-1									43.9	22.9
KL174-1									43.2	22

NOTE: All measurements are in mm.

syrticus hypodigm, suggesting that it might belong to a species other than *Ny. syrticus,* possibly *Ny. australis.* The First Appearances (FAs) and Last Appearances (LAs) of both species, *Ny. syrticus* and *Ny. australis,* are therefore uncertain at Lothagam. The Middle Awash material, on the other hand, clearly indicates that, at least in the Middle Awash, the LA of *Ny. devauxi* and *Ny. syrticus* is no later than 5.5 Ma. The FA of *Ny. australis* is in the Middle Awash succession, and this species not only is larger than its apparent predecessor, *Ny. syrticus,* but also conforms to the major trend in third-molar elongation in the genus in the form of a third cusp pair mesial to the terminal cusp. Therefore, it is plausible to say that *Ny. syrticus* was anagenetically replaced by *Ny. australis,* at least in the Middle Awash, possibly through an intermediate form such as *Ny. waylandi* (discussed next). The latter species has previously been proposed as a possible precursor of *Ny. kanamensis* (Pickford 1989a; White 1995).

Nyanzachoerus waylandi Cooke and Coryndon, 1970

Nyanzachoerus cf. *waylandi*

DESCRIPTION *Nyanzachoerus waylandi* is mainly distinguished from other nyanzachoeres by its small size comparable to *Ny. devauxi,* but with small third and fourth premolars

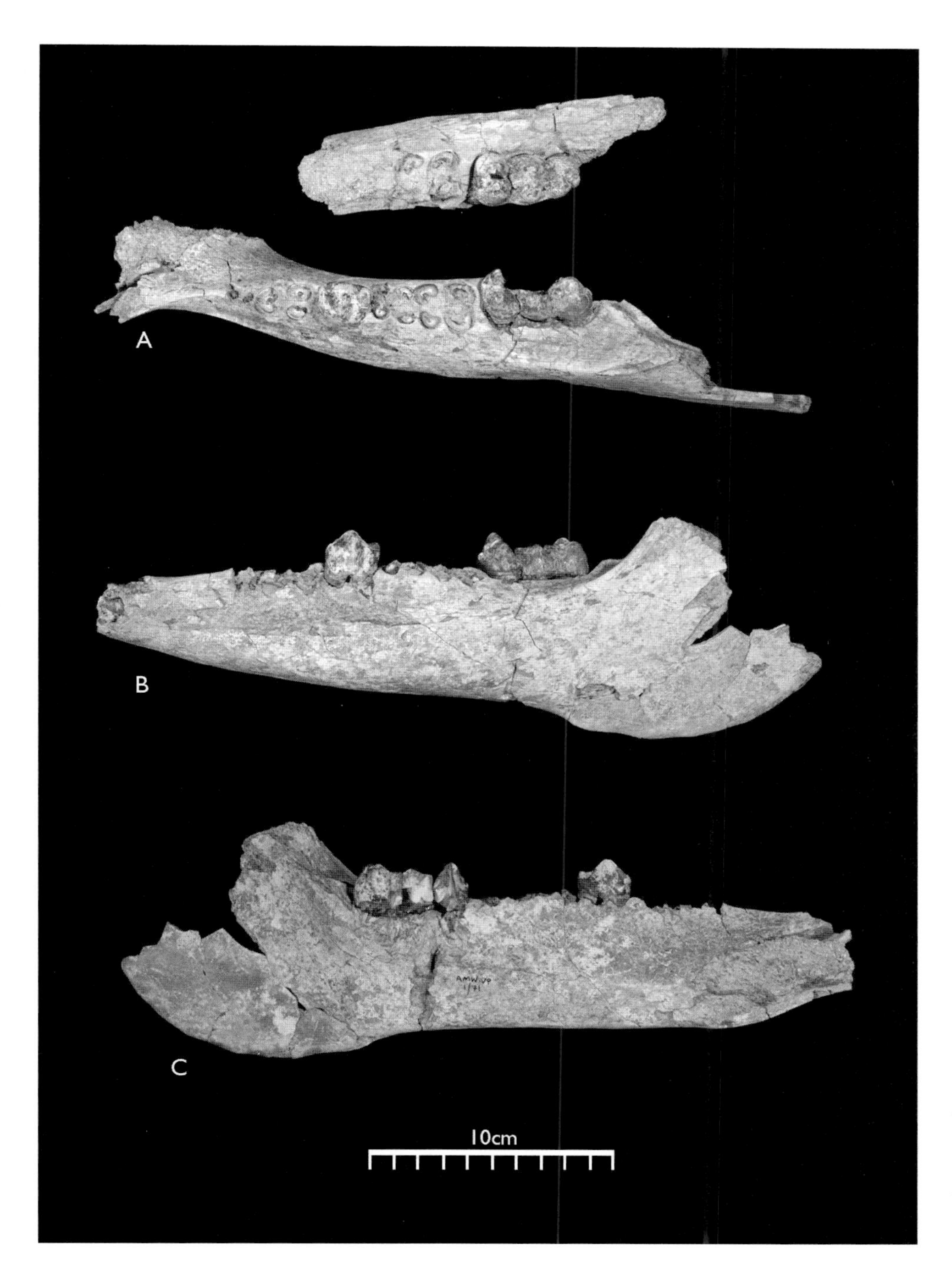

FIGURE 10.9
AMW-VP-1/71, mandibular specimen of *Nyanzachoerus* cf. *waylandi*. A. Occlusal view. B. Lateral view. C. Medial view. P_3 not figured.

and high crowned teeth, unlike *Ny. devauxi* (Pickford 1989a). *Nyanzachoerus waylandi* also has a shallow mandible compared to other nyanzachoeres. Following the naming of the species, a number of specimens have been added to the *Ny. waylandi* hypodigm. All of the specimens initially assigned to this species were from the Nyaburogo site, Toro District, Uganda (Pickford 1989a). Pickford (1993) later included some suid specimens

TABLE 10.3 Comparative Dental Measurements of *Nyanzachoerus waylandi* Specimens

	Middle Awash	*Nyaburogo*
	AMW-VP-1/71	NY 457'87
RP_3 MD	24.1	21.0
BL	15.7	12.0
RP_4 MD	22.3	21.0
BL	16.7	14.0
LP_4 MD	22.1	nm
BL	16.9	nm
RM_3 MD	45.4	44.0
BL	nm	nm

NOTE: Measurements are from specimens AMW-VP-1/71 (Kuseralee Member, Middle Awash) and NY 457'87 (Nyaburogo, Uganda). Dental measurements for NY 457'87 are taken from Pickford (1989). All measurements are in mm; nm = measurement was not possible.

from the Nkondo Formation to this species. Its presence in the Middle Awash, Ethiopia, expands its geographic range further north and east.

The mandibular and dental description of NY 457'87, a mandible from Nyaburogo assigned to the species *Ny. waylandi,* best matches AMW-VP-1/71 (Figure 10.9). This observation is based on the published measurements and images of NY 457'87 in Pickford (1989a). The shallowness of the mandibular body and morphology of the P_3—being high crowned, long, and narrow—are very similar. The symphysis of AMW-VP-1/71 is flat at its base and concave on the lingual side. The canine appears to be horizontal. In terms of individual tooth measurements, the third and fourth premolars of AMW-VP-1/71 are slightly larger than those of NY 457'87 but within the known range of the species (Table 10.3). The P_3–P_4 and the M_1–M_3 lengths (based on alveoli) on the left mandible of AMW-VP-1/71 are 46.4 mm and 91.4 mm, respectively, while those of NY 457'87 are 42 mm and 93 mm, respectively. This indicates that AMW-VP-1/71 has slightly a higher premolar/molar ratio even though the dental row length is almost the same in both specimens. Morphology of the M_3 is also similar in both NY 457'87 and AMW-VP-1/71, with three pairs of cusps arranged transversely and a large terminal cusp with a number of cusplets mesiobuccal to it.

The mandibular depths reported by Pickford (1989a) for NY 457'87 are 18 mm, below the cervix of P_3, and 25 mm, below the cervix of M_2. However, these dimensions, particularly that of the depth below the P_3, are probably underestimated, since the height of the P_3 (18 mm) would be the same as the corpus depth under it. This is very unlikely, since the roots are usually longer than the crown height.

A second specimen, AMW-VP-1/7 (a right mandibular fragment with M_3), is also assigned to this species. Morphology of the M_3 is consistent with AMW-VP-1/71. It lacks a third distal cusp pair but the terminal cusp is large and buccolingually expanded. The mesiodistal and buccolingual demensions of the M_3 are 48.2 mm and 23.6 mm, respectively.

DISCUSSION *Nyanzachoerus waylandi* was initially recognized as *'Sus' waylandi* by Cooke and Coryndon (1970), although they suggested that better material might show that it belonged to *Nyanzachoerus*. With additional material from the Albertine Rift, Pickford (1989:644) diagnosed *Ny. waylandi* as "a small species of *Nyanzachoerus* in which the length of the [lower] third and fourth premolars combined is about 42mm while the length of the molar row is about 93mm." He further stated that the talonids of the lower third molars comprise "one pair of large cusps and several smaller cingular cusps." Pickford stated that there are some dental morphological similarities between *Ny. waylandi* and *Ny. kanamensis* from Kanam West. Van der Made (1999) further argued that *Ny. waylandi* is similar in size and morphology to *Ny. kanamensis* and recognized *Ny. waylandi* as a subspecies of *Ny. kanamensis*.

Prior to its discovery in the Middle Awash, *Nyanzachoerus waylandi* has only been reported from localities in the Nyaburogo Formation outcropping on the eastern flanks of the Nyaburogo Valley, Toro District, Uganda (Pickford, 1989). An age of 5 to 6 Ma has been estimated for the faunal assemblages of the Nyaburogo Formation based on the proboscideans. Although this taxon was recognized in 1970, its remains are always rare in Late Miocene assemblages, so its biogeographic and biochronological ranges are extremely difficult to establish with any accuracy at the present time.

Only two specimens are assigned to *Ny. waylandi* from the entire Middle Awash sequence. The small size of the species alone may not be taxonomically as significant as morphology-based recognition. For example, the size (and to some degree the morphology) of the M_3 of AMW-VP-1/71, and AMW 1/7, is similar to the earlier *Ny. syrticus* M_3s. However, the premolar/molar ratio is different, with the latter having a higher ratio. Mandibular robusticity also distinguishes *Ny. waylandi* from *Ny. syrticus*. Corpus and premolar proportions distinguish AMW-VP-1/71 from *Ny. australis* and *Ny. kanamensis*. The premolars of *Ny. australis* and *Ny. kanamensis* are lower-crowned compared to their breadth. The posterior half of the P_3 in the latter two species is much wider than the mesial half, unlike *Ny. waylandi,* where the mesial and distal breadth measurements are almost the same. The mandibular symphysis of *Ny. waylandi* is also flat compared to the other nyanzachoere species. Therefore, despite the small sample size of *Ny. waylandi* from eastern Africa, there seem to be enough morphological attributes to recognize it as a species distinct from other nyanzachoere species recognized from the Mio-Pliocene of this region.

Nyanzachoerus waylandi appears to be a short-lived, small tetraconodont species. This species is recognized from limited areas in eastern Africa in the late Miocene. It is possible that *Ny. waylandi* was a species that evolved phyletically from *Ny. devauxi* or *Ny. syrticus*. It abruptly went extinct, possibly because of resource competition from a larger tetraconodont *(Ny. australis)* that appeared at about the same time. This could also be one of the reasons for its rarity in the fossil record of eastern Africa. Its distinctiveness from *Ny. devauxi* and *Ny. syrticus,* however, is morphologically justifiable, particularly in the morphology and size of the premolars and the premolar/molar ratio.

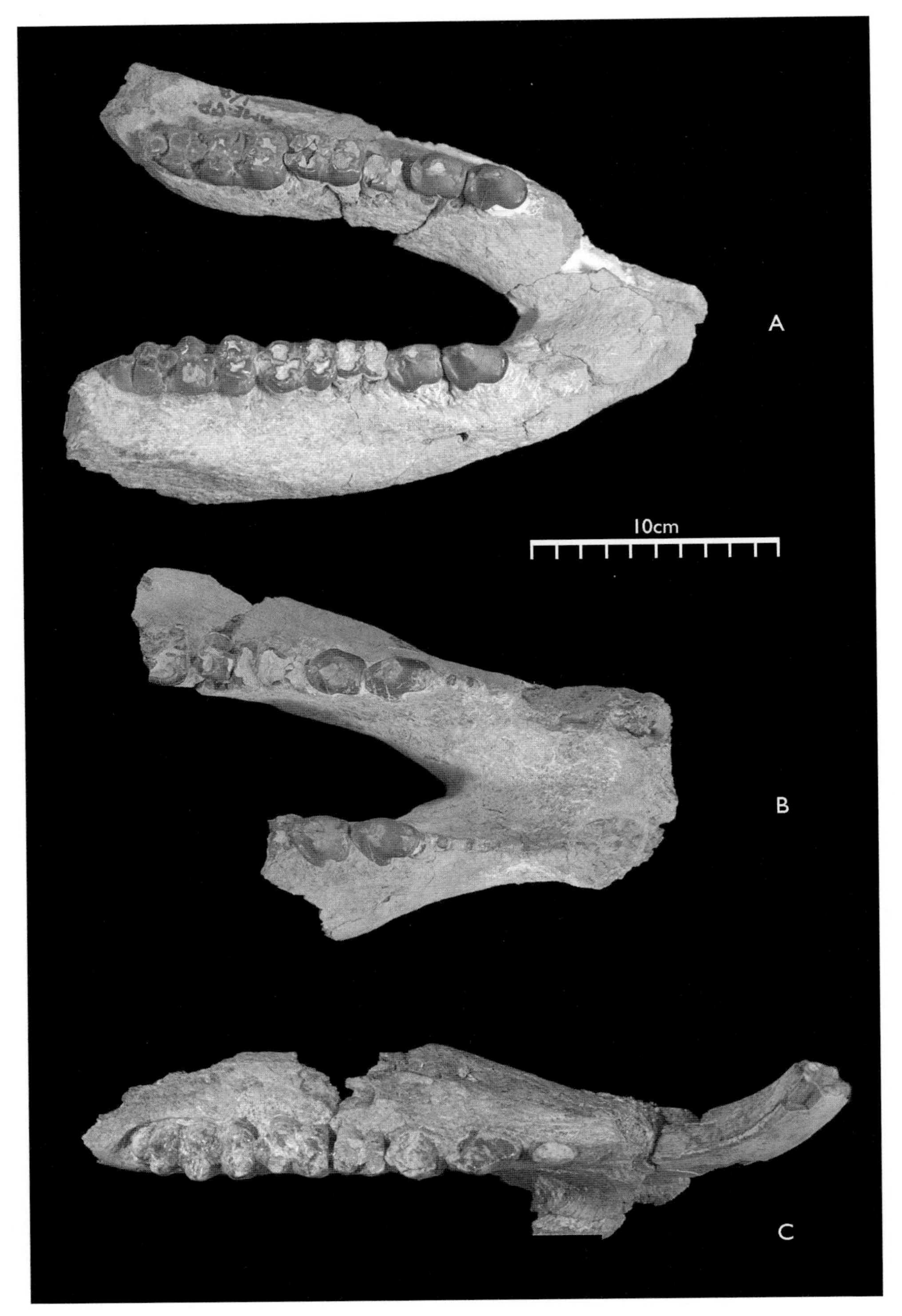

FIGURE 10.10

Occlusal views of mandibular specimens of *Nyanzachoerus australis*.
A. AME-VP-1/8.
B. AME-VP-1/9.
C. AME-VP-1/23.

Nyanzachoerus australis Cooke and Hendey, 1992

(=*Nyanzachoerus kanamensis australis* Cooke and Hendey, 1992; *Sivachoerus australis* Van der Made, 1999)

DESCRIPTION The following description is based largely on the most complete mandibular specimens from the Middle Awash. Most of the isolated upper and lower teeth are assigned to this species based on their morphological and metric similarities with teeth in mandibular specimens described below. Morphology of the upper dentition is also briefly described using specimens KUS-VP-1/30 and KUS-VP-1/32 (not figured).

AME-VP-1/8 is an almost complete mandible with dentition (Figure 10.10A). It lacks the rami and anterior part of the symphysis. All molars and premolars are preserved except for the second premolars and the mesial half of the left M_1. The posterior margin of the symphysis terminates at the level of P_2. The mandible is relatively short and robust. The depth of the mandible at the P_3 level, and also at the M_3 level, is 70 mm. The premolars are not proportionately as large as those of *Ny. syrticus*. The small size of the premolars probably indicates that the mandible belongs to a female individual. The third molars have three pairs of cusps, and the third pair is connected to the terminal cusp without a median cusp. Dental measurements are given in Table 10.4.

AME-VP-1/9 is a mandible very similar to AME-VP-1/8 (Figure 10.10B). The symphyseal region of AME-VP-1/9 is well-preserved. The canine roots are present, but no incisors or their roots are preserved. The canine roots have a verrucose shape. On the left side, the corpus behind the mesial portion of M_3 is missing. On the right side, it is broken at the M_1 level. The premolars are similar in size and morphology to AME-VP-1/8. The P_2 had two roots. The posterior margin of the symphysis terminates at the level of P_2. The mandible shows a slight constriction anterior to P_2. The base of the symphysis is convex, whereas the lingual side is highly concave. The preserved molars on the left side are highly damaged by fracture, wear, and enamel loss. The robusticity of the mandible is comparable to AME-VP-1/8 (see Table 10.4 for dental measurements).

AME-VP-1/23 is a left mandible with the canine and all postcanine teeth (Figure 10.10C). Part of the symphysis is preserved. The corpus base below the molars is damaged, and the ramus is entirely missing. The posterior end of the symphysis extends to the level of the second premolar. The mandibular body is robust, 70.7 mm deep at the P_3 level. The diastema between the canine and the second premolar is 36.8 mm. There is also a small diastema (6.2 mm) between the second and third premolars. The preserved canine is 84 mm in length, and it is almost vertically implanted, is verrucose in cross section, and has enamel only on the inner and outer surfaces. The P_2 is mesiodistally elongate and low-crowned. The distal tubercle of the P_4 is more than half the height of the main cusp. The M_3 has three pairs of cusps and a terminal cusp bifurcated into two subequal halves by a shallow groove. However, there is no median cusp between the third cusp pair and the terminal cusp. This is one of the distinguishing characters seen in all M_3s of *Ny. australis* (Figure 10.11). *Nyanzachoerus syrticus* M_3s lack the third cone pair (see Figure 10.11 for comparisons).

TABLE 10.4 Dental Measurements of Specimens Referred to *Nyanzachoerus australis*

	LP2		*RP2*		*LP3*		*RP3*		*LP4*	
Specimen	MD	BL	MD	BL	MD	BL	MD	BL	MD	BL
Uppers										
AME-VP-1/31					22.8	21.9	23.6	21.1	20.5	26.0
AME-VP-1/35										
AME-VP-1/52										
AME-VP-1/74										
AME-VP-1/76										
AME-VP-1/167										
AME-VP-1/169										
AME-VP-1/196										
AMW-VP-1/13							23.0	18.6		
AMW-VP-1/18					23.6	19.0			17.4	20.9
AMW-VP-1/77									19.0	20.9
AMW-VP-1/89										
KUS-VP-1/10							21.9	20.9		
KUS-VP-1/13					20.7	19.3			16.1	19.4
KUS-VP-1/30	nm	5.7	10.8	6.1	21.5	17.6	18.0	nm	18.0	20.9
KUS-VP-1/32	nm	7.1	12.1	7.5			21.3	nm		
KUS-VP-1/41										
KUS-VP-1/44										
KUS-VP-1/69										
KUS-VP-1/74										
Lowers										
AME-VP-1/7					23.9	16.6			21.2	17.7
AME-VP-1/8					23.7	19.5	25.3	19.3	21.0	18.8
AME-VP-1/13										
AME-VP-1/14										
AME-VP-1/23	16.6	9.4			28.0	20.0			22.7	21.0
AME-VP-1/24										
AME-VP-1/31									24.3	21.2
AME-VP-1/69										
AME-VP-1/132							(27.7)	20.6		
AME-VP-1/139										
AME-VP-1/140					24.5	17.2	23.5	17.4	21.9	18.0
AMW-VP-1/8										
AMW-VP-1/17										
AMW-VP-1/21	16.6	9.1	17.5	9.9	28.1	nm			nm	22.2
AMW-VP-1/37										
AMW-VP-1/39										
AMW-VP-1/68										
AMW-VP-1/78										
KUS-VP-1/9										
KUS-VP-1/13					24.1	17.9	23.9	17.5	21.2	18.4
KUS-VP-1/34										
KUS-VP-1/39										
KUS-VP-1/71										
KUS-VP-1/101			15.3	8.8			26.0	18.5		

NOTE: Measurements in parentheses are estimates. All measurements are given in mm; nm = measurement was not possible.

RP4		*LM1*		*RM1*		*LM2*		*RM2*		*LM3*		*RM3*	
MD	BL	MD	BL	MD	BL	MD	BL	MD	BL	MD	BL	MD	BL
19.4	25.7											53.2	34.4
										42.5	27.8		
												47.6	31.2
										44.3	27.7	45.5	29.0
										54.7	nm		
												49.0	30.6
20.3	23.7			22.7	20.6							49.7	29.3
										54.0	25.9		
												48.0	29.3
		(18.4)	nm			(30.3)	nm			48.5	30.8		
										48.0	(29.2)		
		(22.3)	nm			32.2	26.8			52.2	33.5		
17.6	22.5									50.1	30.7		
										48.5	29.3	48.2	29.3
17.9	20.8			18.8	21.0			27.8	25.3	44.7	31.1	45.1	29.2
		20.0	19.4	21.1	19.8	28.8	27.8	28.7	25.7	50.1	31.1	49.5	30.9
												51.7	33.6
												50.2	30.0
												49.3	28.4
												47.5	28.9
										52.4	22.5		
21.2	18.6			18.6	15.1	31.4	21.0	31.5	21.0	58.4	25.4	56.7	25.3
												53.5	26.1
							(29.0)	nm		56.1	26.6		
		20.4				29.4	20.4			54.7	25.6		
		(22.3)	nm			(32.2)	22.1			53.0	24.8		
24.2	21.8					(31.9)	24.2						
										57.8	30.6		
23.1	20.4												
												53.3	24.5
21.4	17.5	22.9	nm	24.2	15.0								
		(22.0)	(16.2)			28.0	20.3			53.2	25.1		
												50.6	24.9
25.5	21.6					nm	23.6	32.8	23.1	54.5	27.9	54.5	27.5
										(54.0)	(26.8)	(56.0)	(26.4)
												54.6	26.6
										63.4	27.4		
24.3	21.0			21.8	18.1							54.4	23.4
										59.8	27.3		
21.2	17.9									(54.3)	(22.9)	(54.0)	(23.6)
										53.3	23.4		
										54.5	24.6		
										49.2	21.9		
22.3	20.1			19.3	16.9			31.1	22.8			55.1	25.6

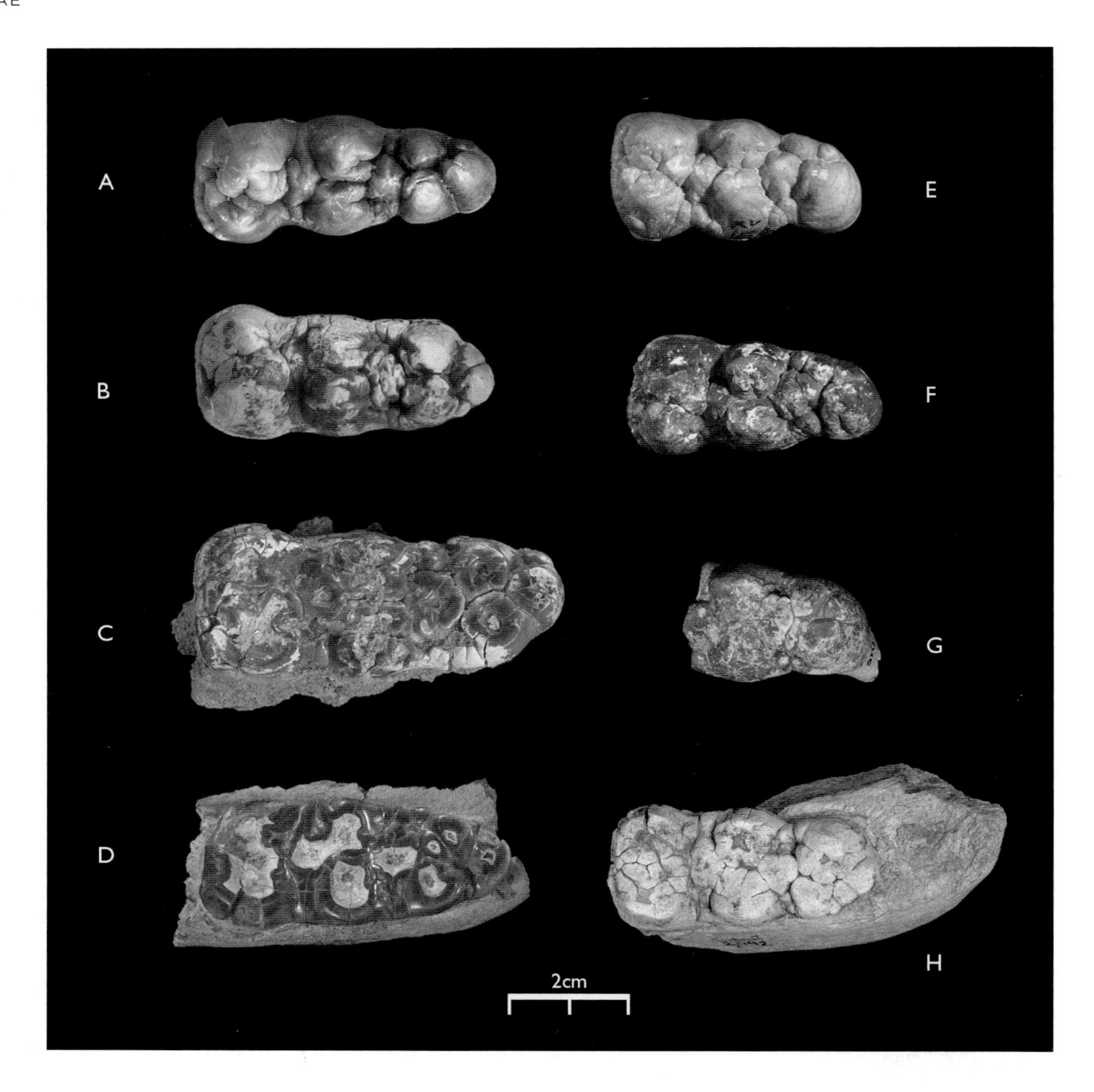

FIGURE 10.11
Comparison of M_3 specimens at various stages of wear. *Nyanzachoerus australis* (A–D): KUS-VP-1/34, AMW-VP-1/17, AMW-VP-1/68, KUS-VP-1/39. *Ny. syrticus* (E–H): KL164-1, KL174-1, STD-VP-1/46, ALA-VP-2/192.

Upper Premolars The P^2 is two-rooted, relatively low-crowned, and mesiodistally long relative to its width. The main cusp is positioned at the center of the crown at about equal height with the distal accessory cusp. The mesial accessory cusp is smaller and lingually connected with its distal counterpart by a strong cingulum. The P^3 is relatively high-crowned, and the mesial accessory cusp extends as high as mid-crown height. The distal accessory cusp is shorter but buccolingually connected with the well-developed hypocone to form a fovea on the distolingual corner of the crown. A lingual cingulum extends from the hypocone to the mesial accessory cusp. The distal edge of the main cusp is steeper than the mesial edge. On the buccal face of the mesial and distal accessory cusps, there are longitudinal grooves extending from the tips of the accessory cusps to almost the crown

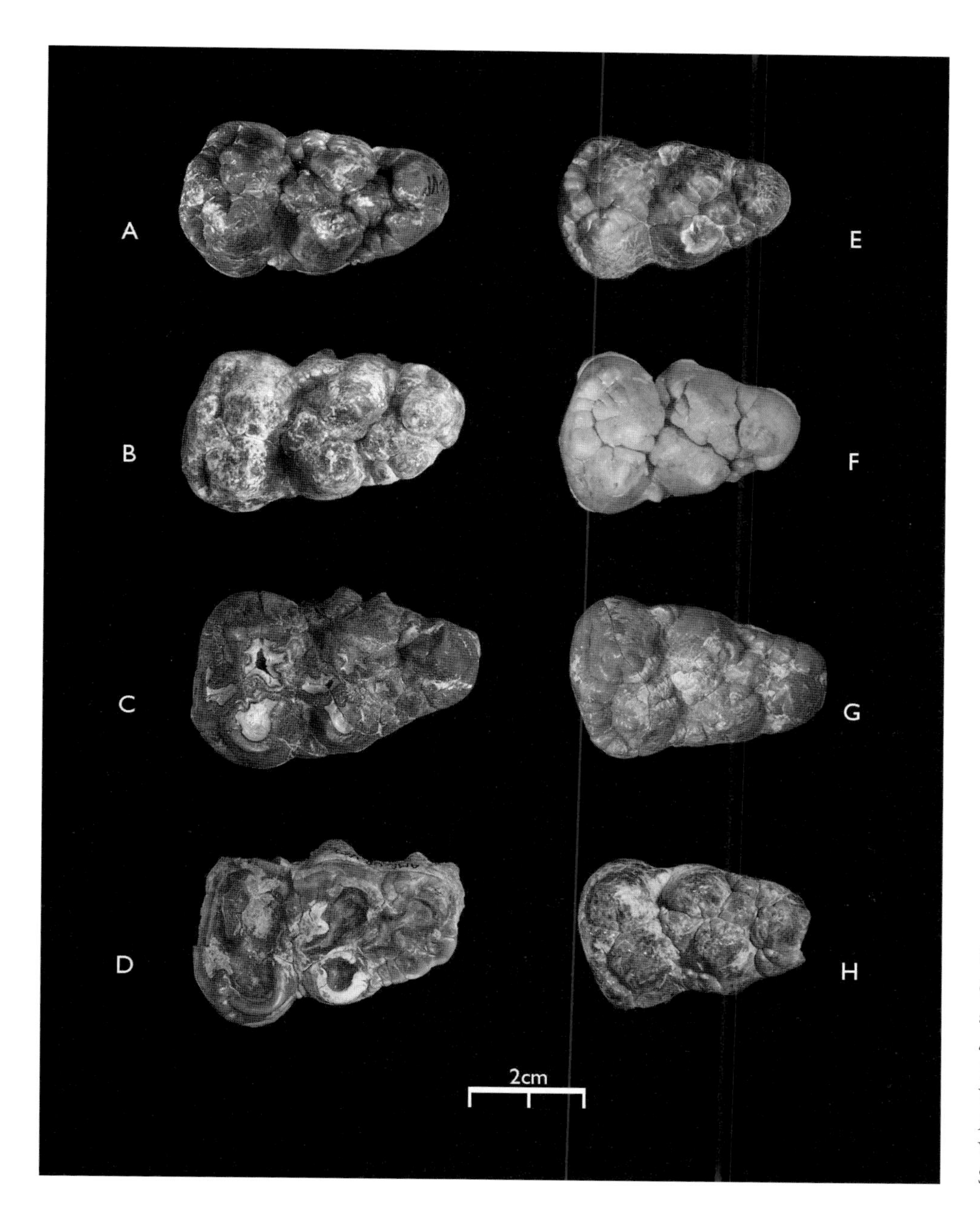

FIGURE 10.12

Comparison of M^3 specimens. *Nyanzachoerus australis* (A–D): KUS-VP-1/13, KUS-VP-1/44, KUS-VP-1/41, AME-VP-1/52. *Ny. syrticus* (E–H): ASK-VP-3/48, ASK-VP-3/538, STD-VP-2/849, KL173-1.

base. The P^4 is buccolingually wider, with its buccal half being wider than its lingual half. The buccal cusp is much larger than the lingual cusp and shows a tendency of bifurcating into two cusps, giving the buccal face a bilobed appearance. It is also higher-crowned than its lingual counterparts and usually shows more wear. The mesial accessory cusp is weakly developed and connects with a weak cingulum on the mesial base of the lingual cusp. The distal accessory cusp is relatively well-developed and positioned posterior to the buccal cusp. The distal cingulum extends from the distal accessory cusp to the distal face of the lingual cusp. A weak cingulum is also present on the distobuccal corner.

Upper Molars The M^1 and M^2 are similar in their occlusal morphology to all nyanzachoere species. However, they have well-developed endostyles and ectostyles usually formed by multiple mammelations. The mesial and distal cinguli are also well-developed, with the mesial cingulum being buccolingually longer. However, they do not extend as far as the ectostyle as is the case in *Ny. syrticus* (Harris and White 1979). The enamel wear is similar to *Ny. syrticus* in being stellate in early wear to more rounded in advanced wear (Harris and White 1979). The M^3s are the same as all other nyanzachoere species in the number of major cusps. However, *Ny. australis* M^3s are distinctly longer than those of *Ny. syrticus* (Figure 10.12). The former species elongates its third molar via increased size of the median cusp at the center of the mesial cingulum and the other two median cusps, and the addition of a usually large fourth cusplet on the buccal side of the talon. *Nyanzachoerus australis* M^3s also tend to have almost equal breadth on the first and second cusp pairs, whereas in *Ny. syrticus* the breadth of the anterior trigon cusp pair is much wider than the breadth of the second cusp pair. *Nyanzachoerus kanamensis* M^3s retain the trend in size increase, and the talon becomes more complex with the addition of numerous cusplets on both the lingual and buccal sides. However, the breadth of the first and second cusp pairs is similar to that of *Ny. australis.*

Nyanzachoerus australis from the Middle Awash and Lothagam are more primitive than the South African form in premolar/molar ratio (if a higher ratio is considered primitive). The premolar to molar ratio of the Langebaanweg suids ranges from 40.5 to 44.9, with a mean of 42.9 (n = 11; Cooke 1987). The Middle Awash form ranges from 41.3 to 52.8 with a mean of 48.1 (n = 9). This evidence is best interpreted as showing a wide range of variation in the premolar/molar ratio of *Ny. australis,* rather than a difference significant at the species level.

DISCUSSION *Nyanzachoerus australis* is a species name coined to refer to specimens from the QSM of Langebaanweg (Cooke and Hendey 1992). The name has been subsequently used by various workers to refer to suid remains older than *Ny. kanamensis.* Historically, however, the South African suid material now referred to *Ny. australis* has been referred to *Ny. pattersoni (= Ny. kanamensis),* and *Ny. kanamensis.* Van der Made (1999) assigned the Langebaanweg suid material to a new species, *"Sivachoerus australis."* He further divided the new species into two subspecies, one of them, *"S. australis australis,"* specifically referring to the material from the QSM, and the other, *"S. australis megadens,"* for the material from the PPM. Van der Made (1999:29) stated that *"S. australis"* is different from other *Sivachoerus* and *Nyanzachoerus* species because it has a "large and complicated M3." This diagnosis is not a strong character with which to differentiate the QSM suid material from *Nyanzachoerus kanamensis sensu stricto,* particularly because *Ny. jaegeri* also has large and complicated third molars. However, *Ny. australis* has unique M_3 morphology that distinguishes it from the earlier *Ny. syrticus* and the younger *Ny. kanamensis.* While *Ny. syrticus* has two cusp pairs and a terminal cusp separated from each other by median cusps, *Ny. kanamensis* has three cusp pairs and a terminal cusp all separated from each other by median cusps. However, *Ny. australis* has no median cusp between the third cusp pair and the terminal cusp. This is one of the strongest characters that can be used to differentiate *Ny. australis* from other nyanzachoere species. *Nyanzachoerus australis* is currently known from Langebaanweg

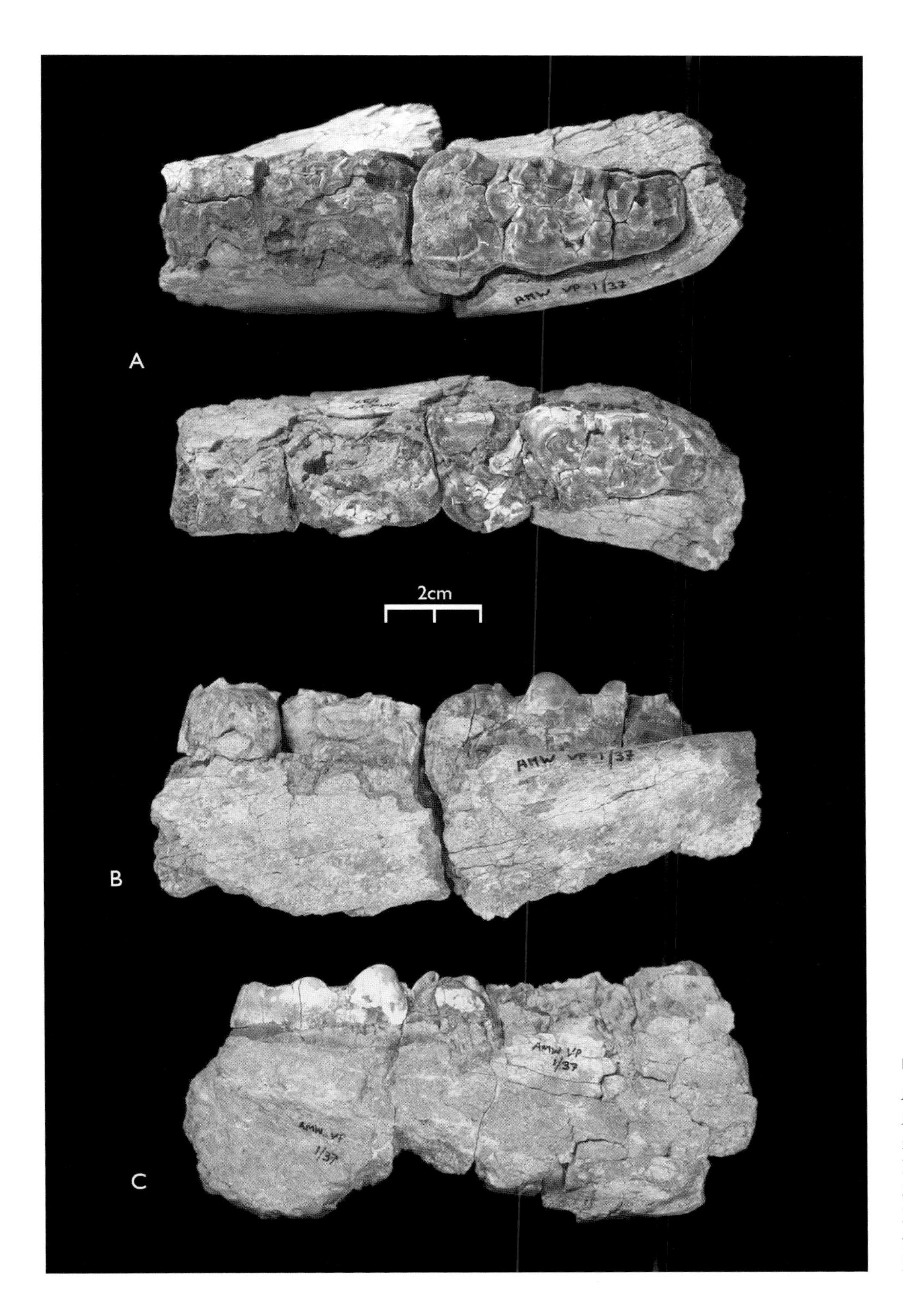

FIGURE 10.13

AMW-VP-1/37, *Nyanzachoerus kanamensis* mandible from the Kuseralee Member of the Sagantole Formation, Middle Awash. A. Occlusal views. B. Lateral view. C. Medial view.

(Cooke and Hendey 1992), Lothagam (Harris and Leakey 2003b), and the Middle Awash (Haile-Selassie 2001a; Haile-Selassie et al. 2004c).

Nyanzachoerus kanamensis Leakey, 1958

(=*Nyanzachoerus pattersoni* Cooke and Ewer, 1972)

Nyanzachoerus cf. *kanamensis*

DESCRIPTION *Nyanzachoerus kanamensis* becomes abundant in the Middle Awash in localities that are slightly younger than 5.2 Ma. However, its FA might predate 5.2 Ma. AMW-VP-1/37 comprises two hemimandibles lacking the area anterior to the first molars on both sides (Figure 10.13). The first and second molars are highly damaged, and no morphological observation can be made. However, the third molars are preserved on both sides. The amount of wear on the molars indicates that the mandible belonged to an older individual. The mesial cusp pairs on the third molars are buccolingually wider than the second and third cusp pairs. These molars are long mesiodistally. The talonid is distally expanded with a third cusp pair and numerous large cusplets on the distolingual corner. Its mesiodistal length is almost half the length on the entire crown. The corpus below the molars is robust, although the corpus base is broken on both sides. The mesiodistal and buccolingual dimensions of the left third molar are 54.5 mm and 27.1 mm, respectively. The right homolog measures 53.3 mm and 27.7 mm, respectively. These dimensions fall in the known range of *Ny. kanamensis* from younger deposits as well as the range of the contemporary *Ny. australis*. However, the occlusal morphology, particularly the talonid morphology, appears to distinguish these two species. The terminal cusp in *Ny. kanamensis* M_3 is separated from the third cusp pair by a distinct median cusp, whereas it is directly connected to the third pair in *Ny. australis*. Because only a single specimen from this stratum bears the derived morphology of *Ny. kanamensis,* its recognition as the FA of the species is tentative.

DISCUSSION Although *Ny. kanamensis, sensu stricto,* has not been documented from the Asa Koma Member of the Adu-Asa Formation or the Kuseralee Member of the Sagantole Formation of the Middle Awash, it is the most abundant suid in the Gawto Member of the Sagantole Formation, dated to between 4.8 Ma and 5.2 Ma (Renne et al. 1999). The suid fauna from the Gawto member is not a subject of discussion in this work. However, its significance in terms of elaborating some evolutionary trends in the dentition of suids can be briefly addressed. It appears that the morphology of the P_2 and orientation of the lower canine are particularly significant characters in distinguishing pre–4.5 Ma *Ny. kanamensis* from the *Ny. kanamensis* at Kanapoi, Hadar, Omo, and other sites younger than 4.5 Ma. In the later forms, the P_2 is a single-rooted bunoid tooth. The lower canines become less vertically implanted. The mandibular body also becomes less robust relative to the length of the dental row. However, even though these characters are stronger than a character such as "narrow-toothed," they do not warrant recognition of two species for the pre– and post–4.5 Ma specimens of *Ny. kanamensis*. This species is one of the longest-lived tetraconodonts in Africa and its LA is documented at 2.7 Ma from the Omo Shungura Formation (White 1995) and 2.5 Ma from the Hatayae Member of the Bouri Formation.

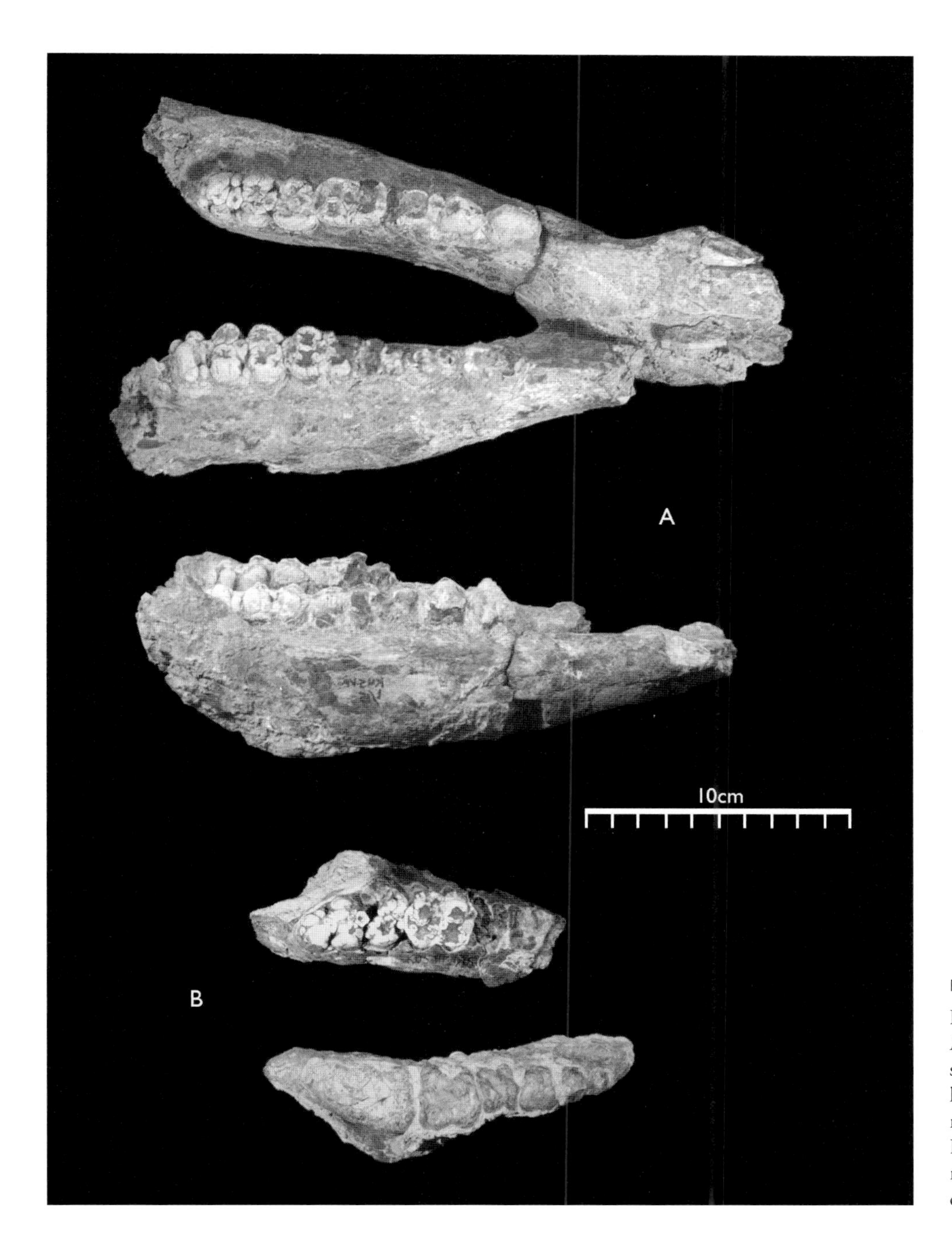

FIGURE 10.14
KUS-VP-1/15, holotype of *Nyanzachoerus kuseralensis* sp. nov. A. Occlusal and lateral views of the type mandible with dentition. B. Associated left and right maxillary fragments with dentition, occlusal view.

Nyanzachoerus kuseralensis sp. nov.

ETYMOLOGY In the local Afar language, Kuseralee Dora is a place where *kusera* trees grow; this locality gave its name to the Kuseralee Member, in which the type specimen was found.

HOLOTYPE KUS-VP-1/15, mandible with dentition and associated maxillary fragments, housed at the National Museum of Ethiopia (Figure 10.14).

PARATYPE AME-VP-1/35, left M^3.

LOCALITIES AND HORIZONS Kuseralee and Ambia East; Kuseralee Member of the Sagantole Formation, radiometrically dated to between 5.18 ± 0.07 and 5.55 ± 0.09 Ma.

DIAGNOSIS A species of *Nyanzachoerus* smaller than all known species of the genus and lacking the first and second mandibular premolars. The M_3 is relatively low crowned, and small, with two pairs of pillars and a terminal median pillar with a cluster of small flanking pillars. The enamel on both lateral and medial molar pillars is strongly folded, forming an H-shaped occlusal wear pattern on each of the major pillars. The P_3 is a mediolaterally compressed. Corpus is highly constricted posterior to the canines. The P_4 has a relatively large posterior accessory cusp. The premolar to molar ratio is higher than all known species of the genus.

DIFFERENTIAL DIAGNOSIS KUS-VP-1/15 is different from all known nyanzachoere mandibles in the lack of both P_1 and P_2, in having a buccolingually compressed P_3, and a robust but shallow mandibular corpus. It is different from *Ny. jaegeri* in being absolutely smaller, in having smaller and bilaterally compressed lower premolars, and an H-shaped wear pattern of the molar pillars. It is different from *Kolpochoerus deheinzelini* (Brunet and White 2001) mandibles in the lack of P_1–P_2 and by having more slender and mesiodistally elongated P_3–P_4, by the unique wear pattern on its molars, and by its overall larger size. *Nyanzachoerus kuseralensis* sp. nov. is different from all other *Nyanzachoerus* and *Notochoerus* species by its relatively smaller overall size (smaller than *Not. clarki;* White and Suwa 2004), small M_3 with two pairs of pillars, and higher premolar/molar ratio.

DESCRIPTION KUS-VP-1/15 is a complete mandible lacking the ascending rami and the posterior corpus base on both sides. All the incisors and canines are also missing, although the root of the left canine is preserved and shows that it was dorsoventrally compressed, having more of an anteriorly pointing triangular shape. The left third and fourth premolars and all the left molars are intact. The roots of the third and fourth premolars, partial M_1, most of the M_2, and a complete M_3 are preserved on the right side. Dental measurements are given in Table 10.5.

The incisor sockets show I_1 and I_2 of comparable size, and a small I_3. There is no diastema between the I_3 and the canine. The distal end of the symphysis terminates in the midline at a level equal to the front of the P_3. The anterior portion of the mandible (anterior to the P_2 position) is long relative to the overall length of the tooth row, and its lingual surface posterior to the incisor sockets is slightly concave. The mandible also shows constriction immediately behind the canines. There is a long diastema between the canine and the P_3 (57.1 mm on the left side). The mandible is relatively robust and shallow. The mental foramen on the left side lies midway between the canine and P_3. On the right side, there are two foramina, one at the same level as the one on the left side, and another one

TABLE 10.5 Dental Measurements of Specimens Referred to *Nyanzachoerus kuseralensis* sp. nov.

	LP3		*RP3*		*LP4*		*RP4*		*LM1*		*RM1*		*LM2*		*RM2*		*LM3*		*RM3*			
Specimen	MD	BL	MD	BL	MD	BL	MD	BL	MD	BL	MD	BL	MD	BL	MD	BL	MD	BL	MD	BL	*P3-P4 Length*	*M1-M3 Length*
Uppers																						
AME-VP-1/35																	42.5	27.2				
KUS-VP-1/15															27.1	21.9			37.5	(23.8)		
Lowers																						
KUS-VP-1/15	19.7	13.4			20.6	14.6			15.7	15.8			27.4	18.1	(26.0)	17.9	43.1	19.3	41.5	19.6	39.7	85.0

NOTE: Values in parentheses are estimates. All measurements are in mm.

TABLE 10.6A Dental Measurements of Permanent Teeth Referred to *Nyanzachoerus* sp.

	LP2		*RP2*		*LP3*		*RP3*		*LP4*	
Specimen	MD	BL	MD	BL	MD	BL	MD	BL	MD	BL
Uppers										
ALA-VP-1/31									19.6	22.4
AME-VP-1/198							21.3	20.3		
Lowers										
AMW-VP-1/55										
AMW-VP-1/83							27.1	18.8	24.3	21.0
KUS-VP-1/106					24.8	20.5			21.7	19.8

NOTE: All measurements are in mm.

at mid-corpus at the level between P_3 and P_4. The P_3 is anteroposteriorly longer than the P_4. It is highly compressed mediolaterally. The P_4 has a well-developed posterior accessory cusp. The M_1 appears to be anteroposteriorly shorter than both premolars across all tetraconodontine species, and this is probably a primitive retention for the subfamily. The associated maxillary fragments are fragmentary and do not yield significant information on the facial morphology of the new species. AME-VP-1/35 is a left M^3.

DISCUSSION The presence of *Ny. kuseralensis* sp. nov in the Middle Awash deposits dated to between 5.2 Ma and 5.55 Ma clearly shows that at least two tetraconodont species emerged to succeed the earlier *Ny. devauxi* and *Ny. syrticus. Nyanzachoerus kuseralensis* sp. nov. is contemporaneous with *Ny. waylandi* in the Middle Awash. This new species shows evolutionary novelties such as the loss of the P_2. The extreme bilateral compression of the P_3 and P_4 is not seen in the earlier nyanzachoeres or in the contemporaneous *Ny. waylandi.* This means that it was specialized and did not have anything to do with the origin of *Ny. australis.* Moreover, it is not currently clear whether this new species evolved from *Ny. devauxi* or *Ny. syrticus,* or an earlier yet unknown stock, since almost all of the dental characters it shares with the two nyanzachoere species that preceded it are plesiomorphic.

Nyanzachoerus sp.

DESCRIPTION The specimens assigned here to *Nyanzachoerus* sp. are isolated teeth from the Adu-Asa Formation that could not be confidently assigned to either *Ny. devauxi* or *Ny. syrticus.* The size and morphology of the incisors, the first and second premolars, and deciduous teeth are similar in both species. Measurements are given in Table 10.6A and B.

Trends in Nyanzachoere Dental Evolution

The trend in nyanzachoere dental evolution can be summarized as an increase in the length of the M_3s, a decrease in the length and breadth of the premolars, and the loss

RP4		*LM1*		*RM1*		*LM2*		*RM2*		*LM3*		*RM3*	
MD	BL	MD	BL	MD	BL	MD	BL	MD	BL	MD	BL	MD	BL
21.7	16.6			21.0	14.5	33.4	19.3	33.0	18.8				
21.8	19.9												

of the anterior premolars. The P_3–P_4/M_3 ratio of various nyanzachoeres clearly shows a decline of this ratio from the earliest *Ny. devauxi* to the youngest *Ny. jaegeri*.

The M_3s became longer by the addition of a cusp pair mesial to the terminal cusp. This was particularly true of nyanzachoere species that evolved after *Ny. syrticus*. The apparently ancestral species had M_3s with two pairs of cusps and a terminal cusp separated from the main cusps by a median cusp (Figure 10.11). *Nyanzachoerus australis* M_3s have three pairs of cusps and a terminal cusp, with the three cone pairs separated by two median cones and the terminal cusp connected to the distal third cusp pair (see Figure 10.11). *Nyanzachoerus kanamensis* further elongated its M_3 by adding a median cusp between the third cusp pair and the terminal cusp. *Nyanzachoerus jaegeri* added a fourth (sometimes more) cusp pair, which is the most derived among the nyanzachoeres.

Other trends in nyanzachoere dentition include an overall increase in crown height, loss of the first premolar and development of bunodonty on an increasingly single-rooted P_2, and a tendency for the lower canines to be implanted more horizontally.

The M^3s in *Nyanzachoerus* tend to be more conservative in their overall morphology. However, there is a tendency on the M^3s in the younger species to be larger in overall size through the addition of more cusplets in the talon region (Figure 10.12).

In general, there are at least three major character changes in the morphology of the tetraconodont M_3s that are taxonomically significant. The first one is the shift from two cusp pairs and a terminal cusp on the M_3 *(Ny. devauxi, Ny. syrticus, Ny. waylandi)* to three cusp pairs and a terminal cusp *(Ny. australis)*. The second character change is the shift from the distal third cone pair being connected to the terminal cusp without a median cusp to the third cone pair being separated from the terminal cusp by a clear median cusp *(Ny. kanamensis)*. This tendency is taken to the extreme in *Ny. jaegeri* with the third character change, the addition of one or more cusp pairs to the M_3. This is all accompanied by a decrease in the premolar/molar ratio. It should be noted here that even though most *Ny. kanamensis* specimens have a median pillar between the third cone pair and the terminal cusp of their M_3s, some individuals of the same species from younger deposits, such as Hadar, tend to secondarily lose the median pillar between

TABLE 10.6B Dental Measurements of Deciduous Teeth Referred to *Nyanzachoerus* sp.

Specimen	*LdP2*		*RdP2*		*LdP3*		*RdP3*		*LdP4*		*RdP4*		*LdM1*		*RdM1*		*LdM2*		*RdM2*	
	MD	BL	MD	BL	MD	BL	MD	BL	MD	BL	MD	BL	MD	BL	MD	BL	MD	BL	MD	BL
Uppers																				
AMW-VP-1/133													17.3	14.3						
Lowers																				
ALA-VP-2/199							11.2	6.5												
ALA-VP-2/202			11.7	6.2																
ALA-VP-2/215					11.2	5.5														
ALA-VP-2/333					10.4	5.6														
AME-VP-1/105																			24.1	14.7
AME-VP-1/146			12.3	7.0			11.2	7.0												
ASK-VP-3/66							11.7	6.3												
ASK-VP-3/383			12.1	6.7																
ASK-VP-3/384			12.9	6.8																
ASK-VP-3/385					12.7	7.0														
ASK-VP-3/386					11.5	6.6														
ASK-VP-3/387							nm	6.9												

NOTE: All measurements are in mm; nm = measurement was not possible.

the third cusp pair and the terminal cusp. However, their premolar/molar ratio remains within the species range.

The morphological changes described in this section reflect not only the trends in the evolution of the third molars in *Nyanzachoerus* but also the mode of evolution in late Miocene and early Pliocene African tetraconodonts. Based on the dental evidence presented in this section, one can argue that the stem African nyanzachoeres—*Ny. devauxi* and *Ny. syrticus*—possibly shared a common ancestor prior to 7–8 Ma. Alternatively, *Ny. syrticus* may be a chronospecies of *Ny. devauxi*. Neither of the two possibilities can be refuted based on the current fossil evidence. However, it is clear that both species share a two-cusp-pair/one-terminal-cusp character in the morphology of their M_3. *Nyanzachoerus australis* is intermediate in its M_3 morphology and size between *Ny. syrticus* and *Ny. kanamensis*. Therefore, the phylogenetic relationship between these three taxa can be interpreted as involving phyletic rather than cladogenetic evolution. The fossil record does not currently support the latter mode of evolution.

Tetraconodont Habitat and Diet

The paleoecology and diet of African tetraconodonts have been approached using dental morphology (Kullmer 1999), skeletal material (Bishop 1994, 1999), comparisons with habitats of extant forms (Cooke 1985), and isotopic analysis (Harris and Cerling 2002). Brachyodonty is often interpreted as an adaptation to browsing and closed habitats, whereas hypsodonty is usually associated with grazing and open habitat. Hypsodonty also tends to correlate with increasing body size among African suids. Earlier nyanzachoeres such as *Ny. devauxi* and *Ny. syrticus* from the Lower Nawata had large premolars and brachyodont molars. There is a tendency for molar crown height increase within *Ny. syrticus* in both the Upper Nawata and the Middle Awash stratigraphic successions. Hypsodonty and mesiodistal elongation of the third molars, coupled with decrease in the premolar size, is the persistent and repeated evolutionary trend in most African suid lineages. The shift in dental proportion is functionally correlated by most workers with a shift in dietary preferences. Some have suggested that this evolution might correspond causally to global climatic changes. Cooke (1985) correlated large premolars with the consumption of abrasive and hard food. However, Kullmer (1999) associated brachyodont molars with a soft and omnivorous diet. The nyanzachoeres with the largest premolars are *Ny. devauxi* and *Ny. syrticus,* which also have relatively brachyodont molars. Hence, Kullmer and Cooke's speculations are based on two closely related variables and nevertheless yield different results. However, studies also show that hypsodonty and brachyodonty, per se, may not be directly correlated with grazing and browsing, respectively (Harris and Cerling 2002). For example, Harris and Cerling's (2002) isotopic analysis suggested that *Ny. devauxi,* albeit the most brachyodont of all nyanzachoeres, showed a higher C_4 signal compared to its less brachyodont descendant *Ny. syrticus* from the Lower Nawata of Lothagam.

Despite the taller, elongate third molars of *Ny. kanamensis,* it is usually found in either more closed (Harris 1983b) or intermediate (Bishop 1999) habitats. Based on their isotopic analysis, Harris and Cerling (2002) suggested that *Ny. pattersoni (=Ny. kanamensis)* was a mixed feeder. The larger size of *Ny. jaegeri* and its elongated and higher third molars would suggest a more open habitat preference and adaptation to a more abrasive, presumably grazing diet (Harris and Cerling 2002). The emergence of *Ny. jaegeri* correlates with the increase in the expansion of C_4 plants in eastern Africa and elsewhere (Cerling et al. 1997a, b; Levin et al. 2004). This was probably induced by a combination of global, regional, and local climatic factors (Levin et al. 2004). *Notochoerus euilus* probably shared the same dietary and habitat preferences as *Ny. jaegeri,* even though some workers have suggested, to the contrary, that *Not. euilus* preferred a more closed and wooded habitat (Cooke 1985; Bishop 1994, 1999). Its sister species, *Not. scotti,* was adapted to drier and more open habitats (Kullmer 1999). Its daughter chronospecies, *Not. clarki,* was probably adapted to similar environments, although it has been associated with wet lake margin habitat at Konso and the Middle Awash in Ethiopia (White and Suwa 2004).

Although the reduction of premolar size among African suids has not yet been fully understood, elongation of the third molars has been correlated with adaptation to a grazing diet (Harris and Cerling 2002). The available paleoenvironmental data also indicates that there was extensive expansion of C_4 biomass between 5 Ma and 7 Ma (Cerling et al. 1997a, b). We also see that at least three tetraconodonts (*Ny. waylandi, Ny. australis,* and *Ny. kuseralensis* sp. nov.) appeared in the fossil record toward the end of the Miocene. This would suggest that availability of new open habitat niches facilitated diversification of taxa, including the suids and bovids, in the late Miocene. The species that succeeded *Ny. syrticus* was *Ny. australis*. It shows reduced premolar size and further elongation of its third molars. This trend culminated in the chronospecies *Ny. kanamensis*. On the other hand, *Ny. waylandi* and *Ny. kuseralensis* sp. nov. were contemporaneous with *Ny. australis,* and were much smaller in size. Their third molars were not as elongate. Based on these observations, it is plausible to infer that the smaller species were adapted to a more closed habitat with a mixed feeding strategy, living in forest/grassland mosaics similar to those inhabited by the extant *Hylochoerus meinertzhageni*.

Cainochoerinae

Cainochoerus Pickford, 1988

GENERIC DIAGNOSIS See Pickford, 1988.

Cainochoerus africanus Hendey, 1976

Cainochoerus cf. *africanus*

DESCRIPTION Specimens identified as *Cainochoerus* cf. *africanus* are shown in Figure 10.15.

Upper Incisors Two lower and three upper incisors are referred to this species. All of the recovered upper incisors are first incisors. The crowns are small and obliquely worn from the mesial tip of the occlusal surface to the distal end of the crown base, giving the crowns a triangular labial shape. They have strong interproximal wear facets positioned on

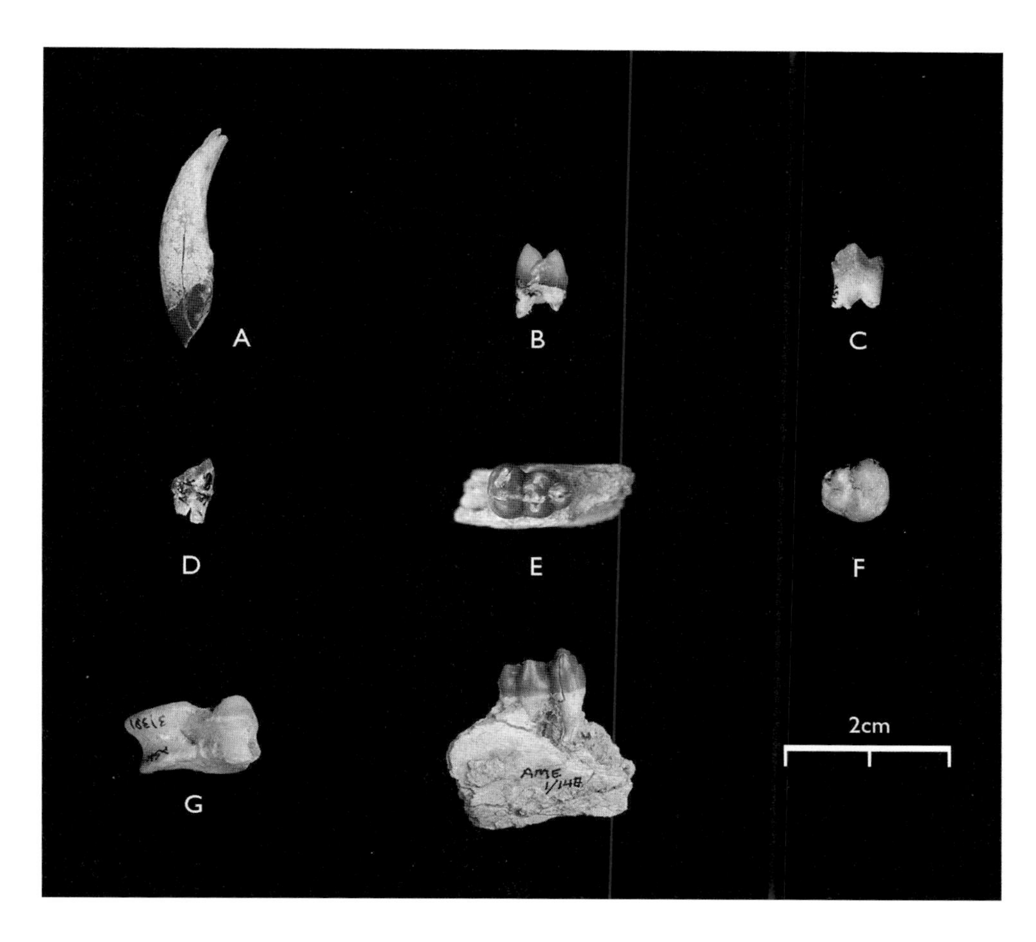

FIGURE 10.15

Middle Awash specimens assigned to *Cainochoerus* cf. *africanus*. A. AME-VP-1/106, left I_1, labial view. B. ASK-VP-3/54, left P^3, mesial view. C. ASK-VP-3/389, left P_4, buccal view. D. ASK-VP-3/65, right P_3, mesial view. E. AME-VP-1/148, right mandible fragment with M_3, occlusal and mesial views. F. ASK-VP-3/398, left M^3, occlusal view. G. ASK-VP-3/391, left astragalus, dorsal view.

the mesial face close to the occlusal tip of the crown. They also have very long, rounded, and curved roots, probably three times as long as the crown itself.

Upper Premolars Specimen ASK-VP-3/54 is an unworn P^3. Specimens ALA-VP-1/28 and ASK-VP-3/517 are P^4s. The P^3 and P^4 are similar in their morphology and number of cusps. They have two main cusps arranged buccolingually, a mesial cingulum, and a distal cingulum. However, the mesial cingulum on the P^4s has at its center a small median cusp, which is absent on the P^3s.

Upper Molars Two specimens, ASK-VP-3/390 and ASK-VP-3/398, are upper molars. Specimen ASK-VP-3/390 is the mesial half of a M^2, and ASK-VP-3/398 is a complete crown of M^3. Both specimens lack their roots. It is possible that these two specimens might belong to the same individual. The M^3 is low-crowned and has four well-defined cusps. The protocone is the largest cusp, followed by the paracone. The hypocone and metacone are equal in their size. The mesial cingulum is well-developed, and its lingual half merges with a small cusplet situated centrally at the mesiolingual corner of the paracone. The cusplet is separated by grooves from the protocone and the buccal half of the

TABLE 10.7 Dental Measurements of *Cainochoerus* cf. *africanus*

	LI1		*RI1*		*LI2*		*RI2*		*LP3*		*RP3*	
Specimen	MD	BL	MD	BL	MD	BL	MD	BL	MD	BL	MD	BL
Uppers												
ASK-VP-3/392			5.04	3.52								
ASK-VP-3/393			6.74	4.1								
ASK-VP-3/54									5.71	6.3		
ASK-VP-3/517									6.4	5.22		
ALA-VP-1/28												
ASK-VP-3/390												
ASK-VP-3/398												
Lowers												
ASK-VP-3/496	3.4	nm										
ASK-VP-3/325					2.36	(3.42)						
KUS-VP-1/89							3.5	2.85				
ASK-VP-3/319											4.6	2.9
ASK-VP-3/320											4.5	2.76
ASK-VP-3/65											5.14	3.6
ASK-VP-3/530									5.36	4.1		
ASK-VP-3/389												
ASK-VP-3/322												
AME-VP-1/148												
GAW-VP-1/47												

NOTE: All measurements are in mm. Values in parentheses are estimates; nm = measurement was not possible.

mesial cingulum. The distal accessory cusp is well-developed and positioned centrally. It merges with a small cingulum on the posterior base of the hypocone. The crown is much wider anteriorly across the paracone and the protocone.

Lower Incisors The lower incisors are represented by one of each of the first and second incisor. I_1 (ASK-VP-3/496) is a long peg-like incisor, mesiodistally narrow relative to its height. Some parts of the base of the crown and the root are missing. The lingual face of the preserved crown shows the presence of a strong ridge running from the tip toward the root, creating vertical grooves on both sides. I_2 (ASK-VP-3/325) is a complete unworn crown with a small portion of the root preserved. It is long and narrow like the I_1 but curves laterally below mid-crown height. The lingual face has a strong vertical ridge positioned more mesially, forming vertical grooves on its left and right sides. The ridge terminates at a strong basal buttressing.

Lower Canine KUS-VP-1/89 is a left lower canine preserving the crown and part of its root. The mesial part of the crown is worn obliquely from contact with the upper canine. The crown has an oval to round cross section. The buccal face of the crown is covered with enamel.

LP4		*RP4*		*LM1*		*RM1*		*LM2*		*RM2*		*LM3*		*RM3*	
MD	BL	MD	BL	MD	BL	MD	BL	MD	BL	MD	BL	MD	BL	MD	BL
		nm	7.83												
								nm	6.9						
												8.31	8.12		
6.45	4.43														
		6.8	4												
														10.36	6.7
										8.5	5.8			9.44	6.1

Lower Premolars The P_3s are unicuspid and sectorial. They tend to lack a distinct mesial accessory cusp and cingulum but have a small posterior accessory cusp, which disappears at a later wear stage. In some specimens (e.g., ASK-VP-3/322), the posterior accessory cusp is larger. When it is worn, it forms a shelf-like structure. The P_4 is also two-rooted and relatively molarized. The mesial accessory cusp is well-developed, and it is connected with a small cingulum on the mesiobuccal corner of the crown. Distal to the mesial accessory cusp is a relatively large median cusp extending almost as high as the main cusp. The latter is bifurcated into buccal and lingual halves by a shallow longitudinal groove. The posterior accessory cusp is buccolingually expanded and is separated from the main cusp by a deep basin.

Lower Molars There are no M_1s and M_2s recovered thus far, and the M_3 is known from one specimen (AME-VP-1/148). It has five well-defined cusps and a strong mesial cingulum. The two mesial cusps (protoconid and metaconid) are taller than all the remaining cusps. The entoconid, hypoconid, and hypoconulid are of equal height. The mesial and distal cusp pairs are separated from each other by a median longitudinal groove. There is a well-defined talonid. The crown is wider on the mesial cusp pair compared to the breadth at the distal cusp pair.

The dental specimens from the Middle Awash are morphologically similar to specimens from Langebaanweg assigned to *Cainochoerus africanus* (Pickford 1988). Dental measurements are given in Table 10.7. However, these specimens metrically fall in the lower range of the South African hypodigm. Craniodental specimens are unknown from Lothagam. These metric differences are inadequate to warrant distinction between the Middle Awash and Langebaanweg samples at a species level. More complete specimens are necessary to determine the relationship between the South African and eastern African *Cainochoerus* hypodigms, and newly recovered specimens from the 4.4 Ma Aramis deposits will be critical in this regard.

DISCUSSION The genus *Cainochoerus* was previously known only from Langebaanweg, South Africa, even though its subfamily was unknown until recently. Pickford (1995) coined the subfamily name Cainochoerinae to accommodate the species. However, its relationship with other suid subfamilies is unknown, largely because of the mosaic nature of craniodental and postcranial characters seen in *Cainochoerus*. The South African material was initially assigned to ?*Pecarichoerus africanus* (Hendey 1976), However, Pickford (1988) later created a new genus name, *Cainochoerus,* for the material, and hence the species name *Cainochoerus africanus.* New material from the Upper Nawata of Lothagam, Kenya, has recently been referred to *Cainochoerus* cf. *africanus* (Harris and Leakey 2003b). The new material from the Middle Awash terminal Miocene deposits is additional evidence for the presence of the genus in eastern Africa.

Conclusions

There are at least two late Miocene and seventeen Plio-Pleistocene suid species known in Africa, three of which are extant today (Bishop 1999). Almost half of these species are known to be endemic to eastern and northern Africa, while the others represent suids distributed more continentally (Bishop 1999). Two *Kolpochoerus* and one *Notochoerus* species have been recently named from Ethiopia and Chad (Brunet and White 2001; White and Suwa 2004) increasing the number of known suid species to 22. A basic understanding of the phylogenetic relationships among these taxa emerged during the early 1970s. Harris and White (1979) recognized three extinct nyanzachoere species and also recognized *Ny. jaegeri* as the parent stock of the *Notochoerus* lineage. Two new nyanzachoere species have been added since then, and this chapter adds a third.

The current fossil record from the Middle Awash late Miocene and early Pliocene deposits allows revision of *Nyanzachoerus* systematics. The earliest deposits, dated at 5.5–6 Ma, contain evidence of only two nyanzachoere species, *Ny. devauxi* and *Ny. syrticus.* These are clearly two distinct species, particularly based on their dental size and premolar/molar ratio, even though an ancestor-descendant relationship cannot be ruled out based on their chronostratigraphic relationships. The deposits of the Central Awash Complex (CAC) represent a subsequent fossil sampling of the Middle Awash record at 5.5 Ma to ca. 5.2 Ma. These deposits contain at least three lineages designated by the species names *Ny. waylandi, Ny. kuseralensis,* and *Ny. australis.* The latter species is known from a much larger sample than the other two, which are each known by a few specimens. These species are discretely separated by the morphology of the M_3, the premolar/molar ratio, the number of premolars, or by a combination of these characters.

In the context of our overall knowledge of the late Neogene of Africa, the new Middle Awash data suggest that *Ny. syrticus* was a widespread, successful late Miocene suid lineage that lasted until ca. 5.5 Ma. There is no compelling evidence that this species gave rise to *Ny. waylandi,* which temporally and morphologically appears to be a transitional form between *Ny. syrticus* and *Ny. australis*. The new Middle Awash species, *Ny. kuseralensis,* is also not a candidate for the ancestry of *Ny. australis* on account of its specialized premolars and loss of the P_2. There are indications in the Middle Awash sequence that *Ny. australis* was a short-lived species replaced by *Ny. kanamensis*. On morphological grounds, *Ny. kanamensis* could have evolved phyletically from *Ny. australis*. However, on chronological grounds, if their contemporaneity is confirmed by further discoveries, these two species could be sister taxa descended from a form like *Ny. waylandi*. This creates the possibility for two equally parsimonious relationships between *Ny. kanamensis* and *Ny. australis*.

The plethora of new suid fossil specimens recovered from the Middle Awash and other eastern African late Miocene sites such as Lothagam have added new tetraconodont taxa and improved our knowledge of their dental evolution. However, these fossils have not yet revealed clear intra- and interspecific phylogenetic relationships among the various nyanzachoere species, particularly among the earlier forms. Moreover, the origin of the genus *Nyanzachoerus* itself remains elusive beyond a hypothesized descent from a form like *Conohyus*. Additional fossils from horizons between 7 and 9 Ma are needed to understand better the origin of African tetraconodonts.

11

Hippopotamidae

JEAN-RENAUD BOISSERIE AND YOHANNES HAILE-SELASSIE

During most of the middle to late Miocene, the first known representatives of the family Hippopotamidae left a limited fossil record. A few fragmentary specimens attributed to the primitive genus *Kenyapotamus* were unearthed in Kenya (Pickford 1983; Behrensmeyer et al. 2002), Tunisia (Pickford 1989b), and possibly Ethiopia (Geraads et al. 2002; Suwa et al. 2007). After 8 Ma, hippopotamids became much more common. They are among the most abundant fossils collected at sites such as Lothagam, Kenya (Weston 2003) and Toros-Menalla, Chad (Boisserie et al. 2005). Their distribution then expanded to southern Europe and southern Asia around 6 Ma (Kahlke 1990). This apparent dramatic increase in hippopotamid abundance is correlated with a significant increase in their species diversity, as shown by recent descriptions of new materials (Weston 2000; Boisserie et al. 2003; Weston 2003; Boisserie 2004; Boisserie and White 2004; Boisserie et al. 2005).

These evolutionary developments have important implications for our knowledge of African late Miocene environments, given that hippopotamids are large, semiaquatic herbivores with a great impact on their habitats. For instance, the most common extant representative, *Hippopotamus amphibius* Linnaeus, 1758, is known for greatly influencing the floral and faunal diversity in African wet ecosystems, for shaping the hydrographic network morphology, and even for generating sedimentary structures (Verheyen 1954; Kingdon 1979; Eltringham 1999; Deocampo 2002). In this regard, the recent discoveries of new late Miocene and early Pliocene hippopotamid material in the Middle Awash study area constitute an important step in understanding this crucial period for Hippopotamidae and, more generally, their evolution in the Afar basin.

Until now, three endemic species have been described in this basin, but from younger deposits. *Trilobophorus afarensis* Gèze, 1985 and *Hexaprotodon coryndoni* Gèze, 1985 are known from between 3.4 Ma and 2.33 Ma at Hadar and Geraru (Gèze 1980; Kimbel et al. 1994). The former species is well documented and has been reported as a new genus on the basis of distinctive morphology of the lacrimal area (Gèze 1985). *Hexaprotodon bruneti* Boisserie and White, 2004 is dated to 2.5 Ma and derives from the Bouri Formation and Maka area of the Middle Awash valley. It is probably an immigrant from Asia, given its affinities with *Hexaprotodon sivalensis* Falconer and Cautley, 1836 from the Siwaliks of India and Pakistan.

The number of hippopotamid genera has always been a disputed issue. The last major phylogenetic assessment was that of Coryndon (1977, 1978), who included most African fossil species in the genus *Hexaprotodon*. Unfortunately, this was done on the basis of primitive features, such as the contact between the frontal and the maxilla in the lachrymal area (Harris 1991a). Subsequent authors recognized the paraphyly of *Hexaprotodon* but generally maintained the *status quo* (Harris 1991a; Harrison 1997b; Weston 2003) or proposed to use only one genus name (Stuenes 1989; Pickford 1995). Recently, a phylogenetic reappraisal of the family was based on the first cladistic analysis ever performed on this group, as well as on a biometrical comparison of various species (Boisserie 2005). This study identified several lineages that have diverged since the late Miocene. According to these results, a new taxonomy was proposed (Boisserie 2005), which is followed in this chapter. Two new genera were created: *Archaeopotamus* for the late Miocene "narrow-muzzled" hippopotamids of Lothagam and Abu Dhabi and *Saotherium* for the early Pliocene hippopotamids from Chad. The genus *Choeropsis* was rehabilitated for the extant Liberian hippopotamid, and the genera *Hippopotamus* and *Hexaprotodon* were preserved. The former continues to be used to designate the same taxa as before; the latter, however, is proposed to be restricted to a lineage mostly known in Asia, for which it was initially created. Finally, the Plio-Pleistocene hippopotamids previously described in the Turkana and Afar basins appeared to be closer to *Hippopotamus* (abbreviated *Hip.* below) than to the Asian *Hexaprotodon* (abbreviated *Hex.* below). However, this material is still in need of a thorough systematic revision. For this reason, Boisserie (2005) proposed a temporary generic designation: aff. *Hippopotamus*. For instance, in this chapter, *Trilobophorus afarensis* from Hadar is called aff. *Hippopotamus afarensis*.

Two assemblages of hippopotamid fossils were collected from the late Miocene–early Pliocene sediments in the Middle Awash. The first one was recovered from the Asa Koma Member of the Adu-Asa Formation, dated to 5.5–5.8 Ma. Those remains are fragmentary and isolated, and they were unearthed from various sediments (sands, silts, and clays) of various localities. Late 2006 discoveries increased the hippopotamid record from the Asa Koma Member (locality STD-VP-1) considerably, but these still need to be prepared prior to study and will therefore not be included in this chapter. The second set came from the Kuseralee Member of the Sagantole Formation, from sandstones just below the 92-15 basalt (dated to 5.2 Ma). This material was recently attributed to a new species (Boisserie 2004). A few additional specimens of the same hippopotamid species were unearthed in the Haradaso Member of the Sangantole Formation. The description of those specimens is also included in this chapter. Those fossils came from the fluvio-lacustrine silty sands that underlie the Abeesa Tuff. The age of this tuff is dated by $^{40}Ar/^{39}Ar$ method to 4.82 ± 0.07 Ma (Renne et al. 1999).

Hippopotamidae

Hippopotaminae Gray, 1821

gen. et sp. indet.

DESCRIPTION All specimens attributed to Hippopotaminae gen. et sp. indet. are from the Asa Koma Member of the Adu Asa Formation.

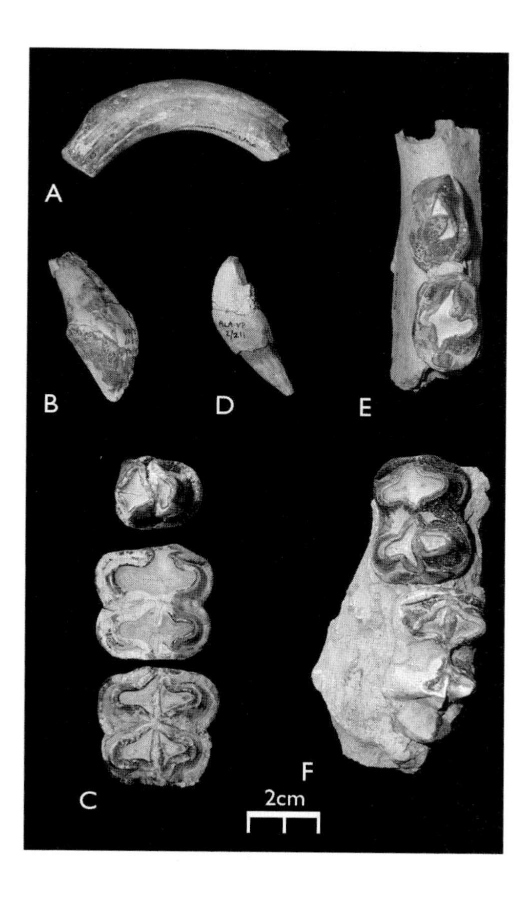

FIGURE 11.1

Dentition of Hippopotaminae indet. from the Adu-Asa Formation. A. Lateral view of DID-VP-1/56, upper incisor. B. Lingual view of ALA-VP-2/210, right P^1. C. Occlusal view of BIK-VP-3/1, right P^4 and left M^2–M^3. D. Lingual view of ALA-VP-2/211, right P_1. E. Occlusal view of DID-VP-1/3, left P_3–P_4. F. Occlusal view of BIK-VP-3/7, left M_2–M_3.

Upper Dentition The only well-preserved anterior tooth is the upper incisor, probably an I^1 (DID-VP-1/56). This curved, open-rooted tooth exhibits a rather square transverse section, no enamel, and an apical transverse wear facet (Figure 11.1A).

Among premolars, ALA-VP-2/210 is a P^1, notably characterized by one strong root, which bears some patches of cement (Figure 11.1B). Although undivided, even apically, this root exhibits two distinct lobes in the occlusal section. The robust crown is roughly triangular in lateral view and slightly kidney-shaped in occlusal view. Its apex leans lingually. A low, small cingulum is present mesially and mesiolingually. On the mesial side of the crown, it connects with two vertical pustulate crests. The rest of the enamel ranges from finely ridged to smooth. The distal side exhibits one simple vertical crest. ALA-VP-2/207 is a left P^2 that shows, in occlusal view, a rectangular outline with medial constriction. The crown of this robust tooth is pyramidal, with its apex leaning lingually. The mesial side is strongly pustulate and ridged, with a vertical crest deviating lingually. The lingual and buccal sides are slightly smoother. Buccodistally, a strong crest bears two small accessory cusps. A pustulate transverse crest is found along the distal side of the crown, just above the cingulum. This cingulum is low and reduced and is found all around the crown, although it is very attenuated on the buccal side. No P^3s were recovered. Two P^4s are known (BIK-VP-3/1 and BIK-VP-3/11). Both exhibit two main cusps of sub-equal size. The lingual cusp is triangle-shaped lingually; the buccal cusp is diamond-shaped.

There is a well-developed cingulum, notably on the distal side, but absent on the buccal side (Figure 11.1C). BIK-VP-3/11 is, however, more buccolingually elongate, with an outline more lingually depressed than in BIK-VP-3/1.

Molars are low-crowned with more or less square outlines in occlusal view. The complete D^4 (ALA-VP-2/208) is a quadrangular tooth with a distal lobe that is wider than the mesial lobe and thin enamel. Its lingual cusps are crescentiform, and its cingulum is moderate but not interrupted. A dental germ identified as possibly an M^1 was reconstituted from various tooth fragments (STD-VP-2/112). Its overall morphology is close to that of the M^2 of BIK-VP-3/1 (Figure 11.1C). The crown of the M^2, in an advanced stage of wear, is completely surrounded by a strong and continuous cingulum. The M^3 from the same specimen (Figure 11.1C) has a similar morphology, except that the cingulum appears to be more pustulate and the distal lobe is narrower than the mesial one. This trait is greatly exaggerated in BIK-VP-1/7, where both the left and right M^3s show strongly reduced metacone and metaconule. The distal lobe is buccally shifted and 21 percent less wide than the mesial lobe (compared to 9 percent in BIK-VP-3/1). This condition is most probably linked to abnormal tooth development. The other M^3 (DID-VP-1/83) is closer to BIK-VP-3/1, although its cingulum is more reduced.

Lower Dentition No lower anterior teeth were collected. One isolated P_1 was identified with some confidence (ALA-VP-2/211, Figure 11.1 D). This tooth is simple, single-rooted, and single-cusped. It has a more or less rounded section at the cervix, and the cusp is conical with no cingulum, but it has one mesial and one distal crest, the latter bearing a small distolingual accessory cusp. The root is robust, with a remaining sleeve of cement below the cervix. No P_2s were collected. A P_3 is present on the mandibular fragment DID-VP-1/3 (Figure 11.1E). This tooth is single-cusped and elongated mesiodistally. A low and thin cingulum is present on all sides but almost disappears mesially. A small accessory cusp is localized on the distal part of the lingual side of the crown, just above the cingulum. The enamel is pustulate and crested on the distal side. The P_4 of the same specimen (Figure 11.1E) exhibits a similar overall shape, but a much larger accessory cusp occupies the center of the lingual side, reaching well above the cingulum. A distal accessory cusp, low but well-marked, is present between the crown and the cingulum and is, overall, stronger on this tooth than on the preceding tooth. A second P_4 (ALA-VP-2/209) differs by not retaining a lingual accessory cusp and having a stronger distal cusp. On this unworn tooth, one vertical crest is seen mesially, and two distally. The enamel on this tooth is finely striated.

No M_1 is known with certainty from these sediments. STD-VP-2/5 is an isolated nonerupted tooth that could be either a M_1 or a M_2. Its morphology is close to that of the M_2 borne by the fragmentary mandible corpus BIK-VP-3/7 (Figure 11.1F). These teeth are elongate, with a cingulum that is well-developed mesially and distally but lacking laterally. Contrary to other cuspids, the entoconid does not show a trilobate wear pattern, and the mesial and distal lobes are reduced. All the M_3s exhibit a similar morphology to M_1/M_2 (Figure 11.1F), except for the presence of the hypoconulid. The latter is well-developed on each tooth, accompanied, at least, by an ectoconulid just distal to the hypoconid.

DISCUSSION In mammalian phylogeny, cheek tooth morphology often plays a crucial role. However, as Coryndon noted in 1977 (p. 63), this is not the case for the Hippopotamidae, "It is unfortunate that, as far as hippopotamids are concerned, molar teeth are very conservative in development and are possibly the least useful element for diagnosis, slight variation in enamel pattern often reflecting slight differences in feeding habits rather than morphogenetic characters." Indeed, hippopotamid cheek teeth show only minor variations, and these variations can be found in most of the known species. Given these conditions, it is not surprising that the above described cheek teeth cannot be morphologically differentiated between aff. *Hip. dulu, A. harvardi,* and other taxa.

These specimens from the Asa Koma Member are all low-crowned and show the moderate trilobate wear pattern that is common to all contemporary hippopotamids. Their measurements largely overlap with those of *A. harvardi* and *Hex. garyam* (Table 11.1). These dimensions seem homogeneous, and these remains could all belong to one form. Although the comparative sample size calls for caution, they also indicate the presence of a form larger than *A. lothagamensis* and aff. *Hip. dulu* (Table 11.1), a view confirmed by preliminary observation of postcranial elements (notably comparison of astragalus sizes). In any case, they are overall not significantly larger than those of *A. harvardi* (see Haile-Selassie et al. 2004c). For these reasons, it is not presently possible to attribute these remains to a particular genus or species of the subfamily Hippopotaminae.

The material collected in 2006 in the Asa Koma Member includes a complete cranium and its associated mandible. Additional late Miocene hippopotamid remains were recently found in a Middle Awash locality (JAB-VP-1) stratigraphically below the Adu-Asa Formation. Future study and comparisons of these remains should lead to more accurate identification and phylogenetic placement of the Afar late Miocene Hippopotaminae.

Hippopotaminae Gray, 1821

aff. *Hippopotamus dulu* Boisserie, 2004

DIAGNOSIS The canine processes are weakly extended externally. The mandible is slender, elongated, with a shallow and long symphysis and a marked anterior projection of the incisor alveolar process above the lower frontal part of the symphysis. The angular processes of the mandible are well-developed but not hook-shaped, and they show a weak to moderate external expansion. The skull displays thin zygomatic arches that expand progressively, a mediolaterally compressed braincase, and large and robust occipital condyles. The canine enamel is finely ridged. The upper canines are bilobate with a deep posterior groove. The cheek tooth rows diverge anteriorly. The first premolars are retained. Diastemas between premolars are short or lacking. Premolars are large, but the length of the P^2–P^4 series is less than that of the M^1–M^3 series. The P^4 has two main cusps of equal size. Molars are low-crowned.

This species shares with *Archaeopotamus harvardi* many dental and cranial features but differs in the following: It is smaller in average size; its premolar rows are short relative to the molar rows; it has a shorter mandibular symphysis relative to its width; it has strong occipital condyles; and its mandibular angular process is slightly to markedly externally shifted. This mandibular morphology recalls that of *Saotherium mingoz*. However, in aff.

TABLE 11.1 Cheek Tooth Measurement Ranges of the Middle Awash Late Miocene–Early Pliocene Hippopotamids, Hippopotaminae indet. and aff. *Hippopotamus dulu*, Compared to Other Hippopotamids

	A-S	*SAG*	*LTh*	*TM*	*KB*		*A-S*	*SAG*	*LTh*	*LTl*	*ABU*	*TM*	*KB*
P^1n	1		4	5		P_1n	2	2					
L	26		21–28	23–28		L	17–19	17–20					
w	19		17–23	19–23		w	13–15	11–12					
P^2n		1	9	14	1	P_2n		1	5	1	1	13	1
L		36	37–41	37–44	37	L		34	31–41	31	39	30–39	30
wd		26	23–35	26–36	28	wd		19	21–25	21	23	19–26	18
P^3n	1	2	10	23		P_3n	3	1	5	1	1	15	1
L	39	39–41	35–48	36–47		L	38–40	34	38–45	31	39	32–48	32
wd	26	30–32	27–37	29–42		wd	23–27	22	23–27	21	23	18–33	19
P^4n	2	4	11	23	1	P_4n	4	3	9		1	19	2
L	29–33	25–33	26–34	27–47	30	L	38–49	33–35	34–45		35	33–43	31–33
wd	33–39	32–37	31–46	32–42	34	wd	19–31	23–27	24–31		24	22–33	21–25
M^1n	1		13	21		M_1n	1	1	10	1		20	2
L	40		31–46	36–49		L	47	39	36–49	37		39–49	37–39
wm	30		32–46	31–43		wm	31	30	26–35	23		27–39	25–25
M^2n	1	2	15	22	1	M_2n	1	5	10	1		26	3
L	45	44–47	37–50	42–52	46	L	50	41–45	41–51	41		42–54	43–46
wm	44	44–44	40–57	41–49	47	wm	35	31–36	34–38	28		31–45	30–33
M^3n	4	5	13	37	1	M_3n	2	14	14	2	1	26	2
L	43–49	43–48	44–52	57–59	48	L	62–70	52–64	58–68	50–51	60	59–72	55–59
wm	39–46	38–46	41–56	41–52	47	wm	38–37	31–39	32–40	28–29	32	34–43	34–35

NOTE: Abbreviations: A-S, Hippopotaminae indet. (Adu-Asa Formation); SAG, aff. *Hip. dulu* (Sagantole Formation); LTh, *A. harvardi* (Lothagam, Kenya); LTl, *A. lothagamensis* (Lothagam, Kenya); ABU, *A.* aff. *lothagamensis* (Abu Dhabi, United Arab Emirates); TM, *Hex. garyam* (Toros-Menalla, Chad); KB, *S.* cf. *mingoz* (Kossom Bougoudi, Chad). Measurements are in mm. Abbreviations: w = width; wm = mesial lobe width; wd = distal lobe width.

Hip. dulu the symphysis is shallower; the corpus is higher; the cheek teeth exhibit greater relative and absolute dimensions; and the braincase is externally compressed with a triangular shape in a transverse section rather than rounded as in *S. mingoz.* The other African and Eurasian hexaprotodont hippopotamids are distinct from aff. *Hip. dulu* in having a much smaller size (*Hex. imagunculus, A. lothagamensis*); less than six lower incisors (aff. *Hip. protamphibius protamphibius,* aff. *Hip. aethiopicus,* aff. *Hip. karumensis,* jointly termed "Turkana Pliocene hippos," and *Hex. crusafonti*); an incisor alveolar process that is less projected ("Turkana Pliocene hippos," aff. *Hip. coryndoni,* aff. *Hip. afarensis*); more differentiated lower incisors (aff. *Hip. afarensis,* aff. *Hip. coryndoni,* aff. *Hip. karumensis*); more expanded canine processes and a relatively shorter mandibular symphysis (all, except for the "narrow-muzzled" hippos); relatively longer symphysis and premolar row (*Hex.* aff. *sahabiensis, A. lothagamensis,* termed as "narrow-muzzled" hippos); a shorter lower premolar row relative to the molar row (aff. *Hip.* cf. *protamphibius,* "Turkana Pliocene hippos," *Hex. sivalensis*); and less robust zygomatic arches (*Hex. sivalensis,* aff. *Hip. afarensis,* "Turkana Pliocene hippos").

DESCRIPTION All specimens attributed to aff. *Hippopotamus dulu* are from the Kuseralee Member of the Sagantole Formation.

Mandible The holotype of aff. *Hippopotamus dulu* (AME-VP-1/33) (Figure 11.2) is the most complete mandible. Its symphysis is well-preserved. However, its anterior border and canine processes are eroded. The corpora are complete. Only the apex of the coronoid process is lacking on the right ramus, but the left ramus is mostly absent. The only remaining anterior tooth is the right canine. The cheek teeth are represented only by the heavily worn M_3, indicating an old individual. In dorsal view, this specimen appears slender and elongated, except for the anteriorly enlarged symphysis. This shallow and long symphysis is hexaprotodont, but, unfortunately, the size and relative position of the incisors cannot be described from these remains. The incisor alveolar process is markedly anteriorly projected above the lower frontal part of the symphysis, and moderately inclined. The symphysis width is not correlated with any extension of the canine process. The canines do not jut out anteriorly past the incisor alveoli. The left canine moderately deviates externally. This position could result from a slight dorsoventral compression, but no distortion marks can be seen on the bone. The symphysis dorsal plane is large, nearly horizontal, and not very curved transversely anterior to P_2–P_3. The ventral part of the symphysis is short and convex in transverse and sagittal planes. The posterior face is high, sagittally rounded, and V-shaped in dorsal view. It extends up to mid-P_3, at the level of the maximal constriction of the mandible. The cheek tooth rows diverge anteriorly. Diastemas are short between premolars. The P_1 size cannot be estimated. The corpus is slender and higher at the M_3 level. Small external mandibular foramina are placed above the P_4. Muscle insertions are reduced, except on the ventral face of the corpus. The M_3s are tilted slightly to the lingual side. The angular processes are thin, well-developed, but not hook-shaped as they are in *Hip. amphibius.* Rather, they show a weak external expansion. The ramus is short and relatively high, showing some shallow muscle imprints, except for the dorsal part of the masseteric fossa. The internal mandibular foramina are of moderate size but are dorsoventrally elongate.

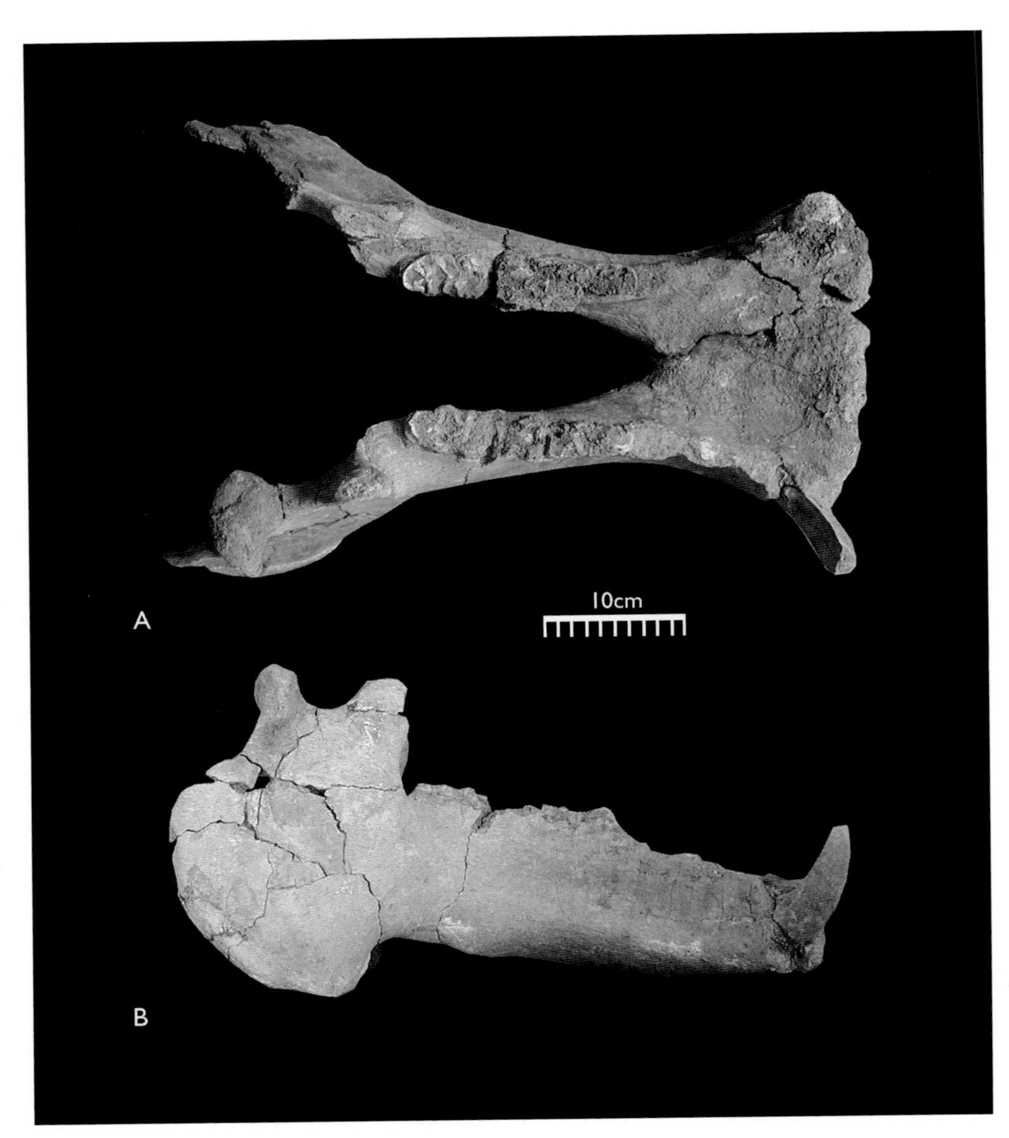

FIGURE 11.2
Mandible AME-VP-1/33, holotype of aff. *Hippopotamus dulu*.
A. Dorsal view.
B. Right lateral view.

The condyle is medially inclined and weakly developed in transverse section, being robust and set on a large neck.

The second incomplete mandible, AME-VP-1/122, is eroded at the incisor alveolar and canine processes. However, the symphysis length may be estimated. P_3–M_3 are conserved and indicate a fully adult individual. The general form of this specimen is close to that of the holotype (i.e., slender and long). The post-symphyseal constriction is, however, more reduced, and the symphysis is slender and shallower at the canine level. Six incisor roots are present. I_1 is somewhat larger than the other two, each of which is equivalent in cross section. The incisor alveolar process is also markedly projected anteriorly, but the lower part is more inclined than that of AME-VP-1/33. In spite of the superficial erosion of the canine process, there does not appear to be any significant extension externally or anteriorly. The symphysis dorsal plane is close to horizontal and more slender and curved than in the holotype. Ventrally, this symphysis is also short but exhibits a weak transverse concavity. The posterior face is somewhat shallower than in AME-VP-1/33 but is identical in other features. No diastema is seen in the cheek tooth rows. The rows begin with a reduced and mono-rooted P_1 (from its alveolus morphology). The mandibular corpora are slender and shallow, similar to those of AME-VP-1/33, and markedly converge anteriorly. The rami are relatively longer, with a less developed angular process, although they are more externally shifted. The muscle insertions are not deep. The condyle and the coronoid process are relatively and absolutely less robust than in the holotype.

AMW-VP-1/104 is a fragmentary mandible that is undistorted and is associated with a cranium. The anterior symphysis is eroded up to P_1. The mandibular corpus is relatively well-preserved but retains only the right M_2–M_3 and left M_3. The only preserved part of the rami is the left angular process. The dorsal part of the symphysis forms a deep transverse curve. It extends posteriorly up to the mid-P_3. The mandible is deeply constricted at this level. The posterior face of the symphysis is dorsoventrally thick and V-shaped in dorsal view. The mandibular corpus displays a fairly constant height and does not show any deep muscle insertions. The cheek tooth rows are anteriorly divergent at the P_2 level. The distal molars lie inclined on their lingual face. The angular process is strong, well-developed, but not hook-shaped. Compared to those in the previously described mandibles, this process exhibits a stronger external shift from the corpus axis.

The fragmentary mandibular symphysis AGG-VP-1/51 shows only the presence of six incisors and is comparable with the above-described specimens in overall size and morphology. The mandible AGG-VP-1/60 consists of two fragmentary mandibular corpora and a posterior fragment of the symphysis. The corpora bear the left P_4–M_3 and right P_4 and M_2–M_3. These teeth are all very worn. Here, also, the morphology and dimensions correspond to the description given for the other specimens.

Lower Dentition The lower incisors are unknown. The only available information comes from the alveoli on mandible AME-VP-1/122 (see preceding paragraphs). The canines (AME-VP-1/33 and AME-VP-1/122) have a kidney-shaped cross section and smooth to finely ridged enamel, without convergent ridges.

Dimensions of the lower cheek teeth are given in Table 11.1. P_1 and P_2 are unknown. The P_3 of AME-VP-1/122 is moderately mesiodistally elongated. The cingulum is mainly distal. The main cusp has one mesial crest turning lingually and two strong pustulate distal crests. It bears a high accessory cusp, somewhat distally shifted. On the P_4, the cingulum is more robust and appears on all sides, although it is very reduced lingually. The main cusp displays the same mesial and distal crests as the P_3. The accessory cusp is larger and more central on the lingual side. A second low accessory cusp occurs distally. This description applies for specimens GAW-VP-3/12 and AMW-VP-1/45 as well.

M_1 and M_2 are closely similar, except for the larger size of M_2. These teeth are wide, without cingulum on their lingual and labial sides. Some crests and conulids are observed at the transverse valleys. The entoconid is reduced, without mesial and distal lobes. The M_3 has the same features: robustness, cingulum restricted to the mesial and distal sides, and reduced entoconid. The hypoconulid is strong in most specimens; it is even divided in GAW-VP-

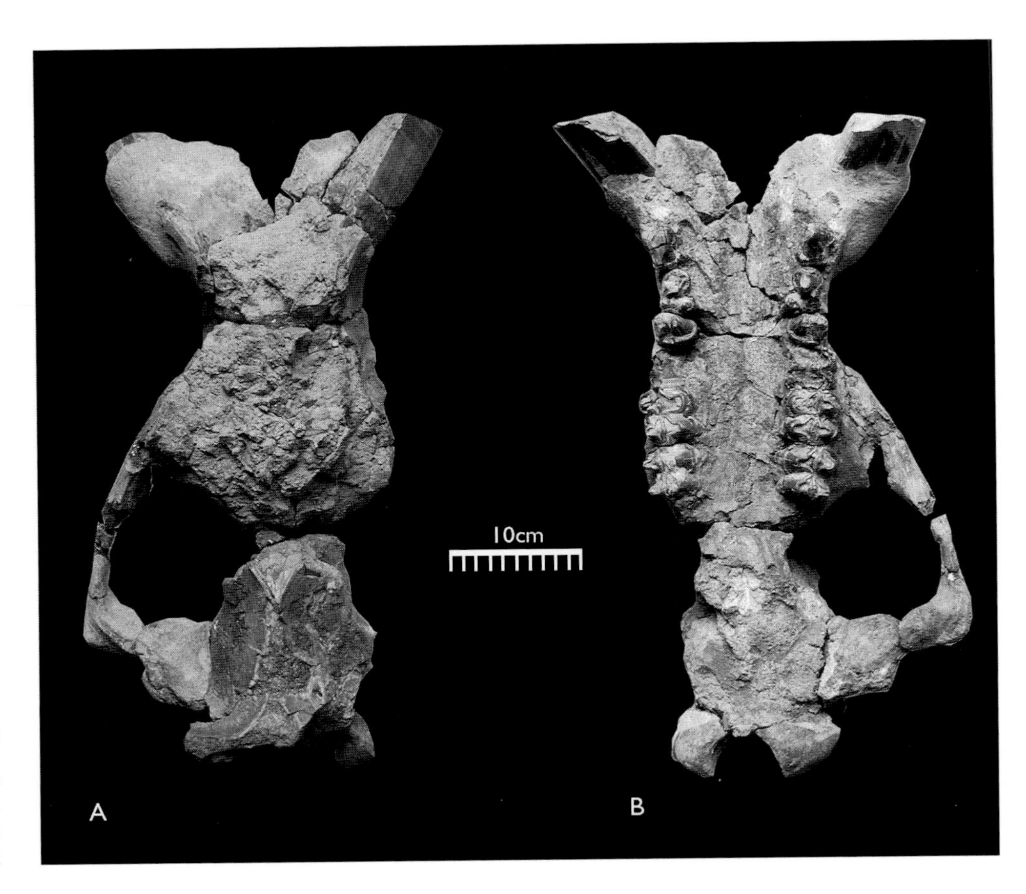

FIGURE 11.3
Cranium AMW-VP-1/104, aff. *Hippopotamus dulu.* A. Dorsal view. B. Ventral view.

3/100. An endoconulid and an ectoconulid, variable in size, often accompany the hypoconulid. These M_3 crowns are low-crowned (hypsodonty index $h = 92.5$ in KUS-VP-1/48).

Cranium The fragmentary skull AMW-VP-1/104 (Figure 11.3) is partially preserved. The premaxillae are lost. The dorsal face is destroyed, except for the anterior part of the sagittal crest. The right zygomatic arch and basicranium are broken and absent. The remaining basicranium is heavily eroded. The skull has abundant breaks and flaking. It required major consolidation prior to extraction from the well-consolidated sandstone matrix. This individual is fully adult, with a fully erupted and partially worn M^3. In ventral view, the maxilla's anterior border is V-shaped and reaches the mesial extremity of the P^2. The canines are strong, although poorly extended externally. The canine processes are robust. Between the canines, the palate shows a shallow groove that is not extended posteriorly. The premolar rows, without a strong diastema, diverge anteriorly at the P^2 level. The P^2–P^4 series is much shorter than the molar row. The anteorbital constriction is relatively long and deep. The palate is rather wide and ends abruptly behind the M^3. The perpendicular laminae of the palatine bone form a weakly open angle. The zygomatic arch starts above the mesial extremity of the M^1. It regularly and quickly deviates from the sagittal plane, without the development of any facial tubercle. The horizontal overhang that it forms above the molars bears a weak crest for the masseter insertion. Posteriorly, the zygomatic arch is considerably thinner, with a minimal thickness of 12 mm just before the glenoid process. This process is medio-externally elongate and externally thin. Its main axis is roughly perpendicular to the sagittal plane. The glenoid fossa is dominated by a weak retroglenoid eminence set on the internal side. Externally, it is limited by a long and thick prominence. Other parts of the basicranium are not preserved well enough for description except for the occipital condyles, which are particularly large and robust and are very prominent around a wide foramen magnum. In external view, a few elements may be observed with some certainty. The canine processes have a weak dorsal extension. The maxillae are depressed above the P^3–M^3 row, but less so under the root of the zygomatic arch. The zygomatic arch does not depart far from the horizontal plane. At its junction with the glenoid process it bears a strong external tubercle. The external acoustic meati are unfortunately not visible. From what is preserved, the sagittal crest seems to be elevated posteriorly. In dorsal view, the robustness of the canine process, the deep muzzle constriction, and the progressive expansion of the zygomatic arches are notable. The braincase is not globular but rather mediolaterally compressed. The temporal fossa is large and deep in relation to a deep postorbital constriction. The dorsal crest is short and probably thin. The supraoccipital tubercle base indicates a small and narrow tubercle. Finally, on the posterior side, the occipital plate is high, with a central plate area separated from the mastoid processes by an abrupt transition.

Upper Dentition Canines are the only anterior teeth to be preserved. They are robust and bilobate with a deep posterior groove, and their enamel is finely ridged.

Dimensions of the upper cheek teeth are given in Table 11.1. The P^1 is unknown, but all the other premolars are present on AMW-VP-104. The P^2 is large, rectangular in occlusal view, and medially constricted, and it bears one main cusp. A low and small cingulum encircles the whole tooth. It is distally thicker and forms a distolingual accessory cusp. The enamel is strongly wrinkled. The larger P^3 displays the same features as the P^2 in exaggerated form. The deeper medial constriction separates two unequal lobes (the distal is wider). A more robust cingulum surrounds the whole tooth. An independent accessory cusp between the main cusp and the cingulum forms a distolingual "heel." The enamel is wrinkled to pustulate. P^4 has two main cusps, equal in size. The labial cusp is four-lobed, the main lobes being mesial and distal. The lingual cusp is crescentiform. This tooth, broadened linguolabially, is surrounded by a strong continuous cingulum, thickened distally. This description applies fully to KUS-VP-1/50 and KUS-VP-2/71.

The only M^1 known (on AMW-VP-1/104) is heavily worn. Its general shape recalls that of the M^2, which is a low-crowned tooth, quadrangular in shape, and surrounded by a continuous cingulum. The protocone and the metaconule are larger than the labial cusps. There is a weak medial endoconule. The two M^2s of KUS-VP-1/50 are similar to this description. The known M^3 looks like the M^2, except for the distal width, which is narrower, and the lack of conules.

DISCUSSION With six lower incisors and poorly extended canine processes, aff. *Hip. dulu* displays a morphology reminiscent of that of other late Miocene and early Pliocene hippopotamids. Among these, *A. harvardi* Coryndon, 1977 (from Lothagam, Kenya) is the most comparable to aff. *Hip. dulu*. Most of the dimensions of aff. *Hip. dulu* are found either at or just beyond the lower end of the range of *A. harvardi* (Tables 11.1 and 11.2). The teeth are similar overall. The cranium AMW-VP-1/104 exhibits the same features as those described in Coryndon (1976, 1977) and Weston (2003) for *A. harvardi*. In both species, the bicanine width is rather small (Table 11.2), the cheek tooth diastemae are short, the thin zygomatic arches form a deep overhang above the maxillae, the braincase is transversally compressed, and the glenoid processes and occipital plate exhibit similar morphologies. However, the occipital condyles are much larger in Sagantole species than in *A. harvardi* (Table 11.2). Moreover, these species differ by the relative proportions of the mandibular symphysis and premolar row, shorter in aff. *Hip. dulu* (Table 11.2), and, by the degree of angular process eversion, nonexistent in *A. harvardi*.

The other representatives of *Archaeopotamus,* denoted "narrow-muzzled" hippopotamids after Gentry (1999a), include the late Miocene *A.* aff. *lothagamensis* from Abu Dhabi, United Arab Emirates (Gentry 1999a) and *A. lothagamensis* (Weston 2000) from the Lower Member of the Nawata Formation, Lothagam, Kenya. They have mandibular symphyses even more elongated than that of *A. harvardi,* and hence more elongated than that of aff. *Hip. dulu* (Table 11.2). They also differ from the Sagantole species by a longer premolar row (Table 11.2). Moreover, *A. lothagamensis* is smaller in size overall (Table 11.2). *Hexaprotodon crusafonti* (Aguirre 1963) from the late Miocene (MN13) of Spain was also included in the "narrow-muzzled" hippopotamids by Weston (2000). Whatever the exact affinities of these specimens are, they are demonstrably tetraprotodont (Lacomba et al. 1986) and thus do not match the anatomy of aff. *Hip. dulu*.

The proportions of the mandibular symphysis of aff. *Hip. dulu* are closer to those of *S. mingoz* Boisserie et al., 2003 from the early Pliocene of Kollé, Djurab, Chad (Brunet et al. 1998). However, the size of the Ethiopian species is slightly, but significantly, larger than that of the Chadian material (Table 11.2; MANOVA on the mesiodistal length and mesial width of the M_3: Wilks's $\lambda = 0.679$; dl (2, 21); $p = 0.017$). Moreover, the mandibular condyles and rami are larger and taller, respectively, in aff. *Hip. dulu,* which exhibits a longer premolar row (Table 11.2). These values correspond better to *S.* cf. *mingoz* from Kossom Bougoudi, Chad, which is slightly older in age than Kollé (Brunet et al. 2000), but both differ from aff. *Hip. dulu* by having a higher mandibular symphysis (Table 11.2), a shallower mandibular corpus, and, above all, a braincase that is not laterally compressed. One of the features that prompted the definition of a new genus for these forms is their peculiar cranial anatomy (Boisserie et al. 2003; Boisserie 2005).

Although poorly known, *Hex. imagunculus* and *Hex.* cf. *imagunculus* from the Mio-Pliocene of the Western Rift (Cooke and Coryndon 1970; Erdbrink and Krommenhoek 1975; Pavlakis 1990; Faure 1994) can be easily distinguished from aff. *Hip. dulu,* being much smaller relative to their respective dental measurements (Boisserie 2004).

Among the late Miocene to early Pliocene African hippopotamids, *Hex. garyam* was recently discovered at Toros-Ménalla, Chad (Boisserie et al. 2005). This species is larger (Tables 11.1 and 11.2) and characterized by a higher, more robust mandibular symphysis

TABLE 11.2 Skull (Cr) and Mandible (Md) Measurement Ranges of aff. *Hippopotamus dulu* Compared to Other Hippopotamids

	Cr1 (1)	*Cr2 (2)*	*100 × (1)/(2)*	*Cr3*	*Md1 (3)*	*Md2 (4)*	*100 × (3)/(4)*	*Md3*	*Md4*	*Md5*
aff. *Hip. dulu*	138	124	111	152	116–155	160–218	71–72	53–66	114–121	77–78
A. harvardi	133–153	105–129	114–127	74–129	170–199	188–213	81–102	58–99	127–127	77–98
A. lothagamensis					124	124	100		91–92	85
A. aff. *lothagamensis*					153	160	96	91	83	100
S. mingoz	115	98	117	111–130	105–142	167–192	63–76	67–87	76–95	65–73
S. cf. *mingoz*		142			139	188–192	74	76–97	81–116	74–83
Hex. garyam	153–154	106–143	115–144	132–151	113–180	184–259	54–79	96–121	101–130	68–90
Hex. bruneti	190	128	106		120	217–228	55	94–106	100–102	70–77
Hex. sivalensis	169–212	117	143	120–150	124–153	210–252	56–73	90–119	100–138	66–82
aff. *Hip. afarensis*	193	117	165	69–136	138–175	271–296	51–59	83–96	143	72–89
aff. *Hip. coryndoni*					121–132	227–238	56–58	55–87	95–111	77
aff. *Hip.* cf. *protamphibius*	216–231	122–137	157–190	107–155	146–147	262–273	53–56	64–72	97–108	67–70
aff. *Hip. protamphibius*	197–245	95–121	203–227	127–147	111–151	231–308	39–60	63–104	103–136	62–79
aff. *Hip. karumensis*	160–252	73–115	217–251	130–159	103–183	231–360	37–52	43–109	103–146	48–78
aff. *Hip. aethiopicus*	160–187	79–93	171–239		62–98	183–248	37–51	49–78	81–104	56–89

NOTE: Abbreviations: Cr1, width between canine alveoli; Cr2, width between mesial P^2s; Cr3, width between lateral tips of occipital condyles; Md1, sagittal length of symphysis; Md2, width between canine process tips; Md3, height of symphysis; Md4, height of ramus at M_3s; Md5, ratio ($100 \times P_2$–P_4 length)/(M_1–M_3 length). Measurements are in mm.

than that of aff. *Hip. dulu* (Table 11.2), which clearly differentiates these two species. This morphology is known for all representatives of the genus *Hexaprotodon,* as redefined by Boisserie (2005). Its type species, *Hex. sivalensis,* has a cranium that is wider at the level of the thick zygomatic arches, a wider rostrum (Table 11.2), a simpler premolar morphology, and relatively larger molars (Colbert 1935; Hooijer 1950). A more recent representative of *Hexaprotodon* was also found in the Middle Awash at Bouri and Maka (ca. 2.5 Ma). *Hexaprotodon bruneti* is characterized by the mandibular morphology just described, as well as by a much enlarged I_3, a feature not seen in aff. *Hip. dulu* (Boisserie and White 2004).

The middle and late Pliocene hippopotamids of the Turkana Basin generally exhibit more advanced characters than those seen in aff. *Hip. dulu.,* aff. *Hip. protamphibius,* aff. *Hip. aethiopicus,* and aff. *Hip. karumensis* (see Arambourg 1944; Coryndon and Coppens 1973; Coryndon 1976, 1977; Gèze 1980, 1985; Harris et al. 1988; Harris 1991a) and differ strongly from aff. *Hip. dulu* in the following characters: tetraprotodont anterior dentition; laterally extended canine processes (Table 11.2); relatively and absolutely shorter mandibular symphysis (Table 11.2); weakly anteriorly projected incisor alveolar processes; and shorter premolar rows relative to the molar row. The earlier taxon aff. *Hip.* cf. *protamphibius* (Weston 2003), including the material from Kanapoi and Allia Bay, exhibits more extended canine processes together with a proportionally shorter mandibular symphysis and premolar rows (Table 11.2). In these, it is more similar to the closely related aff. *Hip. protamphibius turkanensis* Gèze, 1985, and to aff. *Hip. protamphibius protamphibius.* It is, however, more similar to aff. *Hip. dulu* in retaining three incisors and having a more projected lower incisor alveolar process.

Two middle Pliocene species are known from Hadar, Afar Basin. The dominant species, aff. *Hip. afarensis,* is larger than aff. *Hip. dulu,* with a somewhat wider muzzle (Table 11.2) and more robust zygomatic arches. Its lower incisor alveolar process is not prominent as in aff. *Hip. dulu* and most of the late Miocene and early Pliocene hexaprotodont hippopotamids. Its incisors are markedly differentiated. The Sagantole hippotamid is more similar in size to aff. *Hip. coryndoni* (Table 11.2), but aff. *Hip. coryndoni* is closer to aff. *Hip. afarensis* in its mandibular features. However, for both species from Hadar, the relative height of the mandibular symphysis and the poorly differentiated canine processes are not different from those of aff. *Hip. dulu.*

To conclude, the initial generic attribution of aff. *Hip. dulu* to the genus *Hexaprotodon* (Boisserie 2004) was established on the basis of the following primitive features commonly used in the literature (Colbert 1935; Coryndon 1977, 1978; Gèze 1980, 1985; Harris 1991a; Harrison, 1997b; Weston 2003): a primitive deep posterior groove on the upper canines and the presence of fine, parallel ridges on the lower canine enamel not found in *Hippopotamus.* However, by the proportions of its mandible, aff. *Hip. dulu* can be clearly differentiated from the late Miocene "narrow-muzzled" *Archaoeopotamus* from eastern Africa and the Arabian Peninsula, as well as from the mostly Asian *Hexaprotodon,* which exhibits a much more robust morphology. This species does not show derived characters diagnostic of *Saotherium* or *Choeropsis.* Therefore, we choose to refer this species to the provisionary generic attribution used by Boisserie (2005), reflecting the current uncertainty on the phylogenetic position of the Plio-Pleistocene Afar and Turkana hippopotamids. Given the above comparisons, a future revision of these eastern African hippopotamids should constitute a good opportunity to examine carefully the possibility that aff. *Hip. dulu* is a forerunner of the later hippopotamids from Hadar.

Conclusions

Until now, Afar hippopotamids were known only from the middle Pliocene sites of the Hadar area. With the Middle Awash discoveries, the fossil record is substantially enlarged in the Afar Basin (see also Boisserie and White 2004). Specimens of aff. *Hip. dulu* deepen our knowledge of the Hippopotamidae in this basin, and more generally in Ethiopia, as early as the onset of the Pliocene. Specimens of aff. *Hip. dulu* have some patent resemblances to hexaprotodont forms that are generally viewed as primitive, chiefly *A. harvardi*. Nevertheless, its mandibular morphology may indicate a closer relationship with the middle Pliocene hippopotamids from the Afar. For the time being, without more complete cranial remains, it is difficult to confirm its actual phylogenetic relationships. More material from the Mio-Pliocene levels of the Middle Awash Valley is required to decipher the early history of the Hippopotamidae in the Afar basin. In this regard, the Adu-Asa remains, although too fragmentary to allow any accurate identification, indicate the potential of these late Miocene sediments. This was recently confirmed by recent discoveries at locality STD-VP-1, including a complete skull yet to be prepared and analyzed.

The form aff. *Hip. dulu* differs from contemporaneous hippopotamids found in the Turkana Basin (Weston 2003), the Chad Basin (Boisserie et al. 2003), and the Western Rift (Faure 1994). It supports the hypothesis of basin endemism as a major feature of hippopotamid evolution in the African Pliocene. This idea was initially proposed by Gèze (1985) and also evoked by Denys et al. (1987). Such provincialism could well be related to hippopotamid ecology, and hence to the connections of hydrographic networks, or the lack thereof. This interpretation has broad consequences in terms of African paleobiogeography and paleogeography.

12

Giraffidae

YOHANNES HAILE-SELASSIE

The family Giraffidae is represented by two modern genera (*Okapia* and *Giraffa*) in Africa. The family had greater diversity during the Miocene of northern Africa (Churcher 1978). Giraffids first appear in Africa south of the Sahara sometime toward the end of the middle Miocene, at which time *Climacoceras* and *Palaeotragus* are represented at sites such as Fort Ternan and Ngorora (Hamilton 1978). During the late Miocene, however, giraffines, which were also abundant in western Eurasia, dominated the giraffid community of eastern Africa. The earliest genus in the subfamily Palaeotraginae, *Palaeotragus,* is represented in the late Miocene by *P. germaini* in northern Africa (Arambourg 1959) and eastern Africa (Churcher 1979) and by *P. primaevus* in eastern Africa (Hamilton 1978). *Palaeotragus primaevus* is a small giraffid from the middle Miocene that probably lacked ossicones (Churcher 1970). A number of dental and postcranial remains of the species have been recovered from the Ngorora, Chorora, Namurungule, and Ngeringerowa Formations (Sickenberg and Schönfeld 1975; Benefit and Pickford 1986; Nakaya 1994). However, it has not been documented from sediments younger than 7 Ma.

Palaeotragus germaini appears for the first time during the late Miocene. The earliest African record of this species comes from Beni Mellal (Morocco), Oued el Hammam and Bou Hanifia (Algeria), and Bled ed Douarah (Tunisia) (Arambourg 1958, 1959; Lavocat 1961; Robinson and Black 1969). The youngest *P. germaini* thus far known is reported from the Nawata Formation at Lothagam (Patterson et al. 1970; Churcher 1979; Leakey et al. 1996; Harris 2003b; but see Geraads 1986a). A second species of *Palaeotragus* may be represented at Lothagam (Harris 2003b).

Samotherium is a palaeotragine with two paired ossicones, larger in size than *Palaeotragus* species. It is mainly known from northern African localities of middle and early late Miocene age (Stromer 1907; Arambourg 1959). Its record in the rest of Africa is poor. The palaeotragine remains from Sahabi were tentatively assigned to *Samotherium* pending the recovery of more diagnostic material (Harris 1987b). Churcher (1970) assigned ossicones and an atlas from Fort Ternan to *Samotherium africanum,* based on their morphological similarities with those described by Bohlin (1926). Aguirre and Leakey (1974) assigned a single molar and three postcranial elements from the Ngorora Formation to cf. *Samotherium*. The available fossil record of eastern Africa shows that *Samotherium* was limited to

Fort Ternan and Ngorora times in Africa south of the Sahara. However, the sample is too small to establish reliable First Appearance (FA) and Last Appearance (LA) estimates for the genus (if the genus was indeed present in Africa; see Hamilton 1978 for details).

Giraffokeryx is the other giraffid genus reported from the African late Miocene. This genus is only described from Ngorora (Aguirre and Leakey 1974). Hamilton (1978), however, argued that specimens assigned to *Giraffokeryx* by Aguirre and Leakey (1974) were probably misidentified remains of *P. primaevus*. The four-ossiconed *Giraffokeryx* was reported from Nakali by Aguirre and Leakey (1974) and from Lothagam (Churcher 1978). However, Harris (2003b) reported that *Giraffokeryx* was not present in the late Miocene of Lothagam.

The earliest record of the subfamily Giraffinae in Africa is limited to isolated specimens from the Lukeino Formation and possibly the Mpesida Beds of Kenya. These specimens are tentatively assigned to *Giraffa* sp. The earliest *Giraffa jumae* is known from Langebaanweg (Harris 1976b). The genus *Giraffa* is absent from the Nawata Formation of Lothagam (Harris 2003b). The Plio-Pleistocene record of this subfamily is very rich.

Sivatherium maurusium, subfamily Sivatheriinae, is the largest giraffid from Africa (Churcher 1978). It is ubiquitous in Plio-Pleistocene deposits. Its earliest occurrence is in the Adu-Asa Formation in the Middle Awash (Kalb et al. 1982c). Harris (1976b) assigned sivathere remains from the "E" Quarry of Langebaanweg to a new species, *S. hendeyi*. However, since the name was based on a single ossicone, this species remains a probable variant of *S. maurusium* unless additional specimens show further differences between the two groups (Churcher 1978). The Lothagam sivathere is assigned to cf. *Sivatherium* sp. (Harris 2003b). Giraffid remains have also been reported from the Nkondo Formation, although they are identified only at the family level (Geraads 1994a).

Giraffinae

Palaeotragus Gaudry, 1861

Palaeotragus sp.

DESCRIPTION The Middle Awash giraffid teeth referred here to *Palaeotragus* (Figure 12.1) are larger than those of the known palaeotragines from Lothagam (Table 12.1). However, they are much smaller than those of *Sivatherium hendeyi* (Harris 1976b) from Langebaanweg, and hence they are excluded from the genus *Sivatherium*. ALA-VP-2/70 is tentatively identified as a M^2. It lacks most of the protocone and metacone. The preserved portion clearly shows its affinity with *Palaeotragus* species, particularly in the shape of the paracone's labial wall, which has strong labial ribs. The parastyle is well-developed and separated from the rest of the paracone by a deep vertical groove. The mesial edge of the premetaconule crista has enamel folding that has not been documented in upper molars of either *P. primaevus* or *P. germaini*. It lacks an endostyle, which is a character more common in M^2s of *Palaeotragus primaevus* than in *P. germaini*. ALA-VP-2/92 is more damaged but possibly belongs to the same individual as ALA-VP-2/70, based on its size and preservation. This specimen also lacks the endostyle.

The specimens from locality ASK-VP-3 probably belong to a single individual. ASK-VP-3/45 is a right P_4, previously identified (Haile-Selassie 2001a) as a right P_3, lacking the

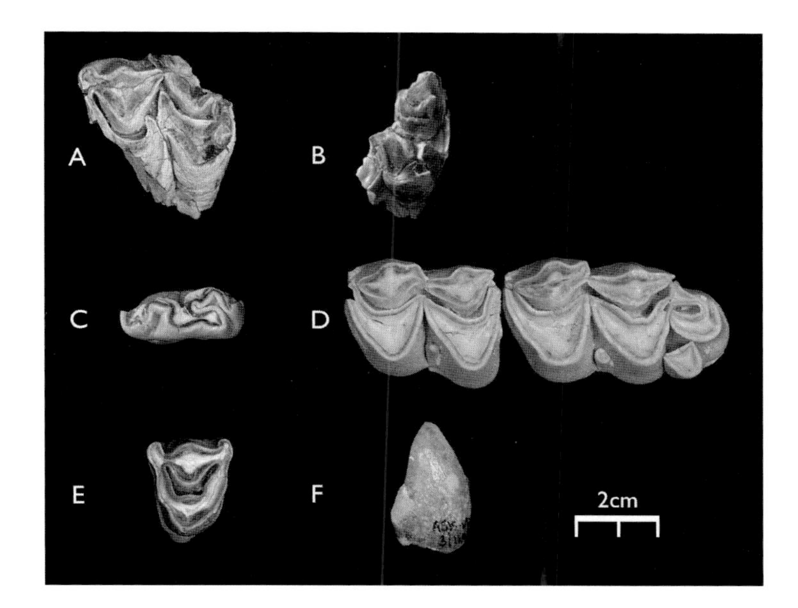

FIGURE 12.1
Palaeotragus sp. specimens from the Adu-Asa Formation (Asa Koma Member), Middle Awash. Occlusal views. A. ALA-VP-2/70, left M^2 fragment. B. ALA-VP-2/92, upper molar fragment. C. ASK-VP-3/45, right P_4. D. ASK-VP-3/46, left M_{2-3}. E. ASK-VP-3/111, P^4. F. ASK-VP-3/114, left I^2.

buccal half of the crown. The preserved portion most closely resembles P_4s of *Palaeotragus primaevus,* a species largely known from the middle Miocene (Churcher 1978). The metaconid and entoconid are fused. The more distal entostylid is separated from the entoconid, with a small enamel island forming between them. The anterior crest and the paraconid are fused by wear. The P^4 (ASK-VP-3/111) also more closely resembles those of *P. primaevus* than *P. germaini* in having a well-developed parastyle and strong labial ribs and being buccolingually elongate.

ASK-VP-3/46 consists of the M_2 and M_3 of a single individual. These molars are low-crowned and have fine striations on their enamel surfaces. The metaconid and entoconid on the M_2 show modest lingual ribbing and highly convex buccal faces. The metaconid has a well-developed cingulum on its mesiolingual corner. The metastylid is positioned more lingually relative to the mesial edge of the pre-entocristid. The pre-protocristid extends lingually to the level of the mesial edge of the pre-metacristid. The post-hypocristid terminates buccally to the post-entocristid and does not close the median valley distally. The pre-hypocristid is separate from the pre-entocristid, and the mesial and distal median valleys are connected at the center of the crown. The M_3 trigonid morphology is similar to that of the M_2. It has an apically worn, low ectostylid between the protoconid and hypoconid. The talonid cusp (hypoconulid) is large and connected with the entoconid via an entoconulid that is as worn as the talonid cusp. The mesial and distal median valleys are connected. What is unique about this molar is the presence of a clearly separated, apically worn extra cusp on the anterobuccal side of the entoconid posterior to the hypoconid (Figure 12.1D). This extra cusp has not been documented in any other described giraffid.

Middle Awash specimens assigned to *Palaeotragus* sp. show a mosaic of dental characters seen in numerous other *Palaeotragus* species. They also show new dental characters, such as the presence of an extra cusp between the hypoconid and the hypoconulid. This might be an indication that the specimens represent yet unnamed *Palaeotragus* species.

TABLE 12.1 Dental Measurements of *Palaeotragus* sp.

Specimen	*I*		*P3*		*P4*		*M2*		*M3*	
	MD	BL	MD	BL	MD	BL	MD	BL	MD	BL
Uppers										
ALA-VP-2/70							32.5	(30.7)		
ALA-VP-2/92										
ASK-VP-3/111					17.8	21.4				
ASK-VP-3/114	17.6	11.4								
Lowers										
ASK-VP-3/45 (L)					24.8	nm				
ASK-VP-3/46 (L)							33.6	25.2 (TR1) 26.7 (TR2)	47.2	26.6 (TR1) 25.5 (TR2)
DID-VP-1/117 (L)			27.0	12.5						

NOTE: Numbers in parentheses are estimates. TR1 is the buccolingual dimension of the anterior lobe, and TR2 is the buccolingual dimension of the second lobe. Abbreviations: MD, mesiodistal; BL, buccolingual; U, upper; L, lower. All measurements are in mm; nm = the measurement was not possible.

However, the small sample size does not allow their assignment to a new species, and therefore they are tentatively assigned here to *Palaeotragus* sp.

Giraffa Brunnich, 1771

Giraffa sp.

DESCRIPTION The assignment of these two isolated teeth and a metacarpal fragment (Figures 12.2 and 12.3) to the genus *Giraffa* is tentative, based only on the observation that they do not belong to a sivathere or palaeotragine species. Their size and morphology are best described by the characters of the genus *Giraffa*. AME-VP-1/34 is a left M_3 with only the posterior lobe and the talonid preserved. The left M^1 (AME-VP-1/128) measures 27 mm mesiodistally and 28.2 mm buccolingually on the posterior lobe. These measurements best fit within the range of *Giraffa stillei* from younger deposits (Harris 1991b: 125).

The metacarpal (ALA-VP-2/67) is tentatively assigned to *Giraffa,* largely because its size is comparable to the size of *Giraffa jumae* and the measurements fall within the range of the Langebaanweg giraffines excluding *Sivatherium*. It is larger than KNM-LT 22865, a metacarpal from Lothagam assigned to *Palaeotragus germaini* (Harris 2003b). It is comparable to, but slightly larger than, KNM-LT 26270, a specimen also from Lothagam and assigned to *Giraffa* sp.

Sivatheriinae

Sivatherium Falconer and Cautley, 1832

Sivatherium sp.

DESCRIPTION KUS-VP-1/110 (Figure 12.4) is a complete left femur of a large sivathere, with a length of 520 mm. The anteroposterior and transverse dimensions of the proximal

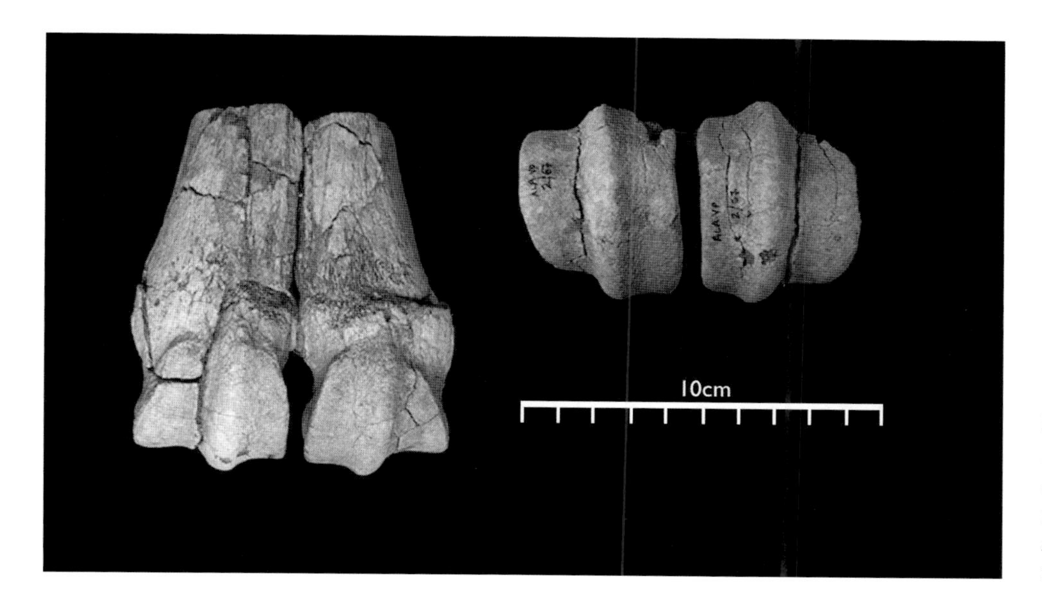

FIGURE 12.2

Giraffa sp. specimen from the Adu-Asa Formation (Asa Koma Member): ALA-VP-2/67, distal metacarpal.

end are 95 mm and 170 mm, respectively. The anteroposterior and transverse dimensions of the distal end are 180 mm and 150 mm, respectively. It is longer than any known *Sivatherium maurusium* femur from the Plio-Pleistocene deposits of Africa (Harris 1991b: 135). Dimensions of the proximal and distal ends, however, slightly overlap with both giraffines and sivatherines. However, there seems to be a morphological difference in the proximal femur between the two subfamilies.

Sivathere and giraffe femora are reported to be virtually identical distally (Harris 1991b). However, the differences or similarities of the proximal ends of both groups have not been documented. Specimen KUS-VP-1/110 documents the proximal femoral morphology of sivatheres, which differs from that of giraffes. The greater trochanter in sivatheres is anteroposteriorly elongated such that the anteroposterior dimension of the femoral head is much smaller than the same dimension in the greater trochanter. Moreover, the greater trochanter is oriented obliquely relative to the main axis of the femoral head. In giraffes, the anteroposterior dimension of the femoral head is the same as the anteroposterior dimension of the greater trochanter. In sivatheres, the lesser trochanter is proximodistally elongated compared to that of giraffes.

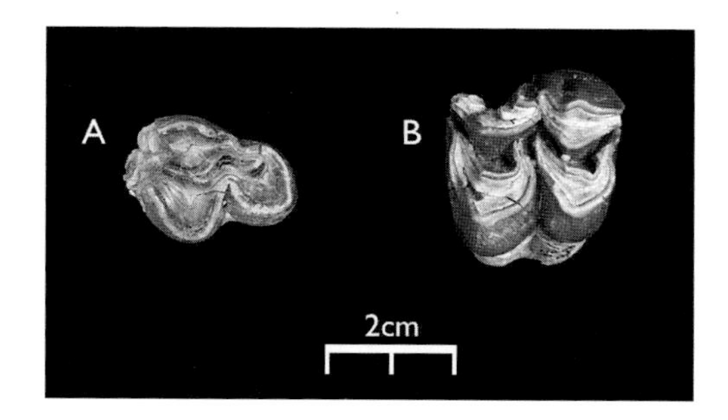

FIGURE 12.3

Giraffa sp. specimens from the Sagantole Formation (Kuseralee Member). Occlusal views: A. AME-VP-1/34, left M_3. B. AME-VP-1/128, M^1.

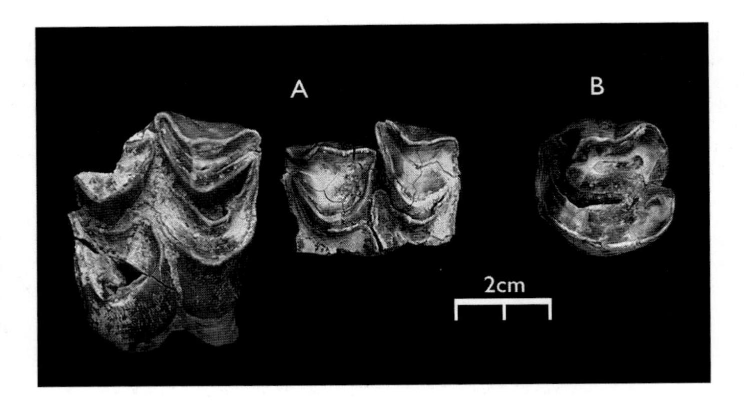

FIGURE 12.4
Sivatherium sp. dental specimens from the Sagantole Formation of the Middle Awash. Occlusal views: A. AME-VP-1/141b-c, left M^1 and M^2. B. KUS-VP-1/1, left P^3.

The teeth assigned to *Sivatherium* sp. (Figure 12.5) are morphologically similar to those of sivatheres described from other localities such as Langebaanweg (Harris 1976b). The dental measurements of the Middle Awash sivathere also fall within the range seen in *Sivatherium maurusium* and *Sivatherium hendeyi*. The incisor measures 21.8 mm mesiodistally and 13 mm buccolingually. The M^1 of AMW-VP-1/141, which is lingually damaged, has an estimated mesiodistal length of 37.7 mm. The M^2, on the other hand, lacks the mesiobuccal corner of the crown and is larger in size, measuring a minimum of 40.3 mm mesiodistally. The buccolingual dimension of this tooth on the posterior lobe is 44 mm. KUS-VP-1/1 (left P^3) measures 28.3 mm mesiodistally and 31.2 mm buccolingually. These dimensions are outside the range seen for *Giraffa* species.

Discussion

The family Giraffidae was divided into two subfamilies (Palaeotraginae and Giraffinae) by Simpson (1945) and into three subfamilies (Palaeotraginae, Sivatheriinae, and Giraffinae) by Churcher (1978). Subsequent work by Geraads (1986a) on the phylogeny and taxonomy of the family subsumes *Palaeotragus* and sivatheres into the subfamily Giraffinae, which includes the extant *Okapia* and *Giraffa*. *Palaeotragus* was first documented from late Miocene localities in northern Africa (Arambourg 1959). Its presence in eastern Africa was documented by Churcher (1979) based on a single specimen from Lothagam. Additional specimens were collected from the Nawata Formation of Lothagam, where it is better known than in the Middle Awash, possibly represented by more than one species (Harris 2003b). *Palaeotragus* appears to be among the rarest taxa in the Middle Awash late Miocene. The genus *Giraffa* is not described from the older Nawata Formation of Lothagam, although *Giraffa stillei* is documented from the younger Apak and Kaiyumung Members (Harris 2003b). If the *Giraffa* designation is correct, then the specimens from the Asa Koma and Kuseralee Members in the Middle Awash may represent the earliest record of the genus in Africa and provide support for the first appearance of the genus in Africa via the Arabian-African land bridge to Ethiopia (Mitchell and Skinner 2003). *Sivatherium* is rare in both the Middle Awash and Lothagam late Miocene deposits. However, it is the most common giraffid at Langebaanweg (Harris 1974) and other younger eastern

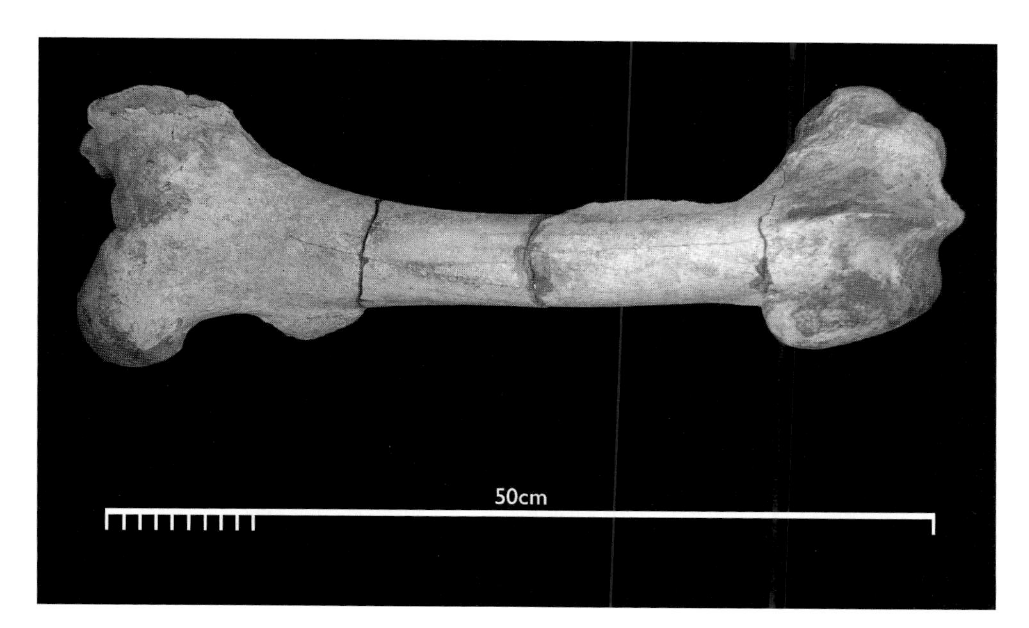

FIGURE 12.5
Sivatherium sp. specimen from the Sagantole Foramation of the Middle Awash. Anterior view of KUS-VP-1/110, left femur.

African sites from the early Pliocene to the middle Pleistocene (Harris 1976b; Churcher 1978; Harris and Cerling, 1998).

Conclusion

Giraffid remains are rare elements in the Middle Awash late Miocene deposits. However, there are possibly three species known from isolated teeth and limited postcrania. A *Palaeotragus* species, better known from other contemporary sites such as Lothagam, is known from the Adu-Asa Formation of the Middle Awash by a limited number of dental remains that appear to belong to a larger species than those represented at Lothagam. *Sivatherium* is present in the Kuseralee Member of the Sagantole Formation of the Middle Awash late Miocene deposits but absent from the earlier Adu-Asa Formation. Its absence in the latter formation is possibly due to sampling bias, since it is known from other contemporaneous deposits in eastern Africa. The genera *Palaeotragus* and *Sivatherium* have also been documented from older deposits in Eurasia. Their presence in the Middle Awash latest Miocene fossil record, and other contemporaneous eastern African deposits, indicates a wider distribution of these two genera in eastern Africa toward the end of the Miocene. The genus *Sivatherium* persists into the early Pliocene of eastern Africa, whereas *Palaeotragus* has not been documented after the late Miocene. Some giraffid remains from the older Asa Koma Member and the younger Kuseralee Member are tentatively referred to the genus *Giraffa* and may represent the earliest record of the genus in eastern Africa. However, discovery of more complete specimens is needed to confirm the first appearance of *Giraffa* in the Middle Awash and whether the palaeotragine dental remains from the Asa Koma Member of the Adu-Asa Formation represent a new species in the Middle Awash late Miocene.

13

Equidae

RAYMOND L. BERNOR AND YOHANNES HAILE-SELASSIE

Hipparionine horses originated in North America ca. 16 Ma and first entered the Old World between 11.1 and 10.7 Ma (Bernor et al. 2004a). The Eurasian late Miocene record is extensive and includes several multispecies superspecific groups. Members of the major clades, including *Hippotherium, Hipparion (sensu stricto),* and *Cremohipparion,* became extinct at the end of the late Miocene (Bernor et al. 1996c). The *"Sivalhippus"* Complex is first recorded in the late Miocene of Indo-Pakistan and eastern Africa and later had ranges that extended across Eurasia and Africa (Bernor and Lipscomb 1995). The Chinese taxa *Plesiohipparion* and *Proboscidipparion* appear to have extended their ranges into Europe and southwestern Asia in the early Pliocene (Bernor et al. 1996c), whereas *Eurygnathohippus* was a vicariant lineage restricted to eastern and southern Africa in the late Miocene (Bernor and Lipscomb 1995; Bernor and Harris 2003; Bernor et al. 2005b) and is known later in the Plio-Pleistocene of northern Africa (Bernor and Armour-Chelu 1999). The only exception known to these observations is the occurrence of an unspecified member of the *"Sivalhippus"* Complex in the latest Miocene northern African locality of Sahabi (Libya; Bernor and Scott 2003). The geographic restriction of species of *Eurygnathohippus* to Africa is thus far an observation and remains a curious aspect of Old World hipparion evolution.

There have been few published studies of Ethiopian hipparions. Eisenmann (1976) provided a brief report on Hadar hipparions, naming a new species, *"Hipparion" afarense.* Bernor et al. (2004a) described a small assemblage of primitive hipparion from Chorora (10.7–10.0 Ma), which they attributed to *"Cormohipparion"* sp. Haile-Selassie (2001a) reported on the Middle Awash hipparion specimens occurring at localities dated between 5.7 and 5.2 Ma. Our study here includes Jara-Borkana (JAB), an older locality dated to 6.0+ Ma located along the western margin of the Middle Awash. The radiometric ages of all these localities are given in Table 13.1.

Bernor et al. (2005b) have provided a detailed morphometric study of Ethiopian metapodial IIIs and 1st phalanges III ranging from 6 to 2.95 Ma, including those from localities listed in Table 13.1, comparing them to those from critical localities in Eurasia and Kenya. The authors proposed that there were two principal lineages during this interval: the elongate and slender *E. feibeli–E. hasumense* lineage and the shorter, more robust *E. turkanense* lineage. The *E. feibeli–E. hasumense* lineage is currently interpreted to have had a chrono-

TABLE 13.1 A List of Middle Awash Late Miocene Collection Areas and Their Radiometric Dates

Locality	*Abbreviation*	*Age*	*Symbol*
Middle Awash Collection Areas			
Jara-Borkana	JAB	6.0+	a
Adu Dora	ADD	5.70	b
Alayla	ALA	5.70	c
Asa Koma	ASK	5.70	d
Bikir Mali Koma	BIK	5.70	e
Gaysale	GAS	5.70	f
Saitune Dora	STD	5.60	g
Bilta	BIL	5.54	h
Kabanawa	KWA	5.40	i
Amba East	AME	5.20	j
Amba West	AMW	5.20	k
Digiba Dora	DID	5.54	l
Agera Gawtu	AGG	5.00	m
Gawto	GAW	4.90	n
Worku Hassan	WKH	4.90	o
Sagantolee	SAG	4.90	p
Kuseralee	KUS	5.20	q
Other Collection Areas			
Lothagam		7.5–5.0	L
Sahabi		6.7–5.3	S

NOTE: Lothagam and Sahabi are included for comparison. Ages are in Ma.

logic range of 6.0+ to 2.95 Ma and may include additional species steps within that range. The *E. turkanense* lineage is best known from the Lower Nawata of the Lothagam sequence and appears to be represented in the Middle Awash sequence at localities younger than those investigated here. *Eurygnathohippus turkanense* has its last documented occurrence at 4.0 Ma, also in the Middle Awash sequence. Postcranial information suggests that the *E. feibeli–E. hasumense* lineage had a postcranial anatomy suitable for open-country running, while the *E. turkanense* lineage had lesser cursorial capabilities.

In this work we will undertake the morphological description and statistical analysis of the Middle Awash hipparions ranging in age from 6.0+ to 5.2 Ma (Renne et al. 1999; WoldeGabriel et al. 2001). We employ the same methodology here as that recently published by Bernor and Harris (2003) and Bernor et al. (2005b). We describe the discrete morphology of skull fragments, mandibular remains, and maxillary and mandibular dentitions by stratigraphic level and locality for each of the recognized taxa. We use the Höwenegg population standard for postcranial metrics and the Eppelsheim population standard for dental metrics in calculating bivariate plots (Bernor et al. 2003; Bernor and Harris 2003). We follow Eisenmann (1995), Bernor and Scott (2003), Bernor et al. (2004a), and Bernor et al. (2005b) in using $\log_{10}$ ratio diagrams for additional analyses of metacarpal III, metatarsal III, and 1st phalanx III. All reported measurements conform to standards established by Eisenmann et al. (1988) and Bernor et al. (1997) and are in mm, rounded to the nearest tenth mm. Anatomical conventions follow Bernor et al. (1997).

Equinae

Eurygnathohippus van Hoepen, 1930

Eurygnathohippus feibeli Bernor and Harris, 2003

DIAGNOSIS (MODIFIED FROM BERNOR AND HARRIS 2003) All African hipparions of the genus *Eurygnathohippus* are united by the synapomorphy of having ectostylids on the permanent cheek teeth. Eurasian and North American hipparions do not have this character, except very rarely in the Dinotheriensande (Germany) and Gaiselberg (Austria) early Vallesian hipparions of the *Hippotherium* clade. *Eurygnathohippus feibeli* is a small hipparionine equid with gracile limbs. The metacarpal III is elongate and slender with its midshaft depth being substantially greater than the width. The anterior 1st phalanx III is elongate with a narrow midshaft width. Maxillary cheek teeth bear a thin parastyle and mesostyle and are labiolingually moderately curved to straight, with a maximum crown height of 50 to 60 mm. There is mostly moderate complexity of the pre- and postfossettes; the posterior wall of the postfossette is mostly separated from the posterior wall of the tooth; the pli caballin is mostly single or a poorly defined double; the hypoglyph varies with wear; the protocone tends to be elongate and compressed; and the protoconal spur is usually absent but may appear as a small, vestigial structure. The premolar and molar protocones are placed lingually to the hypocone. Mandibular cheek teeth have premolar metaconids and metastylids that are mostly rounded; the molar metaconids and metastylids are mostly rounded to elongate; the metastylid spur is absent; the ectoflexid does not separate the metaconid from the metastylid in the premolars and variably separates the metaconid–metastylid junction in the molars; the pli caballinid is mostly absent. When expressed, the protostylid is most often presented as a posteriorly directed, open loop; the ectostylids are variably expressed and, when present, are diminutive structures that do not rise high on the labial side of the tooth. The premolar and molar linguaflexid is a shallow V-shape, the preflexid and postflexid enamel margins are generally simple, and the protoconid enamel band is rounded.

DESCRIPTION The M_3 (JAB-VP-1/3) is in midwear, with a crown height of 33.9 mm. The metaconid has a straight lingual border and is somewhat pointed distally. The metastylid has an elongate-square shape. The entoconid is elongate and has an elongate isthmus continuous with the hypoconulid. The preflexid and postflexid exhibit slight complexity. There is a small, but distinct pli ectoflexid, which very commonly accompanies ectostylids on eastern African *Eurygnathohippus* species. No ectostylid is apparent on the occlusal surface, but the labial surface preserves the wall of an ectostylid rising from the base as high as 17.4 mm on the side of the crown. As with the Lothagam Nawata Formation *Eurygnathohippus* species, the ectostylid evidently did not routinely rise high on the crown. This specimen is best referred to *E. feibeli.*

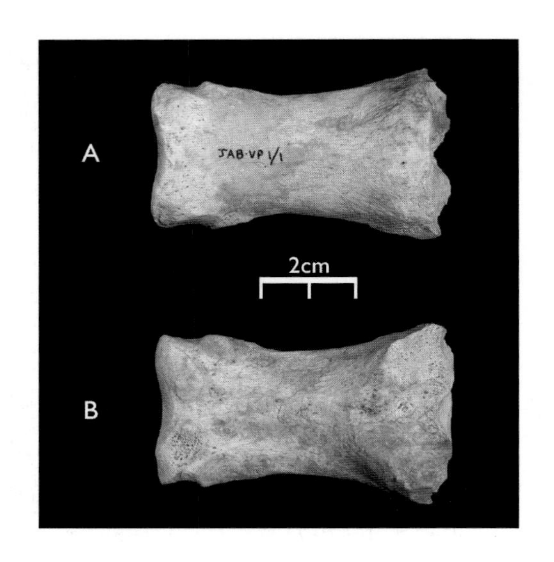

FIGURE 13.1
JAB-VP-1/1, 1st phalanx III. A. Cranial view. B. Caudal view.

Bernor et al. (2005b), and we further below, have undertaken a detailed morphometric analysis of JAB-VP-1/1 (Figure 13.1) by comparing it to a large Ethiopian and other African and Eurasian hipparions and found that it closely compares to the type specimen of *Eurygnathohippus feibeli*. This specimen is relatively elongate and slender and strongly distinct from the late Miocene eastern African robust-limbed form *E. turkanense*.

Eurygnathohippus aff. *feibeli*

DESCRIPTION Most specimens that we refer to *E.* aff. *feibeli* are of single elements. We believe that it is important to describe this material by stratigraphic horizon and, within a stratigraphic horizon, by locality, to preserve the "population unit" character of this sample. Future discoveries from the 5.7–5.2 Ma interval may shed new light that will enable us to taxonomically partition this sample further, but presently we prefer to recognize all of these specimens under this nominal taxon.

5.7 Ma Horizon: Adu Dora, Alayla, Asa Koma, Bikir Mali Koma, Gaysale

The Adu Dora equid fauna includes specimens from localities ADD-VP-1, ADD-VP-3, and ADD-VP-4. ADD-VP-1/14 includes the left P_3–M_3 and the right P_3, M_1–M_3. ADD-VP-1 has also yielded a left P_2 (ADD-VP-1/3), a left P^2 (ADD-VP-1/4), and a left dp^2 (ADD-VP-1/5). ADD-VP-3/3 is a right P^3, and ADD-VP-4/1 is a left astragalus.

The ADD-VP-1/14 mandibular dentition is small and in a middle stage of wear. Figure 13.2A provides a labial view of the left P_3–M_3. The left P_3's (Figure 13.2B) occlusal surface exhibits rounded metaconid and metastylid with a very short isthmus; the ectoflexid does not separate the metaconid and metastylid; preflexid and postflexid are labiolingually restricted and have simple margins; the protostylid forms an open loop that extends labiodistally; the protoconid band is rounded; the ectostylid is a distinct, albeit small rounded structure that rises 20.3 mm on the labial surface. Left P_4 morphology is like that of the P_3. The right M_1 (Figure 13.2C) is similar to the P_3 but differs in its slightly squared metastylid, in having a broad linguaflexid that separates the metaconid

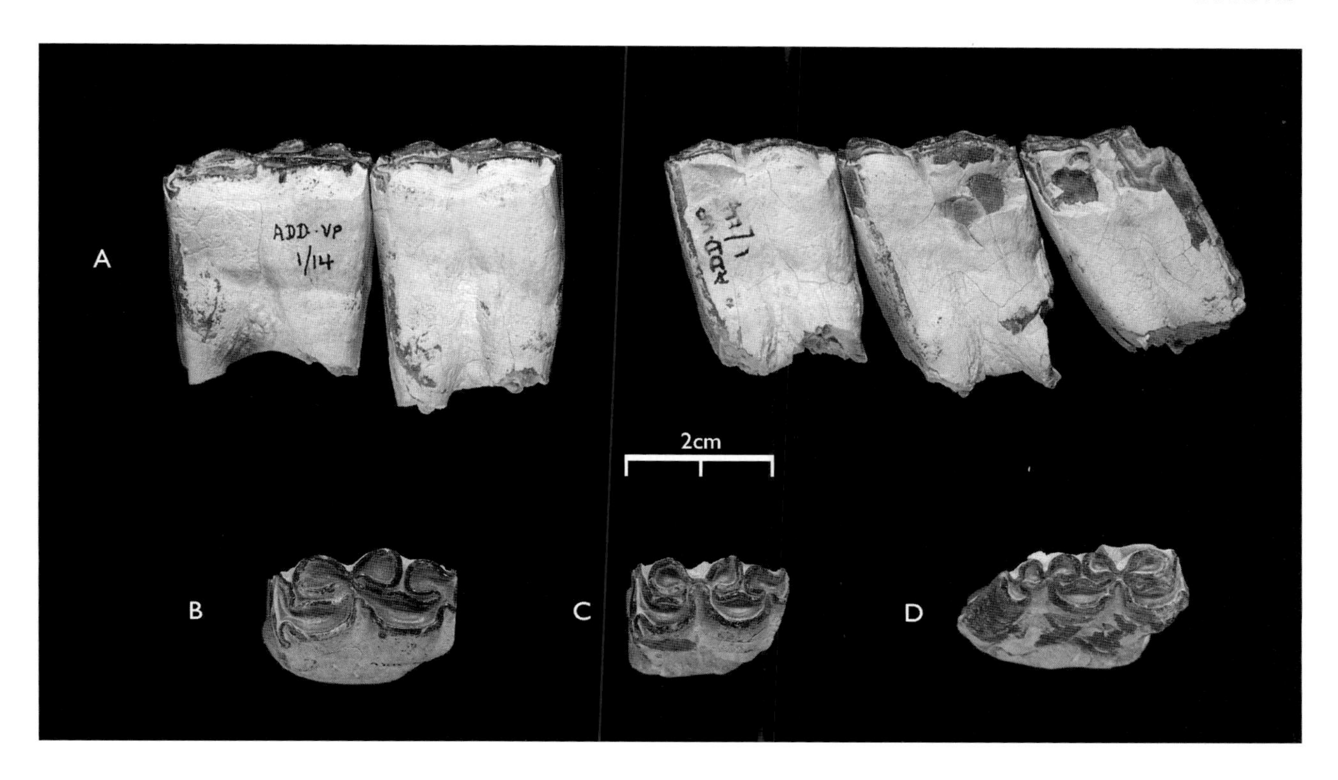

FIGURE 13.2

ADD-VP-1/14. A. Left cheek tooth series (labial view). B–D. Occlusal views of teeth in the tooth series: left P_3 (B); right M_1 (C); right M_3 (D).

and metastylid, and an ectoflexid that strongly separates the metaconid and metastylid. The labial margin is missing, so it is unknown whether this tooth also had an ectostylid. The left M_2 is similar to the P_3 and M_1 in all aspects, as is the left M_3, which has a broken talonid. The dentition of the right side of ADD-VP-1/14 is like those on the left side and accords with it being referred to the same individual. The right M_3 preserves the talonid, and, as in the JAB-VP-1/3 specimen, it is elongate with a distinctly rounded isthmus, with rounded enamel margins, connecting the entoconid to the hypoconulid (Figure 13.2D). There is an additional, distinct enamel ring between the entoconid and hypoconulid.

ADD-VP-1/3 is a very worn left P_2. The most remarkable feature of this specimen is its very elongate paraconid and protoconid enamel band. The metaconid is elongate. The metastylid has a rounded/squared outline. The preflexid and postflexid are simple and labiolingually restricted. The ectostylid is absent.

ADD-VP-1/4 is a very worn left P^2. The anterostyle is elongate. The mesostyle is squared. The preflexid has complex margins. The postflexid is very worn, lacking complicated margins. The protocone is so worn as to be connected to the protoloph; no pli caballin is preserved.

ADD-VP-1/5 (Figure 13.3) is a left dp^2 in an early/middle stage of wear. The occlusal surface preserves the following features: an elongate anterostyle; squared mesostyle; complex margins for the pre- and postfossettes; elongate and double pli caballins; small protocone, rounded with a flat lingual border; hypoglyph deeply incised and encircling the hypocone.

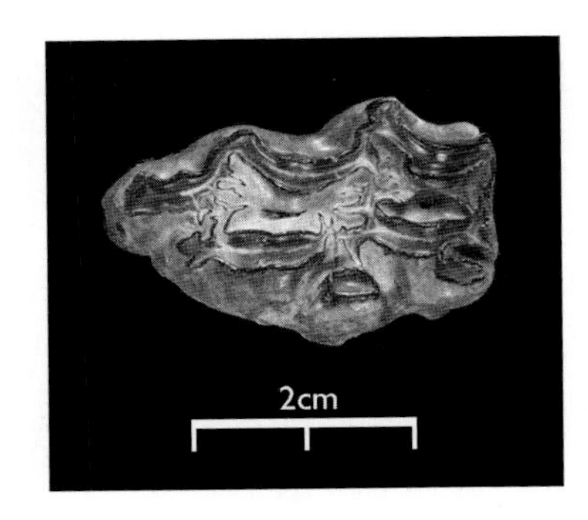

FIGURE 13.3
ADD-VP-1/5, left dp^2, occlusal view.

The Alayla equids include maxillary and mandibular dental and postcranial material. The maxillary material is sparse, including a total of two P^4s, three M^1s, and three M^2s (Table 13.2). ALA-VP-2/14 (right P^4; Figure 13.4A) is in a very late stage of wear, with a crown height of only 11.2 mm. The occlusal surface preserves remarkable ornamentation for this late wear stage: Pre- and postfossettes are very complex, especially on the facing margins; pli caballin is multiple, mesostyle is thin and rounded; protocone is elongate with a flat lingual wall. ALA-VP-2/3 (left M^1) is in a very early stage of wear, with a crown height of 52.2 mm, nearly the maximum for this individual. The entire labial wall, from base to occlusal level, has been broken away. However, the following salient morphological features are maintained: the pre- and postfossette do not yet express plis; the pli caballin is double; the protocone has an elongate lenticular shape, and the hypoglyph is deeply incised, completely encircling the hypocone. ALA-VP-2/4 (Figure 13.4B) is a right M^2 in a relatively early stage of wear and has the following salient features: The mesostyle is slender; the pre- and postflexids are only beginning to come into wear and show negligible ornamentation; the pli caballin is little worn also and presents as a single structure; the protocone is an elongate lenticular structure; the hypoglyph is very deep and completely encircles the hypocone. The measured crown height of this specimen is 47.9 mm, and the very early stage of wear of this specimen suggests that maximum crown height would not have far exceeded 50 mm. ALA-VP-2/8 is a newly erupted, virtually unworn specimen whose lower half has broken away. It contributes little to our understanding here.

ALA-VP-2/39 is a fragmentary right mandible containing very worn dp_2–dp_4; there is no useful morphological information contained in this specimen. The Alayla mandibular cheek tooth assemblage includes a left P_2, a right and left P_3, a right P_4, a right and left M_1, and a right and left M_2. The P_2 (ALA-VP-2/157) is in a later stage of wear and unremarkable. However, the right P_3, ALA-VP-2/155 (Figure 13.5A), is in a middle stage of wear and preserves the following morphological details: metaconid and metastylid are rounded; preflexid and postflexid are labiolingually restricted and have simple margins; ectoflexid does not invade metaconid and metastylid, the ectostylid is a distinct, elongate lenticular structure that rises 23.6 mm on the labial side of the crown.

ALA-VP-2/158 comprises a right P_4 (Figure 13.5B) and a left M_2 (Figure 13.5C), respectively, found in close association and plausibly of a single individual. Both specimens are in very early wear, with the M_2 marginally having the highest crown height, 46.2 mm; the maximum unworn crown height would not have far exceeded 50 mm.

TABLE 13.2 Measurements of Late Miocene Middle Awash Hipparions

Specimen	*Taxon*	*Element*	*M1*	*M2*	*M3*	*M4*	*M5*	*M6*	*M7*	*M8*	*M9*	*M10*	*M11*	*M12*	*M13*	*M14*
JAB-VP-1/1	*Eu. feibeli*	1Ph3	65.2	61.0	26.5	38.0	29.5	33.2	32.7	17.1	18.1	48.5	47.0	13.4	12.1	
JAB-VP-1/13	*Eu. feibeli*	M_3	25.2	25.7	12.0	8.8	9.3	8.1	11.1	11.0	8.0	33.9	2.2		17.4	
ADD-VP-4/1	*Eu.* aff. *feibeli*	Ast	56.2	53.8	25.3		46.1	32.9	40.8							
ADD-VP-1/14B	*Eu.* aff. *feibeli*	M_1	22.2	21.3	13.2	5.0	7.7	12.3	13.4	9.9	10.3	27.7				
ADD-VP-1/14G	*Eu.* aff. *feibeli*	M_1	22.2	21.0	13.3	6.2	7.7	12.3	13.2	11.2	10.9	24.4				
ADD-VP-1/14C	*Eu.* aff. *feibeli*	M_2	23.4	22.4	13.1	5.6	8.3	11.6	12.6	10.1		28.0				
ADD-VP-1/14H	*Eu.* aff. *feibeli*	M_2	23.7	21.4	13.2	5.0	8.3	11.9	12.9	10.9	10.6	28.7				
ADD-VP-1/14D	*Eu.* aff. *feibeli*	M_3	26.1	26.6	11.8	6.3	7.8		11.3	9.4	9.2	30.1				
ADD-VP-1/14I	*Eu.* aff. *feibeli*	M_3			12.4	6.3	10.2	9.5	11.9	9.1		29.2				
ADD-VP-1/3	*Eu.* aff. *feibeli*	P_2	29.3	27.6	12.0	8.0	12.0	10.4	13.5	10.3	12.3	15.0				
ADD-VP-1/14A	*Eu.* aff. *feibeli*	P_3	26.2	26.1	15.2	9.1	12.6	13.4	13.9	13.5	13.4	17.2	1.4	1.3	20.1	
ADD-VP-1/14E	*Eu.* aff. *feibeli*	P_3	26.5	25.3	15.4	9.3	13.0	13.7	14.8	12.4	13.0	18.5	1.3	1.5	16.3	
ADD-VP-3/3	*Eu.* aff. *feibeli*	P_3	26.0	24.2	15.1	9.0	12.8	15.8	13.9	12.4	12.1	32.2				
ADD-VP-1/14F	*Eu.* aff. *feibeli*	P_4	23.9	22.9	13.9	8.0	11.9	12.8	13.9	12.6	12.4	32.2	2.0	1.4	25.2	
ADD-VP-1/5	*Eu.* aff. *feibeli*	dp^2	36.0	34.4	19.8	20.7	16.3	3.0	5.0	3.0	1.0	5.6	3.3			
ADD-VP-1/4	*Eu.* aff. *feibeli*	P^2	29.6	29.2	22.6	20.7	14.3	4.0	5.0	1.0		8.3	6.4			
ALA-VP-2/97	*Eu.* aff. *feibeli*	1Ph3														
ALA-VP-2/110	*Eu.* aff. *feibeli*	Ast	57.3	43.9	27.9			29.3	44.4							
ALA-VP-2/68	*Eu.* aff. *feibeli*	Ast			25.5	52.7	45.8									
ALA-VP-2/39	*Eu.* aff. *feibeli*	Mand											69.4	52.8		
ALA-VP-2/26	*Eu.* aff. *feibeli*	MT III										41.6	38.3	34.8	27.9	30.6
ALA-VP-2/26	*Eu.* aff. *feibeli*	P1Ph3	68.7	55.8	31.6	42.1		38.2	35.9	20.3	15.8	50.0	52.2	21.2	19.0	
ALA-VP-2/26	*Eu.* aff. *feibeli*	P2Ph3	45.1	36.4	32.5	41.2	31.0	35.4								
ALA-VP-2/26	*Eu.* aff. *feibeli*	P3Ph3	62.9	61.9	60.0	38.5	25.0									
ALA-VP-2/39	*Eu.* aff. *feibeli*	dp_2	Broken													
ALA-VP-2/39	*Eu.* aff. *feibeli*	dp_3	26.7		8.0	6.8							9.1	1.9		
ALA-VP-2/39	*Eu.* aff. *feibeli*	dp_4	23.7		16.9	7.4	9.7				10.2		5.2	2.6		
ALA-VP-2/94	Eu. aff. *feibeli*	dp_4	27.3	25.2	15.7	8.0	8.4	13.5	15.1	11.4	9.9	8.5	5.0	2.0	8.2	
ALA-VP-2/156	*Eu.* aff. *feibeli*	M_1	22.7	20.7	13.8	6.8	8.9	13.4	13.9	12.3	11.9	30.4	4.2		14.4	
ALA-VP-2/91	*Eu.* aff. *feibeli*	M_1	25.3	20.1	11.4	6.9	10.3	8.0		7.6	7.3	52.3	4.5		36.0	
ALA-VP-2/154	*Eu.* aff. *feibeli*	M_2	27.5	22.2	12.9	7.7	8.5	9.9	13.1	9.7	8.6	49.9				
ALA-VP-2/88	*Eu.* aff. *feibeli*	M_2										48.4				
ALA-VP-2/157	*Eu.* aff. *feibeli*	P_2	28.3	26.0	11.7	8.5	10.6	14.1	12.8	10.8	12.1	24.8				
ALA-VP-2/155	*Eu.* aff. *feibeli*	P_3	25.1	22.8	14.7	9.2	12.7	13.6	14.0	12.6	12.8	23.6	4.2	1.6	24.8	
ALA-VP-2/158A	*Eu.* aff. *feibeli*	P_4	25.8	21.9	13.7	8.4	12.7	11.2	12.6	10.0	9.9	44.8	1.9		19.6	
ALA-VP-2/158B	*Eu.* aff. *feibeli*	M_2	26.5		12.9	7.9	11.5	11.4	13.1			46.2	2.8		25.5	
ALA-VP-2/139C	*Eu.* aff. *feibeli*	Lower tooth	Juv	Frag												
ALA-VP-2/96	*Eu.* aff. *feibeli*	Lower tooth									13.4					

TABLE 13.2 *(continued)*

Specimen	*Taxon*	*Element*	*M1*	*M2*	*M3*	*M4*	*M5*	*M6*	*M7*	*M8*	*M9*	*M10*	*M11*	*M12*	*M13*	*M14*
ALA-VP-2/95	*Eu.* aff. *feibeli*	Upper tooth?					6.0	6.0	3.0		7.1	4.8				
ALA-VP-2/22	*Eu.* aff. *feibeli*	I^1	14.9	10.1	9.2	10.5	25.0									
ALA-VP-2/139B	*Eu.* aff. *feibeli*	M^1	Juv	Frag												
ALA-VP-2/3	*Eu.* aff. *feibeli*	M^1	26.8	20.0		17.7	52.2	1.0	1.0	0.0	6.6	4.1				
ALA-VP-2/97B	*Eu.* aff. *feibeli*	M^1	Frag													
ALA-VP-2/13	*Eu.* aff. *feibeli*	M^2					54.7									
ALA-VP-2/4	*Eu.* aff. *feibeli*	M^2	26.7	19.9	21.6	22.1	47.9	2.0	2.0	1.0	0.0	6.4				
ALA-VP-2/8	*Eu.* aff. *feibeli*	M^2	24.4	21.4	20.7	21.1	36.9					8.6				
ALA-VP-2/14	*Eu.* aff. *feibeli*	P^4	7.1	25.6	20.7	23.0	11.2	1.0	6.0	4.0	3.0	7.8	4.3			
ALA-VP-2/139A	*Eu.* aff. *feibeli*	P^4	Juv	Frag												
ALA-VP-2/97A	*Eu.* aff. *feibeli*	P^4	Frag													
ALA-VP-2/97C	*Eu.* aff. *feibeli*	Tooth	Frag													
ASK-VP-3/18	*Eu.* aff. *feibeli*	Ast	56.3	56.4	26.1	54.1	42.7	27.9								
ASK-VP-3/9	*Eu.* aff. *feibeli*	Ast	56.9	56.8	26.3	56.9	43.9	32.9	44.4							
ASK-VP-3/110	*Eu.* aff. *feibeli*	MC II			12.4	21.4										
ASK-VP-1/1	*Eu.* aff. *feibeli*	MT III										41.9	35.2	33.7	27.3	30.2
ASK-VP-3/503	*Eu.* aff. *feibeli*	M_1	25.9	12.3	8.0	9.3	11.6		9.6	10.3						
ASK-VP-3/7	*Eu.* aff. *feibeli*	M_1	21.3	19.4	12.9	5.9	7.6	12.9	12.2	11.5	10.5	23.8	1.2	1.6	26.6	
ASK-VP-3/24	*Eu.* aff. *feibeli*	M_2	24.9	22.7	13.1	6.7	10.2	9.6	12.1	8.2	7.9	Ecto	Side	Crown		
ASK-VP-3/44	*Eu.* aff. *feibeli*	M_2	23.5	23.2	11.9	6.2	8.4	11.1	12.3	10.7	9.6	45.5				
ASK-VP-3/502	*Eu.* aff. *feibeli*	M_2	28.1	21.6	13.5	7.8	11.5	11.9	12.5	9.7	10.0	54.4				
ASK-VP-3/23	*Eu.* aff. *feibeli*	M^2	22.7	21.8	21.7	23.1	44.9	6.0	6.0	3.0	2.0	6.7	3.7			
ASK-VP-1/9	*Eu.* aff. *feibeli*	P^4	29.3	29.9			38.0	10.0	7.0			6.1	4.0			
BIK-VP-1/3	*Eu.* aff. *feibeli*	1Ph3	66.6	58.9	31.0	41.7	34.0	36.4			14.2	45.6	51.6	16.2	16.4	
BIK-VP-3/13	*Eu.* aff. *feibeli*	Ast	61.2	62.0	27.6	58.8	45.1	33.8	49.4							
BIK-VP-1/10	*Eu.* aff. *feibeli*	M_1	21.9	20.1	13.5	7.0	10.2	12.0	12.3	10.5	10.5	33.0				
BIK-VP-3/2	*Eu.* aff. *feibeli*	M_1	27.3	22.6	10.6	7.5	8.5	11.2	12.9	7.9	7.9		4.7		34.0	
BIK-VIP-1/11	*Eu.* aff. *feibeli*	P_2	27.9	27.1	11.1	8.9	12.1	13.9		10.3	12.4	31.9				
BIK-VP-1/11	*Eu.* aff. *feibeli*	P_3	26.2	22.5	15.6	9.4	13.2	16.4		12.8	13.3	39.5	2.2	1.6		
BIK-VP-1/12	*Eu.* aff. *feibeli*	M_1	25.6	24.5	15.7	8.5	12.4	15.1	14.8	12.4	12.3	41.2	1.8		31.5	
BIK-VP-1/11	*Eu.* aff. *feibeli*	P_4	24.2	22.1	15.5	7.1	13.3	15.9		12.2	12.7	40.6				
BIK-VP-14/1	*Eu.* aff. *feibeli*	P_4	26.8	26.4	15.8	9.3	13.9	15.4	15.1	13.2	14.7	43.5	2.3	1.7	35.4	
GAS-VP-1/1	*Eu.* aff. *feibeli*	P^4	23.5	22.4	23.8	21.8	41.9	2.0	4.0	2.0	0.0	7.2	4.0			
STD-VP-2/846	*Eu.* aff. *feibeli*	Ast	Frag													
STD-VP-1/45	*Eu.* aff. *feibeli*	M^2	24.9	21.3	15.0	6.6	9.3	13.1	13.5	10.3	10.0	48.5				
STD-VP-9-16	*Eu.* aff. *feibeli*	M^3	26.7		12.2	6.7	11.1	12.2	12.5	11.3	9.9	32.2				
STD-VP-2/38	*Eu.* aff. *feibeli*	P^2	28.0	27.9	16.9	8.3	11.8	12.3	13.9	11.1	12.6	13.5	2.7	1.9	15.7	
STD-VP-2/40	*Eu.* aff. *feibeli*	P^2			7.3	11.5				10.0	11.2	14.6				
STD-VP-2/848	*Eu.* aff. *feibeli*	Upper tooth				13.5	13.8	13.9			10.7	12.0	3.3	2.4	10.5	

STD-VP-1/12	*Eu.* aff. *feibeli*	M^1	22.3	19.3	18.4	19.6	42.4	3.0	3.0	1.0	3.0	6.4	3.0			
STD-VP-1/34	*Eu.* aff. *feibeli*	P^2				37.2	5.0	6.0	6.0	2.0						
STD-VP-2/39	*Eu.* aff. *feibeli*	P^3	23.1	21.0	22.6	24.7	27.9	3.0	8.0	7.0	0.0	7.4	4.7			
STD-VP-2/37	*Eu.* aff. *feibeli*	P^4	21.4	21.0	19.2	20.7	28.7	4.0	6.0	3.0	1.0	8.6	3.9			
STD-VP-2/847	*Eu.* aff. *feibeli*	P^4	26.5	23.2	20.7	23.9	49.6	1.0	6.0	1.0	1.0	9.4	3.8			
BIL-VP-1/1	*Eu.* aff. *feibeli*	Mand		95.7	77.4		151.1		47.8					46.4	74.1	41.9
BIL-VP-1/1	*Eu.* aff. *feibeli*	M_1	21.4	20.0	13.1	7.1	8.5	13.2	13.1	10.3	9.9	31.2				
BIL-VP-1/1	*Eu.* aff. *feibeli*	M_2	22.2	21.6	13.3	7.5	8.4	14.1	13.6	11.7	11.1	35.5				
BIL-VP-1/1	*Eu.* aff. *feibeli*	M_3	25.1	25.8	11.7	7.1	7.9	12.8	11.4	10.1	8.0	39.9	3.3	2.2		
BIL-VP-1/1	*Eu.* aff. *feibeli*	P_2	27.2	29.4	10.7	9.5	12.9	13.1		10.4	11.9	24.5				
BIL-VP-1/1	*Eu.* aff. *feibeli*	P_3	23.6	21.2	16.2	9.1	13.0	15.6		12.8	13.4	27.1	2.0	2.4	24.9	
BIL-VP-1/1	*Eu.* aff. *feibeli*	P_4	25.6	21.2	15.4	8.6	12.1	18.4	15.6	12.8	12.4	31.3	2.2	1.6	32.0	
DID-VP-1/82	*Eu.* aff. *feibeli*	a1ph3	70.4	64.9	30.3	40.9	31.5	34.8	35.0	21.5	16.8	49.7	49.2	14.7	16.1	
DID-VP-1/81	*Eu.* aff. *feibeli*	Ast					33.2									
DID-VP-1/116	*Eu.* aff. *feibeli*	M_1	26.9	23.3	14.0	6.7	9.5	11.4	12.5	9.8	9.8	48.1	2.8		34.0	
DID-VP-1/27	*Eu.* aff. *feibeli*	P_2														
DID-VP-1/4	*Eu.* aff. *feibeli*	P_4	24.3	23.2	13.6	8.3	12.2	13.5	13.5	12.0	12.1	27.5				
KWA-VP-1/2	*Eu.* aff. *feibeli*	P_4	23.9	22.3	14.1	7.8	11.8	16.1	15.9	15.5	12.8	20.2	4.9	2.5	19.3	
AME-VP-1/62	*Eu.* aff. *feibeli*	Ast	55.6	55.5	25.5	56.7	42.7	30.8	39.2							
AME-VP-1/66C	*Eu.* aff. *feibeli*	Frags.														
AME-VP-1/100	*Eu.* aff. *feibeli*	M_1	22.1	21.7	12.7	7.8	8.2		11.8	9.7	9.1	33.2				
AME-VP-1/19	*Eu.* aff. *feibeli*	M_1	26.0		15.4	7.3	9.9	13.6	13.4	11.8		47.1				
AME-VP-1/43D	*Eu.* aff. *feibeli*	M_1	24.4	22.6	14.0	7.4	9.1	10.7	13.6	9.6	10.4	41.3				
AME-VP-1/109	*Eu.* aff. *feibeli*	M_2	22.0	22.7	13.5	5.5	7.9		13.7	11.7	11.3	33.2				
AME-VP-1/110	*Eu.* aff. *feibeli*	M_3					7.7				9.0					
AME-VP-1/130	*Eu.* aff. *feibeli*	M_3	25.3	26.2	10.4				11.9	7.7	6.7	48.7				
AME-VP-1/2	*Eu.* aff. *feibeli*	M_3	24.3		10.3	6.6	6.9	10.3	10.4	8.0	6.8	28.9				
AME-VP-1/25	*Eu.* aff. *feibeli*	M_3	25.9	30.9	13.6	7.9	8.4	9.5	10.9	8.8	8.7	40.9				
AME-VP-1/43E	*Eu.* aff. *feibeli*	M_3	23.0	27.0	11.5	7.5	8.4	9.7		8.9	7.4	43.8				
AME-VP-1/57	*Eu.* aff. *feibeli*	M_3	27.3		12.3	6.7	7.8	12.5		11.9	10.4					
AME-VP-1/66B	*Eu.* aff. *feibeli*	M_3	26.5		12.3	8.4	9.4	13.4		9.1	8.3	50.4				
AME-VP-1/116	*Eu.* aff. *feibeli*	P_2	28.8	26.0	11.9	8.7	13.2	11.7	14.3	11.4	12.3	33.1				
AME-VP-1/120	*Eu.* aff. *feibeli*	P_2	28.1	25.1	11.6	6.9	12.5	12.7	13.2	11.1	11.7	35.5				
AME-VP-1/44	*Eu.* aff. *feibeli*	P_2	27.8		11.6	9.6	11.7	14.1	13.6	11.7	12.8	17.9				
AME-VP-1/103	*Eu.* aff. *feibeli*	P_4	26.0		16.4	9.2	13.0	16.4	13.5	12.5	12.6	33.1				
AME-VP-1/40	*Eu.* aff. *feibeli*	P_4	26.3	24.0	15.5	9.7	13.0	15.7	14.0	13.3	13.9	42.3				
AME-VP-1/41	*Eu.* aff. *feibeli*	P_4	23.4	22.9	13.7	7.9	13.0	15.5	15.5	14.3	13.9	22.6				
AME-VP-1/43C	*Eu.* aff. *feibeli*	P_4	25.0	23.2	14.7	8.4	11.5	13.3	13.7	12.6	12.3	38.7				
AME-VP-1/66A	*Eu.* aff. *feibeli*	P_4/M_1	25.0		14.6	8.2	12.1	16.5		13.0	12.4					
AME-VP-1/3	*Eu.* aff. *feibeli*	M^1	25.8	20.7	21.5	22.6	48.6	2.0	5.0	3.0	3.0	7.6	3.8			
AME-VP-1/43B	*Eu.* aff. *feibeli*	M^1	24.3	22.0	22.1	24.2	39.5	2.0		5.0		5.4	3.6			
AME-VP-1/86	*Eu.* aff. *feibeli*	P^2			22.0	21.5	28.4		3.0	2.0	1.0	6.4	3.8			

TABLE 13.2 *(continued)*

Specimen	*Taxon*	*Element*	*M1*	*M2*	*M3*	*M4*	*M5*	*M6*	*M7*	*M8*	*M9*	*M10*	*M11*	*M12*	*M13*	*M14*
AME-VP-1/43A	*Eu.* aff. *feibeli*	P^4	24.9	23.6	27.2	23.5	39.5	4.0	6.0	2.0	1.0	6.1	3.8			
AMW-VP-1/15	*Eu.* aff. *feibeli*	MC III	235.3	228.2	29.4	23.3	42.9	29.6	38.0	11.3	9.9	39.6	37.8	29.3	25.6	28.3
AMW-VP-1/105	*Eu.* aff. *feibeli*	M_1	24.1	20.4	12.0	7.1	9.4		14.0	8.6	8.3	48.9	2.9		25.0	
AMW-VP-1/117B	*Eu.* aff. *feibeli*	M_1			12.8							32.5				
AMW-VP-1/30	*Eu.* aff. *feibeli*	M_1	23.3	20.0	13.0	5.8	9.4	12.8	12.3		11.6	50.0				
AMW-VP-1/98	*Eu.* aff. *feibeli*	M_1	26.0	21.2	13.4	8.2	10.7	15.1	14.7	10.1	9.5	57.8				
AMW-VP-1/117C	*Eu.* aff. *feibeli*	M_3					9.3				8.7	33.9				
AMW-VP-1/47	*Eu.* aff. *feibeli*	M_3	26.2		9.9				12.5	7.3	6.9	47.6				
AMW-VP-1/11	*Eu.* aff. *feibeli*	P_2	27.7	25.9	10.4	10.1	12.0	12.9	13.5	11.6	13.6	30.0				
AMW-VP-1/117A	*Eu.* aff. *feibeli*	P_2	26.6	25.3	9.5	7.6	11.1	10.5	11.9	8.8	11.3	27.9				
AMW-VP-1/16	*Eu.* aff. *feibeli*	P_4	24.3	22.0	13.6	7.6	11.5	16.1	14.3	12.4	11.5	41.2				
AMW-VP-1/31	*Eu.* aff. *feibeli*	P_4	25.2	22.3	16.3	8.3	11.9	15.0	14.1	13.4	12.4	37.5				
AMW-VP-1/86	*Eu.* aff. *feibeli*	P_4	25.7	24.4	13.8	8.2	14.1	14.5	14.2	12.9	12.9	30.3	3.7	1.4	23.5	
AMW-VP-1/44	*Eu.* aff. *feibeli*	di^1	16.2	13.3	8.3	8.2	10.4									
AMW-VP-1/44	*Eu.* aff. *feibeli*	di^1	16.3	13.8	7.3	8.0	10.5									
AMW-VP-1/44	*Eu.* aff. *feibeli*	di^2	15.8	11.5	6.5	7.0	8.9									
AMW-VP-1/44	*Eu.* aff. *feibeli*	di^2	15.5	13.2	6.5	7.1	9.0									
AMW-VP-1/44	*Eu.* aff. *feibeli*	di^3	14.9	11.3	5.5	5.6	9.4									
AMW-VP-1/44	*Eu.* aff. *feibeli*	di^3	14.9	10.8	5.1	5.9	6.6									
AMW-VP-1/44	*Eu.* aff. *feibeli*	dp^3	25.9	24.0	20.0		7.7	1.0	7.0	2.0	8.0	8.1	6.0			
AMW-VP-1/44	*Eu.* aff. *feibeli*	dp^4	27.7	24.4	21.3	22.5	13.1	1.0	8.0	1.0	2.0	7.9	4.8			
AMW-VP-1/44	*Eu.* aff. *feibeli*	dp^4	27.2	25.8	21.1	22.1	15.0	1.0	8.0	3.0	1.0	7.9	4.8			
AMW-VP-1/44	*Eu.* aff. *feibeli*	I^1		16.6				28.7								
AMW-VP-1/44	*Eu.* aff. *feibeli*	I^1		18.1	12.9	8.9	11.6	32.4								
AMW-VP-1/20	*Eu.* aff. *feibeli*	M^1	23.9	22.7	22.3	24.1	40.2	4.0	6.0	4.0	1.0	7.1	3.8			
AMW-VP-1/44	*Eu.* aff. *feibeli*	M^1	27.5	28.0	20.6		36.2					7.8	3.8			
AMW-VP-1/44	*Eu.* aff. *feibeli*	M^1	27.7	21.2	19.9	23.0	64.1	1.0	3.0	1.0	0.0	7.6	4.0			
AMW-VP-1/9	*Eu.* aff. *feibeli*	M^1	22.1	22.3	19.4	20.9	29.5	3.0	8.0	4.0	1.0	7.5	4.1			
AMW-VP-1/101	*Eu.* aff. *feibeli*	M^2	24.5	18.7	19.3	18.6	57.4	3.0	5.0	2.0		5.4	3.1			
AMW-VP-1/44	*Eu.* aff. *feibeli*	M^2					61.0									
AMW-VP-1/44	*Eu.* aff. *feibeli*	M^2	25.2	20.8	18.8	22.2	60.1					8.3				
AMW-VP-1/22	*Eu.* aff. *feibeli*	M^3	21.5	20.1	20.5	19.4	19.0	1.0	6.0	2.0	1.0	9.7	4.5			
AMW-VP-1/44	*Eu.* aff. *feibeli*	M^3	Frag	Unerup												
AMW-VP-1/53	*Eu.* aff. *feibeli*	M^3	23.0	21.6	20.2	23.2	13.8	0.0	7.0	1.0	0.0	9.1	4.9			
AMW-VP-1/129	*Eu.* aff. *feibeli*	P^2	30.9	30.3	22.6	22.5	37.8	4.0	4.0	5.0	1.0	6.5	3.9			
AMW-VP-1/2	*Eu.* aff. *feibeli*	P^3	26.8	23.9	22.1	21.5	43.3	4.0	5.0	3.0	1.0	9.9	3.9			
AMW-VP-1/23	*Eu.* aff. *feibeli*	P^3	24.0	22.7	22.7	23.0	31.4	4.0	6.0	4.0	1.0	6.2	4.0			
AMW-VP-1/101	*Eu.* aff. *feibeli*	P^4					54.4									
AMW-VP-1/19	*Eu.* aff. *feibeli*	P^4	26.8	22.3	21.6	21.5	50.5	2.0	3.0	3.0	1.0	5.7	3.4			

AMW-VP-1/52	*Eu.* aff. *feibeli*	P^4	27.3	23.1			55.7	3.0	3.0	1.0	0.0		
AMW-VP-1/62	*Eu.* aff. *feibeli*	P^4	27.4	26.4	20.4	19.5	54.7					8.8	
KUS-VP-1/4A	*Eu.* aff. *feibeli*	P^3	25.7	20.4	23.0	23.2	47.5	4.0	5.0	3.0	1.0	8.9	3.4
KUS-VP-1/4B	*Eu.* aff. *feibeli*	P^4	24.1	20.2	20.1	22.5	51.0		7.0	3.0	1.0	7.6	2.8
KUS-VP-1/4C	*Eu.* aff. *feibeli*	M^1	24.5	22.2	22.1	22.8	44.4	5.0	7.0	3.0		8.3	
KUS-VP-1/5	*Eu.* aff. *feibeli*	M^1	22.9	20.8	21.2	21.7	48.4	3.0	5.0	3.0	2.0	6.6	3.5
KUS-VP-1/23	*Eu.* aff. *feibeli*	P_4	28.7	24.9	12.8	10.6	13.5	14.9	13.2	12.2	12.7	48.4	
KUS-VP-1/27	*Eu.* aff. *feibeli*	M^3					40.5						
KUS-VP-1/29	*Eu.* aff. *feibeli*	Lower tooth	Frag										
KUS-VP-1/49	*Eu.* aff. *feibeli*	$P^{3 \text{ or } 4}$	27.7	25.0	21.7	24.0	11.5	5.0	7.0	5.0	3.0	8.5	4.8
KUS-VP-1/51	*Eu.* aff. *feibeli*	Ast	58.1	57.9	27.4	57.9	46.5	32.3	43.6				
KUS-VP-1/95	*Eu.* aff. *feibeli*	Ast	62.0	62.5	26.7	60.4	47.9	33.4	48.2				

NOTE: All measurements in mm. Abbreviations for elements other than teeth: Ast = astragalus, Mand = mandible, MC = metacarpal, MT = metatarsal, 1ph3 = 1st phalanx III, 2ph3 = 2nd phalanx III, 3ph3 = 3rd phalanx III, a = anterior, p = posterior. See text for explanation of measurements M1–M14 for the various elements.

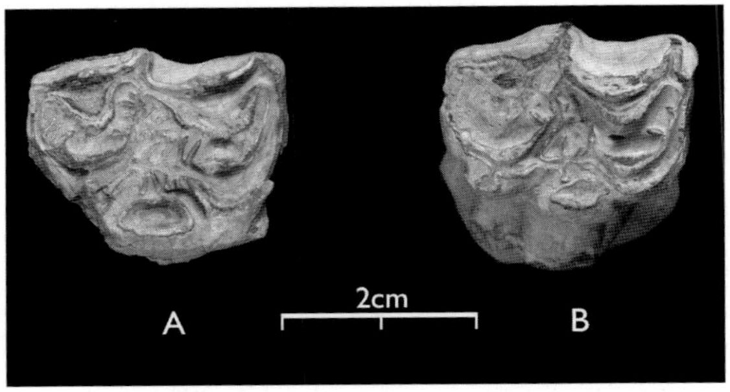

FIGURE 13.4

Occlusal views of maxillary teeth. A. ALA-VP-2/14, right P^4. B. ALA-VP-2-4, right M^2.

Both of these cheek teeth have small ectostylids rising up one-half or less of the total crown height. In both, preflexid and postflexid morphology is poorly developed, and the metaconid and metastylid are very irregularly shaped with long connecting isthmuses. The small, compressed state of the M_2 ectostylid at the base of the tooth is illustrated in Figure 13.5D. The remaining dentition is mostly fragmentary and adds no additional morphologic information.

There are two astragali: ALA-VP-2/68 and ALA-VP-2/110. The former is "rolled" in appearance and was apparently transported, while the latter is well-preserved except that the distomedial surface is broken, disallowing an accurate measurement of the mediolateral dimension. These specimens are similar in size to the Höwenegg population and other Middle Awash astragali from Adu Dora, Amba East (AME), and Bikir Mali Koma (BIK) (discussed later).

The distal hind limb elements of a smaller hipparion ALA-VP-2/26, referred by Haile-Selassie to *Eurygnathohippus* sp. (2001a: figure 5-44, p. 309; Figure 13.6A and B in this chapter), is an important specimen from this site and time interval. The metatarsal III is elongate and slender, as is the 1st phalanx III. The 2nd phalanx III is as in the Höwenegg population, and the 3rd phalanx III is primitive for a hipparionine in having an elongated, distally pointed aspect. Altogether, this limb, as well as the Alayla material described above, is well referable to *Eurygnathohippus* aff. *feibeli,* albeit somewhat more heavily built than the Lothagam type (also, Bernor et al. 2005b).

There are two maxillary cheek teeth from Asa Koma: ASK-VP-1/9, a right P^4, and ASK-VP-3/23, a left M^2. The P^4 is fragmentary but preserves enough detail to exhibit the early middle stage of wear (mesostyle height = 38.0 mm) most useful for characterizing a cheek tooth dentition. In this specimen, the occlusal surface preserves the following characters: a very complex ornamentation of the anterior and posterior borders of the prefossette; the pli caballin is double; the hypoglyph is deeply incised; the protocone is oval. The M^2 is more complete and in an earlier stage of wear with a crown height of 44.9 mm. The labial surface (Figure 13.7A) exhibits the tooth's strong bending distally, and while the ectoloph is chipped, it is clear that it had a flat profile mesially (= "blunt" of Fortelius and Solounias 2000) and a low, round morphology distally, indicative of a grazer. The occlusal surface clearly exhibits a number of important salient features (Figure 13.7B): The mesostyle is thin, being virtually pointed labially; the prefossette is complex on both its anterior and posterior surfaces, while the postfossette is less complex; the protocone is

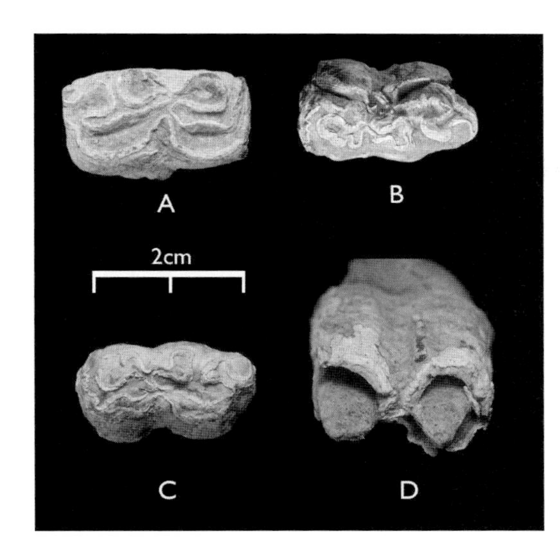

FIGURE 13.5

Occlusal views of mandibular teeth. A. ALA-VP-2/155, right P_3 (with ectostylid). B. ALA-VP-2/158a, right P_4. C. ALA-VP-2/158b, left M_2. D. ALA-VP-2/158b, left M_2 crown base, with close-up of small ectostylid.

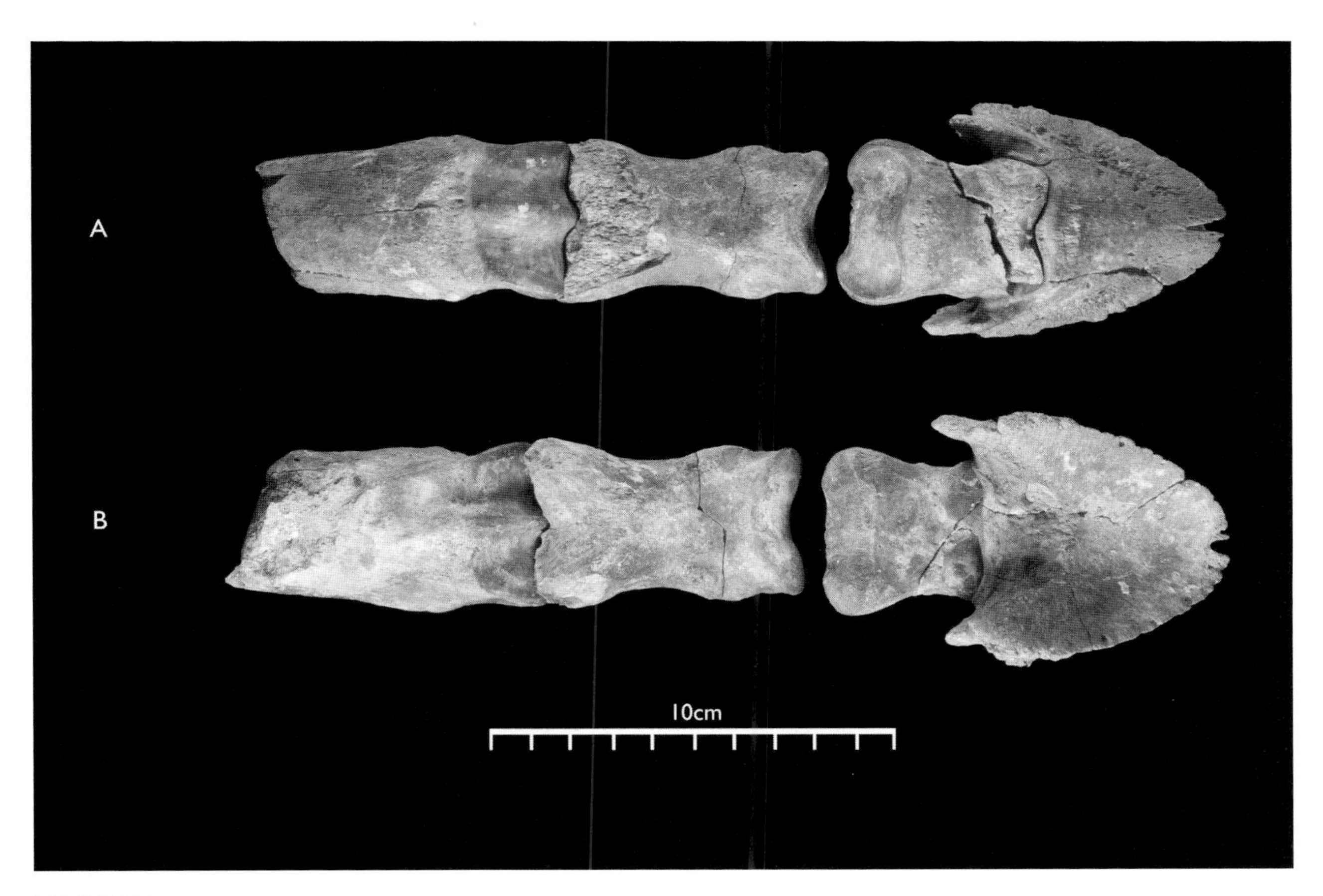

FIGURE 13.6

ALA-VP-2/26, distal hindlimb. A. Cranial view. B. Caudal view.

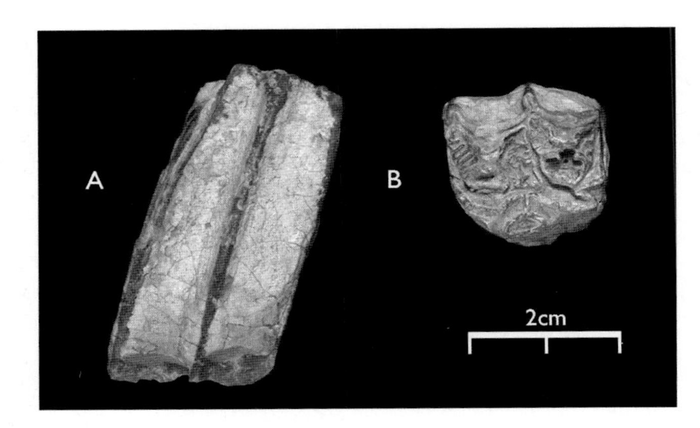

FIGURE 13.7
ASK-VP-3/23, left M^2.
A. Labial view.
B. Occlusal view.

labially rounded and lingually flattened; the hypoglyph is deeply incised. This morphology conforms well with *E. feibeli* from the Nawata Formation, Kenya.

There are two M_1s and three M_2s from Asa Koma. ASK-VP-3/7 is a right M_1 past its middle stage of wear and clearly exhibits a small (L = 1.2 mm; W = 1.6 mm) ectostylid on both its labial wall (Figure 13.8A) and occlusal surface (Figure 13.8). The following are some salient morphological features of the occlusal surface: the metaconid is rounded, while the metastylid is squared; the ectoflexid projects deeply lingually, separating the metaconid from the metastylid, and nearly touching the linguaflexid, which has a very broad, deep U-shape; the pre- and postflexid are labiolingually restricted and have simple margins; the protoconid enamel band is rounded; the ectostylid is a small, rounded structure placed distolabially to the ectoflexid, nearly touching the hypoconid enamel band. ASK-VP-3/502 is a left M_2 with much of the crown from the base upward broken away. However, this specimen also preserves a prominent ectostylid on the labial margin of the tooth. It is in an early stage of wear with a crown height of 54.4 mm, meaning its maximum crown height would have slightly exceeded 55 mm. In this early stage of wear, the metaconid is elongate; the metastylid is somewhat pointed with a long straight mesiolingual edge; there is no ectostylid apparent. ASK-VP-3/24, a left M_2, preserves an ectostylid on the labial surface of the crown.

There are two left astragali preserved in the Asa Koma equid fauna, ASK-VP-3/9, and ASK-VP-3/18. These specimens approximate in size to *E. feibeli* from Lothagam and are somewhat smaller than the Höwenegg population mean.

There is a single right distal metatarsal III, ASK-VP-1/1, that compares closely with specimens from Alaya and is substantially smaller than that of Lothagam *E. turkanense*.

The Asa Koma hipparion assemblage again would appear to be best referable to *E.* aff. *feibeli* and allows us to say, at this stage of evolution, that while maxillary cheek tooth morphology was primitively complex, the maximum crown height was approximately 55+ mm, slightly greater than both Turkish *Cormohipparion sinapensis* (45–50 mm) and central European *Hippotherium primigenium* (50 mm) but not as advanced as Indo-Pakistan *Sivalhippus perimense* (65+ mm; Bernor and Armour-Chelu 1999). The fragmentary distal metatarsal III is very similar to *E.* aff. *feibeli* specimens described above in having a more prominently developed distal sagittal keel than primitive Eurasian hipparions, and

FIGURE 13.8
ASK-VP-3/7, right M_1. A. Labial view with ectostylid. B. Occlusal view with ectostylid.

this, together with the persistent development of ectostylids (albeit small), are hallmarks of *Eurygnathohippus* aff. *feibeli.*

There is a small assemblage from BIK-VP-1, including a few lower cheek teeth, an astragalus, and a 1st phalanx III. BIK-VP-1/11 is composed of associated P_2, P_3, and P_4 with some connecting bone tissue remaining (Figure 13.9). All cheek teeth are in an early middle stage of wear, with P_4 having a crown height of 40.6 mm. Salient morphologic features of these cheek teeth include the following: metaconids and metastylids are essentially sub-rounded; metaconid-metastylid isthmus becomes serially longer from P_2 to P_4; there is a prominent pli ectoflexid on all three premolars; there are several, small rounded protostylids labial to the protoconid enamel band of both P_3 and P_4; a prominent, albeit small ectostylid is found labial to the hypoconid enamel band on the P_3; ectoflexid becomes progressively more deeply projecting from P_2 to P_4 but does not separate the metaconid from the metastylid; linguaflexid is a deep, broad, irregular V-shape on all three teeth; pre- and postfossettes are labiolingually compressed and exhibit little to no enamel complexity.

The single astragalus, BIK-VP-3/13, is the size of that of the Höwenegg population, and somewhat larger than Lothagam *E. feibeli;* it is distinctly smaller than Lothagam *E. turkanense*. BIK-VP-1/3 is a nearly complete 1st phalanx III. This specimen, too, compares most closely with those of the Höwenegg population and *E. feibeli,* being distinctly smaller than that of *E. turkanense*. While the Bikir Mali Koma assemblage is likely referable to *E.* aff. *feibeli,* the premolar dentition includes morphologies not yet documented in this taxon, particularly the elongate metaconid and metastylid isthmuses and multiple protostylids on P_3 and P_4.

Gaysale has produced a single hipparion specimen, GAS-VP-1/1, a right P^4. This is a complete, early middle stage of wear individual with typical *E.* aff. *feibeli* characters: the mesostyle is thin; the prefossettes are just becoming sufficiently worn to exhibit some complexity of the posterior border of the prefossette; the pli caballin is still single (probably because of the early stage of wear); the hypoglyph is deeply incised and virtually encircles the protocone; the protocone is moderately elongate with a flattened lingual border.

5.6 Ma Horizon: Saitune Dora

The left P^2, STD-VP-1/34, is in an early middle stage of wear with a crown height of 37.2 mm. Both the lingual and mesial margins of the tooth are broken, but the specimen retains the following salient morphologic features: although still in a relatively early stage of wear for a P^2, this specimen exhibits complex margins of both fossettes, but in

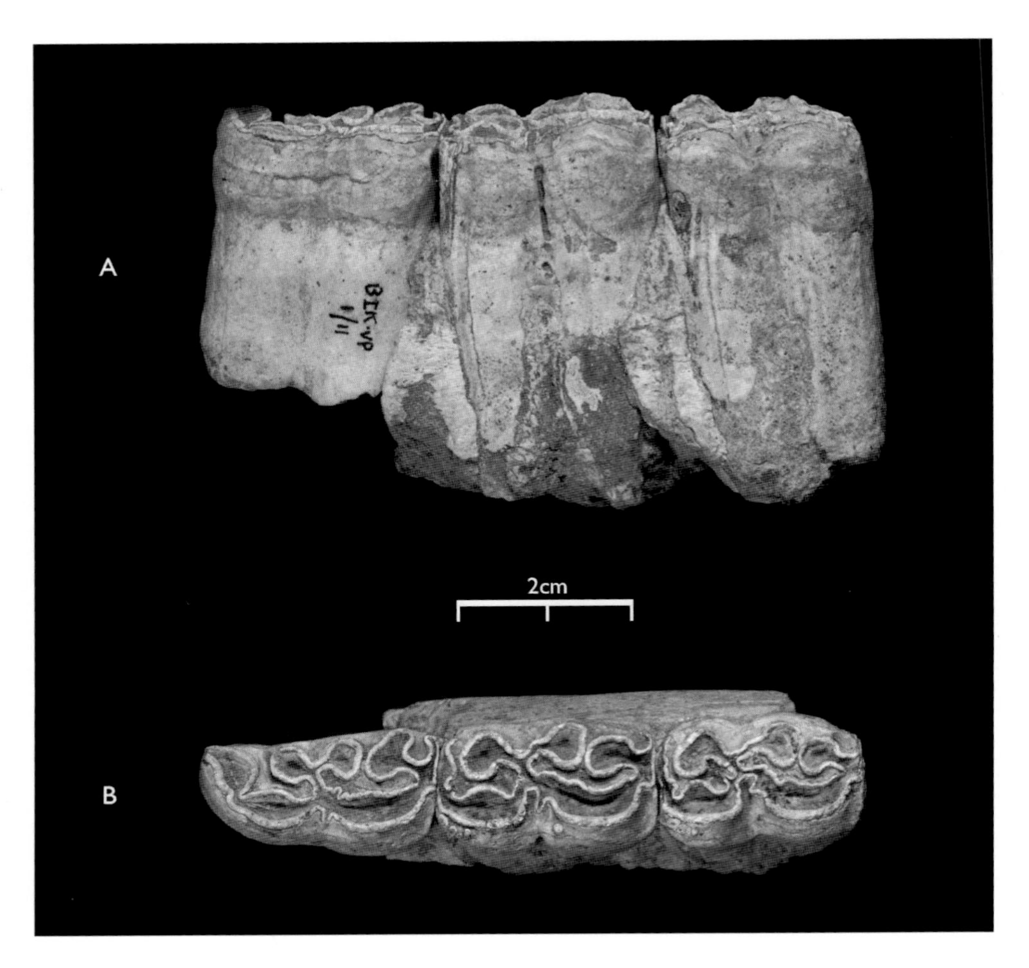

FIGURE 13.9
BIK-VP-1/11, left P_{2-4}.
A. Labial view.
B. Occlusal view.

particular, the two opposing margins in the center of the crown. The mesostyle is a labially rounded and relatively thin loop. The anterostyle, protocone, pli caballins, hypoglyph, and hypocone are all missing.

The right P^3, STD-VP-2/39 (Figure 13.10), is in a middle stage of wear with a crown height of 27.9 mm. It is well-preserved, with the following morphologic characters: the mesial ectoloph is low and rounded, whereas the distal one is virtually blunt (indicative of a grazer); the mesostyle is tightly rounded, with a thin labial aspect occlusally, and with an additional fold displaced slightly mesially; the pre- and postfossette's opposing borders are complex, while the other two borders are simple; the pli caballin is double, with a diminutive additional pli placed distally; the hypoglyph is only moderately incised; the protocone is lingually flattened and labially rounded.

There is a single P^4, STD-VP-2/847, in an early stage of wear (mesostyle height = 49.6 mm). This specimen would not have achieved a crown height much greater than 55 mm.

There are two M^1s: STD-VP-1/12, with a crown height of 42.4 mm, and STD-VP-2/37, with a crown height of 28.7 mm. Considering their different stages of wear, they are remarkably similar to one another. The individual in an earlier stage of wear has more

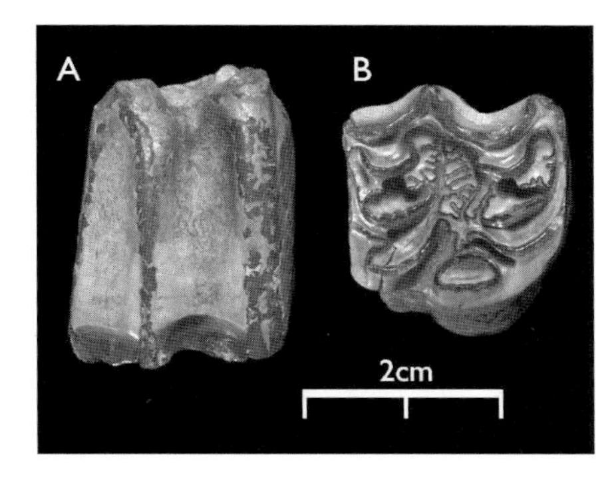

FIGURE 13.10
STD-VP-2/39, right P^4,
A. Labial view.
B. Occlusal view.

complex enamel plications, a more deeply incised hypoglyph, and slightly more elongate-oval protocone. However, they are both very similar in their length and width dimensions, their slender and pointed mesostyle, and complex opposing margins of the fossettes.

There are two P_2s; only STD-VP-2/38 is complete in all its measurements (Figure 13.11A). This specimen is characterized as being in a relatively late stage of wear, having a crown height of 13.5 mm. Occlusally, it preserves a rounded metaconid and squarish metastylid; pre- and postflexids are labiolingually constricted and with relatively simple margins; there is a distinct pli caballinid, and placed slightly labiodistally, there is a small, rounded ectostylid. The ectostylid can be seen clearly on the labial surface of the tooth ascending from the base to the occlusal surface of the crown.

There are two mandibular molars: STD-VP-1/45, a right M_2, and STD-VP-2/916, a right M_3. The M_2 is in an early stage of wear with a crown height of 48.5 mm; the metaconid has an irregular shape and metastlyid is square; the pre- and postflexids are labiolingually constricted and have simple margins; the linguaflexid has a deep and broad U-shape; the ectoflexid is deeply incised, separating the metaconid and metastylid; there is a very strong pli caballinid derived from the distolabial margin of the ectoflexid. The M_3 is in a later stage of wear with a crown height of 32.2 mm. The metaconid is rounded while the metastylid is square-shaped with a pointed distolingual aspect. The entoconid and hypoconulid are separated by an elongate isthmus, and each have distinctly rounded enamel margins.

Altogether, the Saitune Dora assemblage is morphologically very similar to the material described above and, as such, is referable to *E.* aff. *feibeli.*

5.54 Ma Horizon: Bilta, Digiba Dora

There is a single individual recorded from the 5.54 Ma locality of Bilta: BIL-VP-1/1 (Figure 13.12), a fragmentary mandible with right P_2–P_4, M_2 and M_3 and left P_2–M_3. The mandible is from a medium-sized individual. The symphysis is of modest length, with a narrow, rounded incisor arcade. The incisor crowns are broken away. The canine roots are apparent, large, and of a male individual. They are placed close to the I_3s. The premolars are described from the left side and exhibit the following morphologic features: The metaconids are rounded; the metastylids are rounded on P_2, more squarish on P_3 and P_4; the linguaflexid is V-shaped on all premolars; the ectoflexid becomes progressively more deeply incised serially; the pre- and postflexids are labiolingually constricted, with simple margins on all premolars; the protostylid is lacking (broken) on P_2, it is a small circular ring on P_3,

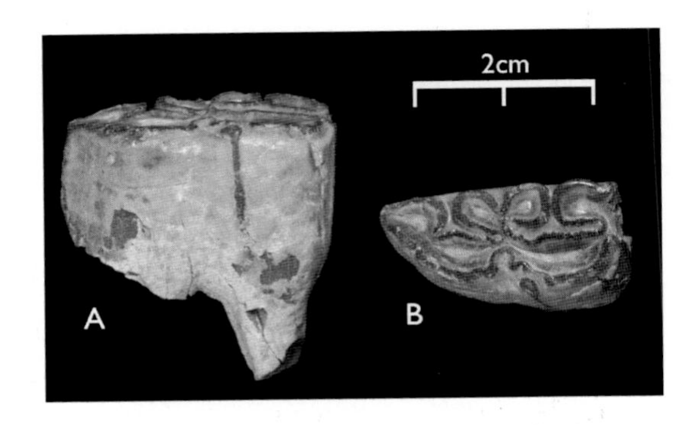

FIGURE 13.11
STD-VP-2/38, left P_2.
A. Labial view.
B. Occlusal view.

and an elongate, labially projecting loop on P_4; the pli caballinid is prominent on P_3 and P_4. The molars, as described on the left side, have the following morphologic features: The metaconids are rounded-elongate, while the metastylids are square-shaped with a pointed distolingual aspect; the linguaflexids have a deep, broad U-shape; the ectoflexids are deep on all molar teeth, clearly separating the metaconid from the metastylid while becoming very closely opposed to the broad linguaflexid; the protostylid is a long, labially projecting loop on M_1 and M_2, and a distinct circular ring on M_3; the preflexids are square-shaped, whereas postflexids are short and elongate, neither having complex margins; the pli caballinid is present only on the M_2; the ectostylid is apparent only on M_3, where it presents as a small, round structure. As with previously described mandibular M_3s, there is a distinctly elongate isthmus separating the entoconid from the hypoconulid.

This specimen is the first mandible reported for what we believe is the *E.* aff. *feibeli* form. The modest length of the symphysis with the accompanying rounded dental arcade suggests that it was a selective feeder (Bernor and Armour-Chelu 1999). The variable protostylids, backwardly pointed molar metastylids, and complicated M_3 talonid, with an accessory isthmus and complex hypoconulid, are all hallmarks of this form-species.

The Digiba Dora equid sample includes three lower cheek teeth: DID-VP-1/27, a right P_2; DID-VP-1/4, a left P_4; and DID-VP-1/116, a right M_1. The P_2 is unerupted and preserves no morphological detail. The P_4 is in a middle stage of wear (crown height = 27.5 mm) and is generally similar to previously described lower cheek teeth in having a rounded metaconid, square and lingually pointed metastylid, compressed pre- and postflexids, a distinct and single pli caballinid, V-shaped linguaflexid, and shallow ectoflexid. The preflexid is remarkable in exhibiting considerable complexity of its enamel margin. There is a small, rounded ectostylid that can be seen both occlusally and on the labial wall of the cheek tooth. The M_1 is in a relatively early stage of wear with a crown height of 48.1 mm. Because of this early wear stage, the metaconid is rather oblong, while the metastylid is square-shaped. The labial wall of this specimen clearly has a well-developed ectostylid, which is not expressed on the occlusal surface, rising 34 mm from the crown's base. The remaining morphology is as in previously described specimens.

Digiba Dora has yielded a fragmentary left astragalus and a complete, left 1st phalanx III, DID-VP-1/82 (Figure 13.13). While the astragalus is of no taxonomic use, the 1st phalanx III is diagnostic. This specimen is at the large end of the Höwenegg population range and is both longer and wider than the Lothagam *E. feibeli* and Middle Awash

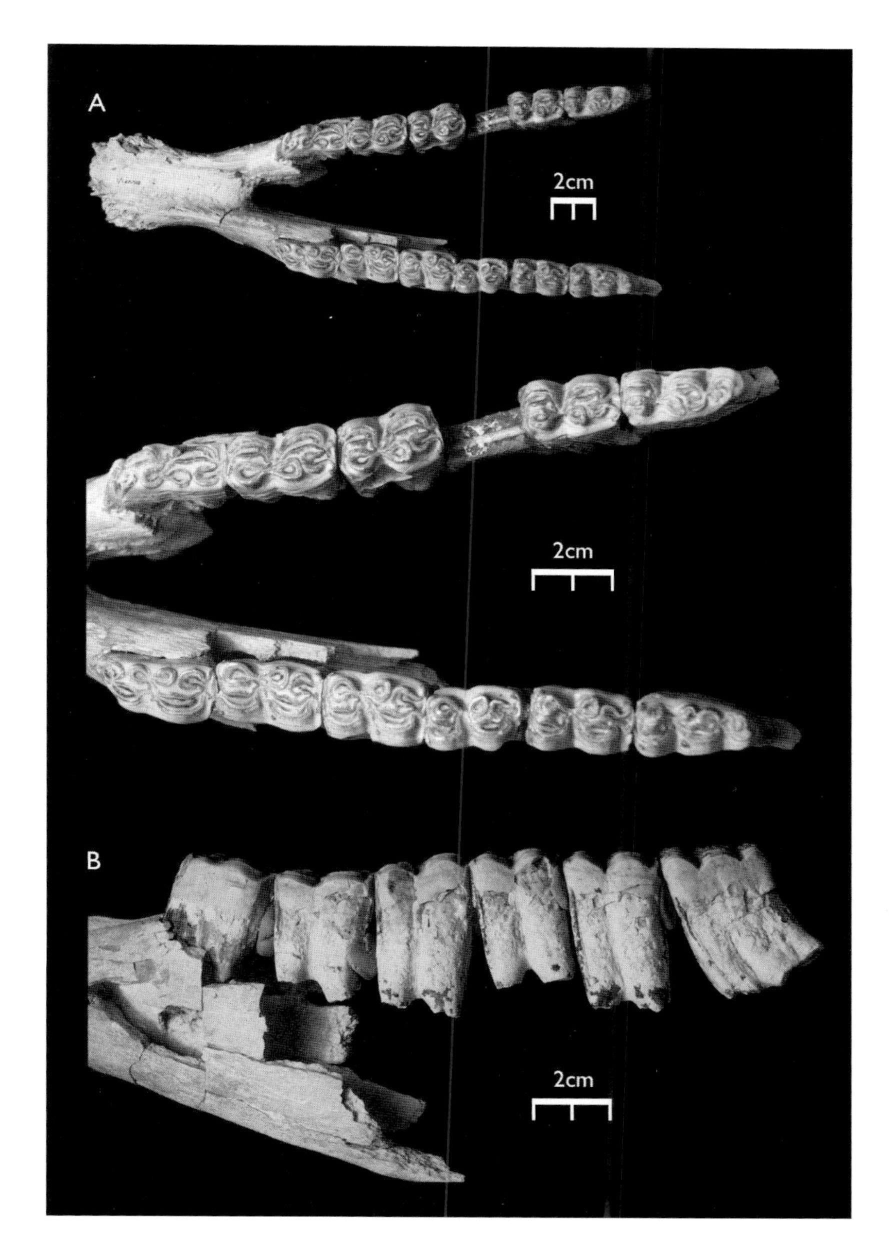

FIGURE 13.12

BIL-VP-1/1, mandible. A. occlusal view. B. Lateral view of left cheek teeth.

specimens from Jara Borkana, Alayla, and Bikir Mali Koma. Yet it certainly groups with this suite of "slender" eastern African forms while contrasting sharply with the three Lothagam specimens of *E. turkanense.*

5.4 Ma Horizon: Kabanawa

The 5.4 Ma interval is represented by the Kabanawa equid fauna, consisting of three isolated teeth. KWA-VP-1/2 is a worn P_4 (crown height = 20.2 mm; Figure 13.14). This specimen is characterized as having a rounded metaconid, distally pointed metastylid, pre-

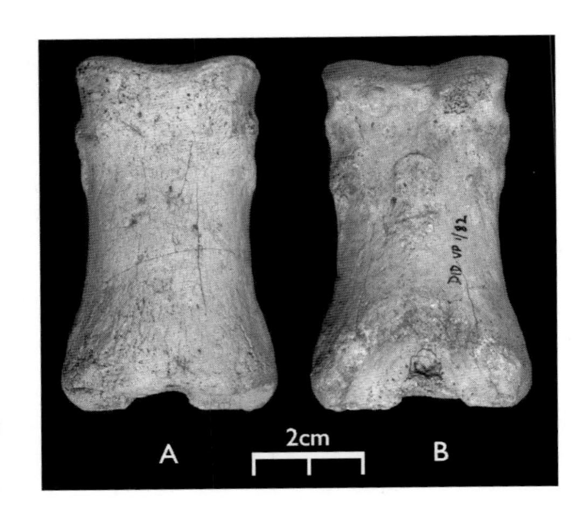

FIGURE 13.13

DID-VP-1/82, 1st phalanx III. A. Cranial view. B. Caudal view.

and postflexid constricted labiolingually and having relatively simple margins, linguaflexid in a deep V-shape, protostylid as a restricted loop extending labially and slightly distally; ectoflexid is deep, but not separating the metaconid and metastylid. The relatively large ectostylid is remarkable, with a length of 4.9 mm and maximum width of 2.5 mm; thus far, in the late Miocene Middle Awash succession, this is the largest measured ectostylid.

5.2 Ma Horizon: Amba East, Amba West, Kuseralee Dora

The 5.2 Ma horizon includes three faunas with hipparionine horses: Amba East (AME), Amba West (AMW), and Kuseralee Dora (KUS). AME-VP-1/86 is a fragmentary left P^2 in a middle stage of wear. This specimen preserves most of the ectoloph, which is worn flat, as in grazing equids. The occlusal surface preserves only modestly plicated fossettes, double pli caballins, oval protocone, a moderately deeply incised hypoglyph, and a pointed mesostyle. AME-VP-1/43 is an associated partial maxillary and mandibular dentition. There are two maxillary teeth: a right and left P^4 and M^1 in a middle stage of wear (both with a crown height = 39.5 mm). These teeth are similar morphologically in virtually all details: the mesostyle is thin and pointed; the pre- and postfosettes show moderate complexity, except for the very complex posterior border of the prefossette; the pli caballin is single or weakly double; the protocone is oval-shaped; the hypoglyph is only moderately incised. The accompanying AME-VP-1/43 mandibular dentition includes a left P_4, right M_1, and left M_3. These specimens are similar in the following morphological details: the metaconids are elongate; the metastylids are rounded to square; the metastylids are connected to one another by a rather elongate isthmus; the pre- and postflexids are labiolingually compressed; the linguaflexids are a deep U-shape; the ectoflexid separates the metaconid and metastylid only in molars; the M_3 talonid is elongate with an elongate isthmus between entoconid and hypoconulid; the labial border of all these teeth is lacking.

AME-VP-1/29 (Figure 13.15) is a well-preserved partial mandible. The mandibular symphysis is not complete, but the dentition includes both right and left incisors and a small, albeit complete, left canine indicating that this individual was a female (Figure 13.15C). The incisors are worn, and each has a distinct, elongate infundibulum. The I_3s are abbreviated with a "pinched" distal aspect typical of chronologically younger east-

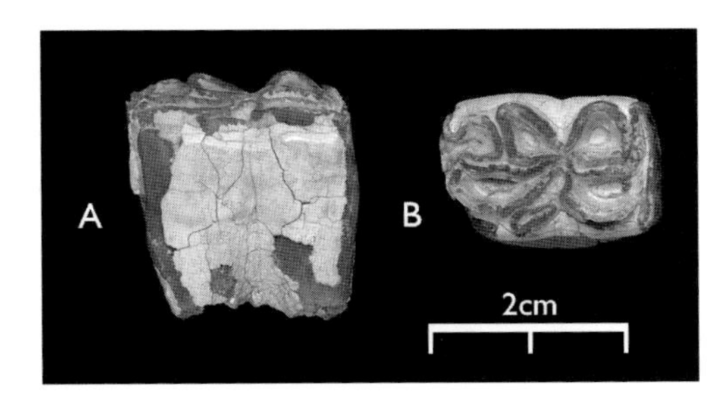

FIGURE 13.14
KWA-VP-1/2, right P_4.
A. Labial view. B. Occlusal view.

ern African hipparions. The incisor arcade is rounded and the incisor teeth are heavily worn, indicating that this was an older adult individual. In the premolars, the metaconids are generally rounded, while the metastylids are squarish and slightly pointed buccally; the linguaflexids are a deep V-shape; the pre- and postflexids are labiolingually restricted; the ectoflexids do not separate the metaconid and metastylid; the protostylid is an elongate, labially projecting loop on the P_3 and a small, rounded structure on the P_4; the protoconid enamel bands are rounded on P_3 and P_4. The molar teeth are virtually the same except for a deeper, more U- shaped linguaflexid; more deeply projecting ectoflexids; elongate pointed metastylids on the M_2 and M_3; a labially projecting protostylid loop preserved on the M_2; an elongate isthmus connecting the entoconid to the hypoconulid on the M_3.

The remaining lower cheek teeth are isolated and add little additional information to this sample, except AME-VP-1/40. This specimen preserves a very strongly developed pli caballinid and an ectostylid buried in the labial wall of the tooth; the ectostylid is not expressed occlusally because of the tooth's relatively early wear (crown height = 42.3 mm).

There is a single postcranial element, an astragalus, AME-VP-1/62. This is the size of that of Lothagam *E. feibeli* and at the smaller end of the Höwenegg population range.

The Amba West assemblage is represented by partial cranial, maxillary, and mandibular teeth and an important, complete metacarpal III. There is a juvenile cranial fragment, AMW-VP-1/44, which includes left and right maxillary fragments (Figure 13.16) and an associated premaxilla (Figure 13.16C). The premaxilla includes right and left di^1–di^3, while the maxillae include right and left dp^3–dp^4, M^1, M^2 (erupting), and evidence of an unerupted left M^3 still in the crypt. The premaxilla is a relatively elongate, slender structure with a rounded dental arcade (Figure 13.16C). There is a distinct tuberosity on the dorsal surface of the premaxilla just posterior to the incisor dentition. The dp^3–dp^4 are worn and exhibit the following morphology: the fossettes show little plication frequency, except for the posterior border of the prefossettes, which are complexly plicated; the mesostyle on the dp^4 is preserved and presents as a rounded loop; the pli caballins are lost with wear; the hypoglyph is moderately deeply incised; the protocones are round to oval. The emerging M^1 is in early wear and preserves few morphologic details, including the following: the mesostyle is pointed; the posterior border of the prefossette is the only border showing any enamel band complexity; there are two weakly developed pli caballins; the protocone is an elongate lenticular structure. The M^2 is just emerging from the crypt and has not yet developed any occlusal details.

AMW-VP-1/129, a P^2, is in a relatively early stage of wear, exhibiting the following morphological details: the anterostyle is an elongate-rounded structure; the prefossette has complex mesial and distal borders, while the postfossette exhibits more moderate plication

FIGURE 13.15
AME-VP-1/29, right and anterior mandible.
A. Ventral view.
B. Occlusal view.
C. Occlusal view.
D. Medial view.

frequency; the mesostyle is relatively thin and pointed; the hypoglyph is deeply incised but does not surround the hypocone; the pli caballin is single; the protocone is labially rounded and lingually flattened. The morphology of P^3 is well-exemplified by AMW-VP-1/23 (left), which is in a later middle stage of wear and exhibits the following morphology: the mesostyle is a rounded loop; pre- and postfossette borders have mostly simple complexity except for the distal border of the prefossette, which is complex; the pli caballin is single; the hypoglyph

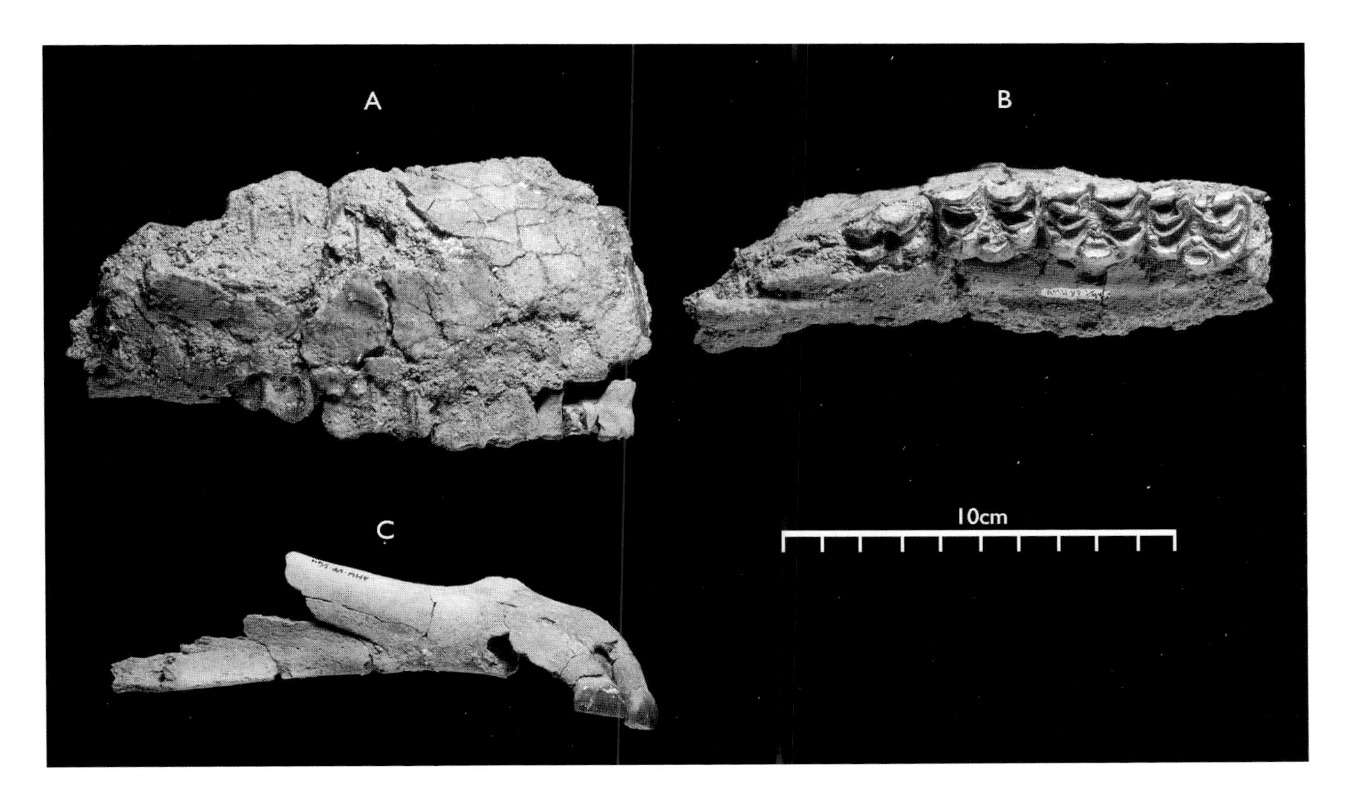

FIGURE 13.16

AMW-VP-1/44. A. Lateral view of maxilla fragment. B. Occlusal view of maxilla fragment. C. Lateral view of premaxilla.

is only moderately deeply incised; the protocone is labially rounded and lingually flattened. There are three P^4s from Amba West Locality 1, of which AMW-VP-1/19 is the only one preserved well enough to have had all measurements taken. AMW-VP-1/19 is in an early stage of wear with a crown height of 50.5 mm. Despite this relatively great crown height, the occlusal surface is sufficiently worn to show some morphological details: pre- and postfossettes are complexly plicated, with enamel ornamentation being particularly developed on the posterior border of the postfossette; the mesostyle is a rounded loop; the parastyle has a particularly broad, mesially oriented, recurved surface; the hypoglyph is only moderately deeply incised; the pli caballin is single; the protocone has an elongate lenticular morphology. Another P^4, AMW-VP-1/52, is important because of its relatively great crown height: 55.7 mm. This specimen is in a very early stage of wear, yet it demonstrates that, at this stage of evolution, crown height most likely exceeded 55 mm.

AMW-VP-1 has yielded four M^1s, three M^2s, and a single erupted M^3. AMW-VP-1/44 has been described above and will not be further referred to here. The two M^1s, AMW-VP-1/9 and AMW-VP-1/20, are very similar in their morphology, sharing the following characters: the parastyle has a distinct mesially directed face; the mesostyle is thin and rounded; the fossettes are mostly moderately complex except for the distal border of the prefossette, which is complex; the hypoglyph is moderately deeply incised; the pli caballin is variable, being single (AMW-VP-1/20) or double (AMW-VP-1/9); the protocone is oval to ovate in shape. The M^2, AMW-VP-1/101, shares all the morphologic details exhibited in the M^1. The M^3, AMW-VP-1/22, is in a late stage of wear (crown height = 19.0

mm) and differs from the previously described cheek teeth in the following regards: the parastyle is a simple, pointed structure; the mesostyle is a broad, flattened structure; the pre- and postfossettes are very simple, except for the distal border of the prefossette, which remains remarkably complex; the hypoglyph is shallow but is met with an unusual pli incisure from its facing surface to nearly enclose the hypocone; the pli caballin is complexly folded, but with very short plis; the protocone is an elongate, lenticular-shaped structure.

The AMW-VP-1 mandibular premolars are represented by two P_2s and three P_4s. AMW-VP-1/117A is in a middle stage of wear and preserves an elongate metaconid and rounded metastylid; a V-shaped linguaflexid; a large, irregularly shaped entoconid; pre- and postflexids lacking any enamel band complexity; a single, yet prominent pli caballinid; a very shallowly incised ectoflexid. The P_4s are very similar to one another in their morphology: the metaconids are mostly rounded; the metastylids are squarish; the pre- and postflexids have simple borders; the linguaflexid has a deep, broad V-shape; the protostyle is a rounded circle on AMW-VP-1/16 and a loop on AMW-VP-1/86; there is a large ectostylid on AMW-VP-1/86 that has an open isthmus connected lingually to the interior of the tooth at the mesial aspect of the hypoconid.

AMW-VP-1 includes three M_1s and two M_3s. AMW-VP-1/105 is in an early stage of wear (crown height = 48.9 mm) and exhibits the following characteristics: the metaconid has an elongate-irregular shape; the metastylid has a squared-pointed lingual aspect; the linguaflexid is a deep U-shape; the entoconid is large and elongate; the pre- and postflexids are labiolingually compressed and with simple margins; the ectoflexid is deep, separating the metaconid and metastylid. AMW-VP-1/98 has an occlusal surface that is well worn and is remarkable for its great crown height (= 57.8 mm); this individual would likely have had a crown height in excess of 60 mm. The M_2 is very similar in morphology to the M_1 and is likewise in an early stage of wear (crown height = 50.0 mm). The two M_3s are poorly preserved and add little to our knowledge of the Amba West lower cheek teeth.

AMW-VP-1/15 is a complete metacarpal III (Figure 13.17A, B). This specimen is slightly longer than the Höwenegg hipparion metacarpal III and is overall larger than that of the Lothagam *E. feibeli,* but it is strikingly more elongate and slender than that of Lothagam *E. turkanense*. We discuss this specimen's proportions, and its taxonomic implications, later in this chapter.

Altogether, the Amba West and Amba East samples would appear to be more closely allied with *E. feibeli* than with *E. turkanense* but are slightly derived over older Middle Awash samples in their increased maximum crown height and larger, although proportionally similar, metacarpal III. These morphological differences do not in themselves warrant species distinction at this time.

The Kuseralee Dora fauna includes the following equid specimens: KUS-VP-1/4, a left P^3–M^1; KUS-VP-1/5, a left M^1; KUS-VP-1/23, a right P_4; KUS-VP-1/27, a left M^3; KUS-VP-1/29, a left M_2; KUS-VP-1/49, a left $P^{3 \text{ or } 4}$; KUS-VP-1/51, an astragalus; KUS-VP-1/95, an astragalus.

KUS-VP-1/4 is the most complete individual including a left P^3, P^4, and M^1. Crown heights for these specimens are P^3, 47.5 mm; P^4, 51.0 mm; M^1, 44.4 mm. The P^3 occlusal surface is just coming into wear enough to exhibit plication development. The most elaborate border is clearly the posterior prefossette, but the anterior border of the prefossette and

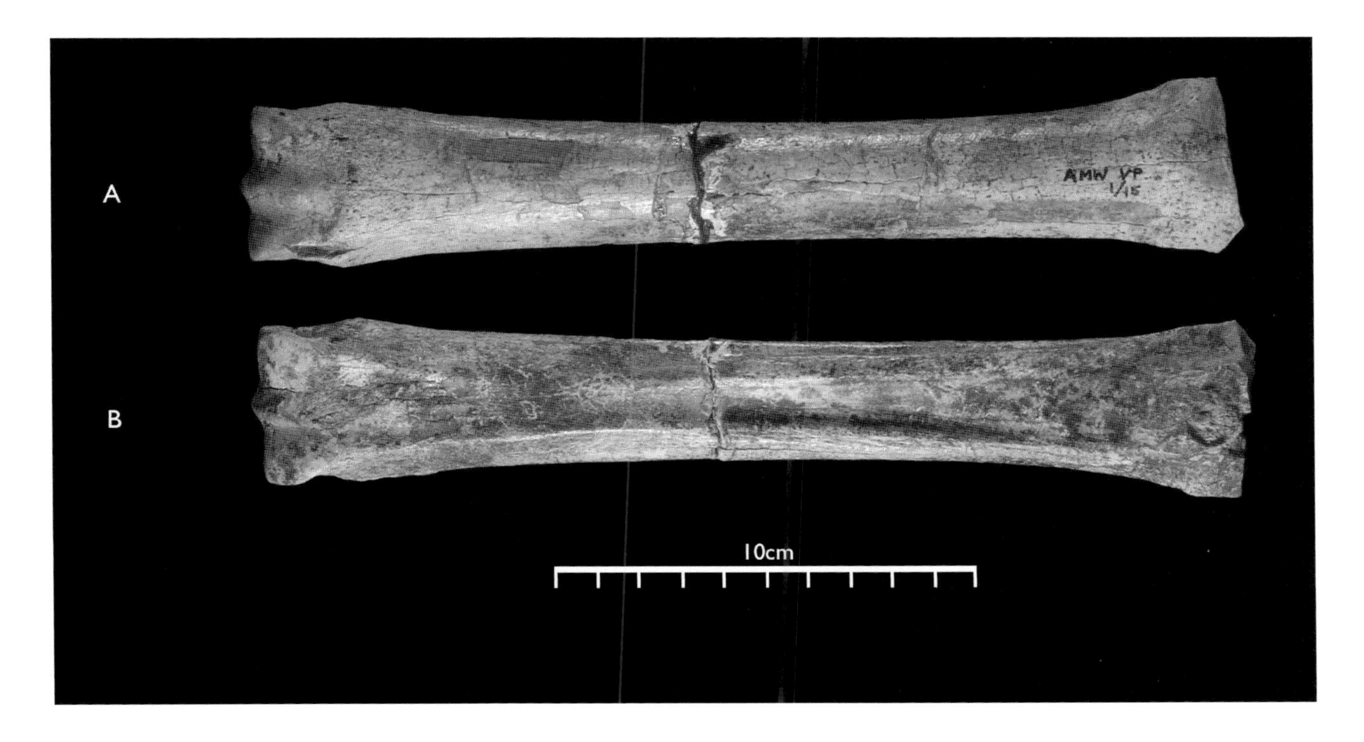

FIGURE 13.17
AMW-VP-1/15, left metacarpal III. A. Cranial view. B. Caudal view.

anterior border of postfossette are moderately complex. The pli caballin is strongly double, with accessory small plis both mesially and distally, giving an overall complex pattern. The protocone has an elongate oval shape with a flat labial border. The area of the hypoglyph is broken. The mesostyle is elongate and slender. The P^4 has very straight, vertical mesial and distal surfaces and is in an earlier stage of wear, so the ornamentations of the pre- and postfossettes are not well expressed. The pli caballin is irregular in shape and with two plis. The protocone is shorter, with a broken mesiolabial border, but having the same basic shape as that found in the P^3. The hypoglyph is preserved in this individual and is very deep, completely encircling the hypocone. The M^1 is more worn than the P^4 and has a broken labial margin. This tooth exhibits well the enamel plication patterns; the posterior border of the prefossette is clearly the most complex; however, the anterior border of the prefossette has four deeply invaginated plis (distally). The pli caballin is double, and the hypoglyph is very deeply incised, virtually completely encircling the hypocone. The protocone has a broken labial margin. The mesostyle is very thin, having a knife-like edge.

KUS-VP-1/5 is a left M^1 that is slightly less worn than the KUS-VP-1/4 M^1, having a crown height of 48.4 mm. As a result of its early wear stage, the fossette plications are not quite as well developed. The pli caballin is a strong double, and the hypoglyph is very deeply incised, engulfing the flat hypocone. The protocone is an elongate, labiolingually compressed oval shape. The mesostyle is thin and blade-like, as in the other Kuseralee Dora specimens.

KUS-VP-1/49 is a very worn posterior premolar, either a P^3 or P^4, having a mesostyle crown height of only 11.5 mm. Despite its late stage of wear, the plication amplitude is relatively complex on all fossette margins. The pli caballin remains stable, having two

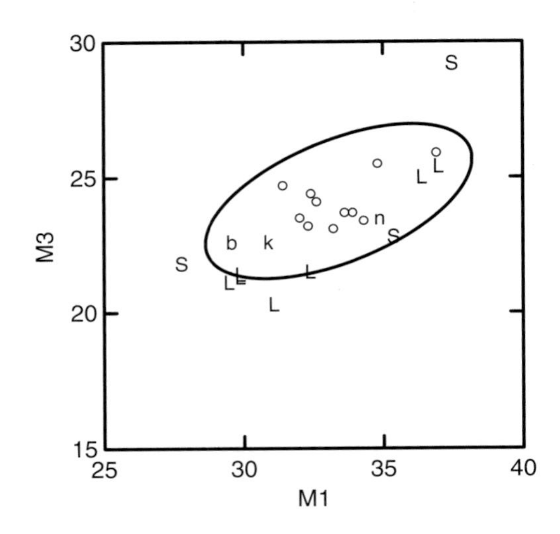

FIGURE 13.18

Bivariate comparison of P^2 from Lothagam, Sahabi, and the Middle Awash. Bivariate plot of M3 (occlusal width) versus M1 (occlusal length), with an Eppelsheim 95 percent confidence ellipse. L = Lothagam; S = Sahabi; b = Adu Dora; k = Amba West; n = Gawto; open circle = Eppelsheim

strong plis. The hypoglyph is moderately deeply incised. The protocone has become shortened but maintains an oval-shaped morphology.

KUS-VP-1/27 is a poorly preserved left M^3 with a mesostyle crown height of 40.5 mm. Occlusal morphological features are not sufficiently well-preserved to be described.

KUS-VP-1/23 is a well-preserved right P_4 with a crown height of 48.4 mm. This specimen is in very early wear, having the following salient features: very irregularly shaped metaconid and metastylid; ectoflexid not yet apparent; pre- and postflexids not yet clearly delineated; ectoflexid inverted with its pli-like structure directed labially. The labial side of this crown has a clearly defined ectostylid, which would appear to increase the total height of the crown, and has a minute circular ring on the distal portion of the protoconid enamel band immediately mesial to the ectoflexid. KUS-VP-1/29 is a fragmentary mandibular cheek tooth too poorly preserved to identify as to tooth position.

There are two complete astragali preserved from Kuseralee Dora: KUS-VP-1/51 (left) and KUS-VP-1/95 (right). Both are very similar in size and consistent with an *E.* aff. *feibeli* attribution and will be discussed further below with the analysis of the astragali.

Statistical Analyses

We provide bivariate and $\log_{10}$ ratio analyses of P^2, metacarpal III, astragalus, metatarsal III, and 1st phalanx III. Measurement numbers (M1, M2, M3, etc.) refer to those published by Eisenmann et al. (1988) and Bernor et al. (1997) for the skulls and postcrania and by Bernor et al. (1997) and Bernor and Harris (2003) for the teeth.

Dentition

Bernor et al.'s (1997, 2003) and Bernor and Armour-Chelu's (1997) analyses of Eurasian and African hipparionines have shown that the P^2 is the best tooth for analyzing potential differences in size (occlusal length [M1 of Table 13.2] versus occlusal width [M3 of Table 13.2]) and length (M10) versus width (M11) of protocone. Figure 13.18 provides comparisons between Middle Awash, Lothagam, and Sahabi P^2s (M1 versus M3). The Eppelsheim sample is characterized by the 95 percent ellipse. The Sahabi sample includes one point above the ellipse and one below, therefore both outside the range of variation recorded for *Hippotherium primigenium* from Eppelsheim. Lothagam includes individuals at the top and at the bottom and slightly lower than the ellipse. The Middle Awash sample includes one individual each from Adu Dora (b), Amba West (k), and Gawto (n), all within the lower portion of the ellipse. This plot provides some size separation, but this separation cannot be characterized as being statistically significant. We plotted other cheek tooth variables including P^2 protocone width (M11) versus length (M10), P_2 width (M8) versus occlusal length (M1), and P_4 width (M8) versus occlusal length (M1) with inconclusive results: there were no differences seen in the sample under consideration. Our plots on cheek teeth confirm Haile-Selassie's (2001a) assertion that one cannot distinguish late Miocene Middle Awash hipparions by tooth morphology and size alone.

Metacarpal III

Figure 13.19A provides a plot of metacarpal III length (M1) versus distal articular width (M11) compared to the Höwenegg 95 percent ellipse. Lothagam plots two individuals, both outside the Höwenegg ellipse: one with an elevated width dimension (= *E. turkanense*) and one with a relatively small width dimension (= *E. feibeli*). Sahabi has a small individual well below the ellipse referred to the Eurasian *Cremohipparion* lineage (Bernor and Scott 2003). There is a single Middle Awash specimen from Amba West (k), larger than the Lothagam species *E. feibeli* (Holotype), and identified as being a more advanced member of the *E. feibeli–E. hasumense* lineage.

Figure 13.19B is a plot of metacarpal III distal sagittal keel diameter (M12) versus distal articular width (M11). Lothagam plots three individuals within the ellipse (*E. feibeli* lineage) and one well above the ellipse (*E. turkanense* lineage). Sahabi plots one individual within the ellipse and two below. The two Sahabi specimens below the ellipse are likely to be members of the *Cremohipparion* lineage, whereas the one within the ellipse is *incertae sedis*. Again, Amba West (k; *E.* aff. *feibeli*) is the only Middle Awash specimen represented here, and it plots in the upper portion of the Höwenegg ellipse.

Figure 13.19C is a $\log_{10}$ ratio diagram of the metacarpal III comparing Amba West, Lothagam, and Sahabi gracile taxa and the type specimen of *Cormohipparion sinapensis* (Sinap, Turkey; 10.135 Ma) to the Höwenegg standard (*Hippotherium primigenium;* 10.3 Ma). Compared to the Höwenegg standard, the Sinap form is primitive and has relatively narrower (M3), and at the same time deeper (M4), midshaft dimensions, which Bernor et al. (2003) have related to its greater adaptation to cursorial locomotion compared to the Höwenegg hipparions: the so-called "Esme Acakoy" effect. The three African metacarpal IIIs all exhibit this "Esme Acakoy" effect, while differing in their size: ISP27P25B (Sahabi) is small, with very small midshaft width (M3), proximal articular (M5 and M6) and distal (M10–14) dimensions. This Sahabi form is distinct from all eastern African hipparions and is best referred to *Cremohipparion* sp. Lothagam *E. feibeli* (KNM-LT 139, Holotype) plots in close parallel to the Sinap specimen, with increased maximum length (M1), distal keel (M12), and distal articular (M13 and M14 only) dimensions. The Amba West specimen

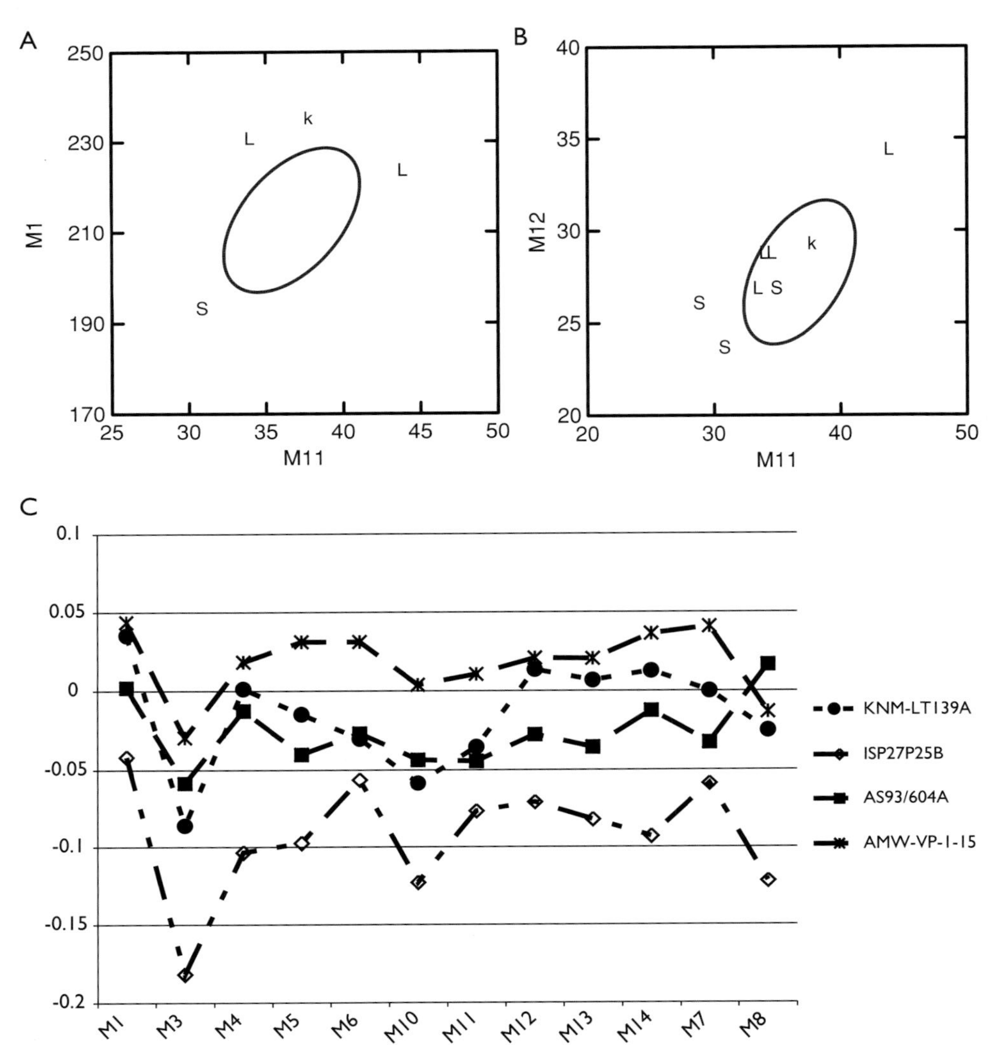

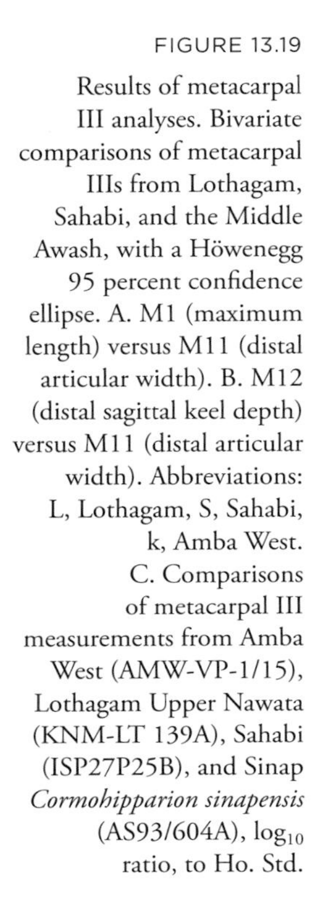
FIGURE 13.19

Results of metacarpal III analyses. Bivariate comparisons of metacarpal IIIs from Lothagam, Sahabi, and the Middle Awash, with a Höwenegg 95 percent confidence ellipse. A. M1 (maximum length) versus M11 (distal articular width). B. M12 (distal sagittal keel depth) versus M11 (distal articular width). Abbreviations: L, Lothagam, S, Sahabi, k, Amba West. C. Comparisons of metacarpal III measurements from Amba West (AMW-VP-1/15), Lothagam Upper Nawata (KNM-LT 139A), Sahabi (ISP27P25B), and Sinap *Cormohipparion sinapensis* (AS93/604A), $\log_{10}$ ratio, to Ho. Std.

(AMW-VP-1/15) plots similarly to KNM-LT 139 but with increased dimensions for proximal articular width (M5) and depth (M6) and distal supra-articular (M10) and distal articular (M11) width dimensions. This data suggests that the Amba West individual has a greater body mass than the Lothagam form and, in fact, is on the trajectory of increased body size reported for the *E. feibeli–E. hasumense* lineage (Bernor et al. 2005b).

Astragalus

Figure 13.20 is a plot of astragalus maximum length (M1) versus distal articular width (M5) and, as such, is a measure of size. Lothagam has three individuals below the Höwenegg ellipse and one just outside the top of the ellipse. This is interpreted as Lothagam having a greater abundance of *E. feibeli*–sized than *E. turkanense*–sized individuals in its astragalus sample. Sahabi has six small individuals at, and below, the smallest Lothagam individuals

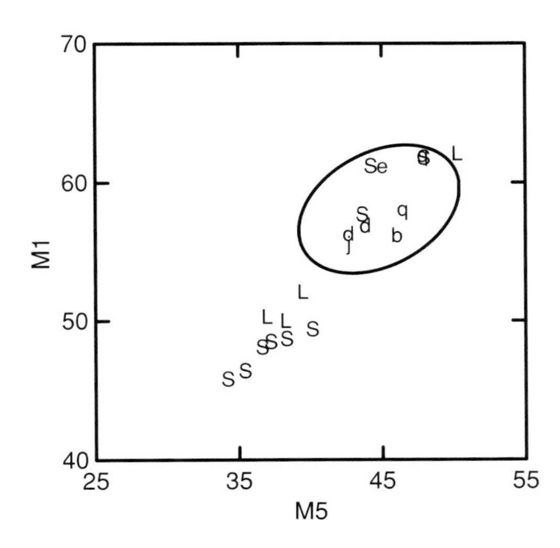

FIGURE 13.20

Bivariate comparison of astragali from Lothagam, Sahabi, and the Middle Awash. Bivariate plot of M1 (maximum length) versus M5 (distal articular width), with a Höwenegg 95 percent confidence ellipse. Abbreviations: L, Lothagam; S, Sahabi; b, Adu Dora; d, Asa Koma; e, Bikir Mali Koma; j, Amba East; q, Kuseralee Dora.

in the sample, while three individuals plot within the Höwenegg ellipse; one is very close in size to the large Lothagam form. The Middle Awash sample includes one individual from Adu Dora (b), two from Asa Koma (d), one from Bikir Mali Koma (e), one from Amba East (j), and two from Kuseralee Dora (q). All of the Middle Awash sample plots within the Höwenegg ellipse, and all of these, except one individual from Asa Koma and one individual from Kuseralee Dora, plot in the lower half of the ellipse. This suggests a likely predominance of the *E. feibeli* lineage at these localities, at least based on the astragalus sample. We also plotted astragalus distal articular depth (M6) versus width (M5) and found a similar result, which we do not figure here.

Metatarsal III

There are no complete Middle Awash specimens of metatarsal IIIs here to plot. However, Figure 13.21A provides an informative plot of metatarsal III distal sagittal keel (M12) versus distal articular width (M11). There is a single Lothagam individual of *E. turkanense*, which plots well above the Höwenegg ellipse and the rest of the sample. Sahabi includes two individuals at the top and just outside the ellipse (*"H". incertae sedis*) and four specimens below the Höwenegg ellipse. The Middle Awash sample has one specimen within the ellipse overlapping with a Sahabi specimen from Agera Gawtu (m). The Middle Awash also includes one individual from Alayla (c) and one individual from Asa Koma (d), just outside and to the left of the Höwenegg ellipse with elevated M12 measurements. All Middle Awash specimens likely belong to a single species of the *E. feibeli* lineage; none are referable to *E. turkanense* from Lothagam.

1st Phalanx III

Figure 13.22A plots 1st phalanx III maximum length (M1) versus proximal width (M4). There are two Lothagam specimens (L) plotted to the right of the Höwenegg ellipse with elevated M4 measurements, two just to the left of the ellipse with reduced M4 measurement,

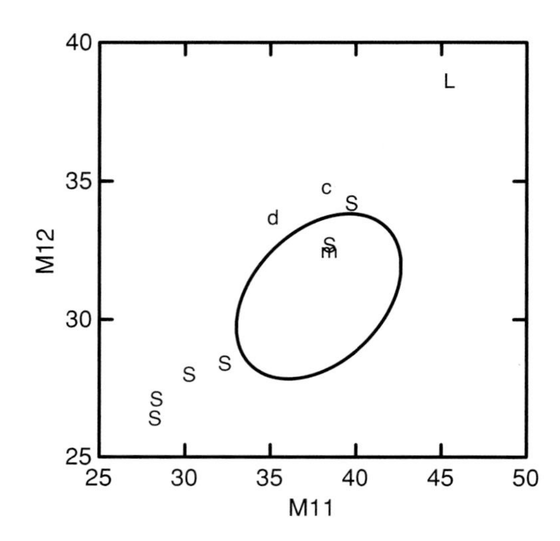

FIGURE 13.21

Bivariate comparison of metatarsal IIIs from Lothagam, Sahabi, and the Middle Awash. Bivariate plot of metatarsal III M12 (distal sagittal keel depth) versus M11 (distal articular width), with a Höwenegg 95 percent confidence ellipse. Abbreviations: L, Lothagam; S, Sahabi; c, Alayla; d, Asa Koma.

and one in the lower right corner of the ellipse. The specimens with elevated M4 measurements are referable to *E. turkanense*, those with reduced M4 measurements are referable to the *E. feibeli* lineage, and the one inside the lower right corner of the Höwenegg ellipse is most likely a small individual of *E. turkanense*. Sahabi has a single individual plotted at the top of the Höwenegg ellipse and one just outside the lower left portion of the ellipse, again indicating two different species. The Middle Awash sample includes a single specimen from Jara-Borkana (a), one from Alayla (c), and one from Bikir Mali Koma (e); all three are referable to the *E. feibeli* lineage, with the Jara-Borkana specimen being essentially identical to the holotype of *E. feibeli* from the Upper Nawata, Lothagam.

Figure 13.22B is a $\log_{10}$ plot of the 1st phalanges III of the *E. feibeli–E. hasumense* lineage between 6.0+ and 5.2 Ma. This plot shows the very close similarity shared between the Type *E. feibeli* (KNM-LT 139) and JAB-VP-1/1 (solid lines with triangles). The Alayla (ALA-VP-2/26), Bikir Mali Koma (BIK-VP-1/3), and Digiba Dora (DID-VP-1/82) specimens (dashed lines) are similar to the Lothagam and Jara-Borkana specimens, but with overall slightly increased dimensions. The Worku Hassan 1st phalanx III is somewhat larger in all relative sizes.

Figure 13.22C is a $\log_{10}$ plot of the 1st phalanges III belonging to robust individuals from Lothagam and Sahabi. These specimens are strikingly similar in their overall shape and contrast particularly with those specimens illustrated in Figure 13.22B in their relatively short length (M1 and M2) versus midshaft (M3), proximal articular (M4 and M5), and distal articular (M6 and M7) dimensions. KNM-LT 25466 was referred to cf. *Hippotherium primigenium* by Bernor and Harris (2003). Referring to the plot by Bernor and Harris (2003: 402, fig. 9.14), one can see that this specimen is smaller than, but has the same shape as, KNM-LT 26294. This relationship is clearly demonstrated in this $\log_{10}$ ratio diagram and suggests a better referral of this specimen to *E.* cf. *turkanense*. Sahabi specimen ISP2P11A clearly plots with all others here and is suggestive that the *E. turkanense* lineage, or a close relative of that lineage, occurs at Sahabi.

Overall, the data presented here suggest that the Middle Awash 1st phalanges III are best referred to the *E. feibeli* lineage, not the *E. tukanense* lineage.

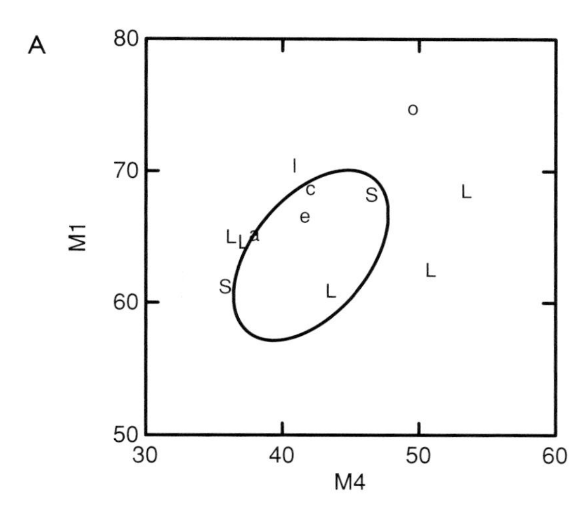

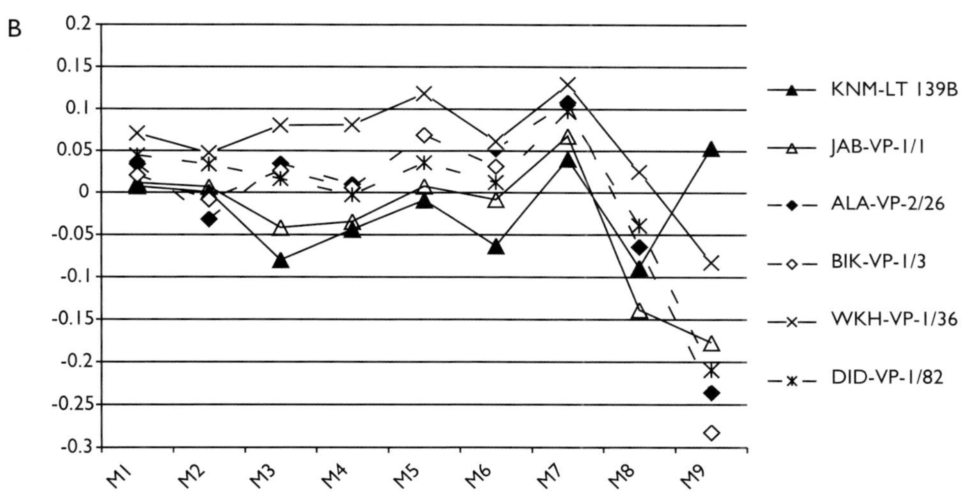

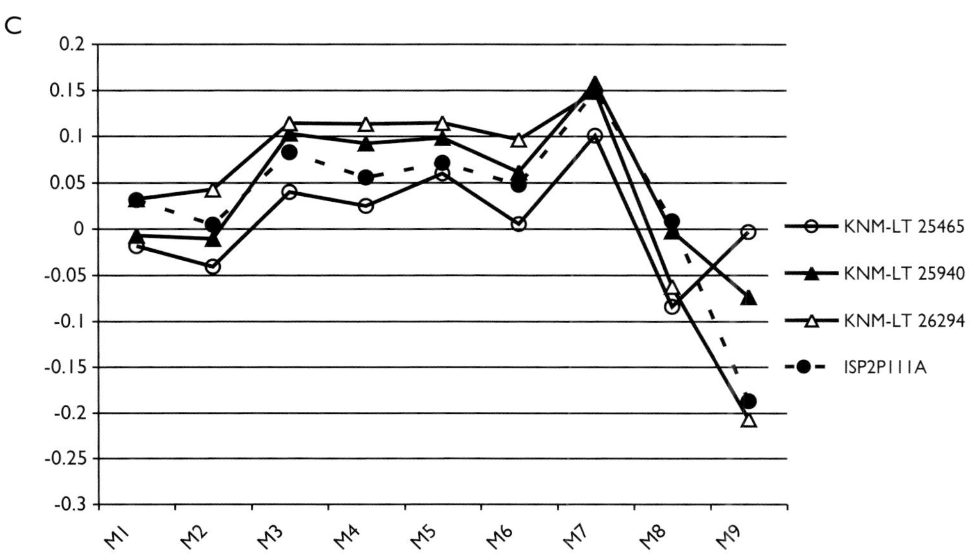

FIGURE 13.22

Results of 1st phalanx III analyses. A. Bivariate comparison of 1st phalanges III from Lothagam, Sahabi, and the Middle Awash, M1 (maximum length) versus M4 (proximal articular width), with a Höwenegg 95 percent confidence ellipse. Abbreviations: L, Lothagam; S, Sahabi; a, Jara-Borkana; c, Alayla; e, Bikir Mali Koma. B. Comparisons of *Eurygnathohippus feibeli* 1st phalanges III from the Middle Awash and Lothagam, $\log_{10}$ ratio, to Ho. Std.. C. Comparisons of the 1st phalanges III of robust taxa from Lothagam and Sahabi, $\log_{10}$ ratio to Ho. Std.

Discussion

The Middle Awash late Miocene sample provides us with about one million years of local hipparionine evolution. This analysis suggests a modest evolutionary pace of the *Eurygnathohippus feibeli* lineage during the 6.0–5.2 Ma interval. The oldest level, Jara-Borkana, includes a 1st phalanx III that is virtually identical to the holotype of *E. feibeli* from the Upper Nawata Member of Lothagam. The remainder of the sample offers no evidence of the *E. turkanense* lineage but rather suggests a gradual progression of morphologic change in the *E. feibeli* lineage. The changes detected include an apparent slight increase in maximum crown height from about 55 mm to about 60 mm; lower molar metastylids become increasingly pointed labially; ectostylids would appear to rise higher on the crown with time; and metacarpal III becomes slightly more robustly built with increased proximal and distal articular dimensions. However, there is insufficient morphologic evidence, both in the holotype *E. feibeli* sample and the 6.0 to 5.2 Ma Middle Awash sample, to provide a convincing case for distinguishing a new species of hipparion in the 6.0 to 5.2 Ma Middle Awash record.

Maxillary cheek tooth morphology exhibits relatively low occlusal relief. Whereas the central European *Hippotherium primigenium* lineage exhibits high relief, such that the ectoloph at the paracone and metacone rises high above the middle section of the tooth (from mesostyle to protocone) with rounded to pointed cusps, the Middle Awash sample is relatively blunt. This suggests an adaptation to eating grass. We believe that this was likely to be a long-term trend in eastern and southern African hipparion evolution, because the 10+ Ma sample of *Cormohipparion* from Chorora, Ethiopia, records the oldest African record of C_4 grazing (Bernor et al. 2004a). The elongate distal limb elements further suggest that the *E. feibeli* lineage was well adapted to open-country running (Bernor et al. 2005b).

Eurygnathohippus feibeli apparently ranged from Ethiopia, through Kenya (Hooijer and Maglio 1974; Bernor and Harris 2003), to Tanzania (Bernor and Armour-Chelu 1997). Curiously, this lineage would appear to be conservative in crown height, achieving a maximum height of about 60 mm. Older late Miocene hipparions from the Siwaliks, ca. 8–7 Ma, achieved crown heights well over 70 mm (Bernor and Armour-Chelu 1999; Bernor and Wolf, unpublished data), and the Langebaanweg *Eurygnathohippus hooijeri* (ca. 5.0 Ma) achieved maximum crown heights of 80 mm (Bernor and Kaiser 2006). This suggests that the East African Rift exhibited some degree of provinciality, at least in late Miocene equid evolution.

Conclusions

We recognize two taxa of Middle Awash late Miocene hipparion, *Eurygnathohippus feibeli*, from the oldest stratigraphic levels of Jara-Borkana and an apparently slowly evolving derivative, *Eurygnathohippus* aff. *feibeli*, from the 5.7–5.2 Ma interval. The *Eurygnathohippus feibeli* lineage would appear to have been restricted to eastern Africa. *Eurygnathohippus,* in turn, would appear to have diverged from southern Asian species of *Sivalhippus* at least by Lower Nawata, or possibly earlier times (>7.4 Ma). As indicated by Bernor and Kaiser (2006), *E. feibeli* may share a close phylogenetic relationship with Langebaanweg *Eurygnathohippus hooijeri*. This requires further study. It would appear that the gracile members of the *Eurygnathohippus* clade were adapted to open-country running and habitually ate C_4 grass (Cerling et al. 2003; Bernor et al. 2005b).

14

Rhinocerotidae

IOANNIS X. GIAOURTSAKIS, CESUR PEHLEVAN, AND YOHANNES HAILE-SELASSIE

During the Miocene, fossil rhinoceroses were diverse and widespread in Africa. At least five different lineages (aceratheres, brachypotheres, iranotheres, dicerorhines, and dicerotines), comprising about eight genera and thirteen species, have been documented (Hooijer 1978; Guérin 2003). However, only one lineage, the dicerotines, managed to survive the biotic turnover event at the Miocene-Pliocene boundary. It persists today with two ecologically differentiated species: the extant black or hook-lipped rhinoceros, *Diceros bicornis,* a browser; and the extant white or square-lipped rhinoceros, *Ceratotherium simum,* a dedicated grazer.

Fossil representatives of the tribe Dicerotini are relatively poorly documented in Miocene Africa. Few African localities have yielded adequate material for a detailed study, and in each case a new species has been described. Despite the number of Miocene species and the more adequate Plio-Pleistocene fossil record, the early evolutionary history of the tribe and the split between the extant black and white rhinoceroses remain tentative and controversial (Thenius 1955; Hooijer 1968; Hooijer and Patterson 1972; Hooijer 1978; Groves 1975; Guérin 1980b , 1982; Groves 1983; Geraads 1988; Guérin 1989; Heissig 1989; Geraads 2005). Therefore, the discovery of a relatively well-preserved Dicerotini skull close to the Miocene-Pliocene boundary in the Middle Awash Valley is of particular interest. A full listing of the institutional, anatomical, and locality abbreviations used herein can be found in the Appendix at the end of this chapter.

Rhinocerotidae

Diceros Gray, 1821

Diceros douariensis Guérin, 1966

RESTRICTED SYNONYMY 2004 *Ceratotherium* cf. *C. praecox* (Haile-Selassie et al. 2004c: 544, figure 5)

2001 *Diceros* sp. (Haile-Selassie 2001a: 318–319)

REVISED DIAGNOSIS *Diceros* of large size; nasal and frontal horns present, nasal bones rostrally rounded with abrupt and broad termination; premaxillary bones reduced;

lower border of orbit sloping laterally downwards; anterior border of orbit above the level between M^1 and M^2; supraorbital process very strong; postorbital process absent; dorsal cranial profile concave; parietal crests widely separate; nuchal crest straight or slightly indented, not extending over the occipital condyles; occipital plane vertically oriented; postglenoid process strong and straight; posttympanic process bending forward, narrowing the external auditory pseudomeatus, but not contacting the postglenoid process. Mandibular symphysis anteriorly abbreviated and narrow, posteriorly extending below the level of P_3; ventral border of mandibular corpus convex without marked angulation at mandibular angle. Lower premolars with open internal valleys, not forming fossetids. Upper and lower incisors absent or vestigial. High-crowned brachydont maxillary dentition with concave occlusal surface, inequalities in enamel thickness, and thin cement coating. Upper premolars with variable persistence of d^1 in adulthood; lingual cingulum strong, crenellated, and continuous; crochet present, crista and medifossette absent; protocone and hypocone not constricted; antecrochet absent; paracone fold present; metacone fold faint or absent. Upper molars (M^1, M^2) with protoloph of M^1 bending slightly distolingually and metaloph vertically oriented; mesial protocone groove present, deep and marked; distal protocone groove absent or faint; lingual protocone groove present; crochet strong, crista and medifossette absent; paracone fold present, moderate; mesostyle swelling developed but weaker than paracone fold; buccal apices of metacone and paracone cusps sharp; M^3 with subtriangular outline and continuous ectometoloph.

DESCRIPTION KUS-VP-1/20 is a moderately well-preserved adult cranium including the complete right and left permanent dentition (Figure 14.1). The specimen has been restored from numerous fragments and bears multiple fractures, postmortem abrasion, and minor dorsoventral crushing. Most affected are the lateral sides, especially the buccinator region and the zygomatic bones, which are poorly preserved. The temporal bones and their zygomatic processes are better-preserved. In dorsal view, the anterior part of the nasal bones is completely preserved. The intervening area, including the posterior part of the nasals, a significant portion of the frontals, and parts of the dorsal border of the maxilla and the lacrimals, is missing. The parietal and interparietal bones are almost intact, but the occipital bone is more fragmentary. In ventral view, the complete dentition and the palatine processes of the maxilla are very well-preserved, including the two small premaxillary bones. The palatine bones are nearly complete, but most of the vomer is lost. The pterygoid bone is better-preserved on the right side. The basisphenoid and the basioccipital of the occipital are moderately well-preserved.

The nasal bones are thick and wide, bearing a very strong nasal horn boss with extensive, rough vascular impressions. In dorsal view, the nasal bones terminate abruptly, and their rostral end is wide and rounded. The internasal groove is deep and marked only at the rostral tip of the nasal dome. The nasal bones are completely fused posteriorly. The ventral surface of the nasals is transversally concave. In lateral view, the nasal incision extends backward to above the mesial half of the P^3. The nasal notch appears to be U-shaped, although this area is fragmentary, especially on the left side. The infraorbital foramen is situated above the distal half of the P^3. The facial morphology of the buccinator region is

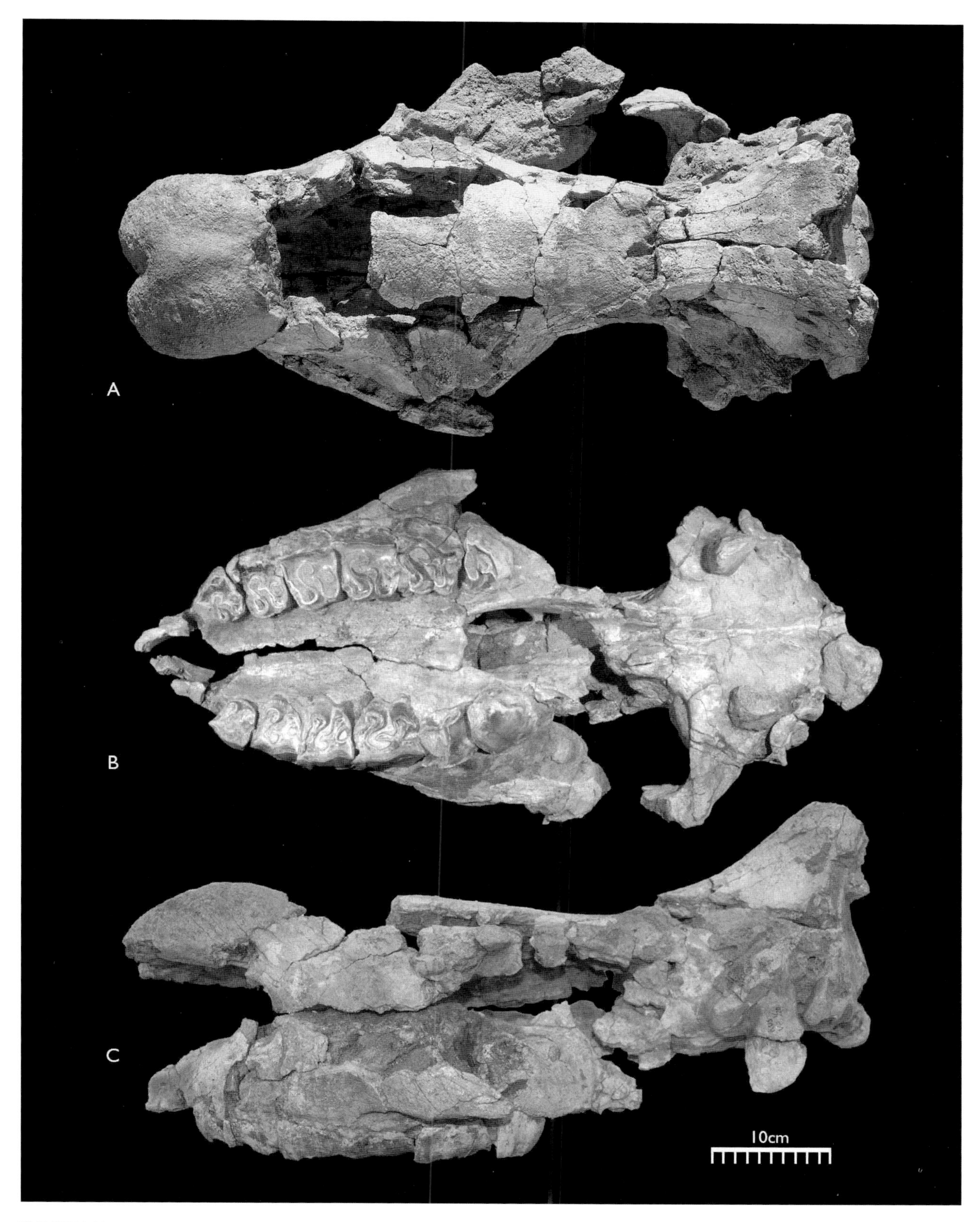

FIGURE 14.1

KUS-VP-1/20, cranium of *Diceros douariensis* with complete right and left permanent dentition. A. Dorsal view. B. Ventral view C. Lateral view.

poorly preserved. A significant portion of the frontal bones is missing, but the remaining fragments bear vascular rugosities at the level between the supraorbital processes, indicating the presence of a smaller frontal horn. The supraorbital process is very strong and prominent. The lacrimal process is weaker and more posteriorly oriented, bearing at least two separated lacrimal foramina at its base. A postorbital process is not developed on the frontals. The anterior border of the orbit is approximately situated above the level between M^1 and M^2. The floor of the orbit (dorsal surface of the zygomatic bone) slopes laterally downwards. The ventral border of the zygomatic bone is low, partly covering the maxillary tuber in lateral view. The temporal process of both zygomatic bones is poorly preserved.

The parietal bones are well-preserved. In lateral view, the dorsal profile of the skull is clearly concave, as in the extant *D. bicornis*. In dorsal view, the two oblique parietal crests are well-separated, and the interparietal bone between them remains wide and slightly convex transversally. Anteriorly, they curve smoothly laterally and are continuous with the temporal lines. Posteriorly, they diverge backward into the nuchal crests. The occipital border of the interparietal bone is damaged at the junction with the squamous part of the occipital bone. The nuchal crest appears dorsally indented (Figure 14.1A). However, it is distorted because of the missing fragments, especially on the left side. Based on the morphology of the more complete right side, we can infer that the nuchal crest was straight or only slightly indented in dorsal view, as in the extant *Diceros*. In occipital view, the squamous part of the occipital is fragmentary. The squamous occipital fossa is deep, and the external occipital protuberance was probably weak. There is no sign of an external median occipital crest. The nuchal tubercle is weak. The foramen magnum appears to be rounded, but it is incomplete. Only the right occipital condyle is preserved; it is kidney-shaped. Despite the incompleteness, the occipital plane appears to be almost vertical in lateral view, as in the extant *D. bicornis,* and not backwardly inclined, as in *C. simum*.

In ventral view, the basioccipital part of the occipital bone is moderately well-preserved. A marked and sharp sagittal crest runs along its middle, extending from the intercondylar incision to the basisphenoid. The paraoccipital process is missing its ventral tip on both sides. Its base is fused with the post-tympanic process, which bends forward, approaching very close to, but not contacting, the postglenoid process. The postglenoid process is strong, long, and straight. The bilateral basilar muscular tubercles are fused and demarcate the junction with the body of the basisphenoid. The pterygoid plates are thin and slope evenly, their posterior margin nearly horizontal. The vomer is poorly preserved. The anterior border of the choanae extends forward to the level between the M^2 and M^3. The palatine processes of the maxilla are fused in the middle, and their rostral border must have been indented. Both premaxillary bones are preserved. The left one is in better condition. They are short, thin, and flattened. They are edentulous, and no alveoli for permanent or persisting deciduous incisors are present. The distance between the rostral tip and the second premolar on the left side is less than 75 mm. A palatine process is not developed. The interincisive fissure is wide. It narrows rostrally as the premaxillary bones bend medially and approach each other but do not come in contact.

The complete left and right permanent dentition of the Kuseralee cranium is well-preserved (Figures 14.1B and 14.2A). The left M^3 is fully erupted and moderately worn, indicating a mature adult individual (Hitchins 1978). The right M^2 is severely deformed and pathologically twisted 90° counterclockwise, so that the metaloph and postfossette are facing the lingual side. This malformation has affected the occlusion of the right M^3, which is completely unworn, and also, to some extent, the right M^1 and P^4. Rhinoceroses chew on one side at a time, and during the occlusal stroke the teeth occlude only on the active side (Fortelius 1985). It is therefore reasonable that the animal tried to avoid the pathological side. We have observed a similar dental malformation in a P^4 of extant *C. simum* (USNM: 164592) and a P^4 of *Dicerorhinus sumatrensis* (BMNH: 1868-4-15-1). Several cases of rare dental malformations have been documented in the dentition of extant and fossil rhinoceroses (Patte 1934; Vialli 1955; Guérin 1980b).

Traces of a thin cement layer are evident in all teeth, especially on the buccal side of the ectoloph and in the medisinus valley of the molars. All premolars are molariform (*sensu* Heissig 1989: figure 21.1): The protoloph and metaloph do not fuse lingually, except at the very late stages of wear, keeping the entrance of the medisinus open. The first premolar (a D^1) has not persisted into adulthood. The P^2 is nearly square-shaped and markedly smaller than the succeeding premolars. As is common in P^2s, the metaloph is slightly longer than the protoloph, and the mesial width is greater than the distal. The hypocone, which bends slightly mesially, is also larger than the protocone. Both are unconstricted, and their lingual sides are rounded. The internal cingulum is strong, continuous, and crenellated. It begins on the medial side, projects lingually, surrounding both lingual cusps and the entrance of the medisinus, and bends again on the distal side, where it is almost completely worn down. A simple crochet is the only secondary fold developed; crista and antecrochet are absent. The postfossette forms a perfect circle in the worn metaloph. The ectoloph is gently convex, but a faint paracone fold can be traced.

P^3 and P^4 are similar in morphology, the last premolar somewhat larger than the third, and its protoloph slightly more oblique. The protocone and hypocone are subequal with a rounded lingual side. The mesial, lingual, and distal (worn down) cingula are strong, crenellated, and continuous. The metaloph is vertically oriented and the postfossette rounded. The crochet is simple and well-developed; crista and medifossette are absent. A mesial protocone groove is not developed on P^3 but is weakly present on P^4. A distal protocone groove and an antecrochet are absent in both. On the ectoloph, a weak but evident paracone fold is developed, as well as a faint metacone fold. A small, crenellated cingular trace is restricted to the distal corner of the ectoloph base.

Because of the abnormal right M^2, descriptions of features refer to the left side except where otherwise noted. The first two molars are morphologically similar. As expected, the M^1 is more rectangular, whereas the M^2 is somewhat longer but distally narrower. The enamel is thicker on the sides of the teeth and thinner around the medisinus. The protoloph of the left M^1 is oblique, bending distolingually, and the mesial protocone groove is deep and marked. The protoloph of the right M^1, which remains less affected by wear as a result of the malformation of M^2, shows better the marked degree of obliqueness (Figure 14.1B). The intensity of the distolingual sloping is more apparent with respect to

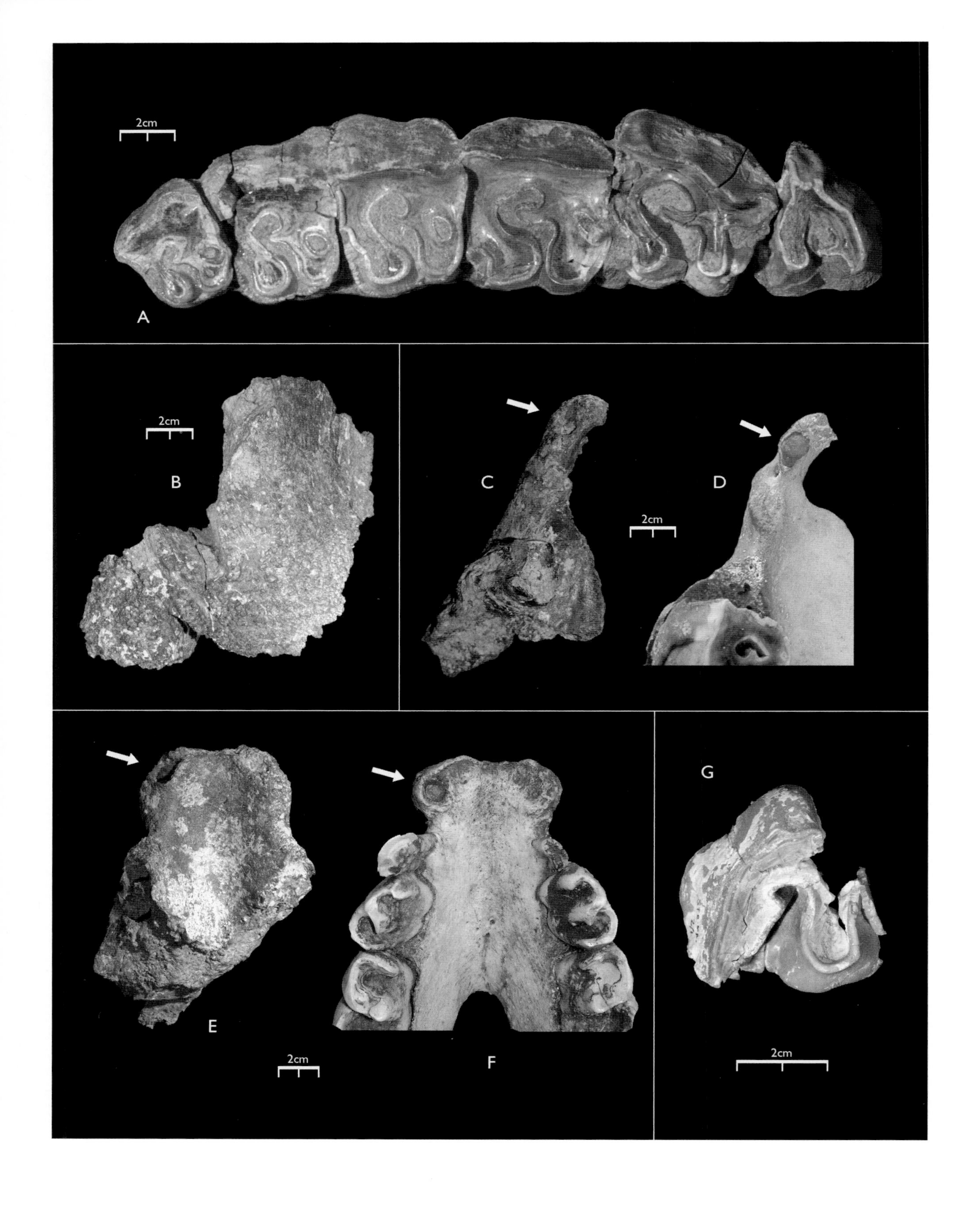

2cm
A
2cm
B
C
2cm
D
E
2cm
F
G
2cm

Plio-Pleistocene and extant *Diceros* skulls examined. The intensity of obliqueness closely resembles the morphology of the Langebaanweg molars but is decisively less marked with respect to the Plio-Pleistocene and extant *Ceratotherium* (see discussion). The metaloph does not bend distolingually but remains almost perpendicular with respect to the longitudinal axis of the tooth. The M^2 follows the same protoloph and metaloph arrangment, but both cross lophs are somewhat more tilted with respect to the ectoloph. A distal protocone groove and an antecrochet are not developed. The lingual wall of the protocone is more flattened than in the premolars. On the less worn right M^1 it is slightly depressed, but no marked groove is actually formed. A lingual cingulum is not developed; the mesial cingulum projects slightly on the lingual side of the protocone, particularly on the left M^2, but does not cross the entrance of the medisinus valley. The latter is V-shaped in the less worn right M^1 and left M^2 and would remain open in the very late stages of wear. A closed postfossette is formed only in the much worn metaloph of the left M^1. The only secondary fold projecting in the medisinus valley is a particularly strong and prominent crochet; a crista is absent. The ectoloph of the M^1 is straight and parallel to the longitudinal axis of the tooth row. The M^2 ectoloph is placed more obliquely, forming a bow with the M^3 ectometaloph. The parastyle is narrow and the parastyle groove flat in both M^1 and M^2. A moderate paracone fold is the most prominent vertical fold on the ectoloph. The mesostyle fold is developed as a broader but less prominent swelling, especially on M^2. A metacone fold is absent on the buccal wall of the ectoloph. However, the coronal tip of the metacone is more prominent than the paracone one. Both are sharp, and the intermediate ectoloph relief is concave. The metastyle is somewhat longer than the parastyle, and the metastyle groove is slightly concave.

The M^3 is subtriangular, bearing a continuous ectometaloph. The paracone fold is better-marked than on the previous two molars, as the tooth is less worn. A mesostyle swelling is evident on the middle of the ectometaloph. The lingual side of the ectometaloph is pointed. The protoloph is vertically oriented, and the mesial groove is very weak. As with the rest of the teeth, a distal groove and an antecrochet are not developed. However, a faint but conspicuous lingual protocone groove is developed on the base of the protocone. A prominent crochet is the only secondary fold developed, as on the two preceding molars; a crista is absent. The crown height of the unworn right M^3 measures 72.6 mm by an ectometaloph length of 62.8 mm; this provides a height/length index of 115.

FIGURE 14.2

A. *Diceros douariensis* from Kuseralee, left upper permanent dentition of the skull KUS-VP1/20 with P^2–M^3 in occlusal view. STD-VP-2/12, *Diceros* sp. from Saitune Dora. B. Nasal bone fragment in dorsal view. C. Right maxillary bone fragment with P^1 and premaxillary bone in ventral view; arrow indicates the presence of rudimentary I^1 alveolus. D. Extant specimen of *Diceros bicornis* (NHMW: 4292), detail of maxillary and premaxillary bone in ventral view; arrow indicates the presence of a rudimentary, unerupted I^1 inside the diminutive alveoli. E. *Diceros* sp. from Saitune Dora, STD-VP-2/12, mandibular symphysis fragment in dorsal view; arrow indicates the presence of a rudimentary I_2 alveolus. F. Extant specimen of *Diceros bicornis* (RMNH: Cat-B), detail of mandibular symphysis in dorsal view; arrow indicates the presence of a rudimentary I_2 alveolus. G. *Diceros* sp. from Saitune Dora, STD-VP-2/1, right M^3 in occlusal view.

Diceros sp.

RESTRICTED SYNONYMY 2004 *Diceros* sp. (Haile-Selassie et al. 2004c: 544, figure 5)
2004 cf. *Brachypotherium lewisi* (Haile-Selassie et al. 2004c: 544, figure 5)
2001 cf. *Brachypotherium lewisi* (Haile-Selassie 2001a: figure 5.45)

DESCRIPTION The remaining identifiable material in the Middle Awash sample under consideration consists mainly of worn upper and lower cheek teeth that cannot be presently assigned with certainty to the species level. An association with *Diceros douariensis* is possible. All specimens recovered from the locality of STD-VP-2 could represent one individual. This material was collected from the same spot during three different field seasons. The dental specimens have been restored from many fragments. The stage of wear and the state of preservation of the recovered teeth correspond perfectly to one another, and there is no repetition of elements. However, for some teeth, a direct contact could not be established, and therefore the different catalog numbers have been preserved (STD-VP-2/1, STD-VP-2/2, STD-VP-2/12, STD-VP-2/113). From the other localities of the Adu-Asa Formation, three dental (STD-VP-1/53, ASK-VP-1/10, ASK-VP-3/71) and two postcranial specimens (STD-VP-1/19, ASK-VP-3/202) have been recovered.

Cranial Elements STD-VP-2/12 comprises the majority of the material from STD-VP-2, including two important cranial fragments (part of the nasal bone and the right premaxillary bone) and the mandibular symphysis.

The nasal fragment retains only the rostral part of the bone, mostly the left side (Figure 14.2B). It is very similar to the nasals of KUS-VP-1/20, displaying the typical Dicerotini morphology. The rostral end is very broad and rounded. Extensive vascular impressions are developed, demonstrating the presence of a strong nasal horn. The internasal groove is deep and marked, but its posterior termination cannot be located because of the fragmentary condition.

The right premaxillary bone is well-preserved, including a small part of the maxilla with the right P^1 (Figure 14.2C). The bone is extremely reduced, and the distance between its anterior tip and the P^1 (probably a persisting d^1) is about 60 mm. The most interesting feature is the presence of a diminutive alveolus for a vestigial I^1. The presence of a rudimentary I^1 (or persisting di^1?) can also be occasionally observed in the extant African species (Figure 14.2D).

The mandibular symphysis is moderately well-preserved (Figure 14.2E). The symphysis is very short and narrow. The lingual face is evenly concave. The labial face is convex and bears several small foramina. There are no marked bilateral ridges developed along the interalveolar margin. The most interesting feature is the presence of a pair of diminutive alveoli, measuring about 11 mm in diameter, for a rudimentary I_2. These are also occasionally observed in the extant species (Figure 14.2F; Hitchins 1978: 72). The posterior part of the symphysis is poorly preserved, and the position of the posterior margin cannot be specified exactly. It must have extended at least to the middle of the P_3 roots, another characteristic feature of the tribe Dicerotini.

Upper Dentition The specimen STD-VP-2/12 comprises the right P^1 and the left P^1–P^3 and M^3. All the teeth are much more worn than those of the KUS-VP-1/20 cranium, indicating a very old adult individual. As described, the right P^1 remains attached on the maxillary fragment (Figure 14.2C). The isolated left P^1 (Figure 14.3C) is a small subtriangular tooth retaining its roots. No particular feature can be observed on the completely worn occlusal surface. A small mesiolingual cingulum appears to have been present. The left P^2 is very well-preserved (Figure 14.3C). In contrast to P^3, its mesial width is smaller than its distal one, and the hypocone is larger than the protocone. Both are unconstricted. The mesial cingulum projects lingually, surrounding the base of the protocone. It terminates at the entrance of the medisinus. The presence of a crochet during this late stage of wear cannot be verified. The same is true for the crista and the medifossette. The postfossette forms a small ring in the metaloph. The ectoloph is slightly sinuous, but no vertical folds can be distinguished. The left P^3 is also well-preserved (Figure 14.3C) and, apart from the differences mentioned, it is morphologically similar with the P^2 but significantly larger. Since it is slightly less worn, the presence of a remnant crochet can be verified. The lingual cingulum is longer, surrounding the base of the hypocone. It is less crenellated than the cingulum of the P^3 from the KUS-VP-1/20 skull. The left M^3 is very poorly preserved. Only a part of the ectometaloph is available. It is a perfect reflection of the more complete right M^3 of STD-VP-2/1 described below, supporting that they probably belong to the same individual.

STD-VP-2/1 consists of a very fragmentary right M^2 and a more complete right M^3. The slightly different color of the teeth with respect to STD-VP-2/12 is a result of more extended surface weathering. Only the metaloph and a small part of the protoloph of the right M^2 are preserved (Figure 14.3E). A worn crochet is present in the medisinus valley. A crista and medifossette are absent. The hypocone is not constricted. The deep postfossette remains distolingually open at this late stage of wear, supporting its identification as M^2. The right M^3 is better preserved; only a small portion of the protoloph is missing (Figure 14.2G). The morphology is similar to the M^3 of KUS-VP-1/20, except for the more worn crochet, which is not as prominent. On the protoloph the mesial protocone groove is marked, and the distal one is absent. An antecrochet is not developed. Crista and medifossette are also absent. The hypocone is angular, nearly V-shaped, and not constricted. The ectometaloph bears a weak paracone fold and is covered by a thin cement layer. Some traces of cement are also observable in the entrance of the medisinus valley. Neither lingual nor buccal cingula are developed.

STD-VP-2/2 is a moderately well-preserved left P^4 missing the mesiobuccal part of the ectoloph (Figure 14.3D). The tooth is very worn. The protocone is larger and more rounded than the hypocone; both are unconstricted. The mesial cingulum projects lingually into a crenellated lingual cingulum surrounding the base of the protocone and the entrance of the medisinus. A crochet was present, but it has been completely worn down. As in the other teeth, the antecrochet, crista, and medifossette are absent. The ectoloph of the tooth is poorly preserved. The coronal apex of the metacone cusp is sharp. A crenellated cingular trace is restricted to the distal corner of the ectoloph base.

STD-VP-2/113 is a bulk specimen including all remaining small dental fragments recovered from STS-VP-2, which could not be restored and securely associated with the larger specimens described.

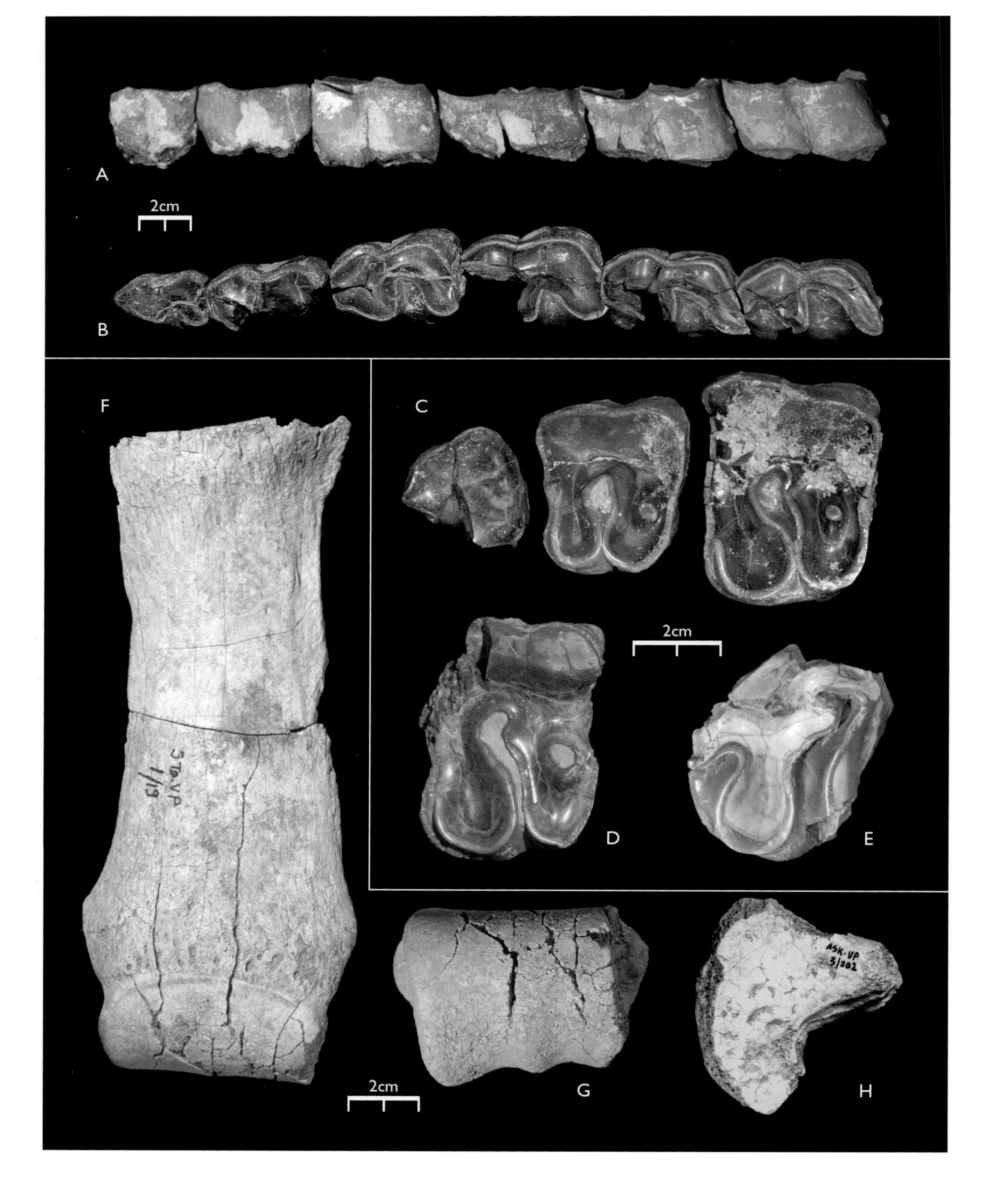
A
2cm
B
F
C
2cm
D
E
G
H
2cm
ASK-VP
3/202

STD-VP-1/53 is the only dental specimen recovered from the locality of STD-VP-1. It represents a right permanent upper molar, probably a M^1. The tooth is fragmentary and completely worn, so that no particular features on the occlusal surface can be described. Although it cannot be accurately measured, the size corresponds well to the described specimens from STD-VP-2.

ASK-VP-1/10 is a fragment of a worn right M^2 from ASK-VP-1. The specimen also includes some small indeterminable fragments of other teeth that belong to the same individual. The assignment to *Diceros* sp. is supported by the absence of a distal protocone groove (no constriction), the absence of an antecrochet, the absence of hypoconone constriction, and the presence of crenellated cingular traces in front of the medisinus entrance. A much worn crochet is also present.

ASK-VP-3/71 is a bulk sample comprising numerous upper and lower dental fragments from at least two individuals. Traces of a thin cement layer are excellently preserved on all ectoloph fragments of the upper dentition. The paracone fold is also well-developed. On the most complete ectoloph fragment, probably a right P^3, neither mesostyle nor metacone fold are developed. Buccal cingula are absent. A medisinus fragment of a much worn premolar shows the presence of a closed mediofossete. Another fragment, probably from a left molar, shows only the typical crochet. Traces of a crenellated lingual cingulum can be observed in some isolated medisinus fragments. Overall, these dental characters are in accordance with the generic morphology of *Diceros*.

Lower Dentition Both lower tooth rows of STD-VP-2/12 are almost completely preserved. They have been restored from many small dental fragments. Most of the teeth are very worn, hindering detailed description of the occlusal morphology (Figure 14.3A and B). The buccal wall of all the teeth is covered by thin cement traces that are more apparent in the ectoflexid groove (sometimes also preserved under the sediment). Cement traces are also preserved in some trigonid and talonid basins of the less worn teeth. Lingual and buccal cingula are not developed, the exception being some crenellated cingular traces at the base of the buccal wall. The mesial and distal cingula are moderately developed. The talonid basin of all teeth is lingually open, even at this late stage of wear (absence of closed fossettids). The P_2 has a reduced trigonid as is the case in advanced rhinocerotids. The paralophid of the P_2 is unconstricted, mesially rounded, and not prominent at this late stage of wear. The ectoflexid is not particularly deep, but better marked than those of the succeeding teeth. A mesial groove on the buccal wall of the trigonid is not developed. The talonid basin is open lingually, even at this late stage of wear. The P_3 and P_4 are very similar in morphology, with the latter being larger in size. Compared to the molars, both

FIGURE 14.3

Specimens of *Diceros* sp. from late Miocene deposits of the Middle Awash. A. STD-VP-2/12, right P_2–M_3, buccal view. B. STD-VP-2/12, right P_2–M_3, occlusal view. C. STD-VP-2/12, left P^1–P^3, occlusal view. D. STD-VP-2/2, left P^4, occlusal view. E. STD-VP-2/1, right M^2, occlusal view. F. STD-VP-1/19, left third metatarsal, dorsal view. G. STD-VP-1/19, left third metatarsal, distal view. H. ASK-VP-3/202, right ectocuneiform.

teeth have a more reduced trigonid and a more angular hypolophid. The M_1 is worn and still retains a rather angular hypolophid. The last two lower molars are morphologically more similar. Their most notable feature is the less angular hypolophid, especially of the M_3. The ectoflexid is moderately marked.

Postcranial Elements STD-VP-1/19 is a nearly complete left third metatarsal (Figures 14.3F and G). Only the morphology of the proximal epiphysis, which bears the articular facets for the adjacent bones, is obscured by abrasion and surface loss. On its medial side, the poorly preserved small dorsal and plantar facets for the second metatarsal are separated. On the proximal side, the facet for the ectocuneiform is very fragmentary, allowing only an estimation of its size. On the lateral side, the dorsal facet for the fourth metatarsal is completely missing. The plantar facet is better-preserved. It is round and separated by a very narrow groove from the proximal ectocuneiform facet. The diaphysis of the bone is rather straight proximally but widens distally. Its dorsal surface is transversally slightly convex. The plantar surface is flattened, bearing longitudinal rugose depressions for the interosseus metatarsal ligaments on either side of the proximal two-thirds of the shaft. The medial and lateral borders are rounded. On the distal part of the shaft, the medial and lateral tubercles for the attachment of the collateral ligaments of the fetlock joint are well-developed, but they do not project dorsally in distal view. The bilateral depressions for the attachments of the collateral sesamoidean ligaments are circular and deep. The distal epiphysis is well-preserved. In dorsal view, the proximal border of the trochlea is slightly convex. In plantar view, the proximal border of the trochlea is slightly sinuous and remains below the level of the bilateral tubercles. There are no deep supratrochlear depressions developed on the diaphysis above this border, only faint traces caused by the sesamoid contact. In distal view, the median sagittal keel of the trochlea is only weakly developed and remains much lower than the medial rim of the trochlea (Figure 14.3G).

ASK-VP-3/202 is a well-preserved right ectocuneiform (Figure 14.3H). The proximal and distal sides are flattened. On the medial side, the dorsal and plantar facets for the second metatarsal are separated, and there is no contact with the small proximal facet for the mesocuneiform. On the lateral side a dorsodistal and a proximoplantar facet are present for the cuboid.

Rhinocerotidae gen. et sp. indet.

DESCRIPTION At ASK-VP-2, an indeterminate ectolophid fragment of a moderately worn left P_3 or P_4 was recovered (ASK-VP-2/1). It has a well-marked and deep ectoflexid, a regularly developed mesial cingulum, and no buccal cingulum.

The presence of a rhinocerotid at Alayla is indicated by an indeterminate small tooth fragment, ALA-VP-2/136.

Discussion and Comparisons

Extant horned rhinoceroses and their fossil relatives are generally classified in three lineages: the dicerotines (includes extant *Diceros bicornis* and *Ceratotherium simum*), the rhinocerotines (includes extant *Rhinoceros unicornis* and *Rhinoceros sondaicus*), and the dicerorhines (includes extant *Dicerorhinus sumatrensis*). Their phylogenetic relationships and suprageneric classification have been highly controversial, with numerous arrangements proposed and debated (Guérin 1980b; Heissig 1981; Guérin 1982; Groves 1983; Prothero et al.1986; Geraads 1988; Heissig 1989; Prothero and Schoch 1989; Cerdeño 1995; McKenna and Bell 1997; Antoine 2002). They are considered here conditionally as three different tribes (Dicerotini Ringström, 1924; Rhinocerotini Owen, 1845; Dicerorhinini Ringström, 1924), forming the subfamily Rhinocerotinae Owen, 1845, of the "true (modern) horned rhinoceroses." Even molecular studies on the five extant species have failed to resolve this trichotomy satisfactorily, resulting in contradicting conclusions (Morales and Melnick 1994; Tougard et al. 2001; Orlando et al. 2003; Hsieh et al. 2003). As a result, molecular clock estimates must be considered cautiously, given that the fossil record for the early radiation of the subfamily is still inadequate. It is generally accepted that the radiation of the three lineages occurred early and rapidly, causing the existing difficulties and disagreements. The monophyly of the dicerotines within all rhinoceroses has been unequivocally supported by all morphological and molecular hypotheses proposed so far.

Because of its key stratigraphic position close to the Miocene-Pliocene boundary, the Kuseralee cranium KUS-VP-1/20 offers valuable indications regarding the potential evolutionary relationships within the tribe Dicerotini and necessitates a detailed and broad discussion. The comparisons begin with the extant and Plio-Pleistocene representatives of the tribe, where material is more abundant and the differences between the craniodental characters can be better analyzed. Then the Miocene African species are discussed, where available material is more limited and the differences more subtle. We finish the comparison with the particular case of the extra-African late Miocene lineage of *"D." neumayri*.

The genus *Diceros* Gray, 1821, as understood here, is paraphyletic. It includes all Dicerotini except the monophyletic lineage of Plio-Pleistocene and extant *Ceratotherium* Gray, 1868, as well as a monophyletic late Miocene extra-African lineage (provisionally referred to as *"Diceros" neumayri*) that arose independently. In this definition, the genus *Diceros* also includes Miocene and early Pliocene species and specimens exhibiting some progressive dental features that apparently represent an ancestral morphology with respect to the true Plio-Pleistocene *Ceratotherium*. These retain an overall craniodental morphology much closer to *Diceros* and do not warrant a generic ascription to *Ceratotherium*. In this aspect, the position of the material from Langebaanweg is left provisionally undecided and to be discussed separately. A complete phylogenetic analysis of the tribe is beyond the scope of the present contribution, since additional fossil evidence is still required, especially from the African Miocene.

Comparison with the Plio-Pleistocene and Extant Ceratotherium *sp.*

The extant white rhinoceros *Ceratotherium simum* (Gray, 1821) has two well-founded subspecies with a strikingly discontinuous range: *C. s. simum* from the southern part of the continent and the critically endangered northern *C. s. cottoni* (Lydekker 1908) from parts of central and eastern Africa. Only 25 animals of *C. s. cottoni* survive today, whereas *C. s. simum* has recovered from a bottleneck of ca. 20 individuals in 1895 to more than

11,000 animals today. A detailed account of their recent and historical status is provided by Emslie and Brooks (1999).

Heller (1913) was the first to point out two important morphological differences between the two subspecies: the length of the tooth row and the depth of the dorsal concavity. Groves (1975) statistically tested and verified these differences using a sample of over 60 skulls. *Ceratotherium simum simum* has a longer toothrow (80 percent joint nonoverlap) and a deeper dorsal profile (95 percent joint nonoverlap) compared to *C. s. cottoni*. Recent molecular studies confirm that the two subspecies are genetically distinct and require separate conservation (George et al. 1983; Merenlender et al. 1989). The genetic difference observed between them is greater than the genetic difference recorded between the various *Diceros* subspecies (Emslie and Brooks 1999). A rare case of a *Ceratotherium–Diceros* hybrid has been documented in captivity (Robinson et al. 2005).

During the middle-late Pliocene and Pleistocene, the white rhinoceros lineage (*Ceratotherium* sp.) was widespread across Africa (Guerin 1980b and references therein), following the expansion of open grasslands and signifying a remarkable example of herbivore adaptation to an exclusively abrasive diet. The distinction from the synchronic Plio-Pleistocene black rhinoceros lineage is easy, since the white rhinoceros lineage had already developed most of its apomorphic craniodental characters. Four fossil (sub)species have been erected based on Pleistocene *Ceratotherium* material: *Rhinoceros mauritanicus* Pomel, 1888; *Rhinoceros simus germanoafricanus* Hilzheimer, 1925; *Rhinoceros scotti* Hopwood, 1926; *Serengeticeros efficax* Dietrich, 1942. Several contradicting arrangements pertaining to their synonymy, specific or subspecific status, and spatiotemporal distribution have been proposed and debated (Arambourg 1938; Dietrich 1945; Arambourg 1948; Dietrich 1945; Cooke 1950; Hooijer 1969; Arambourg 1970; Groves 1972, 1975; Harris 1976a; Guérin 1979, 1980b; Harris 1983a; Guérin 1985, 1987a, 1987b; Geraads 2005). A (sub)specific evolutionary pattern of geographically/ecologically differentiated populations is feasible, but a revision of the Plio-Pleistocene true *Ceratotherium* lineage is beyond the scope of the present contribution. All Pliocene and early Pleistocene specimens assigned to the genus *Ceratotherium* (for a correct generic allocation of the principal cranial material from eastern Africa compare Geraads 2005: table 4) differ from the Middle Awash rhinoceros material described here by the following set of apomorphic features.

The skull is longer and more dolichocephalic with a dorsal cranial profile less concave; the anterior border of the orbit is usually retracted behind the middle of M^2; the occipital plane is inclined backward (posterodorsally), with a strong nuchal crest extending beyond the occipital condyles; the occipital notch of the nuchal crest is deeply concave or forked; the external occipital protuberance is strong with deep bilateral depressions (attachment for the funicular part of the nuchal ligament). The dentition is hypsodont, with constant enamel thickness and thicker cement investment. In the premolars, a crista or medifossette is variably present during the late Pliocene and more frequently in the Pleistocene; the lingual cingulum is progressively reduced. In the upper molars, a crista is usually present, forming in most cases a closed medifossette with the crochet; the protoloph is more oblique, bending markedly distolingually; the metaloph also becomes gradually more oblique; the paracone fold on the ectoloph weakens or disappears; the mesostyle fold becomes stronger than the paracone fold; the occlusal relief of the ectoloph is low or flattened. Furthermore, the middle-late Pleistocene and extant white rhinoceros (*C. simum* ssp.) differ additionally by the following features: the teeth are very high-crowned hypsodont and cement investment is abundant; the protoloph and metaloph fuse after early wear, closing the entrance of the medisinus in the premolars; the lingual cingulum on P^2–P^4 is very reduced or absent; M^1–M^2 with protoloph and metaloph both bending markedly distolingually; a closed medifossette is always present; the paracone fold is completely suppressed by a deep parastyle groove; the mesostyle fold is very prominent; the M^3 is subrectangular with separate ectoloph and metaloph; the mandibular symphysis is anteriorly widened; the lower premolars almost always form closed fossetids after moderate wear, and the lower molars have buccally flattened lophids.

These marked differences do not justify the ascription of the Kuseralee cranium to the genus *Ceratotherium*. All these features reflect the increasing adaptation of the white rhinoceros lineage to an exclusive grass diet: The head is more inclined toward the ground, bearing a more hypsodont and plagiolophodont dentition with abrasion-dominated wear. Effectively, these characters can be also used to distinguish *Ceratotherium* from the synchronic and partly sympatric Plio-Pleistocene and extant *Diceros*.

Review of "Ceratotherium praecox"

A species frequently used to describe Pliocene *Ceratotherium* material was *Ceratotherium praecox*. Hooijer and Patterson (1972) defined *Ceratotherium praecox* based on a fragmentary cranium (KNM-KP 36) with incomplete dentition from Kanapoi, Kenya (~4.2 Ma). The authors complemented their diagnosis with a more complete but crushed cranium from Ekora (KNM-KP 41), estimated to be younger than Kanapoi. They also assigned to the new species a single M^2 (KNM-LT 89) from Lothagam (which indeed bears progressive features). In the same year, Hooijer (1972) described abundant material from Langebaanweg under this name. Subsequently, and based largely on the Langebaanweg sample, the binomen *Ceratotherium praecox* was widely used to refer to the direct ancestor of the extant *Ceratotherium simum* (Hooijer 1973, 1976, 1978; Harris 1976a, 1983a; Guérin 1979, 1980b, 1985, 1987b, 1989; Hooijer and Churcher 1985; Harris and Leakey 2003a; Harris et al. 2003). However, as Geraads (2005) has recently demonstrated, both skulls from Kanapoi and Ekora belong undoubtedly to the Pliocene *Diceros* lineage, and so the former usage has been a source of confusion.

Hooijer and Patterson (1972: 19) themselves underlined the similarities of the type cranium (KNM-KP 36) with the extant *D. bicornis* and its differences with respect to extant *C. simum*: The anterior border of the orbit is placed over the anterior border of the M^2; the posterior elongation of the occipital is missing; the occipital plane appears not to be inclined; and the nuchal crest is not markedly indented. The very incomplete dentition (much worn right P^4–M^2 without ectolophs) bears only *Diceros* characters and is missing all progressive features, not only of the late Pliocene *Ceratotherium* but also of the stratigraphically older Kuseralee cranium and the Langenbaanweg sample (~5 Ma). The protoloph is not bending markedly distolingually (despite accentuation by the very worn rounded protocone); the mesial protocone groove is faint, even at this late wear stage; a lingual protocone groove is absent; a weak crochet is the only secondary fold developed.

The cranium from Ekora that retains a more complete and less worn dentition (P^2–M^2), also shows the same *Diceros* features (Hooijer and Patterson 1972: figures 10A, B).

It is apparent that the combinations *Diceros praecox* or *Diceros bicornis praecox* are available for the Kanapoi *Diceros* population. Geraads (2005) has suggested a broader usage including skulls from Lothgam, Hadar, Laetoli, and Koobi Fora; the Ekora skull was assigned to *D. bicornis*. The revised diagnosis of *D. praecox* provided by Geraads (2005: 455) comprises "a few apomorphic (cranial) features with respect to its likely ancestor *C. neumayri*." However, according to our comparisons, neither is *C. neumayri* its likely ancestor, nor are the suggested cranial features apomorphic. Since a broader evaluation with the Plio-Pleistocene and extant *Diceros* is required, we suggest that more and better-preserved material from the type locality needs to be documented before assessing evolutionary patterns. During recent excavations, the hypodigm of Kanapoi has not increased significantly (Harris et al. 2003).

Unlike the stratigraphically younger Kanapoi and Ekora material, the M^2 from Lothagam (KNM-LT 89) described by Hooijer and Patterson (1972: figures 8c, d) displays indeed several progressive morphological features similar to the Kuseralee cranium and the Langebaanweg sample. The most salient are its large size and relatively high crown, the weak paracone fold combined with a broad and evident mesostyle fold, the strong crochet, the somewhat distolingually bending protoloph, the deep mesial protocone groove, and the presence of a marked lingual protocone groove. A small crista is also developed, as in some teeth of the Langebaanweg sample. Despite its progressive features, isotopic analysis of the tooth indicates a C_3 browsing diet (Harris and Leakey 2003a), and is thus in accordance with our paleoecological inferences for the Kuseralee cranium. According to Harris and Leakey (2003a), the M^2 originates from the Lower Nawata (~6.5–7.5 Ma; McDougall and Feibel 2003) and bridges somewhat the gap between the Douaria and Kuseralee material.

Comparison with the Plio-Pleistocene and Extant Diceros *sp.*

During historical times, the extant black rhinoceros had a nearly continuous distribution throughout most of sub-Saharan Africa. Because of its wide distribution and adaptation to diverse habitats, extant black rhinoceroses show greater variability of locally adapted populations than do white rhinoceroses. This has led to the recognition of several subspecies or ecotypes, whose affinities are still under refinement (Hopwood 1939b; Zukowsky 1965; Mertens 1966; Groves 1967; Du Toit 1986, 1987; Groves 1993; Rookmaaker 1995; Hillman-Smith and Groves 1994). Molecular and biochemical studies generally support the separate management of different subspecies, although their results regarding the degree of genetic variation within populations may vary according to the applied methodology and sample (Ashley et al. 1990; Swart et al. 1994; O'Ryan et al. 1994; Swart and Ferguson 1997; Brown and Houlden 2000). Based on analysis of mtDNA restriction fragment length polymorphism, Ashley et al. (1990) estimated an average divergence of 0.29 percent between the subspecies *D. b. michaeli* and *D. b. minor* and suggested a common ancestry no further than 100,000 years ago. Analysis of the genetic variation in mtDNA control region, which has a higher rate of evolution to detect intraspecific variation than restriction enzymes, found a 2.6 percent nucleotide divergence between the same subspecies and suggested a divergence time of between 0.93 Ma and 1.3 Ma (Brown and Houlden 2000).

During the Plio-Pleistocene, the black rhinoceros lineage had a distribution quite similar to its historic one and is absent from North African localities (Guérin 1980a and references therein). Because of the close morphological resemblance with the extant species and the limitations of the fossil record, all Plio-Pleistocene *Diceros* material has been commonly assigned to the extant species *Diceros bicornis,* with the consideration of a subspecific treatment when more fossil material becomes available (Hooijer 1969, 1973, Harris 1976a; Hooijer 1978; Guérin 1979, 1985, 1987b; Harris and Leakey 2003a).

There are many similarities between the Kuseralee cranium (KUS-VP-1/20) and the Plio-Pleistocene and extant *Diceros bicornis*. The most significant common features include a markedly concave dorsal cranial profile, a straight or only slightly indented nuchal crest that does not extend posteriorly over the occipital condyles, a nearly vertical occipital plane, and anterior border of the orbit not extending behind the middle of M^2. The maxillary dentition is functionally brachyodont, with a concave occlusal surface, irregular enamel thickness, and thin cement coating. The upper premolars have a strong and continuous lingual cingulum, a paracone fold is developed, and a faint metacone fold is occasionally present. The upper molars (M^1, M^2) do not have a closed medifossette; the metaloph is vertically oriented; a paracone fold is present; a mesostyle fold is often developed as a broad swelling but is not stronger than the paracone fold; and the buccal apices of the metacone and paracone cusps are sharp. The M^3 has a subtriangular outline with continuous ectometaloph, lacking crista and medifossette.

All these craniodental similarities and the marked differences with respect to the Plio-Pleistocene and extant true *Ceratotherium* justify the ascription of the Kuseralee cranium to the genus *Diceros*. However, the Kuseralee cranium also displays some derived features with respect to the Plio-Pleistocene and extant *Diceros bicornis*. These include the particularly large size, the relatively high-crowned teeth, the distolingually bending protoloph, the deep mesial protocone groove, and the presence of a faint lingual protocone groove on the molars. The first two need to be carefully evaluated, because they seem to increase independently in some locally adapted *Diceros* populations during the Plio-Pleistocene.

The measurements of the Kuseralee cranium are slightly above or close to the maximum values, and much greater than the mean documented by Guérin (1980b), for ca. 50 skulls of extant *Diceros bicornis* ssp. (Table 14.1). Guérin (1980b: 29, 171) notes that the larger measurements in his sample originate from the "individus vraiment gigantesques" of the Cape black rhinoceros, but a more detailed subspecific analysis was beyond the scope of his study and refers to the work by Groves (1967). Groves (1967) used a larger sample of ca. 84 skulls, but with fewer measurements. He was able to demonstrate subspecific patterns within geographic populations, although the integrity of some of his groups based on a few skulls might be debatable (Du Toit 1987; Groves 1993). The large samples of *D. b. michaeli* (n = 22; occipitonasal length: 532 ± 20.9 mm) and *D. b. minor* (n = 23; occipitonasal length: 576.0 ± 17.0 mm) could indicate that variation within well-founded subspecies might not be much. Nevertheless, intergrades and overlapping with other groups undoubtedly occur (Groves 1967: 274, Tables 1, 2; Groves, 1993). In any case, out of the

TABLE 14.1 Skull Measurements of *Diceros douariensis* from Kuseralee Dora Compared with Other Dicerotini

	Kuseralee Dora	*Omo*	*Koobi Fora*	*Various*	*Africa, Extant*	*Africa, Extant*
	D. DOUARIENSIS KUS-VP-1/19	*DICEROS* SP. SHUNGURA D L.68-1	*DICEROS* SP. (GUÉRIN 1980A)	*CERATOTH.* SP. PLIOCENE-PLEISTOCENE	*D. BICORNIS* (GUÉRIN 1980A)	*C. SIMUM* (GUÉRIN 1980A)
1 L cond.-prmx.	640			720–750	494–619	649–748
n =				5	27 (563)	25 (708)
2 L cond.-nas.	ca. 680		561		519–676	661–786
n =					45 (584)	23 (742)
3 L occ.-nas.	ca. 660	585	537	742–920	480–655	667–836
n =				7	46 (567)	23 (797)
5 W min. of the braincase	128		107	103–154	96–147	94–121
n =				6	53 (116)	26 (112)
7 L nuchal crest-supraorb. proc.	ca. 345	301	308	414–502	285–390	406–454
n =				4	53 (324)	7 (428)
8 L nuchal crest-lacrymal proc.	ca. 405	363	332	470–535	325–424	395–515
n =				6	53 (364)	25 (486)
13 L cond.-M3	ca. 320		236	333–430	235–346	315–430
n =				7	45 (286)	24 (374)
15 W nuchal crest	ca. 208		168	224–280	114–211	181–249
n =				5	53 (186)	(224)
16 W between proc. paraocc.	262		218	250–299	191–264	212–291
n =				5	53 (230)	26 (257)
17 W min. parietal crests	78		76		30–101	30–101
n =					53 (69)	26 (65)
20 W between lacrymal proc.	ca. 305		228		211–312	232–328
n =					51 (255)	25 (290)
21 W bizygomatic	ca. 375		307	337–404	286–363	300–373
n =				5	53 (328)	26 (339)
22 W at nas.inc.	172	ca. 130	129	158–175	127–162	149–178
n =				3	48 (143)	25 (164)

NOTE: All measurements are in mm. Measurement numbers follow Guérin (1980a). Data for extant *D. bicornis* and *C. simum* are after Guérin (1980a). First row of each measurement shows either the value of single specimens or the range for the species. Second row presents the sample size and mean value for species. The Pliocene-Pleistocene material refers to *Ceratotherium* sp. skulls from Hadar, Dikika, Laetoli, Koobi Fora, Chemeron Formation, Olduvai, Rawi, and Ain Hanech. The sample is only used to demonstrate the size difference between the Plio-Pleistocene *Ceratotherium* sp. and the Kuseralee skull and, as such, no mean values are calculated. Data are based on Harris (1983a), Groves (1975), Guérin (1987a), Geraads (2005), and personal observations. Repetition of material is avoided by accepting the minimum and maximum values and the minimum number of skulls for each measurement.

seven subspecies recognized by Groves (1967), six subspecies (representing 79 out of 84 skulls) have maximum values well below that of the Kuseralee cranium. The only subspecies somewhat comparable to the Kuseralee cranium in size is indeed the Cape black rhinoceros, which constitutes the nominate subspecies *D. bicornis bicornis* (Linnaeus 1758), according to Thomas (1911) and Rookmaaker (1998, 2005). The available material of this subspecies was revised by Rookmaker and Groves (1978), including an important amendment of the statistical values of Groves (1967). In Groves (1967: Table 3) the occipitonasal length of *D. b. bicornis* is given as 667.0 ± 37.7 mm based on a sample of five adult skulls. In Rookmaker and Groves (1978: table 1) the same dimension is given as 629–653 mm (mean: 641.3 mm) based on a sample of four skulls. Similar differences occur in other measurements as well. As explained by Rookmaker and Groves (1978), this discrepancy is owed to the fact that Groves (1967) has uncritically included in his calculations the values of the "Groningen skull" as provided by Zukowsky (1965), for example, the occipitonasal length of 732 mm. Zukowsky (1965) had overestimated its dimensions based on the illustrations provided by Camper (1780, 1782). From the remaining four adult skulls of the Cape black rhinoceros (RMNH: cat-A; BMNH: 1838.6.9.101; MNHN: A.7969; SAM: 21383), we have examined the first three and can verify the analysis provided by Rookmaker and Groves (1978: table 1). Compared to the largest extant subspecies, the size of the Kuseralee cranium is slightly above or close to the maximum values.

Unfortunately, very few fossil *Diceros* skulls are available to document any spatiotemporal evolutionary patterns in size and proportions during the Plio-Pleistocene. They are smaller than the Kuseralee cranium and close to the mean values of the extant *Diceros bicornis* ssp. A quite well-preserved late Pliocene subadult skull (KNM-ER 636) was described by Harris (1976a, 1983a) from the KBS Member of Koobi Fora (the KBS tuff at the bottom of the member is dated to 1.88 Ma; McDougall and Brown 2006). Guérin (1980b: 165, table 39) analyzed it statistically and found that it falls within the values of the extant *D. bicornis* ssp. subadult and adult skulls (Table 14.1). A crushed adult skull from Laetoli has an occipitonasal length of 580 mm (Guérin 1987b: table 9.24). Two partial *Diceros* skulls from the Apak Member of Lothagam are reported as comparable in size to the extant species (Harris and Leakey 2003a: 378, figure 9.5), but no measurements are provided. From the ~2 Ma Shungura Member D level of the Omo Valley, a fairly complete but laterally compressed skull (L.68-1) was recovered (Hooijer 1973: Table 6, 1975). Its size is also smaller with respect to the Kuseralee cranium and closer to the mean of the extant *D. bicornis* ssp. (Table 14.1). Other incomplete cranial fragments (NME: Omo-54-2090, Omo-58-2085) from Omo are similar in size to L.68-1. From Hadar, only an incomplete *Diceros* cranium has been recovered so far (Geraads 2005) and no significant measurements can be taken. The holotype cranium (KNM-KP 36) of *"Ceratotherium praecox"* from Kanapoi (~4.2 Ma) is also very incomplete, as well as a second cranium from Kanapoi (KNM-KP 30). The cranium (KNM-KP 41) from the stratigraphically younger Ekora Formation, used by Hooijer and Patterson (1972) to complement the hypodigm of *"C." praecox,* is more complete, but crushed and distorted, and no measurements were provided.

The evolution of dental proportions is somewhat better-documented during the Plio-Pleistocene. Hooijer (1969: 87, 1972: 160, 1973: 165) and Guérin (1980b: 165) have demonstrated that during the early Pliocene, *Diceros bicornis* teeth are less high-crowned and that during the late Pliocene and Pleistocene some specimens achieve crown height similar to the extant species. Dental remains from the Mursi (~4.0 Ma) and the Usno (~3.0 Ma) Formations of the Omo sequence are relatively low-crowned. From Shungura Member D (~2.5 Ma) and younger levels they are similar to the extant form. Compared to the M^3 from the Kuseralee cranium (KUS-VP-1/20), an unworn M^3 from the Usno Formation (White sands W-12) is reported with a height/length index of 100 (Hooijer 1969: 87, 1973: 162). Guérin (1980b: 165, 1985: 81) also reports an index of 100 for a M^3 from the stratigraphically younger Shungura G level. Similar patterns are evident in other eastern African localities (Guérin 1980b: 165). The size of the permanent dentition of fossil specimens of *D. bicornis* falls within the mean values of the extant species (Hooijer 1959; Hooijer and Singer 1960; Hooijer 1969, 1973; Guérin 1980b; Harris 1983a; Guérin 1985; Hooijer and Churcher 1985; Guérin 1987b, 1994). Our observations verify these results, and, although the available record is insufficient for a more detailed analysis, it clearly demonstrates that smaller, medium-sized, lower-crowned *Diceros* populations existed throughout the Plio-Pleistocene. Dentitions from the late Pliocene of Hadar also indicate the presence of large-sized *Diceros* (Geraads 2005: Table 3). Geraads has assigned them, along with some of the Pliocene specimens already discussed, to *Diceros praecox*. However, morphometric comparison between them was not provided, nor were published metrical data from other Plio-Pleistocene localities and extant subspecies considered. The dentition of the Kanapoi cranium is very incomplete (P^4–M^2 without ectolophs) and too worn to allow any usable measurements. The morphology alone falls within the variation observed in the extant species. Certainly, based on its antiquity, we cannot exclude *D. praecox* as the ancestor of the Hadar *Diceros*. But it is equally plausible that it represents an extinct subspecies with no descendants or a subspecies evolving parallel to Hadar and other Plio-Pleistocene populations. We suggest that the Hadar *Diceros,* and eventually some other large-sized Plio-Pleistocene specimens as indicated by Geraads (2005: 457), most likely represent locally adapted populations, similar to the extant large-sized *D. b. bicornis* and *D. b. chobiensis*. They must have temporarily coexisted with the smaller, medium-sized, lower-crowned populations that appear closer to the mean values of the extant species. Groves (1967, 1993) has demonstrated a three-way clinal variation in eastern Africa between the extant subspecies *D. b. minor, D. b. michaeli,* and *D. b. ladoensis,* with geographic intergrades. As Hooijer (1969: 72) has noted, such a pattern of geographic subspeciation must have also existed during the past, but the available material is too limited to evaluate this variability. The same consideration was expressed by all subsequent studies (Hooijer 1973; Harris 1976a; Hooijer 1978; Guérin 1979, 1980b, 1985, 1987a, b, 1994; Harris and Leakey 2003a). Because the fossil record still remains insufficient to establish spatiotemporal subspecific (and perhaps specific) patterns during the Plio-Pleistocene, we also provisionally refer all Plio-Pleistocene black rhinoceroses as *Diceros bicornis* ssp.

The Kuseralee cranium (KUS-VP-1/20) demonstrates more craniodental similarities with the extant *Diceros* than with the extant *Ceratotherium*. However, its particularly large size, the high-crowned teeth, and some progressive dental features indicate that it cannot be the direct ancestor of the Plio-Pleistocene and extant black rhinoceros lineage as a whole. Contrarily, the dental morphology observed in the Kuseralee rhinoceros is further accentuated by even larger-sized and higher-crowned populations during the early Pliocene (notably the Langebaanweg rhinocerotid sample). It appears thus to be closely related to the stock that eventually evolved into the highly specialized *Ceratotherium* lineage during the Pliocene.

Comparison with the Langebaanweg Sample

The abundant rhinoceros sample from the early Pliocene locality of Langebaanweg (Hooijer 1972) has been inaccurately associated with *"Ceratotherium praecox"* and has been used for many years as its flagship reference. Contrary to *Diceros praecox* from Kanapoi, the stratigraphically older Langebaanweg rhinoceros represents indeed a more advanced form with respect to all Miocene African Dicerotini and the Pliocene–extant black rhinoceros lineage. However, it still lacks several key apomorphic features of the true grazing Plio-Pleistocene-extant white rhinoceros, as Hooijer (1972: 153) correctly outlined. A comprehensive revision of the enriched Langebaanweg collection is necessary to further evaluate the variation within this population. In this section, we restrict our discussion to the most essential morphological characters. The age of the Langebaanweg fauna is biochronologically estimated at about 5 Ma (Hendey 1981).

Some of the cranial features cited by Hooijer in his introduction (1972) were reproduced from the descriptions of the Kanapoi and Ekora *Diceros* specimens. From Langebaanweg, however, only some partial skull fragments are recorded (Hooijer 1972). Two large nasofrontal fragments (SAM: L.2520, L.6658) show the typical Dicerotini features: frontal and nasal horn bosses with extensive vascular rugosities, wide and rounded rostral border of nasals, and frontals with strong supraorbital processes. A pair of premaxillaries (SAM: L.13747) shows the presence of rudimentary upper incisors (Hooijer 1972: Plate 28), which sometimes also occur in the extant species (Figure 14.2D). Thus, it is not a diagnostic feature. The occipital morphology can only be observed on the partially reassembled portion of a very fragmentary cranium (SAM: L.31747). It appears to be intermediate between the two extant species, as Hooijer (1972: 158, plate 26) indicated. Compared to the Kuseralee cranium and extant *Diceros,* the occiput seems to be more inclined and the nuchal crest thicker and somewhat more prominent, but still not to the extent achieved by late Pliocene grazing *Ceratotherium* specimens from eastern Africa, or the terminal specimens of the late Miocene extra-African *"D." neumayri*. Important mandibular features of the Langebaanweg rhinocerotid include a symphysis anteriorly abbreviated and narrow (resembling modern *Diceros* and not *Ceratotherium*), posteriorly extending beneath P_3, and a convex ventral border of the mandibular corpus without marked angulation at the mandibular angle (Hooijer 1972: plates 30–33).

The total length of the maxillary toothrows from Langebaanweg (Hooijer 1972: table 1) is greater than that of Kuseralee (Table 14.2). Two unworn M^3s (SAM: L.6696, L.6638) have a height/length index of 120 and 121, respectively (Hooijer 1972), and thus are somewhat larger than the M^3 of the Kuseralee cranium. The following features are similar between the Kuseralee and the Langebaanweg dentitions but are generally more primitive than Plio-Pleistocene *Ceratotherium* (the latter cited in parentheses in the following list): strong continuous cingulum in premolars (progressively reduced during Plio-Pleistocene);

TABLE 14.2 Measurements of the Upper and Lower Permanent Dentition of the Middle Awash Specimens from Kuseralee and Saitune Dora

	Kuseralee Upper Dentition		*Saitune Dora Upper Dentition*		*Saitune Dora Lower Dentition*	
	KUS-VP-1/20		STD-VP-2/1	STD-VP-2/12	STD-VP-2/12	STD-VP-2/12
	DEX.	SIN.	DEX.	SIN.	DEX.	SIN.
P1L				30.2		
Wd				30.5		
P2L	38.5	37.5		37.7	35.2	(35.5)
Wm	41.7	44.0		43.3	23.1	nm
Wd	46.1	47.6		44.5	24.7	nm
P3L	(45.7)	nm		44.6	44.8	43.3
Wm	nm	(62.3)		61.0	nm	27.5
Wd	nm	61.3		59.4	nm	34.1
P4L	47.8	47.8			48.2	(49.7)
Wm	69.9	72.2			32.5	32.3
Wd	65.4	66.1			36.7	39.5
M1L	58.8	56.1			(56.0)	nm
Wm	70.0	69.5			nm	nm
Wd	65.2	66.7			44.5	nm
M2L	nm	61.0			57.6	57.8
Wm	nm	69.2			nm	43.9
Wd	nm	(64.1)			37.1	36.2
M3Lb	63.1	63.4	63.4	nm	nm	54.5
Wm	60.2	60.6	63.9	nm	nm	35.4
La	52.3	54.7	(58.5)	nm	nm	34.0
P2–M3	nm	284.0			nm	ca. 285.0
P2–P4	136.1	134.8			nm	ca. 130.0
P3–P4	97.2	96.8			nm	nm
M1–M3	nm	156.8			nm	ca. 155.0

NOTE: All measurements are in mm. Values in parentheses are estimates; nm = the measurement was not possible.

marked mesial protocone groove in molars (frequently more marked); absence of distal protocone groove and antecrochet in molars and premolars (also absent); inequalities in enamel thickness (enamel thickness remains more equal over the entire tooth); metaloph of M^1 is perpendicular with respect to the ectoloph (progressively more oblique and finally bending distolingually like the protoloph).

The following dental characters, particularly in molars, are further accentuated in the Langebaanweg specimens with respect to the Kuseralee dentition but remain markedly less advanced than in Plio-Pleistocene *Ceratotherium* (the latter cited in parentheses): The teeth are somewhat more high-crowned (but functionally still not hypsodont); the occlusal surface becomes less concave (but not flat); the ectoloph mesowear profile is lower (but not flattened); the cement investment increases (but is still not abundant); the moderate paracone fold persists or weakens, and the wide mesostyle fold is now more evident (structure develops further and the paracone fold gradually disappears); the protoloph bends somewhat more distolingually (obliquity increases further); the lingual protocone groove

TABLE 14.3 Measurements of the Ectocuneiform of ASK-VP-3/202 from Asa Koma and Other Dicerotini Specimens

	Asa Koma	*Langebaanweg*	*Africa, Extant*	*Africa, Extant*
	DICEROS SP. ASK-VP-3/202	DICEROTINI HOOIJER (1972)	*D. BICORNIS* GUÉRIN (1980A)	*C. SIMUM* GUÉRIN (1980A)
DT	62.0	53.0–60.0	43.0–60.0	51.5–62.0
		$n = 8$	$n = 22$	$n = 11$
		$\bar{x} = 56.3$	$\bar{x} = 51.0$	$\bar{x} = 55.9$
DAP	55.8	51.0–59.0	39.0–48.5	46.5–56.0
		$n = 7$	$n = 22$	$n = 11$
		$\bar{x} = 56.7$	$\bar{x} = 43.7$	$\bar{x} = 52.5$
H	(29.0)	27.0–33.0	22.0–22.9	25.5–29.5
		$n = 8$	$n = 22$	$n = 9$
		$\bar{x} = 30.0$	$\bar{x} = 25.0$	$\bar{x} = 27.4$

NOTE: All measurements are in mm. First row of each measurement shows either the value of single specimens or the range for the species. Second row presents the sample size for the species. Third row presents the mean value for the species. Data for extant *D. bicornis* and *C. simum* after Guérin (1980a); for the Langebaanweg sample after Hooijer (1972). Value in parentheses is an estimate.

becomes more evident (frequently present in late Pliocene *Ceratotherium,* but gradually disappearing later as bending of the protoloph increases and the protocone narrows); a small crista appears in some molars, but very rarely a closed medifossette, which occurs only in 3 out of 40 Langebaanweg teeth (a closed medifossette is very frequently present in Pliocene *Ceratotherium* and almost always present in Pleistocene and extant specimens).

A very important feature of the Langebaanweg rhinocerotid is the great length of its relatively slender limb bones, which in most cases significantly exceeds the maximum values recorded in the two extant species (for comparisons of the ectocuneiforms and third metatarsal, see Tables 14.3 and 14.4). Guérin (1979, 1980b, 1987b) analyzed the available postcranial material from the Rift Valley and demonstrated that similar size and proportions to the Langebaanweg limb bones were sustained by the late Pliocene *Ceratotherium* from Hadar-SH and Laetoli (which he referred to as *Ceratotherium praecox,* but the name *C. efficax* might be appropriate). During the early-middle Pleistocene, Guérin (1979, 1980b) recorded a shift toward more massive limb bones in the Olduvai sequence (referred to as *C. simum germanoafricanum*), which was then followed by a size reduction to that of the extant *C. simum* during the middle-late Pleistocene. However, northern African Pleistocene populations (for which the name *Ceratotherium mauritanicum* is available) seem to retain the long and relatively slender metapodials (personal observation: MNHN: 1956-12-109 from Ternifine; MNHN: 1953-21-58 from Ain Hanech). A similar pattern for the Plio-Pleistocene *Ceratotherium* limb bones was suggested by Geraads (2005: 455, figure 4), but using different species names. It is consistent with the craniodental evidence (Hooijer 1969; Groves 1975; Harris 1976a; Hooijer 1978; Guérin 1980b; Harris 1983a; Guérin 1985; Geraads 1987; Guérin 1987b; Likius 2002; Geraads 2005).

TABLE 14.4 Measurements of STD-VP-1/19 (Third Metatarsal) from Saitune Dora and Comparisons with Other Dicerotini Specimens

	Saituna Dora	*Fort Ternan*		*Bou Hanifia*		*Arrisidrift*	*Langebaanweg*	*Hadar*	*E. Mediterranean*	*Africa, Extant*	*Africa, Extant*
	DICEROS SP. STD-VP-1/19	*P. MUKIRII* HOOIJER (1968)		*D. PRIMAEVUS* MNHN: 1951- 9/242	9/144	*D. AUSTRALIS* GUÉRIN (2003)	DICEROTINI HOOIJER (1972)	*CERATOTH.* SP. NME: AL.382-5Y	"*D*". *NEUMAYRI* GIAOURTSAKIS (IN PREP.)	*D. BICORNIS* GUÉRIN (1980A)	*C. SIMUM* GUÉRIN (1980A)
L	188.0	115.0[a]	132.0[a]	161.0	164.0	178.0–190.5 $n = 4$ $\bar{x} = 188.3$	171.0–198.0[a] $n = 20$ $\bar{x} = 183.8$[a]	201.0	160.0–181.0 $n = 16$ $\bar{x} = 170.2$	141.5–178.0 $n = 33$ $\bar{x} = 157.1$	157.0–180.0 $n = 12$ $\bar{x} = 168.4$
DT ep. prox.	(68.0)	43.0	45.0		52.2	54.0–61.0 $n = 3$ $\bar{x} = 57.5$	55.0–70.0 $n = 20$ $\bar{x} = 60.8$	57.3	55.1–67.7 $n = 14$ $\bar{x} = 59.3$	43.5–58.5 $n = 33$ $\bar{x} = 48.9$	51.5–64.5 $n = 12$ $\bar{x} = 55.8$
DAP ep. prox.	(60.0)	36.0	41.0	46.2	48.3	49.0–52.0 $n = 2$ $\bar{x} = 50.5$	51.0–61.0 $n = 16$ $\bar{x} = 55.0$	55.4	49.1–60.5 $n = 15$ $\bar{x} = 52.7$	40.0–56.5 $n = 32$ $\bar{x} = 48.4$	46.0–53.0 $n = 11$ $\bar{x} = 49.4$
DT dia.	(59.5)	31.0	45.0	46.8	41.8	44.0–52.5 $n = 4$ $\bar{x} = 49.5$	47.0–62.0 $n =$ $\bar{x} = 53.5$	49.3	46.7–57.6 $n = 16$ $\bar{x} = 50.2$	37.5–48.0 $n = 33$ $\bar{x} = 42.5$	43.0–52.5 $n = 12$ $\bar{x} = 47.3$
DAP dia.	29.2	20.0	19.0	23.6	23.2	25.5–26.0 $n = 3$ $\bar{x} = 25.7$	25.0–35.0 $n =$ $\bar{x} = 29.3$	25.3	23.7–30.2 $n = 18$ $\bar{x} = 25.7$	17.0–25.5 $n = 30$ $\bar{x} = 21.2$	23.0–28.5 $n = 11$ $\bar{x} = 25.2$
DT max. dist.	77.6	40.0	53.0	59.1	52.3	55.5–61.5 $n = 4$ $\bar{x} = 56.7$	58.0–82.0 $n = 16$ $\bar{x} = 69.1$	69.8	62.0–76.5 $n = 15$ $\bar{x} = 66.2$	50.5–64.0 $n = 33$ $\bar{x} = 56.1$	59.0–72.0 $n = 12$ $\bar{x} = 64.0$
DT ep. dist.	59.8	37.0	42.0	50.3	48.5	51.0–57.0 $n = 4$ $\bar{x} = 53.4$	53.0–69.0 $n = 18$ $\bar{x} = 57.5$	56.8	50.9–61.5 $n = 16$ $\bar{x} = 53.8$	40.0–55.5 $n = 33$ $\bar{x} = 47.3$	48.5–60.0 $n = 12$ $\bar{x} = 52.8$
DAP ep. dist.	53.6	33.0	36.0	42.7	44.3	42.0–47.5 $n = 4$ $\bar{x} = 44.4$	46.0–58.0 $n = 20$ $\bar{x} = 50.8$	41.3	44.2–51.2 $n = 14$ $\bar{x} = 46.2$	32.0–43.5 $n = 27$ $\bar{x} = 40.1$	40.0–49.0 $n = 10$ $\bar{x} = 45.5$

NOTE: All measurements are in mm. Values in parentheses are estimates. First row of each measurement shows either the value of single specimens or the range for the species. Second row presents the sample size for the species. Third row presents the mean value for the species. Data for extant *D. bicornis* and *C. simum* after Guérin (1980a); for *D. australis* after Guérin (2003); for *P. mukirii* after Hooijer (1968); for the Langebaanweg sample after Hooijer (1972). Measurements follow Guérin (1980a).

[a]Hooijer (1968, 1972) provides the median length rather than the total length of the bone. In the case of 3rd metatarsals, this usually results in ca. 2–4 percent smaller values.

The size and proportions of the third metatarsal from Saitune Dora (STD-VP-1/19) corresponds well to the Langebaanweg sample, indicating close relationships. Postcranial elements of *D. douariensis* are still not securely associated, since only fragments have been recovered from the type locality (Guérin 1966).

The Langebaanweg rhinocerotid probably originated from large-sized, high-crowned late Miocene populations similar to the Douaria and Kuseralee rhinocerotids. Its particular morphology shows that it is closely related to the lineage that evolved into the true grazing *Ceratotherium* during the Pliocene, as Hooijer (1972) suggested. This scheme does not imply any concrete migrational evolutionary patterns between northern, eastern, or southern Africa, but rather reflects the inadequate fossil material presently available. The intermediate character of the Langebaanweg rhinocerotid is explained by dietary requirements related to local seasonal environmental conditions.

Comparison with Diceros douariensis

The presence of a fossil rhinoceros in Douaria, Tunisia, was first mentioned by Roman and Solignac (1934) in a "Pontian" fauna list as *Rhinoceros pachygnathus*. The rhinocerotid material was reexamined in detail by Guérin (1966), who assigned all remains to the new species *Diceros douariensis*. The holotype is a partial adult cranium (FSL: 16749), missing the nasal and occipital regions, associated with a fairly complete mandible (FSL: 16750). Guérin (2003) biochronologically estimates the age of the site of Douaria at 9.5 Ma, but based on the associated fauna a younger age cannot be excluded. Some dental remains and an astragalus have been described as cf. *Diceros douariensis* from the Miocene locality of Djebel Krechem el Artsouma, central Tunisia (Geraads 1989). The referred occurrence of the species in Baccinello V3, Italy (Guérin 1980b, 2000) is doubtful.

A partial juvenile cranium (FSL: 16752) was also used by Guérin (1966: figures 2, 6) to complement the hypodigm of *D. douariensis*. The presence of a complete protocone constriction by a mesial and distal protocone groove and a well-developed antecrochet in the M^1 and M^2 of the juvenile skull are atypical for Dicerotini. Dicerotini molars usually have only a mesial (anterior) protocone groove, without a distal (posterior) one. Thus there is no true (complete) protocone constriction and no prominent antecrochet developed, exactly as in the adult skull from Douaria. Moreover, the hypocones of the juvenile specimen also appear to be constricted, at least by a mesial groove. The juvenile specimen represents clearly a different species, probably an acerathere or brachypothere, and is excluded from the comparisons.

The available morphology of the holotype adult cranium (FSL: 16749) is described and illustrated in detail by Guérin (1966). It displays characteristic features of the Dicerotini, including the presence of well-developed nasal and frontal horn bosses, the strong supraorbital process, the laterally sloping lower border of the orbit, and the short edentulous premaxillary bone. Unfortunately, the incomplete condition of the neurocranium obstructs the evaluation of important features related to the development of the posterior cranial region. The associated mandible (FSL: 16750) is also characteristic: The ventral border of the mandibular corpus is convex without marked angulation at mandibular angle; the anterior border of the mandibular symphysis is abbreviated, edentulous, and rather narrow; and the posterior border extends below the level of P_3. One of the diagnostic characters reported by Guérin (1966) is the particularly large size of the skull with respect to extant *D. bicornis*. Because of the incompleteness, the most important measurements had to be estimated. The basal length of the skull is estimated by Guérin (1966) to be about 605 mm, thus larger than the majority of the extant subspecies and comparable only with *D. b. bicornis*. The zygomatic breadth of 260 mm is probably underestimated, because this region is incomplete and apparently distorted (Guérin 1966: figure 5). The length of the dentition (P^2–M^3) measures 264 mm (Guérin 1966: table 1) and is thus somewhat smaller than in the Kuseralee cranium (KUS-VP-1/20).

The teeth of the adult skull were described as high-crowned and, although somewhat worn, the illustrations seem to support this assessment (Guérin 1966: figures 1, 3). However, the hypsodonty indices provided by Guérin (1966) were based on the unworn M^1 and M^2 of the dubious juvenile skull. Hooijer (1973) questioned the "hypsodonty" of *D. douariensis* and recalculated these indices based on greatest ectoloph lengths, concluding that the molars of the juvenile skull are rather lower-crowned compared to the extant *Diceros*. In any case, neither calculation should be considered as applicable for the crown height of the adult skull. The dental morphology of *D. douariensis* follows the unspecialized *Diceros* pattern. The premolars have a strong, continuous crenellated cingulum; a crochet is present, and the crista is absent. Guérin (1966) mentions the presence of a weak antecrochet on P^2–P^3. A paracone fold is developed on the ectoloph. A weak metacone fold is reported on the P^3. The molars, although high-crowned, are functionally brachyodont with concave occlusal surfaces, unequal enamel thickness, and sharp paracone and metacone buccal apices. Cement is present. A crochet is present but no crista or medifossette. The mesial protocone groove appears to be marked, but a distal groove is not developed. The ectoloph bears a paracone fold; a weak mesostyle bulge is apparently also developed (Guérin 1966: figures 5, 8). The M^3 has a continuous ectometaloph. In these morphological features the skull from Douaria is similar to the Kuseralee cranium, as well as to the extant *D. bicornis* (although some subspecies of the latter have secondarily developed a crista or a bifid crochet on premolars [Rookmaaker and Groves 1978; Guérin 1980b; personal observation]).

Besides the reported large size and the apparent high-crowned teeth, some derived dental features also signify a closer relationship between the Douaria and Kusarelee rhinocerotids and distinguish them from Plio-Pleistocene and extant *D. bicornis*. The first one is the obliquity of the protoloph of M^1 reported by Guérin (1966: 30): "Le protolophe est fortement convexe vers l'avant, quelle que soit l'age de l'inividu." The second one is the development of a lingual protocone groove (on the lingual side of the protoloph): "Son extremité linguale est . . . deprimée verticalement en son milieu." This lingual protocone groove is also clearly indicated on the line-drawing illustration (Guérin 1966: figure 8). It is the same feature seen on the M^2 from Lothagam, the molars of the Langebaanweg sample, and the Pliocene true *Ceratotherium*. Some differences can be observed between the Douaria and Kuseralee dentitions. The lingual protocone groove is very marked on the M^1 of the Douaria dentition, and a faint antecrochet seems to be present. Contrarily, the distolingual bending of the protoloph is more conspicuous in the Kuseralee M^1 and the paracone fold weaker. More material is thus nessesary to appreciate the variation of both populations.

Guérin (1966) considered *D. douariensis* as a circum-Mediterranean species showing a mixture of progressive and primitive features with respect to the eastern Mediterranean *"D." neumayri*, a well-established species at the time. Hooijer and Patterson (1972: figure 11) considered *D. douariensis* as a possible ancestor of both extant African lineages. Hooijer (1978: figure 19.1) later deemed its position to be close to the split, on the side of *Diceros*. Heissig (1989) also suggested a placement near the split but on the *Ceratotherium* side. Geraads (2005) considered *D. douariensis* as potentially conspecific with *"D." neumayri* and the latter as the common ancestor of the extant species (all other authors regarded *"D." neumayri* as a separate lineage). The morphological similarities between the Kuseralee cranium and *D. douariensis* support the assignment of the former to the same species/lineage with a position close to the ancestral stock of the *Ceratotherium* clade, as Heissig (1989) suggested. Based on the available material, their precursor could have originated from the earlier *D. primaevus* or *D. australis*.

Comparison with Diceros primaevus

The rhinoceros material from the early late Miocene locality of Bou Hanifia (Oued el Hammam), Algeria, was originally described as *Dicerorhinus primaevus* by Arambourg (1959) and was later allocated to *Diceros* by Geraads (1986b). The Bou Hanifia Tuff, found below the mammal horizon, has provided a radiometric date of 12.18 ± 1.03 Ma (Ameur et al. 1976). The type specimen of *Diceros primaevus* is a partial juvenile cranium with erupting M^1 (MNHN: 1951-9/222; Arambourg 1959: Plate 6, figure 1-3). The rest of the recovered dental material consists of juvenile maxillae and hemimandibles with deciduous dentition. The cranial morphology of the partial juvenile skull displays the typical characters of the tribe Dicerotini: strong nasal and frontal horn boss, laterally sloping lower border of the orbit, well-developed supraorbital process (although broken), and absence of a postorbital process (contra Arambourg 1959). The nasals are mediolaterally crushed and compressed; therefore, the characteristically wide and rounded rostral border is not apparent in Arambourg's illustrations.

The unworn M^1 and M^2 of a juvenile maxilla (MNHN: 1951-9-219) are morphologically identical to extant *Diceros* but somewhat less high-crowned. All deciduous dentitions (five specimens) show some primitive features, such as a weak or absent crista in D^3 and D^4, which are retained by some subspecies of extant *Diceros*. This apparently primitive morphology as well as a misleading comparison with the eastern Mediterranean species *Diceros pachygnathus* (here: *"Diceros" neumayri*) and *Dicerorhinus orientalis (recte: Dihoplus pikermiensis)* have contributed to the initial assignment of the Bou Hanifia rhinocerotid to the genus *Dicerorhinus* by Arambourg (1959). Arambourg (1959: figures 33a, 33b) misidentified and swapped the juvenile maxillae of the two Eastern Mediterranean species from Pikermi and incorrectly associated the Bou Hanifia maxillae with the *"Dicerorhinus"* morphology. Deciduous dentitions of the synchronic and partly sympatric species *"Diceros" neumayri* and *Dihoplus pikermiensis* from the eastern Mediterranean can be easily distinguished based on several unambiguous morphological features (Giaourtsakis et al. 2006: table 3; Geraads 1988).

The postcranial elements found in Bou Hanifia are generally slender for Dicerotini. Their size and morphology fall perfectly within the range and variation documented for the extant *D. bicornis* (Guérin 1980b; personal observation). Compared to the third metatarsal from Saitune Dora, the early Pliocene sample from Lagebaanweg, and the Pliocene *Ceratotherium,* they are significantly shorter (Table 14.4). Compared to the extra-African *"D." neumayri,* they are more slender. A fragmentary atlas (Arambourg 1959: figure 25), showing the presence of an alar incisure lateral to the articular surface of the occipital condyle, does not belong to a rhinoceros but represents probably a short-necked giraffid. This feature has also contributed to the initial assignment to *Dicerorhinus* (in Dicerotini an alar foramen is present instead of an incisure).

D. primaevus preserves an ancestral morphology that essentially persists, with relatively few modifications, in the extant black rhinoceros. Populations similar to *D. primaevus* could have migrated outside Africa, around the middle-late Miocene boundary, and evolved to *"Diceros" neumayri*. In addition, the conservative morphology of *D. primaevus* does not exclude a placement of this species before the split between the extant black and white rhinoceros lineages. This depends, however, on the affinities of the recently discovered early middle Miocene Arrisdrift rhinoceros from Namibia, considered next.

Comparison with Diceros australis

Guérin (2000) described the new species *Diceros australis* based on material discovered at the locality of Arrisdrift in the Orange River Valley of Namibia. A slightly extended version including some additional specimens was presented by Guérin (2003). According to Pickford and Senut (2003), the age of the Arrisdrift fauna is estimated at about 17.5–17 Ma. Besides a small occipital and a few mandibular fragments, the rhinocerotid material from Arrisdrift assigned to *D. australis* comprises several isolated dental and postcranial elements.

The morphology of the upper permanent cheek teeth follows the unspecialized Dicerotini pattern, similar to the extant *Diceros*. Compared to the younger *D. primaevus* from Bou Hanifia, the most notable difference is the size. Guérin (2003) points out a closer resemblance of *D. australis* with the younger (by 8 million years) *D. douariensis,* which shows very similar dimensions. Both still fall within the dental size variation observed in extant *Diceros bicornis* ssp. (Guérin 1980b: table 5), but *D. australis* lacks the few advanced features of *D. douariensis* and the Kuseralee dentition described earlier.

The most prominent feature of *D. australis* (~17 Ma), however, is the significant size of its postcranial elements. The holotype specimen itself is a left third metacarpal (GSN: AD-52'97). The recovered limb bones of *D. australis,* especially the metapodials, are considerably larger than the maximum values recorded for the two extant species, as well as the early late Miocene *D. primaevus* (~12 Ma) and the side branch of the extra-African *"D." neumayri* (Table 14.4). A similar size and morphology can be found in a fourth metatarsal from the much younger Mpesida beds of Kenya (~6.2–6.9 Ma), originally referred to *Ceratotherium praecox* by Hooijer (1973), in the third metatarsal from Saitune Dora (~5.6 Ma) described in this chapter, and in the abundant material from the early

Pliocene of Langebaanweg (~5 Ma) documented by Hooijer (1972). As discussed earlier, this pattern continued thereafter by the true *Ceratotherium* lineage during the Pliocene and early Pleistocene and was followed by a significant size reduction and modification of the osteometric proportions during the middle and late Pleistocene leading to the extant white rhinoceros (Guérin 1979, 1980b, 1987b).

The puzzling discovery of the large *D. australis* as the oldest known representative of the tribe Dicerotini raises important issues regarding the early radiation of the tribe and further perplexes the search for the split between the black and white rhinoceros *(sensu lato),* as it would clearly pose a second center of evolution next to the younger and smaller sized *D. primaevus*. Further fossil evidence is required to shed light on the early evolution of the tribe in Africa.

Revision of Paradiceros mukirii

Hooijer (1968) described the new genus and species *Paradiceros mukirii* from the middle Miocene Fort Ternan Beds in Kenya. Recently, Pickford et al. (2006) refined the age of the Fort Ternan fossiliferous sediments to ca. 13.7 ± 0.3 Ma. Hooijer (1968) originally portrayed *P. mukirii* as a primitive collateral species of the ancestral *Diceros* stock, differing from *Diceros* in a combination of primitive and progressive features. However, our comparisons indicate that the majority of the material, if not all, may belong to the dicerorhine *"Dicerorhinus" leakeyi* Hooijer, 1969. The ascription of the latter species to the genus of the extant Sumatra rhinoceros might be incorrect but shall provisionally be retained, because a more comprehensive comparison with the Eurasian Miocene dicerorhines (*sensu* Guérin 1989) would be required.

There are casts of three important specimens from Fort Ternan in the collections of the BMNH: the juvenile holotype cranium (BMNH: M.29929; original KNM-FT-1962-3113; Hooijer 1968: plate 1), the incomplete adult cranium (BMNH: M.29930; original KNM-FT-1962-3376; Hooijer 1968: plate 2, figures 2, 3) and one mandible (BMNH: M.29931; original KNM-FT-1962-3209). The holotype of *P. mukirii* is a well-preserved juvenile cranium missing the nasal, basioccipital, and premaxillary bones due to incomplete ossification. The available cranial morphology is lacking several important Dicerotini features: The lower border of the orbit is not sloping laterally downwards; the frontal horn boss is developed as a prominent but restricted swelling in the middle of the frontals; and the frontals are only slightly convex at the level of the the supraorbital processes. In contrast, even in juvenile Dicerotini, the supraorbital process is strong; a postorbital process is generally absent, or only faintly developed; the frontal horn boss is more extensive with marked vascular impressions; and the frontals are very convex between the supraorbital processes. In addition, the nasals of the incomplete adult cranium are rather long (length between nasal tip and nasal incision) and do not terminate abruptly rostrally (see also Hooijer 1968: plate 2, figure 3). All these features observed in the Fort Ternan crania are typical of Dicerorhinini (Guérin 1980a; Heissig 1981; Groves 1983; Geraads 1988; Giaourtsakis et al. 2006).

The deciduous dentition of the holotype is markedly smaller than in all fossil and extant Dicerotini examined. Hooijer (1968) describes seven features that distinguish the

Fort Ternan deciduous dentition from extant *D. bicornis*. Although some of them may be variable in Dicerotini (presence and strength of crista, strength of paracone fold, strength of crochet), the combination of all of them is typical Dicerorhinini and characterizes the extant *Dicerorhinus sumatrensis,* as well as *"Dicerorhinus" leakeyi*. In particular, the constriction of the protocone by mesial and distal grooves, the presence of a weak but conspicuous antecrochet, the absent or faint crista on D^3–D^4, and the absence of lingual cingula or cingular pillars in the entrance of the medisinus are distinctive. We can also add the presence of a metacone fold on the ectoloph of D^3–D^4. The deciduous dentition of a juvenile maxilla (BMNH: M.32946) from Rusinga, the type locality of *"Dicerorhinus" leakeyi,* bears the same features and is indistinguishable from the Fort Ternan holotype. The permanent teeth of the Fort Ternan rhinocerotid are very low-crowned brachyodont, and they are also smaller with respect to all fossil and extant Dicerotini examined. The premolars of the partial adult skull do not have a lingual cingulum, which is always well-developed in Dicerotini (except Pleistocene and extant *Ceratotherium*). Hooijer (1968) notes, in some of the isolated permanent teeth, the presence of a weak but conspicuous protocone constriction by a mesial and distal protocone groove, as well as of an antecrochet. A metacone fold appears also to be present (Hooijer 1968: plate 2). These features are more markedly expressed in Miocene dicerorhines but are usually absent or dimly expressed in dicerotines.

The strongest argument of Hooijer was the absence of permanent tusk-like second lower incisors in two mandibles recovered at Fort Ternan. Hooijer (1968: 84) describes the mandible FT-1962-3209 as having "the symphyseal portion complete" and being "edentulous, showing milk incisor alveoli but no traces of permanent canines and incisors" [*sic*]. This mandible was not figured. Our observations of the mandible cast (BMNH: M.29931) suggest that the anterior part of the symphysis is fragmentarily preserved, probably dorsoventrally compressed, and incomplete. The symphysis could have extended further anteriorly, and the "milk incisor alveoli" reported by Hooijer may only represent the distal impression of the roots of the permanent incisors and not the complete alveoli. If the specimen belonged to a female individual, the permanent incisors could have been rather small and their alveoli faded out in front of, and not below, the P_2. For the second mandible (FT-1962-3503; Hooijer 1968: plate 2, figure 1) a close examination of the specimen would be necessary.

The postcranial skeleton also supports an ascription of the Fort Ternan fossils to Dicerorhinini. Hooijer describes an atlas (KNM-FT-1963-3497) and underlines the presence of an alar incisure lateral to the articular surface of the occipital condyle. As Hooijer correctly notes, this is a feature seen in extant *Dicerorhinus* and not in Dicerotini, fossil or extant, where an alar foramen is developed instead (personal observation of ca. 20 skeletons). Detailed description of the Fort Ternan limb bones is not provided by Hooijer (1968). However, their size is significantly smaller than all known Dicerotini (for the third metatarsal comparison, see Table 14.4). To the contrary, it corresponds perfectly to the size reported for the *"Dicerorhinus" leakeyi–Turkanatherium acutirostratum* specimens from Rusinga (Hooijer 1966; Guérin 2003). Hooijer (1966) stated that it was not possible to distinguish the limb bones of these two species, which have similar dimensions, but we agree with Guérin (2003) that revision of the abundant remains preserved in the KNM collection should permit a resolution of this problem.

Rhinoceros findings identified as *Paradiceros mukirii* were reported at Maralal (Hooijer 1968), the Samburu Hills (Nakaya et al. 1984; Tsujikawa 2005) and the Ngorora Formation (Pickford et al. 2006) in Kenya, as well as the Kisegi Formation in Uganda (Guérin 1994) and at Beni Mellal in Morocco (Guérin 1976). The available material from these localities is rather scant, and each case must be revised separately.

Comparison with Diceros neumayri *and Biogeographic Remarks*

"Diceros" neumayri was the first recognized fossil relative of the extant African species (Wagner 1848; Gaudry 1862–1867) and until the 1960s their only Miocene representative. It is a common element of the *Hipparion* faunas of the sub-Paratethyan mammalian province (Bernor 1984) and has been documented in numerous localities from Greece (Gaudry 1862–1867; Weber 1904; Arambourg and Piveteau 1929; Geraads 1988; Geraads and Koufos 1990; Giaourtskis 2003; Giaourtskis et al. 2006) and Turkey (Heissig 1975; Geraads 1994b; Kaya 1994; Heissig 1996; Fortelius et al. 2003a; Antoine and Saraç 2005), as well as from the locality of Maragheh in Iran (Osborn 1900; Thenius 1955) and Eldari-2 in the Caucasus (Tsiskarishvili 1987). The referred occurrence of the species in the Vienna Basin (Thenius 1956) was confirmed as a *Brachypotherium* (Giaourtsakis et al. 2006). Some specimens from Spain reported as *Diceros pachygnathus* by Guérin (1980b) were assigned to *Dihoplus schleiermacheri* (Cerdeño 1989). Specimens referred to as *Rhinoceros pachygnathus* from Mont Léberon, France (Gaudry 1873), and Baltavar, Hungary (Pethő 1884), belong also to *Dihoplus schleiermacheri* (personal observations at MNHN and MAFI). A much worn P^2 from Sahabi, Lybia, reported as *Diceros neumayri* by Bernor et al. (1987: figure 15), can equally belong to *D. douariensis* or another unknown Dicerotini, as it does not bear any diagnostic features.

Although craniodentally very distinct, *"D." neumayri* has been frequently confused and misidentified with the synchronic and partly sympatric *Dihoplus pikermiensis,* a large Dicerorhinini (Heissig 1975; Geraads 1988; Giaourtsakis et al. 2006). Because of its dental similarities with the extant *Diceros,* the taxon has been commonly assigned to this genus (Ringström 1924; Thenius 1955; Hooijer 1972; Heissig 1975; Hooijer 1978; Guérin 1980b, 1982; Tsiskarishvili 1987; Heissig 1989). Geraads (1988) pointed out cranial similarities with *Ceratotherium.* These cranial similarities represent, however, early convergences. Following Geraads (1988), the "common usage" has uncritically changed to *Ceratotherium neumayri* (Geraads and Koufos 1990; Kaya 1994; Heissig 1996; Fortelius et al. 2003a; Giaourtsakis 2003; Antoine and Saraç 2005), although some reservations regarding the preliminary taxonomic status were retained (Guérin 2000, 2003; Giaourtsakis et al. 2006). Because an extensive revision of the taxon is currently in preparation, nomenclatural issues will not be treated further here, but the generic allocation for "common usage" is reconsidered, because it has been used to imply biogeographic, evolutionary, and ecological patterns. The Kuseralee cranium offers new data, in particular because the occipital morphology of Miocene African Dicerotini was practically unknown until now.

The early dispersal and migrational pattern of the Dicerotini outside Africa is not yet well-established. Thomas et al. (1978) described as *Dicerorhinus* cf. *primaevus* an unworn right M^2 from the middle Miocene locality Al Jadidah of the Hofuf Formation in the eastern Province of the Kingdom of Saudi Arabia. A cast of the tooth is housed in the collections of MNHN and is indeed morphologically and metrically very similar with the unworn M^2 of the juvenile specimens from Bou Hanifia (MNHN: 1951-9/219). It bears typical Dicerotini features, such as the absence of a distal protocone constriction and antecrochet. It also has a well-developed paracone fold, a rather prominent crochet, and no crista. It is best referable to as *Diceros* cf. *primaevus* until further material is made available. The oldest occurrence of Dicerotini in the eastern Mediterranean is not yet well-documented. The referred occurrence of a primitive Dicerotini in the middle Miocene of Chios (Heissig 1989) is doubtful, because it has not been followed by evidence. In the well-calibrated Sinap sequence, the lineage of *"Diceros" neumayri* is reported to have a range of ~11.0–6.0 Ma (Fortelius et al. 2003a), and, within this range, its distribution falls more or less in other localities of Turkey (Heissig 1975, 1996), as well as in Greece (Giaourtsakis 2003) and Iran (Bernor et al. 1996d). Practically, this means that the Kuseralee cranium is stratigraphically younger than most specimens of this species.

Cranially, *"D." neumayri* is undoubtedly more specialized than the Kuseralee rhinocerotid, with convergent derived features similar to Pliocene *Ceratotherium,* but expressed to a lesser degree: The skull is usually longer and more dolichocephalic; the anterior border of the orbit is placed either at the same level as in the the Kuseralee cranium (in front of the middle of M^2) or in most specimens more retracted (behind the middle of M^2); the occiput inclines more posterodorsally, and the strong nuchal crest extends in many specimens beyond the level of occipital condyles; the occipital notch of the nuchal crest is deeply concave or forked.

Dentally, *"D." neumayri* follows a different pattern than the Douaria, Kuseralee, and Langebaanweg rhinocerotids. This pattern appears to evolve spatiotemporarily several times during its radiation in the eastern Mediterranean and is probably affected by migrational activities, population exchange, or regional adaptations (Heissig 1975; Fortelius et al. 2003a; personal observation). Increase in size and crown height, development of cristae in premolars and molars (very frequent in Turolian specimens), cingulum reduction, obliquity of the protoloph associated with broadening of the protocone, strengthening of the mesostyle bulge, and broadening or narrowing of the medisinus valley are general trends. Heissig (1975) and Fortelius et al. (2003a) argue that the changes seen (including body size) are of a magnitude that would justify recognition of separate morphospecies or geographic variations. This idea was first put forward by Thenius (1955), but the material he studied was too limited to establish unambiguous characters. Tsiskarishvili (1987) has described the species *Diceros gabuniae* as being a regional variant in the Caucasus, but detailed comparisons are lacking. Our observations (revision in preparation) confirm the radiation of several morphotypes and indicate an evolutionary pattern similar to the numerous, locally adapted extant *Diceros* subspecies. Contrary to the pattern seen in Eurasia, the development of cristae appears to be delayed in Africa, and even in the Langebaanweg sample they are only occasionally developed (less than 40 percent). Instead, the distolingual bending of the protoloph, the fading of the paracone fold, and the flattening of the occlusal surface is favored. The development of a lingual protocone groove, which is so conspicuous in the molars of Douaria, Lothagam (KNM-LT89), Langebaanweg, and Pliocene *Ceratotherium,* is never observed in *"D." neumayri*. The functional interpretation of this groove is not clear, but it might be a structure to gradually enhance the distal curvature of the lingual wall of the protoloph, as Hooijer and Patterson (1972) have suggested.

The postcranial elements of *"D." neumayri* also follow a different specialization pattern than the Langebaanweg sample (which is similar in dimensions to the late Miocene Mpesida and Saitune Dora metapods) and the Pliocene *Ceratotherium*. The Vallesian specimens retain the length of *D. primaevus* but become more robust (personal observations), or "graviportal" as Guérin (1980b, 1982) notes. During the Turolian, several populations increase their size and robustness, but the exact pattern seems to constitute a complicated cline, as other populations retain the smaller dimensions (Heissig 1975; Kaya 1994; Fortelius et al. 2003a; personal observation). In any case, the maximum length values seldom reach the minimum values of the Langebaanweg population or the Mpesida and Saituna Dora metapods (Table 14.4), and the robustness is retained or accentuated. In addition, *"D." neumayri* displays some autapomorphic features. For example, although the proximal epiphysis of the third metacarpal widens, the articular facet for the second metacarpal is shifted more laterally with respect to the medial border of the diaphysis. In this way, the lateral border becomes markedly concave, and the minimum width is shifted closer to the middle of the diaphysis rather than the proximal epiphysis. The anteroposterior diameter of the proximal and distal epiphyses of the bone also increases.

Because of its unique combination of cranial, dental, and postcranial features, *"D." neumayri* has been regarded by the majority of authors as a separate lineage evolving independently from African Dicerotini (Hooijer and Patterson 1972; Heissig 1975; Hooijer 1978; Guérin 1980b, 1982; Heissig 1989).

Geraads (2005) deemed *"Diceros" neumayri* (which he calls *Ceratotherium neumayri*) as the common ancestor of both living species, arguing that it is morphologically and ecologically intermediate between them. In this context, the Miocene African *P. mukirii* and *D. primaevus* were considered as being related to *"D." neumayri,* and *D. douariensis* as potentially conspecific with *"D." neumayri*. However, no arguments were provided to support this grouping. Whereas an ancestry of *D. primaevus* for both *D. douariensis* and the extra-African *"D." neumayri* cannot be excluded (and is, in fact, a feasible option), a lineage of *D. primaevus* –*"D." neumayri* (= *D. douariensis*) splitting then into Pliocene *Diceros praecox* and *Ceratotherium mauritanicum* is inappropriate. Geraads (2005: 455) defines *Diceros praecox* as having "the following apomorphic features with respect to its likely ancestor *C. neumayri*: orbit more anterior to tooth row; skull profile more concave; occipital plane more vertical; nuchal crest less extended posteriorly [*sic*]." Apart from the more concave profile (which we agree is accentuated in Pliocene *Diceros* but not derived from *"D." neumayri*), all other cranial features cited are plesiomorphic (Antoine 2002) with respect to *"D." neumayri*, as they are with respect to the Plio-Pleistocene and the extant *Ceratotherium*. In these cranial features (plus the straight or only slightly indented nuchal crest), the Kuseralee cranium is also more primitive with respect to *"D." neumayri*. Even the occipital morphology of the stratigraphically younger Langebaanweg rhinocerotid (SAM: L-13747) is more conservative than that of Turolian specimens of the extra-African late Miocene *"D." neumayri* (SMNK: Ma 2/15; AMPG: PA 4721/91; MNHN: PIK-971; AUBLA: 18.ÇO-553; MTA: AK4-212; see also Antoine and Saraç 2005: figure 1). An occipital morphology similar to that of the eastern Mediterranean specimens is independently achieved and further developed in Africa during the late Pliocene, by the descendants of the Kuseralee and Langebaanweg populations, that is, the *Ceratotherium* (*sensu stricto*) lineage: Hadar, NME: A.L. 129-25, A.L. 269-4, A.L. 235-3; Dikika, NME: DIK-1-10; Laetoli, LAET-49.

Geraads (2005) suggested that the two extant lineages split soon after the Miocene-Pliocene boundary, leading from an ancestral mixed feeder ("*Ceratotherium neumayri*") to a lineage of grazers (*Ceratotherium*) and a lineage of browsers (*Diceros*). The Kuseralee cranium confirms, however, the scenario that the split of the two extant lineages took place in Africa before the Miocene-Pliocene boundary and that *"D." neumayri* represents a convergent extra-African monophyletic lineage (Hooijer and Patterson 1972; Hooijer 1978; Guérin 1980b, 1982; Heissig 1989). The Kuseralee cranium preserves a cranial morphology more primitive than stratigraphically older skulls of *"D." neumayri* but demonstrates a dentition that develops a different specialization pattern traceable from the late Miocene (Douaria, Lothagam-Nawata) to the Mio-Pliocene boundary (Langebaanweg) before it adapts to an exclusively C_4 grass diet during the course of the Pliocene (*Ceratotherium* sp.). In addition, the dietary inferences of the Kuseralee rhinocerotid show that not only the common ancestor of the two extant lineages should have been a browser, but also that the ancestral stock of the *Ceratotherium* lineage must have favored a browsing diet for as long as available habitats could supply it.

The dispersal of some dicerotine populations outside Africa during the late Miocene was concomitant with the gradual establishment of a unique combination of primitive and derived craniodental features, as well as some autapomorphies, notably in the postcranial elements. Their spatiotemporal expansion in the eastern Mediterranean and adjacent regions must have followed an evolutionary pattern comparable to the numerous, locally adapted, modern *Diceros* subspecies. A separate generic assignment for this monophyletic extra-African Dicerotini lineage is appropriate, an option also considered by Geraads (2005) but in a different phylogenetic context. However, keeping in mind the existing complicated nomenclatural issues concerning the eastern Mediterranean horned rhinoceroses (Heissig 1975; Geraads 1988; Giaourtsakis 2003), which also affect the availability of the generic name *Pliodiceros* Kretzoi, 1945, we suggest that this taxon be preliminarily referred to as *"Diceros" neumayri*.

Paleoecology and Functional Morphology

Inferences about habitat and dietary preferences in fossil mammals are often influenced by the ecological preferences of their extant relatives. However, caution is required, because significant differences may often occur (Solounias and Dawson-Saunders 1988; Solounias et al. 2000). Morphological features that are adapted for particular dietary functions must be carefully compared and evaluated in order to assess accurately the dietary preferences in fossil rhinocerotids (Zeuner 1934; Fortelius 1985; Fortelius and Solounias 2000). Stable isotope analysis of enamel carbonate is another useful tool for understanding the ecology of fossil mammals and the evolution of their habitats during the past (Zazzo et al. 2000 Franz-Odendaal et al. 2002; ; Lee-Thorp and Sponheimer 2005). Microwear dental analysis is also an important method for paleodiet reconstruction (Solounias et al. 2000) but has not yet been sufficiently applied to fossil rhinoceroses and remains a promising source of information for future studies.

Since several misleading interpretations have been uncritically accepted in the past, a brief summary on the ecological and dietary preferences of the two extant species is essential.

The extant *Diceros bicornis* is found in a very wide range of habitats, from montane forest and lowland marginal forest through savanna woodland, bush and thicket, mixed grassland and woodland, scattered tree grassland, to semi-desert and arid desert (Hillmann-Smith and Groves 1994). Distribution ranges from sea level to at least 1,500 m in southern and southeastern Africa, up to 3,000 m in eastern Africa (Guggisberg 1966; Kingdon 1979). Although they are able to deal with patchy vegetation, black rhinoceroses always show a preference for areas with denser cover, especially during the day. In mixed habitats, a direct relationship between the density of black rhinoceroses and the density of habitat has been documented (Hitchins 1969; Mukinya 1973; Goddard 1967). Black rhinoceroses are selective browsers on woody shrubs, young trees, and certain forbs, rejecting generally dry plant material. They are extremely flexible, shifting their food preferences according to circumstances and availability; they are even able to utilize plants that have heavy morphological and chemical defenses against most other herbivores (Hall-Martin et al. 1982; Loutit et al. 1987; ; Oloo et al. 1994). Black rhinoceroses are able to feed on a wide variety of plant species. Goddard (1968, 1970) reported 191 species of plants in Ngorongoro (Tanzania) and 102 in Tsavo (Kenya) browsed by black rhinoceroses, while Leader-Williams (1985) reported 220 species in Luangwa Valley (Zambia) and Hall-Martin et al. (1982) recorded 111 species in Addo (South Africa). Even in the extremely arid Darmaland in Northern Namibia, the desert black rhinoceros utilized 74 out of the 103 plant species encountered (Loutit et al. 1987). Depending on the region and availability, they prefer several *Acacia* species and their relatives, as well as *Grewia similis, Spirostachys* sp., *Phyllanthus fischeri, Euphorbia* sp., and *Hibiscus* spp. Small quantities of grass are taken during the wet season, or together with succulent plants in dry periods when other resources become unavailable (Mukinya 1977; Hall-Martin et al. 1982; Oloo et al. 1994). In very dry seasons, however, excessive consumption of forage with low nutritional value may lead to substantial death rates from malnutrition (Dunham 1985, 1994).

In contrast, the extant *Ceratotherium simum* is a very specialized animal. It does not favor closed forests and mountainous areas and avoids high altitudes. Steep country is only traversed, and relatively flat terrain is preferred. In South Africa, white rhinoceroses occupy open savannas with scattered trees across the open Bushveldt zone. In the Nile region, they inhabit open *Combretum* forest with grassland (Guggisberg 1966; Groves 1972). In both cases, density is of secondary interest because they use the available trees only to provide shade during the hottest parts of the day. White rhinoceroses are entirely grazers, with a preference of feeding on high-quality short grasses. *Panicum maximum, P. coloratum, Urochloa mozambicensis,* and *Digitaria* sp. constitute the bulk of their diet. These grasses occur mostly in shady areas of *Themada triandra* grasslands. However, climax *Themada triandra* is rarely eaten, except for the sprouting green grass regenerating after summer fires, which is an important supplementary food resource during the dry season. An additional 30 species of grasses can be consumed to a lesser extent. Sporadic geophagia, especially around termitaria, has been recorded, presumably to increase mineral content.

When grasses become rare, white rhinoceroses change region, sometimes on a seasonal basis (Foster 1960; Guggisberg 1966; Groves 1972; Owen-Smith 1973, 1988; Skinner and Smithers 1990; Shrader et al. 2006).

Zeuner (1934) and Loose (1975) have demonstrated a close relationship between skull shape and dietary adaptation in the two extant African species. The cranial morphology of the Kuserelee rhinocerotid more closely resembles the morphology of the extant *Diceros,* rather than *Ceratotherium.* In the Kuseralee cranium, as in extant *Diceros,* the occipital plane is nearly vertically oriented, the nuchal crest does not extend beyond the occipital condyles, and the cranial dorsal profile is clearly concave. As a result, the head is held rather horizontally, enabling the animal to browse twigs and leaves from trees and bushes (Mills and Hes 1997: 236, see figure). The skull of *Ceratotherium* is much longer and more dolichocephalic, the occiput inclines strongly backward (posterodorsally), the nuchal crest extends beyond the condyles, and the dorsal profile is only gently concave. As a result, the head is held at an angle close to the ground, enabling the animal to feed on short grasses (Mills and Hes 1997: 233, see figure).

The dental morphology of the Kuseralee cranium, as well as that of the incomplete holotype skull from Douaria, more closely resembles the dentition of extant *Diceros.* Functionally significant morphological similarities related to dietary preferences include the concave occlusal surface of the teeth; the saw-toothed ectoloph wear profile with sharp cusp apices and relatively high intermediate relief; the relatively thin enamel with irregular thickness; the presence of paracone folds that are stronger than the mesostyle folds; the occurrence of thin cement coverage; the presence of a strong, continuous lingual cingulum in the premolors; and the absence of crista and medifossette, particularly in the molars.

Fortelius (1982, 1985) has demonstrated that the concave occlusal surface is a result of two distinct chewing phases during the occlusal stroke: shearing and crushing. During the first phase, only the buccal edges of upper and lower teeth come in contact. The sharp paracone and metacone ridges of the adjacent teeth and the intermediate valleys form the main shearing blades. This configuration corresponds to the profile inducted by the mesowear method, where the buccal cusps are apically sharp with rather high intermediate relief (Fortelius and Solounias 2000). The presence of a paracone fold strengthens the shearing efficiency. During the second phase, the occlusal motion continues lingually and perpendicular to the masticatory force applied, following the inclined occlusal surface of the upper teeth. This results in a high-pressure, low-speed condition, which enables the crushing of the food. Differential wear regulates the shape of the occlusal surface by inequalities in enamel thickness, which is broader on the buccal and lingual sides. The combination of shearing and crushing is an adaptation to the comminution of bulky vegetation, such as soft plants, fruits, and twigs, that forms a typical browsing diet. On the other hand, *Ceratotherium* has a flat occlusal surface and a flat ectoloph mesowear profile. This corresponds to a one-phase, upward-inward, high-pressure occlusal stroke, an adaptation to the rapid comminution of fibrous, thin, and tough vegetation, such as grasses (Fortelius 1982, 1985). Because the occlusal pressure is more uniformly distributed over the entire surface, the enamel thickness remains more equal over the entire tooth. Further, the paracone fold is weakened, and the mesostyle fold is strengthened to maintain the ectoloph pattern.

On the Kuseralee cranium and modern *Diceros,* only a thin cement coat is developed. In fossil and extant *Ceratotherium,* the cement investment is thick and abundant, filling almost completely the inner valleys of the upper and lower dentition in the modern species. This is also an adaptation to a grassy diet, because the cement investment produces a wear-retarding and relief-enhancing structure (Fortelius 1985).

The occurrence of a well-developed, continuous lingual cingulum found in the Kuseralee and extant *Diceros* premolars is related with specific browsing capabilities. Its presence has been interpreted as an adaptation of lower-crowned teeth to protect the gums from injury by thorns and splinters. This necessity would decrease with increasing crown height and disappear with a shift to a diet free of such components, for example, a grassy diet (Fortelius 1982). Indeed, in Plio-Pleistocene *Ceratotherium* the lingual cingulum is gradually reduced, and in the extant species it is absent.

In the Kuseralee and extant *Diceros* molars, a crochet is the only secondary fold developed, the crista is rare, and a closed medifossette is absent. In fossil and extant *Ceratotherium* molars, a closed medifossette is always present, formed by the early junction of crista and crochet. This structure is associated with improved abrasion efficiency, because it increases the effective length (perpendicular to the occlusal motion) of the enamel ridges and may also support the growing demands for additional cement investment in the crown.

In summary, the functional analysis of the available cranial and dental material clearly indicates that the Kuseralee rhinocerotid was, first and foremost, a browser. As discussed, there are also a few advanced morphological features observed with respect to Miocene Dicerotini and Pliocene *Diceros* species, in particular the large size, the relatively higher-crowned teeth, and the tendency of the protoloph to bend distolingually in the molars. Increased crown height is associated with increased resistance to abrasion, resulting directly from the utilization or contamination with extraneous abrasive material, or indirectly from food requiring higher occlusal pressures for comminution (Fortelius 1985). Herbivores that shift to utilize more abrasive forage and exploit more open habitats are frequently larger than their ancestors. The gradual posterodistal deflection of the protoloph is a tendency to increase the effective length of the enamel ridges perpendicular to the occlusal motion (in *C. simum* both protoloph and metaloph are bending markedly distolingually). These features may reflect a slow but gradual adaptation to cope with more open or seasonal environments and their occasionally tougher and nutritionally inferior forage, equivalent to some extant black rhinoceros subspecies. The paleoenviromental reconstruction of the Kuseralee faunal community suggests a riverine woodland habitat with wet grassland (Su et al., Chapter 17) and is thus in accordance with the dietary inferences made here.

In the rapidly changing landscape at the Miocene-Pliocene boundary (Cerling et al. 1997b), being able to adapt to increasingly open or seasonal habitats was an evolutionary advantage (Jernvall and Fortelius 2002). It is not a coincidence that none of the African Miocene rhinoceros lineages other than the dicerotines managed to survive successfully into the Pliocene. The few advanced morphological features observed in the Douaria and Kuseralee rhinocerotids are further accentuated by some large-sized populations at the beginning of the Pliocene. The most adequate and stratigraphically important material is the Langebaanweg sample from the Mio-Pliocene boundary. The teeth from Langebaanweg are somewhat more high-crowned, the occlusal surface less concave (but not flat as in *Ceratotherium* sp.), the ectoloph mesowear profile markedly lower, the cement investment increased, and the lingual protocone groove more evident. The weak protocone fold persists, but the wide mesostyle fold becomes stronger. A small crista appears in some molars, but rarely a closed medifossette as in *Ceratotherium*. This intermediate morphology is in agreement with the stable isotope analyses of enamel carbonate, as well as with the particular local environmental circumstances documented for the locality. Franz-Odendaal et al. (2002) demonstrated that the terrestrial fauna of Langebaanweg existed in a local environment that remained C_3-dominated. The current local Mediterranean climate, controlled by latitudinal, seasonal movement of the South Atlantic high-pressure system, was already established in this region by the early Pliocene, preventing the expansion of C_4 grasses. This climate regime is characterized by wet, rainy winters and markedly arid summers. The rhinoceros of Langebaanweg was probably a mixed feeder, able to browse or graze on C_3 plants, depending on the seasonal conditions in the area.

Isotopic results were also reported for the rhinoceroses of Lothagam (Harris and Leakey 2003a; Cerling et al. 2003), which is geographically closer and temporally overlaps with the Middle Awash succession. All samples from the Lower Nawata (>6.5 Ma), including one referred to as *Ceratotherium praecox,* were C_3 browsers. A tooth from the younger Upper Nawata (>5 Ma) provided a C_4 signal. One sample from the Pliocene Apak Member was from a C_3 browser, and five were C_4 grazers. Harris and Leakey (2003a) tentatively assigned them to *Diceros* and *Ceratotherium,* respectively. Zazzo et al. (2000) applied stable isotope analyses in the Chadian fossil faunas and documented a shift in herbivore paleodiet related to paleoenvironmental changes during the Pliocene. In particular, rhinocerotid material referred to *Ceratotherium praecox* was recorded as a mixed feeder during the early Pliocene and as a pure grazer during the late Pliocene.

Overlooking some of the systematic assessments, which simply followed previously established concepts and might require refinement, these studies clearly indicate that during the late Miocene and early Pliocene some Dicerotini populations gradually increased the abrasive forage in their diet as a response to paleoenviromental changes, including increasing seasonality and expansion of open habitats. Ultimately, the establishment of open grasslands during the Pliocene accelerated the morphological adaptations required for an exclusive grass diet and contributed to the definitive morphogenetic evolutionary step of these populations toward the *Ceratotherium* condition. At the same time, less advanced populations of the *Diceros* lineage continued to survive in more temperate habitats and evolve independently, adapting to new spatiotemporal environmental challenges, probably following an evolutionary pattern analogous to the numerous locally adapted modern subspecies.

Appendix 14.1: Methods and Materials

The fossil rhinoceros material described here was collected by the Middle Awash project and is housed at the National Museum of Ethiopia, Addis Ababa. Skull measurements follow Guérin (1980b). Anatomical conventions follow Getty (1975) and Nickel et al. (1986). Dental measurements and terminology follow Peter (2002). Width measurements include the mesial (Wm) as well as the distal (Wd) width of each tooth. On the first upper and lower deciduous premolar, only the maximal distal width (Wd) is measured. Measurements of M3 include the buccal length of the ectometaloph (Lb), the mesial width (Wm) and the lingual, anatomical length (La) comprising the distal cingular pillar, if present (Guérin 1980b; Peter 2002). Measurements ranging 0–150 mm were taken with a digital caliper to 0.01 mm and rounded to the nearest 0.1 mm. For larger measurements a linear caliper with a precision of 0.1 mm was applied. All measurements are given in millimeters (mm). The terms *low-* and *high-crowned* dentition refer to relative crown height, whereas the terms *brachyodont* and *hypsodont* refer to functionally different types, following Fortelius (1985).

Comparative studies with material from the Plio-Pleistocene localities of Hadar (Guérin 1980a; Geraads 2005) and Dikika (Geraads 2005) have been carried out at the NME; from the Omo Valley (Arambourg 1948; Hooijer 1969, 1972, 1973, 1975; Hooijer and Churcher 1985; Guérin 1985) at NME, RMNH, and MNHN. Pleistocene material from Olduvai Gorge, Laetoli, Kanjera, Kanam West, and Rawi (Hooijer 1969; Groves 1975) has been studied at BMNH. The Kohl-Larsen fossil collection from eastern Africa (Dietrich 1942b, 1945) has been studied at MNHB. Specimens from Bou Hanifia (Arambourg 1959; Geraads 1986b), Ternifine (Pomel 1895), and several Plio-Pleistocene North African localities (Arambourg 1970) have been examined at MNHN. Materials of *"Diceros" neumayri* from Greece (Pikermi, Samos, Axios Valley: Gaudry 1862–1867; Geraads 1988; Geraads and Koufos 1990; Giaourtskis 2003) have been studied at AMPG, LGPUT, MNHN, BMNH, NHMW, IPUW, MAFI, BSPG, SMF, and HLMD; those from Turkey (various localities: Heissig 1975, 1996; Geraads 1994b; Fortelius et al. 2003a) were studied at BSPG, SMNK, AUBLA, MTA and MNHN; and those from Iran (Maragheh: Osborn 1900; Thenius 1956) were examined at NHMW and MNHN. Casts of specimens from Fort Ternan (Hooijer 1968) and Langebaanweg (Hooijer 1972) have been examined at BMNH and BSPG, respectively. Digital images of the dentition of *Diceros douariensis* from Douaria (Guérin 1966) have been kindly provided by C. Guérin and A. Prieur, and of the holotype of *Diceros praecox* from Kanapoi (Hooijer and Patterson 1972) by M. Fortelius. Comparative studies with the extant species have been carried out at the zoological collections of NHMW, IPUW, RMNH, ZMA, SMNK, SMF, BMNH, MNHN, and USNM.

Institutional Abbreviations

AMPG	Athens Museum of Paleontology and Geology, University of Athens
AUABL	Ankara Üniversitesi, Antropoloji Bölümü Laboratuary, Ankara
BMNH	British Museum of Natural History (= Natural History Museum), London
BSPG	Bayerische Staatssammlung für Paläontologie und Geologie, München
FSL	Faculté des Sciences, University of Lyon
GSN	Geological Survey of Namibia, Windhoek
HLMD	Hessisches Landesmuseum, Darmstadt
IPUW	Institut für Paläontologie der Universität, Wien
KNM	Kenya National Museum, Nairobi
LGPUT	Laboratory of Geology and Palaeontology, University of Thessaloniki
MAFI	Magyar Állami Földtani Intézet, Budapest
MNHB	Museum der Naturkunde für Humboldt Universität zu Berlin
MNHN	Muséum National d'Histoire Naturelle, Paris

MTA	Maden Tetkik ve Arama Museum, Ankara
NHMW	Naturhistorisches Museum, Wien
NME	National Museum of Ethiopia, Addis Ababa
NMT	National Museum of Tanzania, Dar-es-Salaam
RMNH	Rijkmuseum van Natuurlijke Historie (Naturalis), Leiden
SAM	South African Museum, Cape Town
SMF	Forschungsinstitut und Naturmuseum Senckenberg, Frankfurt am Main
SMNK	Staatliches Museum für Naturkunde, Karlsruhe
USNM	United States National Museum (Smithsonian), Washington
ZMA	Zoological Museum, Amsterdam

Locality Abbreviations

KUS	Kuseralee
STD	Saitune Dora
ASK	Asa Koma
ALA	Alayla
VP	vertebrate paleontology locality (Middle Awash sample)
AL	Afar locality (Hadar sample)
SH	Sidi Hakoma Member (Hadar sample)
FT	Fort Ternan

Morphology Abbreviations

P, M, d	premolar, molar, deciduous (pre)molar
I, di	incisor, deciduous incisor
prmx.	premaxilla
nas.	nasal(s), nasal tip
nas.inc.	nasal incision
orb.	orbit anterior border
cond.	condyle
occ.	occipital
proc.	process(es)
MC	metacarpal
MT	metatarsal
dia.	diaphysis
ep.	epiphysis
prox.	proximal
dist.	distal
L	length
W	width
H	height
DT	transverse diameter
DAP	anteroposterior diameter

15

Proboscidea

HARUO SAEGUSA
AND
YOHANNES
HAILE-SELASSIE

Late Miocene proboscidean fossils were recovered from the Middle Awash in the 1970s by the Rift Valley Research Mission in Ethiopia (RVRME), led by Jon Kalb. These were described and analyzed in a series of papers published by Kalb, Mebrate, and colleagues (Kalb et al. 1982a, b, c; Mebrate and Kalb 1981, 1985; Kalb and Mebrate 1993; Kalb and Froehlich 1995; Kalb 1995; Kalb et al. 1996). These papers have played a central role in the discussion of late Neogene African proboscidean taxonomy and biochronology (Sanders 1997; Tassy 1994, 1999, 2003). This chapter presents a revision of the late Miocene proboscideans from the Middle Awash. We recognize seven proboscidean taxa and offer interpretations that differ from those of Kalb and Mebrate (1993), who reported the presence of *Stegodon* cf. *S. kaisensis, Stegotetrabelodon orbus,* and cf. *Stegodibelodon schneideri* in the late Miocene of the Middle Awash. However, the available craniodental material from a larger sample clearly shows that *S. orbus, S.* cf. *kaisensis,* and cf. *S. schneideri* are absent from the Middle Awash late Miocene. Their presence was formally inferred based on misidentification of dental elements. We recognize a new subspecies of *Primelephas*. In addition to this new taxon, fragmentary specimens suggest the presence of primitive species of *Mammuthus* and *Loxodonta*. A rich sample of *Anancus* from the Middle Awash documents substantial variation in molar structures among early anancines.

A total revision of the late Neogene African elephantoids and their phylogenetics is beyond the scope of this chapter. However, the species-level taxonomy of the late Miocene proboscideans described here will provide the basis for future work addressing the systematics, biostratigraphy, and biogeography of the late Neogene African elephantoids.

Terminology and Abbreviations

General Dental Terminology of Anancus

We employ the dental terminology used by Tassy (1996a) and Metz-Muller (1995), except for the terms discussed below. In lower molars of *Anancus,* the anterior pretrite central conule is much reduced and is fused with the mesoconelet to form a cusp located mesio-adaxially to the main cusp (e.g., Tassy 1986: 87, 94, pl. XIII, fig. 3). Tassy (1986) called it neither the mesoconelet nor the anterior pretrite central conule, but just the "anterior tubercle," because of the amalgamate nature of the cusp. On the other hand, Metz-Muller (1995) called the same tubercle the "mesoconelet," and we follow her terminology because it is never subdivided. The mesoconelet of the posttrite half-lophid is frequently developed into a large and centrally positioned cusp. It is occasionally identified as a central conule, especially when the molar is heavily worn (e.g., Tassy 1986: 94, pl. XI, fig. 1). It should be called a mesoconelet, regardless of its appearance.

Figure 15.1 is a diagrammatic presentation of the variation seen in adaxial pretrite cusps of upper molars. Because of the complex pattern, the two general terms *mesoconelet* and *central conule* are not always applicable to these cusps, which we collectively call the M-C complex. In fact, a cusp called the anterior central conule by Tassy (1986, pl. XIII, fig. 1) is called the "mesoconelet" by Kalb and Mebrate (1993, fig. 8 and E in fig. 10a). In order to avoid such inconsistency, we redefine these terms according to their positional relationship within the M-C complex. The "mesoconelet" is defined here as a cusp with only a path connecting the pretrite main cusp with the posttrite mesoconelet. The "anterior pretrite central conule" is defined as a cusp that has only the path connecting the pretrite main cusp with the preceding posttrite half-loph. There are cusps that have both types of paths, and one of them is a cusp that was addressed under different names by Tassy (1986) and Kalb and Mebrate (1993). Here, we name these cusps the "junction cusps."

The typology of the M-C complex just described cannot be applied to the first loph, which lacks a preceding loph. There are three morphological types of the first loph (Figure 15.1B). Types A and B are seen in *Tetralophodon,* whereas types B and C are seen in *Anancus*.

General Dental Terminology of Elephants

In this chapter we employ the following terms in the description of elephant molars: lateral pillar (homolog of main cusp); median pillar (homolog of mesoconelet); mesial column (homolog of anterior pretrite central conule); distal column (homolog of posterior pretrite central conule); lateral loop (horizontal section of lateral pillar on wear surface); median loop (horizontal section of median pillar on wear surface); mesial sinus (horizontal section of mesial column on wear surface); distal sinus (horizontal section of distal column on wear surface); digitation (superficial subdivisions of a pillar); Stufenbildung (step-like wear feature of the enamel loop typically seen in stegodonts).

Terminology of Molar Roots

Elephantoid upper molars generally have three major roots (Figure 15.2A, B). Mesial root: transversely wide, mesiodistally narrow, located at the mesiobuccal end of the tooth, supporting most of the buccal part of the first loph. Lingual root: mesiodistally wide and buccolingually narrow, mesiolingually located, supporting lingual sides of the first and second lophs. Distal root: largest, distally located, supporting the buccal side of the second loph

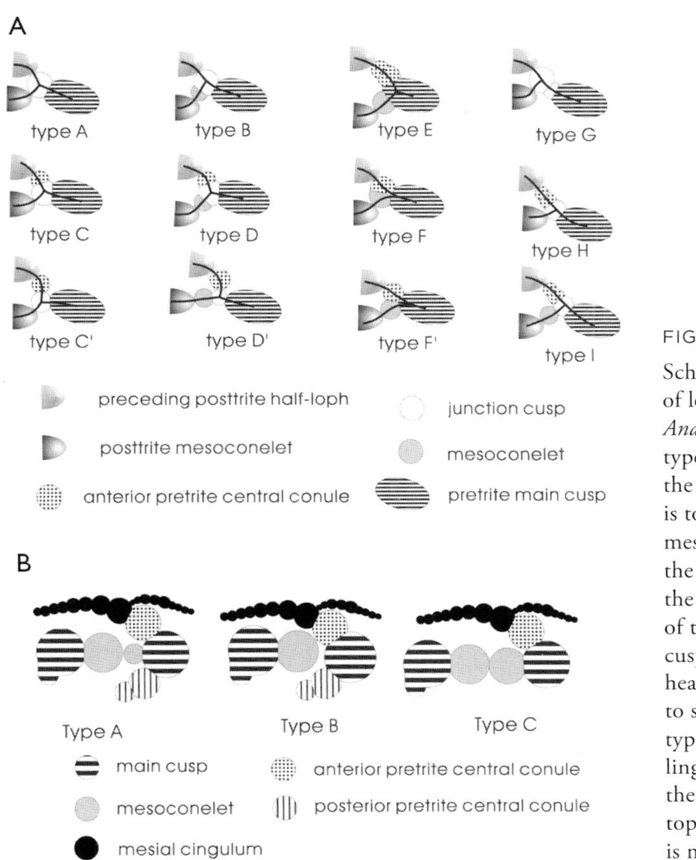

FIGURE 15.1

Schematic occlusal views of loph structures of *Anancus*. A. Morphological types of lophs other than the first loph. The lingual is to the right and the mesial is toward the top of the page. The path across the minimum number of the interfaces between cusps is expressed by a heavy line. Drawing is not to scale. B. Morphological types of the first loph. The lingual is to the right and the mesial is toward the top of the page. Drawing is not to scale.

and the rest of the molar. In addition to these, a tiny accessory root may exist mesiobuccally to the distal root. This basic arrangement of the upper molar root has changed variously in elephants (e.g., the dp^4 of *Primelephas gomphotheroides*).

Lower molars have two roots: mesial and distal (Figure 15.2C–E). In early Elephantoidea, the mesial root supports only the first lophid of the molar. However, a buttress supporting the lingual half of the second lophid is frequently developed on the distolingual surface of the mesial root in gomphotheres. In more derived groups such as advanced elephants and stegodonts, the mesial root supports more than two lophids (Saegusa and Taruno 1991). The rest of the molar is supported by the larger distal root. There may be a tiny accessory root adjacent to the mesiolingual end of the distal root.

Abbreviations

Dental Terminology

The following abbreviations are used: ccprp, posterior pretrite central conule; ccpra, anterior pretrite central conule; ccpop, posterior posttrite central conule; ccpoa, anterior posttrite

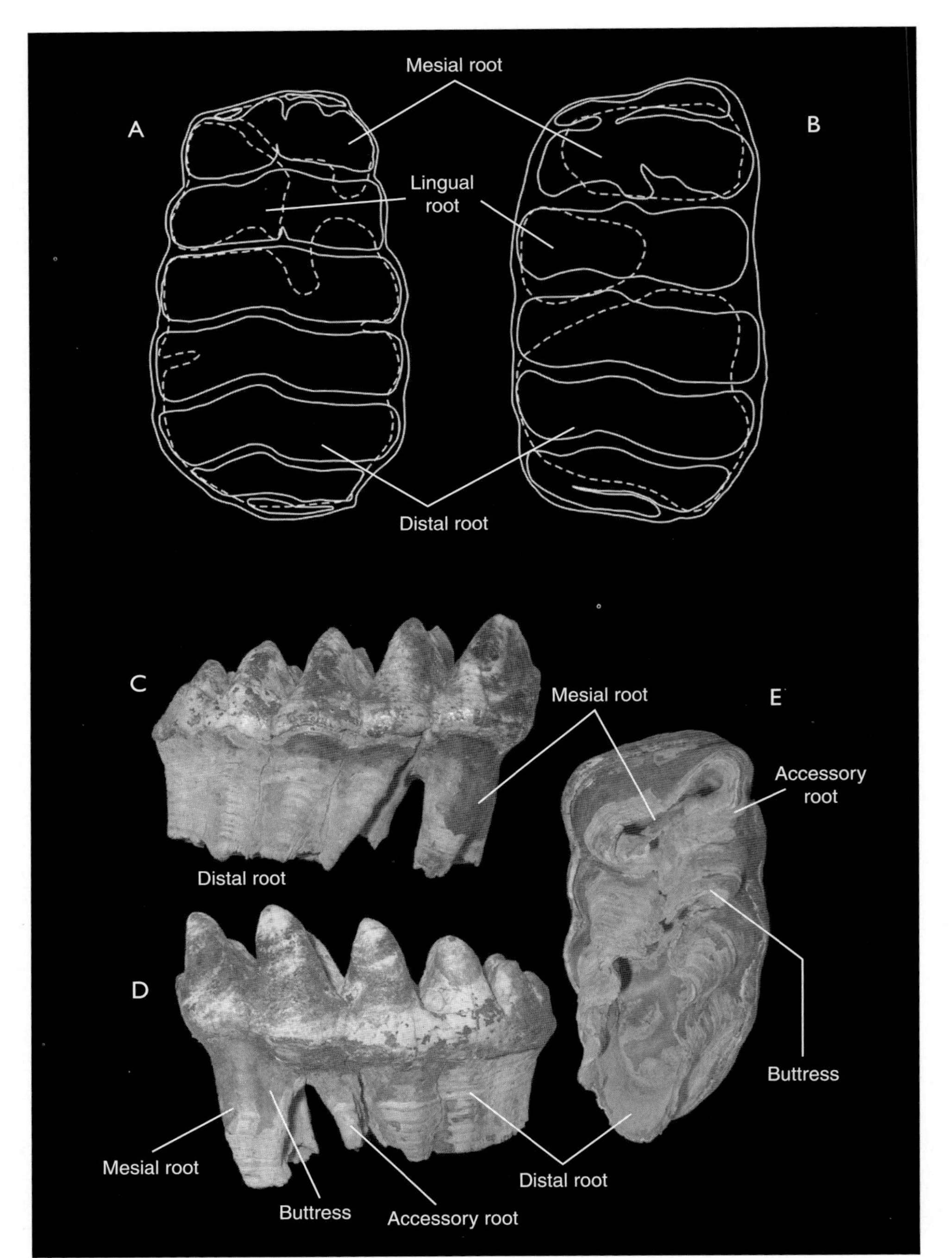

FIGURE 15.2

Line drawings depicting the relationship between crown and root structures. Counters of lophs and cingulum (solid lines) are superimposed on the outlines of the roots (broken lines). A. dp^4 of *Stegodon orientalis* (CBM-PV 814), modified from Taru et al. (2005). B. dp^4 of *Primelephas gomphotheroides saitunensis* (STD-VP-2/3). Right M_3 of *Mammut americanum* (MNHAH DI-000661): C. Buccal view. D. Lingual view. E. Apical view.

central conule; dp, deciduous premolar (dp^2 refers to an upper second deciduous premolar, and dp_2 refers to a lower second deciduous premolar); EDJ, enamel-dentin junction; M, molar (M^2 refers to an upper second molar, and M_2 refers to a lower second molar); P, premolar (P^2 refers to an upper second premolar, and P_2 refers to a lower second premolar).

Locality and Stratigraphic Horizons

The following abbreviations are used for non–Middle Awash collections and institutions: AT, Aterir, Kenya; BC, Baringo, Chemeron Formation, Kenya; KP, Kanapoi, Kenya; LT, Lothagam, Kenya; LU, Lukeino, Kenya; PQ-L, Pelletal Phosphorite and Quartzose Members, Langebaanweg, South Africa; KNM, National Museums of Kenya; MMK, McGregor Museum, Kimberley; NHM, Natural History Museum, London; SAM, South African Museum.

Gomphotheriidae

A number of cladistic analyses have been conducted on the family Gomphotheriidae (Tassy 1982; Tassy and Shoshani 1988; Tassy 1990; Kalb and Mebrate 1993; Kalb et al. 1996; Shoshani 1996; Tassy 1996b; Shoshani et al. 1998). Most of them address the relationship between genera rather than between species. In this respect, analyses conducted by Kalb and Mebrate (1993), Kalb and Froehlich (1995), and Kalb et al. (1996) are exceptional because some species of African *Anancus* are included in their analyses, together with several elephantids. However, their results do not give any insight into the monophyly of African *Anancus* or the relationships between the African anancines and their Eurasian cousins, because neither the Eurasian *Anancus* species nor the North African *Anancus osiris,* described by Arambourg (1945), are included in their analyses. Thus, in this context, Tassy (1986) is still the only published reference that argues species-level phylogenetics within *Anancus,* although more comprehensive cladistic analysis of the genus is conducted in an unpublished doctoral thesis by Metz-Muller (2000). Both Tassy (1986) and Metz-Muller (2000) suggested that *A. kenyensis* and *A. sivalensis* constitute a monophyletic clade, whereas *A. osiris* from North Africa shows a close relationship with Eurasian *A. arvernensis.* The anancine population from the Middle Awash appears to be part of an evolving lineage from eastern and southern Africa that includes the primitive form from Lukeino and the highly derived types from Sagantole, Middle Awash. Morphs have been classified into informal groups, namely, the *A. kenyensis* morph, A. *kenyensis,* and A. *petrocchii* (Tassy 1986), or four successive informal taxa (Kalb and Mebrate 1993). Here we allocate this series to a broadly defined *Anancus kenyensis* species lineage. It should be noted, however, that Sanders (2007) named a new *Anancus* species (*Anancus capensis* sp. nov.) for the anancine material from the Quartzose Sand Member (QSM) of Langebaanweg.

Anancus Aymard in Dhorlac, 1855

Anancus kenyensis (MacInnes, 1942)

DESCRIPTION

Premolars Premolar specimens referred to *Anancus kenyensis* are shown in Figure 15.3, and their measurements are summarized in Table 15.1. The ALA-VP-2/153 dp^2 from the

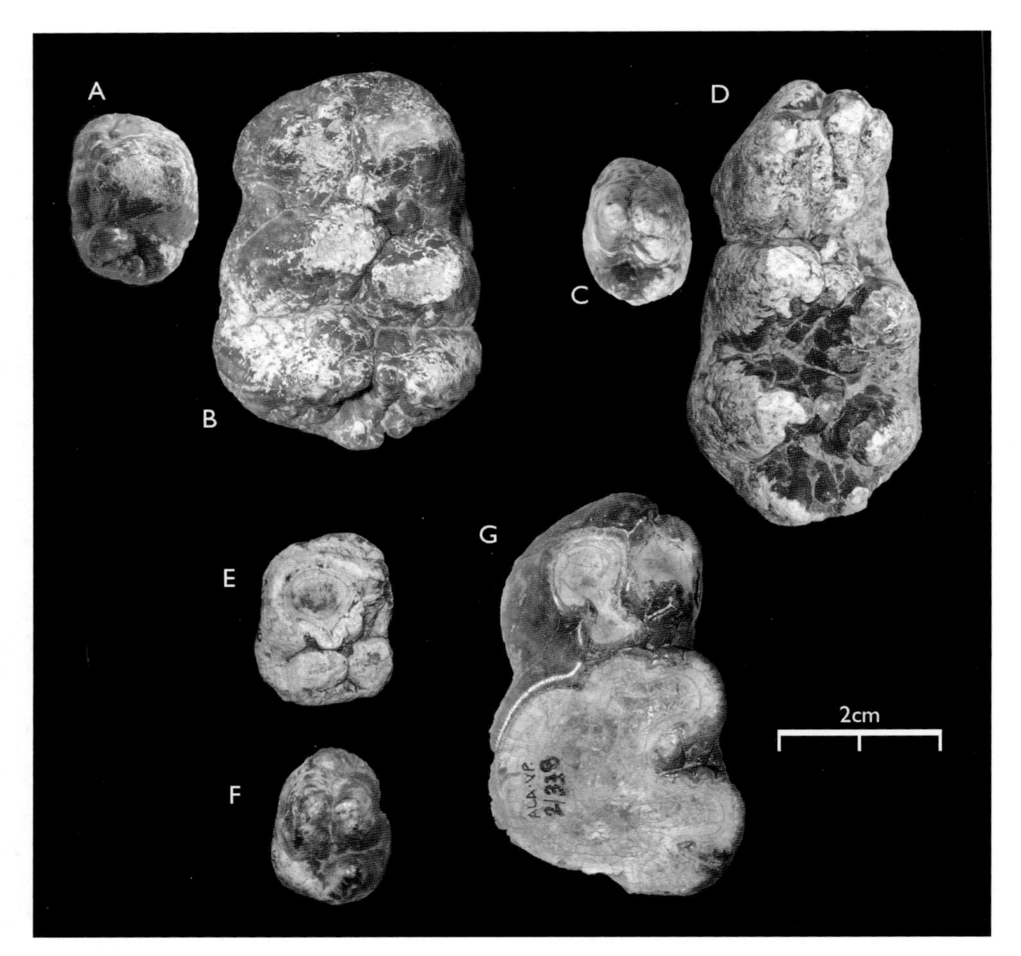

FIGURE 15.3
Occlusal views of deciduous premolars of *Anancus kenyensis:* A. AME-VP-1/91, left dp^2. B. AME-VP-1/91, left dp^3. C. KUS-VP-1/55, left dp_2. D. KUS-VP-1/55, right dp_3. E. ALA-VP-2/153, right dp^2. F. ALA-VP-2/104, right dp_2. G. ALA-VP-2/338, left dp_3.

Asa Koma Member is bilophodont with relatively developed mesial and rudimentary distal cingula. The highest point of the mesial cingulum is located at the lingual side of the mesial margin of the tooth. From this point, a crest goes down to the opposite side of the tooth. The first loph is composed of a large paracone and a rudimentary protocone, connected by a short crest. A much weaker crest runs on the distal flank of the protocone. The second loph is buccolingually wider and lower than the first loph and is composed of a transversely elongate metacone and narrower hypocone. A weak distal cingulum is closely appressed to the second loph. This specimen has no cement. The dp^2 of AME-VP-1/91 from the Kuseralee Member is similar to that of ALA-VP-2/153 apart from following features: The second loph is narrower than the first loph, the protocone is more rudimentary, the mesial cingulum is much reduced, and the hypocone is larger and higher than the metacone.

The dimunition of the protocone and second loph in the dp^2 of *A. kenyensis* appears to be a derived condition, because these features are well-developed in *Gomphotherium angustidens* (e.g., Schlesinger 1917, pl. 2, fig. 2). In *Tetralophodon* (e.g., Schlesinger 1917, pl. 12), the protocone is reduced, thus approaching the condition seen in AME-VP-1/91,

TABLE 15.1 Measurements of Deciduous Premolars of *Anancus kenyensis*

Specimen	*Loph(id) Formula*	*L*	*W*	*H*	*I*	*HI*	*E*
dp^2							
ALA-VP-2/153	x2x	22.7	16.4	nm	72.2	nm	nm
AME-VP-1/91	x2x	22.5	16.4	17.2	72.9	104.9	nm
dp^3							
AME-VP-1/91	x3x	49.4	35.8	22.8	2.57	63.7	1.7–1.9
dp_2							
ALA-VP-2/17	x2	22.1	14.7	19.2	66.5	130.6	nm
ALA-VP-2/104	x2	21	15.8	16	75.2	101.3	nm
KUS-VP-1/46	x2	22.1	14.9	18.1	67.4	121.5	nm
KUS–VP-1/55	2	19.3	14.8	16.5	76.7	111.5	nm
dp_3							
ALA-VP-2/338	x3x	52.8	32.7	nm	61.9	nm	nm
KUS-VP-1/55 l	x3x	51.8	29.1	21.3	56.2	73.2	nm
KUS-VP-1/55 r	x3x	52.5	29.2	22.6	55.6	77.4	nm

NOTE: All measurements in mm. Abbreviations: − = overestimated measurement; + = preserved measurement; E = enamel thickness; H = height; HI = height index (= 100 × H/W); l = left; r = right; L = length; W = width; I = index of robustness (= 100 × W/L); nm = measurement was not possible.

but a strong distal cingulum is still present behind the second loph. The dp^2 of *Anancus arvernensis* is basically similar to that of tetralophodonts, except for several minor differences (Metz-Muller 1996). The dp^2 of *A. kenyensis* is more derived than that of *A. arvernensis* in further degeneration of the protocone and the second loph.

The dp^3 is represented solely by AME-VP-1/91. It is trilophodont and bunolophodont and attains its maximum width at the third loph. The ccpra1 is connected to the mesial cingulum. The ccpop1 is composed of two tubercles and closely appressed to ccpra2. In addition to ccpop1, there is a tubercle located in the middle of the buccal half of the first inter-loph valley. The pretrite half of the first loph consists of a large main cusp and a smaller mesoconelet. On the second and third lophs, pre- and posttrite half-lophs are each composed of a large main cusp and a smaller mesoconelet. Moderately-sized ccpra2, rudimentary ccprp2, and rudimentary ccpop2 are associated with the second loph. The posttrite half-loph of the second loph is slightly displaced distally. The ccpra3 is transversely broad and half embedded in the pretrite half-loph. The distal cingulum is composed of two tubercles. The cusp arrangement and the size of the tooth are within the range of individual variation of the dp^3 of *Anancus arvernensis* described by Metz-Muller (1996). The dp^4 is represented only by AME-VP-1/91 and consists of uninformative small fragments.

The dp_2 is bilophodont. The first lophid is about two times taller than the second lophid and composed of equal-sized protoconid and metaconid that are closely appressed to each other. In KUS-VP-1/46, the apices of the protoconid and the metaconid are further subdivided into small digitations. In KUS-VP-1/55 and KUS-VP-1/46, the mesial cingulum is represented by a slight swelling on the mesial face of the tooth. In ALA-VP-2/104 and

ALA-VP-2/17, the mesial cingulum is composed of a small transversely elongated paraconid and a crest going down to the buccal side of the tooth from the paraconid. On the distal face of the metaconid, a crest blocking the lingual mouth of the interlophid valley develops variably; the crest is composed of either a small but distinct swelling (KUS-VP-1/55, ALA-VP-2/104), two parallel, vertically elongate swellings (ALA-VP-2/17), or two mesiodistally aligned tubercles (KUS-VP-1/46). The second lophid is composed of the entoconid and the hypoconid, and their apices may be superficially subdivided. There is no distal cingulum. The only other African anancine deciduous premolars known are those from the late Pliocene of Ahl al Oughlam, Casablanca, Morocco, whose estimated age is ca. 2.5 Ma (Geraads et al. 1998; Geraads and Metz-Muller 1999). These authors described a dp_2 (AaO-1397) and a mandible with dp_3–dp_4 and M_1 (AaO-3935) from this locality as *Anancus* sp. cf. *A. osiris*. According to them, the dp_2 (AaO-1397) is distinguished from those of *A. arvernensis* by the absence of a mesial cingulum, fusion of the metaconid and the protoconid, a pad-like hypoconid, and the development of the distal cingulum. In the dp_2 from the Middle Awash, the metaconid and the protoconid are appressed to each other, and the distal cingulum is not developed. In two specimens from the Kuseralee Member, the mesial cingulum is rudimentary. Thus, the dp_2 from the Middle Awash exhibits intermediate morphology between those of *A. arvernensis* and AaO-1397.

The dP_3 is represented by KUS-VP-1/55 from the Kuseralee Member and ALA-VP-2/338 from the Asa Koma Member (Figure 15.3D, G). The structure of the dp_3 of *A. kenyensis* from the Middle Awash is mostly identical with that of *A. arvernensis*. It is trilophodont and bunolophodont and attains minimal and maximal width at the first interlophid valley and the third lophid, respectively. Anancoidy is clearly seen on the second and third lophids. The mesial cingulum is lingually placed and slightly appressed to the posttrite half of the first lophid. All the half-lophids are each composed of a single large main cusp and a smaller mesoconelet. The pretrite main cusps of the first and second lophids are placed far apart to form a wide V-shaped interlophid valley. The ccprp1 of KUS-VP-1/55 is flat, small, and separated from the second lophid. In contrast, that of ALA-VP-2/338 is large and contacts with the second lophid, as normally seen in *A. arvernensis*. The ccpoa2 is centrally located and contacts with the third lophid. The ccpop2 is closely appressed to the distally displaced mesoconelet of the second posttrite half-lophid. The ccpra3 is rudimentary but forms a wall blocking the second valley together with the ccpop2.

The dp_4 is represented only by a right and left dp_4 (KUS-VP-1/55), still in the crypt and imperfectly calcified. The mesial three lophids of the right dp_4 are preserved, and strong anancoidy can be assumed. Other morphological features are not clear.

Molars Permanent molars of *Anancus kenyensis* from the Middle Awash (see Table 15.2 for measurements) are characterized by a highly varied loph(id) structure (Figures 15.4–15.5, Tables 15.2–15.6). Features other than those tabulated in Tables 15.3–15.6 are described here.

The M^1 is represented only by KUS-VP-1/73. The molar is tetralophodont, with a small distal cingulum. The second and third lophs are parallel, perpendicular to the tooth axis, without angulations between half-lophs. Posttrite half-lophids are each composed of

TABLE 15.2 Measurements of the Molars of *Anancus kenyensis*

Specimen	*Loph(id) Formula*	*L*	*W*	*H*	*I*	*HI*	*E*
M^1							
KUS-VP-1/73	−4x	88.4+	55		62.2–		3.2–4.1
(M^1 or M^2)							
AMW-VP-1/118 l	−4−	80.6+	62.6+				3.8
AMW-VP-1/118 r	−4−	81.0+					4.6
M^2							
AMW-VP-1/10 l	−4x	102.1+	74.5		73–		4.4–6.2
AMW-VP-1/10 r	−4	116.7+	75		64.3–		5.3–6.4
AMW-VP-1/70	−4x	103.1+	61.9+				3.8–5.4
KUS-VP-1/100	−4x	113.6+	69.3+				4.8–6.4
M^3							
AMW-VP-1/10	x5x	185.5	86.2+	59.4 (3)	46.5+	68.9–	5.1–5.8
AMW-VP-1/10	x5x	194.3	86.8	58.8 (2)	44.7	67.7	
AMW-VP-1/74	x5x	155.8	72.9	50.1+ (3)	46.8	68.7+	4.8–6.7
AMW-VP-1/106 l	x6x	153.8	68.1	43.6 (3)	44.3	64	5.0–5.8
AMW-VP-1/106 r	−4x	105.8+	62.9+	45.7 (2)		72.7−	
AMW-VP-1/147	−5x	147+	73.3				5.4–5.8
DID-VP-1/60	x5x	191.9	88.1	53.6+ (5)	45.9	60.8+	5.7–6.2
DID-VP-1/60	−4x	156.7+	89.7 (2)				5.7–5.9
KUS-VP-1/21	x5x	153.2	73	47.7+	47.7	65.3+	4.2–5.8
KUS-VP-1/64 r	x6x	151.1	63.1	43.7 (4)	41.8	69.3	5.3–5.6
KUS-VP-1/64 l	x6x	148.8	63.1	42.4 (4)	42.4	67.2	
KUS-VP-1/96	x6	153.7	70.9		46.1		5.2–6.0
KUS-VP-1/98	−5	143.6+	69.8+				5.1–6.1
KUS-VP-1/100	x5x	175.2	82.8	53.3 (4)	47.3	64.4	6.2–7
M_1							
ALA-VP-2/151	x3−	60.2+	50.9+				2.9–2.3
AME-VP-1/54	−3−	63.3+	51.1+				4.0–4.7
AME-VP-1/65	−3x	60.3+	60.7				3.3–3.6
AME-VP-1/172	x4x	102.6	46.9+	44.1 (4)	45.7+	94	2.2–3.1
AMW-VP-1/107B	x4x	90.6+	49.2+				3.1–3.8
AMW-VP-1/125	x4x	95.2+	48.7+	37.3+ (3)			3.2–4.3
M_2							
AME-VP-1/65	x4x	139.9	75.6		54.0		5.3–5.6
AME-VP-1/65	x4x	116.3	74.2		63.8		4.8–5.8
AMW-VP-1/107A	x4x	112.9	54.6	40.7 (3)	48.4	74.5	4.5–5.3
AMW-VP-1/120	x4x	132.7	72.8		54.9		4.8–6.3
AMW-VP-1/120	x4x	130.8	71.5		54.7		5.4–6.6
AMW-VP-1/123	x4x	107.1+	63				2.9–4
BIK-VP-3/6	x4−	116.2+	61.6				3.9–4.8
BIK-VP-3/8	x4x	117.7	61.7		52.4		3.8–5.3
M_3							
AME-VP-1/30	−5x	154.3+	71.7				4.2–5.6
AME-VP-1/59	−4x	123.2+	59.7+				4.3–5.2
AME-VP-1/70	−5x	149.7+	78.2				4.9–5.7
AMW-VP-1/97	−6x	211.4+	82.1+	49.9 (3)	38.8–	60.8–	4.2–6.2
AMW-VP-1/63 l	x6x	195.9	80	52.4 (4)	40.8	65.5	5.8–7

TABLE 15.2 *(continued)*

Specimen	*Loph(id) Formula*	*L*	*W*	*H*	*I*	*HI*	*E*
AMW-VP-1/63 r	x6x	184.6	75.2	50.5 (5)	40.7	67.2	5.7–6.3
AMW-VP-1/120 l	x5x	205.9+	90.1	62.2 (4)	43.8−	69.0	
AMW-VP-1/120 r	x5x	198.4+	90.1	57.2 (2)	45.4−	63.5	6.4–6.5
AMW-VP-1/135 r	x5x	211.7	86.8	54.5 (4)	41	62.8	6.1–6.6
AMW-VP-1/135 l	x5x	201.4	87.8	52 (4)	43.6	59.2	
KUS-VP-1/47	−4x	135.8+	70.8+				5.8–6.1

NOTE: All measurements are in mm. For abbreviations, see note to Table 15.1. Numbers in parenthesis indicate the plate number where measurement was taken. "x" before number = anterior platelet present. "x" after number = posterior platelet present.

a main cusp and a slightly smaller mesoconelet. Worn enamel is wrinkled at the second pretrite half-loph.

The M^2 is represented by specimens from the Kuseralee Member (AMW-VP-1/10, KUS-VP-1/100, and AMW-VP-1/70). The M^2 is tetralophodont, with a small distal cingulum. The coronal cement cover is thin. Weak enamel folding is frequently seen on the worn loph. The posttrite half loph is composed of an equally-sized main cusp and mesoconelet. The detail of the M-C complexes of the second and third lophs is not known in M^2s from the Kuseralee Member because of heavy wear. However, the M-C complexes are as large as, or slightly smaller than, the main cusps. Weakly alternating positioning of half-lophs is seen on second and third lophs. Faint chevroning of lophs is seen on the distal half of the molar.

In addition to the M^2s mentioned above, there is a pair of extremely worn left and right intermediate upper molars (AMW-VP-1/118). These were originally tetralophodont and show marked anancoidy. Other crown structure is obliterated by wear.

Six M^3s were recovered from the Kuseralee Member. The loph formula varies from x5x to x6x. The molar generally decreases its width distally. Worn enamel figures are finely folded (e.g., KUS-VP-1/96 and KUS-VP-1/98). Only a small amount of cement is seen in the interloph valleys. In larger molars (e.g., AMW-VP-1/10 and KUS-VP-1/100), the lophs, particularly the first and second, are separated by wide, V-shaped valleys. In buccal view, lophs run fanwise from the cervical rim to the occlusal surface. In contrast, the cone pairs are closely spaced in smaller molars (e.g., KUS-VP-1/64, KUS-VP-1/96). In smaller molars (e.g., KUS-VP-1/98, AMW-VP-1/106), weak chevroning is seen on the middle part of the molar, whereas at the mesial and distal part the lophs run transversally. In larger molars (e.g., AMW-VP-1/10, KUS-VP-1/100), marked chevroning of lophs is seen on the distal part of the molar. Most of posttrite halves of the lophs are composed of an equal-sized mesoconelet and main cusp, except for mesial or distal posttrite half-lophs, where the mesoconelet is subdivided superficially (e.g., second loph of KUS-VP-1/64) or is smaller than the main cusp (e.g., fourth loph of AMW-VP-1/10). The distal cingulum is rudimentary and closely appressed to the last loph in KUS-VP-1/64 and KUS-VP-1/96. The distal cingulum in other molars is a simplified copy of the last loph and is separated from the latter by a deep valley.

Only two M^3s are available from the Adu-Asa Formation. DID-VP-1/60 is a relatively large, pentalophodont molar with a moderately developed distal cingulum. BIK-VP-3/8

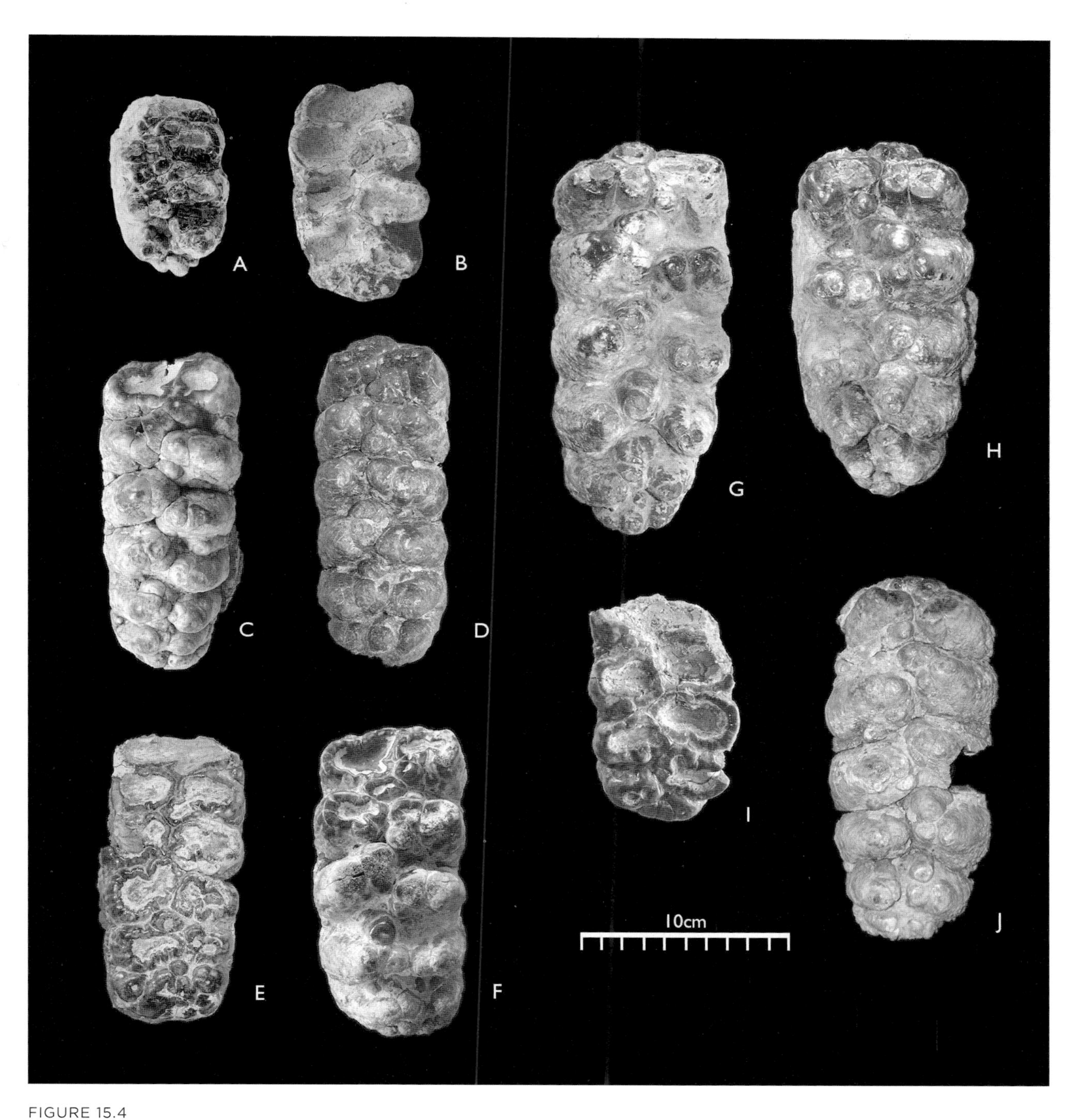

FIGURE 15.4

Occlusal view of upper molars of *Anancus kenyensis*. A. KUS-VP-1/73, right M^1. B. AMW-VP-1/70, left M^2. C. AMW-VP-1/106, left M^3. D. KUS-VP-1/64, left M^3. E. KUS-VP-1/96, left M^3. F. KUS-VP-1/21, left M^3. G. AMW-VP-1/10, left M^3. H. AMW-VP-1/10, right M^3. I. KUS-VP-1/100, right M^2. J. KUS-VP-1/100, right M^3.

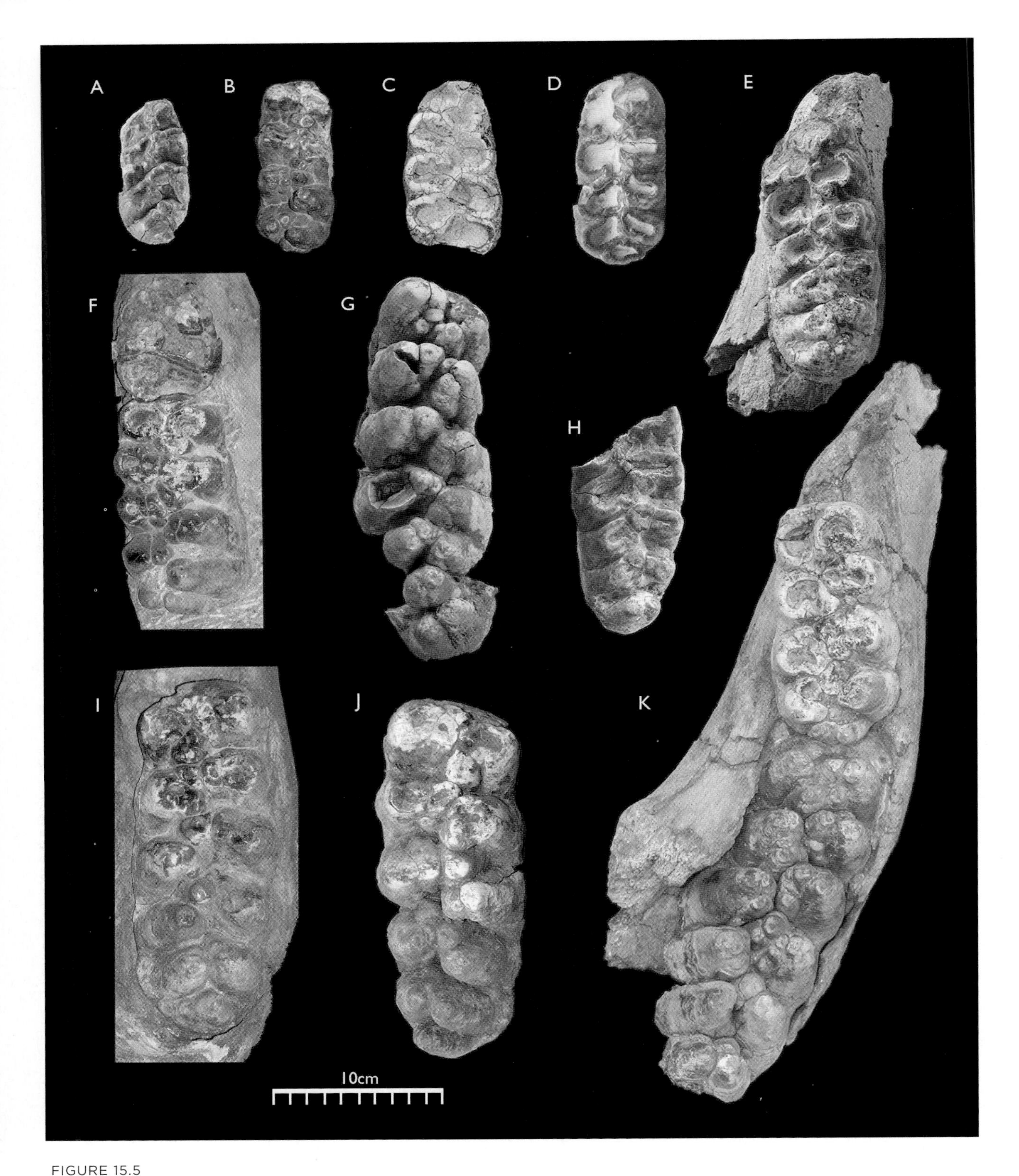

FIGURE 15.5

Occlusal view of lower molars of *Anancus kenyensis*. A. AMW-VP-1/107B, left M_1. B. AME-VP-1/172, right M_1. C. AMW-VP-1/123, right M_2. D. BIK-VP-3/8, left M_2. E. AME-VP-1/30, left M_3. F. AME-VP-1/65, right M_1–M_2. G. AMW-VP-1/97, right M_3. H. KUS-VP-1/47, left M_3. I. AMW-VP-1/135, left M_3. J. AMW-VP-1/135, right M_3. K. AMW-VP-1/120, left mandible with M_2–M_3.

TABLE 15.3 Summary of the Variation of Cusp Structures Seen in the M^1 and M^2 of *Anancus kenyensis* from the Late Miocene of the Middle Awash

	AMW-VP-1/70 *Left* M^2	*AMW-VP-1/10* *Right and Left* M^2	*KUS-VP-1/100* *Right* M^2	*KUS-VP-1/73* *Right* M^1
Loph formula	x4x	–4x	–4x	–4x
Worn loph	x4x	–4x	–4x	–4
Anancoidy	Medium	Medium	Strong	Strong
ccpra2				Large
ccpra3				Large
ccpra4	Large	Large	Large	Large, double
ccprp2	Absent		Absent	Small
ccprp3	Absent		Absent	Absent
ccprp4	Absent	Small	Absent	Absent
ccpoa2			Absent	Absent
ccpoa3		Small	Absent	Small
ccpoa4		Small	Absent	Large
ccpop1	Large		Large?	Large
ccpop2	Large	Medium	Medium	Large
ccpop3	Large	Small	Small	Medium
ccpop4	Absent	Absent	Absent	Absent
ccpos1	Absent		Present	Present
ccpos2	Present	Absent	Present	Present
ccpos3	Absent	Absent	Present	Absent
ccpos4	Absent	Absent	Absent	Absent
Type of first loph				
M-C complex of 2nd loph				F′
M-C complex of 3rd loph				F′
M-C complex of 4th loph	D′	D′	D′	D′
Number of tubercles in distal cingulum	3	1	3	3

NOTE: For central conule (cc) abbreviations and M-C complex type explanations, see the section "Terminology and Abbreviations" in the text and Figure 15.1.

is a fragment of the distal part of M^3. The occlusal crown morphology is simple, with no posttrite conules. The M-C complexes of the first, second, and third lophs of the preserved portion are of type B, C, and F, respectively.

The Kuseralee Member yielded six M_1s and four M_2s; the Adu-Asa formation yielded a mesial fragment of left M_1 and two M_2s (Figure 15.5). Both M_1 and M_2 are tetralophodont and have rudimentary cement in the bottom of the valley. On the second and third lophids, the pretrite half-lophid runs from the median sulcus distobuccally, whereas the posttrite half-lophid runs nearly perpendicular to the lingual margin of the molar. On the last lophid, however, both the pre- and posttrite half-lophids frequently run obliquely to the axis of the molar to form a chevron. The anancoidy is frequently very weak at the second lophid because of poor development of posttrite mesoconelets (Table 15.5). On the other hand, it is always strong at the distal two lophids. Variation of the accessory cusps is tabulated in Table 15.3.

The Kuseralee Member yielded eight M_3s. Lophid formula varies from x5x to x6x. The spacing of the lophids also varies and roughly correlates with the dimension of the molar; the lophids are spaced closely in larger specimens and widely in smaller specimens. A

TABLE 15.4 Summary of the Variation of Cusp Structures Seen in the M^3 of *Anancus kenyensis* from the Late Miocene of Middle Awash

	DID-VP-1/60 Right & Left M³	*AMW-VP-1/10 Right & Left M³*	*KUS-VP-1/100 Right M³*	*AMW-VP-1/147 Left M³*	*AMW-VP-1/74 Left M³*	*KUS-VP-1/21 Left M³*	*KUS-VP-1/98 Right M³*	*KUS-VP-1/96 Left M³*	*AMW-VP-1/106 Right & Left M³*	*KUS-VP-1/64 Right & Left M³*
Loph formula	x5x	x5x	x5x	x5x	x5x	x5x	−6x	x6x	x6x	x6x
Worn loph	x5	x1	x1	x5x	x4	x4	−6x	x6x	x3	x1
Anancoidy	Weak	Medium	Strong	Weak	Weak	Medium	Weak	Weak	Strong	Weak
ccpra2		Double, large	Double, large			Double, large			Large	
ccpra3		Large	Medium			Large			Small	
ccpra4		Large	Large		Large	Large				Large
ccpra5	Large	Large	Large			Double	Medium	Large	Large	Large
ccpra6								Large?	Large	
ccprp2	Absent	Absent	Absent			Medium		Absent	Absent	Medium
ccprp3	Absent	Absent	Absent		Rudimentary	Absent	Small	Absent	Absent	Rudimentary
ccprp4	Absent	Absent	Absent		Absent	Absent	Small	Absent	Absent	Absent
ccprp5	Absent	Absent	Absent		Absent	Absent	Absent	Absent	Absent	Absent
ccpoa2	Absent	Absent	Absent		Absent	Present		Absent	Small	Absent
ccpoa3	Absent	Absent	Absent		Absent	Present		Absent	Absent	Absent
ccpoa4	Absent	Absent	Absent		Absent	Absent	Absent	Absent	Absent	Absent
ccpop1	Double, large	Large	Double, large		Double, large	Double, large		Large	Double, large	Double, large
ccpop2	Large	Small	Large		Double, large	Large	Large	Large	Large	Large
ccpop3	Rudimentary	Rudimentary	Medium		Small	Large	Large	Large	Small	Rudimentary
ccpop4	Absent	Absent	Absent		Absent	Rudimentary	Small	Small	Absent	Rudimentary
ccpop5		Absent	Absent		Absent	Absent	Absent	Rudimentary	Absent	Absent
ccpop6							Absent	Absent	Absent	Absent
ccpos1	Present	Absent	Present		Absent	Absent			Absent	Absent
ccpos2	Present	Absent	Present		Absent	Absent		Absent	Present	Present
ccpos3	Present	Absent	Present		Absent	Absent	Present	Present	Present	Absent
ccpos4	Present	Absent	Absent		Absent	Absent	Present	Present	Absent	Absent
First loph type		C	B			B?			B?	C
M-C complex of 2nd loph	C?	E	E			E			C	A
M-C complex of 3rd loph	F'?	D	D			D			D	A
M-C complex of 4th loph	D?	H/D	F'		D	D			B	D
M-C complex of 5th loph	D'	I/C	C'		A	D	C	D'?	D	D
M-C complex of 6th loph								I?	C	A
Distal end of molar	Round	Pointed	Round	Square	Square	Square	Square	Square	Round	Square

NOTE: For central conule (cc) abbreviations and M-C complex type explanations, see the section "Terminology and Abbreviations" in the text and Figure 15.1.

strong anancoidy is seen from the third lophid through the penultimate lophid, except for AMW-VP-1/97, where anancoidy is exaggerated on the last lophid. The anancoidy on the second lophid varies from specimen to specimen (Table 15.6). The pretrite half-lophids run obliquely to the axis except in the first lophid. The mesial posttrite half-lophids run perpendicular to the axis of the molar. The pretrite and posttrite half-lophids are clearly angled on one another to form anteriorly pointing chevrons on the distal part of the molar. This chevroning of the lophids starts from various positions on the molar: the fifth lophid (AME-VP-1/30 and AMW-VP-1/63), the fourth lophid (AMW-VP-1/97, AME-VP-1/163, and AMW-VP-1/120), and the third lophid (KUS-VP-1/47, AME-VP-1/70, and AMW-VP-1/135). The distal end of the molar is rounded (e.g., AME-VP-1/70) or pointed (e.g., KUS-VP-1/47, AME-VP-1/163). Thin cement cover is seen at the bottom of the interlophid valleys when unworn or little worn.

Mandibles Mandibular morphology is well-preserved in three specimens: AMW-VP-1/120 (Figure 15.5K), AMW-VP-1/63 (Figure 15.6A), and AME-VP-1/65 (Figure 15.6B). The mandibular corpus is robust, and its ventral margin gently curves ventrally as in all *Anancus* species. In dorsal view, the labial outline of the mandibular corpus is widely opened and U-shaped (AMW-VP-1/63) or V-shaped (AMW-VP-1/120). The alveolar part of the corpus is proportionally taller than those of elephants and stegodonts. The masseteric fossa is very deep. The coronoid process flares externally, and the condyles slope mesially. In lateral view the anterior margin of the coronoid process stands nearly perpendicular to the axis of the mandibular corpus. A mediolaterally flattened mandibular angle is located at the posterior margin of the ascending ramus. The mandibular foramina are relatively small, and distinct mylohyoid grooves run anteroinferiorly from these foramina.

DISCUSSION The dental materials of *Anancus* from the Kuseralee Member show great variation in both dimensions and loph(id) structure. The sample includes the smallest known M^3 (KUS-VP-1/64) of *Anancus* ever collected, as well as the second largest M^3 from eastern Africa. However, these specimens show great complexity in M^3 loph structure, necessitating a new terminology for the pretrite cusps of upper molars (see Figure 15.1).

When width and length of third molars are plotted (Figures 15.7 and 15.8), two clusters are seen. One cluster nearly coincides with the previously recorded range of third molar dimensions of eastern African *Anancus,* and the other is smaller. This raises the question of whether these clusters represent one or two species. To address this question, we evaluate variation of the dimensions of the molars, by comparing the coefficient of variation (CV) of width and length of third molars of *Anancus* from the Kuseralee Member with those of some extant and extinct species of Elephantoidea (Lewontin 1966). The CVs of the width and length of *Anancus* third molars from the Kuseralee Member do not significantly differ from those of other species of various elephantoids, except in a few cases (Tables 15.7 and 15.8). This suggests that the variation in the M^3 of Kuseralee Member *Anancus* is best regarded as intraspecific, even though it exceeds the range of metric variation previously known in *Anancus* from eastern Africa.

According to Tassy (1986), there are two morphotypes within *Anancus:* primitive *A. kenyensis* and derived *A. petrocchii*. One morph is represented by the type specimen of *A. kenyensis* from Kanam, Uganda, described by MacInnes (1942). The other is

TABLE 15.5 Summary of Cusp Structure Variation Seen in the M1 and M2 of *Anancus kenyensis* from the Late Miocene of the Middle Awash

	AMW-VP-1/125 Left M_1	*AME-VP-1/172 Right M_1*	*AMW-VP-1/128 Right M_1*	*AME-VP-1/54 Left M_1*
Lophid formula	–4x	x4x	–4x	–3x
Worn lophid	–4	x4	–4	–3
Anancoidy on second lophid	Weak	Weak	Strong	Strong
ccprp1	Large	Large, double	Large, double	
ccprp2	Large, double	Large, double	Large, double	Large
ccprp3		Medium	Medium	Large
ccprp4	Absent	Absent	Absent	Small
ccpop1	Large	Rudimentary	Absent	
ccpop2	Small	Absent	Absent	Medium
ccpop3	Rudimentary	Absent	Absent	Medium
ccpop4	Absent	Absent	Absent	Absent
ccpos1	Absent	Absent		Absent
ccpos2	Absent	Absent		Absent
ccpos3	Absent	Absent	Absent	Absent
ccpos4	Absent	Absent	Absent	Absent
ccpoa2	Absent	Absent	Absent	
ccpoa3	Absent	Absent	Absent	Small
ccpoa4	Absent	Absent	Absent	Absent
Pretrite mesoconelet 2	Large	Large	Small	
Pretrite mesoconelet 3	Rudimentary	Rudimentary	Small	Rudimentary
Pretrite mesoconelet 4	Medium	Small	Rudimentary	Small
Posttrite main cusp 2	Single?	Single?	Single?	
Posttrite main cusp 3	Double	Double	Double	
Posttrite main cusp 4	Single	Single	Single	
Location of posttrite mesoconelet 2	Lingual	Lingual	Central	Central
Location of posttrite mesoconelet 3	Central	Central	Central	Central
Location of posttrite mesoconelet 4	Lingual	Lingual	Central	Central
Tubercles on distal cingulum	2	1	1	1?

NOTE: For central conule (cc) abbreviations and M-C complex types, see the section "Terminology and Abbreviations" in the text and Figure 15.1.

Anancus petrocchii from Sahabi, Libya (Sanders 2008), originally described by Petrocchi (1954) as *Pentalophodon sivalensis*. The *A. petrocchii* morph is distinguished from the *A. kenyensis* morph by the derived traits of its molars, and it is thought to represent a phyletic descendant (chronospecies) of *A. kenyensis*. Recently, Tassy (2003) added new samples from Lothagam to his *A. kenyensis* and *A. petrocchii* morphs.

On the other hand, Kalb and Mebrate (1993) divided sub-Saharan *Anancus* into four successive species, *Anancus kenyensis, Anancus* sp. (Langebaanweg type), *Anancus petrocchii,* and *Anancus* sp. (Sagantole type), mostly based on specimens from the Middle Awash. These authors suggest an evolutionary trend within African *Anancus* towards greater complexity of crown pattern over time (e.g., increase of the number of loph(id)s, enhancement of anancoidy, development of enamel folding, increase in the number and size of posttrite accessory cusps). However, resolution of species-level phylogenetics within African

AMW-VP-1/107A Left M_2	*AMW-VP-1/107B Left M_1*	*AMW-VP-1/123 Right M_2*	*AME-VP-1/65 Right and Left M_2*	*AMW-VP-1/120 Right and Left M_2*	*BIK-VP-3/6 Right M_2*	*BIK-VP-3/8 Left M_2*
x4x	–4x	–4x	x4x	x4x	x4–	x4x
x1	–4	–4x	–4	x4x	x4–	x4x
Strong	Weak	Strong	Weak	Strong	Strong	Weak
Large, double	Large, double	Large	Large, double	Large, double	Large, double	Large
Large	Large	Large	Large, double	Large, double	Large, double?	Large
Medium	Medium	Medium	Large	Large		Medium
	Rudimentary		Small	Medium		Small
Small	Large		Large	Medium	Medium	Large, double
Small	Medium	Small	Rudimentary	Small		Large
Absent	Large	Small	Absent	Large		Large
	Absent	Rudimentary	Absent	Absent		Small
Present		Absent	Absent	Absent	Absent	Absent
Present	Absent	Absent	Rudimentary	Present	Absent	Absent
Absent		Present	Absent	Absent		Absent
		Absent	Absent	Absent		Absent
	Absent	Small	Rudimentary	Medium	Absent	Small
Large	Large	Large	Large	Large		Large
	Small	Small	Absent	Absent		Small
Small	Large		Large	Rudimentary		Large
Small	Small	Rudimentary	Small	Large		Large
Small	Small	Small	Small	Large		Large
				Double		Single?
Double			Double	Double		Single?
			Single	Double		Single?
Central	Lingual	Central?	Lingual	Lingual		Lingual
Central	Central	Central	Lingual	Central		Lingual
Central	Central	Central	Central	Central		Lingual
1?	1		2, Large	2, Large		2?

Anancus has proven difficult because of the extensive overlap of the distribution of derived and primitive characters between available samples from eastern Africa and South Africa.

The large *Anancus* sample described from the Middle Awash will play a role in such resolution. Unfortunately, the other eastern African and South African samples of *Anancus* are still too small or insufficiently described to be compared to the new Ethiopian material. Therefore, the erection of new formal taxonomic units within African *Anancus* is still elusive. However, some samples of *Anancus,* including those from the Kuseralee Member, can be appreciated in evolutionary context based on the presence or absence of derived features in those samples. We allocate them here to *A. kenyensis,* pending a comprehensive revision based on larger samples from other sites, particularly the type sites of already designated species.

The sample from Lukeino, Kenya, exhibits a morphology more primitive than that of the Kuseralee Member. The latter shows more derived dental features in the M_3: the

TABLE 15.6 Summary of the Cusp Structure Variation Seen in the M_3 of *Anancus kenyensis* from the Late Miocene of the Middle Awash

	AME-VP-1/59 *Left M_3*	*AMW-VP-1/63* *Right and Left M_3*	*AME-VP-1/163* *Right M_3*	*AMW-VP-1/135* *Right and Left M_3*
Lophid formula	–4x	x6x	–5	x5x
Worn lophid	–4x	x4	–4	x3
Anancoidy on 2nd lophid		Weak	Weak	Weak/strong
ccprp1		Large, double		Large, double
ccprp2		Large		Large
ccprp3	Small	Large	Large	Small
ccprp4	Small	Absent	Small	Rudimentary
ccprp5	Absent	Small	Absent	Absent
ccpop1		Large, double		Small
ccpop2		Medium		Small
ccpop3	Small	Small/medium	Absent	Rudimentary
ccpop4	Absent	Absent	Small	Rudimentary/absent
ccpop5	Absent	Absent	Absent	Absent
ccpos1		Absent		Absent
ccpos2		Absent		Absent
ccpos3		Absent	Absent	Absent
ccpos4		Absent	Absent	Absent
ccpos5		Absent	Absent	Absent
ccpoa2		Large		Small
ccpoa3		Medium/small	Absent	Large
ccpoa4		Absent	Absent	Large
ccpoa5	Absent	Absent	Absent	Small
Pretrite mesoconelet 2		Large		Medium/large
Pretrite mesoconelet 3		Rudimentary	Large	Small
Pretrite mesoconelet 4	Rudimentary	Small/absent	Absent	Absent
Pretrite mesoconelet 5	Absent	Absent	Absent	Absent
Posttrite main cusp 2		Single		Single
Posttrite main cusp 3		Single	Single	Double
Posttrite main cusp 4		Single	Single	Double
Posttrite main cusp 5		Single	Single	Single
Location of posttrite mesoconelet 2		Lingual		Lingual/central
Location of posttrite mesoconelet 3	Central	Central	Lingual	Central
Location of posttrite mesoconelet 4	Central	Central	Central	Central
Location of posttrite mesoconelet 5	Central	Central/lingual	Central	Lingual/central
Tubercles on distal cingulum	2, Large	1	3, Large	1, Large

NOTE: For central conule (cc) abbreviations and M-C complex types, see the section "Terminology and Abbreviations" in the text and Figure 15.1.

ccpop2 is regularly distinct, the ccpop3 is frequently distinct, and the ccpoa2 is frequently present. Specimens from Mpesida (Tassy 1986), the Nawata Formation of Lothagam (Tassy 2003), and a mandible (KNM-NK 41502) from Lemudong'o, Kenya (Saegusa and Hlusko, 2007) are comparable with those from Lukeino in sharing a primitive molar structure. Specimens from the Adu-Asa Formation are very small and may belong with either those of Lukeino or those from the Kuseralee Member.

Anancus from the "E" Quarry of Langebaanweg, South Africa, based on our observations, shows more derived characters not seen in the Kuseralee Member sample: M^3, the

	AMW-VP-1/97 *Right* M_3	*KUS-VP-1/47* *Left* M_3	*AME-VP-1/30* *Left* M_3	*AME-VP-1/70* *Right* M_3	*AMW-VP-1/120* *Right and Left* M_3
	x6x	–4x	–5x	–5x	x6
	x1	–4x	–5x	–5x	0
	Strong		Strong		Strong
	Large, double		Large, double		Large, double
	Large	Large	Large		Large, double
	Small	Large	Large	Large	Large
	Rudimentary	Medium	Large	Small	Medium
	Rudimentary	Small	Small	Small	Absent
	Large, double		Large		Large
	Small	Medium	Medium		Small
	Small	Large	Medium		Large
	Absent	Absent	Rudimentary	Medium	Medium
	Absent	Absent	Small	Small	Absent
	Absent		Rudimentary	Absent	Absent
	Absent	Absent	Rudimentary	Absent	Absent
	Absent	Absent	Rudimentary	Absent	Absent
	Absent	Absent	Absent	Absent	Absent
	Absent	Absent	Absent	Absent	Absent
	Absent		Medium		Small
	Absent	Absent	Absent	Absent	Medium
	Absent	Absent	Absent	Absent	Absent
	Absent	Absent	Absent	Absent	Absent
	Medium		Large		Small
	Small	Small	Small		Small
	Rudimentary	Rudimentary	Small	Medium	Small
	Absent	Absent	Small	Medium	Rudimentary
	Single		Single?		Single
	Single	Double?	Single	Single?	Single
		Double	Single	Single	Double
	Single	Double	Double?	Single	Single
	Central		Lingual		Lingual
	Central	Central	Central		Central/lingual
	Central	Central	Central	Central	Central
	Central	Central	Central	Lingual	Central
	1, Large	2, Large	2, Large	2, Large	

loph formula varies from x6 to x7x, the supplementary posttrite central conule is always present at anterior interloph valleys, and the ccpop2 is as large as the ccpop 1; M_3, the lophid formula varies from x6 to x7x. The smaller samples from Kanam East, the Apak Member of the Nachukui Formation, Lothagam, and Aterir may be comparable with the morphology seen at Langebaanweg.

Specimens from the Pliocene of the Middle Awash attributed to the "Sagantole type" (KL-150-1 and KL-337-3) by Kalb and Mebrate (1993) are more derived than the Langebaanweg *Anancus* in showing further complication of the molar structure and

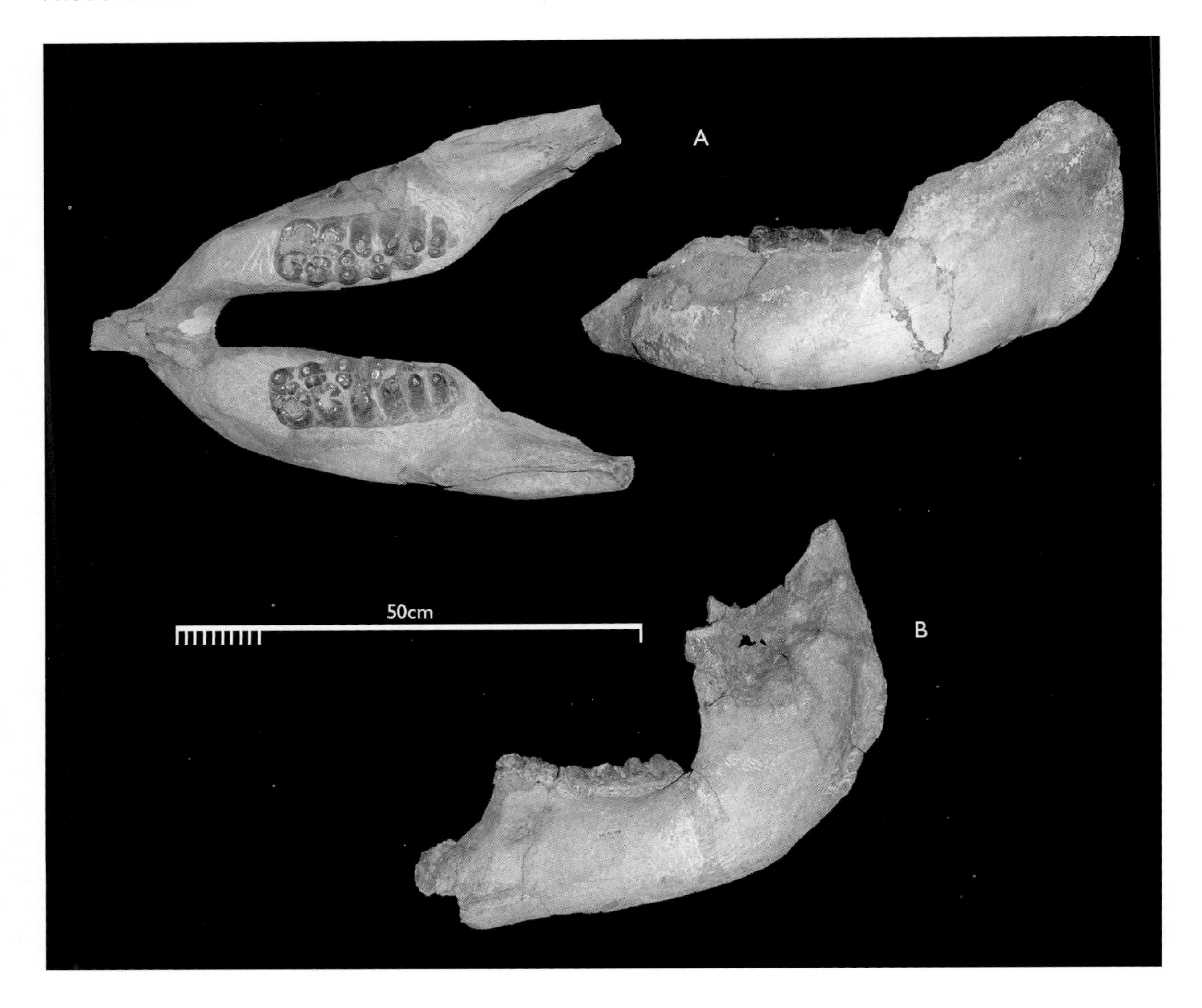

FIGURE 15.6
Mandibles of *Anancus kenyensis:* A. AMW-VP-1/63, mandible with M_3, occlusal and left lateral views. B. AMW-VP-1/65, right mandible with M_1–M_2, right lateral view (reversed).

larger dimensions of the molars. The "Sagantole type" obviously represents the most derived grade of a lineage to which the Lukeino, Kuseralee, and Langebaanweg samples all belong. These geologically younger Sagantole specimens exaggerate such characters as the complicated M-C complex and the ccpos and ccpop in M3 beyond the condition seen in earlier samples of *A. kenyensis*. KNM-KP 384 from Kanapoi, KNM-AT 20 from Aterir, KNM-BC 33 and 1655 from Chemeron, and those from the Nkondo Formation in Uganda reported by Tassy (1994) are comparable with the "Sagantole type" from Ethiopia in the complication of molar structure and large size.

These highly derived eastern African anancine gomphotheres were collectively called the "Petrocchii morph" by Tassy (1986), but the status of *Anancus petrocchii* Coppens, 1965 is problematical. The depository of the only specimens of this species, including the holotype, is currently unknown (Boaz 1987; Bernor and Scott 2003), and the only

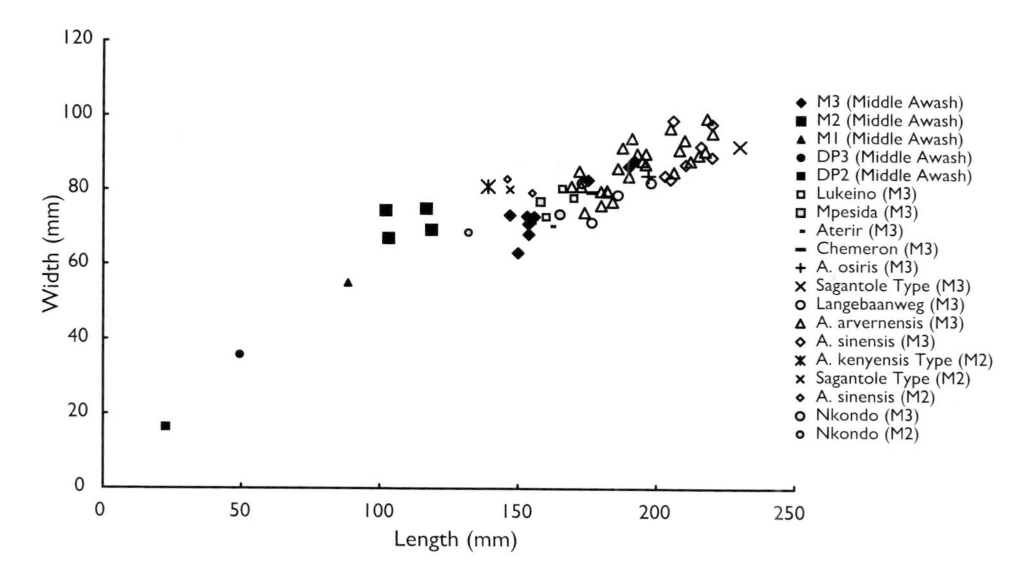

FIGURE 15.7

Scatter diagram of upper cheek teeth of various species of *Anancus* from Africa and Eurasia (width against length). Data for non–Middle Awash fossil specimens are from Lukeino, Mpesida, Aterir, and Chemeron (Tassy 1986); *A. osiris* (Arambourg 1945); Sagantole type (Kalb and Mebrate 1993); *A. arvernensis* (Tobien et al. 1988); *A. sinensis* (Chen 1999); *A. kenyensis* type (MacInnes 1942).

available information on these specimens is the description and figures in Petrocchi (1943, 1954). For this reason, details of the loph(id) structure of *A. petrocchii* are unclear.

The position of the *A. kenyensis* holotype (BMNH M.15400 = KE20 of MacInnes 1942) and associated specimens from Kanam East is also problematical. They are either fragmentary or heavily worn, but they are claimed to be distinguished from Eurasian *Anancus* by the poor development of anancoidy (MacInnes 1942; Tassy 1986). However, there is still no objective means to measure the subtle difference of anancoidy. We tentatively allocate them to the same taxon as other eastern and southern African *Anancus* because a specimen in the Langebaanweg sample (PQ-L 6042) is almost identical in crown dimensions to BMNH M.15400, the type of *A. kenyensis*.

In the above discussion, the samples of *A. kenyensis* are arranged based on the evolutionary grades of the molars. However, this does not necessarily mean that they evolved anagenetically. Although most of *A. kenyensis* evolved highly complicated loph(id) structure, there remains some possibility that some of them headed in another direction: simplification of the crown structure. In the degeneration of the ccpop2, some individuals of *A. kenyensis* from the Kuseralee Member recall the M^3 of *A. osiris* from North Africa, in which ccpop2 is not developed (Arambourg 1945, 1970; Tassy 1986; Metz-Muller 2000). In this context, the dp_2 of *Anancus* from the late Miocene of the Middle Awash is also intriguing because it resembles that of *Anancus* sp. cf. *A. osiris* from Ahl al Oughlam, Morocco (Geraads and Metz-Muller 1999). In recent parsimony analysis of *Anancus, A. osiris* is placed close to *A. arvernensis* and its related forms from Eurasia (Tassy 1986; Metz-Muller 2000). However, the above observations may suggest another possibility: that *A. osiris* may have been derived from *A. kenyensis* or a form close to it with the simplification of the loph(id) structures.

A. arvernensis and its related forms from Eurasia show a less complicated molar loph(id) structure than *A. kenyensis*. In *A. arvernensis,* the decrease in molar size and the increase in hypsodonty index are general tendencies during the period between MN14 and MN17 (Metz-Muller 1995, 2000). At MN17, however, the hypsodonty index markedly increased

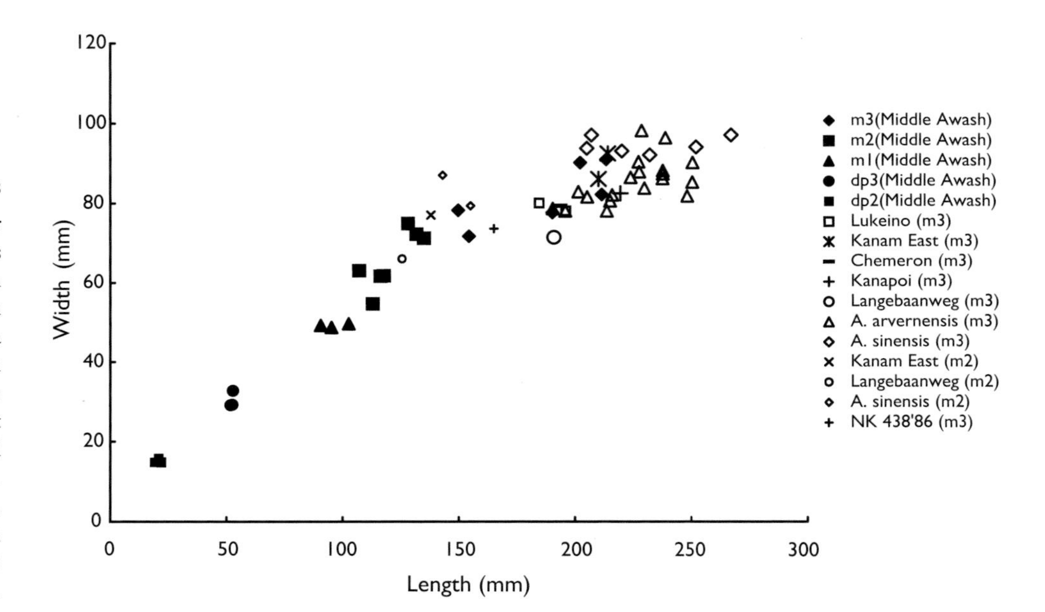

FIGURE 15.8

Scatter diagram of lower cheek teeth of various species of *Anancus* from Africa and Eurasia (width against length). Data for non–Middle Awash fossil specimens are from: Lukeino, Kanam East (M_3), Chemeron, and Kanapoi (Tassy 1986); *A. arvernensis* (Tobien et al. 1988); *A. sinensis* (Chen 1999); Kanam East (M_2) (MacInnes 1940).

at the same time that crown structure became simpler than those of earlier periods (Tobien 1978; Metz-Muller 2000).

A. sinensis and *A. sivalensis* are other major species of Eurasian *Anancus.* Metz-Muller (2000) assigned the former to a subspecies of *A. arvernensis.* The morphology of the M-C complex of *A. sinensis* does not contradict her idea because it is very much simplified, being mostly of Type A and rarely of Type C, E, or F. In contrast, *A. sivalensis* has been considered cladistically close to *A. kenyensis* based on suggested synapomorphies—pentalophodonty of M_2 (Tassy 1986) or high loph(id) frequencies (Metz-Muller 2000). Besides these characters, however, *A. sivalensis* is more similar to *A. arvernensis* than to *A. kenyensis* in its simple M-C complex. In contrast to the crown structure of M^3 of *A. kenyensis,* that of *A. sivalensis* is simple, with poor development of central conules and occasional degeneration of the posttrite mesoconelet at the anterior loph of the M^3 (Metz-Muller 2000). New specimens showing shared derived characters besides those of M_2 are needed in order to resolve this riddle. The complete skull of *Anancus* from Chad pictured in Coppens (1967) is one such specimen. Additional materials from the Middle Awash will also contribute greatly to this issue, especially a skull of *A. kenyensis* (Sagantole type) left in the field by the RVRME (Kalb and Mebrate 1993).

Elephantidae

Recent classifications show that elephantids and stegodontids constitute a monophyletic group that is a sister group of the tetralophodont gomphotheres (Tassy 1982, 1990; Kalb and Mebrate 1993; Kalb et al. 1996; Tassy 1996b; Shoshani 1996; Shoshani et al. 1998). The phylogenetic relationship of stegodonts to elephants still remains rather controversial, and current issues center on the monophyly of Stegodontidae (Tassy 1982; Tassy and Shoshani 1988; Tassy 1990; Kalb and Mebrate 1993; Saegusa 1996; Shoshani 1996; Tassy 1996b; Shoshani et al. 1998). Classification and phylogenetic studies of African elephantids have

TABLE 15.7 Comparison of M^3 Variability of the *Anancus* from the Late Miocene of the Middle Awash against Various Elephantoids

	Length				Width				
	CV	*n*	*F*	FDIST	CV	*n*	*F*	FDIST	*Source*
Mammuthus primigenius	8.1	7	1.1728	0.4309	11.9	17	0.7785	0.6143	Maglio 1973
Elephas maximus	21.6	8	0.1934	0.9772	5.5	8	3.6446	0.0548	Roth and Shoshani 1988
Loxodonta africana (southern Arican combined sample)	12.7	13	0.56	0.7749	6.1	13	2.9629	0.0475	Cooke in Roth 1992
Haplomastodon waringi	6.1	17	2.4254	0.0674	8.9	19	1.3919	0.2681	Simpson and Couto 1957
Palaeoloxodon naumanni (Nojiri)					10.2	20	1.06	0.4253	Takahashi et al. 1991
Mammuthus primigenius (USSR)					11.5	52	0.8336	0.5645	Sher and Garutt (1985), in Roth 1992
Mammuthus primigenius (Predmosti)					10.3	38	1.0392	0.4217	Musil 1968
Anancus from Kuseralee Member	9.5	8			10.5	8			Original data

NOTE: The *F*-value is the ratio of squared coefficients of variation of M^3 width or length. Abbreviations: CV, coefficients of variation; *F*, *F*-value; FDIST, one tailed *p*-value of a given *F*-value; *n*, number of specimens.

TABLE 15.8 Comparison of M_3 Variability of *Anancus* from the Late Miocene of the Middle Awash against Various Elephantoids

	Length				*Width*				
	CV	*n*	*F*	FDIST	CV	*n*	*F*	FDIST	*Source*
Mammuthus primigenius	17.3	5	0.8236	0.5909	12.9	8	0.6375	0.7001	Maglio 1973
Elephas maximus	6.5	6	5.8341	0.0377	8.2	6	1.5778	0.3167	Roth and Shoshani 1988
Loxodonta africana (Uganda)	6.8	12	5.3307	0.0099	6	12	2.9469	0.0575	Elder, in Roth 1992
Loxodonta africana (Zambia)	7	19	5.0304	0.0047	7.6	19	1.8367	0.1482	Laws, in Roth 1992
Loxodonta africana (southern African combined sample)	10.6	17	2.1938	0.1061	6.7	17	2.3633	0.0793	Cooke, in Roth 1992
Haplomastodon waringi	6.7	22	5.491	0.0039	5.9	24	3.0477	0.0242	Simpson and Couto 1957
Mammuthus primigenius (USSR)					9.9	31	1.0824	0.3951	Sher and Garutt (1985), in Roth 1992
Anancus from Kuseralee Member	15.7	6			10.3	7			Original data

NOTE: The *F*-value is the ratio of squared coefficients of variation of M_3 width or length. Abbreviations: CV, coefficients of variation; *F*, *F*-value; FDIST, one tailed *p*-value of a given *F*-value; *n*, number of specimens measured.

been subjects of numerous studies (see Shoshani and Tassy 1996; Todd and Roth 1996; and references therein). The following four extinct elephantid genera have been described from African localities: *Stegotetrabelodon, Primelephas* Maglio, 1970, *Stegodibelodon* Coppens, 1972, and *Selenetherium* Mackaye et al., 2005. Besides these four taxa, fragmentary material from Lothagam (Tassy 2003) and Lemudong'o, Kenya (Saegusa and Hlusko 2007) appear to represent additional primitive elephantid taxa besides these four taxa.

Primelephas Maglio, 1970

Primelephas gomphotheroides Maglio, 1970

Primelephas gomphotheroides gomphotheroides Maglio, 1970

DIAGNOSIS Molars are low-crowned with moderate cover of coronal cement. Distinct central conules are seen on the anterior two or three lophids of the lower intermediate molars. On M_3, the distinct central conule is seen only on the first lophid. M^3 does not have a central conule other than the weak ccprp1. Loph(id) formula of upper and lower third molars varies from x8x to x7x. Enamel is moderately thick.

DESCRIPTION Specimens referred to *Primelephas gomphotheroides gomphotheroides* are shown in Figures 15.10H, 15.11F, G, and 15.12. KL-291-1 is a partial left mandible with M_1–M_2 (Figure 15.11G). The two remaining lophids of the M_1 are deeply worn and damaged, but the remaining loops suggest that they are each composed of a pair of median pillars and elongated buccal and lingual lateral pillars. The penultimate lophid is transversely straight. The buccal half of the last lophid is displaced distally. The distal cingulum is composed of two transversely elongated pillars. There is no central conule on the two preserved lophids. Cement fills the interlophid valley.

The M_2 of KL-291-1 is broken into mesial and distal fragments, but its original lophid formula can be estimated as x6x from the position of its fragments in the alveolus. All of the lophids are blanketed with cement except for the apices of the mesial cingulum and mesial two lophids. The mesial cingulum is large and slightly displaced lingually. The median sulcus of the first lophid is deep. Ccprp1 is centrally located, transversely elongated, and subdivided into two digitations. The second lophid has seven digitations. The surface configuration of the thick cement cover suggests that the third, fifth, and sixth lophids each have four or five digitations and no central conule. There is no marked posterior widening of the crown. The M_2 of KL-291-1 is very similar to that of the holotype of *Elephas nawataensis* Tassy, 2003 (KNM-LT 23783) in having a central conule on the first lophid only, no posterior widening of the crown, and a comparable hypsodonty index (73.8). Its lophid formula is also what is expected for *Elephas nawataensis*.

KL-211-2 was described by Kalb and Mebrate (1993) as a right M^2 of *P. gomphotheroides*. However, it is more derived than M^2s from Saitune Dora in having the median column only behind the first loph and more compressed lophs.

KL-302-2 and KL-302-1 are the right M_3 and M^3 of a single individual. The mesial ends of these molars are damaged, but the location of their roots suggests that their original loph formula was x8x. In the M^3, the only indication of the presence of a central conule is a weak distal sinus on the first loph. The median sulcus is distinct on the first through third lophs. The worn mesial five lophs each appear to have had seven or eight digitations.

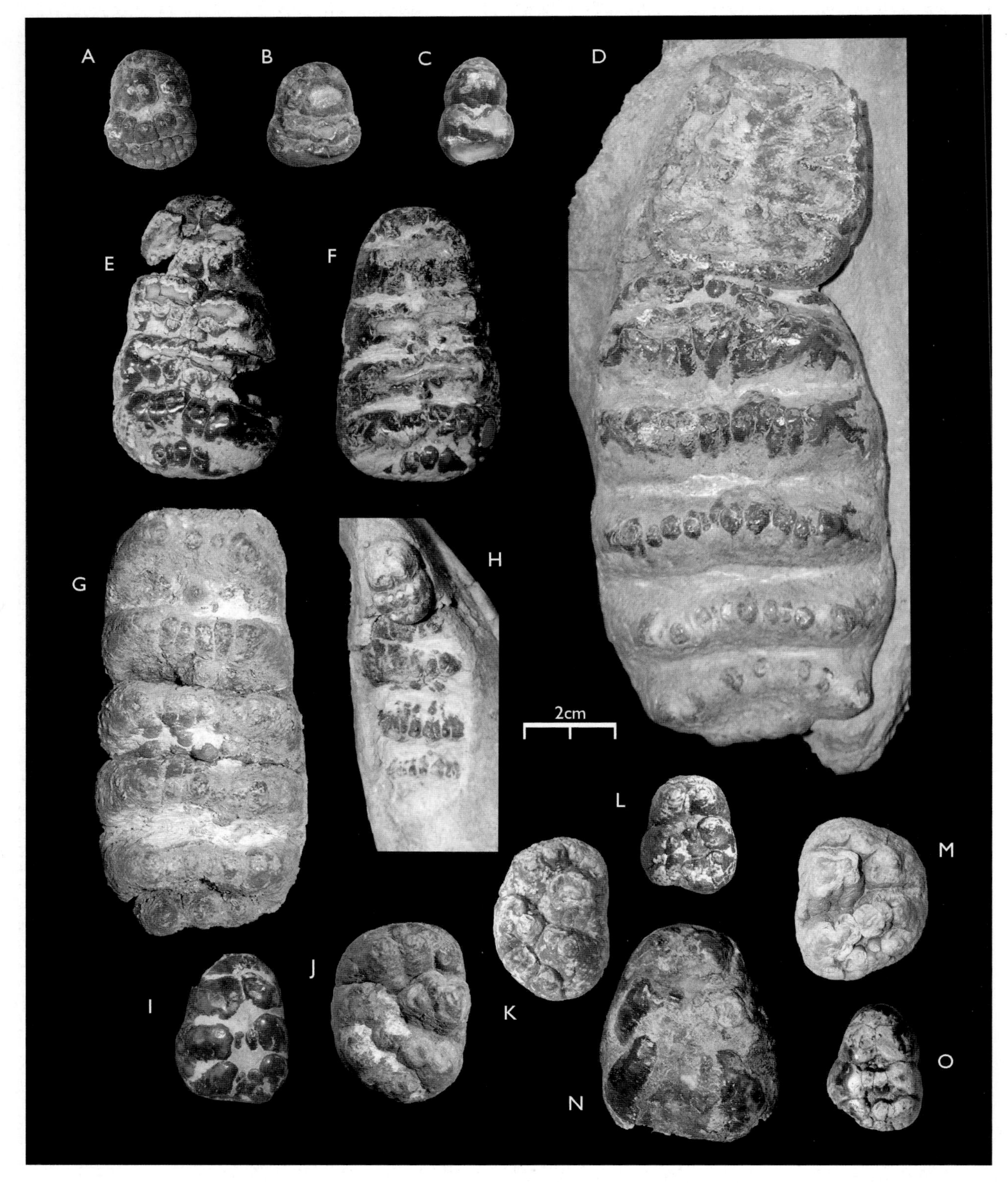

FIGURE 15.9

Occlusal views of deciduous and permanent premolars of *Primelephas gomphotheroides saitunensis*: A. STD-VP-2/89, right dp^2; B. STD-VP-2/60, left dp^2. C. STD-VP-2/59, left dp_2. D. STD-VP-2/897, fragment of right maxilla with dp^3–dp^4. E. STD-VP-2/88, right dp_3. F. STD-VP-2/7, left dp_3. G. STD-VP-2/894, right dp_4. H. STD-VP-2/84, fragment of right mandible with dp_2–dp_3. I. STD-VP-2/64, left P_3. J. STD-VP-2/122, left P^3. K. STD-VP-2/65, left P^3. L. STD-VP-2/6, left P_3. M. ASK-VP-3/50, right P^3. N. STD-VP-2/86, right P_4. O. ALA-VP-2/108, left P_3.

FIGURE 15.10

Occlusal views of dp[4] and upper molars of *Primelephas gomphotheroides*. *Primelephas gomphotheroides saitunensis*: A. KL-113-1, right dp[4]. B. STD-VP-2/3, right dp[4]. C. STD-VP-2/898, right M[1]. D. STD-VP-2/85, fragment of left maxilla with M[1]–M[2]. E. STD-VP-2/895, right M[2]. F. STD-VP-2/896, left M[3]. G. ASK-VP-3/16, left M[1]–M[2]. *Primelephas gomphotheroides gomphotheroides*: H. KL-302-2a, right M[3].

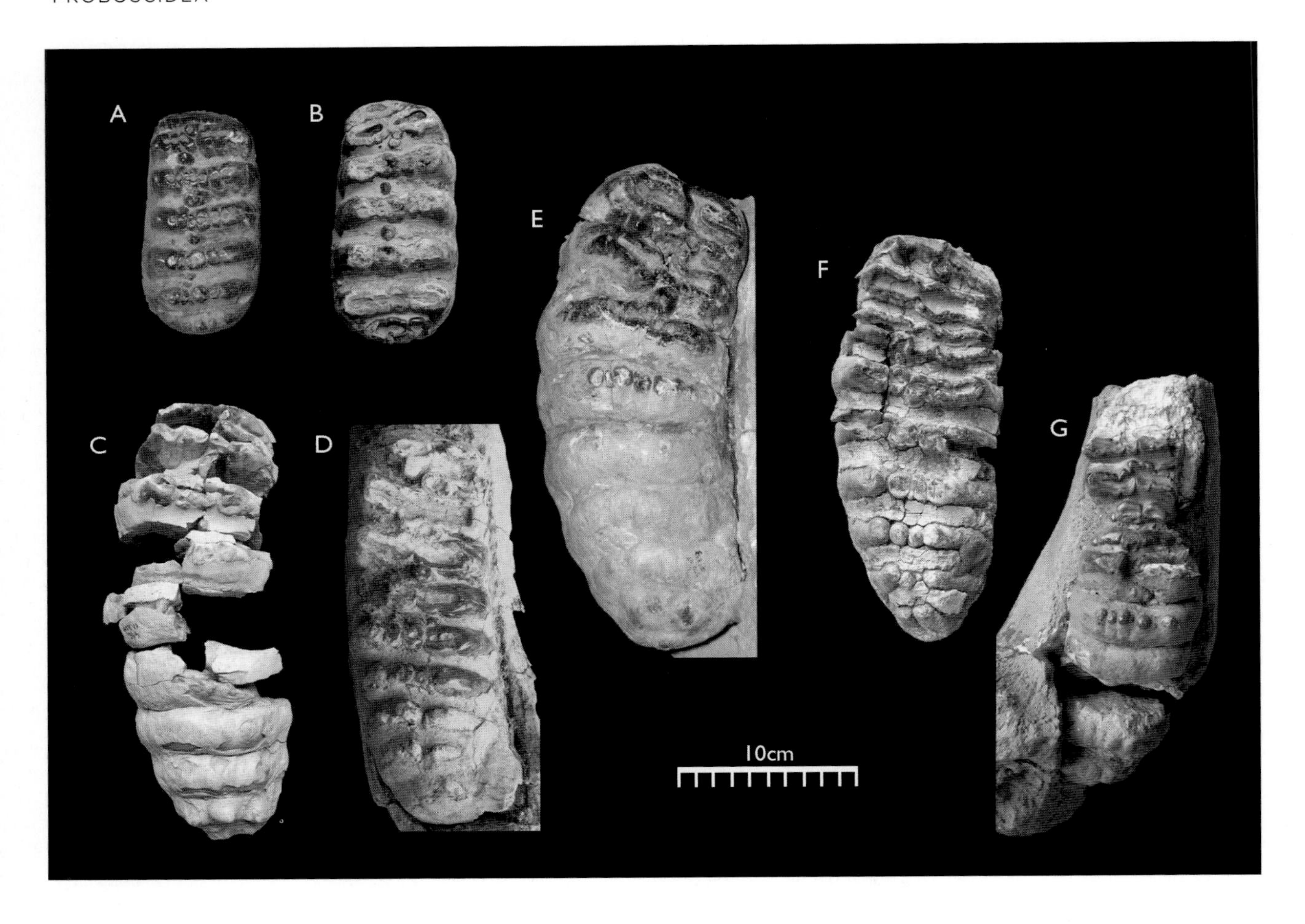

FIGURE 15.11 Occlusal views of lower molars of *Primelephas gomphotheroides*. *Primelephas gomphotheroides saitunensis*: A. STD-VP-2/86, left M_1. B. ALA-VP-2/4, left M_1. C. ASK-VP-3/15, right M_3. D. STD-VP-2/919, right M_3. E. STD-VP-2/833, right M_3. *Primelephas gomphotheroides gomphotheroides*: F. KL-302-1, right M_3. G. KL-291-1, left mandible fragment with M_1–M_2.

Lophs 6, 7, and 8 have four, four, and three digitations, respectively. Interloph valleys are filled with cement. Clear Stufenbildung is seen on the first loph, but the step is formed on the wear surface of the enamel inner layer. Lophs are compressed anteroposteriorly, and their mesial and distal faces are parallel to each other in both occlusal and buccal views.

In the M_3 the median sulcus is deep and clear only on the first lophid. The worn outline of the second and third lophids is propeller-shaped, with a pair of circular median loops and transversely elongated lateral loops. On the fourth through last lophids, the lateral pillars and lingual median pillar are single-cusped, whereas the buccal median pillar is frequently subdivided into two digitations. In occlusal view, the mesial four lophids and other lophids are weakly concave-concave and straight buccolingually, respectively. The molar tapers distally, with a relatively pointed distal profile in occlusal view. The ccprp1 is large and centrally located. A tiny tip of a conule is seen on the worn surface of the cement of the fourth interlophid valley. There is no other indication of the presence of the central conule.

The M^3 of KL-302 is best compared with the M^3 of the holotype of *Primelephas gomphotheroides* (KNM-LT 351; Figure 15.12), not only in the loph structure but also in the loph formula. According to previous authors (Maglio 1970, 1973; Maglio and Ricca 1977; Tassy 2003), the left M^3 of KNM-LT 351 is complete, with seven lophs and the

TABLE 15.9 Measurements of Molars of Elephantids from the Middle Awash

Specimen	*Loph(id) Formula*	*L*	*W*	*H*	*I*	*HI*	*E*	*LF*	*Taxon*
M^1									
STD-VP-2/83	x5x	120.6	66.5		55.1		3.2–3.9	4.6	*P. g. saitunensis*
STD-VP-2/85	–3x	91.2+	64.6				2.9–3.2		*P. g. saitunensis*
STD-VP-2/123	x5x		72.7	45.6 (2)		62.7			*P. g. saitunensis*
STD-VP-2/898	x5x	110.2	71.8	38.6	65.2	53.8	2.8–3.3	4.9	*P. g. saitunensis*
M^2									
KL-211	x6	149.8	80.5	53	53.7	65.8	4.0–4.4	4.2	*P. g. gomphotheroides*
ASK-VP-3/16	x5x	148.5	95.7		64.4		5.1–4.6	3.4	*P. g. saitunensis*
STD-VP-2/85	x5x	141	77.2	50.7	54.8	65.7	3.1–4.0	4	*P. g. saitunensis*
STD-VP-2/895	x5x	161.2	83.4	52.8	51.7	63.1		3.7	*P. g. saitunensis*
M^3									
KL302-2a	–8x	207+	91.8	55.1	–207 (210)		4.5–5.3	4.1	*P. g. gomphotheroides*
ASK-VP-3/16 l	x7x	212.2	104.5	62	49.2	69.3	5.2–6.1	3.3	*P. g. saitunensis*
ASK-VP-3/16 r	x7–	190.9+	102.5	61.1	53.7–	59.6	5.2–6.1		*P. g. saitunensis*
KL-89-7a	–5x	158.7+	88.2				4.1–5.2	3.6	*P. g. saitunensis*
STD-VP-2/896	x7x	197.2	84.3	56.3	42.7	66.8	2.4–5.4	3.4	*P. g. saitunensis*
AME-VP-1/93	–5x	146+	97+				4.0–4.5		cf. *Loxodonta* sp.
M_1									
KL-291-1A	–2x	71+	60				2.5–3.5	4.5	*P. g. gomphotheroides*
ADD-VP-3/3	–5–	96.5+	64.9	42+		64.7+	3.1–3.8	4.7	*P. g. saitunensis*
ALA-VP-2/4	x5x	130.1	70	38+	53.8	54.3+		4.3	*P. g. saitunensis*
STD-VP-2/86	x5x	122.9	66.6	41.7	54.2	62.6	3.5–3.6	5.2	*P. g. saitunensis*
STD-VP-2/87	x5–	124.6+					3.2–3.5		*P. g. saitunensis*
AME-VP-1/159	x5x	134.6	64.3		47.8		3.2–3.6	4.1	cf. *Loxodonta* sp.
M_2									
KL-291-1A	x6x	162	73.8	54.5	45.6	73.9	4.0–5.4	4.2	*P. g. gomphotheroides*
STD-VP-2/829	–2x		75.9	54.4		71.7	5.5	3.8	*P. g. saitunensis*
M_3									
KL-302-1	–8x	237.2+	96.2	55+	40.6–	57.2+	4.2–5.1	3.8	*P. g. gomphotheroides*
ASK-VP-3/15	x8x	249.3	84.9	59.3	34.1	69.9	4.0–6.1	3.8	*P. g. saitunensis*
KL-89-6b & 89-8a	–6–	171.3+	86.9	49.7+	50.7–	57.2+	4.1–5.5	4.3	*P. g. saitunensis*
KL-89-7b & 89-8	–7–	194.7+	81.7	51.2	42–	62.7	4.0–4.6	3.8	*P. g. saitunensis*
STD-VP-2/827	–6–	231.8+	93.6		40.4–		4.8–5.3	3.9	*P. g. saitunensis*
STD-VP-2/829	x6–	136+	75.1+	56.6		75.4–		3.4	*P. g. saitunensis*
STD-VP-2/833	x8x	243.1	98	62.2	40.3	63.5	3.8–6.0	3.4	*P. g. saitunensis*
STD-VP-2/919	x8x	219.8	82.9	54.8	37.7	66.1	4.4–5.2	4.1	*P. g. saitunensis*
ALA-VP-2/309B	–4–5x	223.2+	88.5+				2.9–4.1	4.4	cf. *Mammuthus* sp.
KUS-VP-1/112	–3–3x	235+	87.5		37.3–		3.2–4.9	3.5	cf. *Loxodonta* sp.
KL-210-1	–4–	137.8+	97.6	58+		59.4+	4.2–5.1	3.6	Elephantidae gen. et sp. indet.

NOTE: All measurements in mm. For abbreviations, see note to Table 15.1.

mesial and distal talons. However, the mesial part of the left M^3 of KNM-LT 351, as described, is a mix of fragments from the left and right M^3s. Furthermore, there is enough space to accommodate one more loph between a fragment of the genuine mesial end of the left M^3 and the rest of the molar. Thus, the left M^3 might originally have had a total of eight lophs (problems with the reconstruction of KNM-LT 351 are discussed in detail elsewhere).

Primelephas gomphotheroides saitunensis subsp. nov.

ETYMOLOGY The subspecies name is derived from Saitune Dora, the name of the Middle Awash locality in where most of the specimens assigned to the subspecies have been collected.

HOLOTYPE STD-VP-2/833, right M_3.

REFERRED MATERIAL Adu-Asa Formation (Asa Koma Member): ADD-VP-3/3 (right M_1); ALA-VP-2/4 (left M_1); ALA-VP-2/108 (left P_3); ALA-VP-2/152 (fragment of left dp_3); ALA-VP-2/312 (right dp^3); ASK-VP-3/15 (right M_3); ASK-VP-3/16 (left M^2, left M^3, and right M^3); ASK-VP-3/50 (right P^3); KL-89-5a and b (right and left mandible with right dp_3 and right and left dp_4); KL-89-6b and 89-8a (right mandible fragment with M_3); KL-89-7a (right M^3); KL-89-7b and 89-8 (left mandible fragment with M_3); KL-89-9 (fragment of right mandible with dp_3); KL-113-1 (right dp^4); KL-176-1 (edentulous mandible); STD-VP-2/3 (right dp^4); STD-VP-2/6 (left P_3); STD-VP-2/7 (left dp_3); STD-VP-2/83 (left M^1); STD-VP-2/84 (right mandible with dp_2–dp_3); STD-VP-2/85 (left maxilla fragment with M^1–M^2); STD-VP-2/86 (nearly complete mandible with right P_4 and right and left M_1); STD-VP-2/87 (right M_1); STD-VP-2/88, (right dp_3); STD-VP-2/89 (right dp^2); STD-VP-2/122 (left P^3); STD-VP-2/123 (left M^1); STD-VP-2/59 (left dp_2); STD-VP-2/60 (left dp^2); STD-VP-2/64 (left P_3); STD-VP-2/65 (left P^4); STD-VP-2/827 (right mandible with M_3); STD-VP-2/829 (left mandible fragment with M_2 and M_3); STD-VP-2/833 (right mandible fragment with M_3); STD-VP-2/894 (right dp_4); STD-VP-2/895 (right M^2); STD-VP-2/896 (left M^3); STD-VP-2/897 (right maxilla fragment with dp^3–dp^4); STD-VP-2/898 (right maxilla fragment with M^1); STD-VP-2/899 (left mandible fragment with dp_4); STD-VP-2/919 (right mandible fragment with M_3).

LOCALITIES AND HORIZON

Asa Koma Member, Adu Dora Formation: Saitune Dora VP-2, Ado Dora VP-2, Asa Koma VP-3, Alayla VP-2.

DIAGNOSIS Molars are low-crowned with moderate cover of coronal cement. Distinct posterior central conules are always present on the distal flank of the first and the second lophids of the M_3. Loph(id) formula of M^3 and M_3 is x7x and x8x, respectively. Intermediate molars are pentalophodon. Enamel is moderately thick (about 5 mm at the mid-height of the loph(id)). No enamel folding; no lower tusk; mandible is brevirostrine.

DESCRIPTION More than 90 percent of the *P. g. saitunensis* subsp. nov. fossil specimens come from a single locality, Saitune Dora, and from a single unit that is ca. 1 m thick and

15 m long. This unit is packed with elephant skeletal remains, with postcranial elements representing a minimum of 27 individuals, interspersed with a small number of specimens of other taxa. As studies on the postcranial elements are still under way, only the dentognathic elements are described and discussed here.

Deciduous Premolars and Premolars Measurements of deciduous premolars are summarized in Table 15.10. The dp^2s from Saitune Dora (Figure 15.9) have strong mesial cingula, two lophs, and double (STD-VP-2/89) or single (STD-VP-2/60) distal cingulum. The median sulcus is deep on the first and second lophs. The first loph is much higher than the second. The protocone is flanked with ccpra1 and ccprp1, and the strong mesial cingulum is connected to the ccpra1. The paracone has two digitations and is much larger than the protocone. The second loph is plate-like and has four transversely arranged pillars whose apices are superficially subdivided. The lingual lateral pillar is flanked with anterior and posterior swellings corresponding to the central conules. The distal cingulum consists of seven or eight beady small cusplets with no cement. The two Saitune Dora dp^2s are similar to isolated dp^2s reported from Lukeino and assigned to *Primelephas* by Tassy (1986).

The only dp^3 from the Middle Awash, STD-VP-2/897, is tetralophodont and highly worn, and its lingual and distal enamel rims show fine and somewhat irregular folding. Among three dp^3s reported from the Ibole Member of the Wembere-Manonga Formation of Manonga Valley (Sanders 1997), only WM 181/94 appears to be a true dp^3, because its width closely corresponds with that of the fourth loph of STD-VP-2/897, and it would have been originally tetralophodont. Tassy (2003) identified two dp^3s (KNM-LT 22866 and 26325) from the Lower Nawata member of the Lothagam Formations "*Stegotetrabelodon* or *Primelephas*." However, at least one of them (KNM-LT 22866) is not *Primelephas* because of its trilophodonty.

The dp^4 has five lophs and mesial and distal cingula (Figure 15.10). It is mostly covered with a thin layer of cement. The first loph is divided into four pillars by a strong median sulcus and a weaker lateral sulcus. The transversely elongate ccprp1 extends buccally from the single-cusped lingual lateral pillar. The unsubdivided lingual median pillar is about the same size as the lateral one and is connected to the mesial cingulum by a small ccpra1. The buccal lateral and median pillars are subdivided into three and two cusplets, respectively. On the second loph, the median and lateral sulci are weak and not clear, respectively. The second through fifth lophs each have eight to ten digitations and no distal column. On the distal half of the tooth, lophs are arranged in a shallow convex-convex shape. The distal cingulum is half the width of the last loph and closely appressed to it. In STD-VP-2/3, the first loph, lingual halves of the second and third loph, and the rest of the molar, are supported by the mesial, lingual, and distal roots, respectively. Such root arrangement has never been observed in gomphotheres or stegodonts.

Premolars Measurements of premolars are summarized in Table 15.11. In occlusal view, the P^3 is "kidney-shaped" (STD-VP-2/65) or trapezoid (STD-VP-2/122 and ASK-VP-3/50), with the buccal side much longer than the lingual side (Figure 15.12). It is bilophodont, with strong mesial and distal cingula. The median sulcus is deep on the first

TABLE 15.10 Measurements of Deciduous Premolars of *Primelephas gomphotheroides saitunensis* subsp. nov.

Specimen	*Loph(id) Formula*	*L*	*W*	*H*	*I*	*HI*	*E*
dp^2							
ALA-VP-2/312	x3	25.9+	20.5		79.1−		
STD-VP-2/60	x3	23.9	21		87.9		
STD-VP-2/89	x3x	27.2	20.2	15.5	74.3	76.7	
dp^3							
STD-VP-2/897	−4	52.2+	42.2		80.8−		1.3
dp^4							
KL-113-1	x5x	99.4	55.7		56		1.8–2
STD-VP-2/3	x5x	107	63.1	35.5	59	56.3	
STD-VP-2/897	x5x	99.4	58.3	30.9	58.7	53	
dp_2	x2						
STD-VP-2/84	x2x	20	13.2	11.3+	66	85.6+	
STD-VP-2/59	x3	25	15.9	12.5+	63.6	78.6+	
dp_3							
KL-89-5	−4−	47.7+	35+				
STD-VP-2/7	x4x	58.1	34		58.5		
STD-VP-2/88	x4x	57.7	33.4	18.8+	57.9	56.3+	1.9
dp_4							
KL-89-5 l	x5x	100.9	55.2	33.8	54.7	61.2	2.4
KL-89-5 r	x5−	88+	54.5	32.8		60.2	
STD-VP-2/894	x5x	93.4	46.5	30.5	49.8	65.6	2.4−2.7
STD-VP-2/899	x5x	108.5	56.9	29.8+	52.4	52.4+	2.5−2.9

NOTE: All measurements in mm. For abbreviations, see note to Table 15.1.

loph. The protocone is single-cusped and occasionally displaced mesially (e.g. STD-VP-2/65 and STD-VP-2/122). A strong mesial cingulum extends mesiobuccally from the protocone. A large paracone on the buccal side of the first loph is superficially subdivided into three digitations (STD-VP-2/122, possibly in ASK-VP-3/50). The second loph runs obliquely from the lingual margin to the distobuccal corner of the tooth and is subdivided into four cusps by one median and two lateral sulci. The distal cingulum extends obliquely from the lingual main cusp of the second loph to the distobuccal corner of the tooth. It is made up of five to seven digitations and is closely appressed to the second loph.

Most of the morphological features described for the P^3 of *P. g. saitunensis* match the description of the P^4 of *P. gomphotheroides* from Nkondo (Tassy 1994), the fragmentary P^4 (KNM-LT 23782) of *P. gomphotheroides* from the Lower Nawata Member, Lothagam (Tassy 2003), and the P^4 of *Stegotetrabelodon lybicus* from Sahabi (Gaziry 1987), except that these are much larger than the P^3 from the Adu-Asa Formation. A P^4 of *P. gomphotheroides* (KNM-LU 730) from Lukeino (Tassy 1986) differs from the P^3 of *P. g. saitunensis* in having a thinner and smaller paraconid and a much thicker distal cingulum.

The dp_2 is trilophodont, with a very weak mesial cingulum (Figure 15.9). The first lophid is made up of two cusps and is much taller than the other lophids. A small tubercle (STD-VP-2/59) or strong swelling (STD-VP 2/84) is seen on the distal flank of the

FIGURE 15.12

Upper and lower third molars of *Primelephas gomphotheroides gomphotheroides* Maglio, 1970. KNM-LT 351: A. Fragment of the mesiobuccal corner of right M^3, occlusal view. B. Fragment of the mesiobuccal corner of left M^3, occlusal view. C. Left M^3, lateral view. D. Left M^3, occlusal view. KNM-LU 526: E. Left M_3, occlusal view.

protoconid. The second lophid has two major cusps on the buccal and lingual faces, and two small ones in the middle. The third lophid consists of five or six digitations. The mesial cingulum and the first lophid are supported by the mesial root. The rest of the tooth is supported by the distal root.

Tassy (1986) ascribed a dp_2 (KNM-LU 609) from Lukeino, Kenya, to cf. *Primelephas,* but it may belong to a more derived elephantid because it differs from the Saitune Dora dp_2 in having three tubercles on the second and third lophids. Tassy (2003) also described dp_2–dp_3 associated with a jaw fragment as "*Stegotetrabelodon orbus* or *P. gomphotheroides*." However, the gomphothere-like morphology of the dp_2 clearly refutes the possibility that it belongs to *Primelephas.*

The dp_3 is tetralophodont, with mesial and distal cingula (Figure 15.9). The median sulcus is distinct on all lophids. Buccal half-lophids of the second and third lophids are slightly displaced distally relative to the lingual half-lophids. Each half-lophid is composed of large lateral and smaller median pillars, each of which is further subdivided superficially into two or three digitations. The ccprp is always present on the mesial three lophids, whereas the ccpop is present occasionally. Enamel is finely but somewhat irregularly folded. Thin coronal cement covers the lophids.

A dp_3 (KNM-LU 663) of cf. *Primelephas* from Lukeino, Kenya (Tassy 1986), is comparable with the dP_3 of *Primelephas* from the Middle Awash in crown morphology, although the former is slightly larger. Sanders (1997) described three dp^3s of *P. gomphotheroides* from the Manonga Valley, but two of them, WM 211/94 and WM 1065/92, appear to be dp_3s.

TABLE 15.11 Measurements of Premolars of *Primelephas gomphotheroides saitunensis* subsp. nov.

Specimen	*Loph(id) Formula*	*L*	*W*	*H*	*I*	*HI*
P^3						
ASK-VP-3/50	x3	36.1	29.5		81.7	
STD-VP-2/65	x3	36.8	28.7	15.1	78	52.6
STD-VP-2/122	x3	40.8	31.2	20.4	76.5	65.4
P_3						
ALA-VP-2/108	x3x	28.4+	22.5	12.2	79.3−	54.2
STD-VP-2/64	x3x	33.4	25.2	17.3	75.5	68.7
STD-VP-2/6	x3x	26.4	21.8	14.4	82.5	66.1
P_4						
STD-VP-2/86	3	44.2	40		90.5	

NOTE: All measurements in mm. For abbreviations, see note to Table 15.1.

The P_3 is trilophodont, with a strong mesial cingulum. In lateral view, three lophids lean mesially and converge apically (Figure 15.9). The tooth tapers mesially in occlusal view. The first lophid is higher than the other lophids and is composed of two equal-sized cusps. In STD-VP-2/64 and ALA-VP-2/108, these two cusps are further subdivided superficially into larger lateral and smaller medial digitations. In contrast, they are incompletely subdivided in STD-VP-2/6. The second lophid has four cusps, larger lingual and buccal cusps, and two medially located smaller cusps. The basic structure of the third lophid can be the same as that of the second lophid (e.g., STD-VP-2/64 and ALA-VP-2/108) or differ from that of the second lophid in being made up of five irregularly sized mammillae (e.g., STD-VP-2/6). The distal cingulum is weak, particularly in STD-VP-2/6. Interlophid valleys are filled with cement.

Tassy (2003) allocated two P_3s (KNM-LT 26329 and KNM-LT 26339) from Lothagam to "*Stegotetrabelodon* or *Primelephas*." They closely conform to P_3s of *Primelephas* from the Middle Awash in every morphological detail.

The dp_4 is pentalophodont, is anteriorly narrow, and attains its maximum width at the fourth or fifth lophid. It is blanketed by cement (Figure 15.9). It is occasionally buccolingually constricted behind the second lophid (e.g., STD-VP-2/899). Worn enamel is irregularly and densely folded and exhibits very weak Stufenbildung. Each lophid is subdivided into six to nine digitations. The median sulcus is clear only on the first and second lophids. The posterior central conules are present on first to second (STD-VP-2/899) or first to third (STD-VP-2/894, KL89-5a) lophids. Additionally, small tubercles are present in the first (KL-89-5a) or third (STD-VP-2/894) interlophid valley. The buccal halves of the distal lophids are slightly displaced distally. In buccal and lingual views, the lophids lean mesially, and their interlophid valleys are V-shaped.

Gaziry (1987) described dp_4s of *Stegotetrabelodon syrticus* from Sahabi, Libya. These are comparable with dp_4s of *Primelephas* from the Middle Awash and Lukeino in dimensions of the crown but differ from the latter in lophid formula (x4x) as well as in having a larger central conule. The dP_4 of the holotype mandible (KNM-LT 23783) of *Elephas nawataensis* Tassy, 2003 differs from those of *Primelephas* from the Middle

Awash and Lukeino in lophid formula (x6x) and the attenuated posterior widening of the crown.

Specimen STD-VP-2/86 is the only *Primelephas* P_4 from the Middle Awash (Figure 15.9). It is trilophodont, with rudimentary median cingulum that is mesially higher than distally. Thick cement fills all the interlophid valleys. The first lophid consists of two conical cusps. The second and the third lophids slope mesially and are made up of two large main cusps and two smaller mesoconelets. The P_4 of STD-VP-2/86 is almost identical to KNM-LU 925, which was identified as $P_{?4}$ of *P. gomphotheroides* by Tassy (1986).

Molars Molar specimens referred to *Primelephas gomphotheroides saitunensis* are shown in Figures 15.10C–G and 15.11A–E. Their measurements are presented in Table 15.9. The M^1 is pentalophodont, with mesial and distal cingula (Figure 15.10). The median sulcus is wide and distinct on the first loph, but on the rest of the molar it is thin and less distinct. Each half-loph is further subdivided into median and lateral pillars by a faint lateral sulcus. Each half-loph has three or four digitations. The ccprp1 is always present. A smaller ccprp2 is occasionally present. In occlusal view, the apical part of the loph is parallel-sided and very compressed anteroposteriorly, but at the basal part of the loph the median pillars weakly expand mesiodistally. Distal lophs are slightly concave mesially in occlusal view. Subtle scallop-shaped folding and weak Stufenbildung are seen on worn lophs. Interloph valleys are filled with cement. In STD-VP-2/83, the lingual root supports the buccal margin of the first loph and the lingual half of the second loph, whereas the mesial root supports the rest of the first loph.

The M^2 is pentalophodont, with moderately developed mesial and distal cingula (Figure 15.10). The molar is blanketed with cement. The transverse valleys are V-shaped and open to the base. The median sulcus is distinct on the first through third lophs in all M^2s from the Middle Awash. It is also distinct on the fourth (STD-VP-2/895) or fifth (STD-VP-2/895) loph. Each half-loph is further subdivided by a lateral sulcus into lateral and median pillars. The median pillar is occasionally further subdivided into two to four digitations (e.g., STD-VP-2/895). On the first loph, however, the lateral and median pillars are never subdivided. The median pillars expand mesiodistally slightly more than the lateral pillars at the basal half of the loph. The ccprp1 is large in both STD-VP-2/85 and STD-VP-2/895, whereas the ccprp2 is distinct only in STD-VP-2/85. In contrast, the ccprp3 is rudimentary and distinct in STD-VP-2/85 and -2/895, respectively. In occlusal view, mesial lophs are almost straight transversally, whereas the distal ones are slightly convex-convex shaped. Worn enamel shows weak Stufenbildung and irregular subtle folding in ASK-VP-3/16.

The M^3 has seven lophids and well-developed mesial and distal cingula (Figure 15.10). The median sulcus is very deep on the first loph. On the second through fifth lophs, it is still easily recognizable as a distinct groove. On the sixth and seventh lophs, however, it is obscured by the development of a centrally located large cusp on the median pillar. In STD-VP-2/896 the first loph consists of single-cusped buccal and lingual lateral pillars and apically subdivided median pillars. The adaxial digitation of the lingual median pillar connects with the mesial cingulum. The abaxial digitation of the same pillar is flanked by a large central conule. The first loph of ASK-VP-3/16 is simpler than that of STD-VP-2/896, with a buccal mesoconelet that is not subdivided and a lingual mesoconelet that is less developed. STD-VP-2/896 has a small ccprp2 on the median pillar, whereas ASK-VP-3/16 does not. In the second through fifth lophs of these specimens, distinct median and lateral sulci divide the loph into four pillars. The lateral pillars of these lophs are not subdivided, whereas the median ones are frequently subdivided. The sixth and seventh lophs are each composed of five to six digitations, but neither a median nor a lateral cleft is located on these lophs. In occlusal view, the first through third lophs are each nearly straight transversally, whereas the fourth through seventh lophs are convex-convex shaped in occlusal view. The unerupted M^3 (STD-VP-2/896) does not have cement, whereas the moderately worn ASK-VP-3/16 is mostly covered with cement. The distal margin is rectangular (ASK-VP-3/16) or round (STD-VP-2/896).

Kalb and Mebrate (1993) identified KL89-7 as left M^2 of *Stegotetrabelodon orbus*. However, the lingual and accessory roots preserved beneath the deeply worn mesial end of the molar suggest that it is from the right side and originally had two more lophs in front of the first loph of the preserved portion. Therefore, it is a right M^3, and the original loph formula of the molar can be reconstructed as x7x (not five as published by Kalb and Mebrate 1993).

The M_1 is slightly narrow anteriorly and pentalophodont with mesial and distal cingula (Figure 15.11). All lophids incline mesially, as is usual in lower molars. The enamel loop is extremely smooth, without enamel folding or Stufenbildung. When the molar is moderately worn, the cement cover is exfoliated, except for what remains at the bottom of the interlophid valley. The ccprp is present on the mesial three or four lophids. In addition to ccprp1, ALA-VP-2/4 has a ccpop1, which is larger than the ccprp1 and connected to the first lophid, contrary to the pattern ordinarily seen in elephantoids. The mesial cingulum is transversely asymmetric, with its thickest part situated buccally. The median sulcus is very deep on the first lophid without exception. The median and lateral sulci are also deep on lophids other than the first lophid in ADD-VP-3/3, whereas these sulci are obliterated by the subdivision of the median pillars on distal lophids in ALA-VP-2/4 and STD-VP-2/86. Each lophid is strongly compressed mesiodistally and parallel-sided apically with four to seven apical digitations. In occlusal view they are straight transversely (ALA-VP-2/4) or weakly concave-concave shaped (STD-VP-2/86).

The M_2 is represented only by a distal fragment consisting of two lophids and the distal cingulum (STD-VP-2/829). The preserved part is covered with thick cement except for the broken surface on the mesial end, where thick enamel (ca. 6 mm) is exposed. Other features of this molar are unknown.

The lophid formula of M_3 varies from x7x (STD-VP-2/833) to x9 (STD-VP-2/919) (Figure 15.9). KL-89-8 and KL-89-6b were described by Kalb and Mebrate (1993) as M_3 fragments of *S. orbus*. However, they perfectly join to KL-89-7b and KL-89-8a, respectively. Their original lophid formula can be estimated to be x8x. ASK-VP-3/15 is strongly damaged in the middle, but its lophid formula can be estimated to be x8x. The lophids are weakly concave-concave shaped or straight buccolingually in occlusal view. The lophids have five to six apical digitations except for the most distal ones, which have only three digitations. The molar tapers distally with a relatively pointed distal profile in occlusal view. The mesial two lophids of the M_3 are supported by a mesial root. Unworn lophids are entirely covered with cement except for the tips of the digitations. The cement cover of lophids is rapidly lost by wear except for the very bottom of the valley. Apart from the constrictions marking the boundary of the loops, there is no marked wear feature such as enamel folding or Stufenbildung. The median sulcus is clear on the mesial part of the molar, especially on the first lophid. On the other hand, it is frequently obscured by subdivision and enlargement of the median pillar on the distal part of the molar. Apically, the median and lateral pillars are nearly equal-sized, but at the basal part of the lophid, the former is transversely narrower than the latter. The adaxial median pillar expands mesiodistally to form a gentle distal and mesial projection. The ccprp1 and the ccpop1 are always present. The former is large and occasionally doubled, whereas the latter is small. The ccprp2 is always present. On the other hand, the column on more distal lophids varies greatly; in some molars it is absent on most lophids other than the mesial two lophids, whereas it is present on the distal lophids in other molars. The M_3 of STD-VP-2/829 is still in its crypt, and the deposition of its enamel and cement layer are not completed. The fifth lophid of this molar has a pair of small distal columns. In erupted molars, the presence of such small columns is obscured by thick cement cover.

Mandibles The morphology of the mandible can be observed in STD-VP-2/827 and STD-VP-2/86 (Figure 15.13). Measurements of their molars can be found in Table 15.9. The anterior tip of the symphysis is not preserved in any of the specimens from Saitune Dora. However, the preserved parts of the symphysis do not show any sign of a lower tusk or downturning. The corpus is shallow in STD-VP-2/827, with a height of 140 mm and a width of 150 mm at the mid-M_3 level. In dorsal view, lateral border of the horizontal rami of the mandible is open V-shaped, as in the holotype of *Elephas nawataensis* (KNM-LT 23783) and more derived elephants. The masseteric fossa is deep. The mandibular condyle is elliptical in dorsal view. The distobasal margin of the ascending ramus is very thin, fringed by rugose surface, and flares distally to form a wing-like structure. This rugose surface corresponds to the attachment surface for the sphenomandibular ligament seen in extant *Loxodonta* and *Elephas* (Beden 1979, 1983), but its flaring is a plesiomorphous feature widely seen in gomphotheres. In extant *Loxodonta,* the posteromedian crest connects the rugose surface for the sphenomandibular ligament and the articular condyle of the mandible (Beden 1979, 1983). However, the posteromedian crest between the mandibular condyle and the rugose surface for the sphenomandibular ligament is weak in *P. g. saitunensis,* and in this respect *P. g. saitunensis* is similar to *Elephas* rather than to *Loxodonta*. The lingual margin of the trigonum retromolare forms an acute and distinct projection.

Kalb and Mebrate (1993) reported an edentulous mandible (KL-176-1) from the Adu-Asa Formation as cf. *Stegodibelodon schneideri,* because it shares the "spout-like" short protruding symphysis with a mandible of *S. schneideri* from Chad described by Coppens (1972). The symphysis of the adult *Primelephas* mandibles is poorly preserved in the Saitune Dora sample, but the morphology of the corpus and ascending ramus of preserved specimens is essentially indistinguishable from that of KL-176-1 (Figure 15.13). Taking into account their spatial proximity and similarity, it is more parsimonious to assume that KL176-1 belongs to *Primelephas* rather than to a species not represented by any other material.

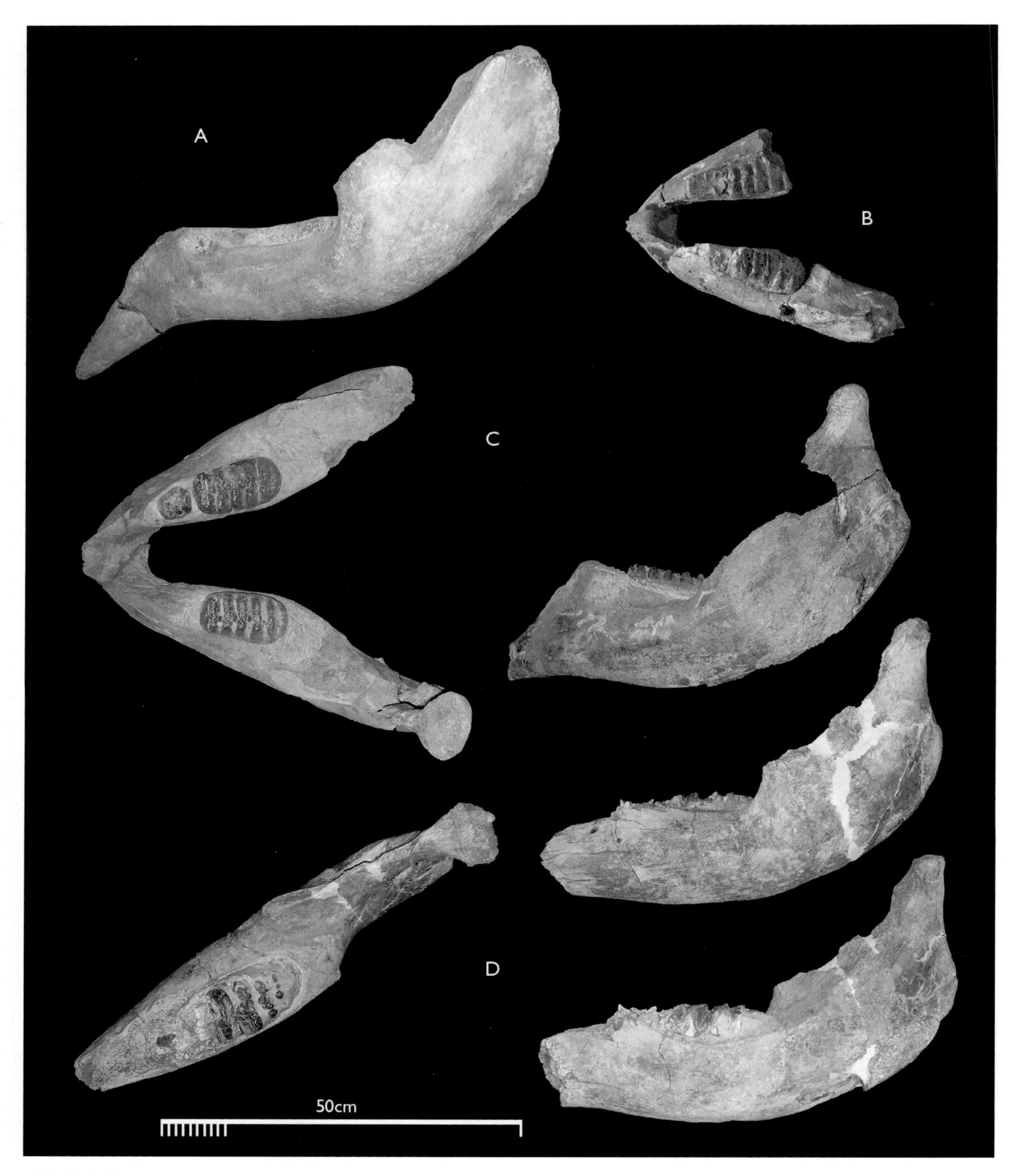

FIGURE 15.13

Mandibles of *Primelephas gomphotheroides saitunensis*: A. KL176-1, edentulous mandible, left lateral view. B. KL89-5, juvenile mandible with dp_3–dp_4, left lateral view. C. STD-VP-2/86, mandible with P_4 and M_1, occlusal and left lateral views. D. STD-VP-2/827, right mandible with M_3, lateral (reversed), medial, and occlusal views.

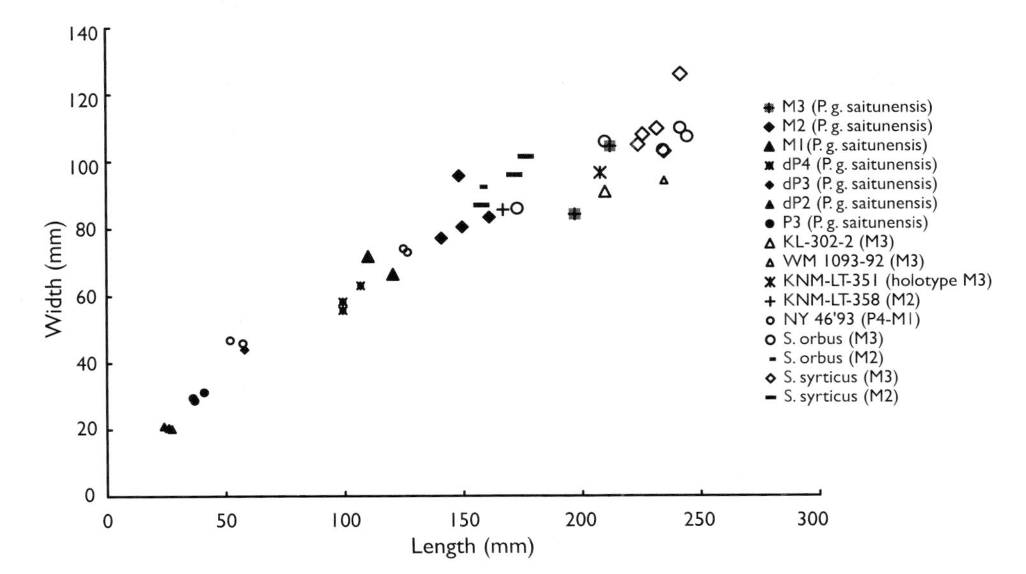

FIGURE 15.14

Scatter diagram of upper cheek teeth of selected specimens of early elephants from various localities in Africa (width against length). Data for non–Middle Awash fossil specimens are from WM 1093-92 (Sanders 1997); KNM-LT 351 and 358 (Tassy 2003); NY 46'93 (Tassy 1994); *S. orbus* (Tassy 2003); *S. syrticus* (Maglio 1973; Gaziry 1987; Tassy 1999).

DISCUSSION It is most parsimonious to assign the elephants from Saitune Dora to *P. gomphotheroides* because the M^3 and M_3 are mostly identical with those of the holotype of *P. gomphotheroides* (KNM-LT 351) in dimensions (Figures 15.14 and 15.15) and number of loph(id)s. However, the elephants from Saitune Dora also differ from KNM-LT 351 in having ccprp2 constantly on the M_3. The difference between the Saitune Dora elephants and KNM-LT 351 can be interpreted in two different ways. The first interpretation would be that KNM-LT 351 represents an exceptional morphological type in a taxon to which Saitune Dora elephants belong. The second interpretation would be that KNM-LT 351 represents a morphological type of a taxon and the M_3s of Saitune Dora represent another taxon. Although no additional specimens of *P. gomphotheroides* have been obtained from Lothagam, even by recent expeditions (Tassy 2003), two specimens found from other contemporaneous localities are nearly identical to KNM-LT 351; one is KL-302 from the Middle Awash, and the other is KNM-LU 526 (see Figure 15.12E) from Lukeino, Kenya, which was reconstructed by one of us (HS) from fragmentary material housed at KNM and will be described elsewhere. The presence of such specimens suggests that the morphology seen in KNM-LT 351 is not exceptional in a taxon to which KNM-LU 526, the KL-302 specimens, and KNM-LT 351 belong. The Saitune Dora elephants are usefully distinguished from this taxon at the subspecific level because the Saitune Dora elephants are identical with *P. gomphotheroides* from Lothagam and Lukeino in most of the dental features, besides the constant presence of the ccprp2 on M_3. *Primelephas gomphotheroides* might therefore be divided into three subspecies according to the degeneration of the central conule of M_3. *Primelaphas g. saitunensis* is a subspecies that still retained distinct central conules on second lophids in M_3, in contrast to that represented by a sample from Manonga Valley, characterized by the total loss of the central conules, approaching the condition seen in stegodonts. *Primelephas g. gomphotheroides,* which includes the type specimen

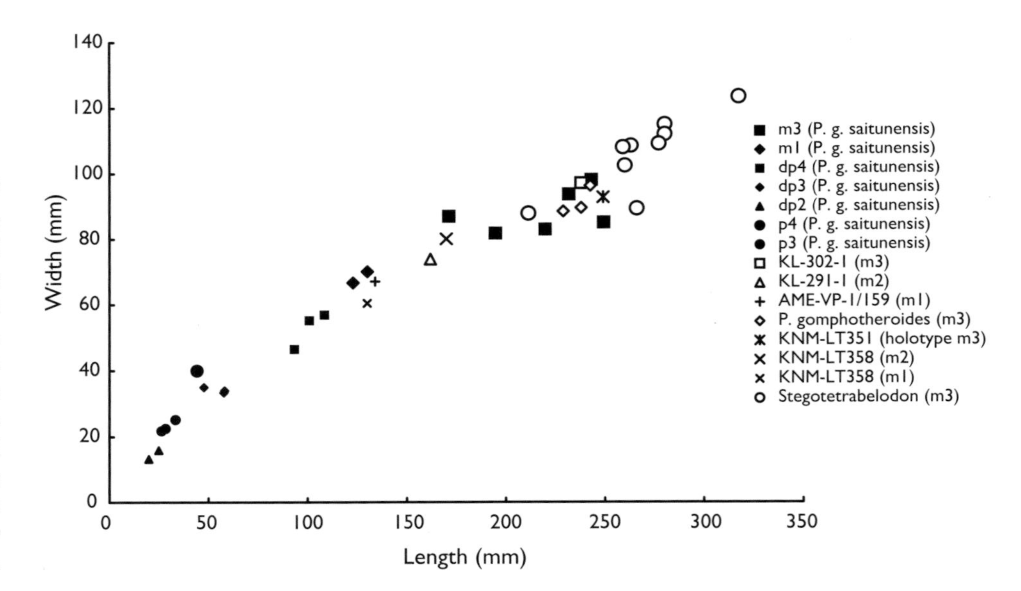

FIGURE 15.15

Scatter diagram of lower cheek teeth of selected specimens of early elephants from various localities in Africa (width against length). Data for non–Middle Awash fossil specimens are from: *P. gomphotheroides* (Sanders 1997); KNM-LT 351 and KNM-LT 358 (Tassy 2003); *Stegotetrabelodon,* Maglio (1973), Gaziry (1987), Tassy (1999, 2003).

of the species, represents the intermediate condition between them in the extent of the reduction of the central conules.

The absence of incisors in the Saitune Dora elephants contradicts the previous idea that *P. gomphotheroides* had lower incisors (Maglio 1973; Maglio and Ricca 1977; Tassy 1986, 2003). Maglio and Ricca (1977) noted that a fragment of lower incisor fits into a gutter on a fragment of the mandibular symphysis of KNM-LT 358, assuming that the gutter is an alveolus of the tusk. However, Tassy (2003) denied such direct contact between mandibular fragment and the incisor, determining that the gutter is part of a mandibular canal, not the alveolus of the lower tusk. Despite this, he argued that the presence of the incisor in *Primelephas* was still evidenced by the fragment of lower incisor, because there is no reason to deny the record that the tusk was found together with other elements of KNM-LT 358. However, it is equally probable the lower tusk actually belongs to another individual (for instance, a female of *Stegotetrabelodon orbus*) and was mistakenly mixed with other elements of KNM-LT 358 during curation, because some (or much) confusion is seen in the numbering of the specimens in the old collection from Lothagam. For example, a molar fragment included in KNM-LT 351 (the holotype of *P. gomphotheroides*) must belong to the same individual as the mandible included in KNM-LT 354 (the holotype of *Stegotetrabelodon orbus*) because the former is morphologically identical with the third and forth lophids of a left M_3 associated with the latter. On the other hand, a right M_3 currently included in KNM-LT 354 differs from the left M_3 of the same accession number in every respect. Another lower tusk of a primitive elephant attributed to *Primelephas* is an isolated tusk fragment from Lukeino (KNM-LU 121). Tassy (1986) tentatively attributed KNM-LU 121 to cf. *Primelephas,* admitting that it could also belong to a young individual of *Stegotetrabelodon*. Although there still remains much ambiguity on the character states of the lower tusks of early elephants, the idea that a lower tusk is not present in *Primelephas* is now well-supported by the new evidence from Saitune Dora, because, as just seen, the instances of counterevidence against this idea are ill-founded, and because of the size of the Saitune Dora sample (absent any incisor fragment).

Primelephas g. saitunensis is thus far known only from the Adu-Asa Formation. In contrast, *P. g. gomphotheroides* is known from several eastern African localities, such as Lothagam, Lukeino, and Mpesida in Kenya (Maglio 1973; Tassy 1986, 2003), Manonga Valley in Tanzania (Sanders 1997), and the Kuseralee Member of the Sagantole Formation of the Middle Awash. A new *Primelephas* species was named from Chad (MačKaye et al. 2008), but comparative studies await. According to the radiometric dating of the Lower Nawata Member of Lothagam (Leakey et al. 1996; McDougall and Feibel 1999), the first appearance of *P. g. gomphotheroides* in the Member predated that of *P. g. saitunensis* in the Middle Awash by ca. 1 Ma, although the former is more derived than the latter. Whatever the explanation for this, *Primelephas* from the Middle Awash is less derived than *Primelephas* from other contemporaneous eastern African sites, whereas the evolutionary level of *Anancus* is roughly concordant with the radiometric ages. This may mean that *Anancus* was less endemic than the elephant was during the late Miocene and perhaps would be more useful for biochronology, provided that a large sample is available for comparison.

Mammuthus Burnet, 1830

cf. *Mammuthus* sp.

DESCRIPTION There are only two specimens referred to cf. *Mammuthus* sp.: an associated tusk fragment and left M_3 (Figure 15.16). The left M_3, ALA-VP-2/309, consists of mesial and distal fragments. Nine lophids and a well-developed distal cingulum are preserved on these fragments. However, the gap between these fragments suggests that the molar had more lophids than what has been recovered and that the molar possibly originally had 10 or more lophids. Each plate has five to six digitations. Distinct Stufenbildung is seen on the wear surface of the enamel. The mesial two lophids are deeply worn and supported by a mesial root. The ccprp1 is transversely elongated and large. There are two distal sinuses abreast with each other at the median part of the second and third lophids. The small, round, buccal sinus contacts the next lophid, but the broader lingual sinus does not contact the succeeding lophid. There is a distal column on the median digitations of the second to fourth lophids of the distal molar fragments. They are free at their apex but fuse with the digitations at one-third the distance from the apex to the crown base. The fused column forms a sharp, distally projecting ridge on the distal wall of the median pillar, which contacts with the mesial surface of the succeeding lophid. The distal end of the molar is round in occlusal view. Similar structures can be observed in the right molar fragment.

A fragment of the distal end of the right M^3 has three lophs and a distal cingulum. Interloph valleys are filled with cement. Two median pillars of the first loph have a vertically running ridge (= fused distal column) on their distal faces. The associated tusk (158 mm long) is twisted, and its preserved diameter is 75 mm. Though its diameter may have been exaggerated by postmortem compression, the cross section of the tusk appears to have originally been slightly elliptical.

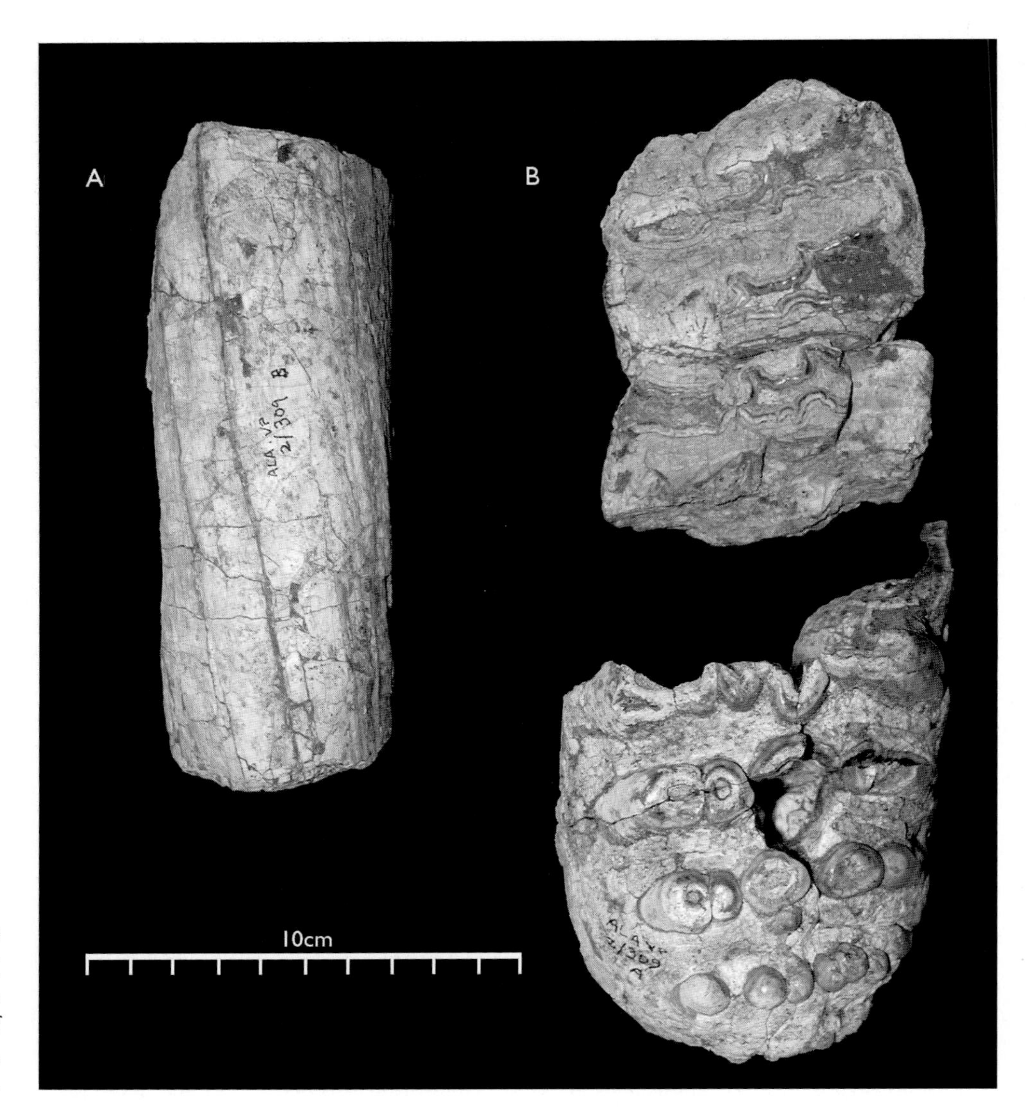

FIGURE 15.16
Tusk and molars of cf. *Mammuthus* sp. A. ALA-VP-2/309b, fragment of tusk. B. ALA-VP-2/309a, left M_3.

DISCUSSION ALA-VP-2/309 from the Asa Koma Member of the Adu-Asa Formation is not distinguishable from the Saitune Dora *Primelephas* in terms of molar dimensions (see Table 15.9). However, it differs from *Primelephas* in the following features: the presence of columns (central conules) on both the mesial and distal flanks of M_3 lophids, a greater number of lophids, a higher lophid frequency, thinner enamel, the absence of distal tapering in M_3, stronger fusion of the distal column, and the presence of a distal ridge (= fused distal column) on the distal face of the median pillar of M^3. Following Maglio (1973) that *Mammuthus* can be distinguished from other genera of elephants by the strong twisting of the upper tusk, ALA-VP-2/309, with associated tusk fragments showing strong twisting, is tentatively assigned to cf. *Mammuthus* sp. Although the molars of ALA-VP-2/309 are more derived than those of *Primelephas,* they are still so primitive that they do not have derived characters exclusively shared with true *Mammuthus* species from Eurasia. KL210-1 and PQ-L12723 have been allocated to *Mammuthus subplanifrons* by previous authors (Maglio and Hendey 1970; Kalb and Mebrate 1993), but ALA-VP-2/309 differs from them in having a mesial column in addition to the distal column and having distinct Stufenbildung on its worn enamel. Thus the validity of the current identification of ALA-VP-2/309 is dependent solely on the morphology of the tusk. The tusk morphology of Plio-Pleistocene African elephants has never been studied comprehensively despite the fact that fossil tusks are not rare in African sites. Further studies on the morphology of these tusks from Africa are necessary to confirm the validity of the criteria of *Mammuthus* tusk set forth by Maglio (1973).

Loxodonta Anonymous, 1827

cf. *Loxodonta* sp.

DESCRIPTION There are four specimens referred to cf. *Loxodonta* sp.; they are shown in Figure 15.17, and their measurements are presented in Table 15.9. AMW-VP-1/67 consists of seven isolated lophs. Though rather fragmentary, the horizontal section of the lophs on the wear surface clearly shows a lozenge pattern typical to loxodonts. The median column (central conule) is half-embedded in the distal wall of the loph, without free apex. The grooves separating the four digitations are deep only at the apex of the loph, so that the enamel loop lacks distinct constructions and is very smooth. Neither enamel folding nor Stufenbildung is present on the thick enamel. The largest plate of AMW-VP-1/67 is 100.8 mm wide and 81.8 mm high.

AME-VP-1/159 is an extremely worn left M_1. Lophid formula of the tooth is x5x. Enamel loops are all parallel-sided. The mesial three lophs show a concave-concave pattern. The median part of the lophid expands mesiodistally to contact adjacent lophids. AME-VP-1/159 is provisionally allocated to cf. *Loxodonta* sp. because of its dissimilarity to STD-VP-2/87.

AME-VP-1/93 is a heavily worn right M^3, with the mesial part damaged. Judging from the preserved distal and accessory roots, there were two more lophs and a mesial cingulum anterior to the preserved five lophs. Therefore, original loph formula can be estimated as x7x. Except for the last loph, all the lophs have a convex-convex arrangement. The completed enamel loop of the third loph is propeller-shaped with a mesiodistally expanded single median pillar. The fourth loph is composed of transversely elongated lateral pillars and a pair of median pillars, of which the buccal one is larger and expanded mesiodistally to contact with the median pillar of the third loph. There is no visible enamel folding. AME-VP-1/93 is similar to the M^3 of *P. g. saitunensis* in its loph formula and low hypsodonty. However, it differs from the latter in the proportionally much wider crown and marked expansion of median pillars associated with the contact between succeeding lophs along the median line of the molar.

KUS-VP-1/112 is composed of the mesial and distal fragments of a right M_3 attached to mandibular fragments. Of these, the distal molar fragment joins a molar fragment (KL-219-4, correctly 291-4?) previously described as M^2 of *Stegotetrabelodon orbus* by Kalb and Mebrate (1993:42–43, fig. 22). A total of seven lophids is

FIGURE 15.17

Molars of cf. *Loxodonta* sp. A. KUS-VP-1/112, right M_3. B. AME-VP-1/159, left M_1. C. AMW-VP-1/67, fragment of a loph. D. AME-VP-1/93, right M^3.

represented by these fragments, but the original lophid formula cannot be established because of the portions missing between these fragments. The molar is fairly worn, but all of the interlophid valleys are still filled with cement. There is a large, centrally located distal column on the first lophid. The enamel loop undulates gently and irregularly, without enamel folding. The second lophid of the distal fragment is composed of two transversely elongated lateral and two round median pillars. The penultimate and last lophids are propeller-shaped, with two lateral pillars and one single median pillar. There is strong distal projection on the median pillar of the penultimate lophid, but it does not contact the last lophid. This molar cannot be the M_3 of *Stegotetrabelodon* because the enamel layer is too thin, the lophid is too compressed anteroposteriorly, and the cement is too thick for the genus. The lophid structure of ALA-VP-2/309 from the Adu-Asa Formation is somewhat similar to KUS-VP-1/112 but differs from the latter in having more closely spaced lophids. We tentatively allocate this specimen to *Loxodonta* sp. because of its similarity to a worn M_3 of *L. exoptata* (Maglio 1969: plate V) in the propeller-shaped lophids and mesiodistal expansion of the median pillar.

Elephantidae gen. and sp. indet

DESCRIPTION KL-210-1 is a right M_3 fragment consisting of second to fifth lophids and a fragment of the posterior column of the first lophid. Median distal columns are present behind the first four plates. KL-210-1 differs from *Primelephas* in the more compressed parallel-sided configuration of the completed enamel loop and weaker development of the median and lateral sulci. Kalb and Mebrate (1993) assigned it to cf. *Mammuthus subplanifrons,* because of its similarity to M_3 on a complete mandible from Langebaanweg (PQ-L12723) described as *M. subplanifrons* by Maglio and Hendey (1970). KL-210-1 is actually very similar to PQ-L12723 in lophid structure except for its dimension. Based on the similarities of molar morphology, it is plausible to conclude that these molars from Ethiopia and South Africa belong to the same group of early elephant, if not conspecific. However, as recently suggested by Mundiger and Sanders (2001), it may not be wise to group them together under the name *M. subplanifrons,* because *M. subplanifrons* is a heterogeneric mixture of primitive elephants from eastern and South Africa. Although the holotype of *"Archidiscodon" subplanifrons* (MMK 3920) is a nondiagnostic fragment of a heavily worn M_3, Maglio and Hendey (1970) and Maglio (1973) allocated several specimens of early elephants from eastern and South Africa, including PQ-L12723, to this taxon, revising the argument set forth by Meiring (1955), Cooke (1960), and Aguirre (1969). Among them, a strongly twisted upper tusk associated with a M_3 and an ulna (A2882) described from Virginia, South Africa, by Meiring (1955) provided the basis for the allocation of *subplanifrons* to *Mammuthus.* Kalb and Mebrate (1993) compared their sample from the Middle Awash with PQ-L12723 rather than with the holotype of the species. However, the molar of PQ-L12723 does not show any particular similarity to the holotype of *A. subplanifrons* (MMK 3920) or A2882. KL-210-1 may belong to the same taxon as PQ-L12723, as Kalb and Mebrate (1993) suggested, but the taxon to which they belong must be a primitive elephant other than *M. subplanifrons.*

Deinotheriidae

Deinotheres are among the rarest elements in the Middle Awash late Miocene fauna, as they are at other contemporaneous sites in Africa such as Lothagam. However, six specimens have been recovered from at least six different localities sampling the time period between 5.2 and 5.8 Ma.

Deinotherium Kaup, 1829

Deinotherium bozasi Arambourg, 1934

DESCRIPTION Specimens referred to *Deinotherium bozasi* are shown in Figure 15.18. Measurements of their premolars are summarized in Table 15.12. AMW-VP-1/60 is a worn left dp^2. It is a mesiodistally elongated trapezoid in occlusal view, with the distal end of the molar wider than the protoloph. The protoloph and ectoloph form an L-shaped enamel loop separated from a deeply worn metacone by a groove. There is a small tubercle at the mouth of the interloph valley. On the distal face, there is a well-developed contact facet with dp^3. Thre is no cement.

AMW-VP-1/110 is a worn dp^3. It is a mesiodistally elongated trapezoid in occlusal view, with the metaloph wider than the protoloph. The protocone and hypocone are deeply excavated by wear. The ridge extending distally from the paracone is interrupted by the interloph valley, failing to form an ectoloph. The tritoloph or the distal cingulum is rudimentary. There is no cement.

AME-VP-1/101 is a worn dp^4. It has proto-, meta-, and tritoloph, and its maximum width (ca. 45 mm) is attained at the metaloph, as is usually the case in deinotheres.

KUS-VP-1/19 is an immature left mandibular fragment with left lower incisor, dp_2–dp_4, and P_3. The dp_2–dp_4 are almost identical (in both size and crown morphology) with specimens from Laetoli (LAET 541, 5452, 81-77) described by Harris (1987a). The lingual wall of the hypo- and tritolophids are damaged on the dp_4, but judging from the width of the apex of these lophids, the width of the latter lophid was slightly greater than that of the former, as in the dp_4 of *D. bozasi* from Laetoli (LAET 81-77, for example). The tip of the incisor is detached from the strongly downturned symphysis. The original direction of the incisor on the symphysis can be reconstructed as shown in Figure 15.18C. The tip of the incisor is covered with a conical enamel cap. The cap is slightly compressed laterally, and the medial and lateral surface of the cap is separated by a blunt, vertical ridge at the anterior end of the cap. Just posterior to the ridge, there are shallow depressions on the medial and lateral surfaces. Compared to the vertical depth of the horizontal ramus (96.6 mm), the vertical distance between the ventral border of the horizontal ramus and the tip of the symphysis is fairly small. This configuration of the symphysis recalls the characteristic sharp downturning of the symphysis seen in adult and subadult mandibles of *D. bozasi*.

The deinothere fossil remains from the Asa Koma and Kuseralee Members of the Middle Awash are limited to deciduous premolars and fragmentary permanent molars. However, close similarity between dp_2–dp_4 from Laetoli and those from the Middle Awash strongly suggests an affinity to *D. bozasi* (Arambourg 1934). This species is an endemic African proboscidean whose earliest record is documented from the

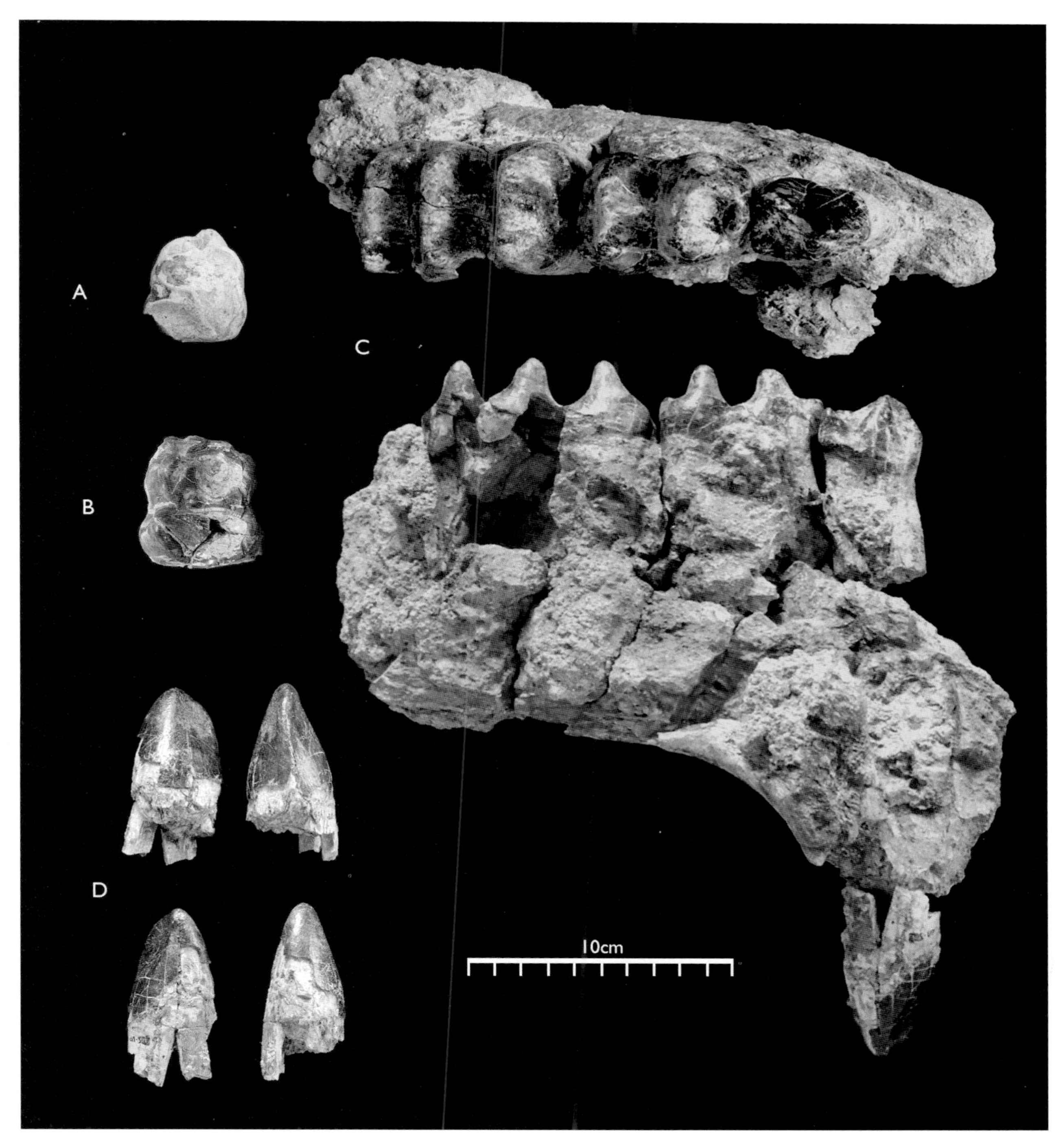

FIGURE 15.18

Specimens of *Deinotherium bozasi*. A. AMW-VP-1/60, left dp[2], occlusal view. B. AMW-VP-1/110, right dp[3], occlusal view. C. KUS-VP-1/19, immature mandibular fragment with left lower incisor, dp_2–dp_4, occlusal and medial views, anterior is to the right. D. KUS-VP-1/19, left lower incisor, lateral, anterior, medial, and posterior views.

TABLE 15.12 Measurements of Deciduous Premolars of *Deinotherum bozasi*

Specimen	*Element*	*L*	*Maximum Width*	*1st Loph(id) Width*	*2nd Loph(id) Width*	*3rd Loph(id) Width*
KUS-VP-1/19	dp_2	36.2	27.2			
KUS-VP-1/19	dp_3	56.9		36.5	41.4	
KUS-VP-1/19	dp_4	79.7		46.3	50+	46.9+
AMW-VP-1/60	dp^2	43.5	38.5			
AMW-VP-1/110	dp^3	51.6		44.2	48.4	

NOTE: All measurements are in mm. For abbreviations, see note to Table 15.1.

Namurungule (Nakaya et al. 1984) and Ngeringerowa Formations (Benefit and Pickford 1986) in Kenya. *D. bozasi* also occurs at terminal Miocene sites such as Lukeino and Lothagam (Leakey et al. 1996; Harris 2003a). The species persisted into the Plio-Pleistocene of eastern Africa until it went extinct sometime between 1.3 and 1.6 Ma (Beden 1987a,b).

Conclusions

The revised list of proboscidean taxa from the late Miocene of the Middle Awash presented here is substantially different from that reported by Kalb and Mebrate (1993). Based on the description and revision of new and old proboscidean material from the Middle Awash late Miocene, seven proboscidean taxa are recognized: *Deinotherium bozasi, Anancus kenyensis, Primelephas gomphotheroides gomphotheroides, P. g. saitunensis,* cf. *Loxodonta* sp., cf. *Mammuthus* sp., Elephantidae gen. et sp. indet. These proboscideans are recognized from two stratigraphic levels: the Asa Koma Member of the Adu-Asa Formation along the western margin and the Kuseralee Member of the Sagantole Formation in the Central Awash Complex (CAC). Radiometric dating has constrained the age of the two members to 5.5–5.8 Ma and 5.2–5.6 Ma, respectively (Haile-Selassie et al. 2004c; Renne et al., Chapter 4). As quantified in Chapter 18, elephants dominate over the gomphothere *Anancus* in the Asa Koma Member, whereas the latter is more abundant in the Kuseralee Member. Deinotheres are relatively rare in both stratigraphic levels. However, the dominance of the elephants over *Anancus* in the Adu-Asa Formation is exaggerated by the specimens from the bone bed of *Primelephas* at Saitune Dora, but even if those from Saitune Dora are removed from the count, the number of elephant specimens is still larger than that of *Anancus* in the other localities of the Adu-Asa Formation.

Primelephas is the dominant proboscidean in the Adu-Asa Formation, whereas *Anancus* dominates in the Kuseralee Member. This difference may be explained by habitat differences. *Anancus kenyensis,* Elephantidae gen. et sp. indet., and cf. *Loxodonta* sp. from the Kuseralee Member can be interpreted as precursors of the proboscideans from Langebaanweg, South Africa. On the other hand, *Primelephas gomphotheroides saitunensis* and cf. *Mammuthus* sp. from the Adu-Asa Formation are currently confined to this formation and do not appear to be ancestral to the elephants from the Kuseralee Member.

16

Tubulidentata

THOMAS LEHMANN

Globally, numerous more or less specialized mammals occupy the anteater ecological niche. The largest African anteaters are the Tubulidentata. These mammals, currently represented by a single extant species—*Orycteropus afer* (Pallas, 1766), the aardvark—are usually scarce and fragmentary in the fossil record. However, Tubulidentata are known in eastern Africa since the early Miocene (MacInnes 1956; Pickford 1975). They are also known from Eurasia since the middle Miocene (Colbert 1933; Fortelius 1990; Tekkaya 1993), but they progressively disappear from that continent until the Pliocene.

The tubulidentate material found in the Middle Awash late Miocene does not depart from the taphonomic rule and is very fragmentary. The present study focuses on the identification and description of these aardvark specimens. Their importance for improving our knowledge of the order Tubulidentata is also discussed.

Orycteropodidae

Orycteropus Geoffroy, 1796

Orycteropus sp. A

DESCRIPTION Two isolated aardvark specimens have been found at Saitune Dora (VP2) in the Asa Koma Member, dated between 5.54 and 5.77 Ma (WoldeGabriel et al. 2001). The sesamoid bone (STD-VP-2/855; Figure 16.1A, B) discovered at Saitune Dora is probably the sesamoid bone for the lateral tendon of the gastrocnemius muscle, which acts on the distal epiphysis of the femur. The dimensions of this bone are 18.7 × 10 × 6.9 mm. In comparison, the dimensions of the same sesamoid bone in the extant specimen AZ 391 (Transvaal Museum, Pretoria, South Africa) are 19.3 × 9.4 × 7.9 mm. This fossil sesamoid bone, like the one in *Orycteropus abundulafus* Lehmann et al., 2005, presents an articular facet more transversely oriented (in respect to the longest length) than in the extant one.

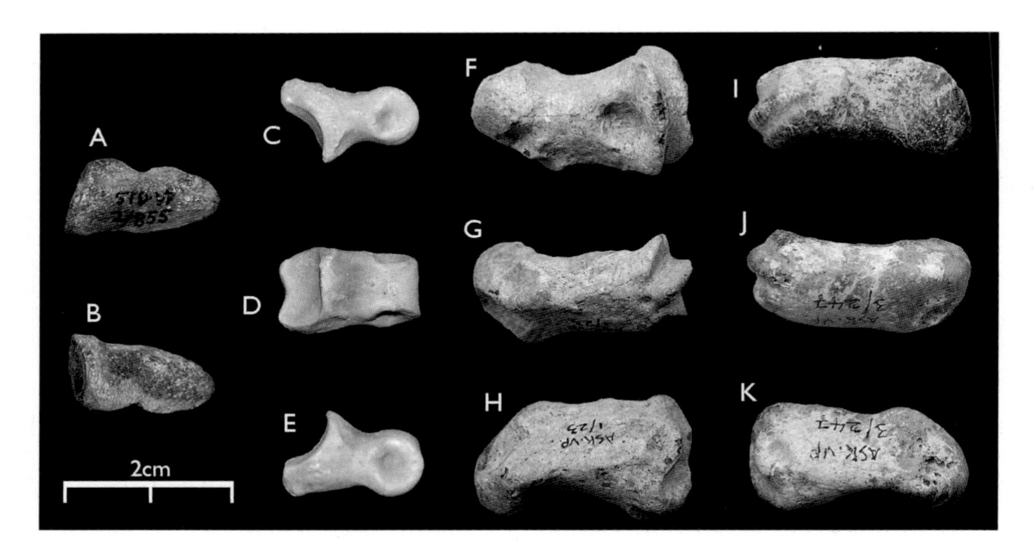

FIGURE 16.1

Aardvark specimens from the western margin sites. STD-VP-2/855: A. lateral view. B. medial view. STD-VP-2/856: C. lateral view. D. dorsal view. E. medial view. ASK-VP-1/23: F. medial view. G. dorsal view. H. lateral view. ASK-VP-3/247: I. ventral view. J. dorsal view. K. lateral view.

The fourth intermediate phalanx of the foot (STD-VP-2/856; Figure 16.1C–E) shows morphological similarities to those of other *Orycteropus* species. For instance, the bone is short and robust, and there is a slight dorsal pinching (proximally) on its diaphysis. However, the dorsal aspect of the proximal epiphysis is less curved than in *O. afer, O. crassidens* MacInnes, 1956, and *O. djourabensis* Lehmann et al., 2004 and is sharper, as in *O. abundulafus* and *O. gaudryi* Major, 1888. The dimensions of this bone, though broader, are close to those of the other Miocene *Orycteropus* but are distinct from those of *Myorycteropus* MacInnes, 1956 (Table 16.1).

DISCUSSION Although very fragmentary, the material from Saitune Dora can be unmistakably attributed to the genus *Orycteropus.* Moreover, it appears to be closer to the late Miocene Chadian *O. abundulafus* and European *O. gaudryi* than to the Plio-Pleistocene West African forms. It is, however, difficult to draw paleobiogeographic conclusions or a species-level determination based on such fragmentary material.

Orycteropus sp. B

DESCRIPTION Two other aardvark specimens have been found at Asa Koma in the Asa Koma Member, dated between 5.54 and 5.77 Ma (WoldeGabriel et al. 2001). The right fifth metacarpal (ASK-VP-1/23; Figure 16.1G, H) is perfectly preserved. Compared to a fifth metacarpal of the extant species, this specimen shows no anatomical differences. The proximal epiphysis presents a triangular outline, whereas the distal, medially twisted epiphysis shows the typical oblique dorsoventral crest that hinders joint dislocation. The left fifth metacarpal (ASK-VP-3/247; Figure 16.1I–K) is more weathered (probably by water) but is also very similar to the extant one. The large dimensions of these bones fall within the range of variation obtained for *O. afer* (Table 16.2). They are

TABLE 16.1 Measurements of the Fourth Intermediate Phalanx of the Foot of Some Orycteropodid Species

Posterior Fourth Intermediate Phalanx	*L*	*Bp*	*Bd*	*Hp*	*Hd*
Plio-Pleistocene Species					
Orycteropus afer ($n = 14$)	20.1 ± 1.1	12.5 ± 1.1	10.3 ± 1	12.4 ± 1.2	9 ± 0.9
O. crassidens (Paratype)	24.9	15.9	13.3	14.4	9.9
O. djourabensis (Holotype)	19	12.1	9.5		
Miocene Species					
STD-VP-2/856	**16.8**	**10.3**	**9**	**9.7**	**7.2**
O. abundulafus (Holotype)	17.6	7.9	6.2		
O. gaudryi (AMNH 22762[a])	16.4	8.7	6.9	8.6	6.2
Myorycteropus africanus (KNM-RU 5968[b])	13.4	8.7	7.7	8.8	6.5

NOTE: All measurements in mm. Abbreviations: L, length; Bp, proximal breadth; Bd, distal breadth; Hp, proximal height; Hd, distal height.

[a]American Museum of Natural History, New York, USA.

[b]National Museums of Kenya, Nairobi, Kenya (RU, Rusinga Island).

also close to the dimensions of the fifth metacarpal of *O. crassidens* but are distinct from those of *O. djourabensis*. The older Miocene aardvarks, in particular *Myorycteropus africanus* MacInnes, 1956, all show significantly smaller fifth metacarpals, with proportionally more slender diaphyses than those of the Ethiopian specimens.

DISCUSSION These Ethiopian late Miocene specimens are, in size, closer to Plio-Pleistocene forms than to other Miocene species. In particular, their lengths are close to the largest individuals of *O.afer* and *O.crassidens*. These metacarpals are thus attributed with confidence to the genus *Orycteropus*. In accordance with the observation made by Colbert (1941: 327), it seems that there is a lengthening of the hand in *Orycteropus* through time, possibly as "a result of the accentuation of its fossorial habits." Nevertheless, these specimens are the first "large" *Orycteropus* specimens found in the Miocene of Africa.

Conclusions

The fossil aardvark specimens found in the Middle Awash late Miocene deposits represent two forms of Tubulidentata. The specimens from Saitune Dora seem to be Miocene-like *Orycteropus* of medium size, like *O. gaudryi* and *O. abundulafus*. Conversely, the specimens from a contemporaneous Asa Koma locality show more affinities with modern Plio-Pleistocene aardvarks such as *O. crassidens* and *O. afer*. It is the first time that "large" Plio-Pleistocene-like aardvarks have been found in Miocene deposits, where only medium- and small-sized forms have been found so far (see Patterson 1975; Lehmann et al. 2005). The presence of these two different forms of Tubulidentata in sites of similar age, and

TABLE 16.2 Measurements of the Fifth Metacarpal of Some Orycteropodid Species

Fifth Metacarpal	*L*	*Bp*	*Bd*	*Hp*	*Hd*
Plio-Pleistocene Species					
Orycteropus afer (*n* = 23)	25.2 ± 1.6	13.2 ± 1.4	11.9 ± 1.2	13.6 ± 1.5	15.2 ± 1.4
O. crassidens (Holotype)	26.0	13.6	12.4	17.2	12.8
O. crassidens (Paratype)	29.1	16.0	13.6	16.0	15.9
O. djourabensis (Holotype)	23.6	11.4	9.7	11.9	13.6
Miocene Species					
ASK-VP-1/23	**26.9**	**13.1**	**12.1**	**14.0**	**15.3**
AK-VP-3/247	**27.2**				
O. abundulafus (Holotype)	20.6	7.4	7.2	7.6	8.0
O. gaudryi (AMNH field No. 4)	20.6	7.2	8.5	8.7	8.9
Myorycteropus africanus (KNM-RU 5968)	16.7	7.6	10.3	10.0	10.6

NOTE: Measurements are in mm. Abbreviations: L, length; Bp, proximal breadth; Bd, distal breadth; Hp, proximal height; Hd, distal height.

geographic proximity, suggests that the larger modern aardvarks may have coexisted for some time with the earlier smaller forms. Thus, the Mio-Pliocene turnover proposed by Lehmann et al. (2005) might not have been as sudden as previously thought. Finally, these Ethiopian specimens are also the first evidence in favor of an African origin for the modern aardvarks (see Lehmann et al. 2004).

17

Paleoenvironment

DENISE F. SU, STANLEY H. AMBROSE, DAVID DEGUSTA, AND YOHANNES HAILE-SELASSIE

The study of paleoenvironments has important implications for questions concerning speciation, extinction, and morphological change and adaptation. The late Miocene is a crucial period for the evolution of African mammals, but until recently there has been a dearth of information from this time period. Recent fossil discoveries in eastern and central Africa have provided new data, including crucial clues about the types of habitats in which early hominids might have evolved (see WoldeGabriel et al. 1994; Leakey et al. 1996; Pickford and Senut 2001; WoldeGabriel et al. 2001; Vignaud et al. 2002; Leakey and Harris 2003a, b; Haile-Selassie et al. 2004c).

The Middle Awash paleoanthropological study area in the Afar Rift of Ethiopia provides a sedimentary record that samples the last six million years. Most of the fossiliferous late Miocene deposits in the study area are radiometrically dated to between 5.2 Ma and 5.8 Ma. These deposits have yielded large assemblages of mammalian fossils. To date, more than 2,500 fossil specimens, representing 53 mammalian genera in 27 families, have been recovered from two basic time slices. The large number of mammalian fossils recovered, in terms of both species richness and abundance, enables an assessment of the kinds of habitats that were locally available in the Middle Awash at the end of the Miocene.

Many paleoenvironmental studies have been constrained by the use of a single taxon or method to infer habitat. The present analysis will avoid those pitfalls by examining the community structure based on locomotor and dietary variables, relative abundances of indicator taxa, and ecomorphology. Independent verification of the results based on large mammalian fauna will be derived from fish, bird, and small mammal faunas and from geological and stable isotopic data.

Materials

All faunal-based analyses presented here are derived from fossil specimens recovered from the Asa Koma Member (5.5–5.8 Ma) of the Adu-Asa Formation and from the

Kuseralee Member (5.2–5.6 Ma) of the Sagantole Formation (see WoldeGabriel et al., Chapter 2, Renne et al., Chapter 4, for detailed discussion of the two members, as well as WoldeGabriel et al. 1994; Renne et al. 1999; WoldeGabriel et al. 2001; Haile-Selassie et al. 2004c). These late Miocene deposits are exposed at localities along the western margin (Asa Koma Member) and in the Central Awash Complex (CAC; Kuseralee Member) (see Figure 1.1) of the Middle Awash study area. Because of temporal and spatial differentiation between the two members, as well as observed taxonomic differences between them (see other chapters in this volume), it is necessary to address the faunal assemblages from the Asa Koma and Kuseralee Members separately.

The member-level assemblages analyzed here are composite assemblages representing different localities. Although the fossils do not all come from a single locality or even a single stratigraphic interval, these time-averaged, composite assemblages can be used to compare and contrast the Asa Koma and Kuseralee Members. The time intervals for these assemblages are relatively constrained, about 0.3 Ma for Asa Koma, and 0.4 Ma for Kuseralee (although for the latter, these radiometric constraints are far looser than the reality that virtually the entire assemblage derives from sands and silts that immediately underlie the Gawto Basalts (see Renne et al. 1999)). Taxonomic homogeneity is observed throughout the stratigraphic section for each member and across different localities within each member. While such broad spatial and temporal composite assemblages have limitations, such as the effects of time averaging, they are buffered against local biases and can provide a general picture of dominant environments through geological time.

Interlocality variation in assemblage composition within members shows some differences in taxonomic composition among the various localities of the Asa Koma and Kuseralee Members. These differences are most likely taphonomic effects of sampling. As locality specimen numbers increase, the number of species sampled increases as well. Asa Koma Vertebrate Locality 3 (ASK-VP-3) yielded the largest number of collected, identifiable specimens (NISP; see Chapter 1 for a discussion of sampling methods) among the localities of the Asa Koma Member (n = 542). It is dated to 5.63–5.57 Ma. There are no significant differences between the faunal list of this locality and that of the overall member. Furthermore, a comparison of the relative proportions of the locality's bovid assemblage with the composite bovid assemblage from all localities within the Asa Koma Member reveals that there are no statistically significant differences between them. In short, there do not appear to be differences between the constrained, local assemblage and the composite assemblages presented in the following analyses.

The faunal lists for the Asa Koma and Kuseralee Members are presented in Table 1.3. Micromammals are excluded from the quantitative analyses presented here due to taphonomic biases uniquely associated with them, differential recovery among localities and between members, and the general scarcity of these taxa at most fossil localities. The quantitative analyses presented in this chapter include (1) community analysis based on locomotor and dietary variables, (2) presence of indicator taxa and their relative abundances, and (3) ecomorphology of the bovids.

Methods and Results

Community Analysis

Method

The goal for this analysis was to determine which modern habitat type is the closest match for the Asa Koma and Kuseralee paleoenvironments. We attempt to compare and contrast the Middle Awash late Miocene faunas to known, extant community structures from different habitat types. The method is based on the observation that differences in community structures reflect differences in habitats. The variables used to characterize community structure include dietary and locomotor adaptations. It has been shown that patterns of community structures based on dietary and locomotor variables are similar for similar habitats, regardless of species composition (e.g., tropical rainforest communities in Asia and South America are similar, even though quite different taxa are represented; Andrews et al. 1979). This is a method for interpreting the paleoecology of fossil communities based on general ecological principles rather than inference through closely related modern taxa (Andrews et al. 1979; Reed 1997). It should be emphasized that the method as presented here is based on faunal list composition only and is thus reliant on presence and absence data and subject to the biases that are associated with this particular type of data.

The modern faunal communities employed in this study are categorized into nine main habitat types: forest, woodland, open woodland, riparian woodland, bushland, shrubland, grassland, floodplain grassland, and desert (Table 17.1). Definition and categorization of the modern communities are modified from those of Lind and Morrison (1974), Pratt and Gwynne (1977), and White (1983). Faunal lists for modern comparative habitats are taken from the published literature (Swynnerton 1958; Ansell 1960; Lamprey 1962; Child 1964; Vesey-FitzGerald 1964; Rahm 1966; Sheppe and Osborne 1971; Smithers 1971; Ansell 1978; Rautenbach 1978a, b; Behrensmeyer et al. 1979; Smithers 1983; Happold 1987; Struhsaker 1997; Coe et al. 1999; Vernon 1999).

One limitation of the analysis as presented here is that modern faunal communities are derived from entire game parks or reserves (sometimes even an entire region of a country or continent), which may encompass many different habitats. The resulting faunal list conflates species that are not actually found in the same area or habitat in nature but artificially "co-occur" within broad geographic confines. Su (2005) has argued that this makes it difficult to compare fossil and modern faunal communities directly, because fossil sites are often much more constrained in size than the area from which modern comparative faunal communities are derived. This can badly degrade the precision of ecological inference even in neontological situations. Furthermore, there is a degree of circularity involved in defining the locomotor and dietary variables used in this type of analysis, because they are themselves dependent on faunal lists. In order to circumvent these problems, it would be necessary to divide modern habitat types clearly within the greater region and examine each individual habitat type separately. Unfortunately, given the fact that most faunal surveys reported do not differentiate between the fauna of different habitat types within the respective greater region,

TABLE 17.1 Modern African Localities and Vegetation Types

Vegetation	*Locality*	*References*
Forest	Congo Rainforest	Rahm 1966
	E. of River Niger	Happold 1987
	W. of River Niger	Happold 1987
	E. of River Cross	Happold 1987
	Kibale	Struhsaker 1997
Riparian Woodland (with swamps and grasslands)	Chobe National Park	Smithers 1971
	Okavango	Smithers 1971
	Moremi Game Reserve	Smithers 1971
	Linyanti Swamp	Smithers 1971
Woodland	Zambia Southern Woodland	Ansell 1978
	Guinea Savanna Woodland	Smithers 1983
	Amboseli National Park	Behrensmeyer et al. 1979
Open Woodland	Tarangire National Park	Lamprey 1962
	Southern Savanna Woodland	Smithers 1983
	Kalahari Thornveld	Rautenbach 1978a,b
	Sudan Savanna	Smithers 1983
	Southwest Arid	Smithers 1983
Bushland	Mkomazi Game Reserve	Coe et al. 1999
	Serengeti Bush	Swynnerton 1958
Shrubland	Sahel Savanna	Smithers 1983
	Karoo-Nama	Vernon 1999
	Karoo-Succulent	Vernon 1999
Floodplain Grassland	Kafue Flats	Sheppe and Osborne 1971
	Makgadikgadi Pan	Smithers 1971
	Rukwa Valley	Vesey-FitzGerald 1964
Grassland	Central Kalahari	Rautenbach 1978a,b
	Serengeti Plains	Swynnerton 1958
	Southern Savanna Grassland	Smithers 1983
Desert	Namib Desert	Rautenbach 1978a

it is not possible to do so, and we can only use the dominant vegetation type as the habitat category (Su 2005).

The variables used to characterize community structure include dietary and locomotor adaptations as outlined in Su (2005). There are nine dietary and five locomotor variables (see Table 17.2 for a list of those locomotor and dietary variables). For the Asa Koma and Kuseralee faunas, these are assigned from published studies on inferred locomotor and dietary behaviors for fossil genera (Lewis 1995; Spencer 1995; Bishop 1999; Cerling et al. 1999; Sponheimer and Lee-Thorp 1999; Zazzo et al. 2000; Werdelin and Lewis 2001a; Harris and Cerling 2002; papers in Leakey and Harris 2003a; chapters in this volume), some unpublished conclusions from specialists for some taxonomic groups, and inferences from their extant relatives (see Table 17.3 for locomotor and dietary variable

TABLE 17.2 Locomotor and Dietary Variable Categories

Code	*Locomotor Adaptations*	*Code*	*Trophic Adaptations*
T	Terrestrial	G	Grazer
T-A	Terrestrial-arboreal	FG	Fresh grass grazer
A	Arboreal	B	Browser
AQ	Aquatic	MF	Mixed feeder
AQ-T	Aquatic-terrestrial	GuI	Gumnivorous-insect
F	Fossorial	Fg	Frugivorous
		FI	Frugivorous-insect
		FL	Fruit and leaf
		C	Carnivorous
		CB	Carnivorous-bone
		CI	Carnivorous-insect
		I	Insectivorous
		O	Omnivorous
		RT	Root and tuber

NOTE: Following Reed 1996; Su 2005.

assignments). Locomotor and dietary variables for fauna from comparative modern communities are taken from published behavioral and carbon isotopic studies (see Su 2005 for a list of references).

Once these data were compiled, the frequency of each locomotor and dietary variable was calculated and comparisons between the Asa Koma and Kuseralee Members and modern faunal communities were made by using hierarchical clustering analysis (Ward's method on Euclidean distances) and principal components analysis (PCA). Before any statistical tests were run, the arcsine transformation was performed on the frequency data in order to normalize the distribution (Zar 1999). This is because percentages form a binomial, rather than normal, distribution and the deviation from normality is great for small or large percentages (Zar 1999).

Results

Relative frequencies of the locomotor and dietary variables indicate that terrestrial and browsing species dominate the faunal lists for both the Asa Koma and Kuseralee Members (Table 17.4). The high proportion of browsing species may indicate a predominance of woody/leafy vegetation for both members. To examine the locomotor and dietary variable frequencies further, they were compared to those of modern communities by using hierarchical clustering and PCA.

In hierarchical clustering analysis, the faunas of the Asa Koma and Kuseralee Members group most closely with each other (see Figure 17.1). While this points to their distinctness, they are also found within a cluster of modern faunal communities that are found in riparian woodland (Chobe, Linyanti Swamp, Moremi, Okavango), woodland (Zambia Southern Woodland, Southern Savanna Woodland, Lake Mweru, Tarangire, Amboseli), bushland (Mkomazi, Serengeti), and floodplain grassland (Kafue Flats, Rukwa). Within this large cluster, the faunas of the Asa Koma and Kuseralee Members are closest to the

TABLE 17.3 Locomotor and Dietary Variable Assignments for the Asa Koma and Kuseralee Fauna Included in the Community Analysis

	Locomotor Adaptations	*Dietary Adaptations*
Primates		
Hominidae		
Ardipithecus kadabba	T	FL
Cercopithecidae		
Colobinae gen. et sp. indet. "large"	A	FL
Kuseracolobus aramisi	A	FL
Pliopapio alemui	T-A	FL
Subfamily indet.	?	?
Carnivora		
Viverridae		
Genetta sp.	T-A	C/CI
Viverra cf. *V. leakeyi*	T	O
Herpestidae		
Herpestes alaylaii sp. nov.	T	CI
Helogale sp.	T	CI
Mustelidae		
Lutrinae gen. et sp. indet.	AQ	C
Sivaonyx cf. *africanus*	AQ	C
Mellivora aff. *M. benfieldi*	F	CI
Plesiogulo botori	T	C
Hyaenidae		
Hyaenictitherium sp.	T	CB
Hyaenictis wehaietu sp. nov.	T	CB
Felidae		
Machairodus sp.	?	C
Dinofelis sp.	T-A	C
Felidae gen. et sp. indet.	?	C
Ursidae		
Agriotherium sp.	T	C
Canidae		
Eucyon sp.	T	C
Artiodactyla		
Bovidae		
Tragoportax abyssinicus sp. nov.	T	MF
Tragoportax sp. "large"	T	MF
Bovini gen. et sp. indet.	T	MF
Ugandax sp.	T	MF
Simatherium aff. *S. demissum*	T	MF
Hippotragini gen. et sp. indet.	T	G
Tragelaphus moroitu sp. nov.	T	B
cf. *Tragelaphus* cf. *moroitu*	T	B
Reduncini gen. et sp. indet.	T	FG
Zephyreduncinus oundagaisus	T	FG
Redunca ambae sp. nov.	T	FG
Kobus cf. *basilcookei*	T	FG
Kobus aff. *oricornis*	T	FG
Kobus cf. *porrecticornis*	T	FG
Gazella sp.	T	G

TABLE 17.3 *(continued)*

	Locomotor Adaptations	*Dietary Adaptations*
Aepyceros cf. *premelampus*	T	MF
Madoqua sp.	T	B
Raphicerus sp.	T	B
Suidae		
Nyanzachoerus syrticus	T	MF
Nyanzachoerus cf. *devauxi*	T	MF
Nyanzachoerus cf. *waylandi*	T	MF
Nyanzachoerus australis	T	G
Nyanzachoerus kuseralensis sp. nov.	T	MF
Cainochoerus aff. *C. africanus*	T	FL
Giraffidae		
Sivatherium sp.	T	B
Palaeotragus sp.	T	B
Giraffa sp.	T	B
Hippopotamidae		
Hippopotamidae gen. et sp. indet.	AQ-T	FG
Hippopotamus aff. *dulu*	AQ-T	FG
Perissodactyla		
Equidae		
Eurygnathohippus feibeli	T	G
Eurygnathohippus aff. *feibeli*	T	G
Rhinocerotidae		
Diceros douariensis	T	B
Proboscidea		
Gomphotheriidae		
Anancus kenyensis	T	G
Elephantidae		
Primelephas gomphotheroides gomphotheroides	T	G
Primelephas gomphotheroides saitunensis subsp. nov.	T	G
cf. *Mammuthus* sp.	T	G
cf. *Loxodonta* sp.	T	G
Deinotheriidae		
Deinotherium bozasi	T	B
Hyracoidea		
Procaviidae		
Procavia sp. indet.	T-A	MF
Tubulidentata		
Orycteropodidae		
Orycteropus sp. A	F	I
Orycteropus sp. B	F	I

NOTE: "?" indicates that the locomotor or dietary adaptation of the taxon is unclear and is not included in the community analysis.

TABLE 17.4 Frequencies of Species Locomotor and Dietary Variables of the Asa Koma and Kuseralee Members

	Asa Koma (%)	*Kuseralee (%)*
Locomotor Adaptations		
T	63.5	73.3
T-A	7.7	6.7
A	7.7	6.7
AQ	3.8	2.2
AQ-T	3.8	4.4
F	13.5	6.7
Dietary Adaptations		
G	14.0	21.0
FG	10.0	9.3
B	18.0	23.3
MF	6.0	7.9
Fg	0.0	0.0
FI	0.0	0.0
FL	12.0	11.6
C	16.0	13.9
CB	4.0	2.3
CI	6.0	4.7
RT	8.0	4.6
I	2.0	0.0
O	4.0	2.3

riparian woodland communities. This indicates that the community structures of these two members are most similar to riparian woodland habitats and other densely wooded habitats. The presence of floodplain grassland habitats within this cluster suggests that there were also some wet grasslands in the Asa Koma and Kuseralee paleoenvironment. Forest communities group together in one cluster, as do riparian woodland communities. This implies that the faunal communities that inhabit these habitats are distinctive, particularly those of forest environments, and may be more easily detectable in the fossil record than those of other habitat types.

The results of the PCA indicate that PC 1 and 2 account for 57.44 percent of the variances seen among the communities (PC 1: 40.46 percent; PC 2: 16.98 percent). Eigenvalues for the PCA are presented in Table 17.5. For PC 1, differences in the proportion of arboreal and frugivorous species had the greatest contribution to the variances observed. For PC 2, differences in the proportion of species that consume mostly insects and edaphic grass contributed most to the variances observed. In order to understand better the results of the PCA, PC 1 was plotted against PC 2 on a two-dimensional graph (Figure 17.2). Both Asa Koma and Kuseralee are in the same quadrant with faunal communities found in riparian woodland, woodland, open woodland, floodplain grassland, and bushland. However, they are most closely associated with riparian woodland, woodland, and open woodland communities. This would indicate a similarity between the faunal communities of the Asa Koma and Kuseralee Members and those of riparian

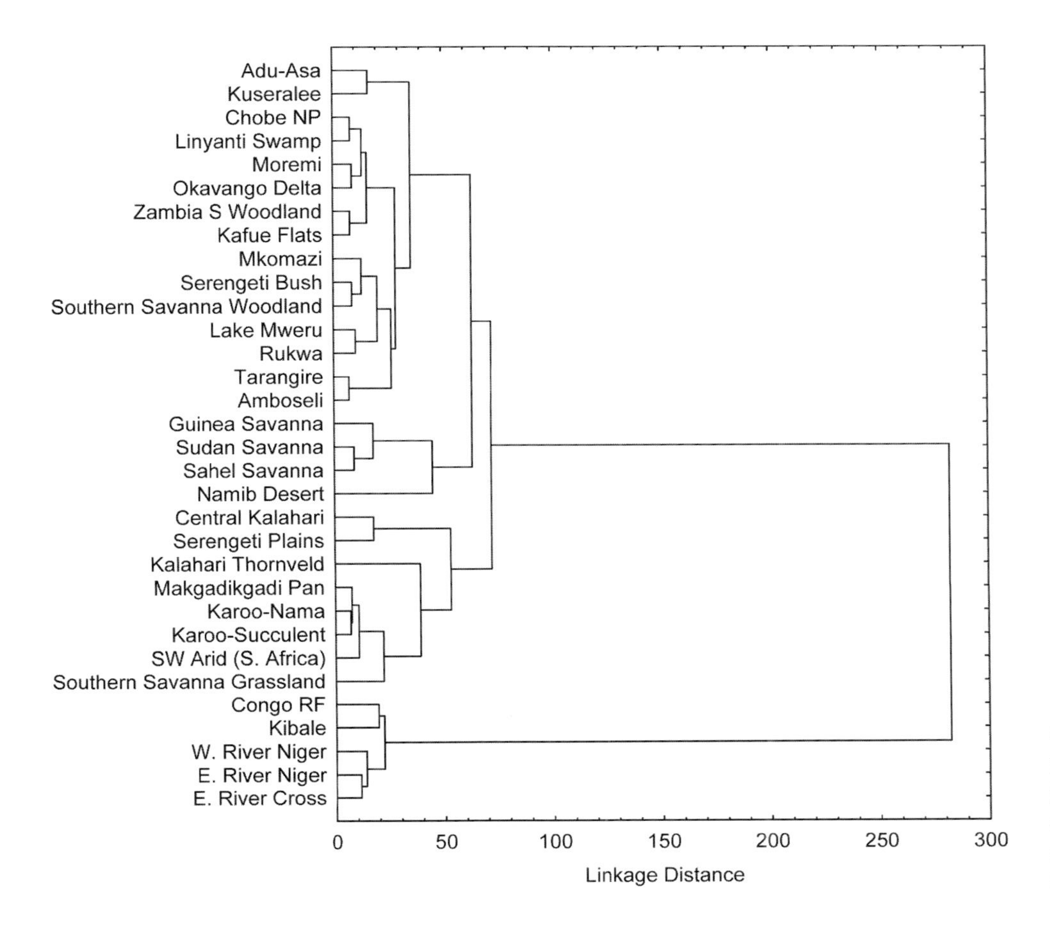

FIGURE 17.1

Dendrogram from the hierarchical clustering analysis (using Ward's method on Euclidean distances).

woodland, woodland, and open woodland habitats. Whereas forest communities cluster together exclusive of other types of faunal communities (similar to the results of the hierarchical clustering analysis), riparian woodland communities do not (unlike the hierarchical clustering results). This points to the distinctiveness of forest communities and suggests that riparian woodland communities share more elements of their fauna with other types of faunal communities.

Indicator Taxa and Their Relative Abundances

Method

In indicator species analysis, the presence of ecologically narrowly constrained species is used to infer paleoenvironment (for example, see Gentry 1970, 1978a; Vrba 1980b; Shipman and Harris 1988). The underlying assumption is that closely related species are behaviorally similar from the past to the present, so that comparisons between extant and extinct groups are possible. In this manner, the habitat preferences of living members of a taxon are extended back to its fossil relatives (Gentry 1970, 1978a; Vrba 1980b; Shipman and Harris 1988). It has been found in recent ecomorphological and stable carbon isotopic

TABLE 17.5 Eigenvalues for Principal Components 1 through 6 and the Percent of Total Variance Each Principal Component Represents

PC	*Eigenvalue*	*Percent of Total Variance*
1	8.495870	40.46
2	3.566238	16.98
3	1.877218	8.94
4	1.592435	7.58
5	1.178071	5.61
6	1.114074	5.31

NOTE: Twenty eigenvalues were extracted, but only the first six are presented because they make up more than 80 percent of the variances observed.

studies that fossil bovids, for example, generally shared similar diets with their extant relatives (Sponheimer et al. 1999). The major advantage of this method is that it is based on presence/absence data, so only a faunal list is required.

However, the indicator taxa method has its limitations. The major advantage, paradoxically, is also a major limitation because the relative abundance of any particular species is not considered. Species presence can be based on single specimens that represent only a small percentage of the total faunal assemblage, often a stratigraphically or ecologically intrusive species (once taphonomic and sampling biases have been accounted for). Without consideration of the relative abundance of fossils representing a species, whether measured by specimen count, NISP, or minimum number of individuals (MNI), such intrusives take on disproportional importance and degrade the precision of inference in simple presence/absence analyses. Relative abundance of ecological indicator species can be important in establishing such precision. For example, it has been suggested that bovid genera, or even tribes, are associated with particular habitat types, defined primarily by the proportions of wood cover (i.e., height and spacing of woody plants and the abundance of open areas) as well as the amount of grass cover in those open areas (Greenacre and Vrba 1984). This approach, however, requires specific conditions involving a host of other taphonomic variables.

Census data of large mammals in Amboseli National Park indicate that there is differential abundance of taxa in different habitats (Western 1973). Species that prefer more open habitats are more abundant in grassland habitats of Amboseli than are those that prefer more wooded habitats (Western 1973). Furthermore, habitat preferences of living species are reflected in surface bone assemblages at Amboseli such that a dominance of a habitat-specific species in the bone assemblage would indicate its predominance in the living community and, in turn, its preferred habitat in the area (Behrensmeyer et al. 1979; Behrensmeyer and Dechant Boaz 1980; Behrensmeyer 1993). Thus, relative abundance data of habitat specific taxa in modern bone assemblages can reflect habitat preference differences in the living community (Behrensmeyer et al. 1979; Behrensmeyer and Dechant Boaz 1980; Behrensmeyer 1993). It is possible to use such data to look at habitat differences in the fossil record. Hence, it is prudent to combine any examination of indicator

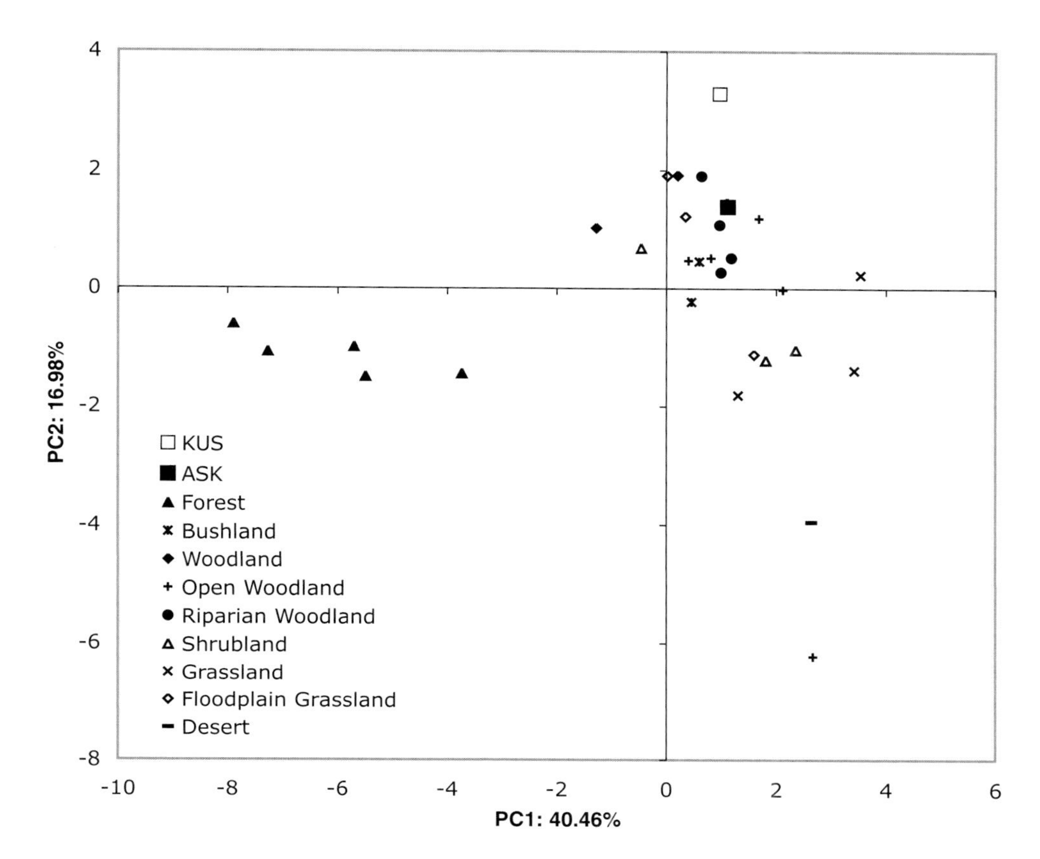

FIGURE 17.2

Results of a principal components analysis (PCA). This is a projection of modern localities and the Asa Koma (ASK) and Kuseralee Members (KUS) on the factor plane (PC 1 × PC 2).

species with analyses of their relative abundances when reconstructing paleoenvironmental conditions. However, before such an analysis can be attempted, a discussion of the biases involved is necessary.

Obviously, the major problem with relative abundance analyses is that the fossil record often only poorly reflects the living biomass and its numerical apportionment at any particular locality. Differential longevity, behaviors, preservational characteristics, time-averaging of surface collections from stratigraphic intervals, and recovery procedures can all distort the relationship between an ecosystem and its fossil residue. It is important to bear these biases in mind when considering relative abundances so that we can be sure that the pattern observed is due to ecological rather than taphonomic factors. Some of these biases can be ameliorated by comparing only mammals within certain size ranges and taxonomic groups and by implementing a collection strategy under which all identifiable elements are collected. For the analysis that follows, we rely on dental specimens to help circumvent the issue of sampling biases and also to avoid differences in element representation due to taphonomic factors.

The bovid family is the most common taxon in the Middle Awash late Miocene deposits and is the focus of the relative abundance analysis that follows. Due to their abundance in the Middle Awash late Miocene fauna and the specificity of their habitat preferences, bovid taxa are ideal for examining relative abundances and using them to infer different habitat types. All locality collection in the geological members analyzed

TABLE 17.6 A List of Comparative Fossil Sites

Locality	*Age (Ma)*	*References*
Lothagam, Kenya		Leakey and Harris 2003a
Lower Nawata Member	7.4–6.5	
Upper Nawata Member	6.5– ~5	
Apak Member	~5–4.2	
Kaiyumung Member	<3.9	
Hadar, Ethiopia		Gray 1980
Sidi Hakoma Member	3.4–3.3	
Denen Dora Member	3.2–3.18	
Laetoli, Tanzania		Su 2005
Upper Laetolil Beds	3.8–3.5	
Omo, Ethiopia		Bobe 1997; Alemseged 2003
Shungura Member B	3.36–2.95	
Shungura Member C	2.95–2.6	

here involved 100 percent recovery of all bovid cheek teeth (even isolated) and identifiable horn core fragments. In the analysis of the relative abundance of bovid tribes that follows, all premolars and molars were counted as identifiable specimens (in NISP). Associated teeth, such as those in a mandible or maxilla, were counted only as a single identifiable specimen each. Horn cores were also included in this analysis because some bovid taxa are represented only by horn cores. Specimens represented by both molars/premolars and associated horn cores were counted as single specimens (= individuals). Associated dental sets or antimeres were counted as single specimens (= individuals). Once all dental and horn core data were tabulated, relative abundances were calculated.

Other taxa that are recognized to be particularly specific in their habitat preferences were also analyzed along with the bovids: giraffids, suids, and primates (see Table 17.4 for a list of the species in each group). In this analysis of indicator taxa (bovids, giraffids, suids, and primates), only M3s (uppers and lowers) were counted to ensure that the NISP value for each taxon was comparable. As before, associated teeth were counted as a single identifiable specimen, so these values probably closely approximate actual MNIs.

Relative abundances of bovid tribes and indicator species from the Asa Koma and Kuseralee Members were compared to those from other Miocene and Plio-Pleistocene sites, including localities from Lothagam (Kenya), Hadar (Ethiopia), Laetoli (Tanzania), and Omo (Ethiopia) (Table 17.6). The count data for comparative Miocene and Plio-Pleistocene sites are taken from published literature (see Table 17.6 for references). The data from different sites are not directly comparable because of different collecting methodologies. However, the use of only dental material and horn cores minimizes this effect, because all of these projects systematically collect these elements.

Results

RELATIVE ABUNDANCES OF BOVID TRIBES Both the Asa Koma and Kuseralee Members have high proportions of reduncines (Asa Koma: 50.7 percent, Kuseralee: 24.4 percent;

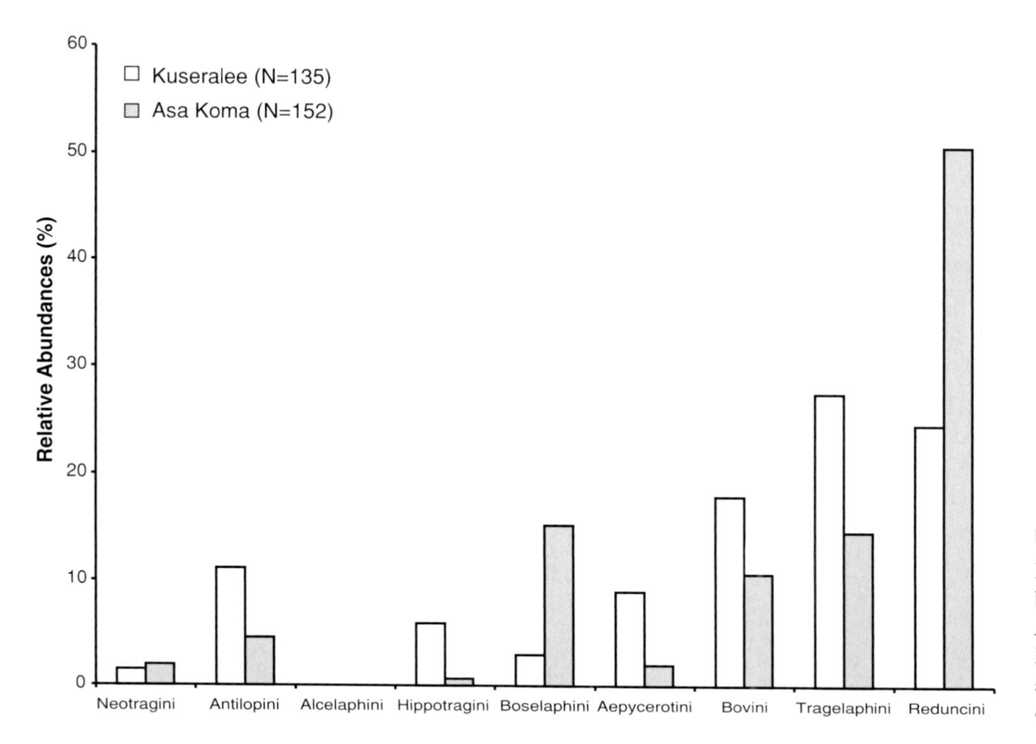

FIGURE 17.3

Relative abundances of bovid tribes from the Asa Koma and Kuseralee Members. Note that there are no alcelaphines from either member.

Figure 17.3). Modern reduncines are water-dependent and are generally found next to permanent sources of water, particularly in floodplain grasslands and/or swampy areas (Mitchell and Uys 1961; Smithers 1971, 1983; Sayer and Van Lavieren 1975; Schuster 1980; Kingdon 1997). The high frequency of reduncines in the late Miocene deposits of Middle Awash indicates a permanent source of water in the paleohabitat, as well as some floodplain grassland or swampy areas.

Relatively high proportions of tragelaphines (Asa Koma: 14.5 percent; Kuseralee: 27.4 percent) and bovines (Asa Koma: 10.5 percent; Kuseralee: 17.8 percent) were also observed in both members (Figure 17.3). This indicates heavily wooded habitats with some grassy glades, and a permanent source of water, in both members. Extant members of both tribes are dependent on water, and whereas tragelaphines prefer densely wooded habitats, bovines are found in areas with dense cover and grass, and some species in both tribes prefer swampy conditions (Smithers 1983; Kingdon 1997).

Bovid tribes associated with drier and less wooded conditions—Neotragini, Antilopini, and Hippotragini (Smithers 1983; Kingdon 1997)—are rare in the Asa Koma and Kuseralee Members. These make up 2.0 percent, 4.6 percent, 0.7 percent in the Asa Koma Member, and 1.5 percent, 11.1 percent, 5.9 percent in the Kuseralee Member, respectively (Figure 17.3). Furthermore, there are no alcelaphines represented in either member, even though, by this time in the late Miocene, alcelaphines were established elements in the bovid community in other parts of eastern Africa. In the Upper Nawata deposits of Lothagam (Kenya) (6.5 to ~5 Ma), alcelaphines make up 17.2 percent of the bovid fauna (Harris 2003c). The low abundance of these taxa suggest that less wooded, arid habitats were not common on the paleolandscape sampled by the Asa Koma and Kuseralee Members.

The emerging picture of the Asa Koma and Kuseralee paleoenvironments based on the bovid relative abundance data is one that is densely wooded with permanent water, such as a river or lake, and some swampy conditions. Areas of the paleohabitat probably also had floodplain grassland, since there is a high proportion of reduncines, which are generally found in floodplain grasslands today.

Paleosol carbonate isotope values and bovid relative abundance data indicate that there might be ecological differences between the Asa Koma and Kuseralee Members. The oxygen isotope data suggest that the Asa Koma Member was relatively humid or cool compared to the Kuseralee Member (Haile-Selassie et al. 2004c). There are higher proportions of bovids associated with arid and open conditions (i.e., Neotragini, Antilopini, Hippotragini) in the Kuseralee Member (18.5 percent) than in the Asa Koma Member (7.3 percent). At the same time, reduncines are most common in the Asa Koma Member and are much more abundant there than in the Kuseralee Member (Figure 17.3). The above evidence suggests that at the time of deposition, the localities of the Kuseralee Member may have been drier or farther away from standing water than those of the Asa Koma Member. However, the relative abundance of certain bovids indicates that the Kuseralee Member was more wooded than the Asa Koma Member. For example, tragelaphines are the most common bovids in the Kuseralee Member and are more abundant there than in the Asa Koma Member (Figure 17.3), suggesting that wooded vegetation was a greater component of the habitat. At the same time, the greater abundance of reduncines recovered from the Asa Koma Member compared to the Kuseralee Member may signal a higher proportion of floodplain grassland in the Asa Koma Member. This is not to say, however, that the bovid assemblage from the Asa Koma Member is indicative of open habitats, only that the proportions of wooded and floodplain grassland habitat may have differed between the two members. It is not possible at this point to determine whether these differences are taphonomic or reflect local or broader regional or continental changes.

To place the relative abundances of the Asa Koma and Kuseralee bovids into context, they were compared to those of other Miocene and Plio-Pleistocene sites (see Table 17.5 for a list of sites). This was accomplished through correspondence analysis, which showed that the first dimension can explain 47.3 percent of the deviation from expected values and the second dimension can explain 22.0 percent of the deviation. The first dimension appears to distinguish mostly between habitat preferences of the different bovid species. In order to more easily visualize this, the first and second dimensions were plotted (Figure 17.4). Taxa that are commonly found in less vegetated and drier habitats (i.e., alcelaphines, neotragines, antilopines, and hippotragines) are in the right quadrants, while those that are found in more wooded and wetter habitats (i.e., tragelaphines, bovines, reduncines, aepycerotines, and boselaphines) are on the left (Figure 17.4). Note that the Asa Koma and Kuseralee Members are closest to taxa that are found in wooded and wet environments, reflecting the fact that those taxa, particularly Tragelaphini and Reduncini, are more abundant in the Asa Koma and Kuseralee bovid assemblages, although not in the same proportions. Most of the comparative fossil localities are also grouping with taxa that are found in wooded or wet habitats, most likely because most of them have a lake or large river on their paleolandscape.

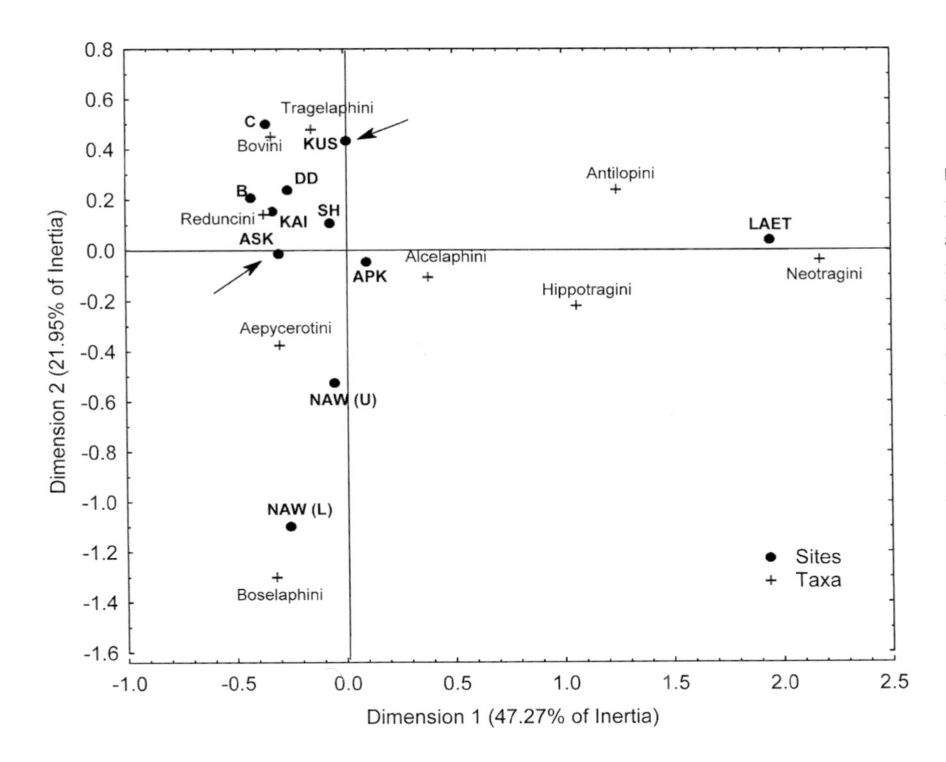

FIGURE 17.4

Results of a correspondence analysis on bovid tribes from the Middle Awash and comparative fossil sites. This is a projection of dimensions 1 and 2 onto a 2D plane. The Asa Koma and Kuseralee Members are highlighted with arrows. Abbreviations: ASK, Asa Koma; KUS, Kuseralee; NAW(L), Lower Nawata; NAW(U), Upper Nawata; APK, Apak; KAI, Kaiyumung; LAET, Upper Laetolil Beds; SH, Sidi Hakoma; DD, Denen Dora; B, Shungura B; C, Shungura C.

RELATIVE ABUNDANCES OF INDICATOR TAXA Relative abundances of indicator taxa other than bovids were also calculated. The indicator taxa used include bovids, giraffids, suids, and cercopithecids. Relative abundance data for these groups are presented in Table 17.7. Correspondence analysis was performed on the count data for indicator taxa from the Asa Koma and Kuseralee Members and those from the comparative Miocene and Plio-Pleistocene sites. Most of the inertia was explained by dimension 1 (27.4 percent) and dimension 2 (25.8 percent). The Asa Koma and Kuseralee Members cluster with other late Miocene localities (i.e., Lower Nawata, Upper Nawata, and Apak), exclusive of any Plio-Pleistocene localities (Figure 17.5). This suggests that Miocene faunal communities may be different from those of the Plio-Pleistocene in terms of the relative abundances of indicator taxa and may point to a shift in ecological conditions from the late Miocene to the Plio-Pleistocene.

To explore this further, the dietary adaptations of indicator taxa were examined (see Table 17.3 for dietary adaptations). In both the Asa Koma and Kuseralee Members, the relative abundance of browsers was higher than that of grazers (Figure 17.6). Mixed feeders were the most abundant animals in both members. There was also a high abundance of grazers that specialize in edaphic grass. Since browsers and grazers are adapted for consuming different types of vegetation (i.e., leaves versus grass) their relative proportions can be used as an indication of the predominant vegetation type. The results suggest that leafy/woody vegetation was more abundant in both the Asa Koma and Kuseralee Members than was grass.

TABLE 17.7 Relative Abundances (Percent) of Indicator Taxa from the Asa Koma and Kuseralee Members

	Asa Koma (%) (n = 54)	*Kuseralee (%) (n = 77)*
Cercopithecinae	7.4	5.2
Colobinae	3.7	6.5
Theropithecus	0.0	0.0
Tragelaphini	7.4	10.4
Bovini	5.6	7.8
Boselaphini	16.7	2.6
Hippotragini	0.0	0.0
Reduncini	11.1	7.8
Alcelaphini	0.0	0.0
Aepycerotini	0.0	1.3
Antilopini	1.9	3.9
Neotragini	3.7	0.0
Giraffa	0.0	0.0
Sivatherium	0.0	0.0
Palaeotragus	0.0	0.0
Nyanzachoerus	40.7	53.2
Notochoerus	0.0	0.0
Potamochoerus	0.0	0.0
Phacochoerus	0.0	0.0
Metridiochoerus	0.0	0.0
Kolpochoerus	0.0	0.0
Cainochoerus	1.9	1.3

NOTE: Indicator taxa include cercopithecids, bovids, giraffids, and suids. Only M3s (uppers and lowers) were included in the counts that resulted in these abundance data.

The relative proportions of browsers and grazers from the Asa Koma and Kuseralee Members were compared to those of the comparative Miocene and Plio-Pleistocene sites (see Figure 17.7). With the exception of the Upper Nawata, late Miocene localities generally have greater abundances of browsers than grazers, whereas Plio-Pleistocene localities have greater abundances of grazers than browsers. This may be indicative of a shift in the relative proportion of vegetation types (wood versus grass) from the late Miocene to the Plio-Pleistocene, evolution of mammals in the ecosystem, or a combination of these and other taphonomic factors. It has been suggested that there was a global replacement of C_3 plants by C_4 plants from the Miocene to the Plio-Pleistocene that was caused by a decrease in global atmospheric CO_2 level (Cerling et al. 1997b). However, recent analyses on topographic effects on eastern African climate suggest that the uplift of the East African Rift System led to changes in atmospheric circulation patterns (i.e., rainfall patterns and amounts) that, in turn, brought about shifts in vegetation patterns, which resulted in the increase of grassland habitats during the Plio-Pleistocene of Africa (Sepulchre et al. 2006). Whatever the cause might be, both analyses agree on the shift from C_3 plants to C_4 plants from the late Miocene to the Plio-Pleistocene of Africa.

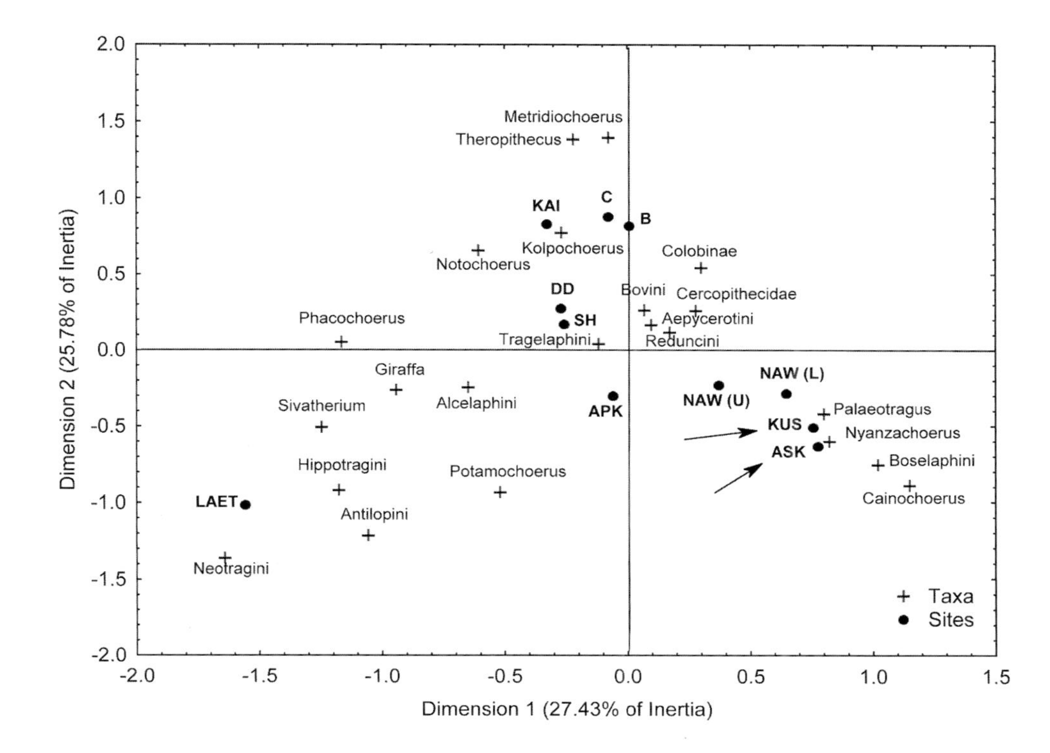

FIGURE 17.5

Results of a correspondence analysis on indicator taxa from the Middle Awash and comparative fossil sites. This is a projection of dimensions 1 and 2 onto a 2D plane. The Asa Koma and Kuseralee Members are highlighted with arrows. Abbreviations: ASK, Asa Koma; KUS, Kuseralee; NAW(L), Lower Nawata; NAW(U), Upper Nawata; APK, Apak; KAI, Kaiyumung; LAET, Upper Laetolil Beds; SH, Sidi Hakoma; DD, Denen Dora; B, Shungura B; C, Shungura C.

Bovid Ecomorphology

Method

Ecomorphology uses functional morphology as a means to understand dietary and locomotor adaptations and the way these adaptations are linked to specific habitats (Kappelman et al. 1997). The assumption is made that the morphology of an animal is related to the habitat in which it lives. Thus, by studying the morphology of living forms, hypotheses about how environment has affected the morphology of the extant organism can be extended to extinct forms. For postcranial remains, an organism's locomotor anatomy is adapted for movement across the substrate. This is particularly true for those animals that rely on running to escape predators, such as bovids. Ecomorphological methods have several advantages over taxon-based methods in reconstructing paleoenvironmental conditions: (1) Habitat preference is not assumed to have stayed the same over time, and (2) taxonomic identification is not necessary. This allows for the use of postcranial material that is often unassigned, or unassignable, to species but that can provide clues to the dietary and locomotor behavioral ecology of fossil animals. However, it is important to keep in mind that taxonomic affinity may influence certain morphological expressions, such that no method, including ecomorphology, is truly "taxon-free."

Bovid astragali and phalanges from the Asa Koma and Kuseralee Members were used for the ecomorphological analysis presented here. These are useful bones for ecomorphological analyses because they interact with the substrate directly and should reflect any differences in the environment. We used measurements and a discriminant function based

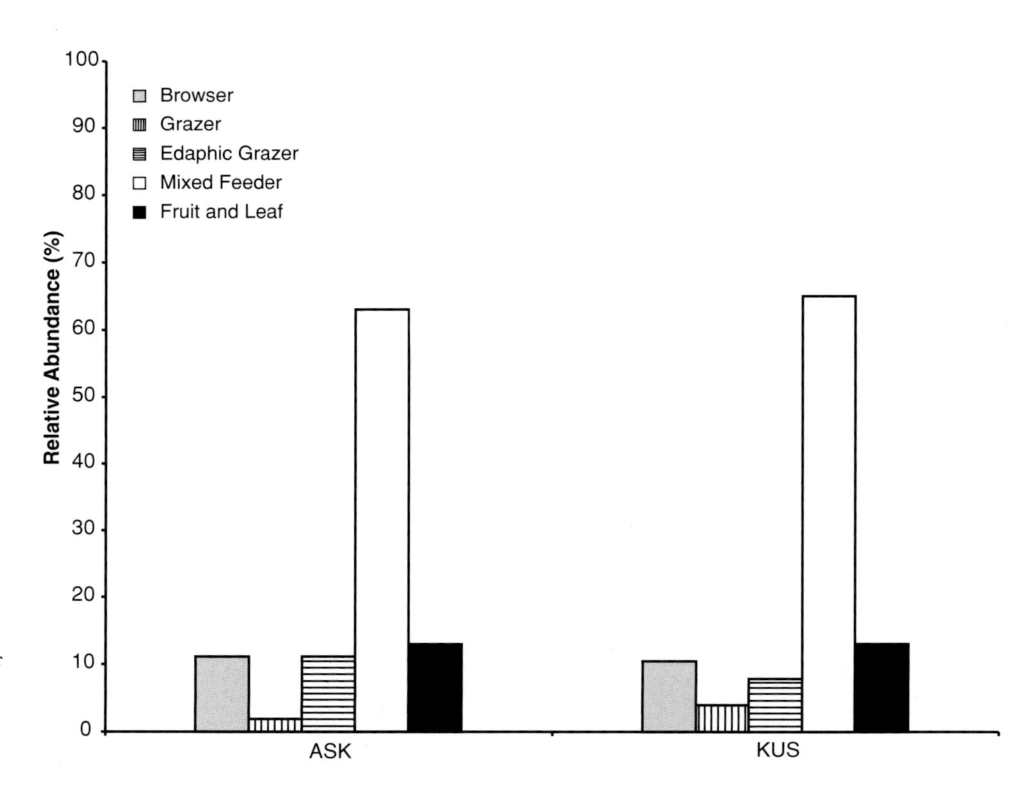

FIGURE 17.6

Frequencies (percent) of the diets of indicator taxa from the Asa Koma and Kuseralee Members.

on a large modern African bovid sample to predict general habitat preferences. Detailed descriptions of the modern bovid sample and the measurements taken, as well as the discriminant function used, can be found in DeGusta and Vrba (2003, 2005a, b).

Results

Ecomorphological analysis was conducted on bovid astragali and phalanges recovered from the Asa Koma and Kuseralee Members. Sixty specimens of astragali and phalanges were measured. Of the specimens analyzed, 29 (48 percent) were classified as "heavy cover," 17 (28 percent) were classified as "forest," and 8 (13 percent) were classified as "open." The confidence level for 37 specimens was 95 percent or higher, and of those 37 specimens, more than half ($n = 25$) were classified as "heavy cover," and the remaining specimens ($n = 10$) were classified as "forest." The results of the ecomorphological analysis suggest that the bovids from Asa Koma and Kuseralee Members were drawn from a predominantly "heavy cover" environment. Since "heavy cover" refers to bush, woodland, swamp, and near-water habitats, this agrees well with the results of the previous analyses.

Other Lines of Evidence

We have explored various aspects of the large mammalian faunal community from the Asa Koma and Kuseralee Members. Our results, based on multiple data sets, suggest that the paleoenvironment of the late Miocene deposits of the Middle Awash was one of densely wooded habitat proximal to permanent sources of water (i.e., lakes or rivers), with swampy

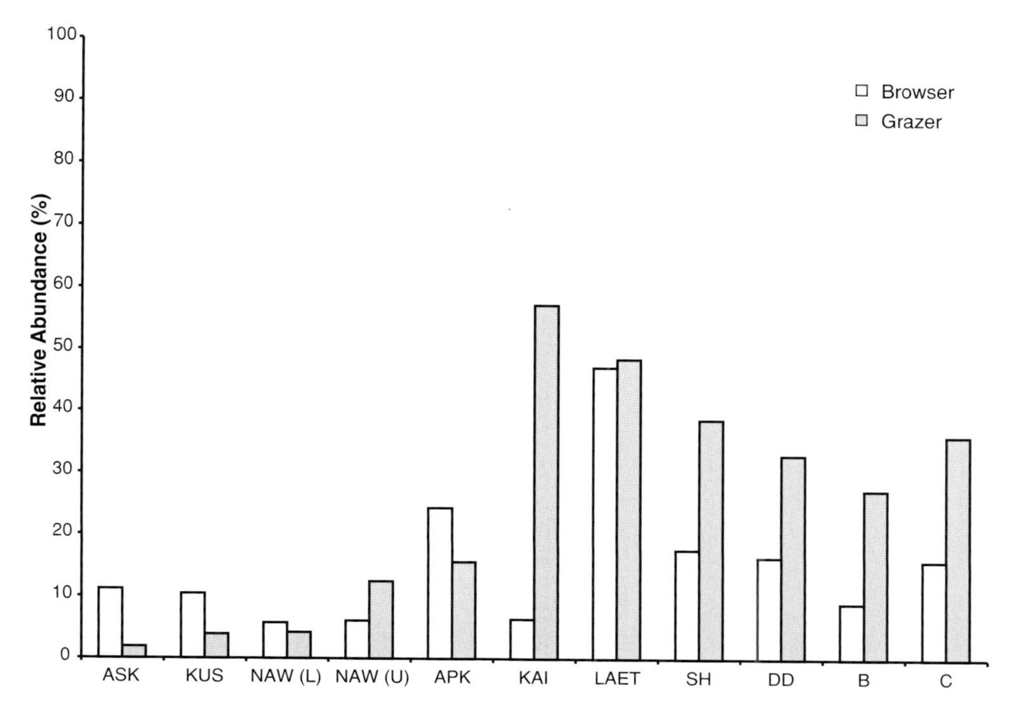

FIGURE 17.7

Frequencies (percent) of browsing versus grazing indicator taxa for Middle Awash localities and comparative fossil sites. Abbreviations: ASK, Asa Koma; KUS, Kuseralee; NAW(L), Lower Nawata; NAW(U), Upper Nawata; APK, Apak; KAI, Kaiyumung; LAET, Upper Laetolil Beds; SH, Sidi Hakoma; DD, Denen Dora; B, Shungura B; C, Shungura C.

conditions and floodplain grasslands. In order to refine and support these conclusions, it is necessary to examine other lines of evidence that would provide information on the paleoenvironment at the Asa Koma and Kuseralee Members. In order to do this, we will present and discuss results from studies on the geology, stable isotopes, and other fauna (small mammals, birds, and fish).

Geology

Geological studies can provide direct evidence for the physical makeup of the paleolandscape. The sediments in the late Miocene deposits of the Middle Awash are composed mostly of lacustrine, freshwater deltaic, fluvial, and terrestrial facies, interbedded with basaltic and silicic ash fallout, diatomites, and lava flows (Haile-Selassie et al. 2004c; WoldeGabriel et al., Chapter 2). The lacustrine and fluvial units indicate that lakes and river systems were the focus of deposition of the late Miocene fossiliferous sediments. The Asa Koma Member is dominated by thick basaltic tuffs interbedded with fluvial sediments (WoldeGabriel et al. 2001; Haile-Selassie et al. 2004c), and the Kuseralee Member is composed of lacustrine and fluvial sediments interbedded with bentonite layers (Renne et al. 1999; Haile-Selassie et al. 2004c). This implies that both rivers and lakes were present during the time of deposition of the Asa Koma and Kuseralee Members. Furthermore, basaltic tephra show evidence of the presence of external water sources prior to and/or during volcanic eruptions, indicating a wet environment with abundant surface water and groundwater (see Hart et al., Chapter 3).

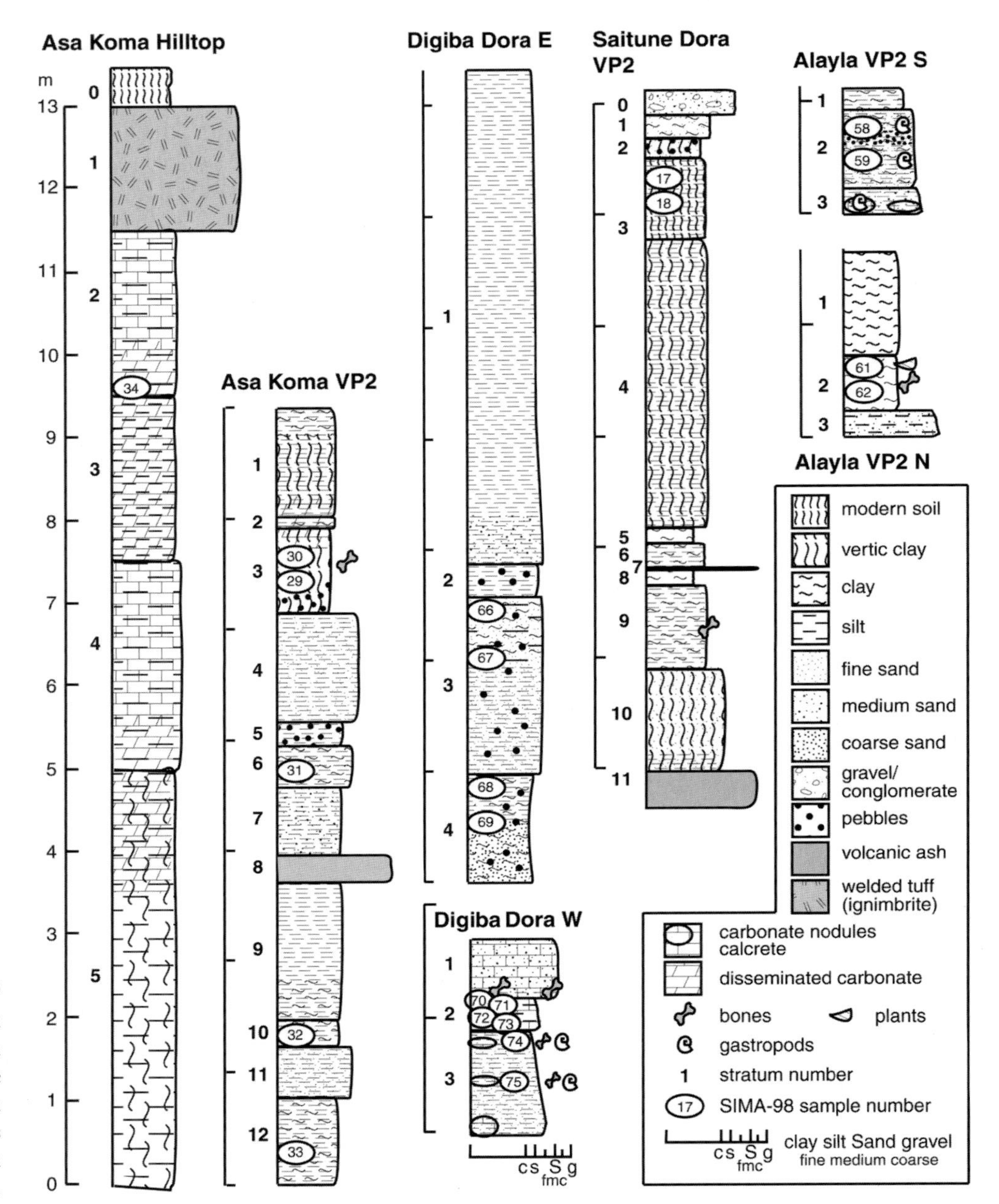

FIGURE 17.8
Stratigraphic sections in the Adu-Asa Formation samples for stable isotope analysis. Numbers in circles show the positions of samples listed in Table 17.7.

Stable Isotopes

Analyses of stable isotopes represent important independent evidence for paleoenvironmental reconstructions (Cerling 1984; Haile-Selassie et al. 2004c; Quade and Levin, in press). Stratigraphic sections were excavated at hominid localities. Samples for isotopic analysis were collected where paleosols, carbonate nodules, or both were observed (see Figure 17.8). At Amba, carbonates adhering to fossils were analyzed as well. Sedimentary matrix samples for isotopic analysis were examined with a stereomicroscope at 10–20× magnification to determine soil micromorphology and the presence of microfossils. Pedogenesis (soil formation) is recognized by root traces, rhizoliths (calcite infillings or replacement of plant roots), and a poorly sorted matrix riddled with tiny voids, often coated with clays or minerals. In clayey soils, polygonal fractures from shrinking and swelling with wetting and drying were present. Most localities in the Asa Koma Member have weakly developed paleosols, and some sampled sections were probably lacustrine to perilacustrine depositional environments. Digiba Dora W levels 2 and 3 contain fish bones and snail opercula, ruling out terrestrial soil formation (Figure 17.8). Saitune Dora, Alayla, and Digiba Dora are mainly nonpedogenic silts and clays. ASK-VP-2 has weak soil development with small carbonate nodules and root traces. The relatively coarse and poorly sorted sediments at ASK-VP-2 were only slightly altered by incipient pedogenesis. Rapid sedimentation may have precluded strong pedogenesis. The Asa Koma Hilltop section has a >6 m thick series of red-brown paleosols with abundant rhizoliths (Figure 17.8), probably reflecting a dry but densely vegetated bushland. Vertebrate fossils are not present in this section. The rarity of pedogenic carbonates suggests mesic to humid climates, because carbonates form mainly under arid to semiarid conditions with rainfall less than 1,000 mm (Retallack 2001). The presence of carbonate skins on fossils at Amba reflects intermittent drying after burial (Pustovoytov 2003).

Soil organic matter occurred in trace amounts (<0.05 percent organic carbon) in these samples, and all were substantially more enriched in ^{13}C than expected from carbonates from the same levels (Table 17.8, Figure 17.9A). Organic carbon isotopes may have been altered by diagenesis, by contamination, or during sample preparation. Our environmental reconstruction thus relies mainly on carbonate carbon and oxygen isotope ratios.

STABLE CARBON ISOTOPES Pedogenic carbonate carbon isotopes are useful in providing an indication of the vegetation type. Low $\delta^{13}C$ values (−10 ‰ to −12 ‰) suggest a habitat with mainly C_3 vegetation (i.e., trees, shrubs, and other dicotyledonous plants). High $\delta^{13}C$ values (>0 ‰) are indicative of a habitat with mainly C_4 vegetation (i.e., tropical grassland) (Cerling 1984; Cerling and Quade 1993). Low $\delta^{13}C$ values at the Asa Koma and Kuseralee localities (−5 ‰ to −7 ‰) suggest the presence of woodland to grassy woodland habitats (Figure 17.9B; Haile-Selassie et al. 2004c). Carbon isotope ratios from the Kuseralee Member at Amba East are slightly higher than those from Asa Koma localities (Figure 17.9B), suggesting a somewhat more open habitat.

STABLE OXYGEN ISOTOPES Pedogenic oxygen isotopes provide evidence of past temperatures and humidity levels. Higher $\delta^{18}O$ values indicate higher temperatures and evaporation rates, as well as lower humidity (Cerling 1984; Cerling and Quade 1993). Asa Koma localities have low $\delta^{18}O$ values (Figure 17.9B), suggesting that the habitat was relatively humid, shady, or cool, with low evaporation rates. This is also consistent with formation at a relatively high altitude. Localities from the Kuseralee Member have higher $\delta^{18}O$ values (Figure 17.9B), indicating that a drier, warmer, low-elevation habitat was present (Haile-Selassie et al. 2004c).

TABLE 17.8 Carbon and Oxygen Isotope Ratios of Pedogenic Carbonates and Organic Matter from Stratigraphic Sections in the Adu-Asa Formation

Site/level	*SIMA Field #*	*MAW Lab #*	*Org* $\delta^{13}C$‰	*MAW Lab #*	*SIMA Field #*	*Carb* $\delta^{13}C$‰	*Carb* $\delta^{18}O$‰	δ^{13} *Ccarb–org*	*%C_4 carb*	*%C_4 org*
Issa Dibo, Level 1	SIMA-98-23	136	–22.69	211	SIMA-98-23	–8.30	–7.24	14.39	19.3	16.5
Issa Dibo, Level 2	SIMA-98-24	137	–21.10	212	SIMA-98-24	1.08	**7.20**	22.18	86.3	27.8
Issa Dibo, Level 2				213	SIMA-98-24	–8.32	–6.79		19.2	
Issa Dibo, Level 3	SIMA-98-25	138	–21.25	214	SIMA-98-25	–7.73	–7.50	13.52	23.4	26.8
Issa Dibo HMS, Level 2	SIMA-98-53	166	–16.84	235	SIMA-98-53	–5.38	–4.37	11.47	40.2	58.3
Issa Dibo HMS, Level 3	SIMA-98-54	167	–15.81	236	SIMA-98-54	–1.41	1.88	14.40	68.5	65.7
Issa Dibo HMS, Level 4	SIMA-98-55	168	–17.61	237	SIMA-98-55	–2.66	–1.37	14.95	59.6	52.8
Issa Dibo HMS, Level 5	SIMA-98-56	169	–17.41	238	SIMA-98-56	–2.18	–0.99	15.23	63	54.2
Saitune Dora VP2, Level 3	SIMA-98-17	130	–16.42							61.3
Saitune Dora VP2, Level 3	SIMA-98-18	131	–14.52							74.9
Asa Koma VP2, Level 3	SIMA-98-29	142	–13.88							79.4
Asa Koma VP2, Level 3	SIMA-98-29	150	–14.13							77.6
Asa Koma VP2, Level 3	SIMA-98-30	143	–14.09							77.9
Asa Koma VP2, Level 6	SIMA-98-31	144	–13.85							79.6
Asa Koma VP2, Level 6	SIMA-98-31	151	–13.62							81.3
Asa Koma VP2, Level 10	SIMA-98-32	145	**–12.06**							92.4
Asa Koma VP2, Level 12	SIMA-98-33	146	–12.93							86.2
Asa Koma Hilltop				217	SIMA-98-34	–7.06	–7.45		28.1	
Asa Koma Hilltop	SIMA-98-34	147	–16.59	216	SIMA-98-34	–5.28	–8.10	11.31	40.9	60.1
Alayla VP2 S, Layer 2	SIMA-98-58	172	–14.03							78.4
Alayla VP2 S, Layer 2	SIMA-98-59	173	–14.93							71.9
Alayla VP2 N, Level 2	SIMA-98-61	174	–18.62							45.6
Alayla VP2 N, Level 2	SIMA-98-62	175	–18.57							45.9
Digiba Dora E, Level 3 up	SIMA-98-66	178	–13.08							85.1
Digiba Dora E, Level 3 mid	SIMA-98-67	179	–13.63							81.3
Digiba Dora E, Level 4 up	SIMA-98-68	180	–16.16							63.1
Digiba Dora E, Level 4 mid	SIMA-98-69	181	–16.45							61.1
Digiba Dora W, Level 2 up	SIMA-98-70	182	–16.41	240	SIMA-98-70	–6.45	–9.06	10.03	32.5	60.9
Digiba Dora W, Level 2 mid	SIMA-98-71	183	–16.02	241	SIMA-98-71	–6.91	–7.17	9.11	29.2	64.2
Digiba Dora W, Level 2 low	SIMA-98-72	184	–16.36	242	SIMA-98-72	–7.12	–8.45	9.24	27.7	61.7
Digiba Dora W, Level 2 base	SIMA-98-73	185	–15.57	243	SIMA-98-73	–7.45	–8.35	8.12	25.4	67.3
Digiba Dora W, Level 3 up	SIMA-98-74	186	–16.81	187	SIMA-98-74	–6.64	–7.06	10.18	31.2	58.5
Digiba Dora W, Level 3 mid	SIMA-98-75	188	–14.80	189	SIMA-98-75	–6.46	–6.08	8.337	32.4	72.9
Digiba Dora W, Level 3 low	SIMA-98-76	190	–15.73	244	SIMA-98-76	–4.07	–7.93	11.66	49.5	66.2
Amba (hominid)				250	SIMA-99-1	–5.17	–4.12		41.6	
Amba (monkey)				251	SIMA-99-2	–6.35	–4.26		33.2	

NOTE: Samples listed in bold type may have modern organic matter contamination, based on high organic carbon content.

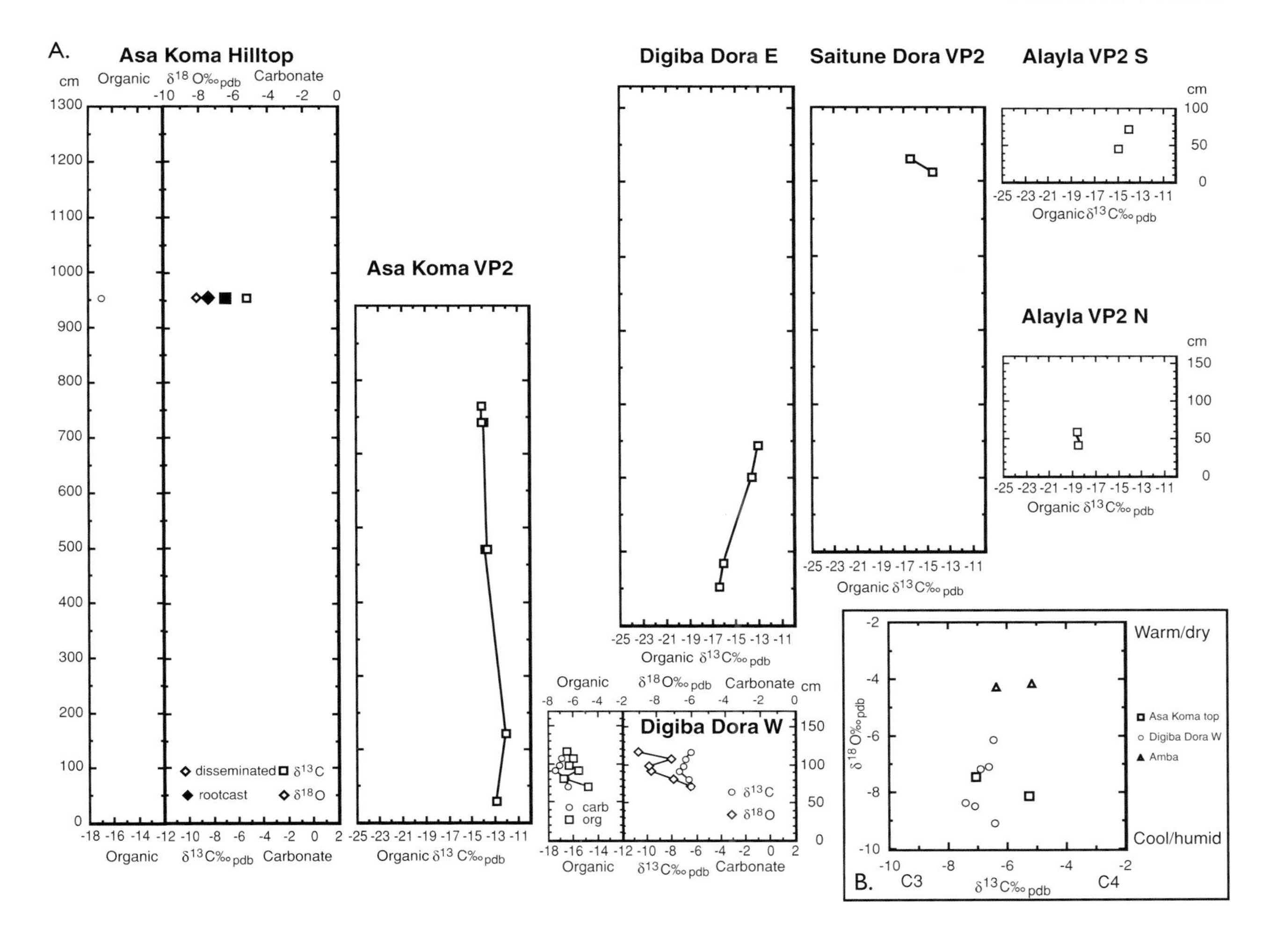

FIGURE 17.9

A. Stable carbon and oxygen isotope ratios of carbonates (carb) and organic matter (org) in stratigraphic sections in the Adu-Asa Formation. B. Bivariate plot of carbonate carbon and oxygen isotope ratios.

Small Mammals

Small mammals are recognized to be particularly useful in paleoecological studies, mostly because many extant small mammals are habitat-specific and tend to spend most of their lives in a small home range (Wesselman 1984, 1995; Denys 1999; Wesselman et al., Chapter 5).

Fourteen genera of micromammals have been recognized from the late Miocene localities of the Middle Awash (see Wesselman et al., Chapter 5). Among them are two species of *Crocidura* (shrew) and a species of *Atherurus* (tree porcupine), both of which may suggest the presence of forests and mesic woodlands supported by an abundance of available water, either in the form of groundwater or rainfall. The presence of *Thryonomys* (cane rat) indicates that mesic, wet long-grass and marshland aquatic environments were also in close proximity (Wesselman et al., Chapter 5). There are, however, indications that drier, less wooded habitats were in the vicinity. Specimens of *Procavia* (hyrax), *Xerus* (squirrel), *Tatera* (gerbil), *Serengetilagus* (hare), and *Lemniscomys* (zebra mouse) suggest that dry acacia savanna, open savanna woodland, and even scrubland habitats were present. Furthermore, the presence of several taxa of small mammals, primarily *Tachyoryctes* (root rat) and *Lophiomys* (crested rat), suggest that the Middle Awash region was at a much higher elevation during the deposition of the late Miocene sediments, possibly above 2,000 meters (Haile-Selassie et al. 2004c; Wesselman et al., Chapter 5).

Since micromammals are particularly habitat-specific and sensitive to habitat changes, relative abundance data for these taxa can further refine our understanding of the paleoenvironment. Small mammals were collected systematically by sieving at both the Asa Koma and Saitune Dora localities, alleviating any sampling biases. The micromammal fauna is dominated by *Tachyoryctes* (40.9 percent) and *Thryonomys* (32.9 percent). Extant species of *Tachyoryctes* are most commonly found in high-altitude open habitats characterized by deep soils (more than 50 cm) that are seasonally waterlogged or flooded (Kingdon 1974; Sillero-Zubiri et al. 1995), such as wet grassland or alluvial plains. The dominance of these two micromammal taxa suggests that there was an abundance of mesic, wet grass and marshland aquatic environments at high elevations during the deposition of the late Miocene sediments. Taxa that indicate drier, less wooded habitats (*Procavia, Xerus, Tatera, Serengetilagus,* and *Lemniscomys*) were more abundant than those that indicate densely wooded habitats (*Crocidura* and *Atherurus*). It was suggested, however, that the dry, open environments they represent were farther away from the site of deposition (Wesselman et al., Chapter 5). This possibility is reinforced by the high proportions of *Thryonomys* specimens in the small mammal assemblage, suggesting that wet environments were abundant during the late Miocene of the Middle Awash.

Birds

The avifauna can help in paleoenvironmental reconstructions, especially when the taxa present have specific habitat preferences. The bird fauna from the late Miocene deposits of the Middle Awash has been studied by Louchart et al. (2008).

Thirteen bird taxa have been described from the Asa Koma Member (see Table 1.3). They comprise mostly aquatic birds that depend on open bodies of water with fish and aquatic vegetation (Louchart et al. 2008). Specimens of *Podiceps* (grebe), *Phalocrocorax* (smaller cormorant), *Anhinga* (darter), anatids (duck), and *Pandion* (osprey) indicate a lake or slow river in the vicinity. *Ardea* sp. (heron), found in swamps, marshes, humid grasslands, and river or lake edges with patches of open vegetation, is also present in the Asa Koma Member (Louchart et al. 2008). The presence of these birds suggests a river or lake with floodplain grassland and swamps or marshes. One species of land bird is represented from the Asa Koma Member, based on a single specimen of *Francolinus* sp. (francolin) recovered from Alayla Vertebrate Locality 2 (ALA-VP-2) (Louchart et al. 2008). This is also the only locality from which species that are not strictly water-adapted were recovered (i.e., falconiformes and ardeids). This may indicate that Alayla may have been sampling parts of the paleoenvironment less dominated by aquatic regimes (Louchart et al. 2008). However, this would have no ecological significance if it were a taphonomic effect of the much larger sample size at ALA-VP-2 ($n = 351$) compared to most Middle Awash late Miocene localities. To examine this more closely, the avian fauna from ALA-VP-2 was compared to that of Asa Koma Vertebrate Locality 3 (ASK-VP-3). The faunal assemblage from ASK-VP-3 (n = 542) was larger than that of ALA-VP-2. Of the eight identified bird specimens from ASK-VP-3, only one was assigned to a taxon that was not strictly water-adapted (i.e., cf. Ardeidae). Of the five identified bird specimens from ALA-VP-2, three were assigned to species that were partially land-adapted (i.e., falconiformes and ardeids), and one was the only land bird identified (i.e., *Francolinus*). It is possible that the prevalence of taxa that are not strictly water-adapted at ALA-VP-2 may indicate the presence of habitats less dominated by aquatic environments at the locality.

Only one bird taxon has been described from the Kuseralee Member. One small Anatinae specimen was recovered from Amba West Vertebrate Locality 1 (AMW-VP-1) (Louchart et al. 2008). Not much ecological information can be gleaned from the presence of this unidentified duck species in the Kuseralee Member, other than that wet conditions must have been present (Louchart et al. 2008).

Fish

We have deduced from the mammalian fauna that a large, permanent body of water was important in the Asa Koma and Kuseralee paleolandscapes. It is possible to speculate on the quality and type of water that was present based on the fish fauna. Specimens of fish were studied only from Asa Koma localities: Adu Dora, Alayla, Asa Koma, Digibi Dora, and Saitune Dora. Preliminary results from studies conducted by Murray and Stewart (unpublished data) indicate that the fish assemblage consists of two taxa of catfish (cf. *Clarias* and cf. *Bagrus*), a minnow (cf. *Barbus*), and a cichlid (Cichlidae, sp. indet., but possibly a member of Tilapiini). Modern representatives of most of these taxa can tolerate all ranges of water conditions and temperatures and are often found in poorly oxygenated waters (Copley 1958; Daget et al. 1986; Lévêque 1997; Murray and Stewart, unpublished data). Therefore, the presence of these fish taxa is not particularly useful as an environmental indicator. However, the absence of taxa that prefer clean, oxygenated open waters may indicate that poor water conditions, such as marshes, were present in the late Miocene waters of the Middle Awash (Murray and Stewart, unpublished data). On the other hand, some extant species of *Barbus* are known to prefer open waters, or waters with a fast current, that are clean and oxygenated (Copley 1958; Daget et al. 1986; Lévêque 1997; Murray and Stewart, unpublished data). This may indicate that better water conditions, such as a large river or inshore waters of lakes, may have been present (Murray and Stewart, unpublished data).

Conclusions

Paleoenvironmental assessment of the vertebrate-bearing late Miocene sediments in the Middle Awash study area was accomplished by examining recovered assemblages of large mammal remains, with particular attention to the overall assemblage compositions, relative abundances of indicator taxa, ecomorphology, and associated contextual evidence.

It was found that the large mammalian fauna from both the Asa Koma and Kuseralee Members was most similar in community structure to modern woodland faunal communities (including riparian, closed, and open woodland communities). To refine this finding further, relative abundances of bovids and other indicator taxa were investigated. The results indicate a greater abundance of bovids associated with densely vegetated habitats (i.e., tragelaphines and bovines) in both members, suggesting that such habitats were dominant in the late Miocene of the Middle Awash area. In addition, both members have high proportions of reduncines, which are often found in floodplain grasslands or swampy areas today. This implies that floodplain grasslands and swampy conditions were also significant. Conversely, the rarity of bovids associated with drier and less wooded conditions suggests that dry, open habitats were not as common in the Asa Koma and Kuseralee Members. Frequencies of browsers and grazers may reflect the proportions of woody/leafy versus grassy vegetation. The greater abundance of browsers and low occurrence of grazers in indicator taxa also suggest that the Asa Koma and Kuseralee Members were dominated by densely vegetated habitats.

Ecomorphological analysis provides another line of paleoenvironmental evidence and the dominance of "heavy cover" bovids implies that heavily wooded habitats, such as riparian and closed woodland, were common in the Asa Koma and Kuseralee Members. Carbon isotope data support these conclusions and suggest that woodland to grassy woodland habitats were present in the Asa Koma and Kuseralee Members (Haile-Selassie et al. 2004c).

As noted above, the high proportions of reduncines in both the Asa Koma and Kuseralee Member assemblages indicate that floodplain grassland or swampy conditions were important aspects of the ecological setting. This implies that a large, permanent water source was present. Results from analyses on the geology, avifauna, and small mammal fauna augment the evidence from the bovid fauna. Geological analyses indicate that a permanent source of water was present at both members in the form of a river system or lake (WoldeGabriel et al. 2001; Haile-Selassie et al. 2004c). An assessment of the bird fauna by Louchart et al. (2008) found a preponderance of aquatic birds in the Middle Awash late Miocene faunal assemblage and therefore emphasizes the significance of water in the region. The presence of *Crocidura* and *Atherurus* in the small mammal fauna (Wesselman et al., Chapter 5) points to a densely wooded habitat that is supported by abundant water. We can speculate on the poor water quality by the lack of fish that inhabit clean, oxygenated open waters (Murray and Stewart, unpublished data), and when combined with the presence of *Ardea* (Louchart et al. 2008) and *Thryonomys* (Wesselman et al., Chapter 5), it implies that marsh- or swamp-like conditions were present.

Differences in the relative abundances of bovid taxa and oxygen isotope compositions between the Asa Koma and Kuseralee Members may reflect subtle ecological distinctions in woodland/floodplain grassland representation and temperature in these two members.

The paleoenvironment of other late Miocene hominid-bearing sites should also be considered. There have been three major discoveries of late Miocene hominids in the last decade: *Ardipithecus kadabba* (Haile-Selassie 2001); *Orrorin tugenensis* (Senut et al. 2001b); and *Sahelanthropus tchadensis* (Brunet et al. 2002). *Orrorin tugenensis* (Tugen Hills, Kenya, 6.2–5.65 Ma) is found with fauna that suggests the presence of open woodland habitats with dense woodland or forest in the vicinity, possibly along lake margins (Pickford and Senut 2001). The presence of hippopotamus, crocodiles, fish, and aquatic snails indicates that a large permanent water source was also in the vicinity (Pickford and Senut 2001).

Sahelanthropus tchadensis (Toros-Menalla, Chad, ~7–6 Ma), is found in association with faunal communities that indicate a mosaic of environments ranging from gallery forest at the edge of a lake to savanna woodland to open grassland (Vignaud et al. 2002). The absence of tragelaphines and boselaphines from the fauna have been interpreted as strong indications of the predominance of open grassland habitats (Vignaud et al. 2002), but the lack of alcelaphines and the presence of reduncines suggest that these were mostly wet grassland habitats. There is also evidence of large and permanent water bodies with some swampy conditions (Vignaud et al. 2002). Detailed paleoenvironmental analyses have not yet been conducted for these two hominid sites, without which it is difficult to assess how their ecological settings might have compared to those documented for the Middle Awash late Miocene. However, the evidence suggests that large, permanent sources of water and wooded environments were in the vicinity of all three hominid sites. The lack of alcelaphines at all three sites may be indicative of the secondary role of arid habitats in contributing to the paleontological assemblages.

One reason to engage in attempts to reconstruct paleoenvironment based on evidence from the fossil record is to establish the kinds of habitats available to early hominids. However, it is only under very rare conditions that hominids can be solidly placed into these various habitats. Given that the small assemblage of *Ardipithecus kadabba* specimens from the Middle Awash was recovered from different localities across the landscape (Alayla Vertebrate Locality 2, Saitune Dora Vertebrate Locality 2, Asa Koma Vertebrate Locality 3, and Digiba Dora Vertebrate Locality 1), conclusions regarding their habitat preferences are difficult to draw. Habitat availability, on the other hand, can be assessed with the data available for the late Miocene of the Middle Awash. The vertebrate faunas as well as the isotopic and geological evidence all indicate that aquatic habitats with large bodies of water were present. These would have been accompanied by riparian woodland and floodplain grassland that supported the terrestrial mammalian communities sampled by the fossil record. Arid savanna environments were unlikely to have been important parts of the overall ecological setting.

18

Paleobiogeography

RAYMOND L. BERNOR, LORENZO ROOK, AND YOHANNES HAILE-SELASSIE

In this study, we conduct comparative biogeographic analyses with the aim of revealing the extent to which the Middle Awash late Miocene localities resemble, at the genus level, well-known eastern and northern African, Arabian, and western Eurasian localities of similar age. The inclusion of Eurasian and African late Miocene localities in this analysis helps us develop knowledge about the first and last occurrences of the taxa considered (see Haile-Selassie et al., Chapter 19), and the geographic areas of these recorded occurrences. It also allows us to consider hypotheses about the timing and direction of paleogeographic connections between western Eurasia and northern and eastern Africa during the late Miocene. Table 18.1 presents a list of all the comparative localities with codes used in the statistical analysis.

Materials and Methods

This research follows recent investigations on evolutionary biogeography of the *Oreopithecus bambolii* "Faunal Zone" by Bernor et al. (2001) and the paleozoogeography of the Rudabánya fauna by Bernor and Rook (2004). As with these previous studies, we follow Fortelius et al. (1996) in undertaking genus-level faunal resemblance index (FRI) studies using both the Simpson's (1943) and the Dice (Sokal and Sneath 1963) algorithms. As previously discussed by Bernor et al. (2001), the Dice FRI is the one most highly recommended by Archer and Maples (1987) and Maples and Archer (1988) and is calculated as $2A / (2A + B + C)$, where A is the number of taxa present in both faunas, B is the number of taxa present in fauna 1 but absent in fauna 2, and C is the number of taxa present in fauna 2 but absent in fauna 1. Simpson's FRI has a long tradition of use (Bernor 1978; Flynn 1986; Bernor and Pavlakis 1987) and is calculated as $A / (A + E)$, where E is the number of taxa in common of the fauna with the fewer number of total genera, B or C. Simpson's FRI is robust and adjusts for differences in sample sizes between pairwise faunas being considered. We follow Bernor et al. (2001) and Bernor and Rook (2004) in plotting the Dice and Simpson's indices together as a heuristic comparison.

Table 18.1 provides a list of Middle Awash localities and selected Eurasian and African localities used in this study. Lukeino (Kenya) and Langebaanweg (South Africa) were not

TABLE 18.1 List of Middle Awash Localities and Selected Eurasian and African Localities Used in This Study

Locality	*Legend*	*Age*
Rudabanya	L1	10.3–10.0
Baltavar	L2	7.5–6.7
Spain MN13	L3	6.7–5.3
OZF zone	L4	9–7.4
Italy MN13	L5	6.7–5.3
Maragheh M&U	L6	8.2–7.4
Pikermi	L7	8.2
Samos Main BB	L8	7.65–7.0
Maramena	L9	6.7–5.3
Baynunah	L10	6.7–5.3
Sahabi	L11	6.7–5.3
Toros Menalla	L12	7.0–6.0
Lothagam Nawata	L13	7.4–5.0
Middle Awash ASK	L14	5.77–5.54
Middle Awash KUS	L15	5.20

included, because of either the small sample size available, the uncertainty in age, or uncertainties in taxonomic identifications. Many of the localities in Table 18.1 are the same analyzed earlier by Bernor et al. (2001) and Bernor and Rook (2004). All faunas have been updated, where applicable, using the current NOW (Neogene Old World) database files, downloaded November 15, 2005 (http://www.helsinki.fi/science/now/). We have further updated these NOW files to reflect our most current understanding of the systematics of all the mammalian groups and faunas used in our analysis (e.g., the Fiume Santo assemblage, Abbazzi et al. 2008). Our analyses here consider only large mammals because small mammal records vary greatly in their size and diversity across the faunas considered as a result of taphonomic and recovery influences.

We have three regional clusters of sites within our analyzed faunal database. Fauna L3 (Spain MN 13 cluster) includes the following assemblages: Venta del Moro, El Arquillo, Las Casiones, and Layna. Fauna L4 (OZF zone) includes faunas of the endemic Tusco-Sardinian paleobioprovince: Baccinello V1, Baccinello V2, Montebamboli, Casteani, and Fiume Santo. The Fauna L5 (Italy MN 13) cluster includes the Baccinello V3, Brisighella, Ciabot Cagna, and Borro Strolla assemblages. For the chronology of European MN (Mammal Neogene) biochronologic units, we follow Steininger et al. (1996). We further incorporate recent updated recommendations and observations made by Bernor et al. (2004, 2005a) for the Vallesian based on recent detailed analyses of the Rudabánya fauna, and Kostopoulos et al. (2003) for the Turolian based on recent magnetostratigraphic correlations of the Samos vertebrate biostratigraphic sequence. For Samos, we incorporate Quarry 4 (White Sands unit) into the Main Bone Bed sample, since there is so little time represented in this sample (7.65–7.0 Ma), and it is so well-documented in the American Museum of Natural History (AMNH) collections. The Lothagam Nawata Member's chronology follows Feibel (2003) and McDougall and Feibel (2003).

In undertaking these analyses, we wish to emphasize that they are entirely dependent on published taxonomic studies. During the last 10 years there have been focused studies relevant to this analysis made by Bernor et al. (1996a,b,d), by Fortelius et al. (2003b), by Leakey and Harris (2003b), and in this volume, as well as the ongoing revisions of the NOW Database (M. Fortelius, Director), which have greatly improved systematic and chronologic resolution of Old World Neogene faunas, including the ones we use in this analysis. Although we have made every effort to recognize viable synonyms, it would be naïve to believe that we have recognized all of them. It would be even more naïve to believe that others will not be recognized with subsequent systematic work.

Analysis

Figure 18.1 provides our analysis of the Middle Awash late Miocene faunas as they compared to the set of 13 other Eurasian and African large mammal faunas: Figure 18.1A is the Asa Koma Member (ASKM) faunal analysis, and Figure 18.1B is the Kuseralee Member (KUSM) faunal analysis.

The ASKM analysis (Figure 18.1A) reveals its closest resemblance to KUSM, with 25 large mammal genera in common, Dice = 0.746 and Simpson's = 0.758. The next closest relationship is clearly to Lothagam Nawata, with 20 taxa in common, Dice = 0.556 and Simpson's = 0.588. The next closest assemblages to ASKM include Toros Menalla, Sahabi, and Baynunah, although there is a sharp drop-off in biogeographic resemblance for each of these faunas. Toros Menalla has seven genera in common, Dice = 0.269 and Simpson's = 0.389. Sahabi has eight genera in common, Dice = 0.258 and Simpson's = 0.286. Baynunah has six genera in common, Dice = 0.250 and Simpson's = 0.429 (high because of Baynunah's small sample size). Spanish and Italian MN 13 have the next highest clusters but are decidedly more distant biogeographically: Spain has six genera in common, Dice = 0.148 and Simpson's = 0.176; Italy has five genera in common, Dice = 0.167 and Simpson's = 0.192. Pikermi, Maragheh M and U, and Samos are the next most distant cluster: Pikermi has five genera in common, Dice = 0.118 and Simpson's = 0.147; Maragheh M and U have five genera in common, Dice = 0.141 and Simpson's = 0.147; Samos also has five genera in common, with Dice = 0.127 and Simpson's = 0.147. Baltavar (genera in common = 2, Dice = 0.080; Simpson's = 0.125), Maramena (genera in common = 1, Dice = 0.040, Simpson's = 0.063) and Rudabánya (genera in common = 1, Dice = 0.032 and Simpson's = 0.034) have weak resemblances. The endemic Tusco-Sardinian Province OZF zone faunas have 0 genera in common and 0.000 indices for both Dice and Simpson's.

In our analyses (Figure 18.1B), the two Middle Awash members, the Kuseralee and the Asa Koma, show the closest resemblances, with 25 large mammal genera in common. For the KUSM assemblages, the next highest resemblance is to the Lothagam Nawata, having 21 large mammal genera in common, with Dice = 0.592 and Simpson's = 0.636. The African localities of Toros Menalla and Sahabi exhibit less resemblance. Toros Menalla shares nine genera in common with KUSM, Dice = 0.353 and Simpson's = 0.500. Sahabi has nine genera in common with KUSM, Dice = 0.295 and Simpson's = 0.321. Baynunah shares four genera in common with the KUSM fauna and has a somewhat

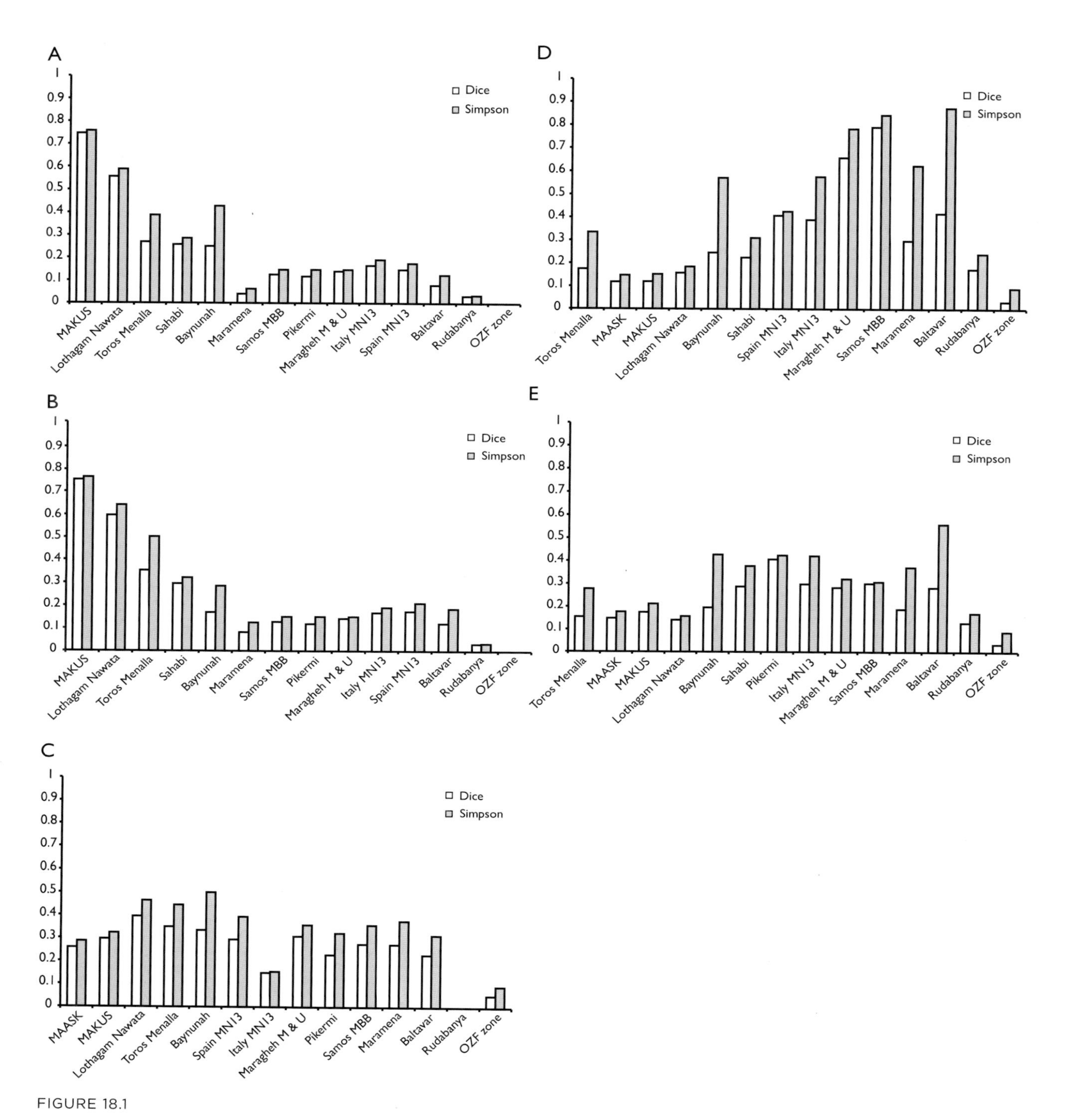

FIGURE 18.1

Genus faunal resemblance sets. A. Comparison of Middle Awash Asa Koma Member with selected Eurasian and African late Miocene large mammal faunas. B. Comparison of Middle Awash Kuseralee Member with selected Eurasian and African late Miocene large mammal faunas. C. Comparison of Sahabi to selected Eurasian and African late Miocene large mammal faunas. D. Comparison of Pikermi to selected Eurasian and African late Miocene large mammal faunas. E. Comparison of Spain MN 13 sites to selected Eurasian and African late Miocene large mammal faunas.

higher Simpson's (0.286) than Dice (0.170) index. Of the Western Eurasian localities, the KUSM fauna exhibits the greatest number of taxa in common with the Spanish MN 13 fauna: total of seven genera in common, Dice = 0.175 and Simpson's = 0.212. The KUSM fauna next closely groups with the Italian MN 13 faunas, with five genera in common, Dice = 0.169 and Simpson's = 0.192. Pikermi has five genera in common with KUSM fauna, Dice = 0.119 and Simpson's = 0.152. Maragheh M and U have five genera in common, Dice = 0.143 and Simpson's = 0.152. Samos also has five genera in common, but with a lower Dice = 0.128 and Simpson's 0.152. Baltavar has 3 genera in common, Dice = 0.122 and Simpson's = 0.188 (relatively high because of Baltavar's low sample size). Rudabánya has a very low resemblance with one genus in common, Dice = 0.032 and Simpson's = 0.034. The Tusco-Sardinian OZF endemic zone has no taxa in common, with Dice and Simpson's both equaling 0.000.

Overall, the Middle Awash latest Miocene faunas exhibit their strongest relationships with each other. The next strongest relationship for both is with Lothagam Nawata, with its relationship to the KUSM fauna slightly greater than with the ASKM fauna. This is likely due to sampling. The next closest relationships are to Toros Menalla and Sahabi, again with KUSM being closer to these than ASKM. Baynunah is slightly more distant still. Maragheh M and U, Italy MN 13, and Spain MN 13 are the next closest related localities. Maramena, the Hungarian faunas, and the OZF Tusco-Sardinian faunas have very distant biogeographic relationships.

We have undertaken three additional plots comparing Sahabi (Figure 18.1C), Pikermi (Figure 18.1D), and the Spanish MN 13 cluster (Figure 18.1E) to the entire sample under consideration to gain further insights into the relationships of northern African and western Eurasian and Arabian late Miocene mammal faunas to the eastern African faunas under consideration.

The Sahabi comparison is significant because it reaffirms the claim by Bernor and Pavlakis (1987) that it is a true biogeographic "crossroads" fauna between Eurasia, Arabia, and eastern Africa. Sahabi exhibits Dice resemblances above 0.300 for KUSM (genera in common = 9; Dice = 0.295; Simpson's = 0.321), Lothagam Nawata (genera in common = 13; Dice = 0.394; Simpson's = 0.464), Toros Menalla (genera in common = 8; Dice = 0.348; Simpson's = 0.444), Baynunah (genera in common = 7; Dice = 0.333; Simpson's = 0.500, high due to small sample size), and Maragheh M and U (genera in common = 10; Dice = 0.308; Simpson's = 0.357). Faunas with Dice between 0.200 and 0.300 include Spain MN 13 (genera in common = 11; Dice = 0.293; Simpson's = 0.393), ASKM (genera in common = 8; Dice = 0.258; Simpson's = 0.286), Pikermi (genera in common also 10; Dice = 0.253; Simpson's = 0.357), Samos (genera in common = 9; Dice = 0.228; Simpson's = 0.321), Maramena (genera in common = 6; Dice = 0.273; Simpson's = 0.375), and Baltavar (genera in common = 5; Dice = 0.227; Simpson's = 0.313). The Italian MN 13 faunas have four genera in common, with Dice (0.148) and Simpson's (0.154) resemblances being low. The OZF zone has one genus in common (Dice = 0.051; Simpson's = 0.091), whereas Rudabánya has no genera in common (Dice and Simpson's both equal 0.000). Our analysis shows that Sahabi's biogeographic relationship is slightly closer to the ASKM-Lothagam Nawata-Toros Menalla-Baynunah-Maragheh M and U cluster (especially for Simpson's faunal resemblance index) than to the KUSM-Spain

MN 13-Pikermian (Pikermi and Samos) MN 11–12 faunas and Maramena (MN 13) fauna. The analysis further suggests that the biogeographic relationship to the western Eurasian faunas is likely embedded in a late MN 11 Turolian dispersion of classical Pikermian faunas circa 8 Ma.

The Pikermi comparison exhibits its highest resemblances with Samos (genera in common = 38; Dice = 0.792; Simpson's = 0.844) and Maragheh M and U (genera in common = 29; Dice = 0.659; Simpson's = 0.784). The next closest relationships are with Spain MN 13 (genera in common = 20; Dice = 0.408; Simpson's = 0.426), Italy MN 13 (genera in common = 15; Dice = 0.390; Simpson's = 0.577), Maramena (genera in common = 10; Dice = 0.299; Simpson's = 0.625, high due to the small sample size), and Baltavar (genera in common = 14; Dice = 0.418; Simpson's = 0.875, high due in part to the low species diversity). Baynunah (genera in common = 8; Dice = 0.246; Simpson's = 0.571) and Sahabi (genera in common = 9; Dice = 0.225; Simpson's = 0.310) are biogeographically less similar. Toros Menalla (genera in common = 6; Dice = 0.174; Simpson's = 0.333), ASKM (genera in common = 5; Dice = 0.118; Simpson's = 0.147), KUSM (genera in common = 5; Dice = 0.119; Simpson's = 0.142), and Lothagam Nawata (genera in common = 7; Dice = 0.157; Simpson's = 0.184) are biogeographically distant from the classical Pikermian faunas. Considering that Rudabánya is both the oldest locality analyzed and biogeographically distinct (Bernor et al. 2004, 2005a), it has a large number of Pikermian taxa (genera in common = 7), with Dice = 0.175 and Simpson's = 0.241. The OZF zone has only one genus in common, and the lowest indices: Dice = 0.032 and Simpson's = 0.091.

In many regards, the Pikermian fauna, characterized most accurately by the Maragheh-Pikermi-Samos triad, contains taxa that were apparently ancestral to many of the northern and eastern African faunas considered here. Pikermi is correlated with Middle Maragheh, ca. 8.2 Ma (Bernor et al. 1996d) and is evidently the time of great intercontinental dispersion of large carnivores and ungulates. A sizable number of genera are shared among western Eurasian Pikermian faunas, European MN 13 faunas, and the African data set. The relationships to Sahabi and Baynunah are less close than those with many of the European and western Asian faunas under consideration, and the relationship to the central and eastern African faunas is even more distant. In most cases we believe that the African faunas under consideration had "Pikermian" elements that were vicariant and evolved independently subsequent to an early-middle Turolian extension. We will develop this aspect of our study later in this chapter.

The Spanish MN 13 cluster comparison did not yield as high resemblance values as did the Pikermi analysis. The highest resemblance for Spanish MN 13 localities is Pikermi, where the number of genera in common = 20, Dice = 0.408, and Simpson's = 0.426. Localities where Dice > 0.200 include Sahabi (number of genera in common = 11; Dice = 0.289; Simpson's = 0.379), Italy MN 13 (genera in common = 11; Dice = 0.301; Simpson's = 0.423), Maragheh M and U (genera in common = 12; Dice = 0.286; Simpson's = 0.324), Samos (genera in common = 14; Dice = 0.304; Simpson's = 0.311). The next grouping with relatively low genus-level FRIs includes Toros Menalla (genera in common = 5; Dice = 0.154; Simpson's = 0.278), ASKM (genera in common = 6; Dice = 0.148; Simpson's = 0.176), KUSM (genera in common = 7;

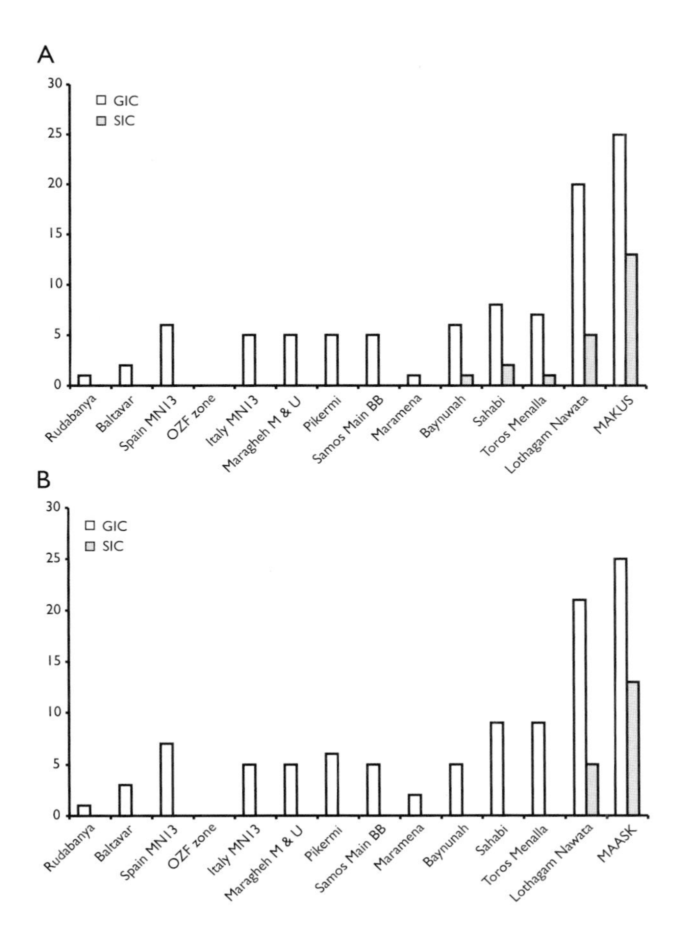

FIGURE 18.2

Number of taxa in common between the Middle Awash late Miocene biochron and selected Eurasian and African late Miocene mammal faunas. A. Genus and species in common between the Asa Koma fauna and Eurasian, Arabian, and African localities. B. Genus and species in common between the Kuseralee fauna and Eurasian, Arabian, and African localities. Abbreviations: GIC = genus in common; SIC = species in common.

Dice = 0.175; Simpson's = 0.212), Lothagam Nawata (genera in common = 6; Dice = 0.141; Simpson's = 0.158), and Baynunah (genera in common = 6; Dice = 0.197; Simpson's = 0.429). Rudabánya has 5 genera in common, Dice is low (0.132) and Simpson's = 0.172. OZF has the weakest biogeographic relationships: One genus is in common, with Dice = 0.034 and Simpson's = 0.091.

Results

Figure 18.2 and the accompanying Table 18.2 provide graphical and numerical data on genus and species identity between the Middle Awash late Miocene fauna and the 13 faunas that have been compared to it. Tables 18.3A (ASKM) and 18.3B (KUSM) summarize this information further by tabulating when there is a genus (G) shared in common and when there is a species (S) shared in common between ASKM and KUSM and the other localities under consideration.

It is clear from these graphics that the closest relationship of the two Middle Awash faunas is between themselves: ASKM and KUSM have 25 genera and 13 species in common. ASKM shows its closest relationships with the Lothagam Nawata fauna, with 19 genera

TABLE 18.2 Genus and Species Homotaxis between Middle Awash Late Miocene Localities (6.0–5.2 Ma) and Selected Eurasian and African Late Miocene Age Localities

Legend	*Locality*	*GIC*	*SIC*
Asa Koma			
L1	Rudabanya	1	0
L2	Baltavar	2	0
L3	Spain MN13	6	0
L4	OZF zone	0	0
L5	Italy MN13	5	0
L6	Maragheh M&U	5	0
L7	Pikermi	5	0
L8	Samos Main BB	5	0
L9	Maramena	1	0
L10	Baynunah	6	1
L11	Sahabi	8	2
L12	Toros Menalla	7	1
L13	Lothagam Nawata	20	5
L15	**MA KUS**	**25**	**13**
Kuseralee			
L1	Rudabanya	1	0
L2	Baltavar	3	0
L3	Spain MN13	7	0
L4	OZF zone	0	0
L5	Italy MN13	5	0
L6	Maragheh M&U	5	0
L7	Pikermi	6	0
L8	Samos Main BB	5	0
L9	Maramena	2	0
L10	Baynunah	5	0
L11	Sahabi	9	0
L12	Toros Menalla	9	0
L13	Lothagam Nawata	21	5
L14	**MA ASK**	**25**	**13**

NOTE: G, genera; S, species; GIC, genera in common; SIC, species in common.

and eight species in common. Three localities share a number of genera and some species in common with ASKM: Sahabi (eight genera and one species), Toros Menalla (seven genera and one species), and Baynunah (six genera, one species). The ASKM fauna shares high genus relationships with a number of western Eurasian faunas, including Spain MN 13 (6) and Pikermi, Samos, Maragheh, and Italy MN 13 (each with 5 genera in common). There is a sharp decline in resemblance between ASKM and Maramena (1), Baltavar (2), and Rudabánya (1).

Other than the ASKM fauna, KUSM shares its closest resemblance to the Lothagam Nawata Formation faunas, with 21 genera in common. The next closest resemblance is

TABLE 18.3 First and Last Occurrences of Taxa in Common between Middle Awash Late Miocene and Selected Eurasian and African Localities

Order	*Family*	*Genus*	*Species*	*L1*	*L2*	*L3*	*L4*	*L5*	*L6*	*L7*	*L8*	*L9*	*L10*	*L11*	*L12*	*L13*	*L15*
Asa Koma																	
Primates	Hominidae	*Ardipithecus*	*kadabba*														S
	Cercopithecidae	*Kuseracolobus*	*aramisi*														S
		Pliopapio	*alemui*														S
		indet.	indet.														
		indet.	indet.														
Carnivora	Viverridae	*Genetta*	sp.													G	G
		Viverra	cf. *leakeyi*					G						G		S	G
		Helogale	sp.														
		Herpestes	*alaylaii* sp. nov.														
	Mustelidae	*Mellivora*	aff. *benfieldi*					G									S
		Lutrinae	indet.														
		Plesiogulo	*botori*			G		G		G			G				
		Sivaonyx	cf. *africanus*														
	Ursidae	*Agriotherium*	sp.			G								G			G
	Felidae	*Machairodus*	sp.		G	G		G	G	G	G		G	G	G		G
		Dinofelis	sp.			G										G	S
	Hyaenidae	*Hyaenictitherium*	sp.											G	G	G	G
		Hyaenictis	*wehaietu* sp. nov.													G	S
Tubulidentata	Orycteropodidae	*Orycteropus*	sp. A & B					G	G	G	G				G	G	G
Hyracoidea	Procaviidae	*Procavia*	sp.														
Proboscidea	Deinotheriidae	*Deinotherium*	*bozasi*	G	G				G	G	G		G		G	S	S
	Gomphotheriidae	*Anancus*	*kenyensis*			G								G	G	S	S
	Elephantidae	*Primelephas*	nov.													G	G
		cf. *Mammuthus*	sp.														
Perissodactyla	Equidae	*Eurygnathohippus*	*feibeli*													S	G
		Eurygnathohippus	aff. *feibeli*														
	Rhinocerotidae	*Diceros*	*douariensis*													G	S
Artiodactyla	Suidae	*Nyanzachoerus*	cf. *devauxi*											S			G
		Nyanzachoerus	*syrticus*										S	S	S		
		Cainochoerus	aff. *africanus*													G	S
	Giraffidae	*Palaeotragus*	sp.						G		G		G			G	
		Giraffa	sp.														G
	Bovidae	*Tragelaphus*	*moroitu* sp. nov														
		Tragelaphus	cf. *moroitu*														

TABLE 18.3 *(continued)*

Order	*Family*	*Genus*	*Species*	*L1*	*L2*	*L3*	*L4*	*L5*	*L6*	*L7*	*L8*	*L9*	*L10*	*L11*	*L12*	*L13*	*L15*
		Tragoportax	*abyssinicus* sp. nov.			G			G	G	G	G	G	G		G	S
		Tragoportax	sp. "large"														
		indet.	indet. (Boselaphini)														
		Aepyceros	cf. *premelampus*													S	S
		Ugandax	sp.														
		indet.	indet. (Bovini)														
		Kobus	cf. *porrecticornis*												G	G	G
		Zephyreduncinus	*oundagaisus*														
		Madoqua	sp.													G	G
		Raphicerus	sp.											G		G	G
Total Number Genera in Common				1	2	6	0	5	5	5	5	1	6	8	7	20	25
Total Number Species in Common				0	0	0	0	0	0	0	0	0	1	2	1	5	13
Kuseralee																	
Primates	Hominidae	*Ardipithecus*	*kadabba*														S
	Cercopithecidae	*Kuseracolobus*	*aramisi*														S
		Pliopapio	*alemui*														S
		indet.	indet.														
		indet.	indet.														
Carnivora	Viverridae	*Genetta*	sp.													G	G
		Viverra	cf. V. *leakeyi*					G						G		S	
	Mustelidae	*Mellivora*	aff. M. *benfieldi*					G									S
		Lutrinae	indet.														
	Ursidae	*Agriotherium*	sp.			G								G			G
	Felidae	*Machairodus*	sp.		G	G		G	G	G	G		G	G	G		G
		Dinofelis	sp.			G										G	G
	Hyaenidae	*Hyaenictitherium*	sp.											G	G	G	G
		Hyaenictis	*wehaietu* sp. nov.													G	S
	Canidae	*Eucyon*	sp.			G		G									
Tubulidentata	Orycteropodidae	*Orycteropus*	sp.					G	G	G	G				G	G	G
Proboscidea	Deinotheriidae	*Deinotherium*	*bozasi*	G	G				G	G	G		G		G	S	S
	Gomphotheriidae	*Anancus*	*kenyensis*			G								G	G	S	S
		Anancus	indet.														
	Elephantidae	*Primelephas*	*gomphotheroides*													S	G
		cf. *Loxodonta*	sp.												G		
Perissodactyla	Equidae	*Eurygnathohippus*	aff. *feibeli*													G	G
	Rhinocerotidae	*Diceros*	*douariensis*													G	S

Artiodactyla	Suidae	*Nyanzachoerus*	*australis*											G	G	G	G
		Nyanzachoerus	*kuseralensis* sp. nov.														
		Nyanzachoerus	*waylandi*														
		Cainochoerus	aff. C. *africanus*													G	S
	Hippopotamidae	aff. *Hippopotamus*	*dulu*														
		Giraffa	sp.														G
	Giraffidae	*Sivatherium*	sp.												G	G	
	Bovidae	*Tragelaphus*	*moroitu* sp. nov.													G	S
		cf. *Tragelaphus*	cf. *moroitu*														S
		Gazella	sp.		G	G			G	G	G	G	G	G		G	
		Tragoportax	*abyssinicus* sp. nov.			G			G	G	G	G	G	G		G	S
		Aepyceros	cf. *premelampus*													S	S
		Kobus	aff. *oricornis*												G	G	G
		Simatherium	aff. *demissum*														
		Madoqua	sp.													G	G
		Raphicerus	sp.											G		G	G
		indet.	indet. (Bovini)														
Total Number Genera in Common				1	3	7	0	5	5	5	5	2	4	9	9	21	25
Total Number Species in Common				0	0	0	0	0	0	0	0	0	0	0	0	5	13

NOTE: See Table 18.1 for locality legend. Abbreviations: G, genus; S, species; GIC, genera in common; SIC, species in common.

to Sahabi and Toros Menalla, both with nine genera in common. A block of localities including Spain MN 13 (seven genera), Pikermi (six genera in common), and Maragheh, Baynunah, Samos, and Italy-MN 13 (each with five genera in common) represent more distant relationships. Baltavar (three genera), Maramena (two genera), and Rudabánya (one genus) have the most distant relationships. The endemic *Oreopithecus* faunas from the Tusco-Sardinian bioprovince have no genera in common with either ASKM or KUSM.

Other than *Dinofelis, Chasmaporthetes, Eucyon,* and *Anancus,* which first occur in the latest Miocene, most of the large mammal taxa shared between the western Eurasian faunas and the Middle Awash and Lothagam Nawata late Miocene faunas first occurred in the early Turolian of Eurasia, such as *Enhydriodon, Plesiogulo, Hyaenictis, Amphimachairodus* (=*Machairodus* of many), *Giraffa* (closely related to *Honanotherium*), *Tragoportax,* and the sister-taxon of *Eurygnathohippus, Sivalhippus.* We believe it most likely that the paleogeographic connection that extended the range of these Eurasian "core" Pikermian lineages into eastern Africa occurred in the earlier Turolian (8.7–8 Ma) and not in the latest Turolian (6.7–5.3 Ma).

Discussion

A variety of relevant biogeographic questions arise from our study. First, to what extent was the Middle Awash fauna derived from a local northern/eastern African Rift source? Second, was the assembly of the Middle Awash fauna influenced by immigrations from more northerly sources and, in particular, sources including central and northern Africa, the circum-Mediterranean region, southwest Asia, and Arabia? Third, does this faunal assembly reflect Vallesian, early-middle Turolian, or late Turolian paleogeographic extensions, and from which definable biogeographic regions?

It is clear from this analysis that ASKM and KUSM, together with the Lothagam Nawata faunas, constitute a provincial large mammal faunal assemblage whose closest resemblance is to Toros Menalla and Sahabi. The relationship to the penecontemporaneous Arabian fauna of Baynunah is more distant. The Pikermian faunas have a more distant temporal relationship yet, but several of their core genera extended their geographic range into northern and eastern Africa in the earlier Turolian. The Spanish MN 13 faunas, and to a lesser extent the Italian MN 13 faunas, likewise share taxa with a Pikermian occurrence including such genera as *Enhydriodon* and *Plesiogulo,* but also include younger first-occurring taxa such as *Agriotherium, Dinofelis, Chasmaporthetes* and *Eucyon,* which do not occur in Pikermian MN 11–12 horizons.

These findings bear on hypotheses regarding the origin and evolution of the African ape-human clade. Considerable controversy has arisen concerning the nexus of this clade's biogeographic origin, and two essentially different hypotheses have emerged on the origins of the Hominini, or African ape-human clade: an "Out of Europe" hypothesis and an Intra-African hypothesis.

Begun has been the prime mover in the "Out of Europe" hypothesis, having published several papers supporting it (Begun 1994; Begun and Kordos 1997; Begun et al. 1997; Begun 2001; Kordos and Begun 2001; Begun 2002; Kordos and Begun 2002; Begun et al. 2003; Begun 2004, 2005). Begun (2001: 235, figure 10.1) published a consensus cladogram (247 characters, LN = 468, CI = 62, RI = 71) proclaiming *Dryopithecus* as the sister taxon of the African ape-human clade. In the same volume, de Bonis and Koufos (2001) suggested that *Ouranopithecus* was more closely related to the human clade and *Dryopithecus* to the African apes. Kordos and Begun (2002: 57) have made the explicit claim that "a relative of the African ape-human clade also probably dispersed from Europe or the eastern Mediterranean into Africa through western Asia some time between about 6 and 9 Ma, based on the fossil and molecular clock evidence, then radiated into the diverse lineages of the African ape and human clade." Begun et al. (2003: 33) have further suggested that a new clade of hominid from the Turolian (MN 11–12) locality of Çorakyerler, Turkey, related to *Ouranopithecus* and *Graecopithecus,* was the sister taxon of *Australopithecus*. Their figure 8 (Begun et al. 2003: 35) suggests the founding of the African ape-human clade in MN 11–12 (Middle Tortonian Stage, early-middle Turolian age) from Turkey.

Begun (2005) has continued to develop the "Out of Europe" hypothesis of hominid origins, suggesting that *Dryopithecus* + *Ouranopithecus* is sister to the African ape-human clade. Moreover, he has made explicit proposals about the evolutionary course of great ape evolution that are briefly summarized as follows: (1) griphopithecines, including European and Turkish *Griphopithecus, Equatorius* (for Begun, a junior synonym for *Griphopithecus*), *Nacholopithecus,* and *Kenyapithecus,* originated in Europe and dispersed into Asia and Africa during the late early/early middle Miocene (MN 5/6); (2) a species of Eurasian griphopith gave rise to a western *Dryopithecus–Ouranopithecus* clade and an eastern *Sivapithecus–Ankarapithecus–Pongo* clade (also including the *Ankarapithecus* clade ca. 12.5 Ma within Eurasia); (3) *Ouranopithecus* likely evolved from a *Dryopithecus* species; (4) a Turolian ape from Turkey immigrated back into Africa due to climatic changes in the higher latitudes of Eurasia sometime between 9 Ma and 7 Ma (Begun 2005: 59, figure 7B).

Bernor et al. (2001, 2004) and Rook and Bernor (2003) have questioned the "Out of Europe" hypothesis on environmental and biogeographic grounds. They noted that *Dryopithecus* is a taxon composed of multiple species all largely restricted to central and western European middle and early late Miocene forested habitats. They argue that these primates were reliant on subtropical–warm temperate forest environments with obligate soft-object frugivory and below-branch locomotion. Bernor and Rook (2004) concluded that they were unlikely candidates for dispersion across the open-country woodland environments resident in the eastern Mediterranean and southwest Asian late Miocene bioprovince (Bernor 1983, 1984; Bernor et al. 1996d, 2001). The de Bonis and Koufos (2001) thesis that latest Vallesian Greek *Ouranopithecus,* or, alternatively, Begun's (2005) thesis that a Turkish early to middle Turolian member of the *Ouranopithecus* (= *Graecopithecus*) clade founded the African ape-human clade seems unlikely given the lack of supporting postcranial material and a postcanine dentition that would appear to be precocious in its convergent development with *Australopithecus*. This viewpoint is strengthened by the earliest African hominids in Chad, Kenya, and Ethiopia (Haile-Selassie et al. Chapter 7), all of which lack these derived parallelisms.

Pickford and Senut (2005) have contested the "Out of Europe" hypothesis on the grounds of newly described fossils from Kenya. They have described an unworn right lower molar from the Ngorora Formation, dated 12.5 Ma. They believe that this specimen, Bar 91 '99, compares closely to European *Dryopithecus* from the Swabian Alb, southwestern

Germany. As a consequence, Pickford and Senut (2005) argue that it is more likely that *Dryopithecus* evolved in Africa and emigrated to Europe than that (Begun's hypothesis) it evolved from *Griphopithecus*. This would be in accord with Andrews and Bernor's (1999) Miocene hominoid vicariance model.

Undoubtedly, the best fossil evidence relevant to the origin of *Dryopithecus* derives from the recent discovery of a partial hominoid skeleton from Spain, *Pierolapithecus catalaunicus* (Moyá-Solá et al. 2004). *Pierolapithecus catalaunicus* has been faunally correlated with MN 7+8, ca. 12.5 Ma, and effectively demonstrated to be near the evolutionary base of the great ape clade. *Pierolapithecus catalaunicus* morphology supports Köhler et al.'s (2001: 209, figure 8.5; cladogram) earlier phylogenetic reconstruction that *Dryopithecus* is a part of the late Miocene great ape radiation and does not share, itself, a sister-taxon relationship to the African ape-human clade (in which they explicitly place *Ardipithecus* as the sister taxon of *Australopithecus*). Moreover, *Pierolapithecus catalaunicus* is fundamentally distinct from griphopithecine primates (*sensu* Begun 2005) in its axial skeleton architecture, more likely being derived from a separate African clade to which early Miocene tropical African *Morotopithecus* may belong. Effectively, cladistic great ape origins are more likely to have an African nexus than a European one, and there may indeed have been multiple hominoid migrations in an Africa-to-Eurasia direction.

We find, as did Bernor et al. (2001), that the Rudabánya mammal community (Bernor et al. 2004, 2005a) has virtually zero faunal resemblance to eastern African latest Miocene faunas: There are no genus-level lineages in common other than the tenuous co-occurrence of *Deinotherium,* which may, in fact, be a paraphyletic taxon. As for *Ouranopithecus,* de Bonis and Koufos (2001) have reported that the Greek specimens are all about the same age (late Vallesian, MN 10) and securely correlated with Chron 4A, circa 9 Ma. The Turkish specimens are, thus far, neither described nor correlated faunally, radioisotopically, or magnetochronologically, so they cannot be legitimately used for supporting a 9–6 Ma extension from Turkey into Africa (*sensu* Begun 2005). As a result, there is no secure evidence of fossil great apes in central or western Europe or, for that matter, Greece and Turkey, after the Vallesian (although the Çorakyerler form may well be Turolian age). Moreover, as demonstrated at Sinap, Turkey (Fortelius et al., 2003b), later Vallesian western Eurasian faunas are very similar in their community structure to regional Turolian faunas. Also, the classic Maragheh faunal sequence, long believed to be late Turolian age, has been demonstrated to range from the late Vallesian (MN 10), ca. 9 Ma, to the Middle Turolian (MN 12), 7.4 Ma, with a very similar faunal aspect throughout. It will be difficult to demonstrate based on faunal correlations alone that Çorakyerler is as young as MN 12.

The results of our biogeographic analysis suggest that a 9–6 Ma immigration of *Dryopithecus* into Africa is highly unlikely. The core Eurasian Pikermian taxa that occur in the Middle Awash late Miocene include (from our database sample) the carnivores *Enhydriodon, Plesiogulo, Agriotherium, Amphimachairodus* (=*Machairodus* of many) and *Hyaenictitherium;* the equid *Eurygnathohippus* (which shares an ancestry with Old World *Cormohipparion* and Siwalik *Sivalhippus perimense*); the rhinoceros *"Diceros" douarensis* (exhibiting convergent characters with *"Ceratotherium" neumayri*); the giraffids *Palaeotragus* and *Sivatherium* (the latter closely related to Pikermian *Helladotherium* and Siwalik *Bramatherium*); and the bovids *Gazella* and *Tragoportax.*

Conclusions

Our analysis reveals that the closest biogeographic connection that the Middle Awash late Miocene localities have are first with the Lothagam Nawata Formation faunas, second to Toros Menalla and Sahabi, and third with the early middle Turolian Pikermian faunas of Greece and Iran and the MN 13 faunas of Baynunah, Spain, and Italy. The Pikermian faunas would appear to represent an important biogeographic and paleoecological source for northern and eastern African latest Miocene faunas, particularly species of carnivores and ungulates. However, the *Nyanzachoerus* species exhibit relationships between the Middle Awash and other localities that are restricted geographically to Africa and Arabia. The truly terminal Miocene biogeographic relationships (MN 13) would appear to include predominantly (but not solely) taxa of African origin that extended their ranges into Eurasia.

We find that eastern African late Miocene faunas were assembled *only in part* by faunal immigrations, most likely dating back to the earliest Turolian, ca. 8.7–8 Ma. Those immigrants were ecologically and faunally of a Pikermian type, not a Rudabánya type. This excludes *Dryopithecus* as a likely immigrant into eastern Africa, but does not exclude the exit of a dryopithecine relative, likely close to *Pierolapithecus* in its stage of evolution, out of Africa and into Eurasia (re: Pickford and Senut 2005). Much of the late Miocene eastern African component was indigenous to tropical, peri-equatorial Africa, and *Ardipithecus* would most likely be derived from a >6 Ma common ancestor with chimpanzees that has yet to be discovered.

19

Biochronology, Faunal Turnover, and Evolution

YOHANNES HAILE-SELASSIE, TIM WHITE, RAYMOND L. BERNOR, LORENZO ROOK, AND ELISABETH S. VRBA

Two available Middle Awash late Miocene faunal assemblages document a number of global first and last appearances of mammalian species in eastern Africa across the Miocene-Pliocene boundary, currently dated to 5.32 Ma (Cande and Kent 1995). The mammalian fauna from the Asa Koma Member (ASKM) of the Adu-Asa Formation is radiometrically dated to between 5.54 Ma and 5.77 Ma, whereas the younger Kuseralee Member (KUSM) of the Sagantole Formation is dated to >5.2 Ma (WoldeGabriel et al. 2001; Renne et al., Chapter 4).

Comparison of time-successive assemblages, particularly those crossing epoch boundaries, may reveal significant evolutionary events, modes, and trends. However, such evaluations work best when there are no sampling biases, when there is sufficient time elapsed between the assemblages, when all speciation is cladogenetic, when the number of specimens sampled from each comparative unit is equivalent, and when there are no other taphonomic or ecological biases (White 1995). The two late Miocene Middle Awash faunal sets do not always conform to these criteria, but they are typical of the assemblages from this part of the African fossil record.

Of the 2,760 specimens collected from the late Miocene deposits of the Middle Awash, 2,209 are from the ASKM. These represent at least 48 mammalian genera. From the KUSM, there are only 551 fossil specimens, assigned to 33 mammalian genera. However, even with such disparities in sample size, there is noticeable change across the approximately 300,000 years separating the ASKM and the KUSM faunas. Differences are particularly evident among the small carnivores, artiodactyls, and micromammals. The late Miocene and early Pliocene saw dramatic changes in African mammalian faunas at localities such as Lothagam in Kenya (Leakey and Harris 2003) and in the Lake Chad basin (Fara et al. 2005).

Chapter 18 in this volume presents a comparative statistical biogeographic analysis with the aim of revealing the extent to which the Middle Awash late Miocene localities resemble eastern and northern African, Arabian, and western Eurasian localities of similar age. However, it should be noted that some sites, such as Langebaanweg and Wembere Manonga (albeit apparently contemporaneous or slightly younger), were not included in that chapter because of the lack of detailed descriptions and comprehensive faunal lists (although the latter were addressed in light of refinement of their biochronological ages).

In the current chapter, we investigate how the two Middle Awash faunal sets, with an age difference of more than 300,000 years, compare at the species level. We start by outlining the materials and methods used. We then present taxon-based biochronological comparisons of the Middle Awash faunas to contemporaneous assemblages from other sites in Africa and Europe. Based on these comparisons, global and local first appearances (FAs) and last appearances (LAs) of the Middle Awash species are presented, with emphasis on what appear to be the most biologically significant FAs and LAs. The discussion section addresses possible causes for intermember, interregional, and intercontinental faunal similarities and differences observed. We believe that this approach (particularly the concept of identifying biologically significant species turnovers), if more broadly applied to other faunal samples, would result in a better understanding of the tempo and mode of mammalian evolution in eastern Africa throughout the Neogene.

Materials and Methods

Table 19.1 provides a combined faunal list for the two Middle Awash members, indicating taxa represented in each member, the number of specimens assigned to each taxon from each member, and local and global FAs and LAs. The table highlights biologically significant LAs and FAs. To assess the significance of Middle Awash FAs and LAs, we used data compiled for African Neogene fauna by one of us (EV). We also consulted data reported in the Neogene Old World (NOW) database (see Bernor et al., Chapter 18, for details).

We examined the evolutionary history of each species represented in both the ASKM and KUSM by examining local (Middle Awash, including unpublished data from deposits younger and older than the KUSM), provincial (eastern African), and intercontinental (Eurasian and African) FAs and LAs, paying special attention to morphological changes in each species lineage through time. We also conducted a comparative bichronological analysis that includes all of the late Miocene sites referred to in Chapter 18. However, faunal assemblages from some African sites (Wembere Manonga, Harrison 1997a; Albertine Rift, Senut and Pickford 1994; Langebaanweg, Hendey 1982) reported as possibly late Miocene in age were not included in the detailed taxon-by-taxon comparison. This is largely because of temporal differences, lack of faunal overlaps with the Middle Awash late Miocene, absence of detailed desciptions, or some combination of these confounding variables.

Intermember Comparison

The ASKM and KUSM are separated by ca. 300 Kyr. However, they share numerous common genera (25) and species (15). There are also taxa that are not shared between them. Twenty-six of the sixty-five taxa from both members are exclusively present in the ASKM. Thirteen taxa are present only in the KUSM. Many of these "turnovers" appear to result from small sample sizes and are thus evolutionarily ambiguous. Others may be

TABLE 19.1 Faunal Lists for the ASKM Assessment of Faunal Turnover within the Middle Awash Late Miocene and Beyond

Order	*Family*	*Genus*	*Species*	*ASKM*	*KUSM*	*SGFA*	*SMAFA*	*BSSFABM*	*BSSLABM*	*SGLA*	*SMALA*
Primates	Hominidae	*Ardipithecus*	*kadabba*	X (17)	? (1)	Y (*)	NA	?	N	Y (*)	Y (P)
Primates	Cercopithecidae	*Kuseracolobus*	*aramisi*	X (18)	X (2)	Y (*)	NA	N	N	N	N
Primates	Cercopithecidae	*Pliopapio*	*alemui*	X (4)	X (6)	Y (*)	NA	N	N	N	N
Carnivora	Viverridae	*Genetta*	sp.	X (3)	X (1)	?	NA	N	N	?	?
Carnivora	Viverridae	*Viverra*	cf. *leakeyi*	X (2)	X (1)	?	NA	N	N	N	N
Carnivora	Viverridae	*Helogale*	sp.	X (1)		?	NA	N		?	?
Carnivora	Viverridae	*Herpestes*	*alaylaii* sp. nov.	X (9)		Y (*)	NA		N	?	?
Carnivora	Canidae	*Eucyon*	sp.		X (1)	?	?	?	?	?	?
Carnivora	Mustelidae	*Mellivora*	aff. *benfieldi*	X (2)	X (1)	?	NA	N	N	?	?
Carnivora	Mustelidae	*Sivaonyx*	cf. *africanus*	X (3)	X (2)	N	NA	N	N	N	N
Carnivora	Mustelidae	*Plesiogulo*	*botori*	X (1)	? (1)	Y (*)	NA	N	N	Y (*, P)	Y (*, P)
Carnivora	Ursidae	*Agriotherium*	sp.	X (6)	X (1)	?	NA	N	N	N	N
Carnivora	Felidae	*Machairodus*	sp.	X (30)	X (7)	?	NA	N	N	N	N
Carnivora	Felidae	*Dinofelis*	sp.	X (4)	X (7)	?	NA	N	N	N	N
Carnivora	Hyaenidae	*Hyaenictitherium*	sp.	X (2)	X (3)	?	NA	N	N	?	?
Carnivora	Hyaenidae	*Hyaenictis*	*wehaietu* sp. nov.	X (1)	X (5)	?	NA	N	N	Y (*)	Y (*)
Insectivora	Soricidae	*Crocidura*	aff. *aethiops*	X (1)		?	NA	N	?	N	N
Insectivora	Soricidae	*Crocidura*	aff. *dolichura*	X (1)		?	NA	?	?	N	?
Rodentia	Muridae	*Lemniscomys*	aff. *striatus*	X (1)		?	NA	N	?	N	N
Rodentia	Muridae	*Tatera*	sp. indet.	X (1)	X (1)	?	NA	?	?	?	?
Rodentia	Lephiomyidae	*Lophiomys*	*daphnae* sp. nov.	X (1)		Y (*)	NA	N	N	?	?
Rodentia	Thryonomyidae	*Thryonomys*	*asakomae* sp. nov.	X (47)		Y (*)	NA	N	N	Y (P)	Y (P)
Rodentia	Thryonomyidae	*Thryonomys*	aff. *gregorianus*		X (2)	Y (P)	Y (P)	N	N	N	N
Rodentia	Sciuridae	*Xerus*	sp. indet.	X (1)		?	NA	?	?	?	?
Rodentia	Hystricidae	*Hystrix*	sp. indet.	X (2)		?	NA	?	?	?	?
Rodentia	Hystricidae	*Xenohystrix*	sp. indet.	X (1)		?	NA	?	?	?	?
Rodentia	Hystricidae	*Atherurus*	*garbo* sp. nov.	X (9)		Y (*)	NA	N	Y (*, P)	Y (*)	Y (*)
Rodentia	Rhyzomyidae	*Tachyoryctes*	*makooka* sp. nov.	X (48)	X (14)	Y (*)	NA	N	N	N	N
Lagomorpha	Leporidae	*Serengetilagus*	*praecapensis*	X (23)		?	NA	?	N	N	?
Tubulidentata	Orycteropodidae	*Orycteropus*	sp. A	X (2)		N	NA	N	?	?	?
Tubulidentata	Orycteropodidae	*Orycteropus*	sp. B	X (2)		Y	NA	Y?	N	N	N
Hyracoidea	Procaviidae	*Procavia*	sp. indet.	X (1)		N	NA	?	?	N	?
Proboscidea	Deinotheriidae	*Deinotherium*	*bozasi*	X (4)	X (8)	N	NA	N	N	N	N
Proboscidea	Gomphotheriidae	*Anancus*	*kenyensis*	X (11)	X (36)	N	NA	N	N	N	N
Proboscidea	Elephantidae	*Primelephas*	*gomphotheroides saitunensis* subsp. nov.	X (36)		?	NA	N	N	?	?
Proboscidea	Elephantidae	*Primelephas*	*gomphotheroides gomphotheroides*		X (5)	N	NA	N	N	N	N

TABLE 19.1 *(continued)*

Order	*Family*	*Genus*	*Species*	*ASKM*	*KUSM*	*SGFA*	*SMAFA*	*BSSFABM*	*BSSLABM*	*SGLA*	*SMALA*
Proboscidea	Elephantidae	cf. *Loxodonta*	sp.		X (4)	N	NA	N	N	N	N
Proboscidea	Elephantidae	cf. *Mammuthus*	sp.	X (1)						N	N
Perissodactyla	Equidae	*Eurygnathohippus*	*feibeli*	X (71)		N	NA	N	N	N	N
Perissodactyla	Rhinocerotidae	*Diceros*	*douariensis*		X (1)	?	?	?	?	?	?
Perissodactyla	Rhinocerotidae	*Diceros*	sp.	X (9)		?	?	?	?	?	?
Artiodactyla	Suidae	*Nyanzachoerus*	cf. *devauxi*	X (8)		N	NA	N	**Y**	Y	Y
Artiodactyla	Suidae	*Nyanzachoerus*	*syrticus*	X (50)		N	NA	N	Y (P)	Y (P)	Y (P)
Artiodactyla	Suidae	*Nyanzachoerus*	*australis*		X (50)	Y	Y (P)	Y (P)	Y (P)	N	N
Artiodactyla	Suidae	*Nyanzachoerus*	*kuseralensis* sp. nov.		X (3)	Y (*)	Y (*)	Y (*)	Y (*)	Y (*)	Y (*)
Artiodactyla	Suidae	*Nyanzachoerus*	*waylandi*		X (3)	N	Y	**Y**	N	Y	Y
Artiodactyla	Suidae	*Cainochoerus*	aff. *africanus*	X (16)	X (3)	N	NA	N	N	N	N
Artiodactyla	Hippopotamidae	*Hippopotamus*	aff. *dulu*		X (33)	?	?	?	?	?	?
Artiodactyla	Giraffidae	*Palaeotragus*	sp.	X (11)		N	NA	?	Y	Y	Y
Artiodactyla	Giraffidae	*Giraffa*	sp.	X (1)	X (2)	?	NA	?	N	N	N
Artiodactyla	Giraffidae	*Sivatherium*	sp.		X (5)	?	NA	**Y**	N	N	N
Artiodactyla	Bovidae	*Tragelaphus*	*moroitu* sp. nov.	X (10)	X (23)	Y (P*)	NA	Y (P*)	N	N	N
Artiodactyla	Bovidae	*Tragoportax*	*abyssinicus* sp. nov.	X (8)	X (2)	Y (*)	NA	N	N	?	?
Artiodactyla	Bovidae	*Tragoportax*	sp. "large"	X (11)		?	?	?	?	?	?
Artiodactyla	Bovidae	Indet.	indet. (Boselaphini)	X (1)		?	?	?	?	?	?
Artiodactyla	Bovidae	*Aepyceros*	cf. *premelampus*	X (3)	X (7)	N	NA	N	N	N	N
Artiodactyla	Bovidae	*Ugandax*	sp.	X (1)		?	?	?	?	?	?
Artiodactyla	Bovidae	*Simatherium*	aff. *demissum*		X (1)	?	Y	Y (?, P)	?	?	?
Artiodactyla	Bovidae	*Redunca*	*ambae* sp nov.		X (4)	Y (*)	Y (*)	Y (P*)	Y (P, *)	Y (P,*)	Y (P, *)
Artiodactyla	Bovidae	*Kobus*	cf. *porrecticornis*	X (13)	X (2)	N	NA	N	N	N	N
Artiodactyla	Bovidae	*Kobus*	aff. *oricornus*		X (5)	Y	Y	**Y**	N	N	N
Artiodactyla	Bovidae	*Zephyreduncinus*	*oundagaisus*	X (7)		Y (*)	NA	N	Y (*)	Y (*)	Y (*)
Artiodactyla	Bovidae	*Gazella*	sp.		X (7)	N	NA	**Y**	N	N	N
Artiodactyla	Bovidae	*Madoqua*	sp.	X (1)	X (1)	N	N	N	N	N	N
Artiodactyla	Bovidae	*Raphicerus*	sp.	X (1)	X (2)	N	N	N	N	N	N

NOTE: Presence of a taxon is denoted by an X. The number after the X is the relevant number of identified specimens belonging to the genus or species. Abbreviations: ASKM, Asa Koma Member; KUSM, Kuseralee Member; SGFA, species global first appearance; SMAFA, species Middle Awash first appearance; BSSFABM, biologically significant, first appearance between members; BSSLABM, biologically significant, last appearance between members; SGLA, species global last appearance; SMALA, species Middle Awash last appearance; N, No; Y, Yes; **?**, not determinable because of unclear taxonomy or insignificant sample size; NA, not applicable because no earlier deposits in the Middle Awash sequence; (P), Phyletic evolutionary pair (chronospecies); **(*)**, first, last, and only occurrence in Middle Awash. Bolded Y's indicate that the first or last occurrences between members are probably biologically significant turnovers, as opposed to reflecting inadequate sampling or uncertain taxonomy.

due to paleoecological differences between members or artificial division of phyletically evolving lineages into separate chronospecies. Some turnovers are, however, apparently significant in evolutionary or biogeographic terms. Some taxa appear to be absent from the KUSM because they went extinct at ca. 5.5 Ma. Other taxa absent from the ASKM but present in the KUSM could have made their first appearance after 5.5 Ma, cladogenetically or via immigration.

Some of the genera absent in the KUSM (e.g., *Helogale, Herpestes, Procavia, Ugandax,* and *Sivaonyx*) have a broad temporal distribution in early Pliocene and younger deposits. This means that their absence in the KUSM is most likely the result of sampling bias, lack of suitable habitat, or both. For example, based on the paleoenvironment inferred for the KUSM (see Su et al., Chapter 17), lack of suitable habitat might be one reason for the absence of *Herpestes* therein. Additionally, because of its high status on the food chain and its small size, its presence or absence may be an artifact of collection, particularly when the total number of specimens collected for this genus is only nine. Similarly, *Ugandax* is present in numerous Pliocene sites of eastern Africa, so its absence in the KUSM assemblage appears to be the result of either paleoenvironmental difference or small sample size (bovine teeth, although present, are difficult to attribute exclusively to this genus). The extreme case of *Procavia* reflects an obvious sampling bias—a single specimen is known from the entire Middle Awash combined assemblage of >17,000 vertebrate fossils!

Middle Awash FAs, LAs, and Faunal Turnovers

There are 18 species or genus global first occurrences in the ASKM (column SGFA in Table 19.1). Fifteen of these are taxa only present in the Middle Awash succession, and one of them is a chronospecies of a lineage represented in earlier sites elsewhere. Because there are inadequate samples/analyses of Middle Awash Fauna from limited sediments antedating the ASKM, the species or genus FAs in the Middle Awash (column SMAFA in Table 19.1) are not reported (NA) in the Local FA column.

There are eight taxa that first appear in the KUSM. Three of the eight species are known exclusively from the Middle Awash KUSM (*Eucyon* sp., *Nyanzachoerus kuseralensis* sp. nov., and *Redunca ambae* sp. nov.). Two species (*Nyanzachoerus australis* and *Thryonomys* aff. *gregorianus*) are phyletic descendants of taxa known from either the ASKM or older deposits elsewhere. *Nyanzachoerus waylandi, Simatherium* aff. *demissum,* and *Kobus* aff. *oricornus* are the only three taxa that appear to have biologically significant first appearancess in the KUSM.

To evaluate faunal turnover during the time interval between the two members, it was necessary to identify the taxa whose appearances or disappearances were biologically significant (BSSFABM and BSSLABM columns of Table 19.1). A taxon's first appearance between members is considered biologically significant when its absence in older deposits is believed to be a "real" absence, and not an artifact of sampling or taxonomic uncertainty. Furthermore, an FA based only on phyletic descent from an earlier chronospecies is only a qualified "turnover." See the legend for Table 19.1 for further details.

Only four species (*Nyanzachoerus waylandi, Sivatherium* sp., *Gazella* sp., and *Kobus* aff. *oricornus*) show "real," biologically significant global FAs relevant to assessing faunal turnover between the two members. These four taxa appear in the African fossil record for the first time in the KUSM. The ASKM might also document the first appearance of *Orycteropus* sp. B. However, its sample size is small, and the taxonomy needs to be refined before it can be used in faunal turnover assessment.

Nyanzachoerus australis appears in the KUSM for the first time, possibly evolving from an earlier *Nyanzachoerus* species. *Nyanzachoerus kuseralensis* sp. nov. also appears in the KUSM, but this species is, so far, known exclusively from the Middle Awash. Moreover, its taxonomic relationship with other *Nyanzachoerus* species is currently unclear. *Tragelaphus moroitu* sp. nov. and *Redunca ambae* sp. nov., are thus far known only from the Middle Awash. The former is sampled from both members, whereas the latter is known only from the KUSM. These two species appear to be phyletic descendants from taxa known in earlier deposits in the Middle Awash or elsewhere (see Haile-Selassie et al., Chapter 9; Vrba 2006). *Simatherium* aff. *demissum* first appears in the KUSM. However, its use in assessing faunal turnovers might be negligible because of the small sample size ($n = 1$). Moreover, a number of specimens from the ASKM, and currently identified only as Bovini indet., might be phylogenetically linked to the *Simatherium* aff. *demissum* of the KUSM.

Turning to disappearances, there are 13 species or genus last local or global occurrences in the ASKM and KUSM (columns SGLA and SMALA in Table 19.1). Eleven of these are either taxa only documented from the two members of the Middle Awash succession, or taxa that "disappeared" only as a result of phyletic evolution (*Plesiogulo botori, Thryonomys asakomae* sp. nov., *Nyanzachoerus syrticus,* and *Redunca ambae* sp. nov.); *Ardipithecus kadabba* "disappears" in the Middle Awash late Miocene, apparently giving rise phyletically to the younger, Pliocene *Ardipithecus ramidus* (White et al. 1994). Determining whether this happened between the ASKM and KUSM requires additional fossil material.

Nyanzachoerus devauxi, Nyanzachoerus waylandi, and *Paleotragus* sp. make their LAs between the two members. Their LAs at the end of the Miocene appear to be real and biologically significant. These three taxa have not been documented anywhere else in deposits younger than 5 Ma. In the Lothagam succession, *Nyanzachoerus devauxi* and *Paleotragus* are not present after Upper Nawata times, whereas *Nyanzachoerus waylandi* was entirely absent from the succession (Harris and Leakey 2003b; Harris 2003b). These three taxa are also absent from slightly younger deposits such as the Haradaso Member of the Middle Awash (unpublished data), the Apak Member of Lothagam (Harris and Leakey 2003b; Harris 2003b), Langebaanweg (Hendey 1982), and Wembere Manonga (Harrison 1997a).

In summary, relative to the overall size of the faunal assemblage, FAs and LAs of mammalian taxa documented between the ASKM and KUSM demonstrate only limited faunal turnover between the two members. However, they document real, and biologically significant, FAs and LAs at the global, regional, and local levels.

Old World Late Miocene Biochronological Comparisons

Chapter 18 presents a formal biogeographic analysis of the late Miocene faunas of the Middle Awash in light of global, continental, and regional taxonomic comparisons. Here, in light of the new data from the Middle Awash, we make broadly based biochronological assessments on major mammalian groups that have previously proven useful in this regard. This section synthesizes information from other chapters in this book and confirms the late Miocene temporal placement of the Middle Awash localities.

Primates

The Middle Awash late Miocene primates appear to be endemic to eastern Africa. At least three cercopithecid genera (one papionine and two colobine species, see Frost et al., Chapter 6) and one hominid genus (see Haile-Selassie et al., Chapter 7) have been reported from the two Middle Awash members. The geographic distribution of these primates is thus far limited to Ethiopia and hence is less significant in a broader biochronological analysis. The papionine *Pliopapio alemui* and the colobine *Kuseracolobus aramisi* first appear in the ASKM (SGFA), and their fossil record continues to Pliocene deposits at Aramis (4.4 Ma), Middle Awash (Frost 2001b). *Kuseracolobus* is also documented from 4.1 Ma deposits at Asa Issie, Middle Awash (Hlusko 2006). It is likely that these genera were also present in similar-aged deposits at Gona, Ethiopia. The hominid *Ardipithecus kadabba* makes its SGFA in the ASKM, and one specimen from the KUSM is assigned to *Ardipithecus* cf. *kadabba* (see Haile-Selassie et al., Chapter 7). A number of specimens from the late Miocene deposits of Gona have also been assigned to *Ar. kadabba* (Simpson et al. 2007). This species evolved to the younger *Ardipithecus ramidus* from Aramis and Gona, both in Ethiopia (White et al. 1994; Semaw et al. 2005).

Carnivora

The Middle Awash carnivoran assemblages share many taxa with other site-based assemblages used in the paleobiogeographic analysis (see Bernor et al., Chapter 18). Nine carnivore genera (out of a total of 13 recorded in the Middle Awash; see Haile-Selassie and Howell, Chapter 8) occur at other Eurasian and African Mio-Pliocene sites. Four species make their first global appearance (SGFA) in the Middle Awash (see the section "Middle Awash FAs, LAs, and Faunal Turnovers" earlier in this chapter).

The family Viverridae is represented by two genera: *Genetta* and *Viverra*. *Genetta* occurs in the latest Miocene of Lothagam (Werdelin 2003b) and Langebaanweg (Hendey 1974) in a form very similar to the Middle Awash taxon (Haile-Selassie et al. 2004c). It is a typically African genus. *Viverra* is also reported in the latest Miocene of Lothagam (Werdelin 2003b) and Langebaanweg (Hendey 1974). However, its geographic range during the latest Miocene was much broader, seen in its representation at Sahabi (Libya; Howell 1987) and Baccinello V3 (Italy; Rook et al. 1991; Rook and Martinez-Navarro 2004). The occurrence of the same species *(Viverra howelli)* both in North Africa (Sahabi) and southern Europe (Baccinello V3) suggests an extension of this African genus into Europe in latest Miocene, MN 13. *Viverra* persists in Europe until the late Pliocene (the LA in Europe of the genus is MN 16 at Triversa).

The family Canidae is represented by the primitive dog *Eucyon* from the KUSM. The genus has its origin in the late Miocene of North America (see Tedford and Qiu 1996) and dispersed across Eurasia and Africa since the latest Miocene (Spassov and Rook 2006). The genus persisted in the Old World until the middle Pliocene, in part overlapping with the arrival of the genus *Canis*.

Three mustelid taxa recognized to the species level are found in the Middle Awash late Miocene sample: *Mellivora, Sivaonyx,* and *Plesiogulo*. Additional mustelid material indicates the presence of an additional one or two lutrine species not currently identified at the specific level due to the scant nature of the sample and/or lack of comparative material.

Mellivora (the extant African honey badger) first occurs (with the same species, *M. benfieldi*) in two latest Miocene age localities: Langebaanweg in South Africa (Hendey 1974), and Brisighella in Central Italy (Rook et al. 1991). *Sivaonyx* is a medium-sized mustelid occurring, in addition to the Middle Awash, at Lukeino and Langebaanweg. This genus is similar to, but smaller than, *Enhydriodon* in its dental morphology. There have been difficulties distinguishing the two genera. *Sivaonyx* first appears in the MN 9 of Europe, where it is reported from Eppelsheim, Germany. It is also known from MN 13 of Spain, which also marks its last occurrence in Europe. In Africa, its LA is the Middle Awash.

Plesiogulo is reported from Pikermi, Spain MN 13 (Venta del Moro, Las Casiones), Italy MN 13 (Baccinello V3), and Baynunah. This genus has a European origin. It is first reported early in MN 6 at Pasalar (Turkey) and is relatively common in Vallesian and Turolian sites of Europe, from Moldova to Spain. The genus apparently extended its geographic range eastward and southward in the latest Miocene, where it has been reported from eastern Kazakhstan (Kalmakpaj), eastern Africa (Middle Awash), and Langebaanweg (South Africa). Its first record in China (Gaozhuang) is in the early Pliocene. The Middle Awash *Plesiogulo botori* makes its first appearance at slightly older deposits of Lemudong'o, Kenya (Haile-Selassie et al. 2004a), and its relationship with the younger *Plesiogulo monspesulanus* is unknown due to a small sample size.

The genus *Agriotherium* includes very large ursids reported from two MN 13 sources cited in Chapter 18 (Bernor et al.): the Spanish MN 13 cluster, and Sahabi. *Agriotherium* is first reported in MN 7–8 at Sofça (Turkey), but this form is not likely to represent the same genus. MN 13 is the likely first occurrence. *Agriotherium* last occurs at MN 16 of Vialette and Les Etouares in France. In Africa, its presence in the Middle Awash marks its first apprearance in sub-Saharan Africa. It expanded its range to South Africa (Langebaanweg) at the end of the Miocene and persisted in Africa into the early Pliocene.

The machairodont cat, *Machairodus* (=*Amphimachairodus* if we restrict *Machairodus aphanistus* to the type specimen of Kaup from Eppelsheim as preferred by Werdelin 1996) had a broad geographic and long chronologic range. There are several localities with this large cat: Baltavar, Spanish MN 13 cluster; Italy MN 13 cluster; Maragheh; Samos; Baynunah; Sahabi; and Toros Menalla. It should be further mentioned that the Lower and Upper Nawata (Lothagam, Kenya) machairodont cat *Lokotunjailurus emageritus* Werdelin, 2003, is smaller than *Machairodus,* and is a form reported to be an evolutionarily intermediate between *Machairodus aphanistus* and *Homotherium*. It can be considered to be a vicariant clade of *Machairodus* (or, alternatively, *Amphimachairodus*) and a sister taxon of *Homotherium* (Werdelin 2003b). This evidence suggests that *Machairodus,* and perhaps specifically *M. (=Amphimachairodus) aphanistus,* made a relatively early Turolian entry into northern and eastern Africa and became vicariant by Lower Nawata times (>7 Ma?). We suspect that the time of entry was not later than the middle Maragheh-Pikermian chron of 8.2 Ma (Bernor et al. 1996a,c).

Dinofelis is reported from the Middle Awash and two other localities: Spanish MN 13 and Lothagam Lower and Upper Nawata and Apak members. The Lothagam material is the oldest known *Dinofelis* from Africa (Werdelin 2003b: 306) and would appear to have exited Africa at the end of the Miocene.

Two hyaenid genera, *Hyaenictitherium* and *Hyaenictis,* are present in the Middle Awash. The first, *Hyaenictitherium,* is also present in four of the sites included in the comparison: Maragheh, Sahabi, Toros Menalla, and Lothagam Nawata. *Hyaenictis* occurs at Pikermi and Lothagam Nawata. Both genera are also present in the latest Miocene of Langebaanweg, and both first appear in Europe. The first record of *Hyaenictis* is in MN 10 of Spain (Villadecavalls). It becomes extinct in Europe in MN 12 (last occurrence at Pikermi in Greece, and San Miguel del Taudell in Spain), but persists in Africa until the end of the Miocene. *Hyaenictitherium* is first recorded in MN 9 of Spain (Los Valles de Fuentidueña) and has a late Miocene record. *Hyaenictitherium hyaenoides* is recorded from MN 11 intervals of Maragheh, Iran (Werdelin and Solounias 1996); is common in the former western USSR and Kazakhstan in MN 12 and MN 13 *(H. parvum)*; and is confined to eastern countries (Ukraine, Moldova, and Kazakhstan). Both *Hyaenictis* and *Hyaenictitherium* seem to have entered Africa in the latest Miocene, and the Middle Awash *Hyaenictis wehaietu* sp. nov. might represent the earliest occurrence of the genus in sub-Saharan Africa. However, the small sample size precludes further comparison with other known *Hyaenictis* species.

Werdelin has recently updated the Langebaanweg (South Africa) carnivore fauna (Langebaanweg faunal symposium, November 2006), which includes *Amphimachairodus* sp. indet. (Werdelin and Sardella 2006). This species is absent in the Middle Awash late Miocene, although *Amphimachairodus* is present at Sahabi (Sardella and Werdelin 2007). At Toros Menalla (Chad), Peigné et al. (2005a) synonymize this genus with *Machairodus.* Werdelin has not yet published his results, but a number of the carnivore taxa reported here from the ASKM and the KUSM are recorded in the Langebaanweg fauna. There is broad agreement that the Langebaanweg mammal fauna lies within the 5.2–4.9 Ma interval, with a bias toward the older part of that range.

Proboscidea

Proboscideans recorded in the Middle Awash late Miocene belong to three families: Deinotheriidae, Gomphotheriidae, and Elephantidae (see Saegusa and Haile-Selassie, Chapter 15). They include both African endemic taxa (e.g., *Primelephas, Loxodonta*) and genera with a broader paleogeographic distribution extending to Europe and Asia (e.g., *Deinotherium, Anancus, Mammuthus*).

Deinotheriidae are first recorded from the late Oligocene of Chilga (Ethiopia; Sanders et al. 2004), with the primitive and still poorly known species *Chilgatherium harrisi*. The more derived *Prodeinotherium* is known from the early Miocene in eastern Africa (Moroto, Uganda; Koru and Rusinga, Kenya) at around 20–18 Ma (species *P. hobleyi;* Tassy 1986). *Prodeinotherium hobleyi,* or a closely related taxon, is also recorded at Adi Ugri (Eritrea;

Vialli 1966) and from several middle Miocene localities in eastern and northern Africa, including Jebel Zelten, Libya (Harris 1973). *Prodeinotherium* extended its range in the late Oligocene to southern Asia (*P. pentapotamiae* in the Bugti Hills, Pakistan) and in the later early Miocene to western Europe, where it is first recorded with the species *P. bavaricum* in a number of MN 4 localities (e.g., La Romieu, France; Pálfy et al. 2007). The occurrence of deinotheres in the Miocene of Israel and Turkey suggests an early Miocene dispersal through the Middle East (Harris 1978). *Prodeinotherium* species are replaced during the middle Miocene by the larger and more cursorial *Deinotherium*.

Deinothere assemblages from Africa, Europe, and Asia have been assigned to three distinct species: *Deinotherium bozasi* (first recorded at Muruyur, Kenya, ca. 12 Ma; Hill et al. 1986), *D. giganteum* (FA MN 6, e.g., Sansan, France), and *D. indicus* (Dhok Pathan), respectively. Phyletic relations among the various deinothere species are unresolved, so it is not possible to accurately reconstruct their dispersal/evolutionary pattern. It is not clear whether the three presently recognized *Deinotherium* species are vicariants derived from a single ancestral *Deinotherium* species (in Africa or Eurasia?), or whether each species evolved *in situ* from a local population of *Prodeinotherium* (which would make *Deinotherium* a polyphyletic taxon). *Deinotherium bozasi* is recorded in Africa as late as the early Pleistocene (ca. 1.3 Ma in the Omo basin; Beden 1987b). *Deinotherium indicus* became extinct at the end of the Miocene, and late representatives of *D. giganteum* (considered by some authors to represent a distinct derived species, *D. gigantissimum*) are known from Ruscinian localities in Moldova (Pripiceni) and Romania (Manazati) (MN 14) and from the northern Caucasus in localities (Kosyakino, Armavir) correlated to MN 15 (Vislobokova 2005).

Unlike the other proboscideans recorded in eastern Africa, the brevirostrine gomphothere genus *Anancus* appears to have originated in Eurasia. A primitive *Anancus* form is known from various late Miocene localities (MN 11 to MN 13) of Europe (e.g., Dorn-Duerkheim, Hohenworth, Alcoy, Venta del Moro, Baltavar) and South Asia (Perim Island; *A. perimensis*). *Anancus kenyensis* is first recorded in Africa (Toros Menalla, Mpesida Beds, Lukeino, Lothagam) in the late Miocene and persists until the early Pliocene (ca. 3.8 Ma; Kanam East; lower Chemeron; Nachukui Formation, Lothagam; earlier Kaiso; Haradaso Member, Sagantole Formation, Middle Awash; Kalb and Mebrate 1993; Sanders 1997; Tassy 2003). *Anancus kenyensis* originated in Africa, giving rise to more progressive forms such as *Anancus* sp. (=*Anancus capensis* sp. nov.; Sanders 2007) from Langebaanweg (Miocene/Pliocene), the pentalophodont *Anancus* Sagantole-type (MA; early Pliocene; Kalb and Mebrate 1993), and the highly derived *A. "petrocchi"* from Sahabi (late Miocene; Petrocchi 1954). The LA of *Anancus* in Africa is slightly younger than 4 Ma at Laetoli (Sanders 2007). The Pliocene pentalophodon *A. sivalensis* from the Siwaliks is closely related to the African *A. kenyensis* lineage (Tassy 1986), perhaps indicating a special paleobiogeographic connection between eastern Africa and South Asia at the Mio-Pliocene transition. Moreover, *A. kenyensis* and *A. sivalensis* may share a common ancestor in the Miocene (Tortonian?) of Eurasia. However, there is no evidence of *A. sivalensis* in the Messinian, whereas *A. kenyensis* is already differentiated in Africa by this time.

In the Pliocene of Europe and North Africa, *Anancus* is represented by the closely related *A. arvernensis* (MN 14, e.g., Alcoy, Spain to MN 17, e.g., Chillac) and *A. osiris* (middle to late Pliocene; Gizeh, Egypt; Lac Ichkeul, Tunisia; Ain Boucherit, Morocco;

Kolinga 1, Chad; Coppens et al. 1978). It has been suggested that the *arvernensis-osiris* clade is the sister group of the *kenyensis-sivalensis* lineage, and that the two groups diverged in the late Miocene from a primitive Eurasian *Anancus* stem group (the Dorn-Duerkheim *Anancus-A. perimensis* assemblage; Tassy 1986; Metz-Muller 2000).

The basal elephantines *(Primelephas)* and the derived elephantines *(Loxodonta)* are strictly endemic to Africa. *Primelephas* was defined on material from the Nawata Formation of Lothagam (*P. gomphotheroides;* Maglio 1973), and has since been recorded in a number of eastern and central African late Miocene–early Pliocene localities: Lukeino, Baringo Basin, Kaiso Formation, Manonga Valley, and Kossom Bougoudi (Sanders 1997; Brunet et al. 2000; Tassy 2003). Although all of the late Miocene African elephants of the genus *Primelephas* are assigned to *Primelephas gomphotheroides,* the new evidence from the Middle Awash shows that there were at least three morphs that can be dentally distinguished at the subspecies level. *Primelephas gomphotheroides saitunensis* from the ASKM and *Primelephas gomphotheroides gomphotheroides* from the younger KUSM appear to be dentally more primitive than the older *Primelephas gomphotheroides* from the Lower Nawata of Lothagam. The material from Wembere-Manonga appears to represent a third subspecies (see Saegusa and Haile-Selassie, Chapter 15). However, their phylogenetic relationship is currently unknown (see Saegusa and Haile-Selassie, Chapter 15).

The genus *Loxodonta* includes the extant African elephant. Two distinct *Loxodonta* lineages have been recognized in the Neogene of Africa: the *L. exoptata-L. atlantica-L. africana* lineage, and the *L. adaurora* lineage. A primitive (still unnamed) member of the first lineage, *Loxodonta* sp., is recorded from Lukeino (Kenya), from Toros Menalla (Chad), and from the Nkondo Formation in Uganda (Tassy 1986; Vignaud et al. 2002). *Loxodonta exoptata* is recorded from Pliocene sites in eastern Africa (Kiloleli Member, Mononga, Nachukui Formation, Lothagam, Laetoli, Koobi Fora, Shungura, Matabaietu, and Hadar; Maglio 1973; Beden 1987b; Kalb and Mebrate 1993; Sanders 1997; Tassy 2003). *Loxodonta exoptata* is replaced in the early Pleistocene by the progressive *L. atlantica*. This species shows a more widespread distribution, ranging from northern (e.g., Ternifine, Algeria) to southern Africa (e.g., Elandsfontein). It has been also recorded in Chad and from the Pliocene Shungura Formation (Coppens et al. 1978; Beden 1987b).

The *L. adaurora* lineage presents a more limited temporal and geographical distribution than the other lineage. Available data indicate that it is limited to the Pliocene of eastern Africa. The oldest occurrence could be that from Wee-ee, in the Sagantole Formation of the Middle Awash at around 4 Ma (Kalb and Mebrate 1993). The species is recorded also from Kanapoi, Koobi Fora, Kanam East, the Kaiso Formation, and Shungura Members B–E.

Revision by Haile-Selassie et al. (2004c) and Saegusa and Haile-Selassie (Chapter 15) has demonstrated that the alleged Middle Awash occurrence of *Stegotetrabelodon* advocated by Kalb and Mebrate (1993) is wrong, being based on misidentifications. The absence of the genus in the Middle Awash is of interest, given the wide distribution of *Stegotetrabelodon* in latest Miocene deposits of Abu Dhabi, Sahabi, Cessaniti, Toros Menalla, Lothagam, Mpesida, Kaperyon, Manonga, and Shuwaihat (Tassy 1999; Ferretti et al. 2003).

Mammuthus is a proboscidean genus that apparently originated within Africa, but is poorly sampled from the African fossil record. It is better-known from Plio-Pleistocene

Eurasia (Lister et al. 2005). The earliest mammoths probably diverged from their last common ancestor with the African elephants ca. 4–6 Ma (Shoshani and Tassy 2005; Todd 2006), or ca. 6–7 Ma (Tassy 2003). The ASKM probably represents the earliest fossil record of the genus at 5.5–5.8 Ma. None of the contemporaneous eastern African sites has yielded remains of *Mammuthus*. Dental specimens from Langebaanweg (South Africa; Maglio and Hendey 1970; Maglio 1973) and the Middle Awash (Ethiopia; Kalb and Mebrate 1993) have been assigned to *Mammuthus subplanifrons*. However, the validity of the species was questioned by Mundiger and Sanders (2001). They argued that *Mammuthus subplanifrons* represents a heterogenetic mixture of eastern African and South African elephants. While some of the specimens addressed by Mundiger and Sanders (2001) belong to *Mammuthus,* the Asa Koma record represents the earliest fossil occurrence of the genus in Africa south of the Sahara. Its presence in the ASKM, despite the small sample size, could also represent a biologically significant FA (BSFABM) for the genus. The genus possibly migrated north immediately after its first appearance, possibly the result of proliferation of other proboscideans in eastern Africa at the end of the Miocene. The absence of *Mammuthus* in the KUSM assemblage, not considering taphonomic bias, is noteworthy, particularly in light of its presence ($n = 1$ of 43 elephantid dental specimens) in the ASKM. It cannot be ruled out that its absence in the KUSM could be a true local absence. It first appears outside Africa in the middle Pliocene (Maglio 1973).

Perissodactyla

The Middle Awash late Miocene equid fauna is currently restricted to *Eurygnathohippus feibeli* Bernor and Harris, 2003, and *Eu.* aff. *feibeli. Eurygnathohippus* is first reported in the Lower Nawata of Lothagam (Bernor and Harris 2003), represented by a large, robust-limbed *Eu. turkanense,* and a smaller, gracile-limbed *Eu. feibeli*. There is no certain evidence of *Eu. turkanense* elsewhere in the late Miocene of Africa, but there are robust metapodials from the Middle Awash early Pliocene that may represent late records of the species (Bernor et al. 2005b). Bernor et al. (2005b) recently evaluated the evolution of the *Eu. feibeli–hasumense* lineage (6.0–2.9 Ma), based on postcranial anatomy, and characterized it as having evolved gradually through that time interval. *Eurygnathohippus* is closely related to the Siwalik horse *Sivalhippus perimense* (*sensu* Bernor and Hussain 1985; Bernor and Lipscomb 1991, 1995). The *Sivalhippus–Eurygnathohippus* clade is distinct from all other Eurasian late Miocene hipparions with the exception of the earliest-occurring members of *Plesiohipparion* from China (Qiu et al. 1987). There are multiple species of *Hippotherium, Hipparion sensu stricto,* and *Cremohipparion* that occurred in Western Eurasia, with *Hipparion sensu stricto* also occurring in China (Bernor et al. 1996a, c), and *Cremohipparion* also occurring in the Siwaliks and at Sahabi (Bernor and Scott 2003). There is very limited evidence that a member of the *Sivalhippus–Eurygnathohippus* clade occurred at Sahabi, but it was rare there, and more fossil material is needed to confirm this occurrence.

Based on current studies of Siwalik and African late Miocene equids, the paleogeographic connection of Siwalik and eastern African equids was active by 8 Ma. The occurrence of an advanced hipparion at Chorora, ca. 10+ Ma (Bernor et al. 2004a), suggests the possibility of an even earlier connection. The equids from Langebaanweg were referred to "*Eurygnathohippus*" *baardi* (Franz-Odendaal et al. 2003), but now Bernor and Kaiser (2006) recognize a new species, *Eurygnathohippus hooijeri*. Their analysis of *Eu. hooijeri* shows a close relationship of this species to the Old World *Cormohipparion* lineage in the postcranial morphology, and at the same time, highly derived cheek tooth crown heights of 80 mm. No known late Miocene or early Pliocene eastern African hipparion has such a high crown. Overall, *Eu. hooijeri* exhibits an interesting assortment of symplesiomorphic and autapomorphic characters, and it may be an early vicariant species of the *Eurygnathohippus* clade (Bernor, unpublished). The *Eurygnathohippus* clade is distinctly African in the late Miocene and exhibits disjunction from Eurasian late Miocene mammal faunas at the beginning of the Lower Nawata. We believe that the paleogeographic connection to southern Asia must have occurred between 10 Ma and 8 Ma (Bernor, unpublished).

At least two rhinoceros taxa occur in the Middle Awash late Miocene (see Giaourtsakis et al. Chapter 14). A *Diceros* very similar to *D. douariensis* is known from the KUSM by a complete cranium. This species was previously known from its type locality in northern Tunisia. Some specimens from the late Miocene of Djebel Krechem el Artsouma, Central Tunisia (Geraads 1989), have also been referred to this species. Additional specimens from Baccinello V3, Italy (Guérin 1980b) and Namibia (Guérin 2000) have been referred to this species (although this is very doubtful). The rhinoceros remains from the western margin of the Middle Awash (5.54–5.8 Ma) are tentatively referred to *Diceros* sp., although it is quite obvious that they are different from *D. douariensis*. There are also isolated remains that are currently not identified to the genus or species level. The black and white rhinoceroses occur in Africa today and are believed to have a close phylogenetic relationship. *"Ceratotherium" neumayri* is a white rhinoceros-like form known from Maragheh, Pikermi, Samos, Sahabi, and Lothagam (Upper Nawata and Kaiyumung Members; Harris and Leakey 2003a). *"Ceratotherium"* is reported as first occurring in late MN 9, ca. 9.7 Ma, of Turkey (Koufos 2003). *Diceros* has a far more restricted reported occurrence in our comparative sample, being reported from the Upper Nawata and Apak of Lothagam (Harris and Leakey 2003a). Heissig (1996) has suggested that *Ceratotherium* may have first occurred in the Vallesian of Anatolia, but this has been recently cast in some doubt (Heissig, personal communication). It remains to be demonstrated that *"Ceratotherium" neumayri* is truly a member of the *Ceratotherium* clade, or a functional convergent on that clade (Giaourtsakis, personal communication). It is evident, however, that *"Ceratotherium" neumayri* was well-established in the Turolian of Western Eurasia and may prove to be an important source for this lineage (see Geraads 2005).

Artiodactyla

Thus far, *Nyanzachoerus* has a strictly African-Arabian distribution. There are several species of *Nyanzachoerus* represented in the Middle Awash (see Haile-Selassie, Chapter 10). *Nyanzachoerus devauxi* is reported from Sahabi and Lothagam Nawata; *Ny. syrticus* is also reported from Baynunah, Toros Menalla, and Lothagam Nawata; *Ny. australis* is also reported at Lothagam Nawata and its type locality at Langebaanweg. The appearance of two new nyanzachoere species in the KUSM indicates that the genus possibly underwent adaptive radiation at the end of the Miocene. Both *Ny. devauxi* and *Ny. syrticus* were replaced by *Ny. australis, Ny. waylandi,* and *Ny. kuseralensis. Nyanzachoerus australis* makes its SGFA in the Middle Awash. Although it is possible to infer from dental morphology that *Ny. australis* directly descended from *Ny. syrticus,* the origin of the other two species is difficult to ascertain at this time. *Nyanzachoerus kuseralensis* is currently limited to the Middle Awash KUSM (SGFA and SMAFA), whereas *Ny. waylandi* (SMAFA) has also been documented (by rare specimens) from Kenya and Uganda (Pickford 1989a).

Nyanzachoerus is an advanced clade of tetraconodont pigs evidently related to the Siwalik *Sivachoerus.* Van der Made (1999) erected the tribe Nyanzachoerini including *Conohyus, Nyanzachoerus,* and *Lophochoerus* that form a sister lineage to the *Sivachoerus* species. Harris and Leakey (2003b) found Van der Made's restricted use of few dental characters in lieu of crania insufficient to support his "non traditional" alignment of *Notochoerus* and *Nyanzachoerus* in a superspecific, distinctly African-Arabian grouping. It is apparent, however, that the Siwaliks and Africa shared tetraconodont pigs and that there was a paleogeographic connection in the late Miocene by, or somewhat before, the Lower Nawata interval.

Cainochoerus cf. *africanus* is a small suid of uncertain subfamilial status (Harris and Leakey 2003b). *Cainochoerus* is reported from the Upper Nawata of Lothagam as well as Langebaanweg (Harris and Leakey 2003b), and it would appear to have been confined to Africa.

Hippopotamids of the Middle Awash late Miocene are represented mostly by isolated dentognathic material from the ASKM and more complete specimens from the KUSM (see Boisserie and Haile-Selassie, Chapter 11). The remains from the younger KUSM are assigned to aff. *Hippopotamus dulu* Boisserie, 2004, and detailed phylogentic analysis of this group has been made by Boisserie (2005). The ASKM specimens are tentatively assigned to Hippopotamidae indet. due to their fragmentary nature (although more complete material recovered in 2007 is currently being cleaned and promises to clarify identity). A number of species from numerous Plio-Pleistocene sites of Africa have been assigned to this provisional genus name, aff. *Hippopotamus,* which appeared in the fossil record of Africa around 6 Ma. This genus, initially known only from Africa, is the sister taxon of *Hexaprotodon* and the stem group of the extant European and Mediterranean hippopotamuses (Boisserie 2005).

There are three giraffid taxa in the Middle Awash late Miocene fauna: *Palaeotragus* sp., *Giraffa* cf. *stillei,* and *Sivatherium* sp. (see Haile-Selassie, Chapter 12). *Paleotragus* was a diverse and broadly distributed genus in the western Eurasian Miocene. Koufos (2003) reports that *Paleotragus* first occurred in North Africa and Spain during the middle Miocene. This, in fact, may not be of the same lineage, instead being related to the late Miocene *Palaeotragus* known from the eastern Mediterranean. Sinap Locality 4, 10.92 Ma, records the presence of *Palaeotragus* (Gentry 2003; Koufos 2003). From then on, *Palaeotragus* had an extensive eastern Mediterranean-southwest Asian, Arabian, and African distribution, including recorded occurrences from Maragheh, Pikermi, Samos, Baynunah, and Lothagam Nawata. However, the genus disappears from the fossil record toward the end of the Miocene. Its presence in the ASKM and its absence in the KUSM further confirm that the genus went extinct before Kuseralee times.

Within our comparative sample, *Giraffa* is only recorded at the Middle Awash. However, other giraffine taxa are reported from Maragheh *(Honanotherium)*, Pikermi (*Bohlinia)*, and Toros Menalla (*Bohlinia*). Koufos (2003) reports a "*Bohlinia* Event," with the first occurrence of this giraffine clade in the Vallesian of Turkey and Greece.

The large sivathere *Sivatherium* sp. is reported from the Middle Awash, Toros Menalla, and Lothagam Nawata. Although *Sivatherium* is not reported in our Eurasian or Arabian sample, the closely related form *Helladotherium* is broadly distributed in our European–western Asian sample, including Baltavar, Italy MN 13, Maragheh, Pikermi, and Samos. *Helladotherium* has its earliest recorded occurrence in the terminal Vallesian locality of Nikiti I, <9.3 Ma, Greece (Koufos 2003). The Siwalik taxon ?*Bramatherium* is believed to occur in the Baynunah fauna (Gentry 1999b). Taken as a whole, the Middle Awash late Miocene giraffid fauna exhibits broad paleogeographic relationships with the western Eurasian earlier Turolian interval, and more proximately, the later Turolian of northern and eastern Africa.

Fourteen bovid taxa are reported from the Middle Awash late Miocene (see Haile-Selassie et al., Chapter 9). Of these, two genera, *Gazella* and *Tragoportax,* are geographically broadly distributed among western Eurasian and African faunas and are believed to have emigrated from Asia (Koufos 2003). *Gazella* first occurs in the Vallesian (MN 9), late Miocene of Turkey (Gentry and Heizmann 1996) and is recorded in our sample from Baltavar, Spanish MN 13, Maragheh, Pikermi, Samos, Maramena, Baynunah, and Lothagam Nawata. *Tragoportax* is a common late Miocene Circum-Mediterranean form. Its first recorded occurrence is Sinap Locality 108, 10.135 Ma (Lunkka et al. 1999; Gentry 2003) correlative with the late Vallesian, MN 10 (Koufos 2003). *Tragoportax* is broadly distributed within our comparative sample, being reported from Spanish MN 13, Maragheh, Pikermi, Samos, Marmena, Baynunah, Sahabi, and Lothagam Nawata. Both *Gazella* and *Tragoportax* exhibit broad paleogeographic relationships between the western Eurasian and African late Miocene.

Among the reduncines, *Zephyreduncinus oundagaisus* (Vrba and Haile-Selassie 2006) and *Redunca ambae* (see Haile-Selassie et al., Chapter 9) make their first appearance (SGFA and SMAFA) in the ASKM and KUSM, respectively. These species have not been sampled from elsewhere. The genus *Kobus* is known in the Middle Awash late Miocene by two species, one of which *(Kobus* aff. *oricornus)* makes its first biologically significant appearance (SGFA and SMAFA) in the KUSM. This species might have given rise to the younger *Kobus basilcookei* sampled from 4.4 Ma deposits at Aramis, Middle Awash (Vrba 2006). The genus is documented from earlier deposits at Toros Menalla and Lothagam Nawata. The Middle Awash taxa *Aepyceros, Madoqua, Raphicerus,* and *Tragelaphus* have a wider temporal and spatial distribution. They occur at Lothagam Nawata Formation and younger Pliocene deposits of eastern and South Africa.

Biochronological Significance

The Middle Awash ASKM and KUSM faunal assemblages are radiometrically dated to between 5.2 Ma and 5.8 Ma (see Renne et al., Chapter 4). The biochronology of various taxa present in these members is consistent with this age range. However, the major biochronological significance of the Middle Awash late Miocene faunal assemblages lies with the establishment of more accurate and reliable biochronological ages for faunal assemblages from other African sites with looser temporal control. The Wembere Manonga Formation (Tanzania; Harrison 1997a), Langebaanweg (South Africa; Hendey 1982), and Toros Menalla (Chad; Vignaud et al. 2002) are good examples in this regard, and they clearly benefit from the well-calibrated Middle Awash late Miocene in terms of biochronologically based age refinement.

The faunal assemblage from the Ibole Member of Wembere Manonga Formation has been biochronologically dated to 5.5–5 Ma (Harrison and Mbago 1997). *Nyanzachoerus kanamensis* has also been documented from the Ibole Member (Bishop 1997). However, this species is thus far known at other African sites in deposits younger than 5 Ma. The Ibole Member fauna in general appears to be more derived than the KUSM fauna, indicating an age younger than 5.2 Ma for the former. The taxa shared between the Middle Awash late Miocene and the Ibole Member are taxa that persist into the Pliocene. In fact, the Ibole Member fauna most closely resembles a yet unpublished faunal assemblage from the Haradaso Member of the Middle Awash and dated to ca. 4.8 Ma (Renne et al. 1999).

The biochronological age of the faunal assemblage from the Quartzose Sand Member (QSM) at Langebaanweg has ranged from ca. 4.5–6 Ma. Cooke and Hendey (1992) reported an age of ca. 4.8–5.5 Ma for the QSM suid remains. We now know that there is substantial faunal overlap between the KUSM and QSM faunas, most likely indicating contemporaneity. These two members share a number of taxa in various families. The dominance of the short-lived suid species *Nyanzachoerus australis* at QSM, combined with its exclusive presence in the KUSM, and possibly in the upper part of the Nawata Member at Lothagam (Harris and Leakey 2003b), indicates at least an age of ca. 5.2 Ma for the QSM. Other QSM taxa such as equids (Franz-Odendaal et al. 2003; Bernor and Kaiser 2006) and proboscideans (Sanders 2007) indicate an age of ca. 5 Ma for the QSM.

The fauna from Toros Menalla (Chad) is diverse and has been biochronologically placed at 6–7 Ma or older (Vignaud et al. 2002). There are not yet any radiometric dates, although attempts are currently being made. Brunet et al. (2005) recently reported that the Toros Menalla fauna and the associated hominid *(Sahelanthropus tchadensis)* from the site are probably as old as 7 Ma. Brunet et al. (2005: 752) wrote, "The TM 266 locality in the Toros-Menalla fossiliferous area yielded a nearly complete cranium (TM 266-01-60-1), a mandible, and several isolated teeth assigned to *Sahelanthropus tchadensis* and biochronologically dated to the late Miocene epoch (about 7 million years ago)."

The biogeographic analysis presented in Chapter 18 (Bernor et al.) shows that there is a substantial amount of faunal overlap between the Middle Awash late Miocene Asa Koma Member assemblage and that from Toros Menalla in Chad. The latter also shows close similarities with the incompletely described assemblages from Lukeino (6.0 Ma). Given the radioisotopic ages of the Middle Awash late Miocene (5.2–5.8 Ma)—and the faunal similarities between Toros Menalla, the Middle Awash, and Lukeino—Toros Menalla may not be more than a million years older than the Middle Awash and Lukeino. The final assessment will, of course, have to await ongoing radioisotopic dating attempts and detailed description of the Toros Menalla fauna, including the description of biochronologically significant taxa such as the proboscideans and suids.

Discussion

The role of the environment in driving, forcing, or influencing human evolution has always played a role in interpretations of human evolution, dating back to the efforts of Darwin and Dart, and becoming particularly prominent following Brain and Vrba's pioneering efforts during the last three decades. A number of environmentally based hypotheses have been proposed following the "turnover pulse" hypothesis (Vrba 1980a; Brain 1981; Vrba 1985b, 1988; Vrba et al. 1989; Vrba 1995a,b; McKee 2001; Hernández Fernández and Vrba 2006), an attempt to examine the relationship between climatic and faunal change. Criticisms of the correlational approach have centered on the adequacy of the available empirical fossil data and on the associated apparent taphonomic and sampling biases (Hill 1987; Kimbel 1995; White 1995; Behrensmeyer et al. 1997; Werdelin and Lewis 2001b, 2005).

The molecular and fossil evidence for human origins converges on the hypothesis that hominids arose during the late Miocene. Chadian *Sahelanthropus* is arguably placed between 7 Ma and 6 Ma (Brunet et al. 2002, 2005; Vignaud et al. 2002). *Orrorin tugenensis* (Senut et al. 2001), from Kenya, is dated between 6.0 Ma and 5.7 Ma; and the Middle Awash *Ardipithecus kadabba,* described in this volume (Chapter 7) is dated as early as 5.8 Ma (Haile-Selassie 2001a,b; Haile-Selassie et al. 2004b,c). Environmental factors that might have triggered the emergence of hominids are currently under study. The Messinian Salinity Crisis (MSC), which extended the last one-third of the Messinian Stage, ca. 5.9–5.3 Ma (Hodell et al. 2001), had enormous impact on the composition of faunal communities of northern Africa, western Asia, and southern Europe. Its impact on Africa south of the Sahara and on the emergence of hominids is not yet understood.

What effects might the Messinian have had on equatorial African faunas? The presence of *Eucyon, Viverra,* cf. *Loxodonta,* aff. *Hippopotamus, Sivatherium, Simatherium, Redunca,* and *Gazella* in the KUSM are possibly their first local and/or regional appearances. *Viverra,* cf. *Loxodonta,* aff. *Hippopotamus, Simatherium,* and *Redunca* are likely derived from African lineages. *Eucyon* has a North American origin. *Gazella* is abundant in earlier late Miocene horizons of Eurasia. *Sivatherium* is likely derived either from the southwest Asian *Helladotherium* or Siwalik *Bramatherium* (also known from the latest Miocene of Arabia). Most of these genera have been documented from slightly older deposits of Lothagam, Nakali, and Lukeino. However, it appears that *Simatherium* appears for the first time in the KUSM. *Eucyon* is a fox-sized dog that emigrated from North America at the end of the Miocene and rapidly extended its range throughout Eurasia and Africa. Its appearance is a legitimate intercontinental FA. Its eastern African first occurrence is now documented in the Middle Awash (Ethiopia), Lukeino (Kenya; Morales et al. 2005), and Langebaanweg (South Africa; Morales et al. 2005), giving *Eucyon* a regional FA of 6.0 Ma at Lukeino.

Nyanzachoerus syrticus and *Ny. devauxi* make their LAs in the ASKM. Their disappearance is consistent with results from other penecontemporaneous African sites. The Middle Awash record indicates that *Nyanzachoerus* underwent cladogenesis during the late Miocene, with at least three new species appearing in the KUSM (*Ny. australis, Ny. waylandi,* and *Ny. kuseralensis* sp. nov.). In some other cases, we see evolutionary changes within apparently evolving lineages. *Nyanzachoerus australis* is more likely to have evolved from *Ny. syrticus*. Also, the *Eurygnathohippus feibeli–hasumense* lineage exhibits a morphological step in the postcranial anatomy in the KUSM. These "turnovers" appear to have occurred between 5.5 Ma and 5.2 Ma, some of them the result of phyletic evolution.

Both the ASKM and the KUSM sample a number of provincial mammalian FAs and LAs. Paleoenvironmental analysis of the KUSM and ASKM shows that the two members sample similar paleoecological settings, with subtle differences in proportions (see Su et al., Chapter 17). Differences in the relative abundances of indicator taxa between the two members may reflect subtle ecological distinctions in woodland and grassland representation. Whether these changes stem from local tectonic influences, global, or more regional climatic changes cannot currently be determined with any confidence. If climatic change is the cause for differences between the two successive faunal samples, are there similar contemporaneous changes in other parts of Africa?

The data are inadequate to answer this question. It is possible that local, rather than regional, continental, or global environmental changes created these taxonomic differences between the ASKM and KUSM faunas. This is also the time when the MSC is believed to have come to an abrupt end (5.32 Ma), causing a major shift to cooler and more humid continental climates and a turnover in European and Asian mammal faunas (Hodell et al. 2001). However, the effect of the end of the MSC on Africa south of the Sahara is currently unknown; nor is there secure evidence that the end of the MSC might have influenced the changes in the Ethiopian faunas. Indeed, given the lack of additional stratigraphic and chronological constraints, it is not possible to determine the exact temporal relationship between the sub–5.2 Ma basalt KUSM fauna and the end of the MSC.

Isotopic studies indicate that toward the end of the Miocene (8–6 Ma), eastern African environments were undergoing an environmental succession from tropical woodlands to expanding grasslands, largely because of the decrease in atmospheric CO_2 (Cerling et al. 1997b). In contrast, beginning with the Pliocene, the expansion of grasslands has been attributed to Indian Ocean sea surface temperature cooling (Cane and Molnar 2001) or the onset of glacial-interglacial cycles (deMenocal 2004). These findings usually attribute tectonic processes as secondary factors for the recurrent eastern African climate changes during the late Neogene. Sepulchre et al. (2006), however, argue that tectonic uplift of eastern Africa is one of the major causes for climate changes in eastern Africa during the late Neogene, and they further suggest that accurate paleonvironmental reconstruction should take topographic history of eastern Africa into account.

The ASKM is located on the western rift margin of the Main Ethiopian Rift (MER), whereas the KUSM is located slightly more toward the rift axis. This suggests a possible topographic and altitudinal difference between the two members. The presence of the micromammals *Tachyoryctes* and *Thryonomys* in the ASKM (and the absence of the former in the KUSM) also suggests that the ASKM was deposited at a higher elevation than it occupies today. Faunal differences between the two members would be expected as a result of this difference. Therefore, altitude appears to be a complicating factor in any assessment of faunal differences between the ASKM and KUSM.

What are the causes for the apparent faunal differences between the ASKM and KUSM? The FAs and LAs documented between the two members are largely attributable to phyletic evolution, a phenomenon seen in other Miocene and Pliocene successions (Werdelin and Lewis 2001b; Badgley et al. 2005). Climatic change may not be the most plausible reason for the differences between the ASKM and the KUSM because they share numerous taxa. Perhaps tectonic processes might better explain the limited differences seen between the two Middle Awash faunal sets. Continued isotopic analysis and results from new research methods might better dissect correlation between climate and faunal changes and the effects of tectonic uplift during the late Neogene of eastern Africa. The Middle Awash will provide a continuing testing ground for such questions and analyses.

Conclusions

The latest Miocene–early Pliocene Middle Awash assemblages show mammalian diversity, slightly different ecological settings, moderate turnover, and biochronological placement fully consistent with available radiometric dates. A total of 65 mammalian taxa are recognized from the Asa Koma Member of the Adu-Asa Formation and the Kuseralee Member of the Sagantole Formation. The two members share a total of 25 genera. These two faunal assemblages also document a number of FAs and LAs. The earlier ASKM documents the first global appearance of 20 species (SGFA in Table 19.1) and a global last appearance of 14 species (SGLA and SMALA in Table 19.1). Between-member faunal differences can be brought to bear on questions of evolutionary tempo and mode in African mammals at the end of the Miocene. The ancestor-descendant relationships documented along several lineages spanning the two members shows that phyletic evolution is one of the causes for apparent differences seen in two time-successive faunal assemblages. The two late Miocene faunal assemblages from the Middle Awash are also significant in terms of presenting more refined biochronologic dates for other African sites, such as Wembere-Manonga and Langebaanweg, establishing a date younger than 5.2 Ma for both sites.

Comparison of the Middle Awash late Miocene faunal assemblages with Eurasian counterparts from the MN 11–13 European biochronologic units clearly indicate the presence of common taxa in the two geographically distinct regions (mostly at the generic level but sometimes at the specific level). These faunal similarities indicate faunal interchange and, in some cases, contemporaneity, whereas the differences indicate temporal and spatial, more so than climatic, differences (see Bernor et al., Chapter 18, for details).

20

Conclusions

YOHANNES HAILE-SELASSIE AND GIDAY WOLDEGABRIEL

This volume presents the work of 27 researcher-authors spanning the fields of paleontology, geology, and paleoecology. The Adu-Asa and lower Sagantole Formations, ranging in age between 5 Ma and 6.1 Ma, provide new information on eastern African Rift geology and African mammal evolution during the transition from the Miocene to the Pliocene. The recovery of 15 hominid specimens from the Asa Koma Member, dated to 5.5–5.8 Ma, boosts the scantily known hominid fossil record prior to 5 Ma and the more than 2,700 total vertebrate specimens add to our understanding of the evolution of various mammals that would come to establish the modern African fauna.

The Middle Awash late Miocene fossil assemblages document the originations and extinctions of numerous mammalian taxa. They indicate the presence of broad biogeographic relationships, both intra-African and between Africa and Eurasia, toward the end of the Miocene. They also document a mosaic of habitats, largely dominated by dense woodland, in eastern Africa at the end of the Miocene long before the expansion of grasslands in the Plio-Pleistocene. The combined results from all aspects of paleontology, geology, biogeography, paleoecology, biochronology, and geochronology of the Middle Awash late Miocene indicate that hominids did not evolve in a savanna setting and that time-successive faunal communities underwent changes for a variety of reasons including local and regional climate changes, differences in elevation induced by tectonic processes, and more global phenomena.

The following sections summarize the geology, paleontology, and paleoenvironment and feature the potential of the Middle Awash late Miocene.

Geology

The project area is in a unique tectonic and volcanic transition zone between the northern sector of the Main Ethiopian Rift (MER) and the Afar Rift. A funnel-shaped rift floor and Quaternary axial rift zone, voluminous hydromagmatic deposits, an arcuate step-faulted broad western margin, a rift-bound domed structure (the Central Awash

Complex), the transverse Bouri horst, and numerous horst and graben structures (mostly confined to the eastern side of the rift floor) characterize the area. Moreover, the western and southeastern rift margins of the transition zone are defined by multiple, densely spaced antithetic fault blocks and half grabens, in contrast to the singular boundary faults of the MER farther to the southwest. The current structural, volcanic, and geomorphic features of the transition zone are products of complex tectonic processes and interactions among the oceanic rifts (i.e., the Red Sea and the Gulf of Aden) and the continental Ethiopian Rift System. The Middle Awash study area is today divided into the east and west sides by the Awash River.

The western margin, the core subject of this volume, is made up of late Miocene (6.3–5.5 Ma) lacustrine and fluvial sedimentary rocks interbedded with generally proximal hydromagmatic mafic and distal silicic tephra deposits. The lithologic units exposed along the frontal fault blocks of the western rift margin of the Middle Awash are assigned to the Adu-Asa Formation, which is subdivided into the Saraitu, Adu Dora, Asa Koma, and Rawa Members in ascending stratigraphic order. The Adu-Asa Formation was deposited following widespread basaltic eruption and a brief volcanic and tectonic quiescence. By the late Miocene, the area currently occupied by the frontal fault blocks of the western rift margin was subjected to intense tectonic activities and subsidence that led to the widespread lacustrine deposits of the Saraitu and Adu Dora Members of the Adu-Asa Formation. Intense and voluminous hydromagmatic eruptions and fluvial sedimentation disrupted the lacustrine environments, resulting in the formation of the terrestrial fossil-rich Asa Koma Member. The Rawa Member was deposited during the waning stages of the hydromagmatic eruptions at the end of the Miocene period and represents diverse units from fluvial sedimentation in a more stable environment characterized by tectonic and volcanic quiescence.

Paleontology

A total of >2,700 vertebrate fossil specimens were collected from the Asa Koma and Kuseralee Members. Twenty percent of the specimens were collected from the Kuseralee Member, and 80 percent (n = 2,209) from the older Asa Koma Member. These fossil assemblages are largely mammalian, although a number of Crocodylia, Chelonia, Aves, and aquatic vertebrates have also been collected. Artiodactyls are the most abundant mammals (49.1 percent; Figure 20.1) in the Middle Awash late Miocene. The diversity of mammalian taxa in the two members is comparable (Figure 20.2), although artiodactyls and perissodactyls appear to be more abundant in the younger Kuseralee Member. The high percentage of proboscideans in the Asa Koma Member is skewed by the "elephant bone bed" at Saitune Dora locality 2 (STD-VP-2), from which 735 proboscidean specimens were collected, representing at least 27 individuals ranging in age from newborns to old individuals (Figure 20.3).

The micromammalian fossil collection is relatively large. At least 13 species are known from 11 genera, most of them deriving from the older Adu Asa Member. Four new species are recognized in the genera *Lophiomys, Thryonomys, Atherurus*, and *Tachyoryctes.* Genera such as *Crocidura, Lemniscomys, Tatera, Hystrix, Xenohystrix, Atherurus,* and *Serengetilagus*

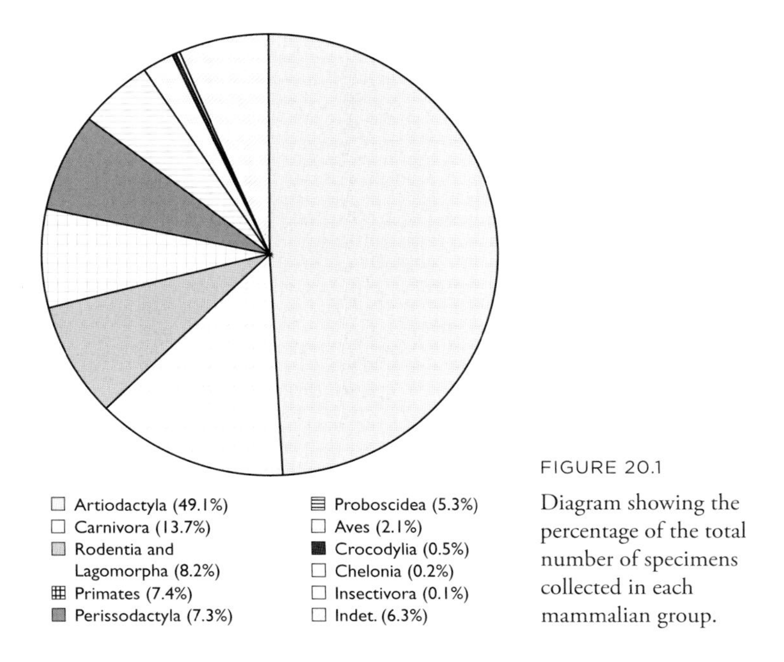

FIGURE 20.1

Diagram showing the percentage of the total number of specimens collected in each mammalian group.

appear to be absent from penecontemporaneous eastern African sites such as Lothagam (Upper Nawata Member; Winkler 2003). However, they are abundant in younger Pliocene deposits (Wesselman 1984), indicating that the Middle Awash may mark their first appearances in eastern Africa. The lack of similarities between Lothagam and the Middle Awash in the micromammalian community is of interest, with possible implications for paleoenvironment.

Primates constitute 7.4 percent of the combined collection. This includes one hominid species, *Ardipithecus kadabba,* and at least four cercopithecid species (*Pliopapio alemui, Kuseracolobus aramisi,* a large colobine, and a small species yet unidentified to the subfamily level). The former two cercopithecid species are also known from 4.4 Ma deposits at Aramis, Middle Awash, Ethiopia (Frost 2001b). *Ardipithecus kadabba, Pliopapio alemui,* and *Kuseracolobus aramisi* mark their first appearance here and have not yet been documented from other sites.

Carnivores are more abundant than the primates and their diversity is much greater, with 13 known species and genera. Most of the genera are immigrants from Eurasia, appearing in sub-Saharan Africa toward the end of the Miocene (co-occurrences within MN 13 European faunal zone). This indicates strong intercontinental biogeographic relationships during the late Miocene. Three new carnivore species are recognized: a mustelid (*Plesiogulo botori* sp. nov.), a herpestid (*Herpestes alaylaii* sp. nov.), and a dog-like hyaenid (*Hyaenictis wehaietu* sp. nov.). Most of these genera have Eurasian origins. The genera *Agriotherium, Machairodus, Deinofelis, Viverra, Genetta, Sivaonyx,* and *Mellivora* are also known from other contemporaneous Eurasian and eastern and South African localities, indicating a wider late Miocene distribution of these genera. The new carnivore species from the Middle Awash increase the diversity of carnivores in eastern Africa at the Mio-Pliocene boundary.

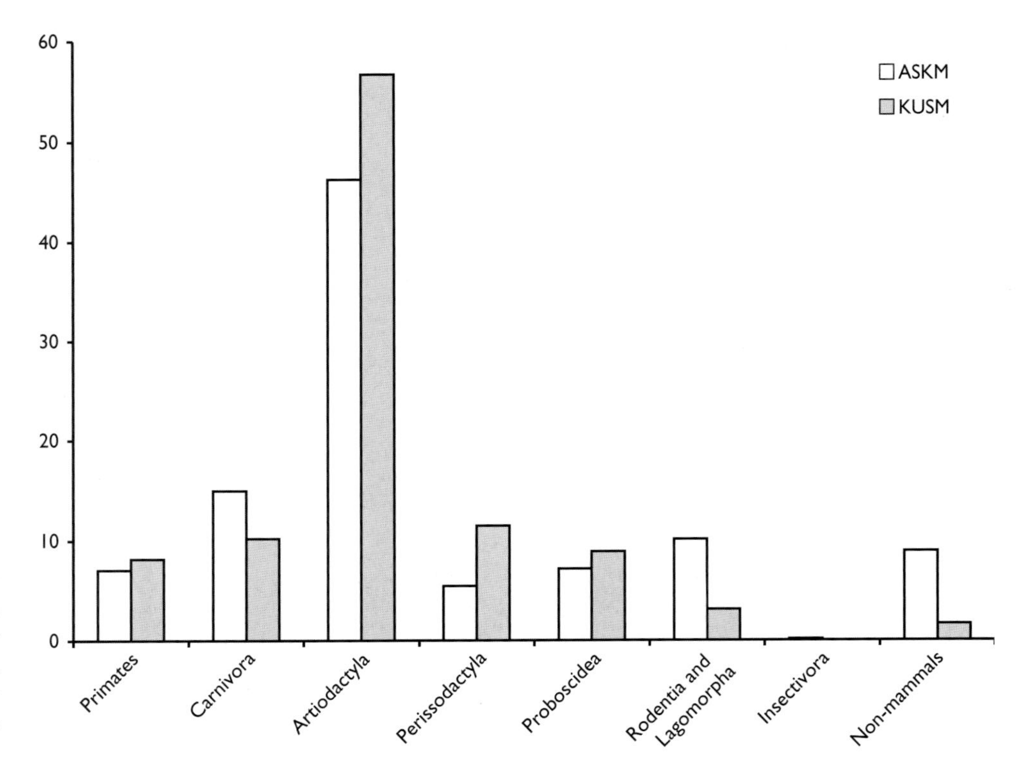

FIGURE 20.2

Bar graph showing the proportions of the major taxonomic groups in the Asa Koma Member (ASKM) and Kuseralee Member (KUSM).

Artiodactyls are the most abundant terrestrial mammals, constituting 49.1 percent of the total number of fossil specimens collected from the two members. Within the artiodactyls, bovids are the most abundant (57.6 percent), followed by suids (29.1 percent) and hippopotamids (10.3 percent). Giraffids are the least abundant (3 percent), as is usually the case in most African Neogene localities.

At least 11 genera and 17 species are recognized in the bovid tribes Tragelaphini, Boselaphini, Reduncini, Bovini, Antilopini, and Aepycerotini. The tragelaphines are represented by a new species, *Tragelaphus moroitu* sp. nov., recovered from both members, and a second species similar to *T. moroitu*. This species is more likely ancestral to *Tragelaphus kyaloae,* which is known from most early Pliocene localities in eastern Africa. Boselaphines are diverse, with at least three known species, one of them a new species, *Tragoportax abyssinicus* sp. nov. A much larger *Tragoportax* species is also known from the older Asa Koma Member. Boselaphines are abundant in the late Miocene of eastern Africa, with three more *Tragoportax* species documented from the Upper and Lower Nawata Members of Lothagam (Harris 2003c), with *Tragoportax* sp. A from Lothagam possibly belonging to *Tragoportax abyssinicus* sp. nov. Reduncines were also relatively abundant, with at least five species known in three or four genera. This includes a new genus and species, *Zephyreduncinus oundagaisus,* and a new *Redunca* species, *Redunca ambae* sp. nov. Reduncines were as diverse in the Middle Awash late Miocene as in other contemporaneous eastern African localities such as

FIGURE 20.3

View to the southeast at Saitune Dora locality STD-VP-2. Proboscidean bones extracted from the colluvial lag below the bone bed (right of photograph at level of excavators) are sorted and prepared for transport. Photograph by Tim White, 1999.

Lothagam, where at least three species are known (Harris 2003b). The bovine *Simatherium,* previously known from Langebaanweg, South Africa (Gentry 1980), makes its first appearance in the Kuseralee Member. It diversifies in the Pliocene and becomes one of the dominant bovine genera at various Pliocene localities of eastern Africa. *Ugandax* makes its first appearance in the Asa Koma Member, even though Thomas (1980) reported *Ugandax* aff. *U. gautieri* from contemporaneous, or slightly older, deposits at Lukeino, Kenya. This genus is not reported from the two Nawata Members of Lothagam. Although present, antilopines, aepycerotines, and hippotragines are rare elements in the Middle Awash late Miocene deposits. Alcelaphines are entirely absent. This could possibly be due to the lack of suitable open habitats to support them. They were abundant in the Upper and Lower Nawata Members of Lothagam (Harris 2003b).

Suids are represented by two species in the Asa Koma Member, *Nyanzachoerus syrticus (= Ny. tulotos)* and *Ny.* cf. *devauxi.* These species are replaced in the Kuseralee Member by the more derived *Ny. australis, Ny. waylandi,* and a new small tetraconodont, *Ny. kuseralensis* sp. nov. The small, peccary-like *Cainochoerus* is present in both members. *Nyanzachoerus syrticus* has been reported from late Miocene localities in Algeria, Libya, Chad, Lukeino, Lothagam, and Lemudong'o. *Nyanzachoerus devauxi* has also been reported from most of these localities. *Nyanzachoerus waylandi* is known from some late Miocene–early Pliocene sites in eastern Africa. *Nyanzachoerus australis* was first reported from Langebaanweg, South Africa. Its presence in the well-calibrated Middle Awash and Lothagam sequences establishes a biochronological placement of ca. 5.2 Ma for the

Quartzose Sand Member (QSM) of Langebaanweg, where the only known suid species is *Ny. australis*. This correlation is supported by the overall faunal similarity between the Kuseralee Member, dated to 5.2 Ma, and the QSM fauna. *Nyanzachoerus kuseralensis* sp. nov. is currently known only from the Kuseralee Member. The diversity of the suid taxa from the Middle Awash late Miocene shows that the family underwent major radiation at the Mio-Pliocene boundary.

In the Asa Koma Member, the Hippopotamidae are mostly represented by isolated teeth of primitive morphology. These are identified as Hippopotamidae indet. In the Kuseralee Member, a species of Hippopotamidae recently described as aff. *Hippopotamus dulu* is a hexaprotodont hippopotamid similar to other Mio-Pliocene forms in its primitive general morphology. Its cranium and dentition display a distinctive association of measurements and features. This species increases the diversity of the hippopotamid fossil record in eastern Africa and strengthens the hypothesis of hippopotamid endemism in each African basin, as early as the terminal Miocene.

Giraffid remains are rare elements as in all other African Neogene localities. However, three species are suggested from isolated teeth and limited postcrania. A *Palaeotragus* species, better known from other contemporary sites such as Lothagam, is represented by a limited number of dental remains. *Sivatherium* is also present, although it is rare. The fossils tentatively referred here to *Giraffa* may represent the earliest record of the genus in eastern Africa.

The equid fossils from the two late Miocene assemblages represent hipparionine horses of the *Eurygnathohippus* clade in the 6–5.2 Ma time frame. *Eurygnathohippus* was restricted to Africa from 7+ to less than 1 Ma (Bernor and Harris 2003). The hipparion remains from the Middle Awash late Miocene mostly consist of isolated elements, but there are cases of more complete associated material. Mandibular cheek teeth confirm the attribution of material from all localities considered as *Eurygnathohippus.* Although species distinction cannot be made by maxillary and mandibular cheek tooth size and morphology alone, mandibular and postcranial material—metapodials and phalanges in particular—confirm the presence of a small to medium-sized, gracile hipparion. The oldest material, from Jara-Borkana (6.0+ Ma), is closely comparable to the Upper Nawata (Lothagam Hill, Kenya) type material of *Eurygnathohippus feibeli* (Bernor and Harris 2003) and is referred to that taxon. The remaining equids are referred to *Eurygnathohippus* aff. *feibeli,* morphologically distinct from the Nawata Formation robust-limbed form *Eurygnathohippus turkanense* (Bernor and Harris 2003). There is some evidence of evolution within the *E.* aff. *feibeli* assemblage, but there is insufficient morphologic information to distinguish a new taxon.

Rhinocerotid remains are limited to a complete cranium from the Kuseralee Member, assigned to the North African *Diceros douariensis.* Isolated dental and mandibular specimens from the Asa Koma Member are assigned to *Diceros* sp. but are distinct from *D. douariensis.* The Kuseralee Member cranium elucidates some of the issues revolving around the origins and phylogenetic relationships between the black rhinoceros and the white rhinoceros. Its morphological similarity with the northern African *Diceros douariensis* puts it within the same species and at a position close to the ancestral stock of the *Ceratotherium* clade. The crania from Douaria (Tunisia) and the Kuseralee Member recall morphological attributes of their precursors, who might have descended from earlier populations of *Paradiceros mukirii, D. primaevus,* and *D. australis.*

Proboscidean fossils document at least four elephantids, two gomphotheres, and one deinothere between 5.2 and 5.8 Ma. The deinotheres are scant, referred to *Deinotherium bozasi,* the only known deinothere from the African late Miocene–early Pliocene and present in both the Asa Koma and Kuseralee Members. The elephantids include the earliest *Primelephas gomphotheroides,* the new subspecies *Primelephas gomphotheroides saitunensis* subsp. nov., cf. *Mammuthus* sp., and cf. *Loxodonta* sp. There might be a fifth elephantid species in the Asa Koma Member whose genus and species could not be determined due to the fragmentary nature of the sample. *Primelephas gomphotheroides saitunensis* subsp. nov. is known only from the Asa Koma Member. The cf. *Mammuthus* sp. is not present in the Kuseralee Member, although its absence from this member could be a sampling bias, since it is documented from other eastern African Pliocene sites. Cf. *Loxodonta* sp. is not present in the Asa Koma Member. However, the elephantids from the Middle Awash late Miocene suggest the need for a major revision in elephantid systematics, descriptions, and serial identification of isolated teeth.

At least two species of *Anancus* are represented between 5.2 and 5.8 Ma. These species appear to represent the tetralophodont lineage, which includes *A. kenyensis,* and a pentalophodont lineage, which includes *Anancus petrochii, Anancus* sp. (Langebaanweg), and *Anancus* sp. (Sagantole, *sensu* Kalb et al. 1996). The larger *Anancus* from the Middle Awash, tentatively referred to *Anancus kenyensis,* is a tetralophodont. However, there also appears to be a small form in the Kuseralee Member with M^3 morphology and cone pair count different from *Anancus kenyensis sensu stricto.* However, it is conservatively included here in *A. kenyensis,* recognizing dental morphological variation in the species. If it is found to be a different species based on additional specimens, it is the most likely candidate for ancestry of the *Anancus* sp. from Langebaanweg and would indicate anagenetic evolution.

The first and last appearance datums (FADs and LADs) of African mammalian species around the Miocene to Pliocene transition provide evidence for extinctions and originations. In the Asa Koma Member, at least 11 genera and 18 large mammal species make their first regional appearance in eastern Africa, whereas four genera and seven species went extinct in the same region. Five of the newcomers *(Agriotherium, Mellivora, Plesiogulo, Diceros,* and *Mammuthus)* appear to have immigrated from Eurasia, whereas the rest evolved within Africa. In the Kuseralee Member, eight new species appear. None of these species is an immigrant; all appear to have evolved from existing African lineages. On the other hand, at least 12 species make their last African appearance in the Kuseralee Member. A precise interpretation of the disparity in the FADs and LADs in the two Middle Awash late Miocene members must await recovery of more specimens from the Kuseralee Member. Meanwhile, the evidence suggests an environmental trend toward increasing aridity from the Asa Koma Member to the Kuseralee Member, accompanied by both faunal turnover and phyletic evolution.

FIGURE 20.4
Aerial view toward the southeast across deposits between the Jara and Borkana rivers at JAB-VP-1, Middle Awash study area. The Central Awash Complex is the low line of dark hills just below the right horizon. Pliocene deposits east of the modern Awash River are seen as a thin white stripe below the left horizon. This >50 m sequence of sediments and interbedded tephras is capped by a weathering basalt dated to >6.0 Ma. Photograph by Tim White, January 11, 2003.

Biogeography and Paleoenvironment

Biogeographic analysis shows that the Middle Awash late Miocene fauna has closest relationships with the Lothagam Upper Nawata Member fauna, followed by Toros Menalla (Chad), Sahabi (Libya), and Baynunah (United Arab Emirates).

The early-middle Turolian Pikermian faunas of Greece and Iran and the MN13 faunas of Spain and Italy also show more limited faunal similarity to the Middle Awash. The northern and eastern African latest Miocene faunas, particularly species of carnivores and ungulates, appear to have been derived from the Pikermian fauna. However, some taxa, such as species of *Nyanzachoerus,* exhibit relationships between the Middle Awash and other localities that are restricted geographically to Africa and Arabia. The eastern African late Miocene faunas were assembled only in part by faunal immigrations, most likely dating back to the earliest Turolian (8 Ma) or possibly somewhat older.

Paleoenvironmental analysis based on the large mammalian taxa from the Kuseralee and Asa Koma members indicates a faunal community structure similar to that of a modern riparian woodland ecosystem. Further analysis based on carbon isotope, ecomorphology, and relative abundance of bovid taxa and other indicator taxa yielded parallel results, suggesting riparian woodland and wet grassland habitats. When combined with other lines of evidence, the results suggest that dry, open habitats were not abundant, whereas large bodies of water were present, providing a variety of riparian habitats for terrestrial mammals in the Middle Awash at the end of the Miocene. The dominance of lacustrine sedimentary deposits in the lower half of the Adu-Asa Formation, followed by successive and voluminous hydromagmatic eruptions, suggests that wetter habitats than those today prevailed in the western margin of the MER during the late Miocene.

FIGURE 20.5

Ground view of eroding sands, silts, and interbedded bentonite at the JAB-VP-1 locality. No primates have yet been found in these, the earliest known deposits in the Middle Awash study area. Photograph by Tim White, December 14, 2003.

Future Prospects

The late Miocene deposits of the Middle Awash study area are far from being exhaustively exploited. There are a number of uninvestigated windows of exposures deep into the western margin and further to the north of designated localities. Foot and vehicle access to these exposures is extremely difficult because of steep-sided fault scarps and heavy vegetation cover. The resistant volcanics usually obscure small sediment patches from ground or aerial/satellite view. Continued survey will undoubtedly result in the recovery of faunal assemblages even older than the Asa Koma Member assemblage. For example, the Jara-Borkana localities (Figures 20.4 and 20.5) have already yielded such fossils. The Middle Awash project plans to continue its survey and exploration of such new areas.

These exposures are targets for future exploration. Several have already been identified west and northwest of the Afar settlement of Dallifage. Figure 20.6 shows the town situated on the rift floor immediately to the east of the major normal fault, which sweeps to the east in the distance. Presumably, late Miocene sediments in this area are intercalated with steeply dipping basalts. No vertebrate localities have been designated in this area, better known for its rich Pleistocene deposits in the Wallia, Halibee, and Talalak catchments. The new all-weather road to Kassagita has allowed access to sedimentary packages previously hidden from the valley floor, and new localities along the road have been identified. This block is the least-explored in the Middle Awash study area.

The project's ongoing site management program (see White 2004) will also augment current collections. It is obvious that the yield from each known locality decreases through time, particularly after intensive collection by a combination of free survey,

FIGURE 20.6

Aerial view of the new Afar town of Dallifage. Photograph by Tim White, January 11, 2003.

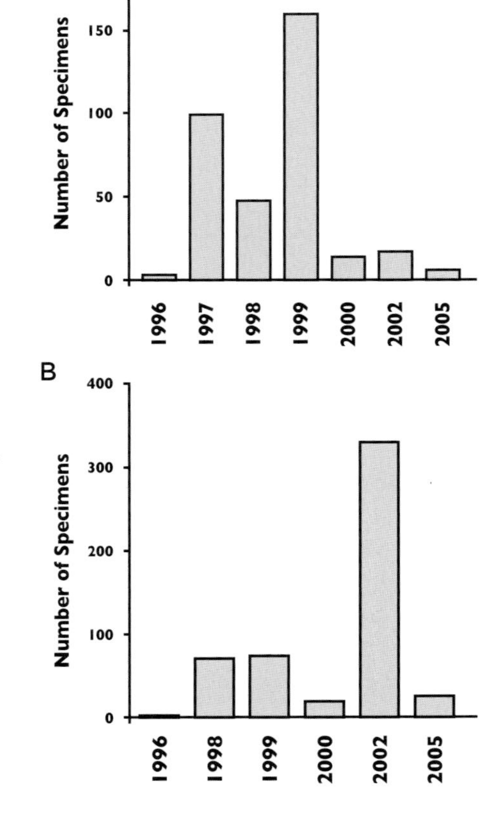

FIGURE 20.7

Bar graphs showing numbers of specimens collected between 1995 and 2005. A. ALA-VP-2. B. ASK-VP-3. The most specimens were collected when a combination of excavation, crawling, and free survey methods was used at the locality. The number of specimens collected after 1999 at ALA-VP-2 and after 2002 at ASK-VP-3 drastically decreased, even though survey intensity and number of collectors increased.

crawling, and excavation. Figure 20.7 shows the numbers of fossil specimens collected from the ALA-VP-2 and ASK-VP-3 localities between 1995 and 2005, respectively. These are two of the three localities that have featured excavation and intensive collection. The number of specimens collected from these localities declined notably after excavations were conducted.

There are still hundreds of fossil specimens on many of these localities. These might be collected in the future to address specific research questions. It is also certain that new specimens will continue to be exposed as a result of erosion. As these processes unfold in years to come, the Middle Awash study area will continue generating new data relevant to the late Miocene evolutionary history and origin of the modern African mammals, including the origins and early evolution of our own family.

Bibliography

Abbazzi, L., M. Delfino, G. Gallai, L. Trebini, and L. Rook. 2008. New data on the vertebrate assemblage of Fiume Santo (northwestern Sardinia, Italy), and overview on the late Miocene Tusco-Sardinian paleobioprovince. *Journal of Palaeontology* 51:425–451.

Aguilar, J.-P., and J. Michaux. 1990. Un *Lophiomys* (Cricetidae, Rodentia) nouveau dans le Pliocène du Maroc: rapport avec les Lophiomyinae fossiles et actuels. *Paleontologia i Evolució* 23:205–211.

Aguirre, E. 1963. *Hippopotamus crusafonti* n. sp. del Plioceno inferior de Arenas del Rey (Granada). *Notas y Comunicaciones del Instituto Geologico y Minero de España* 69:215–230.

Aguirre, E. 1969. Evolutionary history of the elephants. *Science* 164:1366–1376.

Aguirre, E., and P. Leakey. 1974. Nakali, nueva fauna de *Hipparion* del Rift Valley de Kenya. *Estudio Geologia* 30:219–227.

Alcalá, L., P. Montoya, and J. Morales. 1994. New large mustelids from the late Miocene of the Teruel Basin (Spain). *Comptes Rendus de l'Académie des Sciences, Paris* 319:1093–1100.

Alemseged, Z. 2003. An integrated approach to taphonomy and faunal change in the Shungura Formation (Ethiopia) and its implication for hominid evolution. *Journal of Human Evolution* 44:451–478.

Alemseged, Z., F. Spoor, W. H. Kimbel, R. Bobe, D. Geraads, D. Reed, and J. G. Wynn. 2006. A juvenile early hominin skeleton from Dikika, Ethiopia. *Nature* 443:296–301.

Ambrose, S. H., L. J. Hlusko, D. Kyule, A. Deino, M.Williams. 2003. Lemudong'o: A new 6 Ma paleontological site near Narok, Kenya Rift Valley. *Journal of Human Evolution* 44:737–742.

Ameur, R. C., J.-J. Jaeger, and J. Michaux. 1976. Radiometric age of early hipparion fauna in north-west Africa. *Nature* 261:38–39.

Amtmann, E. 1971. Family Sciuridae. In J. Meester and H. W. Setzer (eds.), The Mammals of Africa, an Identification Manual. Part 6.1, pp. 1–12 (not continuously paginated). Smithsonian Institution Press, Washington, DC.

Andrews, P. J. 1978. A revision of the Miocene Hominoidea of East Africa. *Bulletin of the British Museum of Natural History, Geology* 30:85–224.

Andrews, P., and R. L. Bernor. 1999. Vicariance biogeography and paleoecology of European Miocene hominoid primates. In J. Agusti and L. Rook (eds.), Evolutionary History of European Miocene Hominoid Primates, pp. 454–487. Cambridge University Press, Cambridge, UK.

Andrews, P., J. M. Lord, and E. M. Nesbit Evans. 1979. Patterns of ecological diversity in fossil and modern mammalian faunas. *Biological Journal of the Linnean Society* 11:177–205.

Ansell, W. F. H. 1960. Mammals of Northern Rhodesia. Government Printer, Lusaka, Northern Rhodesia.

———. 1978. The Mammals of Zambia. National Park and Wildlife Service, Chilanga, Zambia.

Antoine, P.-O. 2002. Phylogénie et évolution des Elasmotheriina (Mammalia, Rhinocerotidae). *Mémoires du Muséum National d'Histoire Naturelle* 188:1–359.

Antoine, P.-O., and G. Saraç. 2005. Rhinocerotidae (Mammalia, Perissodactyla) from the late Miocene of Akka'dagi, Turkey. *Geodiversitas* 27(4):601–632.

Arambourg, C. 1934. Le *Dinotherium* des gisements de l'Omo. *Comptes Rendus de la Société Géologique de France* 1934:86–87.

———. 1938. Mammifères fossiles du Maroc. *Mémoires de la Société des Sciences Naturelles du Maroc* 46:1–74.

———. 1944. Les hippopotames fossiles d'Afrique. *Comptes Rendus de l'Académie des Sciences* 218:602–604.

———. 1945. *Anancus osiris*, un mastodonte nouveau du Pliocène inférieur d'Egypte. *Bulletin de la Société Géologique de France* 15:479–495.

———. 1948. Contribution à l'étude géologique et paléontologique du bassin du lac Rodolphe et de la basse vallée de l'Omo. In C. Arambourg (ed.), Mission Scientifique de l'Omo. Vol. 1: Géologie et Anthropologie, pp. 231–562. Editions du Muséum, Paris.

———. 1958. La faune de vertébrés Miocènes de l'Oued el Hammam (Oran, Algérie). *Compte-Rendu Sommaire et Bulletin de la Société Géologique de France* 6:116–119.

———. 1959. Vertébrés continentaux du Miocène supérieur de l'Afrique du Nord. *Publications Service Carte Géologique. Algérie (Nouvelle Série). Paléontologie* 4:5–159.

———. 1968. Un suidé fossile nouveau du Miocène supérieure de l'Afrique du Nord. *Bulletin de la Société Géologique de France* 10:110–115.

———. 1970. Les vertébrés du Pléistocène de l'Afrique du Nord. *Archives du Muséum National d'Histoire Naturelle* 10:1–126.

Arambourg, C., and J. Piveteau. 1929. Les Vertébrés du Pontien de Salonique. *Annales de Paléontologie* 18:1–82.

Archer, A. W., and C. G. Maples. 1987. Monte Carlo simulation of selected binomial similarity coefficients (1): Effect of number of variables. *Palaios* 2:609–617.

Arnason, U., A. Gullberg, A. S. Burguete, and A. Janke. 2000. Molecular estimates of primate divergences and new hypotheses for primate dispersal and the origin of modern humans. *Hereditas* 133:217–228.

Arno, V., G. M. Di Paola, and S. M. Berhe. 1981. The Kella Horst: Its origin and significance in crustal attenuation and magmatic processes in the Ethiopian Rift Valley. Proceedings First International Symposium on Crustal Movements in Africa, United Nations, Addis Ababa.

Asfaw, B. 1983. A new hominid parietal from Bodo, Middle Awash Valley, Ethiopia. *American Journal of Physical Anthropology* 61:367–371.

Asfaw, B., T. D. White, C. O. Lovejoy, B. Latimer, S. Simpson, and G. Suwa. 1999. Cladistics and early hominid phylogeny (reply to Strait and Grine). *Science* 285:1210–1211.

Asfaw, B., C. Ebinger, D. Harding, T. White, and G. WoldeGabriel, 1990. Space-based imagery in paleoanthropological research: an Ethiopian example. *National Geographic Research* 6:418–434.

Asfaw, B., W. H. Gilbert, Y. Beyene, W. K. Hart, P. R. Renne, G. WoldeGabriel, E. S. Vrba, and T. D. White. 2002. Remains of *Homo erectus* from Bouri, Middle Awash, Ethiopia. *Nature* 416:317–320.

Ashley, M. V., D. J. Melnick, and D. Western. 1990. Conservation genetics of the black rhinoceros (*Diceros bicornis*). I: Evidence from the mitochondrial DNA of three populations. *Conservation Biology* 4:71–77.

Badgley, C., S. Nelson, J. Barry, A. K. Behrensmeyer, and T. Cerling. 2005. Testing models of faunal turnover with Neogene mammals from Pakistan. In D. E. Lieberman, R. J. Smith, J. Kelley (eds.), Interpreting the Past: Essays on Human, Primate, and Mammal Evolution in Honor of David Pilbeam, pp. 29–46. Brill Academic Publishers, Boston.

Baker, E. W., A. A. Malyango, and T. Harrison. 1998. Phylogenetic relationships and functional morphology of the distal humerus from Kanapoi, Kenya. *American Journal of Physical Anthropology* Supplement 26:66.

Bannert, D., J. Brinckmann, K. C. H. Käding, G. Knetsch, M. Kürsten, and H. Mayrhofer. 1970. Zur geologie der Danakil-Senke (nördliches Afar-Gebiet, NE-Äthiopien). *Geologische Rundschau* 59:409–443.

Barberi, F., and J. Varet. 1975. Nature of the Afar crust. In A. Pilger and A. Rosler (eds.), Afar Depression of Ethiopia, pp. 375–378. Schweizerbart Verlag, Stuttgart.

Barry, J. C. 1987. Large carnivores (Canidae, Hyaenidae, Felidae) from Laetoli. In M. D. Leakey and J. M. Harris (eds.), Laetoli: A Pliocene Site in Northern Tanzania, pp. 235–258. Clarendon Press, Oxford.

———. 1995. Faunal turnover and diversity in the terrestrial Neogene of Pakistan. In E. S. Vrba, G. H. Denton, T. C. Partridge, and L. H. Burckle (eds.), Paleoclimate and Evolution with Emphasis on Human Origins, pp. 115–134. Yale University Press, New Haven, CT.

———. 1999. Late Miocene Carnivora from the Emirate of Abu Dhabi, United Arab Emirates. In P. J. Whybrow and A. Hill (eds.), Fossil Vertebrates of Arabia, with Emphasis on the Late Miocene Faunas, Geology, and Palaeoenvironments of the Emirate of Abu Dhabi, United Arab Emirates, pp. 203–208. Yale University Press, New Haven, CT.

Barry, J. C., M. E. Morgan, L. J. Flynn, D. Pilbeam, A. K. Behrensmeyer, S. M. Raza, I. A. Khan, C. Badgley, J. Hicks, and J. Kelley. 2002. Faunal and environmental change in the late Miocene Siwaliks of northern Pakistan. *Paleobiology* 28:1–71.

Beden, M. 1976. Proboscideans from Omo Group formations. In Y. Coppens, F. C. Howell, G. L. Isaac, and R. Leakey (eds.), Earliest Man and Environments in the Lake Rudolf Basin, pp. 193–208. University of Chicago Press, Chicago.

———. 1979. Les éléphants (*Elephas* et *Loxodonta*) d'Afrique orientale: systématique, phylogénie, intéret biochronologique. Ph.D. thesis, Université de Poitiers, Poitiers, France.

———. 1983. Family Elephantidae. In J. M. Harris (ed.), Koobi Fora Research Project, Vol. 2, pp. 40–129. Clarendon Press, Oxford.

———. 1987a. Fossil Elephantidae from Laetoli. In M. D. Leakey and J. M. Harris (eds.), Laetoli: A Pliocene Site in Northern Tanzania, pp. 259–300. Clarendon Press, Oxford.

———. 1987b. Les Faunes Plio-Pléistocènes de la Basse Vallée de l'Omo (Éthiopie), Tome 2: Les éléphantides (Mammalia, Proboscidea). Cahiers de Paléontologie—Travaux de Paléontologie Est-Africaine. Centre National de la Recherche Scientifique, Paris.

Begun, D. R. 1994. Relations among the great apes and humans: New interpretations based on the fossil great ape *Dryopithecus*. *Yearbook of Physical Anthropology* 37:11–63.

———. 2001. African and Eurasian Miocene hominoids and the origin of Hominidae. In L. de Bonis, G. D. Koufos, and P. Andrews (eds.), Phylogeny of the Neogene Hominoid Primates of Eurasia, pp. 231–253. Cambridge University Press, Cambridge, UK.

———. 2002. European hominoids. In W. Hartwig (ed.), The Primate Fossil Record, pp. 339–369. Cambridge University Press, Cambridge, UK.

———. 2004. The earliest hominins—is less more? *Science* 303:1478–1480.

———. 2005. *Sivapithecus* is east and *Dryopithecus* is west, and never the twain shall meet. *Anthropological Science* 113:53–64.

Begun, D. R., and L. Kordos. 1997. Phyletic affinities and functional convergence in *Dryopithecus* and other Miocene and living hominids. In D. R. Begun, C. V. Ward, and M. D. Rose (eds.), Function, Phylogeny and Fossils: Miocene Hominoid Evolution and Adaptations, pp. 291–316. Plenum Press, New York.

Begun, D. R., C. V. Ward, and M. D. Rose. 1997. Events in hominoid evolution. In D. R. Begun, C. V. Ward, and M. D. Rose (eds.), Function, Phylogeny, and Fossils: Miocene Hominoid Origins and Adaptations, pp. 389–415. Plenum Press, New York.

Begun, D. R., E. Gulec, and D. Geraads. 2003. Dispersal patterns of Eurasian hominoids: Implications from Turkey. In J. W. F. Reumer and W. Wessels (eds.), Distribution and Migration of Tertiary Mammals in Eurasia: A Volume in Honor of Hans de Bruijn, pp. 23–39. Deisea, Rotterdam.

Behrensmeyer, A. K. 1993. The bones of Amboseli. *National Geographic Research and Exploration* 9:402–421.

Behrensmeyer, A. K., and D. E. Dechant-Boaz. 1980. The recent bones of Amboseli National Park, Kenya, in relation to East African paleoecology. In A. K. Behrensmeyer and A. P. Hill (eds.), Fossils in the Making: Vertebrate Taphonomy and Paleoecology, pp. 72–92. University of Chicago Press, Chicago and London.

Behrensmeyer, A. K., D. Western, and D. E. Dechant Boaz. 1979. New perspectives in vertebrate paleoecology from a recent bone assemblage. *Paleobiology* 5:12–21.

Behrensmeyer, A. K., N. E. Todd, R. Potts, and G. E. McBrinn. 1997. Late Pliocene faunal turnover in the Turkana Basin, Kenya and Ethiopia. *Science* 278:1589–1594.

Behrensmeyer, A. K., A. L. Deino, A. Hill, J. D. Kingston, and J. J. Saunders, 2002. Geology and geochronology of the middle Miocene Kipsaramon site complex, Muruyur Beds, Tugen Hills, Kenya. *Journal of Human Evolution* 42:11–38.

Benefit, B. R. 2000. Old World monkey origins and diversification: An evolutionary study of diet and dentition. In P. Whitehead and C. J. Jolly (eds.), Old World Monkeys, pp. 133–179. Cambridge University Press, Cambridge, UK.

Benefit, B. R., and M. McCrossin. 2002. The Victoripithecidae, Cercopithecoidea. In W. Hartwig (ed.), The Primate Fossil Record, pp. 241–254. Cambridge University Press, Cambridge, UK.

Benefit, B. R., and M. Pickford. 1986. Miocene fossil cercopithecoids from Kenya. *American Journal of Physical Anthropology* 69:441–464.

Berhe, S. M., B. Desta, M. Nicoletti, and M. Teferra. 1987. Geology, geochronology, and geodynamic implications of the Cenozoic magmatic province, in West and Southeast Ethiopia. *Journal of the Geological Society of London* 144:213–226.

Bernor, R. L. 1978. The Mammalian Systematics, Biostratigraphy and Biochronology of Maragheh and Its Importance for Understanding Late Miocene Hominoid Zoogeography and Evolution. Ph.D. thesis, University of California, Los Angeles.

———. 1983. Geochronology and zoogeographic relationships of Miocene Hominoidea. In R. L. Ciochon and R. Corruccini (eds.), New Interpretations of Ape and Human Ancestry, pp. 21–64. Plenum Press, New York.

———. 1984. A zoogeographic theater and biochronologic play: The time/biofacies phenomena of Eurasian and African Miocene mammal provinces. *Paléobiologie Continentale* 14: 121–142.

Bernor, R. L., and M. Armour-Chelu. 1997. Later Neogene hipparions from the Manonga Valley, Tanzania. In T. Harrison (ed.), Neogene Paleontology of the Manonga Valley, Tanzania, pp. 219–264. Plenum Press, New York.

Bernor, R. L., and M. Armour-Chelu. 1999. Toward an evolutionary history of African hipparionine horses. In T. Brommage and F. Schrenk (eds.), African Biogeography, Climate Change and Early Hominid Evolution, pp. 189–215. Oxford University Press, Oxford.

Bernor, R. L., and J. M. Harris. 2003. Systematics and evolutionary biology of the Late Miocene and Early Pliocene hipparionine horses from Lothagam, Kenya. In J. M. Harris and M. G. Leakey (eds.), Lothagam: The Dawn of Humanity in Eastern Africa, pp. 387–438. Columbia University Press, New York.

Bernor, R. L., and S. T. Hussain. 1985. An assessment of the systematic, phylogenetic and biogeographic relationships of Siwalik hipparionine horses. *Journal of Vertebrate Paleoontology* 5:32–87.

Bernor, R. L., and T. M. Kaiser. 2006. Systematics and paleoecology of the earliest Pliocene equid *Eurygnathohippus hooijeri* n. sp. from Langebaanweg, South Africa. *Mitteilungen aus dem Hamburgischen Zoologischen Museum und Institut* 103:147–183.

Bernor, R. L., and D. Lipscomb. 1991. The systematic position of "*Plesiohipparion*" aff. *huangheense* (Equidae, Hipparionini) from Gülyazi, Turkey. *Mitteilungen Bayerischen Staatsslammlung für Paläontologie und Historische Geologie* 31:107–123.

———. 1995. A consideration of Old World hipparionine horse phylogeny and global abiotic processes. In E. S. Vrba, G. H. Denton, T. C. Partridge, and L. H. Burckle (eds.), Paleoclimate and Evolution, with Emphasis on Human Origins, pp. 164–177. Yale University Press, New Haven, CT.

Bernor, R. L., and P. P. Pavlakis. 1987. Zoogeographic relationships of the Sahabi large mammal fauna (Early Pliocene, Libya). In N. T. Boaz, A. El-Arnauti, A. W. Gaziry, J. de Heinzelin, and D. D. Boaz (eds.), Neogene Paleontology and Geology of Sahabi, pp. 349–383. Alan R. Liss, New York.

Bernor, R. L., and L. Rook. 2004. Palaeozoogeography of the Rudabánya fauna. *Palaeontographica Italica* 89:21–25.

Bernor, R. L., and R. S. Scott. 2003. New interpretations of the systematics, biogeography and paleoecology of the Sahabi hipparions (latest Miocene), Libya. *Geodiversitas* 25:297–319.

Bernor, R. L., K. Heissig, and H. Tobien. 1987. Early Pliocene Perissodactyla from Sahabi, Libya. In N. T. Boaz, A. El-Arnauti, A. Gaziry, J. de Heinzelin, and D. D. Boaz (eds.), Neogene Paleontology and Geology of Sahabi, pp. 233–254. Alan R. Liss, New York.

Bernor, R. L., V. Fahlbusch, P. Andrews, H. De Bruijn, M. Fortelius, F. Rögl, F. F. Steininger, and L. Werdelin. 1996a. The evolution of European and West Asian later Neogene faunas: A biogeographic and paleoenvironmental synthesis. In R. L. Bernor, V. Fahlbusch, and H.-W. Mittmann (eds.), The Evolution of Western Eurasian Neogene Mammal Faunas, pp. 449–470. Columbia University Press, New York.

Bernor, R. L., V. Fahlbusch, and H.-W. Mittmann. 1996b. The Evolution of Western Eurasian Neogene Mammal Faunas. Columbia University Press, New York.

Bernor, R. L., G. D. Koufos, M. O. Woodbune, and M. Fortelius. 1996c. The evolutionary history and biochronology of European and southwest Asian late Miocene and Pliocene hipparionine horses. In R. L. Bernor, V. Fahlbusch, and H.-W. Mittmann (eds.), The Evolution of Western Eurasian Neogene Mammal Faunas, pp. 307–338. Columbia University Press, New York.

Bernor, R. L., N. Solounias, C. C. Swisher III, and J. A. Van Couvering. 1996d. The correlation of three classical "Pikermian" mammal faunas, Maragheh, Samos and Pikermi, with the European MN unit system. In R. L. Bernor, V. Fahlbusch, and H.-W. Mittmann (eds.), The Evolution of Western Eurasian Neogene Mammal Faunas, pp. 137–156. Columbia University Press, New York.

Bernor, R. L., H. Tobien, L. A. Hayek, and H. W. Mittmann. 1997. The Höwenegg Hipparionine horses: Systematics, stratigraphy, taphonomy and paleoenvironmental context. *Andrias* 10:1–230.

Bernor, R. L., M. Fortelius, and L. Rook. 2001. Evolutionary biogeography and paleoecology of the "*Oreopithecus bambolii* Faunal Zone" (late Miocene, Tusco-Sardinian Province). *Bollettino della Società Paleontologica Italiana* 40:139–148.

Bernor, R. L., R. S. Scott, M. Fortelius, J. Kappelman, and S. Sen. 2003. Systematics and evolution of the late Miocene hipparions from Sinap, Turkey. In M. Fortelius, J. Kappelman, S. Sen, and R. L. Bernor (eds.), The Geology and Paleontology of the Miocene Sinap Formation, Turkey, pp. 220–281. Columbia University Press, New York.

Bernor, R. L., T. M. Kaiser, and S. V. Nelson. 2004a. The oldest Ethiopian hipparion (Equinae, Perissodactyla) from Chorora: Systematics, paleodiet and paleoclimate. *Courier Forschungininstitut Senckenberg* 246:213–226.

Bernor, R. L., L. Kordos, L. Rook, J. Agustí, P. Andrews, M. Armour-Chelu, D. R. Begun, D. W. Cameron, J. Damuth, G. Daxner-Höck, L. de Bonis, O. Fejfar, N. Fessaha, M. Fortelius, J. Franzen, M. Gasparik, A. Gentry, K. Heissig, N. Hernyak, T. Kaiser, G. D. Koufos, E. Krolopp, D. Jánossy, M. Llenas, L. Meszáros, P. Müller, P. Renne, Z. Rocek, S. Sen, R. Scott, Z. Szyndlar, Gy. Topál, P. S. Ungar, T. Utescher, J. A. Van Dam, L. Werdelin, and R. Ziegler. 2004b. Recent advances on multidisciplinary research at Rudabánya, Late Miocene (MN9), Hungary: A compendium. *Palaeontographica Italica* 89:3–36.

Bernor, R. L., L. Kordos, and L. Rook. eds. 2005a. Multidisciplinary research at Rudabánya. *Palaeontographica Italica* 90:1–313.

Bernor, R. L., R. S. Scott, and Y. Haile-Selassie. 2005b. A contribution to the evolutionary history of Ethiopian hipparionine horses: Morphometric evidence from the postcranial skeleton. *Geodiversitas* 27:133–158.

Birchette, M. 1981. Postcranial remains of *Cercopithecoides*. *American Journal of Physical Anthropology* 54:201.

———. 1982. The Postcranial Skeleton of *Paracolobus chemeroni*. Ph.D. thesis, Harvard University, Cambridge, MA.

Bishop, L. C. 1994. Pigs and the Ancestors: Hominids, Suids, and the Environment During the Plio-Pleistocene of East Africa. Ph.D. thesis, Yale University, New Haven, CT.

———. 1997. Fossil suids from the Manonga Valley, Tanzania. In T. Harrison (ed.), Neogene Paleontology of the Manonga Valley, Tanzania, pp. 191–217. Plenum Press, New York.

———. 1999. Suid paleoecology and habitat preferences at African Pliocene and Pleistocene hominid localities. In T. G. Bromage and F. Schrenk (eds.), African Biogeography, Climate Change, and Human Evolution, pp. 216–225. Oxford University Press, Oxford.

Black, C. C. 1972. Review of fossil rodents from the Neogene Siwalik Beds of India and Pakistan. *Paleontology* 15:238–266.

Boaz, N. T. 1987. Introduction. In N. T. Boaz, A. El-Arnauti, A. W. Gaziry, J. de Heinzelin, and D. D. Boaz (eds.), Neogene Paleontology and Geology of Sahabi, pp. xi–xv. Alan R. Liss, New York.

Bobe, R. 1997. Hominid environments in the Pliocene: An analysis of fossil mammals from the Omo Valley, Ethiopia. Ph.D. thesis, University of Washington, Seattle.

Boccaletti, M., M. Bonini, Mazzuoli, R., and T. Trua. 1999. Pliocene-Quaternary volcanism and faulting in the northern Main Ethiopian Rift (with two geological maps at scale 1:50,000). *Acta Vulcanologica* 11:83–97.

Bohlin, B. 1926. Die Familie Giraffidae. *Palaeontologia Sinica, Series C* 4:1–179.

Boisserie, J.-R. 2004. A new species of Hippopotamidae (Mammalia, Artiodactyla) from the Sagantole Formation, Middle Awash, Ethiopia. *Bulletin de la Société Géologique de France* 175:525–533.

———. 2005. The phylogeny and taxonomy of Hippopotamidae (Mammalia: Artiodactyla): A review based on morphology and cladistic analysis. *Zoological Journal of the Linnean Society* 143:1–26.

Boisserie, J.-R., and T. D. White. 2004. A new species of Pliocene Hippopotamidae from the Middle Awash, Ethiopia. *Journal of Vertebrate Paleontology* 24:464–473.

Boisserie, J.-R., M. Brunet, L. Andossa, and P. Vignaud. 2003. Hippopotamids from the Djurab Pliocene faunas, Chad, Central Africa. *Journal of African Earth Sciences* 36:15–27.

Boisserie, J.-R., A. Likius, M. Brunet, and P. Vignaud. 2005. A new late Miocene hippopotamid from Toros-Ménalla, Chad. *Journal of Vertebrate Paleontology* 25:665–673.

Borchardt, G. A., P. J. Aruscavage, and H. T. Millard. 1972. Correlation of the Bishop Ash, a Pleistocene marker bed, using instrumental neutron activation analysis. *Journal of Sedimentary Petrology* 42:301–306.

Bouvrain, G. 1988. Les *Tragoportax* (Bovidae, Mammalia) des gisements du Miocène supérieur de Ditiko (Macedoine, Grece). *Annales de Paléontologie* 74:43–63.

———. 1994. Un Bovide du Turolien inférieur d'Europe orientale; *Tragoportax rugosifrons*. *Annales de Paléontologie* 80:61–87.

Brain, C. K. 1981. The evolution of man in Africa: Was it a consequence of Cainozoic cooling? *Annex of the Transvaal Geological Society of South Africa* 84:1–19.

Brown, F. H. 1995. The potential of the Turkana Basin for paleoclimate reconstruction in East Africa. In E. S. Vrba, G. H. Denton, T. C. Partridge, and L. H. Burckle (eds.), Paleoclimate and Evolution, with Emphasis on Human Origins, pp. 319–330. Yale University Press, New Haven, CT.

Brown, F. H., A. M. Sarna-Wojcicki, C. E. Meyer, and B. Haileab. 1992. Correlation of Pliocene and Pleistocene tephra layers between the Turkana Basin of East Africa and the Gulf of Aden. *Quaternary International* 13/14:55–67.

Brown, S. M., and B. A. Houlden. 2000. Conservation genetics of the black rhinoceros (*Diceros bicornis*). *Conservation Genetics* 1:365–370.

Brunet, M., and T. D. White. 2001. Deux nouvelles espèces de Suini (Mammalia, Suidae) du continent africain (Éthiopie; Tchad). *Comptes Rendus de l'Académie des Sciences, Paris, Sciences de la Terre et des Planètes* 332:51–57.

Brunet, M., A. Beauvilain, D. Geraads, F. Guy, M. Kasser, H. T. Mackaye, L. M. Maclatchy, G. Mouchelin, J. Sudre, and P. Vignaud. 1998. Tchad: découverte d'une faune de mammifères du Pliocène inférieur. *Comptes Rendus de l'Académie des Sciences, Paris* 326:153–158.

Brunet, M., A. Beauvilain, D. Billiou, H. Bocherens, J. R. Boisserie, L. De Bonis, P. Branger, A. Brunet, Y. Coppens, R. Daams, J. Dejax, C. Denys, P. Duringer, V. Eisenmann, F. Fanoné, P. Fronty, M. Gayet, D. Geraads, F. Guy, M. Kasser, G. Koufos, A. Likius, N. Lopez-Martinez, A. Louchart, L. Maclatchy, H. T. Makaye, B. Marandat, G. Mouchelin, C. Mourer-Chauviré, O. Otero, S. Peigné, P. Pelaez Campomanes, D. Pilbeam, J. C. Rage, D. De Ruitter, M. Schuster, J. Sudre, P. Tassy, P. Vignaud, L. Viriot, and A. Zazzo. 2000. Chad: discovery of a Vertebrate fauna close to the Mio–Pliocene boundary. *Journal of Vertebrate Paleontology* 20:205–209.

Brunet, M., F. Guy, D. Pilbeam, H. T. Mackaye, A. Likius, D. Ahounta, A. Beauvilain, C. Blondel, H. Bocherensk, J.-R. Boisserie, L. De Bonis, Y. Coppens, J. Dejax, C. Denys, P. Duringer, V. Eisenmann, G. Fanone, P. Fronty, G. Geraads, T. Lehmann, F. Lihoreau, A. Louchart, A. Mahamat, G. Merceron, G. Mouchelin, O. Otero, P. P. Campomanes, M. P. De Leon, J.-C. Rage, M. Sapanet, M. Schuster, J. Sudre, P. Tassy, X. Valentin, P. Vignaud, L. Viriot, A. Zazzo, and C. Zollikofer. 2002. A new hominid from the Upper Miocene of Chad, Central Africa. *Nature* 418:145–151.

Brunet, M., F. Guy, D. Pilbeam, D. E. Lieberman, A. Likius, H. T. Mackaye, M. S. Ponce de León, C. P. E. Zollikofer, and P. Vignaud. 2005. New material of the earliest hominid from the Upper Miocene of Chad. *Nature* 434:752–755.

Burchell, W. J. 1817. Note sur une nouvelle espèce de rhinoceros. *Bulletin des Sciences* 1817:96–97.

Burgio, E., and M. Fiorio. 1988. *Nesolutra trinacriae* n. sp., lontra quaternaria della Sicilia. *Bollettino della Società Paleontologica Italiana* 27:259–275.

Bush, M. E., C. O. Lovejoy, D. C. Johanson, and Y. Coppens. 1982. Hominid carpal, metacarpal, and phalangeal bones recovered from the Hadar Formation: 1974–1977 collections. *American Journal of Physical Anthropology* 57:651–677.

Camper, P. 1780. Dissertatio de cranio rhinocerotis Africani, cornu gemino. *Acta Academiae Scientiarum Imperialis Petropolitanae* 1(2):193–209.

———. 1782. Natuurkundige Verhandelingen over den Orang outang; en Eenige Andere Aapsoorten; over den Rhinoceros met den Dubbelen Horen, en over het Rendier. Erven P. Meijer en G. Warnars, Amsterdam.

Cande, S. C., and D. V. Kent. 1995. Revised calibration of the geomagnetic polarity time-scale for the Late Cretaceous and Cenozoic. *Journal of Geophysical Research* 100:6093–6095.

Cane, M. A., and P. Molnar. 2001. Closing of the Indian seaway as a precursor to east African aridification around 3–4 million years ago. *Nature* 411:157–162.

Cerdeño, E. 1989. Revisión de la sistemática de los rinocerontes del Neógeno de España. *Colección Tesis Doctorales Universidad Complutense de Madrid* 306/89:1–429.

———. 1995. Cladistic analysis of the Family Rhinocerotidae. *American Museum Novitates* 3143:1–25.

Cerling, T. E. 1984. The stable isotopic composition of modern soil carbonate and its relationship to climate. *Earth and Planetary Science Letters* 71:229–240.

Cerling, T. E., and J. Quade. 1993. Stable carbon and oxygen isotopes in soil carbonates. *American Geophysics Union Monograph* 78:217–231.

Cerling, T. E., J. M. Harris, S. H. Ambrose, M. G. Leakey, and N. Solounias. 1997a. Dietary and environmental reconstruction with stable isotope analyses of herbivore tooth enamel from the Miocene locality of Fort Ternan, Kenya. *Journal of Human Evolution* 33:635–650.

Cerling, T. E., J. M. Harris, B. J. MacFadden, M. G. Leakey, J. Quade, V. Eisenmann, and J. R. Ehleringer. 1997b. Global vegetation change through the Miocene/Pliocene boundary. *Nature* 389:153–158.

Cerling, T. E., J. M. Harris, and M. G. Leakey. 1999. Browsing and grazing in elephants: The isotope record of modern and fossil proboscideans. *Oecologia* 120:364–374.

———. 2003. Isotope paleoecology of the Nawata and Nachukui Formations at Lothagam, Turkana Basin, Kenya. In M. G. Leakey and J. M. Harris (eds.), Lothagam: The Dawn of Humanity in Eastern Africa, pp. 605–624. Columbia University Press, New York.

Charleton, M. D., and G. G. Musser. 1984. Muroid rodents. In S. Anderson and J. K. Jones, Jr. (eds.), Orders and Families of Recent Mammals of the World, pp. 289–379. John Wiley and Sons, New York.

Chen, F. C., and W.-H. Li. 2001. Genomic divergences between humans and other hominoids and the effective population size of the common ancestor of humans and chimpanzees. *American Journal of Human Genetics* 68:444–456.

Chen, G. 1999. The genus *Anancus* Aymard, 1885 (Proboscidea, Mammalia) from the Late Neogene of nothern China. *Vertebrata PalAsiatica* 37:175–189.

Chernet, T., and W. K. Hart. 1999. Petrology and geochemistry of volcanism in the northern Main Ethiopian Rift–southern Afar transition region. *Acta Vulcanologica* 11:21–41.

Chernet, T., W. K. Hart, J. L. Aronson, and R. C. Walter. 1998. New age constraints on the timing of volcanism and tectonism in the northern Main Ethiopian Rift–southern Afar transition zone (Ethiopia). *Journal of Volcanology and Geothermal Research* 80:267–280.

Chessex, R., M. Delaloye, and D. Fontignie. 1980. K-Ar datations on volcanic rocks of the Republic of Djibouti. In Geodynamic Evolution of the Afro-Arabian Rift System, pp. 221–227. Atti Convegni 47. Accademia Nazionale dei Lincei, Rome.

Child, G. S. 1964. Some notes on the mammals of Kilimanjaro. *Tanganyika Notes and Records* 53:77–89.

Churcher, C. S. 1970. Two new upper Miocene giraffids from Fort Ternan, Kenya, East Africa: *Palaeotragus primaevus* sp. nov. and *Samotherium africanum* sp. nov. In L. S. B. Leakey and R. J. G. Savage (eds.), Fossil Vertebrates of Africa II, pp. 1–105. Academic Press, London.

———. 1978. Giraffidae. In V. J. Maglio and H. B. S. Cooke (eds.), Evolution of African Mammals, pp. 509–535. Harvard University Press, Cambridge, MA.

———. 1979. The large palaeotragine giraffid *Palaeotragus germaini*, from late Miocene deposits of Lothagam Hill, Kenya. *Breviora* 453:1–8.

Clark, J. D., and K. D. Schick. 2000. Acheulean archaeology of the eastern Middle Awash. In J. de Heinzelin, J. D. Clark, K. D. Schick, and W. H. Gilbert (eds.), The Acheulean and the Plio-Pleistocene Deposits of the Middle Awash Valley, Ethiopia, pp. 51–121. Annales Sciences Géologiques 104. Royal Museum of Central Africa, Tervuren, Belgium.

Clark, J. D., B. Asfaw, G. Assefa, J. W. K. Harris, H. Kurashina, R. C. Walter, T. D. White, and M. A. J. Williams. 1984. Paleoanthropological discoveries in the Middle Awash valley, Ethiopia. *Nature* 307:423–428.

Clark, J. D., B. Asfaw, R. Blumenschine, J. W. K. Harris, H. Kurashina, C. Sussman, and M. Williams. 1993. Archaeological studies in the Middle Awash, Ethiopia: The 1981 field season. In Proceedings of the 9th Congress of the Pan-African Association of Pre-history and Related Studies, pp. 135–145. Rex Charles Publications, Ibadan, Nigeria.

Clark, J. D., J. de Heinzelin, K. D. Schick, W. K. Hart, T. D. White, G. WoldeGabriel, R. C. Walter, G. Suwa, B. Asfaw, E. Vrba, and Y. Haile-Selassie. 1994. African *Homo erectus*: Old radiometric ages and young Oldowan assemblages in the Middle Awash Valley. *Science* 264:1907–1910.

Clark, J. D., Y. Beyene, G. WoldeGabriel, W. K. Hart, P. R. Renne, W. H. Gilbert, A. Defleur, G. Suwa, S. Katoh, K. R. Ludwig, J.-R. Boisserie, B. Asfaw, and T. D. White. 2003. Stratigraphic, chronological, and behavioural contexts of Pleistocene *Homo sapiens* from Middle Awash, Ethiopia. *Nature* 423:747–752.

Coe, M. 1972. The south Turkana expedition: Ecological studies of small mammals of south Turkana. Scientific Papers, no. 9. *Journal of Geography* 138:316–338.

Coe, M., N. McWilliam, G. Stone, and M. Packer. 1999. Mkomazi: The Ecology, Biodiversity and Conservation of a

Tanzanian Savanna. Royal Geographical Society (with The Institute of British Geographers), London.

Colbert, E. H. 1933. The presence of tubulidentates in the Middle Siwalik Beds of northern India. *American Museum Novitates* 604:1–10.

———. 1935. Siwalik mammals in the American Museum of Natural History. *Transactions of the American Philosophical Society* 26:278–294.

———. 1941. A study of *Orycteropus gaudryi* from the Island of Samos. *Bulletin of the American Museum of Natural History* 78:305–351.

Colbert, E. H., and D. A. Hooijer. 1953. Pleistocene Mammals from the limestone fissures of Szechwan, China. *Bulletin of the American Museum of Natural History* 102:1–134.

Collier, G. E., and S. J. O'Brien. 1985. A molecular phylogeny of the Felidae: Immunological distance. *Evolution* 39:473–487.

Conroy, G. C., C. J. Jolly, D. Cramer, and J. E. Kalb. 1978. Newly discovered fossil hominid skull from the Afar Depression, Ethiopia. *Nature* 276:67–70.

Conroy, G. C., M. Pickford, B. Senut, J. van Couvering, and P. Mein. 1992. *Otavipithecus namibiensis*, first Miocene hominoid from southern Africa. *Nature* 356:144–148.

Cooke, H. B. S. 1950. A critical revision of the Quaternary Perissodactyla of southern Africa. *Annals of the South African Museum* 31:393–479.

———. 1960. Further revision of the fossil Elephantidae of southern Africa. *Palaeontologia Africana* 7:46–58.

———. 1978. Suid evolution and correlation of African hominid localities: An alternative taxonomy. *Science* 201:460–463.

———. 1982. A preliminary appraisal of fossil Suidae from Sahabi, Libya. *Garyounis Scientific Bulletin* 4:71–82.

———. 1985. Plio-Pleistocene Suidae in relation to African hominid deposits. In Y. Coppens (ed.), L'Environnement des Hominidés au Plio-Pléistocène, pp. 101–117. Masson, Paris.

———. 1987. Fossil Suidae from Sahabi, Libya. In N. T. Boaz, A. El-Arnauti, A. W. Gaziry, J. de Heinzelin, and D. D. Boaz (eds.), Neogene Paleontology and Geology of Sahabi, pp. 255–266. Alan R. Liss, New York.

Cooke, H. B. S., and S. C. Coryndon. 1970. Pleistocene mammals from the Kaiso Formation and other related deposits in Uganda. In L. S. B. Leakey and R. G. J. Savage (eds.), Fossil Vertebrates of Africa, Vol. 2, pp. 107–224. Academic Press, London.

Cooke, H. B. S., and R. F. Ewer. 1972. Fossil Suidae from Kanapoi and Lothagam, Northwestern Kenya. *Bulletin of the Museum of Comparative Zoology* 143:149–295.

Cooke, H. B. S., and Q. B. Hendey. 1992. *Nyanzachoerus* (Mammalia: Suidae: Tetraconodontinae) from Langebaanweg, South Africa. *Durban Museum Novitates* 17:1–20.

Cooke, H. B. S., and A. F. Wilkinson. 1978. Suidae and Tayassuidae. In V. J. Maglio and H. B. S. Cooke (eds.), Evolution of African Mammals, pp. 435–482. Harvard University Press, Cambridge, MA.

Copley, H. 1958. Common Freshwater Fishes of East Africa. H. F. and G. Witherby Ltd., London.

Coppens, Y. 1967. Les faunes de vertébrés quaternaires du Tchad. In W. W. Bishop and J. A. Miller (eds.), Background to Evolution in Africa, pp. 89–97. University of Chicago Press, Chicago.

———. 1971. Une nouvelle espèce de Suidé du Villafranchien du Tunisie, *Nyanzachoerus jaegeri* nov. sp. *Comptes Rendus Hebdomadaires des Séances de l'Académie des Sciences, Paris* 272:3264–3267.

———. 1972. Un nouveau Proboscidien du Pliocène du Tchad, *Stegodibelodon schneideri* nov. gen., nov. sp., et le phylum des Stegotetrabelodontinae. *Comptes Rendus Hebdomadaires des Séances de l'Academie des Sciences, Série D, Sciences Naturelles* 274:2962–2695.

Coppens, Y., V. J. Maglio, C. T. Madden, and M. Beden. 1978. Proboscidea. In V. J. Maglio and H. B. S. Cooke (eds.), Evolution of African Mammals, pp. 336–367. Harvard University Press, Cambridge, MA.

Coryndon, S. C. 1976. Fossil Hippopotamidae from Plio-Pleistocene successions of the Rudolf Basin. In Y. Coppens, F. C. Howell, G. L. Isaac, and R. E. F. Leakey (eds.), Earliest Man and Environments in the Lake Rudolf Basin, pp. 238–250. University of Chicago Press, Chicago.

———. 1977. The taxonomy and nomenclature of the Hippopotamidae (Mammalia, Artiodactyla) and a description of two new fossil species. *Proceedings of the Koninklijke Nederlandse Akademie van Wetenschappen*, 80:61–88.

———. 1978. Hippopotamidae. In V. J. Maglio and H. B. S. Cooke (eds.), Evolution of African Mammals, pp. 483–495. Harvard University Press, Cambridge, MA.

Coryndon, S. C., and Y. Coppens. 1973. Preliminary report on Hippopotamidae (Mammalia, Artiodactyla) from the Plio/Pleistocene of the Lower Omo Basin, Ethiopia. In R. J. G. Savage and L. B. S. Leakey (eds.), Fossil Vertebrates of Africa, pp. 139–157. Academic Press, London.

Daget, J., J.-P. Gosse, and D. F. E. Thys van den Audenaerde. 1986. Check-List of the Freshwater Fishes of Africa. Volume 2. Institut Royal des Sciences Naturelles de Belgique, Tervuren.

Davies, C. 1987. Note on the fossil Lagomorpha from Laetoli. In M. D. Leakey and J. M. Harris, (eds.), Laetoli: A Pliocene Site in Northern Tanzania, pp. 190–193. Clarendon Press, Oxford.

de Bonis, L., and G. D. Koufos. 2001. Phylogenetic relationships of *Ouranopithecus macedoniensis* (Mammalia, Primates, Hominoidea, Hominidae) of the late Miocene deposits of Central Macedonia (Greece). In L. de Bonis, G. D. Koufos, and P. Andrews (eds.), Phylogeny of the Neogene Hominoid

Primates of Eurasia, pp. 231–253. Cambridge University Press, Cambridge, UK.

de Bonis, L., S. Peigné, A. Likius, H. T. Mackaye, P. Vignaud, and M. Brunet. 2005. *Hyaenictitherium minimum*, a new ictithere (Mammalia, Carnivora, Hyaenidae) from the Late Miocene of Toros-Menalla, Chad. *Comptes Rendus Palevol*, 4:671–679.

DeGusta, D., and E. Vrba. 2003. A method for inferring paleohabitats from the functional morphology of bovid astragali. *Journal of Archaeological Science* 30:1009–1022.

———. 2005a. Methods for inferring paleohabitats from the functional morphology of bovid phalanges. *Journal of Archaeological Science* 32:1099–1113.

———. 2005b. Methods for inferring paleohabitats from discrete traits of the bovid postcranial skeleton. *Journal of Archaeological Science* 32:1115–1123.

de Heinzelin, J. 2000. Stratigraphy. In J. de Heinzelin, J. D. Clark, K. D. Schick, and W. H. Gilbert (eds.), The Acheulean and the Plio-Pleistocene Deposits of the Middle Awash Valley, Ethiopia, pp. 11–25. Annales Sciences Géologiques 104. Royal Museum of Central Africa, Tervuren, Belgium.

de Heinzelin, J., J. D. Clark, T. D. White, W. K. Hart, P. R. Renne, G. WoldeGabriel, Y. Beyene, and E. S. Vrba. 1999. Environment and behavior of 2.5-million-year-old Bouri hominids. *Nature* 284:625–629.

de Heinzelin, J., J. D. Clark, and T. D. White. 2000a. Chapter 1: History and introduction. In J. de Heinzelin, J. D. Clark, K. D. Schick, and W. H. Gilbert (eds.), The Acheulean and the Plio-Pleistocene deposits of the Middle Awash Valley, Ethiopia, pp. 1–4. Annales Sciences Géologiques 104. Royal Museum of Central Africa, Tervuren, Belgium.

———. 2000b. Chapter 2: Geography, mapping, and nomenclature. In J. de Heinzelin, J. D. Clark, K. D. Schick, and W. H. Gilbert (eds.), The Acheulean and the Plio-Pleistocene deposits of the Middle Awash Valley, Ethiopia, pp. 5–10. Annales Sciences Géologiques 104. Royal Museum of Central Africa, Tervuren.

Deinard, A., and K. Kidd. 1999. Evolution of a *HOXB6* intergenic region within the great apes and humans. *Journal of Human Evolution* 36:687–703.

Deino, A., and R. Potts. 1990. Single-crystal ^{40}Ar/^{39}Ar dating of the Olorgesailie Formation, southern Kenya Rift. *Journal of Geophysical Research* 95:8453–8470.

———. 1992. Age-probability spectra for examination of single-crystal ^{40}Ar/^{39}Ar dating results: examples from Olorgesaillie, southern Kenya Rift. *Quaternary International* 13/14:47–53.

Deino, A. L., P. R. Renne, and C. C. Swisher. 1997. ^{40}Ar/^{39}Ar dating in paleoanthropology and archeology. *Evolutionary Anthropology* 2:63–75.

Deino, A. L., L. Tauxe, M. Monaghan, and A. Hill. 2002. Ar-40/Ar-30 geochronology and paleomagnetic stratigraphy of the Lukeino and lower Chemeron Formations at Tabarin and Kapcheberek, Tugen Hills, Kenya. *Journal of Human Evolution* 42:117–140.

Delson, E. 1973. Fossil Colobine Monkeys of the Circum-Mediterranean Region and the Evolutionary History of the Cercopithecidae (Primates: Mammalia). Ph.D. thesis, Columbia University, New York.

———. 1980. Fossil macaques, phyletic relationships and a scenario of deployment. In D. G. Lindburg (ed.), The Macaques: Studies in Ecology, Behavior and Evolution, pp. 10–30. Van Nostrand Reinhold, New York.

———. 1994. Evolutionary history of the colobine monkeys in paleoenvironmental perspective. In A. G. Davies and J. F. Oates (eds.), Colobine Monkeys: Their Ecology, Behavior, and Evolution, pp. 11–43. Cambridge University Press, Cambridge, UK.

Delson, E., C. J. Terranova, W. L. Jungers, E. J. Sargis, N. G. Jablonski, and P. C. Dechow. 2000. Body mass in Cercopithecidae (Primates, Mammalia): estimation and scaling in extinct and extant taxa. *Anthropological Papers of the American Museum of Natural History* 83:1–159.

deMenocal, P. B. 2004. African climate change and faunal evolution during the Pliocene-Pleistocene. *Earth and Planetary Science Letters Frontiers* 6976:1–22.

Denys, C. 1987. Fossil rodents (other than Pedetidae) from Laetoli. In M. D. Leakey and J. M. Harris (eds.), Laetoli: A Pliocene Site in Northern Tanzania, pp. 118–170. Clarendon Press, Oxford.

———. 1989. Paleoecological and paleobiogeographical implications of jerboa presence (Mammalia, Rodentia) in the East-African rift system during middle Pleistocene times. *Comptes Rendus de l'Academie des Sciences, Paris, Série II* 309:1261–1266.

———. 1999. Of mice and men: Evolution in East and South Africa during Plio-Pleistocene times. in T. G. Bromage and F. Schrenk (eds.), African Biogeography, Climate Change, and Human Evolution, pp. 226–252. Oxford University Press, Oxford, London, and New York.

Denys, C., and J.-J. Jaeger. 1986. A biostratigraphic problem: The case of the East African rodent faunas. *Modern Geology* 10:215–233.

Denys, C., J. Chorowicz, and J. J. Tiercelin. 1987. Tectonic and environmental control on rodent diversity in the Plio-Pleistocene sediments of the African Rift system. In L. E. Frostick, R. W. Renaut, I. Reid, and J. J. Tiercelin (eds.), Sedimentation in the African Rifts, pp. 363–372. Blackwell, Oxford.

Denys, C., L. Viriot, R. Daams, P. Pelaez-Campomanes, P. Vignaud, L. Andossa [A. Likius], and M. Brunet. 2003. A new Pliocene xerine sciurid (Rodentia) from Kossom Bougoudi, Chad. *Journal of Vertebrate Paleontology* 23:676–687.

Deocampo, D. M. 2002. Sedimentary structures generated by *Hippopotamus amphibius* in a lake-margin wetland, Ngorongoro Crater, Tanzania. *Palaïos* 17:212–217.

Dietrich, W. O. 1941. Die saugetierpaläoontologische Ergebnisse der Kohl-Larsen'schen Expedition, 1937–1939 im nordlichen Deutsch-Ostafrika. *Centralblatt für Minerologie, Geologie, und Paläontologie (Stuttgart)* 1941 B:217–223.

———. 1942a. Altestquartäre Saugetiere aus der südlichen Serengeti, Deutsch-Ostafrika. *Paläontographica (Stuttgart)* 94 A:43–133.

———. 1942b. Zur Entwicklungsmechanik des Gebisses der afrikanischen Nashörner. *Zentralblatt für Mineralogie* 1942:297–300.

———. 1945. Nashornreste aus dem Quartär Deutsch-Ostafrikas. *Palaeontographica* 96:45–90.

Disotell, T. R. 1996. The phylogeny of Old World monkeys. *Evolutionary Anthropology* 5:18–24.

———. 2000. Molecular systematics of Old World monkeys. In P. F. Whitehead and C. J. Jolly (eds.), Old World Monkeys, pp. 29–56. Cambridge University Press, Cambridge, UK.

Drapeau, M. S. M., C. V. Ward, W. H. Kimbel, D. C. Johanson, and Y. Rak. 2005. Associated cranial and forelimb remains attributed to *Australopithecus afarensis* from Hadar, Ethiopia. *Journal of Human Evolution* 48:593–642.

Dunham, K. M. 1985. Ages of black rhinos killed by drought and poaching in Zimbabwe. *Pachyderm* 5:12–13.

———. 1994. The effect of drought on the large mammal populations of the Zambezi riverine woodlands. *Journal of Zoology* 234:489–526.

Du Toit, R. 1986. Re-appraisal of black rhinoceros subspecies. *Pachyderm* 6:5–9.

———. 1987. The existing basis for subspecies classification of black and white rhinos. *Pachyderm* 9:3–5.

Ebert, S. W., and R. O. Rye. 1997. Secondary precious metal enrichment by steam-heated fluids in the Crowfoot-Lewis Hot Spring gold-silver deposit and relation to paleoclimate. *Economic Geology* 92:578–600.

Eisenmann, V. 1976. Nouveaux cranes d'hipparions (Mammalia, Perissodactyla) Plio-Pléistocène d'Afrique orientale (Ethiopie et Kenya): *Hipparion* sp., *Hipparion* cf. *ethiopicum* et *Hipparion afarense* nov. sp. *Geobios* 9:577–605.

———. 1995. What metapodial morphometry has to say about some Miocene hipparions. In E. S. Vrba, G. H. Denton, T. C. Partridge, and L. H. Burckle (eds.), Paleoclimate and Evolution, with Emphasis on Human Origins, pp. 148–163. Yale University Press, New Haven, CT.

Eisenmann, V., M. T. Alberdi, C. De Giuli, and U. Staesche. 1988. Studying fossil horses Volume I: Methodology. In M. O. Woodburne and P. Y. Sondaar (eds.), Collected Papers after the "New York International Hipparion Conference, 1981," pp. 1–71. Brill, Leiden.

Ellerman, J. R. 1941. Families and Genera of Living Rodents, Part 2. British Museum (Natural History), London.

Eltringham, S. K. 1999. The Hippos. Academic Press, London.

Emslie, R., and M. Brooks. 1999. African Rhino. Status Survey and Conservation Action Plan. IUCN/SSC African Rhino Specialist Group. IUCN, Gland, Switzerland, and Cambridge, UK.

Erdbrink, D., and W. Krommenhoek, 1975. Contribution to the knowledge of the fossil Hippopotamidae from the Kazinga Channel area (Uganda). *Säugetierkundliche Mitteilungen* 23:258–294.

Ewer, R. F. 1973. The Carnivores. The World Naturalist. Weidenfeld and Nicolson, London.

Fara, E., A. Likius, H. T. Mackaye, P. Vignaud, and M. Brunet. 2005. Pliocene large-mammal assemblages from northern Chad: sampling and ecological structure. *Naturwissenschaften* 92: 537–541.

Faure, M. 1994. Les Hippopotamidae (Mammalia, Artiodactyla) du rift occidental (bassin du lac Albert, Ouganda). Etude préliminaire. In B. Senut and M. Pickford (eds.), Geology and Paleobiology of the Albertine Rift Valley, Uganda-Zaïre. II: Paleobiology, pp. 321–337. CIFEG, Orléans, France.

Feibel, C. S. 2003. Stratigraphy and depositional history of the Lothagam sequence. In M. G. Leakey and J. M. Harris (eds.), Lothagam: The Dawn of Humanity in Eastern Africa, pp. 17–30. Columbia University Press, New York.

Ferretti, M. P., L. Rook, and D. Torre. 2003. *Stegotetrabelodon* cf. *syrticus* (Proboscidea, Elephantidae) from the Upper Miocene of Cessaniti (Calabria, southern Italy) and its bearing on Late Miocene paleogeography of central Mediterranean. *Journal of Vertebrate Paleontology* 23:659–666.

Fessaha, N. 1999. Systematics of Hadar (Afar, Ethiopia) Suidae. Ph.D. thesis, Howard University, Washington, DC.

Fischer, G. 1817. Adversaria Zoologica, fasciculus primus. *Mémoires de la Société des Naturalistes de Moscou* 5:357–446.

Fischer, M. 1986. Die Stellung der Schliefer (Hyracoidea) um phylogenetischen system der Eutheria. *Courier Forschungsinstitut Senckenberg* 84:1–32.

Fleagle, J. G., and W. S. McGraw. 2002. Skeletal and dental morphology of African papionins: unmasking a cryptic clade. *Journal of Human Evolution* 43:267–292.

Flynn, J. J., N. A. Neff, and R. H. Tedford. 1988. Phylogeny of the Carnivora. In M. J. Benton (ed.), The Phylogeny and Classification of the Tetrapods, pp. 73–116. The Systematics Association Special Volume No. 35B. Oxford University Press, New York.

Flynn, J. J., J. A. Finarelli, S. Zehr, J. Hsu, and M. A. Nedbal. 2005. Molecular phylogeny of the Carnivora (Mammalia): Assessing the impact of increased sampling on resolving enigmatic relationships. *Systematic Biology* 54:317–337.

Flynn, L. 1986. Faunal provinces and the Simpson coefficient. *Contributions to Geology, University of Wyoming Special Paper* 3:317–338.

Flynn, L. J., and M. Sabatier. 1984. A muroid rodent of Asian affinity from the Miocene of Kenya. *Journal of Vertebrate Paleontology* 3:160–165.

Flynn, L. J., and A. J. Winkler. 1994. Dispersalist implications of *Paraulacodus indicus:* A South Asian rodent of African affinity. *Historical Biology* 9:223–235.

Flynn, L. J., W. Wu, and W. R. Downs. 1997. Dating vertebrate microfaunas in the late Neogene record of Northern China. *Palaeogeography, Palaeoclimatology, Palaeoecology* 133:227–242.

Fortelius, M. 1982. Ecological aspects of dental functional morphology in the Plio-Pleistocene rhinoceroses of Europe. In B. Kurtén (ed.), Teeth: Form, Function And Evolution, pp. 163–181. Columbia University Press, New York.

———. 1985. Ungulate cheek teeth: Developmental, functional, and evolutionary interrelations. *Acta Zoologica Fennica* 180:1–76.

———. 1990. Less common ungulate species from Paşalar, middle Miocene of Anatolia (Turkey). *Journal of Human Evolution* 19:479–487.

Fortelius, M., and N. Solounias. 2000. Functional characterization of ungulate molars using the abrasion-attrition wear gradient: A new method for reconstructing paleodiets. *American Museum Novitates* 3301:1–36.

Fortelius M., L. Werdelin, P. Andrews, R. L. Bernor, A. Gentry, L. Humphrey, H.-W. Mittmann, and S. Viratana. 1996. Provinciality, diversity, turnover, and paleoecology in land mammal faunas of the later Miocene of western Eurasia. In R. L. Bernor, V. Fahlbusch, and H.-W. Mittmann (eds.), The Evolution of Western Eurasian Neogene Mammal Faunas, pp. 414–448. Columbia University Press, New York.

Fortelius, M., K. Heissig, G. Sarac, and S. Sen. 2003a. Rhinocerotidae (Perissodactyla). In M. Fortelius, J. Kappelman, S. Sen, and R. L. Bernor (eds.), Geology and Paleontology of the Miocene Sinap Formation, Turkey, pp. 282–307. Columbia University Press, New York.

Fortelius, M., J. Kappelman, S. Sen, and R. L. Bernor. eds. 2003b. Geology and Paleontology of the Miocene Sinap Formation, Turkey. Columbia University Press, New York.

Foster, W. E. 1960. The square-lipped rhinoceros. *Lammergeyer* 1:25–35.

Franz-Odendaal, T. A., J. A. Lee-Thorp, and A. Chinsamy. 2002. New evidence for the lack of C_4 grassland expansions during the early Pliocene at Langebaanweg, South Africa. *Paleobiology* 28(3):378–388.

Franz-Odendaal, T. A., T. M. Kaiser, and R. L. Bernor. 2003. Systematics and dietary evaluation of a fossil equid from South Africa. *South African Journal of Science* 99:453–459.

Freedman, L. 1957. Fossil Cercopithecoidea of South Africa. *Annals of the Transvaal Museum* 23:121–262.

Froehlich, D. J., and J. E. Kalb. 1995. Internal reconstruction of elephantid molars: Applications for functional anatomy and systematics. *Paleobiology* 21:379–392.

Frost, S. R. 2001a. Fossil Cercopithecidae of the Afar Depression, Ethiopia: Species Systematics and Comparison to the Turkana Basin. Ph.D. thesis, City University of New York, New York.

———. 2001b. New early Pliocene Cercopithecidae (Mammalia: Primates) from Aramis, Middle Awash Valley, Ethiopia. *American Museum Novitates* 3350:1–36.

Frost, S. R., and E. Delson. 2002. Fossil Cercopithecidae from the Hadar Formation and surrounding areas, Ethiopia. *Journal of Human Evolution* 43:687–748.

Gatesy, J., and Arctander, P. 2000. Molecular evidence for the phylogenetic affinities of Ruminantia. In E. S. Vrba and G. B. Schaller (eds.), Antelopes, Deer, and Relatives: Fossil Record, Behavioral Ecology, Systematics, and Conservation, pp. 143–155. Yale University Press, New Haven, CT.

Gatesy, J., D. Yelon, R. Desalle, and E. S. Vrba. 1992. Phylogeny of the Bovidae (Artiodactyla, Mammalia) based on mitochondrial ribosomal DNA-sequences. *Molecular Biology and Evolution* 9:433–446.

Gaudry, A. 1862–1867. Animaux Fossiles et Géologie de l'Attique. F. Savy éditions, Paris.

———. 1873. Animaux Fossiles du Mont Léberon (Vaucluse): Étude sur les Vertébrés. F. Savy éditions, Paris.

Gaziry, A. W. 1987. Remains of Proboscidea from the early Pliocene of Sahabi, Libya. In N. T. Boaz, A. El-Arnauti, A. W. Gaziry, J. de Heinzelin, and D. D. Boaz (eds.), Neogene Paleontology and Geology of Sahabi, pp. 183–203. Alan R. Liss, New York.

Gentry, A. W. 1970. Revised classification for *Makapania broomi* Wells and Cooke (Bovidae, Mammalia) from South Africa. *Palaeontologia Africana* 13:63–67.

———. 1974. A new genus and species of Pliocene boselaphine (Bovidae, Mammalia) from South Africa. *Annals of the South African Museum* 65:145–188.

———. 1978a. Bovidae. In V. Maglio and H. B. S. Cooke (eds.), Evolution of African Mammals, pp. 540–572. Harvard University Press, Cambridge, MA.

———. 1978b. The fossil Bovidae of the Baringo area, Kenya. In W.W. Bishop (ed.), Geological Background to Fossil Man, pp. 293–308. Geological Society Special Publication Number 6. Scottish Academic Press, Edinburgh.

———. 1980. Fossil Bovidae from Langebaanweg, South Africa. *Annals of the South African Museum* 79:213–337.

———. 1981. Notes on Bovidae from the Hadar Formation, Ethiopia. *Kirtlandia* 33:1–30.

———. 1985. The Bovidae of the Omo Group deposits, Ethiopia. In Y. Coppens and F. C. Howell (eds.), Les Faunes Plio-Pléistocènes de la Basse Vallée de l'Omo (Ethiopie). Volume 1. Perissodactyles, Artiodactyles (Bovidae), pp. 119–191. Centre National de la Recherche Scientifique, Paris.

———. 1987. Pliocene Bovidae from Laetoli. In M. D. Leakey and J. M. Harris (eds.), Laetoli: A Pliocene Site in Northern Tanzania, pp. 378–408. Clarendon Press, Oxford.

———. 1990. Evolution and dispersal of African Bovidae. In G. A. Bubenik and A. B. Bubenik (eds.), Horns, Pronghorns, and Antlers, pp. 195–233. Springer-Verlag, New York.

———. 1992. The subfamilies and tribes of the family Bovidae. *Mammal Review* 22:1–32.

———. 1997. Fossil ruminants (Mammalia) from the Manonga Valley, Tanzania. In T. Harrison (ed.), Neogene Paleontology of the Manonga Valley, Tanzania, pp. 107–135. Plenum Press, New York.

———. 1999a. A fossil hippopotamus from the Emirate of Abu Dhabi, United Arab Emirates. In P. J. Whybrow and A. Hill (eds.), Fossil Vertebrates of Arabia, with Emphasis on the Late Miocene Faunas, Geology, and Palaeoenvironments of the Emirate of Abu Dhabi, United Arab Emirates, pp. 271–289. Yale University Press, New Haven, CT.

———. 1999b. Fossil pecorans from the Baynunah Formation, Emirate of Abu Dhabi, United Arab Emirates. In P. J. Whybrow and A. P. Hill (eds.), Fossil Vertebrates of Arabia, with Emphasis on the Late Miocene Faunas, Geology, and Palaeoenvironments of the Emirate of Abu Dhabi, United Arab Emirates, pp. 290–316. Yale University Press, New Haven, CT.

———. 2003. Ruminantia (Artiodactyla). In M. Fortelius, J. Kappelman, S. Sen, and R.L. Bernor (eds.), Geology and Paleontology of the Miocene Sinap Formation, Turkey, pp. 332–379. Columbia University Press, New York.

Gentry, A. W., and Gentry, A. 1978. Fossil Bovidae of Olduvai Gorge, Tanzania. *Bulletin of the British Museum of Natural History (Geology). London. Part I* 29:289–445. Part II. 30:1–83.

Gentry, A., and E. P. J. Heizmann. 1996. Miocene ruminants of the Central and eastern Paratethys. In R. L. Bernor, V. Fahlbusch, and H.-W. Mittmann (eds.), The Evolution of Western Eurasian Neogene Mammal Faunas, pp. 378–391. Columbia University Press, New York.

George, M., L. A. Puentes, and O. A. Ryder. 1983. Genetische Unterschiede zwischen den Unterarten des Breitmaulnashorns. In W. Klos (ed.), International Studbook of African Rhinoceroses, Vol. 2, pp. 60–67. Berlin Zoologischer Garten, Berlin.

Geraads, D. 1986a. Remarques sur la systematique et la phylogénie des Giraffidae (Artiodactyla, Mammalia). *Geobios* 19:465–477.

———. 1986b. Sur les relations phylétiques de *Dicerorhinus primaevus* Arambourg, 1959, rhinocéros du Vallésien d'Algérie. *Comptes Rendus de l'Académie des Sciences, Paris* 302:835–837.

———. 1987. La faune des dépôts pléistocènes de l'Ouest du lac Natron (Tanzanie); interprétation biostratigraphique. *Sciences Géologiques, Bulletin* 40:167–184.

———. 1988. Révision des Rhinocerotidae (Mammalia) du Turolien de Pikermi. Comparaison avec les formes voisines. *Annales de Paléontologie* 74:13–41.

———. 1989. Vertébrés du Miocène supérieur du Djebel Krechem el Artsouma (Tunisie centrale). Comparaisons biostratigraphiques. *Géobios* 22:777–801.

———. 1992. Phylogenetic analysis of the tribe Bovini. *Zoological Journal of the Linnean Society* 104:193–207.

———. 1994a. Girafes Fossiles D'Ouganda. In Geology and Paleobiology of the Albertine Rift Valley, Uganda-Zaire, Vol II: Palaeobiology, pp. 375–381. CIFEG Occasional Publication 29. CIFEG, Orléans, France.

———. 1994b. Les gisements de Mammifères du Miocène supérieur de Kemiklitepe, Turquie: 4. Rhinocerotidae. *Bulletin du Muséum National d'Histoire Naturelle 4ème sér., C* 16: 81–95.

———. 1995. *Simatherium shungurense* n. sp., un noveau Bovini (Artiodactyla, Mammalia) du Pliocène terminal de l'Omo (Éthiopie). *Annales de Paléontologie* 81:87–96.

———. 2005. Pliocene Rhinocerotidae (Mammalia) from Hadar and Dikika (Lower Awash, Ethiopia), and a revision of the origin of modern African rhinos. *Journal of Vertebrate Paleontology* 25(2):451–461.

Geraads, D., and G. Koufos. 1990. Upper Miocene Rhinocerotidae (Mammalia) from Pentalophos-1, Macedonia, Greece. *Palaeontographica* 210:151–168.

Geraads, D., and F. Metz-Muller. 1999. Proboscidea (Mammalia) du Pliocene final d'Ahl al Oughlam (Casablanca, Maroc). *Neues Jahrbuch für Geologie und Palaeontologie Monatshefte* 1999:52–64.

Geraads, D., and H. Thomas. 1994. Bovidés du Plio-Pléistocène d'Ouganda. In B. Senut and M. Pickford (eds.), The Geology and Palaeobiology of the Albertine Rift Valley, Uganda-Zaire, pp. 383–407. CIFEG, Orléans, France.

Geraads, D., F. Amani, J. P. Raynal, and F. Z. Sbihi-Alaoui. 1998. La faune de Mammifères du Pliocène terminal d'Ahl al Oughlam, Casablanca, Maroc. *Comptes Rendus de l'Academie des Sciences, Paris, Sciences de la Terre et des Planètes* 326:671–676.

Geraads, D., Z. Alemseged, and H. Bellon, 2002. The late Miocene fauna of Chorora, Awash Basin, Ethiopia: systematics, biochronology and the ^{40}K-^{40}Ar ages of the associated volcanics. *Tertiary Research* 21:113–122.

Getty, R. 1975. Sisson and Grossman's: The Anatomy of the Domestic Animals, Volume 1. W.B. Saunders, Philadelphia.

Gèze, R. 1980. Les Hippopotamidae (Mammalia, Artiodactyla) du Plio-Pléistocène de l'Éthiopie. Ph.D. thesis, Université Pierre et Marie Curie, Paris.

———. 1985. Répartition paléoécologique *et* relations phylogénétiques des Hippopotamidae (Mammalia, Artiodactyla) du Néogène d'Afrique Orientale. In M. Beden, A. K. Berhensmeyer, N. T. Boaz, R. Bonnefille, C. K. Brain, B. Cooke, Y. Coppens, R. Dechamps, V. Eisenmann, A. Gentry, D. Geraads, R. Gèze, C. Guérin, J. Harris, J. Koeniguer, F. Letouzey, G. Petter, A. Vincens, and E. Vrba (eds.), L'Environnement des Hominidés au Plio-Pléistocène, pp. 81–100. Fondation Singer-Polignac Masson, Paris.

Giaourtsakis, I. X. 2003. Late Neogene Rhinocerotidae of Greece: distribution, diversity and stratigraphical range. In J. W. F. Reumer and W. Wessels (eds.), Distribution and Migration of Tertiary Mammals in Eurasia. A Volume in Honour of Hans de Bruijn, pp. 235–253. Deinsea, Annual of the Natural History Museum of Rotterdam 10. Natural History Museum, Rotterdam.

Giaourtsakis, I. X., G. Theodorou, S. Roussiakis, A. Athanassiou, and G. Iliopoulos. 2006. Late Miocene horned rhinoceroses (Rhinocerotinae, Mammalia) from Kerassia (Euboea, Greece). *Neues Jahrbuch für Geologie und Paläontologie, Abhandlungen* 239:367–398.

Gill, T. 1872. Arrangement of the families of mammals with analytical tables. *Smithsonian Miscellaneous Collections* 11:1–98.

Ginsburg, L., and J. Morales. 1998. Hemicyoninae (Ursidae, Carnivora, Mammalia) and the related taxa from Early and Middle Miocene of Western Europe. *Annales de Paléontologie* 84:71–123.

Glazko, G. V., and M. Nei. 2003. Estimation of divergence times for major lineages of primate species. *Molecular Biology and Evolution* 20:424–434.

Goddard, J. 1967. Home range, behaviour, and recruitment rates of two black rhinoceros populations. *East African Wildlife Journal* 6:133–150.

———. 1968. Food preferences of two black rhinoceros populations. *East African Wildlife Journal* 6:1–18.

———. 1970. Food preferences of black rhinoceros in the Tsavo National Park. *East African Wildlife Journal* 8:145–161.

Gortani, M., and A. Bianchi. 1973. Missione Geologica dell'Azienda Generale Italiana Petroli (A.G.I.P.) nella Dancalia Meridonale e sugli Altipiani Hararini (1936–1939). Accademia Nazionale dei Lincei, Rome.

Gray, B. T. 1980. Environmental reconstruction of the Hadar Formation (Afar, Ethiopia). Ph.D. thesis, Case Western Reserve University, Cleveland, Ohio.

Gray, J. E. 1821. On the natural arrangement of vertebrose animals. *London Medical Repository* 15:296–310.

———. 1868. Observations on the preserved specimens and skeletons of Rhinocerotidae in the collection of the British Museum and Royal College of Surgeons, including the description of three new species. *Proceedings of the Zoological Society of London* Year 1867:1003–1032.

Greenacre, M. J., and E. S. Vrba. 1984. Graphical display and interpretation of antelope census data in African wildlife areas, using correspondence analysis. *Ecology* 65:984–997.

Grine, F. E., and Q. B. Hendy. 1981. Earliest primate remains from South Africa. *South African Journal of Science* 77:374–376.

Groves, C. P. 1967. Geographic variation in the black rhinoceros *Diceros bicornis* (Linnaeus, 1758). *Zeitschrift für Säugetierkunde* 32:267–276.

———. 1972. *Ceratotherium simum*. *Mammalian Species* 8:1–6.

———. 1975. Taxonomic notes on the white rhinoceros *Ceratotherium simum* (Burchell, 1817). *Säugetierkundliche Mitteilungen* 23(3):200–212.

———. 1983. Phylogeny of the living species of rhinoceros. *Zeitschrift für Zoologische Systematik und Evolutionsforschung* 21:293–313.

———. 1993. Testing rhinoceros subspecies by multivariate analysis. In O. A. Ryder (ed.), Rhinoceros Biology and Conservation, pp. 92–100. Publications of the Zoological Society of San Diego, California.

Grubb, P. 1993. Family Bovidae. In D. E. Wilson and D. M. Reeder (eds.), Mammal Species of the World: A Taxonomic and Geographic Reference, pp. 393–414. Smithsonian Institution Press, Washington, DC.

Guérin, C. 1966. *Diceros douariensis* nov. sp., un rhinocéros du Mio-Pliocène de Tunisie du Nord. *Documents des Laboratoires de Géologie de la Faculté des Sciences de Lyon* 16:1–50.

———. 1976. Les restes de rhinocéros du gisement miocène de Beni Mellal, Maroc. *Géologie Méditerranéenne* 3:105–108.

———. 1979. Chalicotheriidae et Rhinocerotidae (Mammalia, Perissodactyla) du Miocène au Pléistocène de la Rift Valley (Afrique orientale). Un exemple d'évolution: Le squelette post-crânien des *Diceros* et *Ceratotherium* plio-pléistocènes. *Bulletin de la Société Géologique de France* 21:283–288.

———. 1980a. À propos des rhinocéros (Mammalia, Rhinocerotidae) néogenènes et quaternaires d'Afrique: Essai de synthèse sur les espèces et sur les gisements. In R. E. Leakey and B. A. Ogot (eds.), Proceedings of the 8th Panafrican Congress of Prehistory and Quaternary Studies Nairobi 1977, pp. 58–63. International Louis Leakey Memorial Institution for African Prehistory, Nairobi.

———. 1980b. Les rhinocéros (Mammalia, Perissodactyla) du Miocène terminal au Pléistocène supérieur en Europe occidentale. Comparaison avec les espèces actuelles. *Documents des Laboratoires de Géologie de la Faculté des Sciences de Lyon* 79:1–1185.

———. 1982. Les Rhinocerotidae (Mammalia, Perissodactyla) du Miocène terminal au Pléistocène supérieur d'Europe occidentale comparés aux espèces actuelles: Tendances évolutives et relations phylogénétiques. *Geobios* 15:599–605.

———. 1985. Les rhinocéros et les chalicothères (Mammalia, Perissodactyla) des gisements de la vallée de l'Omo en Ethiopie. In Y. Coppens and F. C. Howell (eds.), Les Faunes Plio-Pléistocènes de la Basse Vallée de l'Omo (Ethiopie), pp. 67–89. Cahiers de Paléontologie, Travaux de Paléontologie Est-Africaine. Centre National de la Recherche Scientifique, Paris.

———. 1987a. A brief paleontological history and comparative anatomical study of the Recent rhinos of Africa. *Pachyderm* 9:5.

———. 1987b. Fossil Rhinocerotidae (Mammalia, Perissodactyla) from Laetoli. In M. D. Leakey and J. M. Harris (eds.),

Laetoli, a Pliocene Site in Northern Tanzania, pp. 320–348. Clarendon Press, Oxford.

———. 1989. La famille des Rhinocerotidae (Mammalia, Perissodactyla): Systématique, histoire, évolution, paléoécologie. *Cranium* 6(2):3–14.

———. 1994. Les rhinocéros (Mammalia, Perissodactyla) du Néogène de l'Ouganda. In B. Senut and M. Pickford (eds.), Geology and Palaeobiology of the Albertine Rift Valley, Uganda-Zaire, pp. 263–280. Publication occasionnelle. CIFEG, Orléans, France.

———. 2000. The Neogene rhinoceroses of Namibia. *Palaeontologia Africana* 36:119–138.

———. 2003. Miocene Rhinocerotidae of the Orange River Valley, Namibia. In M. Pickford and B. Senut (eds.), Geology and Palaeobiology of the Central and Southern Namib Desert, Southwestern Africa. Volume 2: Palaeontology, p. 257–281. Memoirs of the Geological Survey of Namibia 19. Geological Survey of Namibia, Windhoek.

Guggisberg, C. A. W. 1966. SOS Rhino. Andre Deutsch Publications, London.

Gundling, T., and A. Hill. 2000. Geological context of fossil Cercopithecoidea from eastern Africa. In P. F. Whitehead and C. J. Jolly (eds.), Old World Monkeys, pp. 180–213. Cambridge University Press, Cambridge, UK.

Haile-Selassie, Y. 2001a. Late Miocene Mammalian Fauna from the Middle Awash Valley, Ethiopia. Ph.D. thesis, University of California, Berkeley.

———. 2001b. Late Miocene hominids from the Middle Awash, Ethiopia. *Nature* 412:178–181.

Haile-Selassie, Y., L. J. Hlusko, and F. C. Howell. 2004a. A new species of *Plesiogulo* (Mustelidae: Carnivora) from the late Miocene of Africa. *Palaeontologica Africana* 40:85–88.

Haile-Selassie, Y., G. Suwa, and T. D. White. 2004b. Late Miocene teeth from Middle Awash, Ethiopia, and early hominid dental evolution. *Science* 303:1503–1505.

Haile-Selassie, Y., G. WoldeGabriel, T. D. White, R. L. Bernor, D. L. DeGusta, P. Renne, W. K. Hart, E. Vrba, S. Ambrose, and F. C. Howell. 2004c. Mio-Pliocene mammals from the Middle Awash, Ethiopia. *Geobios* 37:536–552.

Hall, C. M., R. C. Walter, J. A. Westgate, and D. York. 1984. Geochronology, stratigraphy, and geochemistry of Cindery Tuff in Pliocene hominid-bearing sediments of the Middle Awash, Ethiopia. *Nature* 308:26–31.

Hall-Martin, A. J., T. Erasmus, and B. P. Botha. 1982. Seasonal variation of diet and faeces composition of black rhinoceros (*Diceros bicornis*) in the Addo Elephant National Park. *Koedoe* 25:63–82.

Hamilton, W. R. 1978. Fossil giraffes from the Miocene of Africa and a revision of the phylogeny of the Giraffoidea. *Philosophical Transactions of the Royal Society of London B* 283:165–229.

Happold, D. C. D. 1987. The Mammals of Nigeria. Clarendon Press, Oxford.

———. 2001. Ecology of African small mammals: Recent research and perspectives. In C. Denys, L. Granjon, and A. Poulet (eds.), African Small Mammals, pp. 377–414. Institut du Recherche pour le Developpement Editions, Paris.

Harris, E. E., and T. R. Disotell. 1998. Nuclear gene trees and the phylogenetic relationships of the mangabeys (Primates: Papionini). *Molecular Biology and Evolution* 15:892–900.

Harris, J. M. 1973. *Prodeinotherium* from Gebel Zelten, Libya. *Bulletin of the British Museum of Natural History, Geology* 23:285–350.

———. 1974. Orientation and variability in the ossicones of African Sivatheriinae (Mammalia; Giraffidae). *Annals of the South African Museum* 65:189–198.

———. 1976a. Fossil Rhinocerotidae (Mammalia, Perissodactyla) from East Rudolf, Kenya. *Fossil Vertebrates of Africa* 4:147–172.

———. 1976b. Pliocene Giraffoidea (Mammalia; Giraffidae) from the Cape Province. *Annals of the South African Museum* 69:325–353.

———. 1978. Deinotherioidea and Barytherioidea. In V. J. Maglio and H. B. S. Cooke (eds.), Evolution of African Mammals, pp. 315–332. Harvard University Press, Cambridge, MA.

———. 1983a. Family Rhinocerotidae. In J. M. Harris (ed.), Koobi Fora Research Project, Vol 2: The Fossil Ungulates: Proboscidea, Perissodactyla and Suidae, pp. 130–156. Clarendon Press, Oxford.

———. 1983b. Family Suidae. In J. M. Harris (ed.), Koobi Fora Research Project, Vol. 2: The Fossil Ungulates: Proboscidea, Perissodactyla and Suidae, pp. 215–300. Clarendon Press, Oxford.

———. 1987a. Fossil Deinotheriidae from Laetoli. In M. D. Leakey and J. M. Harris (eds.), Laetoli, a Pliocene Site in Northern Tanzania, pp. 294–297. Clarendon Press, Oxford.

———. 1987b. Fossil Giraffidae from Sahabi, Libya. In N. T. Boaz, A. El-Arnauti, A. W. Gaziry, J. de Heinzelin, and D. D. Boaz (eds.), Neogene Paleontology and Geology of Sahabi, pp. 317–321. Alan R. Liss, New York.

———. 1991a. Family Hippopotamidae. In J. M. Harris (ed.), Koobi Fora Research Project Vol. 3: The Fossil Ungulates: Geology, Fossil Artiodactyls, and Palaeoenvironments, pp. 31–85. Clarendon Press, Oxford.

———. 1991b. Family Giraffidae. In J. M. Harris (ed.), Koobi Fora Research Project Vol. 3: The Fossil Ungulates: Geology, Fossil Artiodactyls, and Palaeoenvironments, pp. 93–138. Clarendon Press, Oxford.

———. 1991c. Family Bovidae. In J. M. Harris (ed.), Koobi Fora Research Project Vol. 3: The Fossil Ungulates: Geology, Fossil Artiodactyls, and Palaeoenvironments, pp. 139–320. Clarendon Press, Oxford.

———. 2003a. Deinotheres from the Lothagam succession. In M. G. Leakey and J. M. Harris (eds.), Lothagam: The Dawn of Humanity in Eastern Africa, pp. 359–361. Columbia University Press, New York.

———. 2003b. Lothagam Giraffids. In M. G. Leakey and J. M. Harris (eds.), Lothagam: The Dawn of Humanity in Eastern Africa, pp. 523–530. Columbia University Press: New York.

———. 2003c. Bovidae from the Lothagam Succession. In J. M. Harris and M. G. Leakey (eds.), Lothagam: The Dawn of Humanity in Eastern Africa, pp. 531–579. Columbia University Press, New York.

Harris, J. M., and T. E. Cerling. 1998. Isotopic changes in the diet of giraffids. *Journal of Vertebrate Paleontology, Supplement* 18:49A.

———. 2002. Dietary adaptations of extant and Neogene African suids. *Journal of Zoology, London* 256:45–54.

Harris, J. M, and M. G. Leakey. 2003a. Lothagam Rhinocerotidae. In M. G. Leakey and J. M. Harris (eds.), Lothagam: the Dawn of Humanity in Eastern Africa, pp. 371–385. Columbia University Press, New York.

———. 2003b. Lothagam Suidae. In M. G. Leakey and J. M. Harris (eds.), Lothagam: The Dawn of Humanity in Eastern Africa, pp. 485–522. Columbia University Press, New York.

Harris, J. M., and T. D. White. 1979. Evolution of the Plio-Pleistocene African Suidae. *Transactions of the American Philosophical Society* 69:1–128.

Harris, J. M., F. H. Brown, and M. G. Leakey, 1988. Stratigraphy and paleontology of Pliocene and Pleistocene localities west of Lake Turkana, Kenya. *Contributions in Science* 399:128.

Harris, J. M., M. G. Leakey, and T. E. Cerling. 2003. Early Pliocene tetrapod remains from Kanapoi, Lake Turkana Basin, Kenya. In J. M. Harris and M. G. Leakey (eds.), Geology and Vertebrate Paleontology of the Early Pliocene Site of Kanapoi, Northern Kenya, pp. 39–113. Natural History Museum of Los Angeles County.

Harris, J. W. K. 1983. Cultural beginnings: Plio-Pleistocene archaeological occurrences from the Afar, Ethiopia. *African Archaeological Review* 1:3–31.

Harrison, T. 1989. New postcranial remains of *Victoriapithecus* from the middle Miocene of Kenya. *Journal of Human Evolution* 18:3–54.

———. 1997a. Neogene Paleontology of the Manonga Valley, Tanzania: A Window into East African Evolutionary History. Plenum Press, New York.

———. 1997b. The anatomy, paleobiology, and phylogenetic relationships of the Hippopotamidae (Mammalia, Artiodactyla) from the Manonga Valley, Tanzania. In T. Harrison (ed.), Neogene Paleontology of the Manonga Valley, Tanzania, pp. 137–190. Plenum Press, New York.

———. 2002. Late Oligocene to middle Miocene catarrhines from Afro-Arabia. In W. Hartwig (ed.), The Primate Fossil Record, pp. 311–338. Cambridge University Press, Cambridge, UK.

Harrison, T., and M. L. Mbago. 1997. Introduction: paleontological and geological research in the Manonga Valley, Tanzania. In T. Harrison (ed.), Neogene Paleontology of the Manonga Valley, Tanzania, pp. 1–32. Plenum Press, New York.

Hart, W. K., and R. C. Walter. 1983. Geochemical investigation of volcanism in the west-central Afar, Ethiopia. *Carnegie Institution of Washington Year Book* 82:491–497.

Hart, W. K., R. C. Walter, and G. WoldeGabriel. 1992. Tephra sources and correlations in Ethiopia: application of elemental and Nd isotope data. *Quaternary International* 13/14:77–86.

Hart, W. K., G. WoldeGabriel, S. Katoh, P. R. Renne, G. Suwa, B. Asfaw, and T. D. White. 2003. Reply to Dating of the Herto Hominin Fossils. *Nature* 426:622.

Hart, W. K., G. WoldeGabriel, P. R. Renne, and T. D. White. 2004. Bimodal volcanism and rift basin development in the Middle Awash Region, Ethiopia. *Geological Society of America* 36:485.

Hassanin, A., and E. J. P. Douzery. 1999. The tribal radiation of the family Bovidae (Artiodactyla) and the evolution of the mitochondrial cytochrome *b* gene. *Molecular Phylogenetics and Evolution* 13:227–243.

Hayward, N. J., and C. J. Ebinger. 1996. Variations in the along-axis segmentation of the Afar Rift. *Tectonics* 15:244–257.

Hedberg, H. D. 1976. International Stratigraphic Code. Wiley, New York.

Heiken, G. and K. Wohletz. 1985. Volcanic Ash. University of California Press, Berkeley.

Heim de Balsac, H., and J.-J. Barloy. 1966. Revision des Crocidures du groupe *-flavescens, -occidentalis, -manni. Mammalia* 30:601–633.

Heim de Balsac, H., and J. Meester. 1977. Order Insectivora. In J. Meester and H. W. Setzer (eds.), The Mammals of Africa: An Identification Manual. Part 1, pp. 1–29 (not continuously paginated). Smithsonian Institution Press, Washington, DC.

Heissig, K. 1975. Rhinocerotidae aus dem Jungtertiär Anatoliens. *Geologisches Jahrbuch (B)* 15:145–151.

———. 1981. Probleme bei der cladistischen Analyse einer Gruppe mit wenigen eindeutigen Apomorphien: Rhinocerotidae. *Paläontologisches Zeitschrift* 55:117–123.

———. 1989. The Rhinocerotidae. In D. R. Prothero and R. M. Schoch (eds.), The Evolution of Perissodactyls, p. 399–417. Oxford Monographs on Geology and Geophysics 15. Oxford University Press, Oxford.

———. 1996. The stratigraphical range of fossil rhinoceroses in the late Neogene of Europe and the Eastern Mediterranean. In R. L. Bernor, V. Fahlbusch, and H.-W. Mittmann (eds.), The Evolution of Western Eurasian Neogene Mammal Faunas, pp. 339–347. Columbia University Press, New York.

Heller, E. 1913. The white rhinoceros. *Smithsonian Miscellaneous Collections* 61(1):1–77.

Hendey, Q. B. 1974. The late Cenozoic Carnivora of the southwestern Cape province. *Annals of the South African Museum* 63:1–369.

———. 1976. Fossil peccary from the Pliocene of South Africa. *Science* 192:787–789.

———. 1977. Fossil bear from South Africa. *South African Journal of Science* 73:112–116.

———. 1978a. Late Tertiary Hyaenidae from Langebaanweg, South Africa, and their relevance to the phylogeny of the family. *Annals of the South African Museum* 76:265–297.

———. 1978b. Late Tertiary Mustelidae (Mammalia, Carnivora) from Langebaanweg, South Africa. *Annals of the South African Museum* 76:329–357.

———. 1980. *Agriotherium* (Mammalia, Ursidae) from Langebaanweg, South Africa, and relationships of the genus. *Annals of the South African Museum* 81:1–109.

———. 1981. Geological succession at Langebaanweg, Cape Province, and global events of the late Tertiary. *South African Journal of Science* 77:33–38.

———. 1982. Langebaanweg: A Record of Past Life. South African Museum, The Rustica Press, Cape Town.

Hernández Fernández, M., and E. S. Vrba. 2005. A complete estimate of the phylogenetic relationships in Ruminantia: A dated species-level supertree of the extant ruminants. *Biological Reviews* 80:69–302.

———. 2006. Plio-Pleistocene climatic change in the Turkana Basin (East Africa): Evidence from large mammal faunas. *Journal of Human Evolution* 50:595–626.

Hill, A. 1987. Causes of perceived faunal change in the later Neogene of East Africa. *Journal of Human Evolution* 16:583–596.

———. 1999. The Baringo Basin, Kenya: From Bill Bishop to BPRP. In P. Andrews and P. Banham (eds.), Late Cenozoic Environments and Hominid Evolution: A Tribute to Bill Bishop, pp. 85–97. Geographical Society, London.

———. 2002. Paleoanthropological research in the Tugen Hills, Kenya. *Journal of Human Evolution* 42:1–10.

Hill, A., and S. Ward. 1988. Origin of the Hominidae: The record of African large hominoid evolution between 14 My and 4 My. *Yearbook of Physical Anthropology* 31:49–83.

Hill, A., G. Curtis, and R. Drake. 1986. Sedimentary stratigraphy of the Tugen Hills, Baringo, Kenya. In L.E. Frostick, R. W. Renault, I. Reid, and J.-J. Tiercelin (eds.), Sedimentation in the African Rifts, pp. 285–295. Geological Society of London Special Publication 25. Blackwell, Oxford.

Hill, A., S. Ward, and B. Brown. 1992. Anatomy and age of the Lothagam mandible. *Journal of Human Evolution* 22:439–451.

Hillman-Smith, A. K. K., and C. P. Groves. 1994. *Diceros bicornis*. *Mammalian Species* 445:1–8.

Hilzheimer, M. 1925. *Rhinoceros simus germano-africanus* n. subsp. aus Oldoway. In H. Reck (ed.), Wissenschaftliche Ergebnisse der Oldoway-Expedition 1913, Neue Folge, Heft 2, pp. 47–79. Bornträger, Leipzig.

Hinton, M. A. C. 1933. Diagnoses of new genera and species of rodents from Indian Tertiary deposits. *Annals and Magazine of Natural History* 72:620–622.

Hitchins, P. M. 1969. Influence of vegetation types on sizes of home ranges of black rhinoceroses in Hluhluwel Game Reserve, Zululand. *Lammergeyer* 10:81–86.

———. 1978. Age determination of the black rhinoceros (*Diceros bicornis* Linn.) in Zululand. *South African Journal of Wildlife Research* 8:71–80.

Hlusko, L. J. 2002. Expression types for two cercopithecoid dental traits (interconulus and interconulid) and their variation in a modern baboon population. *International Journal of Primatology* 23:1309–1318.

———. 2006. A new large Pliocene colobine species (Mammalia: Primates) from Asa Issie, Ethiopia. *Geobios* 39:57–69.

———. 2007a. A new late Miocene species of *Paracolobus* and other cercopithecoidea (Mammalia: Primates) fossils from Lemudong'o, Kenya. *Kirtlandia* 56:72–85.

———. 2007b. Earliest evidence for *Atherurus* and *Xenohystrix* (Hystricidae, Rodentia) from the Late Miocene site of Lemudong'o, Kenya. *Kirtlandia* 56:86–91.

Hodell, D. A., J. H. Curtis, F. J. Sierro, and M.E.Raymo. 2001. Correlation of Late Miocene to Early Pliocene sequences between the Mediterranean and North Atlantic. *Paleoceanography* 16:164–178.

Hollister, N. 1919. East African Mammals in the United States Museum, Part 2: Rodentia, Lagomorpha, and Tubulidentata. Bulletin 99. Smithsonian Institution, United States National Museum, Washington, DC.

Hooijer, D. A. 1950. The fossil Hippopotamidae of Asia, with notes on the recent species. *Zoologische Verhandelingen* 8:123.

———. 1955. Fossil Proboscidea from the Malay Archipelago and the Punjab. *Zoologische Verhandelingen* 28:1–146.

———. 1959. Fossil rhinoceroses from the Limeworks Cave, Makapansgat. *Paleontologia Africana* 6:1–13.

———. 1963. Miocene mammalia of Congo. *Annals of the Musée Royal de l'Afrique Centrale. Sciences Géologiques* 46:1–71.

———. 1966. Miocene rhinoceroses of East Africa. Fossil Mammals of Africa No. 21. *Bulletin of the British Museum (Natural History)* 12:120–190.

———. 1968. A rhinoceros from the late Miocene of Fort Ternan. *Zoologische Mededelingen* 43:77–92.

———. 1969. Pleistocene East African rhinos. *Fossil Vertebrates of Africa* 1:71–98.

———. 1970. Miocene mammalia of Congo, correction. *Annals of the Musée Royal de l'Afrique Centrale. Sciences Géologiques* 67:163–167.

———. 1972. A Late Pliocene rhinoceros from Langebaanweg, Cape Province. *Annals of the South African Museum* 59: 151–191.

———. 1973. Additional Miocene to Pleistocene rhinoceroses of Africa. *Zoologische Mededelingen* 46:149–178.

———. 1975. Note on some newly found perissodactyl teeth from the Omo Group deposits, Ethiopia. *Proceedings of the Koninklijke Nederlandse Akademie van Wetenschappen, B* 78: 188–190.

———. 1976. Phylogeny of the rhinocerotids of Africa. *Annals of the South African Museum* 71:167–168.

———. 1978. Rhinocerotidae. In V. J. Maglio and H. B. S. Cooke (eds.), Evolution of African Mammals, pp. 371–378. Harvard University Press, Cambridge, MA.

Hooijer, D. A., and C. S. Churcher. 1985. Perissodactyla of the Omo Group deposits, American Collections. In Y. Coppens and F. C. Howell (eds.), Les Faunes Plio-Pléistocènes de la Basse Vallée de l'Omo (Éthiopie), p. 99–117. Cahiers de Paléontologie, Travaux de Paléontologie Est-Africaine. Centre National de la Recherche Scientifique, Paris.

Hooijer, D. A., and V. J. Maglio. 1974. Hipparions from the late Miocene and Pliocene of northwestern Kenya. *Zoologische Verhandelingen (Leiden)* 134:3–34.

Hooijer, D. A., and B. Patterson. 1972. Rhinoceroses from the Pliocene of northwestern Kenya. *Bulletin of the Museum of Comparative Zoology* 144:1–26.

Hooijer, D. A., and R. Singer. 1960. Fossil rhinoceroses from Hopefield, South Africa. *Zoologische Mededelingen* 37:113–128.

Hopwood, A. T. 1926. Fossil Mammalia. In E. J. Wood (ed.), The geology and palaeontology of the Kaiso Bone Beds II: Paleontology, 13–36. Occasional Papers of the Geological Survey of the Uganda Protectorate 2.

———. 1939a. The mammalian fossils. In T. P. O'Brien (ed.), The Prehistory of Uganda Protectorate, p. 308–316. Cambridge University Press, Cambridge, UK.

———. 1939b. The subspecies of black rhinoceros, *Diceros bicornis* (Linnaeus), defined by the proportions of the skull. *Journal of the Linnean Society of London* 40:447–457.

Houghton, B. F., C. J. N. Wilson, and D. M. Pyle. 2000. Pyroclastic fall deposits. In H. Sigurdsson, B. F. Houghton, S. R. McNutt, H. Rymer, and J. Stix (eds.), Encyclopedia of Volcanoes, pp. 555–570. Academic Press, San Diego.

Howell, F. C. 1982. Preliminary observations on Carnivora. *Garyounis Scientific Bulletin, Garyounis University* 4:49–62.

———. 1987. Preliminary observations on Carnivora from the Sahabi Formation (Libya). In N. T. Boaz, A. El-Arnauti, A. W. Gaziry, J. de Heinzelin, and D. D. Boaz (eds.), Neogene Paleontology and Geology of Sahabi, pp. 153–181. Alan R. Liss, New York.

Howell, F. C., and G. Petter. 1976. Carnivora from Omo Group Formations, Southern Ethiopia. In Y. Coppens, F. C. Howell, G. Isaac, and R. Leakey (eds.), Earliest Man and Environments in the Lake Rudolf Basin, pp. 314–332. University of Chicago Press, Chicago.

———. 1985. Comparative observations on some Middle and Upper Miocene hyaenids: Genera: *Percrocuta* Kretzoi, *Allohyaena* Kretzoi, *Adcrocuta* Kretzoi (Mammalia, Carnivora, Hyaenidae). *Geobios* 18:419–476.

Hsieh, H.-M., L.-H. Huang, L.-C. Tsai, Y.-C. Kuo, H.-H. Menga, A. Linacre, and J. C.-I. Lee. 2003. Species identification of rhinoceros horns using the cytochrome *b* gene. *Forensic Science International* 136:1–11.

Hunt, R. M., Jr. 1987. Evolution of the aeluroid Carnivora: Significance of auditory structure in the nimravid cat *Dinictis*. *American Museum Novitates* 2886:1–74.

———. 1996. Biogeography of the order Carnivora. In J. L. Gittleman (ed.), Carnivore Behavior, Ecology and Evolution, Vol. 2, pp. 451–485. Cornell University Press, Ithaca, NY.

———. 1998. Evolution of the aeluroid Carnivora: Diversity of the earliest aeluroids from Eurasia (Quercy, Hsanda-Gol) and the origin of felids. *American Museum Novitates* 3252:1–65.

Hunt, R. M., Jr., and R. H. Tedford. 1993. Phylogenetic relationships within the aeluroid Carnivora and implications of their temporal and geographic distribution. In F. S. Szalay, M. J. Novacek, and M. C. McKenna (eds.), Mammal Phylogeny (Placentals), pp. 53–73. Springer-Verlag, New York.

Ishida, H., Y. Kunimatsu, M. Nakatsukasa, and Y. Nakano. 1999. New hominoid genus from the middle Miocene of Nachola, Kenya. *Anthropological Sciences* 107:189–191.

Ishida, H., Y. Kunimatsu, T. Takano, Y. Nakano, and M. Nakatsukasa. 2004. *Nacholapithecus* skeleton from the Middle Miocene of Kenya. *Journal of Human Evolution* 46:67–101.

Izett, G. A. 1981. Volcanic ash beds: recorders of upper Cenozoic silicic pyroclastic volcanism in the western United States. *Journal of Geophysical Research* 86:10200–10222.

Jablonski, N. 2002. Fossil Old World monkeys: The late Neogene radiation. In W. Hartwig (ed.), The Primate Fossil Record, pp. 255–299. Cambridge University Press, Cambridge, UK.

Jacobs, L. L. 1978. Fossil rodents (Rhizomyidae and Muridae) from Neogene Siwalik deposits, Pakistan. *Museum of Northern Arizona Press, Bulletin Series* 52:1–103.

Jacobs, L. L., L. J. Flynn, and W. R. Downs. 1989. Neogene rodents of southern Asia. In C. C. Black and M. R. Dawson (eds.), Papers on Fossil Rodents in Honor of Albert Elmer Wood, pp. 157–177. Special Publication 33. Natural History Museum of Los Angeles County, Los Angeles.

Jaeger, J.-J. 1975. Les Muridae (Mammalia, Rodentia) du Pliocène et du Pléistocène du Maghreb: Origine, evolution, données biogeographiques et paleoclimatiques. Ph.D. thesis, Université des Sciences et Techniques du Langedoc, Montpellier, France.

Jaeger, J.-J., J. Michaux, and M. Sabatier. 1980. Premières données sur les rongeurs de la formation de Ch'orora (Éthiopie) d'âge Miocène supérieur. I: Thryonomyides. In *Palaeovertebrata*: Mémoire Jubilaire R. Lavocat, pp. 365–374.

James, G. T., and B. H. Slaughter. 1974. A primitive new Middle Pliocene murid from Wadi El Natrun, Egypt. *Annals of the Geological Survey of Egypt*, IV:333–362.

Jenkins, P. 2001. Other cavy-like rodents. In D. Macdonald (ed.), The New Encyclopedia of Mammals, pp. 682–685. Oxford University Press, Oxford, London, and New York.

Jernvall, J., and M. Fortelius. 2002. Common mammals drive the evolutionary increase of hypsodonty in the Neogene. *Nature* 417:538–540.

Johanson, D., C. O. Lovejoy, W. H. Kimbel, T. D. White, S. C. Ward, M. E. Bush, B. M. Latimer, and Y. Coppens. 1982a. Morphology of the Pliocene partial hominid skeleton (AL 288-1) from the Hadar formation, Ethiopia. *American Journal of Physical Anthropology* 57:403–451.

Johanson, D. C., M. Taieb, and Y. Coppens. 1982b. Pliocene hominids from the Hadar Formation, Ethiopia (1973–1977): Stratigraphic, chronologic, and paleoenvironmental contexts, with notes on hominid morphology and systematics. *American Journal of Physical Anthropology* 57:373–402.

Jolly, C. J. 1972. The classification and natural history of *Theropithecus* (*Simopithecus*) (Andrews, 1916), baboons of the African Plio-Pleistocene. *Bulletin of the British Museum of Natural History* 22:1–122.

Juch, D. 1975. Geology of the southeastern escarpment of Ethiopia between 39° and 42° longitude east. In A. Pilger, and Rosler (eds.), Afar Depression of Ethiopia, pp. 310–315. Aschweizerbart Verlag, Stuttgart.

———. 1980. Tectonics of the southeastern escarpment of Ethiopia. In Geodynamic Evolution of the Afro-Arabian Rift System, pp. 407–418. Atti Convegni 47. Accademia Nazionale Dei Lincei, Rome.

Justin-Visentin, E., and B. Zanettin. 1974. Dike swarms, volcanism and tectonics of the western Afar margin along the Kombolcha-Eloa traverse (Ethiopia). *Bulletin Volcanologique* 38:187–205.

Kahlke, R. D. 1990. Zum Stand der Erforschung fossiler Hippopotamiden (Mammalia, Artiodactyla). Eine Übersicht. *Quartärpaläontologie* 8:107–118.

Kalb, J. E. 1978. Miocene to Pleistocene deposits in the Afar depression, Ethiopia. *Sinet (Ethiopian Journal of Science)* 1:87–98.

———. 1993. Refined stratigraphy of the hominid-bearing Awash Group, Middle Awash Valley, Afar Depression, Ethiopia. *Newsletter in Stratigraphy* 29:21–62.

———. 1995. Fossil elephantoids, Awash paleolake basins, and the Afar triple junction, Ethiopia. *Palaeogeography, Palaeoclimatology, Palaeoecology* 114:357–368.

———. 2000. Adventures in the Bone Trade: The Race to Discover Human Ancestors in Ethiopia's Afar Depression. Copernicus Books, New York.

Kalb, J. E., and D. J. Froehlich. 1995. Interrelationships of Late Neogene elephantoids: New evidence from the Middle Awash Valley, Afar, Ethiopia. *Geobios* 28:727–736.

Kalb, J. E., and A. Mebrate. 1993. Fossil elephantoids from the hominid-bearing Awash Group, Middle Awash valley, Afar depression, Ethiopia. *Transactions of the American Philosophical Society, New Series* 83:1–114.

Kalb, J. E., C. B. Wood, C. Smart, E. B. Oswald, A. Mebrate, S. Tebedge, and P. F. Whitehead. 1980. Preliminary geology and paleontology of the Bodo D'Ar hominid site, Afar, Ethiopia. *Palaeogeography, Palaeoclimatology, Palaeoecology* 30:107–120.

Kalb, J. E., M. Jaeger, C. J. Jolly, and B. Kana. 1982a. Preliminary geology, paleontology and paleoecology of a Sangoan site at Andalee, Middle Awash Valley, Ethiopia. *Journal of Archaeological Science* 9:349–363.

Kalb, J. E., C. J. Jolly, A. Mebrate, S. Tebedge, C. Smart, E. B. Oswald, D. Cramer, P. F. Whitehead, C. B. Wood, G. C. Conroy, T. Adefris, L. Sperling, and B. Kana. 1982b. Fossil mammals and artefacts from the Middle Awash Valley, Ethiopia. *Nature* 298:25–29.

Kalb, J. E., C. J. Jolly, S. Tebedge, A. Mebrate, C. Smart, E. B. Oswald, P. F. Whitehead, C. Wood, T. Adefris, and V. Rawn-Schatzinger. 1982c. Vertebrate faunas from the Awash group, Middle Awash Valley, Afar, Ethiopia. *Journal of Vertebrate Paleontology* 2:237–258.

Kalb, J. E., E. B. Oswald, A. Mebrate, S. Tebedge, and C. J. Jolly. 1982d. Stratigraphy of the Awash Group, Middle Awash Valley, Afar, Ethiopia. *Newsletter on Stratigraphy* 11:95–127.

Kalb, J. E., E. B. Oswald, S. Tebedge, A. Mebrate, E. Tola, and D. Peak. 1982e. Geology and stratigraphy of Neogene deposits, Middle Awash Valley, Ethiopia. *Nature* 298:17–25.

Kalb, J. E., D. J. Froehlich, and G. L. Bell. 1996. Phylogeny of African and Eurasian Elephantidae of the late Neogene. In J. Shoshani and P. Tassy (eds.), The Proboscidea: Evolution and Palaeoecology of Elephants and Their Relatives, pp. 101–116. Oxford University Press, Oxford.

Kappelman, J., T. Plummer, L. Bishop, A. Duncan, S. Appleton. 1997. Bovids as indicators of Plio-Pleistocene paleoenvironments in East Africa. *Journal of Human Evolution* 32:229–256.

Katoh, S., T. Danhara, W. K. Hart, and G. WoldeGabriel. 1999. Use of sodium polytungstate solution in the purification of volcanic glass shards for bulk chemical analysis. *Nature and Human Activities* 4:45–54.

Kaya, T. 1994. *Ceratotherium neumayri* (Rhinocerotidae, Mammalia) in the Upper Miocene of Western Anatolia. *Turkish Journal of Earth Sciences* 3:13–22.

Kazmin, V. and S. M. Berhe. 1978. Geology and development of the Nazret Area, northern Ethiopian Rift, Sheet NC37-15. *Ethiopian Geological Survey Note* 100:1–26.

Kazmin, V., S. M. Berhe, M. Nicoletti, and C. Petrucciani. 1980. Evolution of the northern part of the Ethiopian rift.

In Geodynamic Evolution of the Afro-Arabian Rift System, pp. 275–292. Atti Convegni 47. Accademia Nazionale Dei Lincei, Rome.

Kilburn, C. R. J. 2000. Lava flows and flow fields. In H. Sigurdsson, B. F. Houghton, S. R. McNutt, H. Rymer, and J. Stix (eds.), Encyclopedia of Volcanoes, pp. 291–305. Academic Press, San Diego.

Kimbel, W. H. 1995. Hominid speciation and Pliocene climatic change. In E. Vrba, G. Denton, L. Burckle, and T. Partridge (eds.), Paleoclimate and Evolution with Emphasis on Human Origins, pp. 425–437. Yale University Press, New Haven, CT.

Kimbel, W. H., D. C. Johanson, and Y. Rak, 1994. The first skull and other new discoveries of *Australopithecus afarensis* at Hadar, Ethiopia. *Nature* 368:449–451.

Kimbel, W. H., Y. Rak, and D. C. Johanson, 2004. The skull of *Australopithecus afarensis*. Oxford University Press, Oxford, England.

Kingdon, J. 1971. Hyraxes. In East African Mammals: An Atlas to Evolution in Africa, Vol. I, pp. 329–351. Academic Press, London and New York.

———. 1974. Hares and Rodents. In East African Mammals: An Atlas of Evolution in Africa, Vol. II, part B, pp. 343–703. Academic Press, London and New York.

———. 1979. East African Mammals: An Atlas of Evolution in Africa, Volume III, Part B. Academic Press, New York.

———. 1997. The Kingdon Field Guide to African Mammals. Academic Press, San Diego.

Köhler, M., S. Moyá-Solá, and D. M. Alba. 2001. Eurasian hominoid evolution in the light of recent *Dryopithecus* findings. In L. de Bonis, G. D. Koufos, and P. Andrews (eds.), Phylogeny of the Neogene Hominoid Primates of Eurasia, pp. 192–212. Cambridge University Press, Cambridge, UK.

Kono, R. T. 2004. Molar enamel thickness and distribution patterns in extant great apes and humans: New insights based on a 3-dimensional whole crown perspective. *Anthropological Sciences* 112:121–146.

Kordos, L., and D. R. Begun. 2001. A new cranium of *Dryopithecus* from Rudabánya, Hungary. *Journal of Human Evolution* 41:689–700.

———. 2002. Rudabánya: A late Miocene subtropical swamp deposit with evidence of the origin of the African apes and humans. *Evolutionary Anthropology* 11:45–57.

Kostopoulos, D. S. 2005. The Bovidae (Mammalia, Artiodactyla) from the late Miocene of Akkasdagi. *Geodiversitas* 27:747–791.

Kostopoulos, D. S., S. Sen, and G. D. Koufos. 2003. Magnetostratigraphy and revised chronology of the late Miocene mammal localities of Samos, Greece. *International Journal of Earth Sciences* 92:779–794.

Kotsakis, T., and S. Ingino. 1980. Osservazioni sui *Nyanzachoerus* (Suidae, Artiodactyla) del Terziaro superiori di Sahabi (Cirenaica, Libya). *Bollettino Servizio di Geologia d'Italia* C:391–408.

Koufos, G. D. 1982. *Plesiogulo crassa* from the upper Miocene (Lower Turolian) of northern Greece. *Annales Zoologici Fennici* 19:193–197.

———. 1993. A mandible of *Ouranopithecus macedoniensis* from a new late Miocene locality of Macedonia (Greece). *American Journal of Physical Anthropology* 91:225–234.

———. 2003. Late Miocene mammal events and biostratigraphy in the Eastern Mediterranean. In J. W. F. Reumer and W. Wessels (eds.), Distribution and Migration of Tertiary Mammals in Eurasia. *Dinsea* 10:343–372.

Kovarovic, K., P. Andrews, and L. Aiello. 2002. The paleoecology of the Upper Ndolanya Beds at Laetoli, Tanzania. *Journal of Human Evolution* 43:395–418.

Kramer, A. 1986. Hominid-pongid distinctiveness in the Miocene-Pliocene fossil record: The Lothagam mandible. *American Journal of Physical Anthropology* 70:457–473.

Kretzoi, M. 1942. Bemerkungen zum System der nachmiozänen Nashorn Gattungen. *Földtani Közlöny* 72(4–12):309–318.

Kullmer, O. 1999. Evolution of African Plio-Pleistocene suids (Artiodactyla: Suidae) based on tooth pattern analysis. *Kaupia Darmstädter Beiträge zur Naturgeschichte* 9:1–34.

Kumar, S., A. Filipski, V. Swarna, A. Walker, S. B. Hedges. 2005. Placing confidence limits on the molecular age of the human-chimpanzee divergence. *Proceedings of the National Academy of Sciences of the United States of America* 102:18842–18847.

Kuntz, K., H. Kreutzer, and P. Muller. 1975. Potassium-argon age determination of the trap basalt of the southeastern part of the Afar Rift. In A. Pilger and A. Rosler (eds.), Afar Depression of Ethiopia, pp. 370–374. Schweizerbart Verlag, Stuttgart.

Kurtén, B. 1966. Pleistocene bear of North America. 1. Genus *Tremarctos*, spectacled bears. *Acta Zoologica Fennica* 115:1–120.

———. 1970. The Neogene wolverine *Plesiogulo* and the origin of *Gulo* (Carnivora, Mammalia). *Acta Zoologica Fennica* 131:1–22.

Lacomba, J. I., J. Morales, F. Robles, C. Santisteban, and M. T. Alberdi. 1986. Sedimentologia y paleontologia del yacimiento finimioceno de La Portera (Valencia). *Estudios Geologicos* 42: 167–180.

Lamprey, H. F. 1962. The Tarangire Game Reserve. *Tanganyika Notes and Records* 60:10–22.

Larson, T. L., W. K. Hart, and G. WoldeGabriel. 1997. Heterogeneous magmatic systems: Evidence from a mixed tephra deposit, Middle Awash region, southern Afar Rift, Ethiopia. *Geological Society of America* 29:393.

Latimer, B. M., and O. C. Lovejoy. 1990. Hallucal tarsometatarsal joint in *Australopithecus afarensis*. *American Journal of Physical Anthropology* 82:125–133.

Latimer, B. M., C. O. Lovejoy, D. C. Johanson, and Y. Coppens. 1982. Hominid tarsal, metatarsal and phalangeal bones recovered from the Hadar Formation: 1974–1977 collections. *American Journal of Physical Anthropology* 57:701–719.

Latimer, B. M., J. C. Ohman, and C. O. Lovejoy. 1987. Talocrural joint in African hominoids: implications for *Australopithecus afarensis. American Journal of Physical Anthropology* 74:155–175.

Lavocat, R. 1957. Sur l'âge des faunes de rongeurs des grottes a Australopithèques. In J. D. Clark (ed.), Proceedings of the 3rd Pan-African Congress on Prehistory, pp. 133–134. Chatto and Windus, London.

———. 1961. Le gisement de vertébrés Miocène de Beni Mellal (Maroc). Etude systématique de la faune de Mammifères et conclusions générales. *Notes et Mémoires du Service Géologique Maroc* 155:29–94,109–145.

———. 1973. Les rongeurs du Miocène d'Afrique orientale. *Memoires et Traveaux de l'Institute de Montpellier* 1:1–284.

Leader-Williams, N. 1985. Black rhino in South Luangwa National Park: Their distribution and future protection. *Oryx* 19:27–33.

Leakey, L. S. B. 1958. Some East African Fossil Suidae. Fossil Mammals of Africa No. 14. British Museum of Natural History, London.

———. 1962. A new lower Pliocene fossil primate from Kenya. *Annals and Magazines of Natural History* 4:689–696.

Leakey, M. D., C. S. Feibel, I. McDougall, and A. C. Walker. 1995. New four-million-year-old-hominid species from Kanapoi and Allia Bay, Kenya. *Nature* 376:565–571.

Leakey, M. G. 1982. Extant large colobines from the Plio-Pleistocene of Africa. *American Journal of Physical Anthropology* 58:153–172.

Leakey, M. G., and E. Delson. 1987. Fossil Cercopithecidae from the Laetoli Beds, Tanzania. In M. D. Leakey and J. M. Harris (eds.), The Pliocene Site of Laetoli, Northern Tanzania, pp. 91–107. Oxford University Press, Oxford.

Leakey, M. G., and J. M. Harris. 2003a. Lothagam: The Dawn of Humanity in Eastern Africa. Columbia University Press, New York.

———. 2003b. Lothagam: Its significance and contributions. In M. G. Leakey and J. M. Harris (eds.), Lothagam: The Dawn of Humanity in Eastern Africa, pp. 625–660. Columbia University Press, New York.

Leakey, M. G., and A. Walker. 2003. The Lothagam hominids. In M. G. Leakey and J. M. Harris (eds.), Lothagam: The Dawn of Humanity in Eastern Africa, pp. 249–256. Columbia University Press, New York.

Leakey, M. G., C. S. Feibel, R. L. Bernor, J. M. Harris, T. E. Cerling, K. M. Stewart, G. W. Storrs, A. Walker, L. Werdelin, and A. J. Winkler. 1996. Lothagam: A record of faunal change in the Late Miocene of East Africa. *Journal of Vertebrate Paleontology* 16:556–570.

Leakey, M. G., C. S. Feibel, I. McDougall, C. Ward, and A. Walker. 1998. New specimens and confirmation of an early age for *Australopithecus anamensis. Nature* 393:62–66.

Leakey, M. G., M. F. Teaford, and C. V. Ward. 2003. Cercopithecidae from Lothagam. In M. G. Leakey and J. M. Harris (eds.), Lothagam: The Dawn of Humanity in Eastern Africa, pp. 201–248. Columbia University Press, New York.

Lee-Thorp, J. A., and M. Sponheimer. 2005. Opportunities and constraints for reconstructing palaeoenvironments from stable light isotope ratios in fossils. *Geological Quarterly* 49(2):195–204.

Legendre, S., S. Montuire, O. Maridet, and G. Escarguel. 2005. Rodents and climate: A new model for estimating past temperatures. *Earth and Planetary Letters* 235:408–420.

Le Gros Clark, W. E., and L. S. B. Leakey. 1951. The Miocene Hominoidea of East Africa. *Fossil Mammals of Africa* 1:1–117.

Lehmann, T., P. Vignaud, A. Likius, and M. Brunet. 2004. A fossil aardvark (Mammalia, Tubulidentata) from the lower Pliocene of Chad. *African Journal of Earth Science* 40:201–217.

Lehmann, T., P. Vignaud, H. T. Mackaye, and M. Brunet. 2005. A new species of Orycteropodidae (Mammalia, Tubulidentata) in the Mio-Pliocene of Northern Chad. *Zoological Journal of the Linnean Society* 143:109–131.

Lehmann, U., and H. Thomas. 1987. Fossil Bovidae from the Mio-Pliocene of Sahabi (Libya). In N. T. Boaz, A. El-Arnauti, A. W. Gaziry, J. de Heinzelin, and D. D. Boaz (eds.), Neogene Paleontology and Geology of Sahabi, pp. 323–335. Alan R. Liss, New York.

Leonardi, P. 1952. I suidi di Sahabi nella Sirtica (Africa Settentrionale). *Rendiconti dell' Accademia Nazional XL, Ser. IV.* 4:75–88.

Lévêque, C. 1997. Biodiversity Dynamics and Conservation: The Freshwater Fish of Tropical Africa. Cambridge University Press, Cambridge, UK.

Levin, N. E., J. Quade, S. W. Simpson, S. Semaw, M. Rogers. 2004. Isotopic evidence for Plio-Pleistocene environmental change at Gona, Ethiopia. *Earth and Planetary Science Letters* 219:93–110.

Lewis, G. E. 1934. Notice of the discovery of *Plesiogulo brachygnathus* from the Siwalik measures of India. *American Journal of Science* 26:80.

Lewis, M. E. 1995. Plio-Pleistocene carnivoran guilds: Implications for hominid paleoecology. Ph.D. thesis, State University of New York, Stony Brook.

Lewontin, R. C. 1966. On the measurement of relative variability. *Systematic Zoology* 15:141–142.

Likius, A. 2002. Les Grands Ongulés du Mio-Pliocène du Tchad (Rhinocerotidae, Giraffidae, Camelidae): Systématique,

Implications Paléogéographiques et Paléoenvironnementales. Ph.D. thesis, Université de Poitiers, Poitiers, France.

Lind, E. M., and M. E. S. Morrison. 1974. East African Vegetation. Longman, Bristol.

Linnaeus, C. 1758. Systema Naturae I: Regnum Animale. Editio Decima, Reformata. Laurentii Salvii, Holmiae (Stockholm).

Lister, A. M., A. V. Sher, H. van Essen, and G. Wei. 2005. The pattern and process of mammoth evolution in Eurasia. *Quaternary International* 126–128:49–64.

Loose, H. 1975. Pleistocene Rhinocerotidae of W. Europe with reference to the recent two-horned species of Africa and S. E. Asia. *Scripta Geologica* 33:1–59.

López-Antoñanzas, R. L., S. Sen, and P. Mein. 2004. Systematics and phylogeny of the cane rats (Rodentia: Thryonomyidae). *Zoological Journal of the Linnean Society* 142:423–444.

Louchart, A., P. Vignaud, Y. Haile-Selassie, A. Likius, and M. Brunet. 2008. Fossil birds from the late Miocene of Chad and Ethiopia and zoogeographical implications. In J. Le Loeuff (ed.), Proceedings of the 6th International Meeting of the Society of Avian Paleontology and Evolution. *Oryctos* 7:147–167.

Loutit, B. D., G. N. Louw, and M. K. Seely. 1987. First approximation of food preference and the chemical composition of the diet of the desert-dwelling black rhinoceros (*Diceros bicornis*). *Madoqua* 15:35–54.

Lovejoy, C. O., D. C. Johanson, and Y. Coppens. 1982. Hominid upper limb bones recovered from the Hadar Formation: 1974–1977 collections. *American Journal of Physical Anthropology* 57:637–649.

Lunkka, J. P., M. Fortelius, J. Kappelman, and S. Sen. 1999. Chronology and mammal faunas of the Miocene Sinap Formation, Turkey. In J. Agusti, L. Rook, and P. Andrews (eds.), Climate and Environmental Change in the Neogene of Europe, pp. 238–264. Cambridge University Press: Cambridge, UK.

Lydekker, R. 1880. Siwalik and Narbada Proboscidea. Memoirs of the Geological Survey of India. *Palaeontologia Indica* 1:182–294.

———. 1908. The white rhinoceros. *The Field, London* 111:319.

MacFadden, B. J., and B. J. Shockey. 1997. Ancient feeding ecology and niche differentiation of Pleistocene mammalian herbivores from Tarija, Bolivia: Morphological and isotopic evidence. *Paleobiology* 23:77–100.

MacInnes, D. G. 1942. Miocene and post-Miocene Proboscidea from East Africa. *Transactions of Zoological Society of London* 25:33–106.

———. 1956. Fossil Tubulidentata from East Africa. *Fossil Mammals of Africa* 10:1–38.

Maglio, V. J. 1969. The status of the east African elephant "*Archidiskodon exoptatus*" Dietrich 1942. *Breviora* 336:1–25.

———. 1970. Four new species of Elephantidae from the Plio-Pleistocene of northwestern Kenya. *Breviora* 341:1–43.

———. 1973. Origin and evolution of the Elephantidae. *Transactions of the American Philosophical Society* 63:1–149.

Maglio, V. J., and Q. B. Hendey. 1970. New evidence relating to the supposed stegolophodont ancestry of the Elephantidae. *South African Archaeological Bulletin, Cape Town* 25:85–87.

Maglio, V. J., and A. B. Ricca. 1977. Dental and skeletal morphology of the earliest elephants. *Verhandelingen der Koninklijke Nederlandse Akademie van Wetenschappen, Afdeling Natuurkunde, Eerste Reeks* 29:1–51.

Malatesta, A. 1977. The skeleton of *Nesolutra ichnusae* sp. n., a Quaternary otter discovered in Sardinia. *Geologica Romana* 16:173–210.

Maples, C. G., and A. W. Archer. 1988. Monte Carlo simulation of selected binomial similarity coefficients (2): Effect of sparse data. *Palaios* 3:95–103.

Markov, G. N., and N. Spassov. 2003. Primitive mammoths from Northeast Bulgaria in the context of the earliest mammoth migrations in Europe. In A. Petculescu and E. Ştiucă (eds.), Advances in Vertebrate Paleontology, "Hen to Panta," Volume in Honor of Constantin Radulescu and Petre Mihai Samson, pp. 53–58. Institute of Speleology, Bucharest.

Martin, L. D. 1989. Fossil history of the terrestrial Carnivora. In J. L. Gittleman (ed.), Carnivore Behavior, Ecology, and Evolution, pp. 536–568. Cornell University Press, Ithaca, NY.

Mazzarini, F., T. Abebe, F. Innocenti, P. Manetti, and M. T. Pareschi. 1999. Geology of the Debre Zeyt area (Ethiopia) (with a geological map at scale 1:100,000). *Acta Vulcanologica* 11:131–141.

McCrossin, M. L. 1994. The phylogenetic relationships, adaptations, and ecology of *Kenyapithecus*. Ph.D. thesis, University of California, Berkeley.

McDougall, I., and F. H. Brown. 2006. Precise $^{40}Ar/^{39}Ar$ geochronology for the upper Koobi Fora Formation, Turkana Basin, northern Kenya. *Journal of the Geological Society, London* 163:205–220.

McDougall, I., and C. S. Feibel. 1999. Numerical age control for the Miocene-Pliocene succession at Lothagam, a hominoid-bearing sequence in the northern Kenya Rift. *Journal of the Geological Society, London* 56:731–745.

———. 2003. Numerical age control for the Miocene-Pliocene succession at Lothagam, a hominoid-bearing sequence in the northern Kenya Rift. In M. G. Leakey and J. M. Harris (eds.), Lothagam: The Dawn of Humanity in Eastern Africa, pp. 43–66. Columbia University Press, New York.

McKee, J. 2001. Faunal turnover rates and mammalian biodiversity of the late Pliocene and Pleistocene of eastern Africa. *Paleobiology* 27:500–511.

McKenna, M. C., and S. K. Bell. 1997. Classification of Mammals above the Species Level. Columbia University Press, New York.

Mebrate, A., and J. Kalb. 1981. A primitive elephantid from the Middle Awash Valley, Afar Depression, Ethiopia. *Sinet (Ethiopian Journal of Science)* 4:45–55.

Megrue, G. H., E. Norton, and D. W. Strangway. 1972. Tectonic history of the Ethiopian rift as deduced by K-Ar ages and paleomagnetic measurements of basaltic dikes. *Journal of Geophysical Research* 77:5744–5754.

Meikle, E. 1987. Fossil cercopithecidae from the Sahabi Formation. In N. T. Boaz, A. El-Arnouti, and A. W. Gaziry (eds.), Neogene Paleontology and Geology of Sahabi, pp. 119–127. Alan R. Liss, New York.

Meiring, A. J. D. 1955. Fossil proboscidean teeth and ulna from Virginia, O. F. S. *Researches of the National Museum, Bloemfontein* 1:187–202.

Merenlender, A. M., D. S. Woodruff, O. A. Ryder, R. Kock, and J. Váhala. 1989. Allozyme variation and differentiation in African and Indian rhinoceroses. *Journal of Heredity* 80: 377–382.

Merriam, J. C. 1911. Carnivora from the Tertiary formations of the John Day Region. *University of California Publication in Geology* 5:479–484.

Mertens, R. 1966. Zur Typenterminologie und Nomenklatur einiger Nashörner der Gattung *Diceros. Zoologischer Garten Leipzig* 32:116–117.

Metz-Muller, F. 1995. Mise en évidence d'une variation intraspécifique des caractères dentaires chez *Anancus arvernensis* (Proboscidea, Mammalia) du gisement de Dorkovo (Pliocène ancien de Bulgarie, biozone MN14). *Geobios* 28:737–743.

———. 1996. A propos des spécimens-types d'*Anancus arvernensis* (Proboscidea, Mammalia): Caractéristiques des deux premières dents de lait (D2-D3) chez les gomphothères tétralophodontes. *Annales de Paléontologie (Vertébré-Invertébré)*, 82:27–52.

———. 2000. La Population d'*Anancus arvernensis* (Proboscidea, Mammalia) du Pliocène de Dorkovo (Bulgarie); Étude des Modalités Évolutives d'*Anancus arvernensis* et Phylogénie du Genre *Anancus*. Ph.D. thesis, Muséum National d'Histoire Naturelle, Paris.

Mills, M. G. L., and L. Hes. 1997. The Complete Book of Southern African Mammals. Struik New Holland Publishers, Cape Town.

Misonne, X. 1974. Order Rodentia. In J. Meester and H.W. Setzer (eds.), The Mammals of Africa: An Identification Manual, Part 6, pp. 1–39 (not continuously paginated). Smithsonian Institution Press, Washington, DC.

Mitchell, B. L., and J. M. C. Uys. 1961. The problem of the lechwe (*Kobus leche*) on the Kafue Flats. *Oryx* 6:171–183.

Mitchell, G., and J. D. Skinner. 2003. On the origin, evolution and phylogeny of giraffes *Giraffa camelopardalis*. *Transactions of the Royal Society of South Africa* 58:51–73.

Mohr, P. A. 1967. The Ethiopian Rift System. *Bulletin Geophysical Observatory, Addis Ababa* 12:27–56.

———. 1987. Pattern of faulting in the Ethiopian rift valley. *Tectonophysics* 143:169–179.

Morales, J. 1984. Venta del Moro: Su Macrofauna de Mamíferos Biostratigraphía Continental del Neógeno Terminal Mediterráneo. Ph.D. thesis, Université Complutense de Madrid, Madrid.

Morales, J. C., and D. J. Melnick. 1994. Molecular systematics of the living rhinoceros. *Molecular Phylogenetics and Evolution* 3:128–134.

Morales, J., and M. Pickford. 2005. Giant bunodont Lutrinae from the Mio-Pliocene of Kenya and Uganda. *Estudios Geológicos* 61:233–246.

Morales, J., M. Pickford, and D. Soria. 2005. Carnivores from the late Miocene and basal Pliocene of the Tugen Hills, Kenya. *Revista de la Sociedad Geológica de España* 18:39–61.

Morton, W. H., and R. Black. 1975. Crustal attenuation in Afar. In A. Pilger and A. Rosler (eds.), Afar Depression of Ethiopia, pp. 362–369. Schweizerbart Verlag, Stuttgart.

Moyà-Solà, S. 1983. Los Boselaphini (Bovidae Mammalia) del Neogeno de la Península Ibérica. Publicaciones de Geología 18. Universidad Autònoma de Barcelona.

Moyá-Solá, S., M. Köhler, D. M. Alba, I. Casanovas-Vilar, and J. Galindo. 2004. *Pierolapithecus catalaunicus*, a new Middle Miocene great ape from Spain. *Science* 306:1339–1344.

Mukinya, J. G. 1973. Density, distribution, population structure and social organisation of the black rhinoceros in Masai Mara Game Reserve. *East African Wildlife Journal* 11:385–400.

———. 1977. Feeding and drinking habits of the black rhinoceros in the Masai Mara Game Reserve. *East African Wildlife Journal* 15:135–148.

Mundiger, G. S., and W. J. Sanders. 2001. Taxonomic and systematic re-assessment of the Mio-Pliocene elephant *Primelephas gomphotheroides*. In 8th International Theriological Congress, p. 429. Sun City, South Africa..

Munthe, J. 1987. Small mammal fossils from the Pliocene Sahabi Formation of Libya. In N. T. Boaz, A. El-Arnauti, A. W. Gaziry, J. de Heinzelin, and D. Dechant-Boaz (eds.), Neogene Paleontology and Geology of Sahabi, pp. 135–144. Alan R. Liss, New York.

Musil, R. 1968. Die Mammutmolaren von Předmostí (ČSSR). *Paläntologische Abhandlungen A* 3:1–192.

Musser, G. G., and M. D. Carleton. 1993. Family Muridae. In D. E. Wilson and D. M. Reeder (eds.), Mammal Species of the World: A Taxonomic and Geographic Reference (2nd Edition), pp. 501–755. Smithsonian Institution Press, Washington, D.C.

Nakatsukasa, M. 2004. Acquisition of bipedalism: the Miocene homonoid record and modern analogues for bipedal protohominids. *Journal of Anatomy* 204:385–402.

Nakatsukasa, M., A. Yamanaka, Y. Kunimatsu, D. Shimizu, and H. Ishida. 1998. A newly discovered *Kenyapithecus* skeleton and its implications for the evolution of positional behavior in

Miocene East African hominoids. *Journal of Human Evolution* 34:657–664.

Nakaya, H. 1994. Faunal change of Late Miocene Africa and Eurasia: Mammalian fauna from the Namurungule Formation, Samburu Hills, northern Kenya. *African Study Monographs* Supplementary Issue 20:1–112.

Nakaya, H., M. Pickford, Y. Nakano, and H. Ishida. 1984. The late Miocene large mammal fauna from the Namurungule Formation, Samburu Hills, northern Kenya. *African Study Monographs* Supplementary Issue 2:87–132.

Nakaya, H., M. Pickford, K. Yasui, and H. Ishida. 1987. Additional large mammalian fauna from the Namurungule Formation, Samburu Hills, Northern Kenya. *African Study Monographs* Supplementary Issue 5:79–129.

Nei, M., and G. V. Glazko. 2002. Estimation of divergence times for a few mammalian and several primate species. *The American Genetic Association* 93:157–164.

Nesbitt, L. M. 1935. Hell-Hole of Creation, The Exploration of the Abyssinian Danakil. Alfred A. Knopf, New York.

Nickel, R., A. Schummer, E. Seiferle, H. Wilkens, K.-H. Wille, and J. Frewein. 1986. The Anatomy of the Domestic Animals, Volume I: The Locomotor System. Verlag Paul Parey-Springer Verlag, Berlin.

Nowak, R. M. 1991. Walker's Mammals of the World, Fifth Edition: I and II. Johns Hopkins University Press, Baltimore.

Odintzov, I. A. 1967. New species of Pliocene Carnivora, *Vulpes odessanus* sp. nov., from the karstic caves of Odessa. *Paleontologichesky Sbornik, Lwow* 4:130–137.

Oloo, T. W., R. Brett, and T. P. Young. 1994. Seasonal variation in the feeding ecology of black rhinoceros (*Diceros bicornis*) in Laikipia, Kenya. *African Journal of Ecology* 32:142–157.

Orlando, L., J. A. Leonard, A. Thenot, V. Laudet, C. Guérin, and C. Hännia. 2003. Ancient DNA analysis reveals woolly rhino evolutionary relationships. *Molecular Phylogenetics and Evolution* 28:485–499.

O'Ryan, C., J. R. B. Flamand, and E. H. Harley. 1994. Mitochondrial DNA variation in black rhinoceros (*Diceros bicornis*): Conservation management implications. *Conservation Biology* 8:495–500.

Osborn, H. F. 1900. Phylogeny of the rhinoceroses of Europe. *Bulletin of the American Museum of Natural History* 12:229–267.

Owen-Smith, R. N. 1973. The behavioural ecology of the white rhinoceros. Ph.D. thesis, University of Wisconsin, Madison.

———. 1988. Megaherbivores: The Influence of Very Large Body Size on Ecology. Cambridge University Press, Cambridge, UK.

Pálfy, J., R. Mundil, P. R. Renne, R. L. Bernor, L. Kordos, and M. Gesparik. 2007. Uranium-lead and $^{40}Ar/^{39}Ar$ dating of the Miocene fossil track site at Ipolytarnoc, Hungary, and its implications. *Earth and Planetary Sciences Letters* 258(2007):160–174.

Patte, E. 1934. Anomalies dentaires de quelques Ongulés fossiles; remarques sur le cingulum et le tubercule de Carabelli. *Bulletin de la Société Géologique de France* 4(5):777–796.

Patterson, B. 1975. The fossil aardvarks (Mammalia: Tubulidentata). *Bulletin of the Museum of Comparative Zoology Cambridge* 147:185–237.

Patterson, B., A. K. Behrensmeyer, and W. D. Sill. 1970. Geology and fauna of a new Pliocene locality in northwestern Kenya. *Nature* 226:918–921.

Patterson, N., D. J. Richter, S. Gnerre, E. S. Lander, and D. Reich. 2006. Genetic evidence for complex speciation of humans and chimpanzees. *Nature* 441:1103–1108.

Pavlakis, P. P. 1990. Plio-Pleistocene Hippopotamidae from the Upper Semliki. In N. T. Boaz (eds.), Results from the Semliki Research Expedition, pp. 203–223. Virginia Museum of Natural History Memoir, Martinsville.

Peigné, S., L. de Bonis, A. Likius, H. T. Mackaye, P. Vignaud, and M. Brunet. 2005a. A new machairodontine (Carnivora, Felidae) from the Late Miocene hominid locality of TM 266, Toros-Menalla, Chad. *Comptes Rendus Palevol* 4:243–253.

———. 2005b. The earliest modern mongoose (Carnivora, Herpestidae) from Africa (late Miocene of Chad). *Naturwissenschaften* 92:287–292.

Perkins, M. E., W. P. Nash, F. H. Brown, and R. J. Fleck. 1995. Fallout tuffs of Trapper Creek, Idaho: A record of Miocene explosive volcanism in the Snake River Plain volcanic province. *Geological Society of America Bulletin* 107:1485–1506.

Peter, K. 2002. Odontologie der Nashornverwandten (Rhinocerotidae) aus dem Miozän (MN 5) von Sandelzhausen (Bayern). *Zitteliana* 22:3–168.

Pethő, G. 1884. Über die Fossilien Säugethier-Überreste von Baltavár. *Földtani Intézet Évi Jelentése* 1884:63–73.

Petrocchi, C. 1943. Il giacimento fossilifero di Sahabi. Parte 2, Paleontologia. *Collezione Scientifica e Documentaria dell'Africa Italiana, Ministero dell'Africa Italiana* 12:69-162.

———. 1954. Paleontologia di Sahabi (Cirenaica). I. Proboscidati di Sahabi. *Rendiconti dell' Accademia Nazionale dei Quaranta* 4/5:1–66.

Petter, F. 1971. Subfamily Gerbillinae. In J. Meester and H. W. Setzer (eds.), The Mammals of Africa, an Identification Manual. Part 6.3, pp. 1–7 (not continuously paginated). Smithsonian Institution Press, Washington, DC.

———. 1963. Etudes de quelque viverridés (Mammifères, Carnivores) du Pléistocène inférieur du Tanganyika (Afrique Orientale). *Bulletin de la Société Géologique de France* 5:265-274.

———. 1974. Rapports phylétiques des Viverridés (Carnivores, Fissipèdes). Les formes de Madagascar. *Mammalia* 134:605–636.

———. 1987. Small Carnivores (Viverridae, Mustelidae, Canidae) from Laetoli. In M. D. Leakey and J. M. Harris (eds.),

Laetoli: A Pliocene Site in Northern Tanzania, pp. 194–234. Clarendon Press, Oxford.

Petter, G., and H. Thomas. 1986. Les Agriotheriinae (Mammalia, Carnivora) néogènes de l'ancien monde: Présence du genre *Indarctos* dans la faune de Menacer (ex-Marceau), Algérie. *Geobios* 19:573–586.

Petter, G., M. Pickford, and F. C. Howell. 1991. The Pliocene piscivorous otter of Nyaburogo and Nkondo (Uganda, East Africa): *Torolutra ougandensis* n.g., n.sp. (Mammalia, Carnivora). *Comptes Rendus de l'Academie des Sciences, Paris, Série II* 312:949–955.

Petter, G., M. Pickford, and B. Senut. 1994. Presence du genre *Agriotherium* (Mammalia, Carnivora, Ursidae) dans le Miocène terminal de la Formation de Nkondo (Ouganda, Afrique orientale). *Comptes Rendus de l'Academie des Sciences, Paris, Série II* 319:713–717.

Pickford, M. 1975. New fossil Orycteropodidae (Mammalia, Tubulidentata) from East Africa. *Netherlands Journal of Zoology* 25:57–88.

———. 1983. On the origins of Hippopotamidae together with descriptions of two species, a new genus and a new subfamily from the Miocene of Kenya. *Géobios* 16:193–217.

———. 1985. A new look at *Kenyapithecus* based on recent discoveries in western Kenya. *Journal of Human Evolution* 14:113–143.

———. 1988. Un étrange suide nain du Nèogene supérieur de Langebaanweg (Afrique du Sud). *Annales de Paléontologie (Vertébré-Invertébré)* 74:229–250.

———. 1989a. New specimens of *Nyanzachoerus waylandi* (Mammalia, Suidae, Tetraconodontinae) from the type area, Nyaburogo (upper Miocene), Lake Albert Rift, Uganda. *Géobios* 22:641–651.

———. 1989b. Update on hippo origins. *Comptes Rendus de l'Académie des Sciences, Paris, Série II* 309:163–168.

———. 1993. Fossil Suidae of the Albertine Rift, Uganda-Zaire. In B. Senut and M. Pickford (eds.), The Geology and Palaeobiology of the Albertine Rift Valley, Uganda-Zaire, Vol. 2, pp. 339–374. CIFEG, Orléans, France.

———. 1995. Old World suoid systematics, phylogeny, biogeography and biostratigraphy. *Paleontologia i Evolució* 26/27:237–269.

———. 2005. Fossil hyraxes (Hyracoidea: Mammalia) from the late Miocene and Plio-Pleistocene of Africa, and the phylogeny of the Procaviidae. *Palaeontologia Africana* 41:141–161.

Pickford, M., and B. Senut. 2001. The geological and faunal context of Late Miocene hominid remains from Lukeino, Kenya. *Comptes Rendus de l'Académie des Sciences, Sciences de la Terre et des Planètes* 332:145–152.

———. 2003. Miocene palaeobiology of the Orange River Valley, Namibia. In M. Pickford and B. Senut (eds.), Geology and Palaeobiology of the Central and Southern Namib Desert, Southwestern Africa, Volume 2: Palaeontology, p. 1–22. Memoirs of the Geological Survey of Namibia 19. Geological Survey of Namibia, Windhoek.

———. 2005. Hominoid teeth with chimpanzee- and gorilla-like features from the Miocene of Kenya: Implications for the chronology of ape-human divergence and biogeography of Miocene hominoids. *Anthropological Science* 113:95–102.

Pickford, M., B. Senut, and D. Hadoto. 1993. Geology and Palaeobiology of the Albertine Rift Valley, Uganda-Zaire, Vol. I: Geology. CIFEG, Orléans, France.

Pickford, M., B. Senut, D. Gommery, and J. Treil. 2002. Bipedalism in *Orrorin tugenensis* revealed by its femora. *Comptes Rendus Palevol* 1:1–13.

Pickford, M., Y. Sawada, R. Tayama, Y. Matsuda, T. Itaya, H. Hyodo, and B. Senut. 2006. Refinement of the age of the Middle Miocene Fort Ternan Beds, Western Kenya, and its implications for Old World biochronology. *Comptes Rendus Geoscience* 338:545–555.

Pilbeam, D., M. D. Rose, C. Badgley, and B. Lipschutz. 1980. Miocene hominoids from Pakistan. *Postilla* 181:1–94.

Pilgrim, G. E. 1932. The fossil Carnivora of India. *Palaeontologia Indica* 18:173–180.

———. 1937. Siwalik antelopes and oxen in the American Museum of Natural History. *Bulletin of the American Museum of Natural History* 72:729–874.

———. 1939. The fossil Bovidae of India. *Paleontologia Indica* 26:1–356.

———. 1947. The evolution of the buffaloes, oxen, sheep and goats. *Journal of the Linnean Society, Zoology* 41:272–286.

Pomel, A. 1888. Visite faite à la station préhistorique de Ternifine (Palikao) par le groupe excursionniste D. *Association Française pour l'Avancement des Sciences, Comptes Rendus* 17:208–212.

———. 1895. Les Rhinocéros Quaternaires. Carte géologique de l'Algérie, Paléontologie, Monographies, Alger.

Pratt, D. J., and M. D. Gwynne. 1977. Rangeland Management and Ecology in East Africa. Hodder and Stoughton, London.

Preece, S. J., J. A. Westgate, and M. P. Gorton. 1992. Compositional variation and provenance of late Cenozoic distal tephra beds, Fairbanks area, Alaska. *Quaternary International* 13/14:97–101.

Prothero, D. R., and R. M. Schoch. 1989. Classification of the Perissodactyla. In D. R. Prothero and R. M. Schoch (eds.), The Evolution of Perissodactyls, pp. 530–537. Oxford Monographs on Geology and Geophysics 15. Oxford University Press, Oxford.

Prothero, D. R., E. Manning, and B. C. Hanson. 1986. The phylogeny of the Rhinocerotoidea (Mammalia, Perissodactyla). *Zoological Journal of the Linnean Society* 87:341–366.

Pustovoytov, K. 2003. Growth rates of pedogenic carbonate coatings on clasts. *Quaternary International* 106–107:131–140.

Pyle, D. M. 2000. Sizes of volcanic eruptions. In H. Sigurdsson, B. F. Houghton, S. R. McNutt, H. Rymer, and J. Stix (eds.), Encyclopedia of Volcanoes, pp. 263–269. Academic Press, San Diego.

Qiu, Z.-X., and N. Schmidt-Kittler. 1983. *Agriotherium intermedium* (Stach 1957) from the Pliocene fissure fillings of Xiaoxian county (Anhuei province, China) and the phylogenetic position of the genus. *Palaeovertebrata, Montpellier* 13:65–81.

Qiu, Z., H. Weilong, and G. Zhihui. 1987. Chinese hipparionines from the Yushe Basin. *Palaeontologica Sinica, Series C* 175:1–250.

Qiu, Z.-X., W. Wu, and Z. Qiu. 1999. Miocene mammal faunal sequence of China: Palaeozoogeography and Eurasian relationships. In G. E. Rossner and K. Heissig (eds.), The Miocene Land Mammals of Europe, pp. 443–455. Fredrich Pfiel, Munich.

Quade, J., and N. Levin. In press. East African hominid paleoecology: isotopic evidence from paleosols. In M. Sponheimer, K. Reed, J. Lee-Thorp, and P. Ungar (eds.), Early Hominin Paleoecology. University of Colorado Press, Boulder.

Quade, J., N. Levin, S. Semaw, D. Stout, P. Renne, M. Rogers, and S. Simpson. 2004. Paleoenvironments of the earliest stone toolmakers, Gona, Ethiopia. *Geological Society of America Bulletin* 116:1529–1544.

Rahm, U. 1966. Les mammifères de la forêt équatoriale de l'est du Congo. *Musée Royal de l'Afrique Centrale Annales, Serie 8* 149:9–121.

Rasmussen, D. T., M. Pickford, P. Mein, B. Senut, and G. Conroy. 1996. Earliest known procaviid hyracoid from the late Miocene of Namibia. *Journal of Mammalogy* 77:745–754.

Rautenbach, I. L. 1978a. A numerical re-appraisal of southern African biotic zones. *Bulletin of the Carnegie Museum of Natural History* 6:175–187.

———. 1978b. Ecological distribution of the mammals of the Transvaal (Vertebrata: Mammalia). *Annals of the Transvaal Museum* 31:131–153.

Raynal, J.-P., D. Lefevre, D. Geraads, and M. El Graoui. 1999. Contribution du site paléontologique de Lissafa (Casablanca, Maroc) à une nouvelle interpretation du Mio-Pliocène de la Meseta. *Comptes Rendus de l'Academie des Sciences, Sciences de la Terre et des Planètes* 329:617–622.

Reed, K. E. 1996. The paleoecology of Makapansgat and other African Pliocene hominid localities. Ph.D. thesis, State University of New York, Stony Brook.

———. 1997. Early hominid evolution and ecological change through the African Plio-Pleistocene. *Journal of Human Evolution* 32:289–322.

Renne, P. R. 1995. Excess ^{40}Ar in biotite and hornblende from the Noril'sk 1 intrusion, Siberia: Implications for the age of the Siberian Traps. *Earth and Planetary Science Letters* 131:165–176.

Renne, P. R., K. Deckart, M. Ernesto, G. Féraud, and E. M. Piccirillo. 1996. Age of the Ponta Grossa dike swarm (Brazil) and implications to Paraná flood volcanism. *Earth and Planetary Science Letters* 144:199–212.

Renne, P. R., C. C. Swisher, A. L. Deino, T. Owens, D. J. DePaolo, and D. B. Karner. 1998. Intercalibration of standards, absolute ages and uncertainties in ^{40}Ar/^{39}Ar dating. *Chemical Geology (Isotope Geoscience Section)* 145:117–152.

Renne, P. R., G. WoldeGabriel, W. K. Hart, G. Heiken, and T. D. White. 1999. Chronostratigraphy of the Miocene-Pliocene Sagantole Formation, Middle Awash Valley, Afar Rift, Ethiopia. *Geological Society of America Bulletin* 111:869–885.

Retallack, G. J. 2001. Soils of the Past. 2nd edition. Blackwell Science, Oxford.

Ringström, T. 1924. Nashörner der Hipparion-Fauna Nord-Chinas. *Palaeontologia Sinica C* 1(4):1–156.

Robinson, P., and C. C. Black. 1969. Note préliminaire sur les vertébrés fossiles du Vindobonien (Formation Béglia) du Bled Douarah, Gouvernorat de Gafsa, Tunisie. *Service Géologique de Tunisie, Notes* 31:67–70.

Robinson, T. J., V. Trifonov, I. Espie, and E. H. Harley. 2005. Interspecific hybridisation in rhinoceroses: Confirmation of a black _ white rhinoceros hybrid by karyotype, fluorescence in situ hybridisation (FISH) and microsatellite analysis. *Conservation Genetics* 6:141–145.

Roman, F., and M. Solignac. 1934. Découverte d'un gisement de mammifères pontiens à Douaria (Tunisie septentrionale). *Comptes Rendus de l'Académie des Sciences Paris* 199:1649–1650.

Rook, L. 1992. "*Canis*" *monticinensis* sp. nov., a New Canidae (Carnivora, Mammalia) from the Late Messinian of Italy. *Bollettino della Societá Paleontológica Italiana* 31:151–156.

Rook, L., and R. L. Bernor. 2003. Ancestry of the African ape-human clade. *Palaeontographica Italica* 89:30–31.

Rook, L., and B. Martinez-Navarro. 2004. *Viverra howelli* n.sp., a new viverrid (Carnivora, Mammalia) from the Baccinello-Cinigiano basin (latest Miocene, Italy). *Rivista Italiana di Paleontologia e Stratigrafia* 110:719–723.

Rook, L., G. Ficcarelli, and D. Torre. 1991. Messinian carnivores from Italy. *Bolletino della Società Paleontologica Italiana* 30:7–22.

Rookmaaker, L. C. 1995. Subspecies and ecotypes of the black rhinoceros. *Pachyderm* 20:39–40.

———. 1998. The sources of Linnaeus on the rhinoceros. *Svenska Linnesallskapets Arsskrift* 1996–97:61–80.

———. 2005. Review of the European perception of the African rhinoceros. *Journal of Zoology, London* 265:365–376.

Rookmaaker, L. C., and C. P. Groves. 1978. The extinct Cape rhinoceros, *Diceros bicornis bicornis* (Linnaeus, 1758). *Säugetierkundliche Mitteilungen* 26(2):117–126.

Rose, M. D. 1986. Further hominoid postcranial specimens from the Late Miocene Nagri Formation of Pakistan. *Journal of Human Evolution* 15:333–367.

Roth, J., and A. Wagner. 1954. Die fossilen Knochen-Ueberreste von Pikermi in Griechenland. *Abhandlungen der Bayerischen Akademie der Wissenschaften* 7:371–464.

Roth, V. L. 1992. Quantitative variation in elephant dentitions: Implications for the delimitation of fossil species. *Paleobiology* 18:84–202.

Roth, V. L., and J. Shoshani. 1988. Dental identification and age determination in *Elephas maximus*. *Journal of Zoology* 214:567–588.

Roussiakis, S. J. 2002. Musteloids and Feloids (Mammalia, Carnivora) from the late Miocene Locality of Pikermi (Attica, Greece). *Geobios* 35:699–719.

Ruvolo, M. 1997. Genetic diversity in hominoid primates. *Annual Review of Anthropology* 26:515–540.

Sabatier, M. 1978. Un nouveau *Tachyoryctes* (Mammalia, Rodentia) du bassin pliocène de Hadar (Éthiopie). *Geobios* 11:95–99.

———. 1982. Les rongeurs du site Pliocène à hominidés de Hadar (Éthiopie). *Palaeovertebrata* 12:1–56.

Saegusa, H. 1996. Stegodontidae: Evolutionary relationships. In J. Shoshani and P. Tassy (eds.), The Proboscidea: Evolution and Palaeoecology of Elephants and Their Relatives, pp. 178–190. Oxford University Press, Oxford.

Saegusa, H., and L. J. Hlusko. 2007. New late Miocene elephantoid (Mammalia: Proboscidea) fossils from Lemudong'o, Kenya. *Kirtlandia* 56:140–147.

Saegusa, H., and H. Taruno. 1991. Stegodonts. In T. Kamei (ed.), Japanese Proboscidean Fossils, pp. 68–110. Tsukiji-Shokan, Tokyo.

Saegusa, H., Y. Thasod, and B. Ratanasthien. 2005. Notes on Asian stegodontids. *Quaternary International* 126–128:31–48.

Salles, L. O. 1992. Felid phylogenetics: Extant taxa and skull morphology (Felidae, Aeluroidea). *American Museum Novitates* 3047:1–67.

Salvador, A. 1994. International Stratigraphic Guide: A Guide to Stratigraphic Classification, Terminology, and Procedure. International Union of Geological Sciences and the Geological Society of America, Trondheim, Norway, and Boulder, CO.

Sanders, W. J. 1997. Fossil Proboscidea from the Wembere-Manonga Formation, Manonga Valley, Tanzania. In T. Harrison (ed.), Neogene Paleontology of the Manonga Valley, Tanzania, pp. 265–310. Plenum Press, New York.

———. 2007. Taxonomic review of fossil Proboscidea (Mammalia) from Langebaanweg, South Africa. *Transactions of the Royal Society of South Africa* 62:1–16.

Sanders, W. J., J. Kappelman, and D. T. Rasmussen. 2004. New large-bodied mammals from the late Oligocene site of Chilga, Ethiopia. *Acta Palaeontologica Polonica* 49:365–392.

Sardella, R. 1993. Sistematica e Distribuzione Stratigrafica dei Macairodontini dal Miocene Superiore al Pleistocene. Ph.D. thesis, Universita Consorziate, Modena, Bologna, Firenze, Roma.

Sardella, R., and L. Werdelin. 2007. *Amphimachairodus* (Felidae, Mammalia) from Sahabi (latest Miocene–earliest Pliocene, Libya), with a review of African Miocene Machairodontinae. *Rivista Italiana di Paleontologia e Stratigrafia* 113:67–77.

Sarich, V., and A. Wilson. 1967. Immunological time scale for hominid evolution. *Science* 158:1200–1203.

Sarna-Wojcicki, A. M. 1976. Correlation of Late Cenozoic Tuffs in the Central Coast Ranges of California by Means of Trace- and Minor-Element Chemistry. U.S. Geological Survey, Boulder, CO.

———. 2000. Tephrochronology. In J. S. Noller, J. M. Sowers, and W. R. Lettis (eds.), Quaternary Geochronology: Methods and Applications, pp. 77–100. American Geophysical Union, Washington, DC.

Sayer, J. A., and L. P. van Lavieren. 1975. The ecology of the Kafue lechwe population of Zambia before the operation of hydroelectric dams of the Kafue River. *East African Wildlife Journal* 13:9–37.

Schlesinger, G. 1917. Die Mastodonten des K. K. naturhistorischen Hofmuseums. *Denkschriften des Kaiserlich-Königlichen Naturhistorischen Hofmuseums 1, Geologisch-Paläontologische Reihe* 1:1–230.

Schlosser, M. 1903. Die fossilen Säugethiere Chinas nebst einer Odontographie der recenten Antilopen. *Abhandlungen der Bayerische Akademie der Wissenschaften (II C1)* 22:1–221.

Schmidt-Kittler, N. 1987. Zur Stammegeschichte der marderverwandten Raubtiergruppen (Musteloidea, Carnivora). *Eclogae Geologicae Helvetiae* 74:753–801.

Schuster, R. H. 1980. Will the Kafue lechwe survive the Kafue dams? *Oryx* 15:476–489.

Semaw, S., P. R. Renne, J. W. K. Harris, C. S. Feibel, R. L. Bernor, N. Fessaha, and K. Mowbray. 1997. 2.5 million-year-old stone tools from Gona, Ethiopia. *Nature* 385:333–336.

Semaw, S., S. W. Simpson, J. Quade, P. R. Renne, R. F. Butler, W. C. McIntosh, N. Levin, M. Dominiguez-Rodrigo, and M. J. Rogers. 2005. Early Pliocene hominids from Gona, Ethiopia. *Nature* 433:301–305.

Senut, B. 1978. Contribution à l'Étude de l'Humérus et de Ses Articulations chez les Hominidés Plio-Pléistocenes. Ph.D. thesis, Université Pierre et Marie Curie, Paris.

———. 1981. Humeral outlines in some hominoid primates and in Plio-Pleistocene hominids. *American Journal of Physical Anthropology* 56:275–282.

———. 1994. Cercopithecoidea Néogènes et Quaternaires du rift occidental (Ouganda). In B. Senut, and M. Pickford (eds.), Geology and Palaeobiology of the Albertine Rift Valley, Uganda-Zaire, Volume II: Palaeobiology, pp. 195–204. CIFEG, Orléans, France.

Senut, B., and M. Pickford, eds. 1994. Geology and Paleobiology of the Albertine Rift Valley, Uganda-Zaire. Volume II. CIFEG, Orléans, France.

———. 2004. La dichotomie grands singes–homme revisitée. *Comptes Rendus Palevol* 3:263–274.

Senut, B., and C. Tardieu. 1985. Functional aspects of Plio-Pleistocene hominid limb bones: Implications for taxonomy and phylogeny. In E. Delson (ed.), Ancestors: The Hard Evidence, pp. 193–201. Alan R. Liss, New York.

Senut, B., M. Pickford., D. Gommery, P. Mein, K. Cheboi, and Y. Coppens. 2001. First hominid from the Miocene (Lukeino Formation, Kenya). *Comptes Rendus de l'Academie des Sciences, Sciences de la Terre et des Planètes* 332:137–144.

Sepulchre, P., G. Ramstein, F. Fluteau, M. Schuster, J.-J. Tiercelin, and M. Brunet. 2006. Tectonic uplift and Eastern Africa aridification. *Science* 313:1419–1423.

Sharp, W. D., B. D. Turrin, P. R. Renne, and M. A. Lanphere. 1996. The ^{40}Ar/^{39}Ar and K/Ar dating of core from the Hilo 1-km core hole, Hawaiian Scientific Drilling Project. *Journal of Geophysical Research* 101:11,607–11,616.

Sheppe, W., and T. Osborne. 1971. Patterns of use of a flood plain by Zambian mammals. *Ecological Monographs* 41:179–205.

Shipman, P., and Harris, J. M. 1988. Habitat preference and paleoecology of *Australopithecus boisei* in eastern Africa. In F. E. Grine (ed.), Evolutionary History of the "Robust" Australopithecines, pp. 343–382. Aldine de Gruyter, New York.

Shoshani, J. 1996. Para- or monophyly of the gomphotheres and their position within Proboscidea. In J. Shoshani and P. Tassy (eds.), The Proboscidea: Evolution and Palaeoecology of Elephants and Their Relatives, pp. 149–177. Oxford University Press, Oxford.

Shoshani, J., and P. Tassy. 1996. Summary, conclusions, and a glimpse into the future. In J. Shoshani and P. Tassy (eds.), The Proboscidea: Evolution and Palaeoecology of Elephants and Their Relatives, pp. 335–348. Oxford University Press, Oxford.

———. 2005. Advances in proboscidean taxonomy and classification, anatomy and physiology, and ecology and behavior. *Quaternary International* 126–128:5–20.

Shoshani, J., E. M. Golenberg, and H. Young. 1998. Elephantidae phylogeny: Morphological versus molecular results. *Acta Theriologica* 5:89–122.

Shrader, A. M., N. Owen-Smith, and J. O. Ogutu. 2006. How a mega-grazer copes with the dry season: Food and nutrient intake rates by white rhinoceros in the wild. *Functional Ecology* 20:376–384.

Sickenberg, O., and M. Schonfeld. 1975. The Chorora Formation: Lower Pliocene limnical sediments in the southern Afar. In A. Pilger and A. Rosler (eds.), Afar Depression of Ethiopia, pp. 277–284. Schweizerbart Verlag, Stuttgart.

Sillero-Zubiri, C., F. H. Tattersall, and D. W. Macdonald. 1995. Habitat selection and daily activity of giant molerats *Tachyoryctes macrocephalus*: Significance to the Ethiopian wolf *Canis simensis* in the Afroalpine ecosystem. *Biological Conservation* 72:77–84.

Simpson, G. G. 1943. Mammals and the nature of continents. *American Journal of Science* 241:1–31.

———. 1945. The principles of classification and a classification of the mammals. *Bulletin of the American Museum of Natural History* 85:1–350.

Simpson, G. G., and C. de P. Couto. 1957. The mastodons of Brazil. *Bulletin of the American Museum of Natural History* 112:125–190.

Simpson, S. W., J. Quade, N. Levin, P. Renne, R. F. Butler, W. C. Macintosh, and S. Semaw. 2004. Early Pliocene hominids and their environments from Gona, Ethiopia. *American Journal of Physical Anthropology* 123 S38:182.

Simpson, S. W., J. Quade, L. Kleinsasser, N. Levin, W. MacIntosh, N. Dunbar, S. Semaw. 2007. Late Miocene hominid teeth from Gona Project Area, Ethiopia. *American Journal of Physical Anthropology* 132, S44:219.

Skinner, J. D., and R. H. N. Smithers. 1990. The Mammals of the Southern African Subregion, 2nd edition. University of Pretoria Press, Pretoria, South Africa.

Smart, C. L. 1976. The Lothagam 1 fauna: Its phylogenetic, ecological and biogeographic significance. In Y. Coppens, F.C. Howell, G. Ll. Issac, and R. E. F. Leakey (eds.), Earliest Man and Environments in the Lake Rudolf Basin, pp. 361–369. University of Chicago Press, Chicago.

Smithers, R. H. N. 1971. The mammals of Botswana. *Museum Memoirs of the National Museums and Monuments of Rhodesia* 4:1–340.

———. 1983. The Mammals of the Southern African Subregion. University of Pretoria Press, Pretoria, South Africa.

Sokal, R. R., and P. H. A. Sneath. 1963. The Principles of Numerical Taxonomy. Freeman and Company, San Francisco.

Solounias, N. 1981. The Turolian fauna from the Island of Samos, Greece; with special emphasis on the hyaenids and the bovids. *Contributions to Vertebrate Evolution* 6:1–248.

Solounias, N., and B. Dawson-Saunders. 1988. Dietary adaptations and paleoecology of the Late Miocene ruminants from Pikermi and Samos in Greece. *Palaeogeography, Palaeoclimatology, Palaeoecology* 65(3–4):149–172.

Solounias, N., W. S. McGraw, L. Hayek, and L. Werdelin. 2000. The paleodiet of Giraffidae. In E. S. Vrba and G. B. Schaller (eds.), Antelopes, Deer and Relatives: Fossil Record, Behavioral Ecology, Systematics, and Conservation, pp. 83–95. Yale University Press, New Haven, CT.

Sotnikova, M. V., and N. G. Noskova. 2004. The History of *Machairodus* in Eurasia. In 18th International Senckenberg Conference, VI International Palaeontological Colloquium in Weimar, Germany, p. 238. Terra Nova.

Spassov, N., and D. Geraads. 2004. *Tragoportax* Pilgrim, 1937 and *Miotragocerus* Stromer, 1928 (Mammalia, Bovidae) from the Turolian of Hadjidimovo, Bulgaria, and a revision of the late Miocene Mediterranean Boselaphini. *Geodiversitas* 26:339–370.

Spassov, N., and L. Rook. 2006. *Eucyon marinae* sp. nov. (Mammalia, Carnivore) a new canid species from the Pliocene of Mongolia with a review of forms referable to the genus. *Rivista Italiana di Paleontologia e Stratigrafia* 112:123–133.

Spencer, L. M. 1995. Antelopes and Grasslands: Reconstructing African Hominid Environments. Ph.D. thesis, State University of New York, Stony Brook.

Sponheimer, M., and J. A. Lee-Thorp. 1999. Isotopic evidence for the diet of an early hominid, *Australopithecus africanus*. *Science* 283:368–370.

Sponheimer, M., K. E. Reed, and J. A. Lee-Thorp. 1999. Combining isotopic and ecomorphological data to refine bovid paleodietary reconstruction: a case study from the Makapansgat Limeworks hominin locality. *Journal of Human Evolution* 36:705–718.

Steininger, F. F., W. A. Berggren, D. V. Kent, R. L. Bernor, S. Sen, and J. Agusti. 1996. Circum-Mediterranean Neogene (Miocene and Pliocene) marine-continental choronologic correlations of European mammalian units. In R. L. Bernor, V. Fahlbusch, and H.-W. Mittmann (eds.), The Evolution of Western Eurasian Neogene Mammal Faunas, pp. 7–46. Columbia University Press, New York.

Stern, J. T. 2000. Climbing to the top: A personal memoir of *Australopithecus afarensis*. *Evolutionary Anthropology* 9:113–133.

Stevens, M. S., and J. B. Stevens. 2003. Carnivora (Mammalia, Felidae, Canidae, and Mustelidae) from the earliest Hemphillian Screw Bean Local Fauna, Big Bend National Park, Brewster County, Texas. *Bulletin American Museum of Natural History* 279:177–211.

Stokes, S., D. J. Lowe, and P.C. Froggatt. 1992. Discriminant function analysis and correlation of late Quaternary rhyolitic tephra deposits from Taupo and Okataina volcanoes, New Zealand, using glass shard major element compositions. *Quaternary International* 13/14:103–117.

Strasser, E. 1988. Pedal evidence for the origin and diversification of cercopithecid clades. *Journal of Human Evolution* 17:225–245.

Strasser, E., and E. Delson. 1987. Cladistic analysis of cercopithecid relationships. *Journal of Human Evolution* 16:81–99.

Stromer, E. 1907. Fossile Wilbeltiere-Reste aus dem Uadi Faregh and Uadi Natrun in Ägypten. *Abhandlungen der Senckenbergischen Naturforschenden Gesellschaft* 29:97–132.

Struhsaker, T. T. 1997. Ecology of an African Rain Forest. University Press of Florida, Gainesville.

Stuenes, S. 1989. Taxonomy, habits, and relationships of the subfossil Madagascan hippopotami *Hippopotamus lemerlei* and *H. madagascariensis*. *Journal of Vertebrate Paleontology* 9:241–268.

Su, D. 2005. The Paleoecology of Laetoli, Tanzania: Evidence from the Mammalian Fauna of the Upper Laetolil Beds. Ph.D. thesis, New York University, New York.

Suwa, G. 1990. A Comparative Analysis of Hominid Dental Remains from the Shungura and Usno Formations, Omo Valley, Ethiopia. Ph.D. thesis, University of California, Berkeley.

Suwa, G., and R. T. Kono. 2005. A micro-CT based study of linear enamel thickness in the mesial cusp section of human molars: Reevaluation of methodology and assessment of within-tooth, serial, and individual variation. *Anthropological Science* 113:273–289.

Suwa. G., R. T. Kono, S. Katoh, B. Asfaw, and Y. Beyene. 2007. A new species of great ape from the late Miocene epoch in Ethiopia. *Nature* 448:921–924.

Swart, M. K. J., and J. W. H. Ferguson. 1997. Conservation implications of genetic differentiation in southern African populations of black rhinoceros (*Diceros bicornis*). *Conservation Biology* 11:79–83.

Swart, M. K. J., J. W. H. Ferguson, R. du Toit, and J. R. B. Flamand. 1994. Substantial genetic variation in southern African black rhinoceros. *Journal of Heredity* 85:261–266.

Swynnerton, G. H. 1958. Fauna of the Serengeti National Park. *Mammalia* 22:435–450.

Szalay, F. S., and E. Delson. 1979. Evolutionary History of the Primates. Academic Press, New York.

Taieb, M. 1974. Evolution Quaternaire du Bassin de l'Awash. Ph.D. thesis, Université de Paris VI, Paris.

Taieb, M., D. C. Johanson, Y. Coppens, and J. L. Aronson. 1976. Geological and paleontological background of Hadar hominid site, Afar, Ethiopia. *Nature* 260:289–293.

Takano, T., M. Nakatsukasa, Y. Kunimatsu, Y. Nakano, and H. Ishida. 2003. Functional morphology of the *Nacholapithecus* forelimb long bones. *American Journal of Physical Anthropology* 120 S36:205–206.

Takahashi, K., N. Mazima, and Fossil Mammal Research Group for Nojiri-ko Excavation. 1991. [Morphological description and its variation in the molars of the Naumann's elephant (*Palaeoloxodon naumanni* (Makiyama)) from Lake Nojiri, Nagano Prefecture, Central Japan.] *Journal of Fossil Research* 24:7–32.

Taru, H., H. Okazaki, S. Isaji, and T. Yanagisawa. 2005. The fourth deciduous premolar of *Stegodon orientalis* Owen 1870 (Mammalia: Proboscidea) from the Middle Pleistocene Mandano Formation in Ichihara, Chiba Prefecture, Central Japan. *Journal of the Natural History Museum and Institute, Chiba* 8:1–10.

Tassy, P. 1977. Le plus ancien squelette de gomphothère (Proboscidea, Mammalia) dans la Formation burdigalienne des Sables de l'Orléanais, France. *Bulletin du Muséum National d'Histoire Naturelle, New Series C* 37:1–51.

———. 1982. Les principales dichotomies dans l'histoire des Proboscidea (Mammalia): Une approche phylogénétique. *Geobios* 6:225–245.

———. 1986. Nouveaux Elephantoidea (Mammalia) dans le Miocène du Kenya. Cahiers de Paléontologie, Travaux de Paléontologie Est-Africaine. Centre National de la Recherche Scientifique, Paris.

———. 1990. Phylogénie et classification des Proboscidea (Mammalia): Historique et actualité. *Annales de Paléontologie (Vertébré-Invertébré)* 76:159–224.

———. 1994. Les proboscidiens (Mammalia) fossiles du Rift Occidental, Ouganda. In B. Senut and M. Pickford (eds.), Geology and Palaeobiology of the Albertine Rift Valley, Uganda-Zaire, Vol. II: Palaeobiology, pp. 217–257. CIFEG, Orléans, France.

———. 1996a. Dental homologies and nomenclature in the Proboscidea. In J. Shoshani and P. Tassy (eds.), The Proboscidea: Evolution and Palaeoecology of Elephants and Their Relatives, pp. 21–25. Oxford University Press, Oxford.

———. 1996b. Who is who among the Proboscidea? In J. Shoshani and P. Tassy (eds.), The Proboscidea: Evolution and Palaeoecology of Elephants and Their Relatives, pp. 39–48. Oxford University Press, Oxford.

———. 1999. Miocene elephantids (Mammalia) from the Emirate of Abu Dhabi, United Arab Emirates: Palaeobiogeographic implications. In J. Whybrow and A. Hill (eds.), Fossil Vertebrates of Arabia, with Emphasis on the Late Miocene Fauna, Geology, and Palaeoenvironments of the Emirate of Abu Dhabi, United Arab Emirates, pp. 209–233. Yale University Press, New Haven, CT.

———. 2003. Elephantoidea from Lothagam. In M. G. Leakey and J. M. Harris (eds.), Lothagam: The Dawn of Humanity in Eastern Africa, pp. 331–358. Columbia University Press, New York.

Tassy, P., and J. Shoshani. 1988. The Tethytheria: Elephants and their relatives. In M. J. Benton (ed.), The Phylogeny and Classification of the Tetrapods, Vol. 2: Mammals, pp. 238–315. The Systematics Association, Special Volume No. 35B. Clarendon Press, Oxford.

Taylor, M. E., and C. A. Goldman. 1993. The taxonomic status of the African mongooses, *Herpestes sanguineus, H. nigratus, H. pulverulentus* and *H. ochraceus* (Carnivora: Viverridae). *Mammalia* 57:375–391.

Taylor, M. E., and J. Matheson. 1999. A craniometric comparison of the African and Asian mongooses in the genus *Herpestes* (Carnivora : Herpestidae). *Mammalia* 63:449–463.

Tedford, R. H., and Z.-X. Qiu. 1996. A new canid genus from the Pliocene of Yushe, Shanxi Province. *Vertebrata PalAsiatica* 34:27–40.

Teilhard de Chardin, P. 1945. Les formes fossiles. In P. Teilhard de Chardin and P. Leroy (eds.), Les Mustélidés de Chine, pp. 36. Institut de Géobiologie, Peking.

Tekkaya, I. 1993. Türkiye fosil Orycteropodidae'leri. *T. C. Kültür Bakanlığı Anıtlar ve Müzeler Genel Müdürlüğü Arkeometri Sonuçları Toplantısı* VIII:275–289.

Tesfaye, S., D. J. Harding, and T. M. Kusky. 2003. Early continental breakup and migration of the Afar triple junction, Ethiopia. *Bulletin of the Geological Society of America* 115:1053–1067.

Thenius, E. 1955. Zur Kenntnis der unterpliozänen *Diceros* Arten. *Annalen der Naturhistorisches Museum Wien* 60:202–209.

———. 1956. Über das Vorkommen von *Diceros pachygnathus* (Wagner) im Pannon (Unter-Pliozaen) des Wiener Beckens. *Neues Jahrbuch für Geologie und Palaeontologie, Monatshefte* 1:35–39.

———. 1959. *Indarctos arctoides* (Carnivora, Mammalia) aus dem Pliozän Österreichs nebst einer Revision der Gattung. *Neues Jahrbuch für Geologie und Paläontologie, Abhandlung* 108(3):270–295.

———. 1965. Lebende Fossilien-Zeugen Vergangener Welten. Franck'sche Ver lagsbuchhandlung W. Keller, Stuttgart, Kosmos-Bibliothek 246.

———. 1979. Die taxonomische und stammesgeschichtliche Position des Bambusbären (Carnivora, Mammalia). *Anzeiger Österreichische Akademie der Wissenschaften, Mathematisch-Naturwissenschaftliche Klasse* 3:67.

———. 1982. Zur Stammes und Verbreitungs geschichtlichen Herkunft des Bambusbären (Carnivora, Mammalia). *Zeitschrift fur Geologische Wissenschafte Berlin* 10:1029–1042.

Thomas, H. 1979. *Miotragocerus cyrenaicus* sp. nov. (Bovidae, Artiodactyla, Mammalia) du Miocène supérieur de Sahabi (Lybie) et ses rapports avec les autres *Miotragocerus. Géobios* 12: 267–281.

———. 1980. Les bovidés du Miocène supérieur des couches de Mpesida et de la formation de Lukeino (district de Baringo, Kenya). In R. E. F. Leakey and B. A. Ogot (eds.), Proceedings of the 8th Panafrican Congress of Prehistory and Quaternary Studies, pp. 82–91. International Louis Leakey Memorial Institute for African Prehistory, Nairobi, Kenya.

———. 1984. Les Bovidae (Artiodactyla; Mammalia) du Miocène du sous-continent Indien, de la peninsule Arabique et de l'Afrique; biostratigraphie, biogeographie et ecologie. *Palaeogeography, Palaeoclimatology, Palaeoecology* 45:251–299.

Thomas, H., P. Taquet, G. Ligabue, and C. Del'Agnola. 1978. Découverte d'un gisement de vertébrés dans les dépôts continentaux du Miocène moyen du Hasa (Arabie saoudite). *Compte Rendu Sommaire des Séances de la Société Géologique de France* 1978(2):69–72.

Thomas, O. 1911. The mammals of the tenth edition of Linnaeus; an attempt to fix the types of the genera and the exact bases and localities of the species. *Proceedings of the Zoological Society of London* 1911:120–158.

Tiercelin, J. J., M. Taieb, and H. Faure. 1980. Continental sedimentary basins and volcano-tectonic evolution of the Afar Rift. *Accademia Nazionale dei Lincei, Rome* 47:491–504.

Tobien, H. 1973. The structure of the mastodont molar (Proboscidea, Mammalia), Part 1: The bunodont pattern. *Mainzer Geowissenschaften Mitteilungen* 2:115–147.

———. 1976. Zur paläontologischen Geschichte der Mastodonten (Proboscidea, Mammalia). *Mainzer Geowissenschaften Mitteilungen* 5:143–225.

———. 1978. On the evolution of mastodonts (Proboscidea, Mammalia). Part 2: The bunodont tetralophodont groups. *Geologisches Jahrbuch Hessen* 106:159–208.

Tobien, H., G. Chen, and Y. Li. 1988. Mastodonts (Proboscidea, Mammalia) from the Late Neogene and Early Pleistocene of the People's Republic of China. Part 2: The genera *Tetralophodon*, *Anancus*, *Stegotetrabelodon*, *Zygolophodon*, *Mammut*, *Stegolophodon*; some generalities on the Chinese mastodonts. *Mainzner Geowissenschaften Mitteilungen* 17:95–220.

Todd, N.E. 2006. Trends in proboscidean diversity in the African Cenozoic. *Journal of Mammalian Evolution* 13:1–10.

Todd, N., and V. L. Roth. 1996. Origin and radiation of the Elephantidae. In J. Shoshani and P. Tassy (eds.), The Proboscidea: Evolution and Palaeoecology of Elephants and Their Relatives, pp. 193–202. Oxford University Press, Oxford.

Topachevskii, V. A., and A. F. Skorik. 1984. Pervaya nakodka iskopaemykh ostatkov kosmatykh khomyakov—Lophiomyinae (Rodentia, Cricetidae) [The first find of fossil remains of a maned hamster—Lophiomyinae (Rodentia, Cricetidae)]. *Vestnik Zoologii* 2:57–60.

Tosi, A. J., J. C. Morales, and D. J. Melnick. 2000. Comparison of Y chromosome and mtDNA phylogenies leads to unique inferences of macaque evolutionary history. *Molecular Phylogenetics and Evolution* 17:133–144.

———. Y-chromosome and mitochondrial markers in *Macaca fascicularis* indicate introgression with Indochinese *M. mulatta* and a biogeographic barrier in the isthmus of Kra. *International Journal of Primatology* 23:161–178.

Tosi, A. J., T. R. Disotell, J. C. Morales, and D. J. Melnick. 2003. Cercopithecine Y chromosome data provide a test of competing morphological evolutionary hypotheses. *Molecular Phylogenetics and Evolution* 27:510–521.

Tougard, C., T. Delefosse, C. Hänni, and C. Montgelard. 2001. Phylogenetic relationships of the five extant rhinoceros species (Rhinocerotidae, Perissodactyla) based on mitochondrial cytochrome *b* and 12S rRNA genes. *Molecular Phylogenetics and Evolution* 19:34–44.

Tsiskarishvili, G. V. 1987. Pozdnetvetichnye Nosorogi (Rhinocerotidae) Kavkaza [Late Tertiary Rhinoceroses (Rhinocerotidae) of the Caucasus]. Georgian SSR, Gosudarstvennyy Muzey Gruzii, Izdatel'stvo Metsnierba, Tbilisi.

Tsujikawa, H. 2005. The updated Late Miocene large mammal fauna from Samburu Nills, northern Kenya. *African Study Monographs* Supplementary Issue 32:1–50.

Turner, A. 1990. Late Neogene/Lower Pleistocene Felidae of Africa: Evolution and dispersal. *Quartärpaläontologie, Berlin* 8:247–256.

Ukstins, I., P. R. Renne, E. Wolfenden, J. Baker, and M. Menzies. 2002. Matching conjugate volcanic rifted margins: $^{40}Ar/^{39}Ar$ chrono-stratigraphy of pre- and syn-rift bimodal flood volcanism in Ethiopia and Yemen. *Earth and Planetary Science Letters* 198:289–306.

van der Made, J. 1999. Biometrical trends in the Tetraconodontinae, a subfamily of pigs. *Transactions of the Royal Society of Edinburgh: Earth Sciences* 89:199–225.

van Hoepen, E. C. N. 1930. Fossiele pferde van Cornelia. *O. V. S. Paleontologiese Navorsing van die Nasionale Museum, Bloemfontein* 11:13–24.

Van Weers, D. J. 2002. *Atherurus karnuliensis* Lydekker, 1886, a Pleistocene brush-tailed porcupine from India, China, and Vietnam. *Paläontologische Zeitschrift* 76:29–33.

Verheyen, R. 1954. Monographie Éthologique de l'Hippopotame (*Hippopotamus amphibius* Linné), Exploration Parc National Albert. Institut des Parcs Nationaux du Congo Belge, Morges.

Vernon, C. J. 1999. Biogeography, endemism and diversity of animals in the karoo. In W. R. J. Dean and S. J. Milton (eds.), The Karoo: Ecological Patterns and Processes, pp. 57–85. Cambridge University Press, Cambridge, UK.

Veron, G., and S. Heard. 2000. Molecular systematics of the Asiatic Viverridae (Carnivora) inferred from mitochondrial cytochrome *b* sequence analysis. *Journal of Zoological Systematics and Evolutionary Research* 38:209–217.

Veron, G., M. Colyn, A. E. Dunham, P. Taylor, and P. Gauberta. 2004. Molecular systematics and origin of sociality in mongooses (Herpestidae, Carnivora). *Molecular Phylogenetics and Evolution* 30:582–598.

Vesey-Fitzgerald, D. F. 1964. Mammals of the Rukwa Valley. *Tanganyika Notes and Records* 62:61–72.

Vialli, V. 1955. Su una anomalia nella dentatura di un rinoceronte africano. *Natura, Rivista di Scienze Naturali, Milano* 46(3):131–134.

———. 1966. Sul rinvenimento di Dinoterio (*Deinotherium* cf. *hobleyi* Andrews) nelle ligniti di Adi Ugri (Eritrea). *Giornale di Geologia (Bologna)* 33:447–458.

Vignaud, P., P. Duringer, H. T. Mackaye, A. Likius, C. Blondel, J.-R. Boisserie, L. de Bonis, V. Eisenmann, M. E. Etienne, D. Geraads, F. Guy, T. Lehmann, F. Lihoreau, N. Lopez-Martinez, C. Mourer-Chauviré, O. Otero, J. C. Rage, M. Schuster, L. Viriot, A. Zazzo, and M. Brunet. 2002. Geology and palaeontology of the Upper Miocene Toros-Menalla hominid locality, Chad. *Nature* 418:152–155.

Viret, J. 1939. Monographie paléontologique de la faune de vertébrés des Sable de Montpellier III, Carnivora, Fissipedia. *Travaux de Laboratoire de Géologie de la Faculté des Sciences de Lyon* 37:7–26.

Vislobokova, I. 2005. On Pliocene faunas with Proboscideans in the territory of the former Soviet Union. *Quaternary International* 126–128:93–105.

Vrba, E. S., 1980a. Evolution, species and fossils: How did life evolve? *South African Journal of Science* 76:61–84.

———. 1980b. The significance of bovid remains as indicators of environment and prediction patterns. In A. K. Behrensmeyer and A. P. Hill (eds.), Fossils in the Making, pp. 247–271. University of Chicago Press, Chicago.

———. 1984. Evolutionary pattern and process in the sister-group Alcelaphini-Aepycerotini (Mammalia: Bovidae). In N. Eldredge and S. M. Stanley (eds.), Living Fossils, pp. 62–79. Springer-Verlag, New York.

———. 1985a. African Bovidae: Evolutionary events since the Miocene. *South African Journal of Science* 81:263–266.

———. 1985b. Environment and evolution: Alternative causes of the temporal distribution of evolutionary events. *South African Journal of Science* 81:229–236.

———. 1987. A revision of the Bovini (Bovidae) and a preliminary revised checklist of Bovidae from Makapansgat. *Palaeontologia Africana* 26:33–46.

———. 1988. Late Pliocene climatic events and hominid evolution. In F. E. Grine, (ed.), Evolutionary History of the "Robust" Australopithecines, pp. 405–426. Aldine de Gruyter, New York.

———. 1995a. On the connections between paleoclimate and evolution. In E. S. Vrba, G. H. Denton, T. C. Partridge, and L. H. Burckle (eds.), Paleoclimate and Evolution with Emphasis on Human Origins, pp. 24–45. Yale University Press, New Haven, CT.

———. 1995b. The fossil record of African antelopes (Mammalia, Bovidae) in relation to human evolution and paleoclimate. In E. S. Vrba, G. H. Denton, T. C. Partridge, and L. H. Burckle (eds.), Paleoclimate and Evolution with Emphasis on Human Origins, pp. 385–424. Yale University Press, New Haven, CT.

———. 2000. Major features of Neogene mammalian evolution in Africa. In T. C. Partridge and R. R. Maud (eds.), The Cenozoic of Southern Africa, pp. 277–304. Oxford University Press, Oxford.

———. 2006. A possible ancestor of the living waterbuck and lechwes: *Kobus basilcookei* sp. nov. (Reduncini, Bovidae, Artiodactyla) from the early Pliocene of the Middle Awash, Ethiopia. *Transactions of the Royal Society of South Africa* 62:63–74.

Vrba, E. S., and J. E. Gatesy. 1994. New fossils of hippotragine antelopes from the Middle Awash deposits, Ethiopia, in the context of a phylogenetic analysis of Hippotragini (Bovidae, Mammalia). *Palaeontologia Africana* 31:1–18.

Vrba, E. S., and Haile-Selassie, Y. 2006. *Zephyreduncinus oundagaisus* (Reduncini, Artiodactyla, Bovidae) from the late Miocene of the Middle Awash, Afar Rift, Ethiopia. *Journal of Vertebrate Paleontology* 26:213–218.

Vrba, E. S., and Schaller, G. B. 2000. Phylogeny of Bovidae (Mammalia) based on behavior, glands and skull morphology. In E. S. Vrba and G. B. Schaller (eds.), Antelopes, Deer, and Relatives: Fossil Record, Behavioral Ecology, Systematics, and Conservation, pp. 203–222. Yale University Press, New Haven, CT.

Vrba, E. S., G. H. Denton, and M. L. Prentice. 1989. Climatic influences on early hominid behavior. *Ossa* 14:127–156.

Vrba, E. S., J. R. Vaisnys, J. E. Gatesy, R. Desalle, and K. Y. Wei. 1994. Analysis of pedomorphosis using allometric characters—the example of Reduncini antelopes (Bovidae, Mammalia). *Systematic Biology* 43:92–116.

Wagner, A. 1848. Urweltliche Säugetiere-Überreste aus Griechenland. *Abhandlungen der Bayerischen Akademie der Wissenschaften* 5:335–378.

Wahlert, J. H. 1984. Relationships of the extinct rodent *Cricetops* to *Lophiomys* and the Cricetinae (Rodentia: Cricetidae). *American Museum Novitates* 2784:1–15.

Wallace, A. R. 2003. Geology of the Ivanhoe Hg-Au district, northern Nevada: Influence of Miocene volcanism, lakes, and active faulting on epithermal mineralization. *Economic Geology* 98:409–424.

Walter, R. C., and J. L. Aronson. 1982. Revsions of K/Ar ages for the Hadar site, Ethiopia. *Nature* 296:122–127.

Walter, R. C., W. K. Hart, and J. A. Westgate. 1987. Petrogenesis of a basalt-rhyolite tephra from the west-central Afar, Ethiopia. *Contributions to Mineralogy and Petrology* 95:462–480.

Ward, C.V., M. G. Leakey, and A. Walker. 2001. Morphology of *Australopithecus anamensis* from Kanapoi and Allia Bay, Kenya. *Journal of Human Evolution* 41:255–368.

Ward, S., and A. Hill. 1987. Pliocene hominid partial mandible from Tabarin, Baringo, Kenya. *American Journal of Physical Anthropology* 72:21–37.

Weber, M. 1904. Über tertiäre Rhinocerotiden von der Insel Samos. *Bulletin de la Société Impériale des Naturalistes de Moscou* 17:477–501.

Wells, L. H., and H. B. S. Cooke. 1956. Fossil Bovidae from the Limeworks Quarry, Makapansgat, Potgietersrus. *Palaeontologia Africana* 4:1–55.

Werdelin, L. 1996. Carnivores, exclusive of Hyaenidae, from the later Miocene of Europe and western Asia. In R. L. Bernor, V. Fahlbusch, and H.-W. Mittmann (eds.), The Evolution of Western Eurasian Miocene Mammal Faunas, pp. 271–289. Columbia University Press, New York.

———. 2003a. Carnivora from the Kanapoi Hominid Site, Turkana Basin, Northern Kenya. In J. M. Harris and M. E. Leakey (eds.), Geology and Vertebrate Paleontology of the Early Pliocene Site of Kanapoi, Northern Kenya, pp. 115–132. Natural History Museum of Los Angeles County, Los Angeles.

———. 2003b. Mio-Pliocene Carnivora from Lothagam, Kenya. In M.G. Leakey and J.M. Harris (eds.), Lothagam: The Dawn of Humanity in Eastern Africa, pp. 261–238. Columbia University Press, New York.

Werdelin, L., and M. E. Lewis. 2000. Carnivora from the southern Turkwell hominid site, northern Kenya. *Journal of Paleontology* 74:1173–1180.

———. 2001a. A revision of the genus *Dinofelis* (Mammalia, Felidae). *Zoological Journal of the Linnean Society* 132:47–258.

———. 2001b. Diversity and turnover in eastern African Plio-Pleistocene carnivora. *Journal of Vertebrate Paleontology* 21:112–113.

———. 2005. Plio-Pleistocene Carnivora of eastern Africa: Species richness and turnover patterns. *Zoological Journal of the Linnean Society* 144:121–141.

Werdelin, L., and N. Solounias. 1990. Studies of fossil hyaenids: The genus *Adcrocuta* and the interrelationships of some hyaenid taxa. *Zoological Journal of the Linnean Society* 98:363–386.

———. 1991. The Hyaenidae: Taxonomy, systematics and evolution. *Fossils and Strata* 30:1–104.

———. 1996. The evolutionary history of hyaenas in Europe and western Asia during the Miocene. In R. L. Bernor, V. Fahlbusch, and H.-W. Mittmann (eds.), The Evolution of Western Eurasian Late Neogene Mammal Faunas, pp. 290–306. Columbia University Press, New York.

Werdelin, L., and R. Sardella. 2006. The "*Homotherium*" from Langebaanweg, South Africa and the origin of *Homotherium*. *Palaeontographia Abteilung A*, 277:123–130.

Werdelin, L., and A. Turner. 1996. The fossil and living Hyaenidae of Africa: Present status. In K. Stewart and K. Seymour (eds.), Palaeoecology and Palaeoenvironments of Late Cenozoic Mammals. Tributes to the Career of C. S. (Rufus) Churcher, pp. 637–659. Toronto University Press, Toronto.

Werdelin, L., A. Turner, and N. Solounias. 1994. Studies of fossil hyaenids: The genera *Hyaenictis* Gaudry and *Chasmaporthetes* Hay, with a reconsideration of the Hyaenidae of Langebaanweg, South Africa. *Zoological Journal of the Linnean Society* 111:197–217.

Wesselman, H. B. 1984. The Omo micromammals: Systematics and paleoecology of early man sites from Ethiopia. *Contributions to Vertebrate Evolution* 7:1–219.

———. 1985. Fossil micromammals as indicators of climatic change about 2.4 Myr ago in the Omo Valley, Ethiopia. *South African Journal of Science* 81:260–261.

———. 1995. Of mice and almost men: Regional paleoecology and human evolution in the Turkana Basin. In E. S. Vrba, G. H. Denton, T. C. Partridge, and L. H. Burckle (eds.), Paleoclimate and Evolution, with Emphasis on Human Origins, pp. 356–368. Yale University Press, New Haven, CT.

Western, D. 1973. The structure, dynamics and changes of the Amboseli ecosystem. Ph.D. thesis, University of Nairobi, Nairobi.

Westgate, J. A., and M. P. Gorton. 1981. Correlation techniques in tephra studies. In S. Self and R. S. J. Sparks (eds.), Tephra Studies, pp. 73–94. D. Reidel Publishing Co., Dordrecht, The Netherlands, and Boston.

Weston, E. M. 2000. A new species of hippopotamus *Hexaprotodon lothagamensis* (Mammalia: Hippopotamidae) from the late Miocene of Kenya. *Journal of Vertebrate Paleontology* 20: 177–185.

———. 2003. Fossil Hippopotamidae from Lothagam. In J. M. Harris and M. G. Leakey (eds.), Lothagam: The Dawn of Humanity in Eastern Africa, pp. 380–410. Columbia University Press, New York.

White, F. 1983. The Vegetation of Africa: A Descriptive Memoir to Accompany UNESCO/AETFAT/UNSO Vegetation Maps of Africa. UNESCO, Paris.

White, J. A. 1991. North American leporinae (Mammalia: Lagomorpha) from Late Miocene (Clarendonian) to latest Pliocene (Blancan). *Journal of Vertebrate Paleontology* 11:67–89.

White, T. D. 1977. New fossil hominids from Laetolil, Tanzania. *American Journal of Physical Anthropology* 46:197–230.

———. 1984. Pliocene hominids from the Middle Awash, Ethiopia. *Courier Forschunginstitut Senckenberg* 69:57–68.

———. 1986. *Australopithecus afarensis* and the Lothagam mandible. *Anthropos (Brno)* 23:79–90.

———. 1995. African omnivores: Global climatic change and Plio-Pleistocene hominids and suids. In E. S. Vrba, G. H. Denton, T. C. Partridge, and L. H. Burckle (eds.), Paleoclimate and Evolution with Emphasis on Human Origins, pp. 369–384. Yale University Press, New Haven, CT.

———. 2002. Earliest hominids. In W. C. Hartwig (ed.), The Primate Fossil Record, pp. 407–417. Cambridge University Press, Cambridge, UK.

———. 2003. Early hominids—diversity or distortion? *Science* 299:1994–1996.

———. 2004. Managing paleoanthropology's nonrenewable resources: A view from Afar. *Comptes Rendus Palevol* 3:341–351.

———. 2006. Early hominid femora: The inside story. *Comptes Rendus Palevol* 5:99–108.

White, T. D., and J. M. Harris. 1977. Suid evolution and correlation of African hominid localities. *Science* 198:13–21.

White, T. D., and D. C. Johanson. 1982. Pliocene hominid mandibles from the Hadar Formation, Ethiopia: 1974–1977 collections. *American Journal of Physical Anthropology* 57:501–544.

White, T. D., and G. Suwa. 2004. A new species of *Notochoerus* (Artiodactyla, Suidae) from the Pliocene of Ethiopia. *Journal of Vertebrate Paleontology* 24:474–480.

White, T. D., G. Suwa, W. K. Hart, R. C. Walter, G. WoldeGabriel, J. de Heinzelin, J. D. Clark, B. Asfaw, and E. Vrba. 1993. New discoveries of *Australopithecus* at Maka in Ethiopia. *Nature* 366:261–265.

White, T. D., G. Suwa, and B. Asfaw. 1994. *Australopithecus ramidus*, a new species of early hominid from Aramis, Ethiopia. *Nature* 371:306–312.

———. 1995. Corrigendum: *Australopithecus ramidus*, a new species of early hominid from Aramis, Ethiopia. *Nature* 375:88.

White, T. D., G. Suwa, S. Simpson, and B. Asfaw. 2000. Jaws and teeth of *Australopithecus afarensis* from Maka, Middle Awash, Ethiopia. *American Journal of Physical Anthropology* 111:45–68.

White, T. D., B. Asfaw, D. DeGusta, W. H. Gilbert, G. D. Richards, G. Suwa, and F. C. Howell. 2003. Pleistocene *Homo sapiens* from Middle Awash, Ethiopia. *Nature* 423:742–747.

White, T. D., F. C. Howell, and H. Gilbert. 2006a. The earliest *Metridiochoerus* (Artiodactyla: Suidae) from the Usno Formation, Ethiopia. *Transactions of the Royal Society of South Africa* 61:75–79.

White, T. D., G. WoldeGabriel, B. Asfaw, S. Ambrose, Y. Beyene, R. Bernor, J.-R. Boisserie, B. Currie, W. H. Gilbert, Y. Haile-Selassie, W. K. Hart, L. Hlusko, F. C. Howell, R. T. Kono, A. Louchart, C. O. Lovejoy, P. R. Renne, H. Saegusa, E. Vrba, H. Wesselman, and G. Suwa. 2006b. Asa Issie, Aramis, and the origin of *Australopithecus*. *Nature* 440:883–889.

Whybrow, P. J., and A. Hill (eds.). 1999. Fossil Vertebrates of Arabia, with Emphasis on the Late Miocene Faunas, Geology, and Palaeoenvironments of the Emirate of Abu Dhabi, United Arab Emirates. Yale University Press, New Haven, CT.

Wildman, D. E., M. Uddin, G. Liu, L. I. Grossman, and M. Goodman. 2003. Implications of natural selection in shaping 99.4% nonsynonymous DNA identity between humans and chimpanzees: Enlarging genus *Homo*. *Proceedings of the National Academy of Sciences of the United States of America* 100:7181–7188.

Willemsen, G. F. 1992. A revision of the Pliocene and Quaternary Lutrinae from Europe. *Scripta Geologica* 101:1–115.

Williams, M. A. J., G. Assefa, and D. A. Adamson. 1986. Depositional context of Plio-Pleistocene hominid-bearing formations in the Middle Awash Valley, southern Afar Rift, Ethiopia. In L. E. Frostick, R. W. Renaut, J. Reid, and J. J. Tiercelin (eds.), Sedimentation in the African Rifts, pp. 241–251. Blackwell Scientific, Palo Alto, CA.

Willows-Munro, S., T. J. Robinson, and C. A. Matthee. 2005. Utility of nuclear DNA intron markers at lower taxonomic levels: Phylogenetic resolution among nine *Tragelaphus* spp. *Molecular Phylogenetics and Evolution* 35:624–636.

Wilson, D. E., and D. H. M. Reeder. 2005. Mammal Species of the World: A Taxonomic and Geographic Reference. Johns Hopkins University Press, Baltimore.

Winkler, A. J. 1990. Systematics and biogeography of Neogene rodents from the Baringo District, Kenya. Ph.D. thesis, Southern Methodist University, Dallas.

———. 1997. Systematics, paleobiogeography, and paleoenvironmental significance of rodents from the Ibole Member, Manonga Valley, Tanzania. In T. Harrison (ed.), Neogene Paleontology of the Manonga Valley, Tanzania, pp. 311–332. Plenum Press, New York.

———. 2002. Neogene paleobiogeography and East African paeloenvironments: Contributions from the Tugen Hills rodents and lagomorphs. *Journal of Human Evolution* 42:237–256.

———. 2003. Rodents and lagomorphs from the Miocene and Pliocene of Lothagam, northern Kenya. In M. G. Leakey and J. M. Harris (eds.), Lothagam: The Dawn of Humanity in Eastern Africa, pp. 169–198. Columbia University Press, New York.

WoldeGabriel, G. 1987. Volcanotectonic history of the central sector of the Main Ethiopian Rift. Ph.D. thesis, Case Western Reserve University, Cleveland, Ohio.

WoldeGabriel, G., J. L. Aronson, and R. C. Walter. 1990. Geology, geochronology, and rift basin development in the central sector of the Main Ethiopian Rift. *Geological Society of America Bulletin* 102:439–458.

WoldeGabriel, G., R. C. Walter, J. L. Aronson, and W. K. Hart. 1992a. Geochronology and distribution of silicic volcanic rocks of Plio-Pleistocene age from the central sector of the Main Ethiopian Rift. *Quaternary International* 13/14:69–76.

WoldeGabriel, G., T. White, G. Suwa, S. Semaw, Y. Beyene, B. Asfaw, and R. Walter. 1992b. Kesem-Kebena: A newly discovered paleoanthropology research area in Ethiopia. *Journal of Field Archaeology* 19:471–493.

WoldeGabriel, G., T. D. White, G. Suwa, P. R. Renne, J. de Heinzelin, W. K. Hart, and G. Heiken. 1994. Ecological and temporal placement of early Pliocene hominids at Aramis, Ethiopia. *Nature* 371:330–333.

WoldeGabriel, G., P. R. Renne, T. D. White, G. Suwa, J. de Heinzelin, W. K. Hart, and G. Heiken. 1995. Age of early hominids. *Nature* 376:559.

WoldeGabriel, G., R. C. Walter, W. K. Hart, S. A. Mertzman, and J. L. Aronson. 1999. Temporal relations and geochemical features of felsic volcanism in the central sector of the Main Ethiopian Rift. *Acta Vulcanologica* 11:53–67.

WoldeGabriel, G., Y. Haile-Selassie, P. R. Renne, W. K. Hart, S. H. Ambrose, B. Asfaw, G. Heiken, and T. D. White. 2001. Geology and paleontology of the late Miocene Middle Awash valley, Afar Rift, Ethiopia. *Nature* 412:175–178.

WoldeGabriel, G., P. R. Renne, W. K. Hart, S. H. Ambrose, B. Asfaw, and T. D. White. 2004. Geoscience methods lead to paleo-anthropological discoveries in Afar Rift, Ethiopia. *Eos, Transactions, American Geophysical Union* 273:276–277.

WoldeGabriel, G., W. K. Hart, S. Katoh, Y. Beyene, and G. Suwa. 2005. Correlation of Plio-Pleistocene tephra in Ethiopian and Kenyan rift basins: Temporal calibration of geological features and hominid fossil records. *Journal of Volcanology and Geothermal Research* 147:81–108.

Wolfenden, E., C. Ebinger, G. Yirgu, A. Deino, and D. Ayalew. 2004. Evolution of the northern Main Ethiopian Rift: Birth of a triple junction. *Earth and Planetary Science Letters* 224:213–228.

Wozencraft, W. C. 1984. A phylogenetic reappraisal of the Viverridae and its relationship to other Carnivora. Ph.D. thesis, University of Kansas, Lawrence.

———. 1989. Classification of the Recent Carnivora. In J. Gittleman (ed.), Carnivore Behavior, Ecology, and Evolution, pp. 569–593. Cornell University Press, Ithaca, NY.

Wynn, J. G., Z. Alemseged, R. Bobe, D. Geraads, D. Reed, and D. C. Roman. 2006. Geological and palaeontological context of a Pliocene juvenile hominin at Dikika, Ethiopia. *Nature* 443:332–335.

Wyss, A., and J. H. Flynn. 1993. A phylogenetic analysis and definition of the Carnivora. In S. F. Szalay, M. J. Novacek, and M. C. McKenna (eds.), Mammal Phylogeny (Placentals), pp. 32–52. Springer-Verlag, New York.

Yalden, D. W. 1975. Some observations on the giant mole-rat *Tachyoryctes macrocephalus* (Rüppell, 1842) (Mammalia, Rhizomyidae) of Ethiopia. *Monitore Zoologico Italiano, N. S., Supplemento* 6:275–303.

Yalden, D. W., and M. J. Largen. 1992. The endemic mammals of Ethiopia. *Mammal Review* 22:115–150.

Yalden, D. W., M. J. Largen, and D. Kock. 1976. Catalogue of the mammals of Ethiopia. *Monitore Zoologico Italiano, N. S., Supplemento* 8:1–118.

Yellen, J., A. Brooks, D. Helgren, M. Tappen, S. Ambrose, R. Bonnefille, J. Feathers, G. Goodfriend, K. Ludwig, P. Renne, and K. Stewart. 2005. The archaeology of Aduma Middle Stone Age sites in the Awash Valley, Ethiopia. *PaleoAnthropology* 10:25–100.

Zanettin, B., and E. Justin-Visentin. 1975. Tectonical and volcanological evolution of the western Afar margin (Ethiopia). In A. Pilger and A. Roesler (eds.), Afar Depression of Ethiopia, pp. 300–309. Schweizerbart Verlag, Stuttgart.

Zar, J. H. 1999. Biostatistical Analysis. Prentice Hall, Upper Saddle River, NJ.

Zazzo, A., H. Bocherens, M. Brunet, A. Beauvilain, D. Biliou, H. T. Mackaye, P. Vignaud, and A. Mariotti. 2000. Herbivore paleodiet and paleoenvironmental changes in Chad during the Pliocene using stable isotope ratios of tooth enamel carbonate. *Paleobiology* 26:294–309.

Zeuner, F. 1934. Die Beziehungen zwischen Schädelform und Lebensweise bei den rezenten und fossilen Nashörnern. *Berichte der Naturforschenden Gesellschaft zu Freiburg* 34:21–79.

Zukowsky, L. 1965. Die Systematik der Gattung *Diceros* Gray, 1821. *Zoologischer Garten Leipzig* 30:1–178.

Index

Note: page numbers followed by *f* or *t* indicate that the citation may be found in a figure or table.